Geomatics Applied to Civil Engineering

Irineu da Silva • Paulo C. L. Segantine

Geomatics Applied to Civil Engineering

Second Edition

 Springer

Irineu da Silva
São Carlos School of Engineering
University of São Paulo
São Carlos, SP, Brazil

Paulo C. L. Segantine
São Carlos School of Engineering
University of São Paulo
São Carlos, SP, Brazil

ISBN 978-3-031-75736-5 ISBN 978-3-031-75737-2 (eBook)
https://doi.org/10.1007/978-3-031-75737-2

1st edition: © Elsevier 2015

This Springer imprint is published by the registered company Springer Nature Switzerland AG
The registered company address is: Gewerbestrasse 11, 6330 Cham, Switzerland

If disposing of this product, please recycle the paper.

To
Our families

Preface

At a time when digital information, Internet communications, satellite navigation and computer applications for urban mobility are globally accessible, people's perception of space has changed radically, allowing them to understand and use the environment in which they live much more comprehensively than ever before. Navigating on digital maps and browsing geospatial information of our cities is now something that most people understand perfectly.

Although most users do not realise it, this new interpretation of space translates into access to geospatial information related to the environment in which they live. For the general public, this means access to online digital maps or geocoded information, while for the professional sector, it means the mining and availability of relevant, comprehensive, and computationally organised geospatial data for use in various engineering fields. However, to make these data usable, it must be collected, stored, systematised and made available in a consistent way. Hence the importance of systematising a new discipline of geospatial data management, given its range of applications in the most diverse domains of human activity, which is now called Geomatics Engineering.

On the other hand, Civil Engineers as well as other engineering professionals are facing new challenges in their profession due to new technologies, computational methods and market forces that have redefined the role of engineers towards a new productive vision. Concepts such as Industry 4.0, big data, artificial intelligence and the Internet of Things (IoT) are examples of some of these contemporary drivers of innovation.

From this point of view, Geomatics, one of the oldest activities related to Civil Engineering and still an indispensable tool for any design and construction work, has become an important discipline in the teaching of Civil Engineering, especially with regard to the synergy between them. At the same time, like all technical fields, Geomatics has made important technological advances over the years, incorporating new tools, technologies and working methods that must be understood by Civil Engineers so that they can continue to develop complementary work seamlessly. The aim of this book is, therefore, to present new Geomatics technologies and their applications in Civil Engineering projects and field operations, so that geomatics and Civil Engineers can collaborate and make efficient use of the new technologies at their disposal.

Although several books partially addres the concepts of Geomatics applied to Civil Engineering, few of them cover the theoretical and practical concepts related to the full understanding of the use of Geomatics in Civil Engineering. This is the merit of this book: it deals with several topics from a very specific point of view, since it has been written by Civil Engineers with a professional background in Geomatics. In other words, to the extent of their efforts, the authors sought to (i) analyse geodetic references and their coordinate systems in order to enable their conversion and application, considering the local ground-based horizontal plane, where Civil Engineering projects are designed; (ii) present an overview of the adjustment computation theory with a view to its use in point positioning and error reduction when redundant data are available; (iii) present in detail most of the coordinate transformation methods useful for Civil Engineering projects and field work; (iv) discuss the practical use of modern surveying instruments in the measurement of directions, angles and distances with reference to their use in Civil Engineering works; (v) describe in detail the main components of surveying instruments used in Geomatics, highlighting robotic total stations, digital and laser levels, laser scanners and GNSS instruments, according to their Civil Engineering applications; (vi) describe instrumental errors and their implications in Civil Engineering projects; (vii) present a complete study of the UTM Projection System and its variants, equations and practical use in Civil Engineering projects, as well as its coordinate transformations to allow combining GNSS observations and total station measurements; (viii) present the theories and practical applications of Digital Terrain Modelling with a view to its application in Civil Engineering projects; (ix) present a full chapter on conventional and UAV-based digital photogrammetry, highlighting the basic concepts so that Civil Engineers can use these technologies in their engineering projects; (x) present an updated theory of adjustment computation, coordinate geometry, traversing, detail surveying and setting-out works using adjustment computation models; and finally, (xi) give an updated overview of volume calculation, horizontal and vertical curve geometry, and BIM concepts for Geomatics applied to Civil Engineering.

The format of the book is based on several years of experience of Professors Irineu da Silva and Paulo C. L. Segantine in teaching Geomatics to Civil Engineers at undergraduate and postgraduate levels at the School of Civil Engineering of São Carlos, University of São Paulo, Brazil. The book is not intended to turn Civil Engineers into surveyors. On the contrary, it is intended to give civil engineers a broad view of Geomatics so they can understand and discuss surveying techniques when applied to Civil Engineering projects.

For the benefit of students and teachers, the topics in the book are organised to cover most aspects of Surveying and Geomatics in Civil Engineering courses. In order to facilitate the learning and practical use of modern surveying methods and innovative technologies, the chapters have been carefully organised in an appropriate learning sequence. In addition, application examples are presented at the end of each section to aid understanding of the theories developed in the chapters.

The book was edited by Dr. Irineu da Silva (Associate Professor at the University of São Paulo – Brazil) with substantial contributions from Dr. Paulo Cesar Lima Segantine (Associate Professor at the University of São Paulo – Brazil), Dr. Guilherme Poleszuk dos Santos Rosa (PhD Cartographic Engineer), Dr. Marcelo Monari (Assistant Professor at the Federal University of São Carlos – Brazil) and Julio Franco (MSc Civil Engineer).

São Carlos, SP, Brazil   Irineu da Silva

Contents

1 **Geomatics: An Introduction** 1
Irineu da Silva

2 **Basic Concepts of Geomatics** 15
Irineu da Silva and Paulo C. L. Segantine

3 **Introduction to Adjustment Computation** 37
Irineu da Silva and Guilherme Poleszuk dos Santos Rosa

4 **Reference Systems** 97
Irineu da Silva and Marcelo Monari

5 **Coordinate Reference System** 119
Irineu da Silva and Guilherme Poleszuk dos Santos Rosa

6 **Directions and Angles** 177
Irineu da Silva and Guilherme Poleszuk dos Santos Rosa

7 **Distances** 207
Irineu da Silva

8 **Distance Measurement** 237
Irineu da Silva and Paulo C. L. Segantine

9 **Surveying Instruments** 265
Irineu da Silva

10 **Instrumental and Operational Errors** 315
Irineu da Silva and Guilherme Poleszuk dos Santos Rosa

11 **Coordinate Geometry for Point Positioning** 355
Irineu da Silva and Guilherme Poleszuk dos Santos Rosa

12 **Horizontal Control Survey: Traversing** 439
Irineu da Silva and Guilherme Poleszuk dos Santos Rosa

13 **Vertical Control Survey: Levelling** 499
    Irineu da Silva and Marcelo Monari

14 **Levelling Applications for Relief Representations** 569
    Irineu da Silva and Paulo C. L. Segantine

15 **Digital Terrain Modelling** 589
    Irineu da Silva

16 **Map Projections** 619
    Irineu da Silva and Paulo C. L. Segantine

17 **Global Navigation Satellite System: GNSS** 677
    Irineu da Silva, Paulo C. L. Segantine,
    and Guilherme Poleszuk dos Santos Rosa

18 **Terrestrial Laser Scanning** 739
    Irineu da Silva

19 **Airborne Photogrammetry** 769
    Irineu da Silva

20 **Detail Survey** 841
    Irineu da Silva and Paulo C. L. Segantine

21 **Setting Out** 861
    Irineu da Silva

22 **Areas and Volumes** 887
    Irineu da Silva, Paulo C. L. Segantine, and Marcelo Monari

23 **Horizontal and Vertical Curves** 943
    Irineu da Silva, Paulo C. L. Segantine, and Marcelo Monari

24 **Basics of Building Information Modelling for Geomatics** 989
    Júlio Franco and Irineu da Silva

Index 1007

Abbreviations

AAGS	American Association for Geodetic Surveying
AASHTO	American Association of State Highway and Transportation Officials
ACSM	American Association on Surveying and Mapping
AEC	Architecture, Engineering and Construction
ALS	Airborne Laser Scanning
APOS	Australian Positioning Service
ARP	Antennas´ Geometric Centre
AS	Anti-Spoofing
ASCE	American Society of Civil Engineers
ASI	Agenzia Spatiale Italiana
ASPRS	American Society for Photogrammetry and Remote Sensing
ASTER	Advanced Spaceborne Thermal Emission and Reflection Radiometer
ATR	Automatic Target Recognition
BCF	BIM Collaboration Format
BEP	BIM Execution Plan
BIM	Building Information Modelling
BM	Benchmarks
BS	Backsight
C/A	Course Acquisition
CAD	Computer-Aided Design
CCD	Charged Coupled Device
CDGPS	Canada-Wide Differential GPS
CGCS	China Geodetic Coordinate System
CGPM	General Conference on Weights and Measures
CHA	Channel of High Accuracy
CMOS	Complementary Metal Oxide Semiconductor
CNSS	Compass Navigation Satellite System
COGO	Coordinate Geometry

CORS	Continuously Operating Reference Station
CS	Commercial Access Service
CSA	Channel of Standard Accuracy
DEM	Digital Elevation Model
DEM	Digital Elevation Modelling
DGPS	Differential Global Positioning System
DLR	German Aerospace Center
DMA	US Department Mapping Agency
DORIS	Doppler Orbitography and Radiopositioning Integrated by Satellite
DPW	Digital Photogrammetric Workstation
DSM	Digital Surface Model
DTM	Digital Terrain Modelling
EDM	Electronic Distance Measurement
EGM	Earth Gravitational Modell
EGNOS	European Geostationary Navigation Overlay Service
EO	Exterior Orientation
EPP	Expanded Poly Propylene
ESA	European Space Agency
FGDC	Federal Geographic Data Committee
FIG	International Federation of Surveyors
FOC	Full Operational Capability
FS	Foresight
GA	Ground Antenna
GAGAN	Indian GPS-Aided Geo Augmented System
GALILEO	European Global Navigation Satellite System
GBAS	Ground-based Augmentation System
GCP	Ground Control Point
GCS	Geographic Coordinate System
GDOP	Geometric Dilution of Precision
GEO	Geosynchronous Equatorial Orbit
GIS	Geographic Information Systems
GLONASS	Globalnaya Navigatsionnaya Sputnikovaya Systema
GNSS CORS	GNSS Continuously Operating Reference Stations
GNSS RTK	GNSS Real-Time Kinematic
GNSS	Global Navigation Satellite systems
GPS	Global Positioning System
GRS80	Geodetic Reference System 1980
GSD	Ground Sample Distance
GSM	Global System for Mobile Communications
GTRF	GALILEO Terrestrial Reference Frame
HTTP	Hypertext Transfer Protocol
IATS	Image-Assisted Total Station
ICA	International Cartography Association

IEC	International Electrotechnical Commission
IERS	International Earth Rotation and Reference Systems Service
IGS	International GNSS Service
IGSO	Inclined Geosynchronous Orbit
IMU	Inertial Measurement Unit
INS	Inertial Navigation System
IO	Internal Orientation
IOV	In-Orbit Validation
IP	Protection Rating
IRNSS	Indian Regional Navigation Satellite System
ISO	International Organization for Standardization
ISPRS	International Society for Photogrammetry and Remote Sensing
ITRF	International Terrestrial Reference Frame
ITRS	International Terrestrial Reference System
JPEG	Joint Photographic Experts Group
JPL	Jet Propulsory Laboratory
JPO	Joint Program Office
KASS	Korean Augmentation Satellite System GPS
LASS	Local Area GPS Enhancement
LCD	Liquid Crystal Display
LDP	Low Distortion Projection
LED	Light-Emitting Diode
LiDAR	Light Detection and Ranging
LLR	Lunar Laser Ranging
LoD	Level of Development
LRS	Linear Referencing System
LTM	Local Transverse Mercator
MAC	Master-Auxiliary Concept
MCS	Master Control Station
MEO	Median Earth Orbit
MS	Monitor Station
MSAS	Japan´s Spaced-Based Augmentation System
MTOW	Maximum Take-Off Weight
MVS	Multi-View Stereo
N, E	UTM coordinates
NASA	National Aeronautics and Space Administration
NAVCEN	US Coast Guard Navigation Center
NaviC	Navigation Indian Constellation Satellite System
NAVSTAR	Navigation Satellite Timing and Ranging
NDS	Detection System Payload
NGS	National Geodetic Survey
NMEA	National Marine Electronics Association
NNSS	Navy Navigational Satellite System
NRTK	Network Real-Time Kinematic Positioning

NTRIP	Networked Transport of RTCM via Internet Protocol
NUDET	Nuclear Detonation
OCS	Operational Control Segment
OPUS	Online Positioning User Service
OS GALILEO	Open Access Service GALILEO
OS	Ordnance Survey
PDOP	Position Dilution of Precision
PPK	Post-Processed Kinematic
PPP	Precise Point Positioning
PPP-RTK	PPP Real-Time Kinematic Positioning
PPS	Precise Positioning Services
PRS	Public Regulated Service
QZSS	Quasi-Zenith Satellite System
RINEX	Receiver Independent Exchange Format
RMS	Root Mean Square
RMSE	Root Mean Square Error
RPA	Remotely Piloted Aircraft
RPAS	Remotely Piloted Aircraft Systems
RS	Restricted Service
RTCM	Radio Technical Commission for Maritime Services
RTK	Real-Time Kinematic
RTS	Robotic Total Station
SAD69	South American Datum 1969
SAPOS	German Satellite Positioning Service
SAR	Search and Rescue Service
SAR	Synthetic Aperture Radar
SBAS	Space-Based Augmentation Systems
SDCN	System for Differential Correction and Monitoring
SfM	Structure from Motion
SI	International System of Units
SIFT	Scale-Invariant Feature Transform
SIRGAS	Geodetic Reference System for the Americas
SIRGAS-CON	Geodetic Reference System for the Americas – Continuously Operating Network
SLR	Satellite Laser Ranging
SNR	Signal-to-Noise Ratio
SoL	Safety-of-Life Service
SPS	Standard Positioning Services
SQL	Structured Query Language
SRTM	Shuttle Radar Topography Mission
STS	Scanning Total Station
TCP/IP	Transfer Control Protocol/Internet Protocol
TDI	Time Delay Integrator
TIN	Triangular Irregular Network

TLS	Terrestrial Laser Scanning
TM	Transverse Mercator
UAS	Unmanned Aircraft System
UAV	Unmanned Aerial Vehicle
UHF	Ultra High Frequency
USACE	US Army Corps of Engineers
UTM	Universal Transverse Mercator
VGA	Video Graphics Array
VIM	International Vocabulary of Metrology
VLBI	Very Long Baseline Interferometry
VRINEX	Virtual RINEX
VRS	Virtual Reference Stations
WAAS	Wide Area Augmentation System
WAGE	Wide Area GPS Enhancement
WFD	Wave Form Digitiser
WGS84	World Geodetic System 1984
XML	Extensible Markup Language

Chapter 1
Geomatics: An Introduction

Irineu da Silva

1.1 Introduction

As with many technical professions, today's Civil Engineering professionals face significant challenges in responding to new technological advances and market forces that require them to design increasingly complex and diversified projects. Depending on the size of the project, today's design team must include many experts from different disciplines who will bring their expertise to each stage of project development and construction site. One of the most important of these experts is the Geomatics Engineer.

Along with Civil Engineering, Geomatics is concerned with measuring the existing features of the natural and built environment and presenting the data in a format suitable for use by architects and Civil Engineers in their construction projects. By providing the position, shape and nature of geographic features using topographic and geodetic instruments, geomatics lies between Civil Engineering and the real world, playing an essential role from the early stages of design to final as-built mapping and structural monitoring.

In this context, Civil Engineers have increasingly recognised that a narrow focus on Engineering Surveying work is no longer sufficient to meet the real and current needs of Civil Engineering and that a broader view of geospatial data[1] management is required. Thus, considering that Geomatics is a discipline that encompasses the sciences, techniques and methods that deal with the acquisition, integration, storage, distribution, analysis, processing, modelling, georeferencing, presentation,

---

[1] Information that identifies the geographic location and characteristics of natural or built features of the earth.

I. da Silva (✉)
São Carlos School of Engineering, University of São Paulo, São Carlos, SP, Brazil
e-mail: irineu@sc.usp.br

© The Author(s), under exclusive license to Springer Nature Switzerland AG 2025
I. da Silva, P. C. L. Segantine, *Geomatics Applied to Civil Engineering*,
https://doi.org/10.1007/978-3-031-75737-2_1

distribution, setting out and management of spatially referenced data, its concepts are of great interest to Civil Engineers.

Geomatics as a discipline of study is a recent idea that emerged in Canada in the 1980s. Today, Geomatics Engineering is taught as a worldwide career path, complementing or replacing the teaching of Surveying Engineering. From the civil engineer's point of view, however, the concepts and techniques of geomatics should be seen as additional tools in the development of Civil Engineering works. Civil engineers are users of geomatics. They are generally expected to have sufficient knowledge to determine the best practices and appropriate tools to apply to their projects. They must be able to communicate conveniently with Geomatics Engineers, clearly understand the surveying documents, have sufficient knowledge to work with the different coordinate systems and map projections involved in civil engineering, be able to perform basic topographic calculations and have the knowledge necessary to use geomatics software applied to civil engineering and to evaluate the quality of the results obtained. In this sense, the Civil Engineer must include in his education the fundamental study of some sciences and techniques related to geomatics, such as Engineering Surveying techniques, the basics of Geodesy, Adjustment Computation, the basics of Cartography, Photogrammetry, Remote Sensing, Cadastral Management, Database Management, Geographic Information Systems (GIS), Global Navigation Satellite System (GNSS), Building Information Modelling (BIM) and others. To help the reader understand the terms mentioned, a brief description of each and its relevance to Civil Engineering is given below.

1.1.1 Engineering Surveying Techniques

The American Society of Civil Engineers (ASCE) defines *Engineering Surveying* as the activities involved in the planning and execution of surveys for the planning, design, construction, operation and maintenance of engineered projects.[2] In terms of Geomatics applied to Civil Engineering, Engineering Surveying should be understood as the discipline that relies on the fundamental points of the geodetic system of a country or region to generate geospatial data that allow the description, management and representation of objects and elements of construction sites and buildings. Therefore, the concepts related to this discipline must be fully mastered by Civil Engineers, which is why a significant part of this book is dedicated to them.

---

[2] https://www.asce.org/advocacy/policy-statements/ps333%2D%2D-engineering-surveying-definition

1.1.2 Geodesy

Among the disciplines involved with geomatics, *Geodesy* stands out, as it provides the mathematical foundations for studying the others. In terms of Geomatics applied to Civil Engineering, the main objective of geodetic surveying is to determine the location of precise reference points on the Earth's surface, which is necessary for georeferencing geospatial data on large Civil Engineering projects. It is essential, however, to emphasise that the importance and scope of Geodesy go far beyond the determination of control points; therefore, in this book, only the essential concepts for understanding the mathematical models studied throughout the chapters will be presented.

1.1.3 Cartography

According to the International Cartography Association's (ICA) Strategic Plan 2003–2011, *Cartography* is the discipline dealing with the art, science and technology of making and using maps. In this case, a map is defined as a symbolised representation of geographical reality, representing selected features or characteristics resulting from the creative effort of its author's execution of choices. It is designed for use when spatial relationships are of primary relevance.

Despite its importance to Civil Engineering, this book will cover only the basics of Cartography, focusing on methods and fieldwork related to digital mapping production for Civil Engineering projects.

1.1.4 Adjustment Computation

Measurements made using surveying instruments contain unavoidable inaccuracies or errors inherent to the measuring process, regardless of the quality of the instruments and the operator's skill. Understanding the causes and effects of these errors is essential to validate the results obtained according to allowable tolerances and technical specifications. There is, therefore, a discipline of geomatics that takes particular care of the analysis of the reliability and errors made during field measurements, intending to know if they are statistically acceptable and if their magnitudes are lower than certain thresholds imposed by the technical and specific regulations so that they can then be mathematically adjusted to indicate the most probable values that satisfy the mathematical model adopted. As the reader will have the opportunity to observe, several of the mathematical models applied throughout this book use the error theory and adjustment computation for their solution; therefore, an entire chapter is included in the study of the main principles of this discipline.

1.1.5   Photogrammetry

Photogrammetry is a technique that allows the collection and analysis of geospatial data on the Earth's surface based on measurements performed through digital images. The main characteristic of this technique is that it allows the almost instantaneous recording of the state of geospatial data, enabling further exploration of physical and geometric attributes under conditions that are generally more favourable than those observed in field surveys. As it is a science and a technique with wide applications in Civil Engineering, especially in projects involving large areas, a chapter in this book is dedicated exclusively to photogrammetry for Civil Engineering.

1.1.6   Remote Sensing

Remote Sensing is a technique of remote observation by measuring and processing electromagnetic signals emitted or reflected by objects on the Earth's surface to obtain information concerning their nature, properties and state. It is primarily based on measuring electromagnetic energy variations captured by photoelectric sensors installed in artificial satellites or airborne platforms. The energy thus captured is recorded in an appropriate digital medium and made available as an image for various uses in engineering projects. The names of the images are generally related to the names of the satellites carrying the sensor, such as *LANDSAT, SPOT, IKONOS, QUICKBIRD, AVIRIS and CBERS.*

It is important to note that although Remote Sensing can be used to obtain geometric information about objects on the Earth's surface, in topographic terms, it is primarily used to produce thematic maps. This information can be used with significant advantages in the preliminary design phase of Civil Engineering works, such as in drainage, ports, airports and highways projects or land use planning. Furthermore, Remote Sensing allows the spectral study of images for the analysis of the matter, such as heat or chemical energy, which can enable the identification of objects on the ground, the recognition of varieties of vegetation cover, diseases in crops and other characteristics of phenomena related to the Earth's surface. Under these circumstances, although it is a science with many applications in civil engineering, it is not within the scope of this book.

1.1.7   Cadastral Management

According to the International Federation of Surveyors (FIG), a cadastre is usually a parcel-based and up-to-date land information system containing a record of interests in land (e.g., rights, restrictions and responsibilities). It usually includes a geometric description of parcels of land linked to other records describing the nature of the interests, the ownership or control of those interests and often the parcel's value and improvements. It may be established for tax purposes (e.g., valuation and equitable taxation), legal purposes (conveyancing), to assist in the management of land and land use (e.g., for planning and other administrative purposes) and to enable sustainable development and environmental protection.[3]

Generally, the information in a cadastre exceeds that relative to the geospatial data, covering complementary information such as ownership and boundaries, possession (ownership rights and rents), improvements and land values, among others. In addition, depending on the purpose of the cadastre, other information may be included, such as:

- Real estate data relating to buildings.
- Agricultural information (land capacity classification and land use).
- Forest information.
- Infrastructure (drinking water, sewerage, electricity and web services).
- Environmental quality data.
- Demographic data (population statistics).

More recently, there has been a growing interest in Cadastral Management to include space below and above the surface (e.g., underground structures and infrastructures, such as cables and pipes). Therefore, 3D information has become an essential extension of cadastral registration, characterising the so-called 3D cadastre.

Cadastral Management is, therefore, a multidisciplinary activity involving geomatics and Civil Engineering professionals, as well as professionals from different areas, such as cadastral survey authorities, lawyers and municipal managers. In most cases, professionals in the field of geomatics are responsible for the cadastral surveying, updating and maintaining the cadastre. In Civil Engineering, a cadastral survey provides digital maps and alphanumeric information related to the cadastre, which, in most cases, are used as auxiliary reference data in the elaboration of projects in general. Therefore, although this book does not contain an exclusive chapter on Cadastral Management, most of the topics covered are used as technical support for cadastral surveys.

1.1.8   Database Management

A database is essentially software designed to compose and manipulate datasets stored in digital data structures. It consists of two distinct systems: the data manipulation system and the data storage system, which communicate with each other to generate answers to its users' questions. By its nature, it is designed to ensure the integrity, accuracy and availability of data, regardless of periodic updates.

---

[3]FIG Statement on the Cadastre, FIG Publication N° 11, 1995.

As an element of data management, databases are essential resources for geomatics and civil engineering, storing information collected in the field and managing spatial data. Due to the increasing amount of spatial data provided by surveying instruments, geomatics and Civil Engineering professionals need to have adequate knowledge of these systems, both in terms of structuring and using them. Examples of databases are computational structures based on query languages (e.g., SQL—*Structured Query Language*), managed by specialised systems (Database system), such as Oracle Database, Microsoft Access and MySQL. Due to the wide range of applications and purposes, this subject is beyond the scope of this book.

1.1.9 Geographic Information System

The first stage of handling spatial data through computers occurred through Computer-Aided Design (CAD) techniques in the 1980s. Due to its ease of use and efficiency, it quickly replaced the old methods of technical drawing. Since then, the constant technological evolution, both in terms of computer systems and surveying instruments, has allowed the development of numerous applications in a CAD environment, which have automated everything from data acquisition to the final mapping edition, making CAD the primary technical drawing tool used in civil engineering for a long time.

Although CAD is still widely used in Civil Engineering, it is gradually being replaced by intelligent data management systems, such as Geographic Information Systems (GIS) and BIM.

In the specific case of Geographic Information Systems, their great advantage is that, in addition to establishing spatial relationships between the graphic elements of a drawing, they allow them to be related to alphanumeric information stored in a database, thus constituting an intelligent surveying map through which the user can obtain descriptive and geometric information to assist in their decision making. Therefore, as they are characterised as a management tool, Geographic Information Systems have become an essential part of Civil Engineering projects that require geographic information. Although important for Civil Engineering, this subject is not within the scope of this book.

1.1.10 Global Navigation Satellite System (GNSS)

The *Global Navigation Satellite System* (GNSS) refers to a constellation of satellites providing signals from space that allow for determining the geographic position, speed and time transfer of a GNSS antenna located at any point on the Earth's surface or close to it. Because of these characteristics, it has been used in the most diverse areas of human activity, ranging from simple pedestrian and vehicle navigation to the precise location of points in space. In Civil Engineering, it has been used with great success in topographic and geodetic surveys, setting-out works and special applications such as machine control and geodetic structural monitoring. Few Civil Engineering projects do not use GNSS technology at some stage of their development. A whole chapter of this book is therefore devoted to the main principles of this technology as applied to Civil Engineering.

1.1.11 Building Information Modelling (BIM)

Building Information Modelling (BIM) is a Civil Engineering project management model. It comprises a set of technologies, practices and collaborative processes for developing and managing projects based on geometric and semantic data of parametric graphical representation orientated to three-dimensional objects and progressively developed and integrated throughout the lifecycle of a construction project.

Although introduced in the 1980s, the concept of BIM has only gained prominence in recent decades due to the technological evolution of design tools and the need for digital transformation in the Civil Engineering construction industry. It aims to facilitate the coordination and compatibility of disciplines, anticipating the problem-solving for the stages where the ability to influence quality and cost is greatest and the financial impact of changes is least. These features improve information management and reduce waste and rework throughout the construction lifecycle.

As an integration and management tool, BIM technology allows approaches to n-dimensions by incorporating factors such as time (allowing to portray the state of construction at different times), cost and others, depending on the project's objectives. Due to its managerial characteristics, it has been systematically adopted and institutionalised by public and private entities worldwide. To help readers understand the basic concepts of and their integration with geomatics, a chapter in this book is dedicated exclusively to the subject.

1.2 Geomatics Engineering Assignments for Civil Engineering

Applying Geomatics Engineering technologies to the Civil Engineering production workflow requires Civil Engineers to clearly understand Geomatics Engineering assignments. To facilitate this understanding, the following is a brief description of each of these assignments from the point of view of a Civil Engineer. Figure 1.1 shows a conjectural view of the Geomatics Engineering assignments in the production chain of a Civil Engineering project.

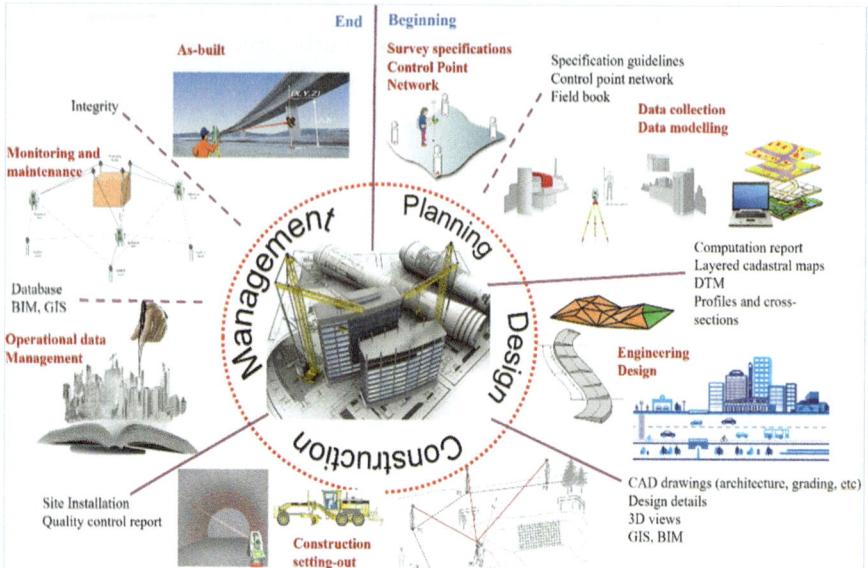

Fig. 1.1 Geomatics engineering assignments in the production chain of a Civil Engineering project. (Adapted from Silva 2020)

1.2.1 Planning

No significant civil engineering project can be developed without detailed information from the construction site. Furthermore, no civil engineering project can be executed without a network of geodetic control points. The first group of activities in this work phase includes the preparation of surveying and mapping specifications, database standardisation and measurement tolerances. In this case, the objective is to comply with international norms and standards to ensure that the same database can be used for different applications and data exchange in the same project. Failure to comply with these standards entails reworking risks, data loss and wasted time in manual data formatting. Preventing this from happening is the responsibility of project managers, who should involve both Geomatics and Civil Engineers.

The second group of activities includes field surveying planning and data collection. Surveying planning involves a detailed study of the surveying area and selecting surveying instruments and methods. Spatial data collection, however, means the execution and handling of field data collection. A critical point to highlight in this phase of activities is that raw data can be collected through different types of sensors in the same project, and each sensor presents its data typology and format. It is therefore essential to consider using a data management system that would adequately manage these different data, enabling the user to retrieve and apply them conveniently, at any time, easily and quickly.

Once information has been collected from the field, it must be modelled and referenced using mathematical models and appropriate coordinate systems to produce georeferenced geospatial data.

Data modelling is performed by coordinate geometry calculations depending on the measurement method used in the field. The primary data modelling is based on plane surveying because it is widely used in engineering and surveying practice. Based on the fundamentals of measuring distance, angle, direction and elevation or directly from GNSS positioning or scanning data, data modelling will provide point location, geospatial data dimensioning and orientation throughout azimuths, slopes, areas and volumes.

Another point to be considered in the data modelling concept is the increasing use of redundant measurements. This leads to using adjustment techniques to resolve the inconsistency between the observations and the model. It is therefore recommended that Civil Engineers have at least a basic understanding of adjustment computation to avoid misuse.

Data modelling also involves a data referencing process that links the modelled data to geodetic reference systems. This can be done using map projections or local ground-based plane coordinates. In both cases, it is essential to understand that distortion will result from this geometric assumption. Failure to understand or ignore its effects can be problematic for the development of many engineering projects.

It is important to note that to use the measurement technologies currently available properly, Civil Engineers must be aware of the characteristics of the use and applications of each instrument, its accuracy and its limitations. It is also essential to ensure that the instruments selected are correctly calibrated and in good working order and that the specified procedures are rigorously followed.

The Geomatics products provided at this stage include specification guidelines, horizontal and vertical control point networks, measurement field books, calculation reports and layered cadastral plans, including digital orthophotos, landmarks, point meshes, terrain profiles, cross-sections, Digital Terrain Modelling (DTM) or Digital Elevation Modelling (DEM) and levelling benchmarks. It can also include quantities for inventory, database management, economic assessment and cost accounting.

1.2.2   Design

Upon completion of the planning phase, engineers should have at their disposal a complete set of data needed to develop engineering projects. This typically includes CAD drawings of surveying measurements on specified map projections and digital terrain models, complemented by specific Geomatics products described in the planning phase. As a result, Civil Engineers will gather all available information from a variety of sources to develop and manage infrastructure projects, including architectural and construction designs, parcelling, intermediate construction, 3D visualisation, grading, earth movement, volume computation, mass haul plans, machine control guidance and machine control automation, environmental mitigation plans and many others, depending on the type of project and its lifetime. GIS and Building Information Modelling (BIM) methodologies are also initiated at this stage. At this stage the greater the interaction between Geomatics and Civil Engineering, the greater the success of the project.

1.2.3 Construction

During the construction phase, the amount of information needs to be increased considerably, and the topographic information must be fine enough to follow the construction progress on a daily basis. Real-time spatial data measurement and automated procedures are required to manage the difference between the daily construction progress and the scheduled value. At the same time, setting-out processes are implemented to enable stakes, templates or other markers to be set out to control the construction work and ensure that each element of the work is built in the correct position and level.

The setting-out operations are generally performed using total stations or GNSS instruments operating in Real-Time Kinematic (RTK) mode. These instruments already have internal application programs for the work to be implemented, which indicate the direction and distance to be measured for positioning points on the ground, as described in detail in *Chap. 17—Global Navigation Satellite System— GNSS.*

However, earthworks tend to use machine control systems that allow dimensional control of earthworks movements in Civil Engineering construction areas through stakeout techniques based on machinery automation systems, as described in *Chap. 21—Setting Out.*

The execution and certification of quality control and production reports are critical tasks performed at this project stage.

1.2.4 Management

As shown in the previous sections, as in many disciplines, Geomatics and Civil Engineering professionals face the challenge of big data management and the extensive use of engineering software requiring high rates of speed, variety, variability and complexity in database management and analysis of the results. In this context, it becomes necessary to standardise this kind of dataset's storage structures and data management to provide mass storage and easy access to the users, mainly in cloud services. The complexity of the problem is even more remarkable when the geospatial data is connected to already deployed engineering structures on the ground, connecting the geospatial databases with engineering design databases. The solution to these cases has been using GIS structures for Civil Engineering projects and BIM methodology for building data management. Real-time measurement sensors, digital cameras, Unmanned Aerial Vehicles (UAV), and web-based data transfer are technologies currently available to ensure the quality and effectiveness of construction and structural monitoring. Needless to say, meta-data, coding and standardisation of the communication protocol are essential for integrating information at this stage.

Finally, the completion of the work is mapped by an as-built survey to represent the construction's status quo. The purpose of an as-built survey is to prove to the project owner or government authorities that the designed construction details have been carried out in accordance with the specifications established during the planning phase and shown in the work plan, or eventually to show precisely what has been completed to date or modified during the operational work. Considering this, as-built work is carried out in the same way as any surveying mapping work. Given the amount of data to be measured, the level of automation of the measurement and data management is critical to enable easy connectivity between field books and processing/mapping software. For this application, it is important to note that GNSS and terrestrial scanning instruments are also of interest.

Finally, depending on the type and size of the project, it may be necessary to check the safety of the works or to monitor their structural behaviour once they have been completed. Several monitoring methods are available for this quality control process, and geodetic monitoring is one of the most important, depending on the type of project. The ultimate goal of such monitoring is to periodically determine the spatial coordinates of specific points on the structure, from which structural deformations or displacements are calculated. For this purpose, current geodetic monitoring systems operate through Robotic Total Stations, GNSS instruments working in static or RTK measurement mode and, more recently, laser scanning instruments.

1.3   The Future of Geomatics Engineering

In the foreseeable future, the field of Geomatics Engineering will continue to undergo significant changes due to technological developments in digital image sensors, inertial measurement systems, laser sensing, artificial intelligence, computer vision, Big Data, Machine Learning and Database Management, augmented reality, as well as other related technologies. These ongoing and other emerging technologies are expected to change and expand the scope of applications of Geomatics Engineering and impact the synergy between Geomatics and Civil Engineering. On the other hand, automation and 3D-based projects will be increasingly incorporated into the operational flow of Civil Engineering works, which indicates the need for civil and Geomatics Engineers to keep up to date with these new technologies and their advances.

1.4 Important Institutions and Organisations for Geomatics

Due to its scientific scope and the diversity of its applications, there are several international institutions and organisations seeking to rationalise the development and use of the various disciplines involved with geomatics. They provide a wide range of information through websites, including mathematical models, computerised applications, images, maps, literature, discussion forums, etc. To guide the reader in their search on the Internet, the following is a list of websites of important institutions and organisations (governmental and civil) for Geomatics.

- www.fig.net—Fédération Internationale des Géomètres (FIG)
- www.aagsmo.org—The American Association for Geodetic Surveying (AAGS)
- www.navcen.uscg.gov—U.S. Coast Guard Navigation Center (NAVCEN)
- www.asprs.org—American Society for Photogrammetry and Remote Sensing (ASPRS)
- www.asce.org—American Society of Civil Engineers (ASCE)
- www.isprs.org—International Society for Photogrammetry and Remote Sensing (ISPRS)
- www.ordnancesurvey.co.uk—The Ordnance Survey
- www.ngs.noaa.gov—National Geodetic Survey (NGS)
- www.iers.org—International—Earth Rotation and Reference Systems Service
- itrf.ign.fr—International Terrestrial Reference Frame
- www.igs.org—International GNSS Service
- www.jpl.nasa.gov—Jet Propulsor Laboratory—NASA/USA
- www.iso.org/committee/53732/x/catalogue—ISO/TC 172/SC 6 Secrétariat
- www.din.de/en—DIN standards
- http://www.transportation.org—American Association of State Highway and Transportation Officials (AASHTO)
- http://europa.eu.int/comm/dgs/energy_transport/galileo/index_en.htm—Galileo
- http://www.glonass-iac.ru—GLONASS home page

1.5 Review Questions

1. Explain what Engineering Surveying is.
2. Explain what Geomatics is.
3. Explain the difference between plane and geodetic surveys.
4. Explain what are Photogrammetry and Remote Sensing and their applications in Civil Engineering.
5. Explain the difference between GIS and BIM.
6. Explain the difference between topographic and cadastral surveys.
7. Explain the difference between GPS and GNSS.
8. Explain why Geomatics is an essential study area for Civil Engineering.
9. Explain the objective of as-built surveying for Civil Engineering.

References

Bill R, Blankenbach J, Breunig M, et al (2022). *Geospatial Information Research: State of the Art, Case Studies and Future Perspectives*. PFG – J Photogramm Remote Sens Geoinf Sci 90:349–389.

Gagnon, P., Coleman, D. J. (1990). *La géomatique-une approche systématique intégrée pour répondre aux besoins d'information sur le territoire*. CISM Journal ASCGC. Vol 44, N° 4.

Ghosh, J.K., Silva, I. (2020). *Applications of Geomatics in Civil Engineering*. Select Proceedings of ICGCE 2018. Springer Nature Singapore.

Lam, S. Y. W., Tang, H. W. C. (2002). *Role of surveyor under ISO 9000 in the construction industry*. Journal of Surveying Engineering. USA.

Silva, I. (2020). *Geomatics Applied to Civil Engineering State of the Art*. In: Ghosh, J., da Silva, I. (eds) Applications of Geomatics in Civil Engineering. Lecture Notes in Civil Engineering, vol 33. Springer, Singapore. https://doi.org/10.1007/978-981-13-7067-0_2.

Chapter 2
Basic Concepts of Geomatics

Irineu da Silva and Paulo C. L. Segantine

2.1 Introduction

As described in the first chapter of this book, Geomatics should be understood as the discipline that groups mathematical concepts and methodological and technological procedures that can be used to manage spatial data. For Civil Engineers, this means understanding the concepts and techniques for collecting field data using surveying instruments and determining geospatial data through the algebraic relationship between the measured values of quantities. In this context, it is necessary to understand the basic concepts underlying such procedures before starting any measurement or data modelling process. Therefore, this chapter presents the basic mathematical concepts that the readers should know when collecting geospatial data for their Civil Engineering projects.

2.2 Measurement

Collecting data for geomatics purposes means carrying out a measurement, defined by the International Vocabulary of Metrology (VIM) as *experimentally obtaining one or more values that can reasonably be attributed to an individual quantity.* This process must be based on a *measurement procedure*, which is a detailed description of a measurement according to one or more *measurement principles* and a given *measurement method*, and on a *measurement model,* including any calculations to obtain a *measurement result.* For more details on these highlighted terms, readers are suggested to refer to the VIM3 version on the BIMP website.

I. da Silva (✉) · P. C. L. Segantine
São Carlos School of Engineering, University of São Paulo, São Carlos, SP, Brazil
e-mail: irineu@sc.usp.br; pclsegantine@usp.br

© The Author(s), under exclusive license to Springer Nature Switzerland AG 2025
I. da Silva, P. C. L. Segantine, *Geomatics Applied to Civil Engineering*,
https://doi.org/10.1007/978-3-031-75737-2_2

The term *quantity*, in turn, should be understood as the property of a phenomenon, body or substance, where the property has a magnitude that can be expressed as a number and a reference, generically called the *value of a quantity*. The *measured value of a quantity* is expressed as the ratio of the magnitude of the property being measured to a fundamental scalar quantity of the same kind adopted by convention and called the *unit of measurement*. The individual entity intended to be measured is called a *measurand*.

In addition to the above, it is also crucial that professionals involved with Geomatics are familiar with other essential definitions, such as those described below:

- *Measuring instrument*: a device used for carrying out measurements, alone or in conjunction with one or more supplementary devices. It is referred to as *measuring equipment* or *measuring device* when used alone. When composing a set of measuring instruments, it is referred to as a *measuring system*.
- *Calibration*: a set of operations to determine systematic errors in a measuring instrument by comparing its measurements with markings or measurements made with a working standard, that is, another measuring device found to be correct and which has itself been calibrated. The calibration result is usually recorded in a specific *calibration certificate report*. This certificate must provide information on the metrological performance of the measuring instrument and clearly describe the procedures performed.
- *Verification (or check)*: simplified periodic comparison of the quality of a *measuring instrument* against a standard to verify that there are no changes in the measurement results compared with the specified tolerances or the results of the last calibrations. In summary, verification can be understood as the set of actions performed to confirm that the characteristics related to the performance or technical requirements of an instrument (usually specified by the manufacturer) are met.
- *Instrument adjustment*: a set of operations carried out to bring a *measuring instrument* to a state of a quality adequate to perform the measurements for which it was designed. In practice, it is carried out after a first calibration to determine the approximate adjustment amount required, followed by a second calibration to certify its performance.

2.3 Unit of Measurement

According to the VIM, a unit of measurement is the *fundamental scalar individual quantity with which any other quantity of the same kind can be compared by ratio, resulting in a number*. It is an abstract concept used to express the unit value of a quantity, which is independent of physical conditions and to which name and symbol are conventionally assigned. The set of measurement units of different classes, grouped to allow their rational use, is called a *System of Units*.

The units of measurement used in most countries are those specified by the *International System of Units*, universally abbreviated as SI,[1] which was adopted by the 11th General Conference on Weights and Measures (CGPM), held in Paris in 1960.

The SI is scientifically composed of seven base units chosen so that one can, in principle, measure all the physical quantities known today. Derived units, in turn, are formed from a combination of base units without needing any conversion factor. In summary, the system is based on three classes of units:

1. *Base units*, which are seven well-defined units and considered dimensionally independent: *the metre, the kilogram, the second, the ampere, the kelvin, the mole* and *the candela.*
2. *Derived units*, which can be formed by combining the base units using the algebraic relationships between the corresponding quantities. These units are given special names and symbols.
3. *Supplementary units*, which include some special units.

The main SI units of measurement used in geomatics are described below.

2.3.1 Unit of Length Measurement

Geomatics uses the SI base unit called *metre* (symbol m) as a unit of measurement for length, whose multiples and submultiples most commonly used in geomatics and Civil Engineering, are listed in Table 2.1.

According to the seventh General Conference for Weights and Measures (CGPM), on 20 October 1983, the *metre* is defined as follows:

One metre is the length of the path travelled by light in a vacuum during a time interval of 1/299,792,458 s.

Table 2.1 Multiples and submultiples of the metre

Dimensional relationships	Prefix—Name	Symbol
$1000 \text{ m} = 10^3 \text{ m}$	kilo—kilometre	km
$100 \text{ m} = 10^2 \text{ m}$	hecto—hectometre	hm
$10 \text{ m} = 10^1 \text{ m}$	deca—decametre	dam
$0.1 \text{ m} = 10^{-1} \text{ m}$	deci—decimetre	dm
$0.01 \text{ m} = 10^{-2} \text{ m}$	centi—centimetre	cm
$0.001 \text{ m} = 10^{-3} \text{ m}$	milli—millimetre	mm
$0.001 \text{ mm} = 10^{-6} \text{ m}$	micro—micron	μm
$0.001 \text{ μm} = 10^{-9} \text{ m}$	nano—nanometre	nm
$0.001 \text{ μm} = 10^{-12} \text{ m}$	pico—picometer	pm

[1]For more details on the SI, consult the page of *International Bureau of Weights and Measures (BIPM)*. https://www.bipm.org/en/measurement-units

2.3.2 Units of Angular Measurement

In the case of angular measurements in the plane, SI defines three units: the *radian*, the *degree* and the *gon*.

The *radian* (symbol rad) is a dimensionless angular unit defined as follows:

> One *radian* is the angle subtended at the centre of a circle by an arc that is equal in length to the radius.

The algebraic relationship between the length of an arc and the radius of a circle is given by Eq. (2.1).

$$s = d * \alpha \tag{2.1}$$

where

s = arc length
d = radius of the circle
α = central angle of the arc, in radians

The conversion from radians to other angular units is based on the constant *pi* (symbol π), which defines the ratio of the circumference of a circle to its diameter. As shown in Fig. 2.1, a complete circle has an angle equivalent to 2π rad, approximately 6.283185307 rad, which allows calculating the arc value of a given angle in any angular unit.

Fig. 2.1 Representation of the four quadrants of the radian angular unit on the circle

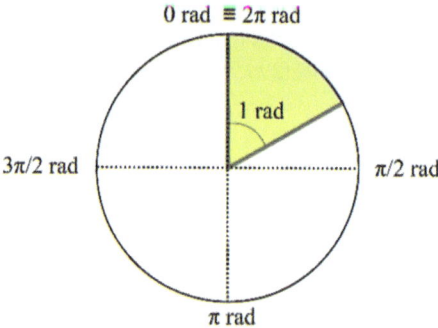

The value of *pi* is an irrational number with an infinite sequence. For calculations in Geomatics, however, its value with nine decimal places is accepted, that is, $1\pi = 3.141592654$.

The *degree* is a *sexagesimal* angular unit in which the circle is divided into 360 equal parts, as shown in Fig. 2.2. Each part corresponds to an angle of $1°$ (one degree). Each degree is subdivided into 60 equal parts, each part corresponding to an angle of 1′ (one minute). Each minute is subdivided into 60 equal parts, each part corresponding to an angle of 1″ (one second). So, by definition:

> 1 *degree* is equal to the central angle that intersects, on a circle, an arc of length equal to 1/360 of the circle's perimeter.

Fig. 2.2 Representation of the four quadrants of the sexagesimal angular unit on the circle

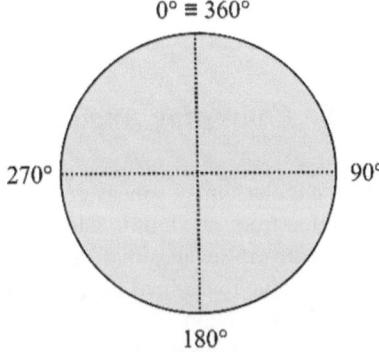

For the *sexagesimal degree*, the degrees, minutes and seconds are indicated by the degree symbol, an acute accent and quotation marks, all placed to the upper right of the corresponding number. For example, 135° 26′ 42″ reads as 135°, 26 min and 42 s.

An alternative to using the sexagesimal degree is its decimal form, wherein minutes and seconds are given as decimal fractions, as described in Sect. 2.3.3.

The *gon*[2] is a decimal angular unit defined as follows:

> 1 *gon* is equal to the central angle that intersects, on a circle, an arc of length equal to 1/400 of the circle's perimeter.

In this case, the circle is divided into 400 equal parts, as shown in Fig. 2.3, where each part is equal to an angle of 1^g (one *gon*). Each gon is subdivided into 100 equal parts, each part corresponding to an angle of 1 *centigon*. Each centigon is subdivided into 100 equal parts, each part corresponding to an angle of 1 *miligon*. Thus, 135.6342^g corresponds to 135 *gons* (135^g), 63 *centigons* (63^c) and 42 *miligons* (42^{cc}).

---

[2] Also known as gradian, grad or grade.

Fig. 2.3 Representation of
the four quadrants of the
angular unit gon on the
circle

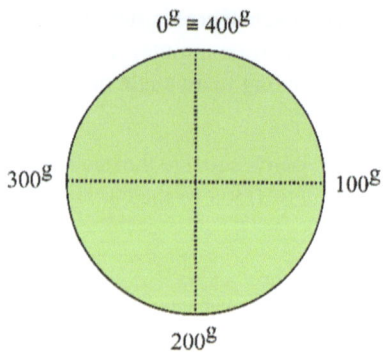

2.3.3   Converting Angular Units

Since angular units of measurement are related to the circle, angular values can be converted from one unit to another, as shown in Table 2.2.

To convert sexagesimal degrees to other units, it is preferable to convert the original value to decimal degrees first, as shown in the sequence, and then carry out the final conversion.

To comply with the designations found in scientific calculators, decimal degrees are indicated as $dd.dddddddd$ and sexagesimal degrees as $dd\ mm\ ss.ss$. Thus,

$$dd.dddddddd = dd + mm/60 + ss.ss/3600$$

For the reverse case, it follows:

$$dd = \text{integer of } dd.dddddddd$$

$$mm = \text{integer of } [(dd.dddddddd{-}dd) * 60]$$

$$ss.ss = dd.dddddddd * 3600 - dd * 3600 - mm * 60$$

Table 2.2 Conversion of
angles

Sexagesimal degree	to	Decimal degree
Decimal degree		Sexagesimal degree
Degree		Gon
Gon		Degree
Radian		Degree
Degree		Radian
Radian		Gon
Gon		Radian

Conversions between the other angular units can be made using the "rule of three" principle, which gives the algebraic relationships shown below:

1. To convert *degrees* to *gons*, one must divide the angle in degrees by 0.9.
2. To convert *gons* to *degrees*, one must multiply the angle in gons by 0.9.
3. To convert *radians* to *degrees* or *gons*, consider the algebraic relationships in Table 2.3.

Table 2.3 Converting radians to degrees or gons

2π rad $= 360°$	2π rad $= 400^g$
$1\,\text{rad} = \frac{180°}{\pi} = 57°\,17'\,44.8062''$	$1\,\text{rad} = \frac{200^g}{\pi} = 63.661977237^g$

Usually, the angle 1 rad, expressed in another angular unit, is designated by the Greek letter ρ *(rho)*, as shown in Table 2.4.

Table 2.4 Corresponding values of radian units to other angular units

1 rad =				1 rad $= 63.661977237^g = \rho^g$
	$57.295779513°$	$=$	$\rho°$	
	$3437.7467708'$	$=$	ρ'	
	$206264.8062''$	$=$	ρ''	

To convert an angle in *radians* to *degrees* or *gons*, it is necessary to multiply it by the corresponding value of ρ. To convert an angle in *degrees* or *gons* to the *radian* unit, it is necessary to divide it by the corresponding value of ρ.

Notes
1. To avoid errors, it is recommended to use leading zeros to fill the decimal spaces without units in the minutes and seconds values when specifying sexagesimal degrees, i.e., $4°\,00'\,07.1234''$.
2. When expressing latitudes and longitudes in sexagesimal degrees, four decimal places must be considered for the seconds because, as seen in Example 2.2, on the Earth's surface, a central angle of $1''$ corresponds to an arc length of about 31 m. An arc of $0.0001''$, therefore, corresponds to 0.3 mm.

Example 2.1
Calculate the length s of an arc intersected by an angle α at a distance d for different values of d and α, as shown in Fig. 2.4.

Fig. 2.4 Geometric representation of the arc

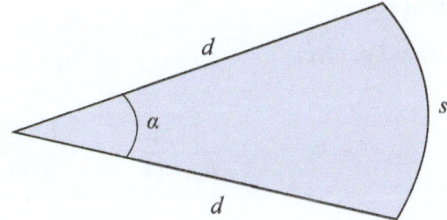

Solution:

Considering Eq. 2.1 and the values indicated for α and d:

$$\alpha = 1° \quad d = 10.000 \text{ m} \quad s = d * \frac{\alpha°}{\rho°} = 10.000 * \frac{1°}{57.295779513°} = 0.175 \text{ m}$$

$$\alpha = 1' \quad d = 100.000 \text{ m} \quad s = d * \frac{\alpha'}{\rho'} = 100.000 * \frac{1'}{3437.7467708'} = 2.9 \text{ cm}$$

$$\alpha = 1'' \quad d = 1000.000 \text{ m} \quad s = d * \frac{\alpha''}{\rho''} = 1000.000 * \frac{1''}{206264.80622''} \cong 5 \text{ mm}$$

According to the results of Example 2.1, the following approximate values are obtained:

- 1 s of arc (1") intercepts 5 mm of arc at 1 km distance
- 1 s of arc (1") intercepts 1 mm of arc at 206.265 m distance
- 1 min of arc (1') intercepts 3 cm of arc at 103.132 m distance
- 1 min of arc (1') intercepts 1 cm of arc at a 34.377 m distance.

Example 2.2
Calculate the angle at the centre of the Earth, in seconds of arc, that intersects, on the surface, an arc of a circle 31 m in length at a place where the mean radius R_0 of the Earth, considered as a sphere, is equal to 6371 km.

Solution:

Using Eq. (2.1).

$$\alpha'' = \frac{s[\text{m}]}{R_0[\text{m}]} * \rho'' = \frac{31}{6371.000} * 206264.8062'' = 1''$$

Example 2.3
Calculate the angle at the centre of the Earth, in minutes of arc, that intersects, on the surface, an arc of a circle 1.852 km long at a place where the mean radius R_0 of the Earth, considered as a sphere, is equal to 6371 km.

Solution:

Using Eq. (2.1).

$$\alpha' = \frac{s[\text{km}]}{R_0[\text{km}]} * \rho' = \frac{1.852}{6371} * 3437.7467708 = 1'$$

The value of 1.852 km corresponds to a nautical mile.

Example 2.4

Convert 65° 12′46″ to decimal degrees.

Solution:

According to the conversion rule given in Sect. 2.3.3.

$$65°\,12'\,46'' = 65 + \left(\frac{12}{60}\right) + \left(\frac{46}{3600}\right) = 65.2128°$$

Example 2.5

Convert 45.3215° to sexagesimal degrees.

Solution:

According to the sequence of calculations given in Sect. 2.3.3.

$$dd = integer\ of\ 45.3215° = 45°$$

$$mm = integer\ of\ [(45.3215 - 45) * 60] = 19'$$

$$ss.ss = 45.3215 * 3600 - 45 * 3600 - 19 * 60 = 17.4''$$

$$Therefore,\ \ 45.3215° = 45°\ 19'17.4''$$

Example 2.6

Convert 32° 22′ 30″ to *gons.*

Solution:

For this conversion, it is necessary to first convert the sexagesimal angular unit to the decimal angular unit and then convert it to gon.

$$32°\ \ 22'30'' = 32 + \left(\frac{22}{60}\right) + \left(\frac{30}{3600}\right) = 32.3750° \rightarrow \alpha^g = \frac{32.3750°}{0.9} = 35.9722^g$$

Example 2.7

Convert 103.6368^g to *sexagesimal degrees.*

Solution:

For this conversion, it is necessary first to convert the angle given in gon into decimal degrees.

$$103.6368^g * 0.9 = 93.2731°$$

Then, the decimal degrees are converted to sexagesimal degrees, as shown in Example 2.5.

$$103.6368^g = 93.2731^\circ = 93^\circ\,16'\,23.2''$$

Example 2.8

Convert $127^\circ\,18'\,54''$ to radians.

Solution:

For this conversion, it is necessary first to convert the sexagesimal degrees into decimal degrees and then into radians.

$$127^\circ\,18'\,54'' = 127^\circ + \frac{18'}{60} + \frac{54''}{3,600} = 127.3150^\circ$$

$$\frac{127.3150^\circ}{57.295779513^\circ} = 2.222065937 \ \text{rad} = 0.707305556\pi\,\text{rad}$$

Example 2.9

Convert 2.7535481 rad to *sexagesimal degrees*.

Solution:

For this conversion, it is necessary first to convert the radians into decimal degrees and then into sexagesimal degrees.

$$2.7535481 * 57.295779513^\circ = 157.7666848^\circ$$

To convert decimal to sexagesimal degrees, it is necessary to follow the sequence of calculations given in Sect. 2.3.3, *resulting in* 2.7535481 rad $= 157.7666848^\circ = 157^\circ46'01''$.

2.3.4 Units of Area

For area measurement, i.e., the extent of a surface or an appropriately defined part of it, the SI-derived units adopted in geomatics are given in Table 2.5.

The units *are* and *hectare* are not part of the SI, but their use as special units is permitted.

Table 2.5 Units of area

1 centiare (*ca*)	$1\,\text{m}^2$
1 are (*a*)	$100\ \text{m}^2$
1 hectare (*ha*)	$10,000\ \text{m}^2$

2.3.5   Units of Volume

According to the SI, the cubic metre (m^3) is the derived unit of volume measurement. Its submultiples are the cubic decimetres (dm^3) and the cubic centimetres (cm^3). In Geomatics, it is the unit of measure used in engineering projects to calculate earthwork volume, the volume of reservoirs, the volume of concrete in Civil Engineering and others.

In some countries, the unit *litre* and its submultiples are also used. They are not units of volume but capacity and are not part of the SI. Generally, they indicate the amount of liquid held by a container. By definition, 1 litre is the amount of liquid required to fill a cube with a volume of 1 dm^3.

2.4   Important Geometric Figures in Geomatics

Important geometric figures in Geomatics are those defined by Euclidean geometry, such as points, lines, polygons, planes, cubes, spheres, etc. Generally, they have some dimension, i.e., they define a space. However, they can be dimensionless as well. Below is a brief description of the most important ones.

2.4.1   Point

The point is an essential geometric primitive for Geomatics. It is defined as a dimensionless geometric figure that occupies a unique location in space, represented by a set of coordinates related to a predefined coordinate system. The point is, therefore, not an object but a location. In practice, it is only possible to define a point with a dimension because it must be visual. Thus, physical marks on stakes, whose position on Earth has been determined by surveying, temporary pins used as a reference for angular and linear measurements, and others are practical examples of materialised points in Geomatics. The same can be said of a point on a map corresponding to an identifiable element in the real world. In a database, however, a point is represented by an identifier and its coordinates.

2.4.2   Line

A line is a one-dimensional geometric figure physically defined as a series of points. It has no thickness and can be infinite or have a specific length. It can be straight or curved. In Geomatics, it is used to represent the edges of objects or geographic elements of linear features, such as centrelines, watercourses, etc. In GIS, it is often referred to as an arc. In a database, it is characterised by an identifier and the coordinates of its parameters.

2.4.3 Polygon

A polygon is a closed geometric figure formed by straight line segments whose start and end points coincide. In Geomatics, it is used to represent geographic objects that comprise an area, such as a parcel of land, a building, etc.

2.4.4 Straight Line

A straight line is an endless one-dimensional figure that has no width. From it derives the term *semi-straight line*, which is a straight line with a beginning but without an end, and *line segment*, which is the part of a straight line between two points. Algebraically, it is defined by the linear Eq. (2.2) in slope-intercept form.

$$y = ax + b \qquad (2.2)$$

where

x, y = coordinates of any point on the line
a = slope of the line
b = y-intercept

2.4.5 Plane

A plane is geometrically a two-dimensional surface on which a line segment connecting any two points on the surface lies entirely on that surface. Of the infinite planes in Euclidean space, the horizontal and the vertical are of particular interest in Geomatics because they are the projection planes used in engineering projects. A horizontal plane at a given point is perpendicular to the direction of gravity at that point. On the other hand, a vertical plane is any plane passing through a point on the Earth and containing the zenith (and nadir) of that point. Algebraically, it is defined by the parametric Eq. (2.3).

$$z = a_0 + a_1 x + a_2 y \qquad (2.3)$$

where

$x, y, z =$ Three-dimensional Cartesian coordinates
$a_i =$ coefficients of the equation

According to Eq. (2.3), three non-colinear points define a plane, i.e., they allow determining the values of the coefficients a_i.

2.4.6 Ellipse

An ellipse is a two-dimensional figure formed by a set of points whose sum of the distances (r_1 and r_2) between any of them and the two focal points (F_1) and (F_2) of the ellipse is equal to $2a$. See the illustration in Fig. 2.5. Algebraically, it is defined by Eq. (2.4).

Fig. 2.5 Ellipse

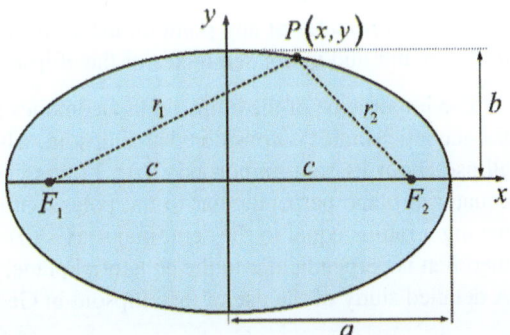

$$\frac{x^2}{a^2} + \frac{y^2}{b^2} = 1 \qquad (2.4)$$

where

$x, y =$ coordinates of any point on the ellipse
a and $b =$ sizes of the semi-axes of the ellipse on the coordinate axes (x) and (y), respectively, where a is the semi-major axis, and b is the semi-minor axis

2.4.7 Ellipsoid of Revolution

An ellipsoid of revolution or ellipsoid of rotation is a three-dimensional figure formed by rotating an ellipse about one of its semi-axes (a, b), as shown in Fig. 2.6. Algebraically, it is defined by Eq. (2.5).

Fig. 2.6 Ellipsoid

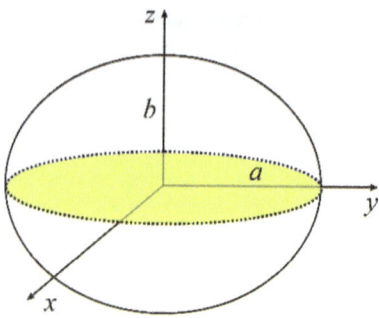

$$\frac{x^2}{a^2} + \frac{y^2}{a^2} + \frac{z^2}{b^2} = 1 \tag{2.5}$$

where

x, y, z = coordinates of any point on the ellipsoid
a, b = major and minor semi-axes of the ellipse

The importance of the ellipsoid in Geomatics is that the Earth's geometric shape for geodetic studies is considered an ellipsoid, which is formed by the rotation of the ellipse about its semi-minor axis (b).[3] In this case, the figure of the Earth has the equatorial plane perpendicular to the poles, with the circumference of the equator having a radius equal to the semi-major axis (a) of the ellipse. The section of each meridian is perpendicular to the equatorial plane, as defined in the previous section. A detailed study of the use of the ellipsoid in Geomatics is presented in Sect. 4.2.4.

2.4.8   Sphere

A sphere is a uniformly curved Three-dimensional figure formed by a set of points equidistant from its centre. See the illustration in Fig. 2.7. Algebraically, it is defined by Eq. (2.6).

$$(x-a)^2 + (y-b)^2 + (z-c)^2 = R^2 \tag{2.6}$$

where

x, y, z = coordinates of any point of the sphere
a, b, c = coordinates of the centre of the sphere
R = radius of the sphere

---

[3] Because the ellipsoid of revolution is very close to a sphere in its representation of the shape of the earth, it may also be called a spheroid.

Fig. 2.7 Sphere

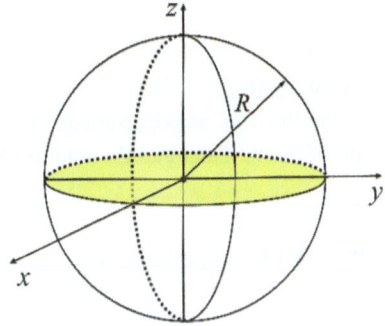

The sphere is essential for geomatics because the representation of the geometric shape of the Earth, in some cases, is considered a sphere, as presented in Sect. 4.2.5.

2.5 Map Scale

Upon completion of field data collection and subsequent data modelling, Civil Engineers often need to present measured geospatial data in printed form. It is, therefore, necessary to use the map scale concept.

A map scale m is defined as the factor by which a distance i obtained from a map by calculation or measurement is multiplied to give the actual distance o at the map datum, as given by Eq. (2.7).

$$m = \frac{i}{o} \tag{2.7}$$

For example, the ratio of one to thousand is expressed as 1:1000 or 1/1000. In other words, it is represented by a fraction of the type $1/M$, where M is called the *scale modulus*. A map scale is considered small when the value of M is large and vice versa. So, comparatively, the map scale of 1:50,000 is small, and a map scale of 1:1000 is large.

Using Eq. (2.7).

$$m = \frac{1}{M} = \frac{i}{o} \tag{2.8}$$

From where

$$o = i * M \tag{2.9}$$

Equation (2.9) allows calculating the actual size of a mapped object by knowing the scale modulus of this map and the size of the object it represents.

Example 2.10

The length of an airport runway was measured on a 1/100,000 scale map, given a value of 3 cm. Calculate the length of this runway on the ground.

Solution:

Using Eq. (2.9).

$$o = 3 * 100{,}000 = 300{,}000 \text{ cm} = 3000.00 \text{ m}$$

Example 2.11

Consider a distance of 453.279 metres measured on the ground. Calculate its length on the drawing if it is to be drawn at a scale of 1/2000.

Solution:

Using Eq. (2.9).

$$i = \frac{453.279}{2000} = 0.227 \text{ m} = 22.7 \text{ cm}$$

2.6   Significant Figures

As already described, measuring is a process that may benefit from redundant observations, resulting in different values for the measurand, suggesting indecision about its true value. In this case, each measured value consists of a sequence of exact figures and a doubtful one. The sum of the number of exact figures plus one (the doubtful number) gives the number of significant figures in a number. For example, consider measuring a bar using a graduated ruler, as shown in Fig. 2.8.

In this figure, it is noted that the first reading is easily defined as being equal to 10.6 measurement units. The uncertainty is about the value of the second decimal, which, in this case, could be 4, 5 or 6. Thus, the first three digits are considered exact, and the fourth is an estimate but still a significant figure. The measured values, therefore, have four significant figures.

In this sense, significant figures are defined as the number of figures in a number necessary to give physical meaning to a measured value of a quantity.

Fig. 2.8 Definition of significant figures in a measurement

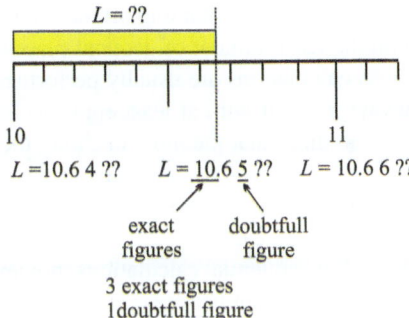

To find the number of significant figures in a number, the following practical rules can be followed:

1. Any number other than zero is a significant figure.
2. Zeros between non-zeros are always significant figures.
3. Zeros before non-zeros are never significant figures.
4. Trailing zeros to the right of non-zeros and located after the decimal point are significant figures.
5. In the scientific notation of the type $N*10^x$, the digits of N are significant, but the numbers 10 and (x) are not significant.
6. Zeros that complete numbers multiples of powers of 10 are ambiguous. The notation does not allow to say whether they are significant or not. The number 200, for example, can have one significant figure (2), two significant figures (20) or three significant figures (200). This ambiguity can be corrected by specifying the significant figures or using scientific notation. For example, for the value 200, the number of significant figures must be specified, or the result should be presented in scientific notation; i.e., $2*10^2$ has one significant figure, $2.0*10^2$ has two significant figures, and $2.00*10^2$ has three significant figures.

Table 2.6 presents examples of significant figures.

Table 2.6 Examples of significant figures

Rule	Measurement	Significant figures
1	124.386	6
2	102.004	6
3	0.000249	3
4	0.20	2
5	$3.102*10^4$	4
6	520.000	6

Obviously, significant figures cannot be created arbitrarily. Therefore, to ensure the compatibility of the number of significant figures in mathematical operations, the calculations must be carried out with the total number of digits involved in the numbers and then rounding off the final value of the operation to the number of digits with the least number of decimal places.

As calculations are usually performed using calculators or spreadsheets, there is always a result with at least eight decimal places. The less experienced technician tends to think that the more decimal, the more accurate the calculated value. This is not true. The results cannot have more decimals than the input data. For example, if the coordinates of two points are given in centimetres, the calculated distance between them cannot be given in millimetres. It would even be a lack of common sense. For sequential calculations, however, storing the results with at least one digit more than the number of significant figures is advisable to round off the value of the last calculation. Some rules to follow are given below to clarify this matter.

When calculating sums and subtractions, all significant figures of the added or subtracted values must be considered. The final result should be rounded off to the number of decimal places that are least in the numbers involved in the calculation.

Example 2.12

Calculate the operation result in the sequence and indicate the number of significant figures of the final value.

$$302.568 + 1.23 - 147.250 - 32.589 = ???$$

Solution:

According to the rule of addition and subtraction, the result of the mathematical operation equals 123.959. *Due to the least number of decimal places of the values involved, the correct answer is* 123.96 m.

For multiplication and division, the number of significant figures in the overall result must be the same as that in the value with the least number of significant figures. When multiplying or dividing by an exact constant, the number of significant figures is not affected.

Example 2.13

Calculate the volume of a cylindrical reservoir, given that its internal radius is 2.52 m, and its height is 34.214 m. Indicate how many significant figures the result of the operation has.

Solution:

The calculation volume is given as follows.

$$V = \pi * R2 * h = 3.141592653 * 2.522 * 34.214 = 682.582 \text{ m}^3$$

As explained, the final result must be rounded off to 682.58 m^3 *based on the number of significant figures of the number* 2.52.

In addition to being careful with the number of significant figures, it is also essential to analyse how rounding should be performed to present the result of the operation. The general rule is to round down the decimals between 1 and 4 and round up those between 6 and 9. The problem occurs when the decimal is the number 5. In this case, some professionals adopt the practice of rounding to the above, which can cause a systematic bias in the results. To prevent this, it is proposed to round up when the decimal before 5 is odd and round down when it is even. Therefore, using this criterion, both values 32.835 and 32.845 must be rounded to 32.84.

2.7   Standards and Regulations

Engineers must know that technical decisions must be guided by standards or technical instructions. Technical standards do not have the force of law but guarantee professional, technical, and legal support. In this sense, a series of relevant international technical standards for Geomatics are presented below.

- ASPRS Accuracy standards for large-scale maps. American Society for Photogrammetry and Remote Sensing, 1990.
- ASPRS Guidelines: Vertical Accuracy Reporting for Lidar Data. American Society for Photogrammetry and Remote Sensing, 2004.
- ASPRS Positional Accuracy Standards for Digital Geospatial Data. American Society for Photogrammetry and Remote Sensing 2014.
- ISO 19157:2013—Geographic Information—Data quality.
- ISO 19111:2019—Geographic Information—Referencing by Coordinates.
- ISO 9849:2017—Optics and optical instruments. Geodetic and surveying instruments. Vocabulary.
- ISO 7078:2020—Buildings and civil Engineering works—Procedures for setting out, measurement and surveying.
- FHWA. Geometric Design. US Federal Highway Administration. https://www.fhwa.dot.gov/programadmin/standards.cfm.
- USGS. NMAS: United States National Map Accuracy Standards. United States Geological Survey. Reston, 1947.
- USGS. Part 1: General, Standards for Digital Elevation Models. United States Geological Survey. Reston, 1997.
- USGS. Part 2: Specifications, Standards for Digital Elevation Models. United States Geological Survey. Reston, 1998.
- USGS. Part 3: Quality Control, Standards for Digital Elevation Models. United States Geological Survey. Reston, 1997.
- USGS Standards and Specifications. United States Geological Survey. https://www.usgs.gov/ngp-standards-and-specifications/standards-and-specifications.

2.8   Problems

1. Convert the angle 0.6° to sexagesimal minutes.
2. Convert the angle 0.016° to sexagesimal seconds.
3. Convert the angle 95.1245° to sexagesimal degrees.
4. Convert the angle 2.5 rad to decimal degrees.
5. Convert the angle 2.5 rad to sexagesimal minutes.
6. Convert the angle 2.5 rad to sexagesimal seconds.
7. Convert the angle 95.1245° to radians.
8. Convert the angle 49° 18′37″ to decimal degrees.
9. Convert the angle 5° 46′12″ to radians.
10. Convert the angle 125° 32′45″ to gons.
11. Convert the angle 66. 4941^g to sexagesimal degrees.
12. Calculate 35° 42′28″ + 57° 31′59″.
13. Calculate 28° 13′28″ - 25° 33′42″.
14. Calculate 13° 28′35″ * 2.7354.
15. Calculate 73° 32′10″/2.831.
16. Given the angles of a triangle measured by a theodolite, as shown in Table 2.7, check if there is an error of closure. If so, distribute it equally over the angles measured.

Table 2.7 Angle values

Vertice	Angle
1	42° 55′ 27″
2	65° 12′ 13″
3	71° 52′ 14″

17. Given the polygon shown in Fig. 2.9 and the values of the angles in Table 2.8, calculate the value of the angle α_4.

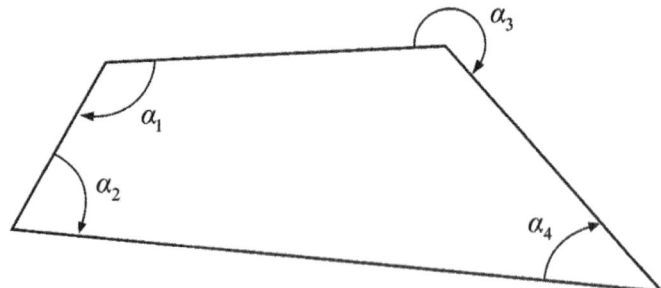

Fig. 2.9 Polygon geometry

Table 2.8 Angle values

Angle	Value
α_1	112° 35′ 33″
α_2	81° 52′ 20″
α_3	263° 22′ 11″

18. Using the geometric elements in Fig. 2.10 and the values in Table 2.9, calculate the value of the radius R_2.

Fig. 2.10 Geometric elements of the problem

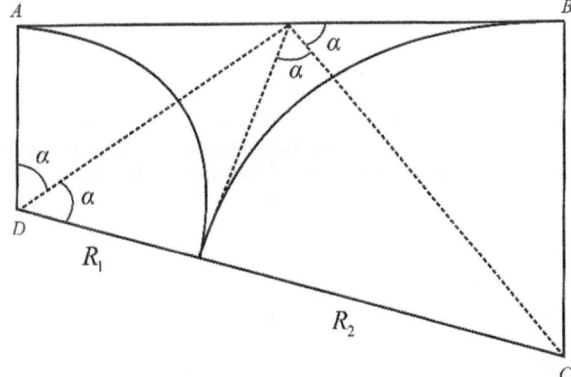

Table 2.9 Geometric elements

$d_{AB} = 1800.000$ m	$R_1 = 600.000$ m

19. Using the geometric elements in Fig. 2.11 and the values in Table 2.10, calculate the perimeter of the figure.

Fig. 2.11 Geometric elements of the problem

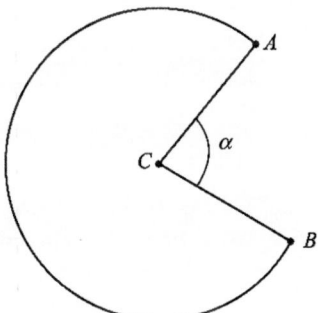

Table 2.10 Geometric elements

$d_{AC} = 12.000$ m	$\alpha = 60°\ 00′\ 00″$

20. Knowing that the distance between two points on a topographic map is 80 cm and that the scale of the map is 1:250, calculate the distance between the two points on the ground.
21. Which scale represents the ratio 1 cm = 500 m?
22. Which scale represents the ratio 10 cm = 200 m?

References

JCGM 200:2012. International vocabulary of metrology—*Basic and general concepts and associated terms (VIM)*. 3rd edition.

Savvaidis. 2009. *Standardization of equipment and techniques in surveying engineering and deformation measurements*. Laboratory of Geodesy and Geomatics, Department of Civil Engineering, Aristotle University of Thessaloniki, Greece.

Chapter 3
Introduction to Adjustment Computation

Irineu da Silva and Guilherme Poleszuk dos Santos Rosa

3.1 Introduction

Adjustment Computation, formerly known as the *Theory of Errors*, is an essential part of the study of Geomatics and, more recently, the qualification of Civil Engineers. This is partly because the mathematical modelling of physical phenomena is now very sophisticated, surveying instruments are highly automated and the results of surveying measurements are more reliable and accurate than ever before. For these reasons, Civil Engineering also requires more sophisticated data processing and statistically coherent results to assess the accuracy of measurements and to estimate confidence intervals of calculated values.

Statistically coherent means that the amount of data available must be greater than what is strictly necessary to obtain a unique solution. In addition, the errors made in each measurement are considered random to allow the application of an adjustment computation model, as presented later in this chapter.

A thorough study of the concepts involved in adjustment computation is beyond the scope of this book. Therefore, this chapter presents a synthesis of the theories and methods of adjustment computation that are essential for Civil Engineers to achieve the objectives mentioned above.

In this context, to assist readers in studying the theories presented, it is essential, first of all, that they understand two fundamental properties of a surveying measurement, which are:

I. da Silva (✉)
São Carlos School of Engineering, University of São Paulo, São Carlos, SP, Brazil
e-mail: irineu@sc.usp.br

G. Poleszuk dos Santos Rosa
Cartographic Engineer, Researcher at the Faculty of Science and Technology of the São Paulo State University (UNESP), São Paulo, Brazil

© The Author(s), under exclusive license to Springer Nature Switzerland AG 2025
I. da Silva, P. C. L. Segantine, *Geomatics Applied to Civil Engineering*,
https://doi.org/10.1007/978-3-031-75737-2_3

- Measuring always means performing a physical operation using a physical instrument; therefore, the resulting values will always have some uncertainties, called *observational errors.*[1]
- As described later in this chapter, measuring always refers to a theoretical abstraction of a physical situation or a set of real-world events represented by a mathematical model.

In summary, the adjustment calculation aims to determine the best possible value of the measured quantity and to estimate the precision of the results obtained.

3.2   Measurement Errors

In Geomatics, measurement errors can be classified in two ways: according to causes and according to the effects, as described in the following subsections.

3.2.1   *According to Causes*

Measurement errors can be classified in three ways according to causes, i.e., the environment in which the measurement is made, the measuring instrument and the observer.

Regarding the environment, measurement errors are related to natural causes such as unstable ground, varying weather conditions, effects of atmospheric refraction and the curvature of the Earth, among others.

Regarding the measuring instrument, measurement errors are related to instrumental errors, which vary from instrument to instrument. They usually result from mechanical errors, such as instrument axis and angular circle reading errors, or electronic errors, such as electronic distance measurement (EDM) or GNSS antenna phase centre offsets and variation.

Regarding the observer, the sources of measurement errors are related to personal skills, such as experience, fatigue, sighting inaccuracy, optical parallax effects, incorrect readings and others.

3.2.2   *According to the Effects*

According to the effects, measurement errors can be classified as *blunders, systematic and random errors* (or *accidental errors*). It is also possible to establish a fourth class of error, intermediate between the last two, called *correlated error.*

---

[1] Also referred to as measurement errors.

Blunders (*gross errors*, *mistakes*) are those caused by the observer's lack of attention or by malfunction of the measuring instrument. From a statistical point of view, these errors are considered outliers and should be rejected whenever detected. They can only be avoided by applying measurement procedures that allow them to be detected, such as redundant measurements from different stations or using different techniques, equipment performance verification, geometric checks to ensure consistency, redundancies to cross-verify results, etc.

Systematic errors occur when a critical physical phenomenon is neglected in the mathematical model, such as temperature effects, atmospheric refraction, the curvature of the Earth and others, or due to poor instrument calibration. Under identical conditions, they repeat with the same magnitude and the same algebraic sign, which means that they cannot be detected even with redundant data. To avoid or correct them, it is necessary to apply accurate mathematical models and to use calibrated instruments or other field measurement methods to measure the influencing conditions and to apply appropriate corrections based on these measurements.

Finally, random errors will remain, even after detecting blunders and systematic errors. These are observational errors that do not follow any deterministic mathematical law and can, therefore, only be treated by statistical means. In surveying measurements, experience and probability calculations have shown that when many such errors are observed, they follow a frequency law, which suggests that small errors occur more frequently than large ones and that there are both negative and positive errors about the mean. They are generally assumed to be random and conform to a normal distribution curve (Gaussian distribution). The mathematical analysis of this type of error is the aim of the adjustment computation, which helps to estimate and mitigate their impact on measurements.

On the other hand, given that the systematic observations are repeated identically under the same conditions, reflecting consistent biases or errors in the measurement process, they can be interpreted as highly correlated, whereas the random observations are considered to be statistically independent. Thus, there are correlated observations between the two, which occur when systematic errors can only be partially eliminated or modelled, even when mixed with random errors. They thus indicate the existence of an interrelationship between the measured values, caused by some external phenomenon, whose influence on the results is measured by a correlation coefficient, which must be taken into account in the adjustment computation to ensure that the impact of systematic and random errors on the final data is properly managed and corrected, as described later in this chapter.

3.3   True and Residual Errors

As explained in Chap. 2, carrying out measurement is the act of measuring something, which in adjustment computation is commonly referred to as performing an *observation*. In Geomatics applied to Civil Engineering, observations refer to lengths (distances), horizontal directions, angles and coordinate differences. In this book, individual observations are represented by the symbol l_i, and a group of observations is represented by the vector l.

In terms of observational errors in the context of Geomatics or any measurement-based discipline, there are, by definition, two primary types of error: *true error* and *residual error*, as described in the following subsections.

3.3.1   True Error

By definition, an error ε_i is the difference between the true value $\tilde{l}$ and the measured value l_i. Such an error is often referred to as the *true error*, written mathematically as given by Eq. (3.1).

$$\text{Observation} + \text{true error} = \text{true value}$$

$$l_i + \varepsilon_i = \tilde{l} \tag{3.1}$$

where

$i = 1, 2, 3, \ldots, n$

The concept of true error assumes that the true value $\tilde{l}$ of an observation is known. In theory, this is impossible. In practice, however, it is always possible to assume a measurement process capable of producing a more accurate result than the observation being analysed. Consider, for example, the measurement of a difference in elevation to be carried through a trigonometric levelling whose accuracy is of the order of ± 2 cm, compared with the result of a corresponding differential levelling, whose accuracy is of the order of ± 1 mm. In this case, if all systematic and gross errors have been eliminated, the difference in elevation obtained from differential levelling can be considered a 'quasi-true' value.

3.3.2   Residual Error

In practice, if the same individual quantity is measured n times using the same measurement procedure, they are considered to be of the same quality. Thus, if they are free of blunders and systematic errors, the arithmetic mean of these measured values, as given by Eq. (3.3), is the quantity closest to the true value. In this sense, *true errors* are replaced by *residual errors* (or simply *residuals*), defined as the difference between the mean and the measured value given by the Eq. (3.2).

$$\text{Observation} + \text{residual error} = \text{arithmetic mean}$$

$$l_i + v_i = \bar{l} \tag{3.2}$$

where

$i = 1, 2, 3, \ldots, n$
l_i = observation value
v_i = residual error

$$\bar{l} = \frac{e^T l}{e^T e} = \text{arithmetic mean} \tag{3.3}$$

$$e^T = [1 \ \ 1 \ \ 1 \ \ \cdots \ \ 1] \tag{3.4}$$

$$l = \begin{bmatrix} l_1 \\ \vdots \\ l_n \end{bmatrix} = \text{observation vector}$$

From the above, considering

$$v = \begin{bmatrix} v_1 \\ \vdots \\ v_n \end{bmatrix} = \text{residual error vector}$$

$$e^T v = 0 \tag{3.5}$$

If there is a possible systematic error among the observations, the calculated mean will be affected and should be considered biased.

In the case of an adjustment computation, as presented in Sect. 3.9, the difference between the observed and the adjusted values is referred to as the residual.

3.4   Measures of Quality

Several quantitative concepts are involved with measurement quality for Geomatics applied to Civil Engineering. The most important for the context of this book are as follows:

- Accuracy
- Precision
- Resolution
- Redundancy
- Tolerance
- Standard Error and Standard Deviation
- Root Mean Square (RMS) and
- Root Mean Square Error (RMSE)

3.4.1   Accuracy

Accuracy is defined as the degree of closeness of a set of measurements to the true value. It indicates how close the measurement results are to this true value. However, since the true value of a quantity can never be determined without error, accuracy is theoretically an unknown value.

3.4.2 Precision

By definition, *precision* is the degree of agreement between multiple observations of the same quantity or their degree of reproducibility. In other words, it represents how close the observations are to each other, rather than how close they are to the true value. It is usually expressed as the deviation of a set of results from the arithmetic mean of the set (to be discussed later in this section). It is the term currently used to describe the quality of a measurement result, the quality of an instrument, or even the skill of an observer. It should be noted that the higher the scalar value of the precision calculated from a set of observations, the less precise the measurements are, as a higher value indicates a greater spread or deviation among the observed values.

At this point, it is important for the reader to understand that precise does not necessarily mean accurate, as undetected systematic errors can lead to biased results.

3.4.3 Resolution

The *resolution*, in Geomatics, should be understood as the smallest incremental change that a measuring instrument can provide through its readings. It is, therefore, not a direct indicator of quality, as observational errors will still degrade the quality of the measurement. In this sense, it is worth noting that the resolution value must be small enough to ensure that the instrument achieves or exceeds the expected accuracy of the measurement.

3.4.4 Redundancy

For an adjustment computation model to be applied, there must be more observations than is strictly necessary. The number of measurements beyond what is needed to calculate a unique result for a set of parameters to be determined is called measurement *redundancy (r)*. It is calculated as the difference between the number of observations n and the number of parameters u.

3.4.5 Tolerance

When it is necessary to decide whether or not to accept the result of an adjustment, it is convenient to deal with a range (interval) of values that sets the acceptable limits of error in the measurements or the values estimated by the adjustment computation. This range is called *tolerance*. Typical tolerance values are usually based on the percentage of the values to be retained. For example, assuming that a set of measurements has normally distributed errors, normalised to have a population mean of zero and a population variance of 1, the 95% error is the most commonly accepted range in Geomatics applications, meaning that 95% of all results should be retained. It is calculated from 1.960σ and, therefore, called the two-sigma error. Another tolerance interval sometimes specified is 99.7%, calculated from 3σ. Note that when the 1σ interval is considered, only 68.3% of the values are retained. See the next section for an explanation of the meaning of σ. To better understand the use of percentage errors, it is suggested that the reader takes a brief review of statistical concepts related to the Normal Distribution. The reader may also be interested in research on the difference between tolerance and confidence interval (not discussed here).

3.4.6 Standard Error and Standard Deviation

Given that random errors can be positive and negative, with a mean equal to zero, the mean value of the squared terms can be a parameter to indicate the dispersion of the observed dataset. Statisticians call this the *population variance* when applied to the entire population. However, since the variance is a squared term, a more practical value to represent the measurement quality of the true errors is the *standard error σ*, given by Eq. (3.6).

$$\sigma = \pm \sqrt{\frac{\varepsilon^T \varepsilon}{n}} \qquad (3.6)$$

where

$$\varepsilon = \begin{bmatrix} \varepsilon_1 \\ \vdots \\ \varepsilon_n \end{bmatrix} = \text{true error vector}$$

$n \to \infty$

The σ value given in Eq. (3.6) is also referred to by some authors as the *population standard deviation* when the error vector is formed by comparing the measured value with the population mean. However, it is also important to note that the reader will also find the standard error term used to represent the standard deviation of the mean, as given in Eq. (3.56).

On the other hand, if the variance is calculated from the mean of a small number n of measurements, statisticians can show that a good approximation to the population standard deviation is the *sample standard deviation s* (or simply standard deviation) given by Eq. (3.7).

$$s = \pm \sqrt{\frac{v^T v}{n - 1}} \tag{3.7}$$

where

v = residual error vector
$n < \infty$

Since residual errors are involved in Eq. (3.7), for a series of n measurements of the same measurand, the standard deviation s characterises the dispersion of the results around the arithmetic mean of the n results considered. A small standard deviation means that the values are all close together and therefore more precise. A large standard deviation means that the values are not very similar and therefore less precise. The standard deviation is, therefore, a useful and common measure of internal precision.

3.4.7 Root Mean Square (RMS)

In statistics, the RMS is the square root of the arithmetic mean of the squares of a set of observations (quadratic mean), as given by Eq. (3.8). It is mainly used for noisy quantities to obtain an integrated and filtered value free from fluctuations. It is, therefore, of little use in Geomatics applied to Civil Engineering.

$$\text{RMS} = \sqrt{\frac{l^T l}{n}} \tag{3.8}$$

where

n = number of observations

3.4.8 Root Mean Square Error (RMSE)

By definition, the RMSE is a statistical quantity that measures the difference between a random variable (observation) and a reference or accepted value of higher quality, as given by the Eq. (3.9).

$$b_i = a_i - l_i \qquad\qquad (3.9)$$

where

b_i= difference between the reference and the observed values
l_i= random variable (observation i)
a_i= reference value

Taking into account Eq. (3.9), the RMSE value is determined by Eq. (3.10).

$$\text{RMSE} = \pm \sqrt{\frac{b^{\mathrm{T}} b}{n}} \qquad\qquad (3.10)$$

where

$$b = \begin{bmatrix} b_1 \\ \vdots \\ b_n \end{bmatrix} = \text{error vector}$$

n = number of observations

Note that the RMSE is similar to calculating the population standard deviation, except that it indicates how much the observed value deviates from the reference value, rather than the population mean.

RMSE is used in Geomatics in various situations, for example, to estimate the positional accuracy of geospatial data by comparing the coordinates predicted from a data model with the coordinates from an independent source of higher accuracy for identical points. In this case, the data model can be photogrammetric, a DTM, a vectorised map and others.

Compared to the sample standard deviation (Eq. 3.7), the RMSE is considered a better parameter for estimating accuracy because it includes both random and systematic errors. In contrast, the standard deviation comprises only random errors. See Example 3.2 for a comparative analysis.

3.5 Covariance and Correlation

In the same way that the variance is defined for one random variable, it is also possible to determine the covariance between two random variables. In other words, while the variance expresses the variation of the distribution around the true value or around the mean, the covariance describes the mutual variation between two random variables. Thus, by definition, for two random variables x and y, the covariance is given by Eq. (3.11).

$$\sigma_{xy} = \frac{\boldsymbol{\varepsilon}_x^T \boldsymbol{\varepsilon}_y}{n} \tag{3.11}$$

where

σ_{xy} = covariance between x and y
$\boldsymbol{\varepsilon}_x, \boldsymbol{\varepsilon}_y$ = true error vectors of variables x and y
n = number of observations

However, in practical applications, the covariance can also be determined using residual errors, as given by Eq. (3.12).

$$s_{xy} = \frac{\boldsymbol{v}_x^T \boldsymbol{v}_y}{n-1} \tag{3.12}$$

where

s_{xy} = *covariance* between x and y
$\boldsymbol{v}_x, \boldsymbol{v}_y$ = residual error vectors of variables x and y

The correlation coefficients ρ_{xy} and r_{xy} are the factors that indicate the degree of interdependence between two variables. They are determined using the variances and covariance for the observations, as shown in Eqs. (3.13) or (3.14).

$$\rho_{xy} = \frac{\sigma_{xy}}{\sigma_x \sigma_y} \tag{3.13}$$

or

$$r_{xy} = \frac{s_{xy}}{s_x s_y} \tag{3.14}$$

where

$$\sigma_x = \pm \sqrt{\frac{\boldsymbol{\varepsilon}_x^T \boldsymbol{\varepsilon}_x}{n}} \tag{3.15}$$

$$\sigma_y = \pm \sqrt{\frac{\boldsymbol{\varepsilon}_y^T \boldsymbol{\varepsilon}_y}{n}} \tag{3.16}$$

or

$$s_x = \pm \sqrt{\frac{\boldsymbol{v}_x^T \boldsymbol{v}_x}{n-1}} \tag{3.17}$$

$$s_y = \pm \sqrt{\frac{v_y^T v_y}{n-1}} \qquad (3.18)$$

The correlation coefficient is a non-dimensional value that can vary between -1 and $+1$. The closer it is to ± 1, the greater the correlation. If it is equal to zero, it is assumed that there is no correlation between the observed quantities, i.e., they are independent. As mentioned in Sect. 3.2.2, interdependence is due to the persistence of systematic errors that could not be corrected and remain in the random model and must, therefore, be considered in the adjustment computation.

Example 3.1

Consider the surveying measurement shown in Fig. 3.1. From point (P), repeated observations of the horizontal direction (L_i) and horizontal distance (d_i) to points (A) and (B)[2] were made every half hour using a total station. Each direction observation consists of two readings taken from the direct and reversed positions of the telescope.[3] Given the observation in Table 3.1, calculate the following:

(a) the mean, the residual errors, the standard deviation of the observed direction PA and PB and the correlation coefficient between the observations to point (A) and to point (B).
(b) the mean, the residual errors, the standard deviation of the observed distances d_A and d_B and the correlation coefficient between the observations to point (A) and to point (B).

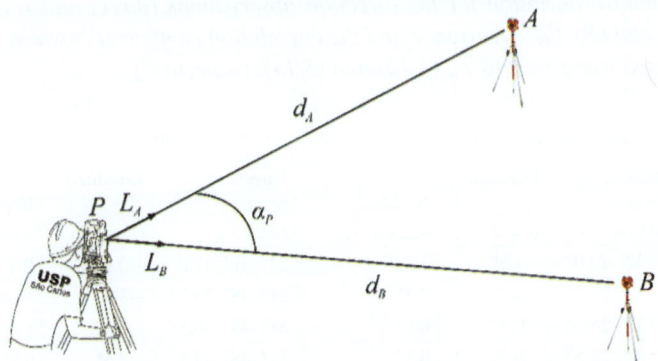

Fig. 3.1 Measurement geometry

---

[2] For further information on horizontal direction and distance observation, read Chaps. 6 and 7.
[3] See Sect. 6.2.1.1.

Table 3.1 Field measurements

Station	n	Sighting point	Horizontal direction (direct)	Horizontal direction (reversed)	Horizontal distance [m]
P	1	A	20° 32′ 25″	200° 32′ 23″	1465.123
		B	51° 48′ 44″	231° 48′ 38″	1423.111
	2	A	20° 32′ 30″	200° 32′ 27″	1465.124
		B	51° 48′ 42″	231° 48′ 35″	1423.115
	3	A	20° 32′ 29″	200° 32′ 28″	1465.120
		B	51° 48′ 42″	231° 48′ 39″	1423.119
	4	A	20° 32′ 32″	200° 32′ 25″	1465.121
		B	51° 48′ 45″	231° 48′ 44″	1423.112
	5	A	20° 32′ 26″	200° 32′ 29″	1465.122
		B	51° 48′ 48″	231° 48′ 45″	1423.109
	6	A	20° 32′ 24″	200° 32′ 31″	1465.127
		B	51° 48′ 43″	231° 48′ 44″	1423.108
	7	A	20° 32′ 32″	200° 32′ 31″	1465.123
		B	51° 48′ 40″	231° 48′ 39″	1423.117
	8	A	20° 32′ 23″	200° 32′ 30″	1465.131
		B	51° 48′ 39″	231° 48′ 43″	1423.110

Solution:

Table 3.2 *gives the mean and residual values of each set of angular observations taken in the direct and reversed positions of the telescope.*

The standard deviation for the direction observations (direct and reversed) to points (A) and (B), the covariance and the correlation coefficient between them can be calculated using Eqs. (3.7), (3.12) and (3.14), respectively.

Table 3.2 Mean value and residuals of horizontal direction observations

n	Direct/ reversed result L_A	Residual v_{L_A} (Eq. 3.2)	Residual squared $v_{L_A}^2$	Direct/ reversed result L_B	Residual v_{L_B} (Eq. 3.2)	Residual squared $v_{L_B}^2$
1	20° 32′ 24.0″	−3.8″	14.5″	51° 48′ 41.0″	−0.9″	0.8″
2	20° 32′ 28.5″	0.7″	0.5″	51° 48′ 38.5″	−3.4″	11.4″
3	20° 32′ 28.5″	0.7″	0.5″	51° 48′ 40.5″	−1.4″	1.9″
4	20° 32′ 28.5″	0.7″	0.5″	51° 48′ 44.5″	2.6″	6.9″
5	20° 32′ 27.5″	−0.3″	0.1″	51° 48′ 46.5″	4.6″	21.4″
6	20° 32′ 27.5″	−0.3″	0.1″	51° 48′ 43.5″	1.6″	2.6″
7	20° 32′ 31.5″	3.7″	13.6″	51° 48′ 39.5″	−2.4″	5.6″
8	20° 32′ 26.5″	−1.3″	1.7″	51° 48′ 41.0″	−0.9″	0.8″
Mean	20° 32′ 27.8″		Sum = 31.5″	51° 48′ 41.9″		Sum = 51.4″

$$s_{L_A} = \pm 2.1'' \qquad s_{L_B} = \pm 2.7''$$
$$s_{L_A L_B} = -1.1''^2 \quad r_{LALB} = -0.191 = -19.1\%$$

From Table 3.2, *the mean horizontal direction value for directions L_A and L_B are*

$$\overline{L}_A = 20° 32' 27.8'' \quad \overline{L}_B = 51° 48' 41.9''$$

The same reasoning and the same sequence of equations can be applied to the horizontal distances measured to points (A) and (B), as shown in Table 3.3.

The standard deviation for distance observations to points (A) and (B), the covariance and the correlation coefficient between them can be calculated by using Eqs. (3.7), (3.12) and (3.14), respectively.

$$s_{d_A} = \pm 3.6 \ \ \text{mm} \qquad s_{d_B} = \pm 4.0 \ \ \text{mm}$$
$$s_{d_A d_B} = -7.3 \ \ \text{mm}^2 \quad r_{dAdB} = -0.52 = -52.0\%$$

From Table 3.3, *the mean horizontal distances are as follows:*

$$\overline{d}_A = 1465.124 \ \text{m} \quad \overline{d}_B = 1423.113 \ \text{m}$$

Table 3.3 Mean and residual values for distance observations

n	Horizontal distance d_A [m]	Residual v_{d_A} [mm] (Eq. 3.2)	Residual squared $v_{d_A}^2$ [mm^2]	Horizontal distance d_B [m]	Residual v_{d_B} [mm] (Eq. 3.2)	Residual squared $v_{d_B}^2$ [mm^2]
1	1465.123	−0.9	0.8	1423.111	−1.6	2.6
2	1465.124	0.1	0.0	1423.115	2.4	5.6
3	1465.120	−3.9	15.0	1423.119	6.4	40.6
4	1465.121	−2.9	8.3	1423.112	−0.6	0.4
5	1465.122	−1.9	3.5	1423.109	−3.6	13.1
6	1465.127	3.1	9.8	1423.108	−4.6	21.4
7	1465.123	−0.9	0.8	1423.117	4.4	19.1
8	1465.131	7.1	50.8	1423.110	−2.6	6.9
	Mean = 1465.124		Sum = 88.9	Mean = 1423.113		Sum = 109.9

Example 3.2
Table 3.4 shows the elevation values obtained from the *Shuttle Radar Topography Mission* (SRTM) DEM[4] for a set of points and the homologous elevations determined in the field by GNSS relative kinematic positioning measurements.[5] Using the

---

[4] For further information, consult Sect. 15.1.
[5] For further information, consult Sect. 17.6.2.6.

GNSS values as the reference values, calculate the mean error, the standard deviation and the RMSE of the altimetric assessment.

Table 3.4 Homologous elevations

Point ID	H_{DEM} [m]	H_{GNSS} [m]	Point ID	H_{DEM} [m]	H_{GNSS} [m]
1	428.427	427.653	11	427.582	428.346
2	439.437	439.346	12	451.375	451.656
3	438.390	437.468	13	452.120	452.585
4	450.458	450.830	14	434.322	433.715
5	439.063	438.719	15	434.848	434.561
6	437.022	436.637	16	430.772	431.608
7	436.123	436.820	17	431.890	431.782
8	439.409	439.008	18	451.282	451.279
9	456.356	455.818	19	443.011	442.622
or	448.184	448.619	20	444.882	445.020

Solution:

Table 3.5 *shows the parameter values used to calculate the standard deviation and RMSE for the DEM altimetric evaluation.*

Table 3.5 Parameter values for standard deviation and RMSE calculation

Point ID	H_{DEM} [m]	H_{GNSS} [m]	Error b_i [m] (Eq. 3.9)	Squared error b_i^2 [m^2]	Residual v_i [m] $(v_i = b_i - \bar{b})$	Squared residual v_i^2 [m^2]
1	428.427	427.653	0.774	0.599	0.731	0.534
2	439.437	439.346	0.091	0.008	0.048	0.002
3	438.390	437.468	0.922	0.850	0.879	0.773
4	450.458	450.830	−0.372	0.138	−0.415	0.172
5	439.063	438.719	0.344	0.118	0.301	0.091
6	437.022	436.637	0.385	0.148	0.342	0.117
7	436.123	436.820	−0.697	0.486	−0.740	0.548
8	439.409	439.008	0.401	0.161	0.358	0.128
9	456.356	455.818	0.538	0.289	0.495	0.245
10	448.184	448.619	−0.435	0.189	−0.478	0.229
11	427.582	428.346	−0.764	0.584	−0.807	0.651
12	451.375	451.656	−0.281	0.079	−0.324	0.105
13	452.120	452.585	−0.465	0.216	−0.508	0.258
14	434.322	433.715	0.607	0.368	0.564	0.318
15	434.848	434.561	0.287	0.082	0.244	0.060
16	430.772	431.608	−0.836	0.699	−0.879	0.773
17	431.890	431.782	0.108	0.012	0.065	0.004
18	451.282	451.279	0.003	0.000	−0.040	0.002
19	443.011	442.622	0.389	0.151	0.346	0.120
20	444.882	445.020	−0.138	0.019	−0.181	0.033
			Sum = 0.861	Sum = 5.198		Sum = 5.161

Using Eq. (3.3),

$$\bar{b} = \frac{e^T b}{e^T e} = \frac{0.861}{20} = 0.043 \text{ m} = 4.3 \text{ cm}$$

Then, considering

$$v_i = b_i - \bar{b}$$

The standard deviation can be calculated using Eq. (3.7).

$$s = \pm \sqrt{\frac{v^T v}{n-1}} = \pm \sqrt{\frac{5.161}{19}} = \pm \ 0.521 \text{ m} \ = \ \pm 52.1 \text{ cm}$$

Finally, the RMSE can be calculated using Eq. (3.10).

$$\text{RMSE} = \pm \sqrt{\frac{b^T b}{n}} = \pm \sqrt{\frac{5.198}{20}} = \pm \ 0.510 \text{ m} \ = \ \pm 51.0 \text{ cm}$$

3.6   Error Propagation

In several situations of mathematical modelling in Geomatics, it is sought to determine the values of quantities (unknowns) that cannot be measured directly with a measuring instrument, such as, for example, the values of the coordinates of a point. These values cannot be measured but rather calculated through mathematical relationships with values from direct measurements. Thus, assuming that the measurements have only random errors and knowing the standard deviations with which they were measured, it is possible to determine the precisions (standard deviations) of the derived quantities. This leads to the error propagation problem, which is discussed in the following subsection.

For a better understanding of the error propagation theory presented below, it is divided into linear and non-linear functions.

3.6.1   Error Propagation for Linear Functions

Consider the case of two observations, l_1 e l_2, with standard deviations σ_1 and σ_2 and covariance σ_{12}. Consider also two linear functions, y and z, of these observations, as given in Eqs. (3.19).

$$y = a_0 + a_1 l_1 + a_2 l_2$$
$$z = b_0 + b_1 l_1 + b_2 l_2 \qquad (3.19)$$

where

$a_i, \ b_i$ = equation parameters (real numbers)

The true values of these two observations are $\tilde{l}_1$ and $\tilde{l}_2$. In these circumstances, inserting the true values and the true errors into Eq. (3.19) and considering Eq. (3.1) gives Eqs. (3.20).

$$\tilde{y} = y + \varepsilon_y = a_0 + a_1 \tilde{l}_1 + a_2 \tilde{l}_2 = a_0 + a_1(l_1 + \varepsilon_1) + a_2(l_2 + \varepsilon_2)$$
$$\tilde{z} = z + \varepsilon_z = b_0 + b_1 \tilde{l}_1 + b_2 \tilde{l}_2 = b_0 + b_1(l_1 + \varepsilon_1) + b_2(l_2 + \varepsilon_2) \qquad (3.20)$$

From this follows the *law of propagation of true errors* for linear functions, as shown as follows:

$$\varepsilon_y = a_1 \varepsilon_1 + a_2 \varepsilon_2$$
$$\varepsilon_z = b_1 \varepsilon_1 + b_2 \varepsilon_2 \qquad (3.21)$$

Consider, temporarily, that l_1 and l_2 are just two measurements from a set of n pairs of measurements. So, there are n pairs of equations of the type (3.21), formulated in matrix notation as in Eqs. (3.22).

$$\boldsymbol{\varepsilon}_y = a_1 \boldsymbol{\varepsilon}_1 + a_2 \boldsymbol{\varepsilon}_2$$
$$\boldsymbol{\varepsilon}_z = b_1 \boldsymbol{\varepsilon}_1 + b_2 \boldsymbol{\varepsilon}_2 \qquad (3.22)$$

where $\boldsymbol{\varepsilon}_y$ and $\boldsymbol{\varepsilon}_z$ are the true error vectors for the functions y and z and $\boldsymbol{\varepsilon}_1$ and $\boldsymbol{\varepsilon}_2$ are the true error vectors of the observation sets l_1 and l_2.

Forming the corresponding matrix products, it follows that

$$\boldsymbol{\varepsilon}_y^T \boldsymbol{\varepsilon}_y = a_1^2 \boldsymbol{\varepsilon}_1^T \boldsymbol{\varepsilon}_1 + 2a_1 a_2 \boldsymbol{\varepsilon}_1^T \boldsymbol{\varepsilon}_2 + a_2^2 \boldsymbol{\varepsilon}_2^T \boldsymbol{\varepsilon}_2$$
$$\boldsymbol{\varepsilon}_z^T \boldsymbol{\varepsilon}_z = b_1^2 \boldsymbol{\varepsilon}_1^T \boldsymbol{\varepsilon}_1 + 2b_1 b_2 \boldsymbol{\varepsilon}_1^T \boldsymbol{\varepsilon}_2 + b_2^2 \boldsymbol{\varepsilon}_2^T \boldsymbol{\varepsilon}_2 \qquad (3.23)$$
$$\boldsymbol{\varepsilon}_y^T \boldsymbol{\varepsilon}_z = a_1 b_1 \boldsymbol{\varepsilon}_1^T \boldsymbol{\varepsilon}_1 + (a_1 b_2 + a_2 b_1) \boldsymbol{\varepsilon}_1^T \boldsymbol{\varepsilon}_2 + a_2 b_2 \boldsymbol{\varepsilon}_2^T \boldsymbol{\varepsilon}_2$$

Dividing each term of Eqs. (3.23) by n and considering Eqs. (3.24) and (3.25).

$$\sigma_i^2 = \frac{\boldsymbol{\varepsilon}_i^T \boldsymbol{\varepsilon}_i}{n} \qquad (3.24)$$

$$\sigma_{ij} = \frac{\boldsymbol{\varepsilon}_i^T \boldsymbol{\varepsilon}_j}{n} = \frac{\boldsymbol{\varepsilon}_j^T \boldsymbol{\varepsilon}_i}{n} \qquad (3.25)$$

gives the *law of propagation of variances* as follows:

$$\sigma_y^2 = a_1^2\sigma_1^2 + 2a_1a_2\sigma_{12} + a_2^2\sigma_2^2$$
$$\sigma_z^2 = b_1^2\sigma_1^2 + 2b_1b_2\sigma_{12} + b_2^2\sigma_2^2 \qquad (3.26)$$
$$\sigma_{yz} = a_1b_1\sigma_1^2 + (a_1b_2 + a_2b_1)\sigma_{12} + a_2b_2\sigma_2^2$$

Considering that in practical applications the population standard deviations σ and the standard deviations s can be interchanged, the following mathematical relationships can be written in matrix form for a set of m functions with n observations.

(a) For only *one* function y with n observations l_i,

$$y = a_0 + a^T l$$
$$s_y^2 = a^T \Sigma_{ll} a \qquad (3.27)$$

(b) For m functions f with n observations l_i,

$$f = f_0 + F^T l$$
$$\Sigma_{ff} = F^T \Sigma_{ll} F \qquad (3.28)$$

The second parts of Eqs. (3.27) and (3.28) are known as the *general law of propagation of variance-covariances*[6] so that

$$a_0 = \text{constant of } y$$

$$a = \begin{bmatrix} a_1 \\ \vdots \\ a_n \end{bmatrix} \quad \text{and} \quad b = \begin{bmatrix} b_1 \\ \vdots \\ b_n \end{bmatrix} = \text{coefficient vectors} \qquad (3.29)$$

$$l = \begin{bmatrix} l_1 \\ \vdots \\ l_n \end{bmatrix} = \text{observation vector} \qquad (3.30)$$

$$f_0 = \begin{bmatrix} a_0 \\ \vdots \\ m_n \end{bmatrix} = \text{constant coefficient of } f \qquad (3.31)$$

---

[6]This term is usually written as "general law of variance propagation" or simply "covariance propagation".

$$f = \begin{bmatrix} y \\ \vdots \\ m \end{bmatrix} = \text{function vector} \qquad (3.32)$$

$$F = \begin{bmatrix} a_1 & & m_1 \\ & \ddots & \\ a_n & & m_n \end{bmatrix} = \text{coefficient matrix for the function } f \qquad (3.33)$$

$$\Sigma_{ll} = \begin{bmatrix} s_1^2 & & s_{1n} \\ & \ddots & \\ s_{n1} & & s_n^2 \end{bmatrix}$$

$$= \text{variance-covariance matrix}^7 \text{ for the observation } l \qquad (3.34)$$

$$\Sigma_{ff} = \begin{bmatrix} s_y^2 & & s_{ym} \\ & \ddots & \\ s_{my} & & s_m^2 \end{bmatrix} = \text{variance-covariance matrix for the function } f \quad (3.35)$$

Example 3.3

Suppose that two sets of horizontal distance measurements have been carried out with a total station installed over points (P_1) and (P_2) to determine the horizontal distance between points (A) and (B), as shown in Fig. 3.2. Field observations are given in Table 3.6. Assuming that there is a correlation coefficient of 0.75 for observations taken from the same station, calculate the two horizontal distances measured between points (A) and (B) and their precisions (standard deviations).

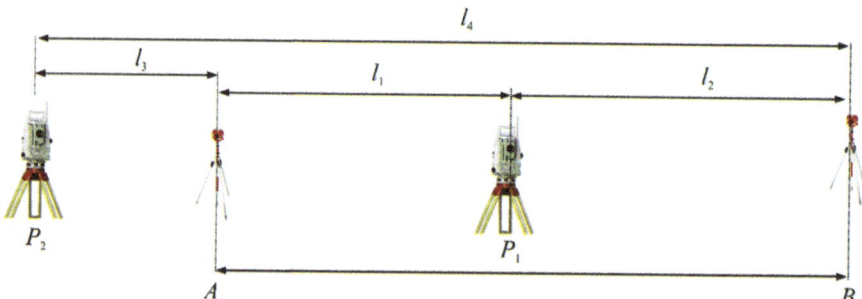

Fig. 3.2 Distance measurement configuration

---

[7] Also, simply covariance matrix.

Table 3.6 Field observations

Horizontal distance	Length (m)	Precision (mm)
l_1	283.9967	±2.6
l_2	370.1201	±2.7
l_3	165.1925	±2.3
l_4	819.3078	±3.6

Solution:

The first step in solving this problem is to determine the variance-covariance matrix for the observations. This can be done by taking into account the precision of the measured distances given in Table 3.6 *and the assumed correlation factor.*

Using Eq. (3.14),

$$s_{12} = s_{21} = 0.75 * s_1 * s_2 = 5.28 \text{ mm}^2$$

$$s_{34} = s_{43} = 0.75 * s_3 * s_4 = 6.36 \text{ mm}^2$$

$$s_{13} = s_{31} = s_{23} = s_{32} = s_{14} = s_{41} = s_{24} = s_{42} = 0$$

According to Eq. (3.34), the variance-covariance matrix Σ_{ll} for the observation l_i is given as follows:

$$\Sigma_{ll} = \begin{bmatrix} s_1^2 & \cdots & s_{1n} \\ \vdots & \ddots & \vdots \\ s_{n1} & \cdots & s_n^2 \end{bmatrix} = \begin{bmatrix} 6.59 & 5.28 & 0 & 0 \\ 5.28 & 7.51 & 0 & 0 \\ 0 & 0 & 5.43 & 6.36 \\ 0 & 0 & 6.36 & 13.24 \end{bmatrix}_{\text{mm}^2}$$

Considering the distance AB measured from point (P1) as d_{AB_1} and from point (P2) as d_{AB_2}, the following observation equations can be written for calculating d_{AB_1} and d_{AB_2}:

$$d_{AB_1} = 1l_1 + 1l_2 + 0l_3 + 0l_4 = 654.1168 \text{ m}$$

$$d_{AB_2} = 0l_1 + 0l_2 - 1l_3 + 1l_4 = 654.1153 \text{ m}$$

Considering the observation equations above, the following coefficient matrix for the functions can be written as

$$F^T = \begin{bmatrix} 1 & 1 & 0 & 0 \\ 0 & 0 & -1 & 1 \end{bmatrix}$$

The variance-covariance matrix Σ_{ff} for the function f can then be calculated using Eq. (3.28).

$$\Sigma_{ff} = \begin{bmatrix} s^2_{dAB_1} & s_{dAB_1 dAB_2} \\ s_{dAB_2 dAB_1} & s^2_{dAB_2} \end{bmatrix} = \begin{bmatrix} 1 & 1 & 0 & 0 \\ 0 & 0 & -1 & 1 \end{bmatrix}$$

$$* \begin{bmatrix} 6.59 & 5.28 & 0 & 0 \\ 5.28 & 7.51 & 0 & 0 \\ 0 & 0 & 5.43 & 6.36 \\ 0 & 0 & 6.36 & 13.24 \end{bmatrix} * \begin{bmatrix} 1 & 0 \\ 1 & 0 \\ 0 & -1 \\ 0 & 1 \end{bmatrix} = \begin{bmatrix} 24.66 & 0.00 \\ 0.00 & 5.95 \end{bmatrix}_{mm^2}$$

From which

$$d_{AB_1} = 654.1168 \text{ m} \pm 4.97 \text{ mm}$$

$$d_{AB_2} = 654.1153 \text{ m} \pm 2.44 \text{ mm}$$

There is no correlation between d_{AB_1} and d_{AB_2}.

3.6.2   Error Propagation for Non-linear Functions

Linear functions are only sometimes available in Civil Engineering applications. In most cases, data modelling requires knowledge of the variances and covariances for non-linear functions of several observations, as given by Eq. (3.36).

$$y = f(l_1, l_2, \cdots, l_n) \tag{3.36}$$

As shown in the previous section, the true value of y is given by Eq. (3.37).

$$\tilde{y} = f\left(\tilde{l}_1, \tilde{l}_2, \cdots, \tilde{l}_n\right) \tag{3.37}$$

Given the hypothesis that the function y is differentiable for all possible observations l_i, a first-order Taylor series expansion can be used to linearise it. Therefore, after linearisation, Eq. (3.37) can be rewritten as given by Eq. (3.38).

$$\tilde{y} = y + \varepsilon_y = f(l_1 + \varepsilon_1, l_2 + \varepsilon_2, \cdots, l_n + \varepsilon_n) = f(l_1, l_2, \cdots, l_n) + \left(\frac{\partial y}{\partial l_1}\right)_0 \varepsilon_1$$

$$+ \left(\frac{\partial y}{\partial l_2}\right)_0 \varepsilon_2 + \cdots + \left(\frac{\partial y}{\partial l_n}\right)_0 \varepsilon_n \tag{3.38}$$

Using the same reasoning as for the law of propagation of true errors for linear functions results

$$\varepsilon_y = \left(\frac{\partial y}{\partial l_1}\right)_0 \varepsilon_1 + \left(\frac{\partial y}{\partial l_2}\right)_0 \varepsilon_2 + \cdots + \left(\frac{\partial y}{\partial l_n}\right)_0 \varepsilon_n \tag{3.39}$$

where $\left(\frac{\partial y}{\partial l_i}\right)_0$ are the partial derivatives of y, concerning the measurements l_i, calculated from approximate values of the observations.

From Eq. (3.39) follows the vector f'^{T} of the partial derivatives of the function y, as given by Eq. (3.40).

$$f'^{\mathrm{T}} = \left[\left(\frac{\partial y}{\partial l_1}\right)_0 \left(\frac{\partial y}{\partial l_2}\right)_0 \cdots \left(\frac{\partial y}{\partial l_n}\right)_0\right] \tag{3.40}$$

Thus, for a function y with n observations l_i, the variance for y is given by Eq. (3.41).

$$\begin{aligned} y &= f(l_1, l_2, \ldots\ldots, l_n) \\ s_y^2 &= f'^{\mathrm{T}} \boldsymbol{\Sigma}_{ll} f' \end{aligned} \tag{3.41}$$

The reader should note that, compared to the case of the linear function, in the same reasoning, the vector f'^{T} replaces the vector a^{T} in Eq. (3.27).

For a set of m functions f with n observations, the variance-covariance matrix for the function f is given by Eq. (3.42).

$$\boldsymbol{\Sigma}_{ff} = J^{\mathrm{T}} \boldsymbol{\Sigma}_{ll} J \tag{3.42}$$

whose values are given as follows:

$$J^{\mathrm{T}} = \begin{bmatrix} \left(\frac{\partial y_1}{\partial l_1}\right)_0 & \cdots & \left(\frac{\partial y_m}{\partial l_1}\right)_0 \\ \vdots & \ddots & \vdots \\ \left(\frac{\partial y_1}{\partial l_n}\right)_0 & \cdots & \left(\frac{\partial y_m}{\partial l_n}\right)_0 \end{bmatrix}$$

$= $ matrix of the partial derivatives (Jacobian matrix) $\tag{3.43}$

$$\boldsymbol{\Sigma}_{ll} = \begin{bmatrix} s_1^2 & & s_{1n} \\ & \ddots & \\ s_{n1} & & s_n^2 \end{bmatrix} = \text{variance-covariance matrix for the observation } l$$

3.6.3 Standard Deviation Propagation for Some Classical Functions

3.6.3.1 General Functions

Consider a series of n **independent observations** $l_1, l_2, \ldots, l_n$ with standard deviations $s_1, s_2, \ldots, s_n$, respectively. Consider also a function f of these observations. For the functions given by Eqs. (3.44) to (3.50), the following standard deviations can be written for f.

$$f = kl_1 \quad \rightarrow \quad s_f = \pm ks_1 \tag{3.44}$$

$$f = l_1 + l_2 \quad \rightarrow \quad s_f = \pm \sqrt{s_1^2 + s_2^2} \tag{3.45}$$

$$f = l_1 - l_2 \quad \rightarrow \quad s_f = \pm \sqrt{s_1^2 + s_2^2} \tag{3.46}$$

$$f = l_1 + l_2 + \cdots + l_n \quad \rightarrow \quad s_f = \pm \sqrt{s_1^2 + s_2^2 + \cdots + s_n^2} \tag{3.47}$$

$$f = al_1 + bl_2 \quad \rightarrow \quad s_f = \pm \sqrt{a^2 s_1^2 + b^2 s_2^2} \tag{3.48}$$

$$f = l^2 \quad \rightarrow \quad s_f = \pm 2ls \tag{3.49}$$

$$f = l_1 l_2 \quad \rightarrow \quad s_f = \pm \sqrt{l_2^2 s_1^2 + l_1^2 s_2^2} \tag{3.50}$$

3.6.3.2   Standard Deviation of the Arithmetic Mean

Consider the arithmetic mean $\bar{l}$ of n observations l_i, each with the same standard deviation s and a correlation coefficient r. According to Eq. (3.3), the arithmetic mean is expressed by Eq. (3.51) and as a result, the standard deviation of the mean can be calculated as follows:

$$\bar{l} = \frac{e^T l}{e^T e} = \frac{1}{n} l_1 + \frac{1}{n} l_2 + \cdots \frac{1}{n} l_n \tag{3.51}$$

As $\bar{l}$ is a linear function, its standard deviation is given by Eq. (3.52).

$$s_{\bar{l}}^2 = a^T \Sigma_{ll} a \tag{3.52}$$

whose values are given by Eqs. (3.53) and (3.54).

$$a^T = \begin{bmatrix} \dfrac{1}{n} & \cdots & \dfrac{1}{n} \end{bmatrix} \tag{3.53}$$

$$\Sigma_{ll} = s^2 \begin{bmatrix} 1 & r & \cdots & r \\ r & 1 & \cdots & \vdots \\ \vdots & \vdots & \ddots & r \\ r & \cdots & r & 1 \end{bmatrix} \tag{3.54}$$

where r is the correlation coefficient.

Finally,

$$s_{\bar{i}}^2 = \begin{bmatrix} \frac{1}{n} & \cdots & \frac{1}{n} \end{bmatrix} * s^2 * \begin{bmatrix} 1 & r & \cdots & r \\ r & 1 & \cdots & \vdots \\ \vdots & \vdots & \ddots & r \\ r & \cdots & r & 1 \end{bmatrix} * \begin{bmatrix} \frac{1}{n} \\ \vdots \\ \frac{1}{n} \end{bmatrix}$$

$$= s^2 * \frac{1 + (n-1) * r}{n} \tag{3.55}$$

If observations are considered independent, i.e., $r = 0$.

$$s_{\bar{i}}^2 = \frac{s^2}{n} \rightarrow s_{\bar{i}} = \frac{s}{\sqrt{n}} \tag{3.56}$$

It is important to emphasise that if the observations are effectively correlated, Eq. (3.56) will give a very small standard deviation value for the arithmetic mean, leading to the assumption of an accuracy that has not been achieved.

Example 3.4
Using the measurements and results obtained in Example 3.1, determine the distance between points (A) and (B) and its standard deviation.

Solution:

According to Example 3.1.

$$s_{d_A} = \pm 3.6 \text{ mm} \qquad s_{d_B} = \pm 4.0 \text{ mm} \qquad r_{dAdB} = -0.52$$
$$\bar{d}_A = 1465.124 \text{ m} \qquad \bar{d}_B = 1423.113 \text{ m}$$
$$s_{L_A} = \pm 2.1'' \qquad s_{LB} = \pm 2.7'' \qquad r_{LALB} = -0.19$$
$$\bar{L}_A = 20°32'27.8'' \qquad \bar{L}_B = 51°48'41.9''$$

Using Eq. (3.56).

$$s_{\bar{L}_A} = \pm \frac{2.1''}{\sqrt{8}} = \pm 0.74'' \qquad\qquad s_{\bar{L}_B} = \pm \frac{2.1''}{\sqrt{8}} = \pm 0.95''$$

$$\alpha_P = 51°48'41.9'' - 20°32'27.8'' = 31°16'14.1''$$

$$s_{\bar{d}_A} = \pm \frac{3.6}{\sqrt{8}} = \pm 1.27 \text{ mm} \qquad\qquad s_{\bar{d}_B} = \pm \frac{4.0}{\sqrt{8}} = \pm 1.41 \text{ mm}$$

Using Eq. (3.27).

$$s_{\alpha P}^2 = \begin{bmatrix} 1 & 1 \end{bmatrix} * \begin{bmatrix} 0.74^2 & 0.13 \\ 0.19*0.74*0.95 & 0.95^2 \end{bmatrix} * \begin{bmatrix} 1 \\ 1 \end{bmatrix} \quad \rightarrow \quad s_{\alpha P} = \pm 1.3''$$

The distance d_{AB} can be calculated using the cosine law.

$$d_{AB} = \sqrt{d_A^2 + d_B^2 - 2d_A d_B \cos(\alpha_P)} = 779.468 \text{ m}$$

According to Eq. (3.41), the standard deviation of the distance AB is given by

$$s_{d_{AB}}^2 = f'^{\mathrm{T}} \Sigma_{ll} f'$$

where

$$f'^{\mathrm{T}} = \left[\frac{\partial d_{AB}}{\partial d_A} \ \frac{\partial d_{AB}}{\partial d_B} \ \frac{\partial d_{AB}}{\partial \alpha_A} \right] = \left[\frac{d_A - d_B \cos(\alpha_A)}{d_{AB}} \ \frac{d_B - d_A \cos(\alpha_A)}{d_{AB}} \ \frac{d_A d_B \sin(\alpha_A)}{d_{AB}} \right]$$

$$f'^{\mathrm{T}} = [0.319132 \ 0.219168 \ 1,388.513145]$$

$$\Sigma_{ll} = \begin{bmatrix} s_{d_A}^2 & s_{d_A d_B} & 0 \\ s_{d_B d_A} & s_{d_B}^2 & 0 \\ 0 & 0 & s_{\alpha_P}^2 [\text{rad}] \end{bmatrix} = \begin{bmatrix} 1.62 * 10^{-6} & 9.36 * 10^{-7} & 0 \\ 9.36 * 10^{-7} & 2.00 * 10^{-6} & 0 \\ 0 & 0 & 4.07 * 10^{-11} \end{bmatrix}$$

Then,

$$s_{d_{AB}} = \pm 8.9 \text{ mm}$$

Finally,

$$d_{AB} = 779.468 \text{ m} \pm 8.9 \text{ mm}$$

3.7   Weights of Observations

Measurements performed with surveying instruments are not always of the same quality. This difference in quality must be considered in the adjustment models so that the adjusted results are consistent with the measurements made. Therefore, a quality parameter related to the standard deviation of the measured values must be adopted to account for this assumption. Such a parameter is called *weight*. By definition, high-precision measurements should have a high weight, and low precision measurements should have a low weight.[8] Therefore, to consider this statement, the weight p_i is mathematically defined as given by Eq. (3.57).

---

[8] The reader should remember that high precision means a small standard deviation value.

$$p_i = \frac{\text{scale factor}}{s_i^2} \tag{3.57}$$

From Eq. (3.57), the weight value can be set to one if the *scale factor* is assumed to be equal s_i^2. The standard deviation, in this case, is classically denoted as s_0 and is known as the *standard deviation of unit weight* so that

$$p_i = 1 = \frac{s_0^2}{s_0^2} \tag{3.58}$$

Finally, to account for the variation in measurement quality, Eq. (3.58) becomes

$$p_i = \frac{s_0^2}{s_i^2} \tag{3.59}$$

which leads to Eq. (3.60).

$$s_i = \frac{s_0}{\sqrt{p_i}} \tag{3.60}$$

When applied to a series of observations with different standard deviations, the concept of weight is related to the diagonal matrix as given in Eq. (3.61), which is restricted to the case of independent observations.

$$\boldsymbol{P} = \begin{bmatrix} p_1 & 0 & \cdots & 0 \\ 0 & p_2 & \cdots & 0 \\ \vdots & \vdots & \ddots & \vdots \\ 0 & 0 & \cdots & p_n \end{bmatrix} = s_0^2 \begin{bmatrix} s_1^2 & 0 & \cdots & 0 \\ 0 & s_2^2 & \cdots & 0 \\ \vdots & \vdots & \ddots & \vdots \\ 0 & 0 & \cdots & s_n^2 \end{bmatrix}^{-1} \tag{3.61}$$

3.7.1   Weighted Mean

If weights are assigned to observations, they must be considered when calculating the mean of a set of observations l, resulting in a *weighted mean* $\bar{l}$, as given by Eq. (3.62) in matrix form.

$$\bar{l} = \frac{e^{\mathrm{T}} \boldsymbol{P} l}{e^{\mathrm{T}} \boldsymbol{P} e} \tag{3.62}$$

where the matrix $\boldsymbol{P}$ is given by Eq. (3.61) and e is the unit vector.

For example, consider the case of three observations, the first of which has a weight of 1, the second of which has a weight of 2 and the third of which has a weight of 3. The weighted mean, in this case, should be calculated as follows:

$$\bar{l} = \frac{1l_1 + 2l_2 + 3l_3}{1 + 2 + 3} \qquad (3.63)$$

Since the first observation has a weight of 1, its standard deviation is equal to s_0, given by Eq. (3.59), and the weighted mean can be considered equivalent to six standard observations l_1. Consequently, its precision is given by Eq. (3.64).

$$s_{\bar{l}} = \sqrt{\frac{s_0^2}{e^T P e}} \qquad (3.64)$$

Example 3.5
The same angle was calculated with three theodolites of different precision. The results are shown in Table 3.7. Calculate the weighted mean and its precision by considering the calculated values and the corresponding precisions.

Table 3.7 Field observations

Observation	Angle	Precision
l_1	192°34′17″	±7″
l_2	192°34′26″	±5″
l_3	192°34′15″	±2″

Solution:

Considering the value of 7″ as the standard deviation of unit weight s_0, the following weights are assigned to each observation:

$$p_1 = \frac{7^2}{7^2} = 1 \qquad\qquad p_2 = \frac{7^2}{5^2} = 1.96 \qquad\qquad p_3 = \frac{7^2}{2^2} = 12.25$$

Then,

$$l = \begin{bmatrix} 192°\ 34'17'' \\ 192°\ 34'26'' \\ 192°\ 34'15'' \end{bmatrix} \qquad P = \begin{bmatrix} 1 & 0 & 0 \\ 0 & 1.96 & 0 \\ 0 & 0 & 12.25 \end{bmatrix} \qquad e = \begin{bmatrix} 1 \\ 1 \\ 1 \end{bmatrix}$$

$$e^{T}Pl = \begin{bmatrix} 1 & 1 & 1 \end{bmatrix} \begin{bmatrix} 1 & 0 & 0 \\ 0 & 1.96 & 0 \\ 0 & 0 & 12.25 \end{bmatrix} \begin{bmatrix} 192° & 34'17'' \\ 192° & 34'26'' \\ 192° & 34'15'' \end{bmatrix} = 2,929° \ 00' 32.1''$$

$$e^{T}Pe = \begin{bmatrix} 1 & 1 & 1 \end{bmatrix} \begin{bmatrix} 1 & 0 & 0 \\ 0 & 1.96 & 0 \\ 0 & 0 & 12.25 \end{bmatrix} \begin{bmatrix} 1 \\ 1 \\ 1 \end{bmatrix} = 15.21$$

Using Eq. (3.62).

$$\bar{l} = \frac{2,929° \ 00' \ 32.1''}{15.21} = 192° \ 34'16.5''$$

Using Eq. (3.64).

$$s_{\bar{l}} = \sqrt{\frac{7^2}{15.21}} = \pm 1.8''$$

3.8 Cofactors

The concept of weights for independent observations was introduced by C. F. Gauss (1777–1855) in the nineteenth century. However, for cases of correlated observations, it is recommended to use the concept of cofactors q_{ij}, which, by definition, are related to covariances by Eq. (3.65) and variance by Eq. (3.66).

$$q_{ij} = \frac{s_{ij}}{s_0^2} \tag{3.65}$$

$$q_i = \frac{s_i^2}{s_0^2} \tag{3.66}$$

Applying Eqs. (3.65) and (3.66) to the variance-covariance matrix Σ_{ll} gives the cofactor matrix Q_{ll} as follows.

$$Q_{ll} = \frac{1}{s_0^2} \Sigma_{ll} \tag{3.67}$$

where

$$\boldsymbol{Q}_{ll}=\boldsymbol{Q}=\frac{1}{s_0^2}\begin{bmatrix} s_1^2 & s_{12} & \cdots & s_{1n} \\ s_{21} & s_2^2 & \cdots & s_{2n} \\ \vdots & \vdots & \ddots & \vdots \\ s_{n1} & s_{n2} & \cdots & s_n^2 \end{bmatrix}=\begin{bmatrix} q_{11} & q_{12} & \cdots & q_{1n} \\ q_{21} & q_{22} & \cdots & q_{2n} \\ \vdots & \vdots & \ddots & \vdots \\ q_{n1} & q_{n2} & \cdots & q_{nn} \end{bmatrix} \tag{3.68}$$

With Eq. (3.68) it is always possible to switch from cofactors to covariances and vice versa. From Eqs. (3.60) and (3.67), it can also be seen that the inverse of the cofactor matrix is the weight matrix $\boldsymbol{P}$, as given by Eq. (3.69).

$$\boldsymbol{P}=\boldsymbol{Q}^{-1} \tag{3.69}$$

However, Eq. (3.69) is only valid in the case of independent observations. In the general case of correlated observations, the concept of cofactors, as given by Eq. (3.67), must be used for weighting procedures.

3.9   Least Squares Adjustment Method

The Least Squares method is the most common of the various adjustment methods available in the literature. It was first introduced in the nineteenth century by Karl Friedrich Gauss. Since then, it has been widely used for adjusting all basic measurements in Geomatics where the number of observations is redundant. As already mentioned, a redundant number of observations means that there is more information than necessary to have a unique mathematical model solution, thus allowing the Least Squares Adjustment method to fit the data to the model. This has two major advantages: the availability of precision indicators and the possibility of checking reliability by detecting and eliminating outliers.

The Least Squares Adjustment method is based on the assumption that for a set of redundant observations, represented by the vector l, where blunders and systematic errors have been eliminated, it is possible, after the adjustment, to obtain a set of *adjusted values* (most probable values) represented by the vector $\tilde{l}$. The difference between the two data sets gives a set of values called *residuals*, as given by Eq. (3.70).

$$v=\tilde{l}-l \tag{3.70}$$

where

$$v = \begin{bmatrix} v_1 \\ \vdots \\ v_n \end{bmatrix} = \text{residual vector} \qquad l = \begin{bmatrix} l_1 \\ \vdots \\ l_n \end{bmatrix} = \text{observation vector}$$

$$\tilde{l} = \begin{bmatrix} \tilde{l}_1 \\ \vdots \\ \tilde{l}_n \end{bmatrix} = \text{adjusted observation vector}$$

n = number of observations

As will be seen later in this chapter, residuals and weights play an important role in the adjustment computation. Residuals indicate how close the adjusted values are to the observations, and weights allow the stochastic properties of the data set to be considered in the mathematical model. In addition, after the adjustment, both make it possible to assess the adequacy of the mathematical model adopted.

Based on the above criterion, mathematicians can easily prove that the most probable results are given when

$$v^T P v = v^T Q^{-1} v = \text{minimum} \tag{3.71}$$

where P is the weight matrix of the observations, and Q is the cofactor matrix.

At this point, it is important to emphasise that to solve an adjustment computation problem, it is necessary to adopt a mathematical model that best describes the physical phenomenon represented. This mathematical model is divided in two other models: the *functional* and *stochastic* models.

The functional model establishes the mathematical relationships between the observations and, therefore, the relationships between the unknowns of the equations. In practice, it must describe the physical and geometrical characteristics of the measurement environment.

The stochastic model is associated with the statistical properties of all elements involved in the functional model, described by the matrix of variance-covariance assigned to the observations. It defines the precision of the measurement and the a priori correlations that influence the adjustment results. Such elements must be verified after adjustment to ensure that the values assumed a priori are compatible with those obtained a posteriori. If they are incompatible, the functional and stochastic models and their hypotheses must be carefully re-examined.

3.9.1 Adjustment Models

Three models of observation adjustment are presented in this book: the *parametric adjustment model*, the *Conditional adjustment model* and the *general adjustment model*, the latter also known as the *Gauss-Helmert adjustment model*.

3.9.1.1   Parametric Adjustment Model

The Parametric adjustment model consists of relating the observations to the unknown parameters of the functional model through error equations as given by Eq. (3.72).

$$\text{Observation} + \text{Residual error} = \text{Linear function of the unknowns}$$

$$l_i \; + \; v_i \; = \; a_i x + b_i y + c_i z + \text{constant value} \tag{3.72}$$

where a_i, b_i, c_i are coefficients of the function and x, y, z are the unknown parameters.

An error equation must be written for each measurement. Thus, if redundant observations have been made, there will be more error equations than needed for a unique solution, and the Least Squares principle can determine the most probable values of the unknowns. The reader should also bear in mind that Least Squares calculations are lengthy and best performed on a computer, mainly because of the need for matrix inversion, as indicated in Eq. (3.87).

In most cases, the initial mathematical relationship between observations and unknowns is non-linear, and it is necessary to linearise the function to obtain linear error equations. At this stage, reduced values for the observations l_{0_i} and approximate values for the unknowns x_{k_0} are introduced into the model. As a result, the corrections to be applied to the unknowns will be small, legitimising the linearisation of the equations by series expansion.

The generalisation of Eq. (3.72) for n observations and u unknowns, where $n > u$, gives

$$\tilde{l}_i = l_i + v_i = f_i(\tilde{x}_1, \tilde{x}_2, ..., \tilde{x}_u) \tag{3.73}$$

$i = 1, 2, ..., n$

The above equation shows that each adjusted observation is expressed as a function, linear or not, of the adjusted values of the unknowns. Therefore, for the adjustment to be processed, it is necessary to assume approximate values for the unknowns, as given by Eq. (3.74).

$$\tilde{x}_k = x_{k_0} + x_k \tag{3.74}$$

where

$\tilde{x}_k = $ adjusted value
$x_{k_0} = $ approximate value
$x_k = $ correction value (unknown)
$k = 1, 2, ..., u.$

The linearised form of Eq. (3.73) is given as follows:

$$f_i(\tilde{x}_1, \tilde{x}_2, \ldots, \tilde{x}_u) = f_i(x_{1_0}, x_{2_0}, \ldots, x_{u_0}) + \left(\frac{\partial f_i}{\partial x_1}\right)_0 x_1 + \left(\frac{\partial f_i}{\partial x_2}\right)_0 x_2 + \ldots$$

$$= f_i(x_{1_0}, x_{2_0}, \ldots, x_{u_0}) + a_i x_1 + b_i x_2 + \ldots$$

$$(3.75)$$

where

$$a_i = \left(\frac{\partial f_i}{\partial x_1}\right)_0, \quad b_i = \left(\frac{\partial f_i}{\partial x_2}\right)_0.$$

Substituting Eq. (3.75) into (3.73) and using a reduced value l_{0_i} to account for the scalar values, as shown in Eq. (3.76), gives the generalised error Eq. (3.77).

$$l_{0_i} = l_i - f_i(x_{1_0}, x_{2_0}, \ldots, x_{u_0}) \qquad (3.76)$$

$$v_i = a_i x_1 + b_i x_2 + \ldots - l_{0_i} \qquad (3.77)$$

As mentioned, an adjustment model by the Least Squares principle can only be applied if the number n of error equations is greater than the number u of unknowns, i.e., there is a redundancy of observations (degrees of freedom). Therefore, to facilitate the development of the equations, consider the case where $u = 3$, i.e., unknowns equal to x, y, z and coefficients of the equations equal to a_i, b_i, c_i. Under these conditions, the error equation (3.77) becomes

$$v_1 = a_1 x + b_1 y + c_1 z \ldots \ldots -l_{0_1}$$

$$\cdot$$

$$(3.78)$$

$$\cdot$$

$$v_n = a_n x + b_n y + c_n z \ldots \ldots -l_{0_n}$$

Then, considering

$$A = \begin{bmatrix} a_1 & b_1 & c_1 \\ a_2 & b_2 & c_2 \\ & \ddots & \\ a_n & b_n & c_n \end{bmatrix} \quad v = \begin{bmatrix} v_1 \\ v_2 \\ \cdot \\ \cdot \\ v_n \end{bmatrix} \quad l_0 = \begin{bmatrix} l_{0_1} \\ l_{0_2} \\ \cdot \\ \cdot \\ l_{0_n} \end{bmatrix} \quad x = \begin{bmatrix} x \\ y \\ z \end{bmatrix}$$

Equation (3.78) can be written in the matrix form as follows.

$$\underset{(n \times 1)}{v} = \underset{(n \times u)(u \times 1)}{A \ x} - \underset{(n \times 1)}{l_0} \qquad (3.79)$$

where

$$\underset{(u\times 1)}{\tilde{x}} = \underset{(u\times 1)}{x_0} + \underset{(u\times 1)}{x} \tag{3.80}$$

Assuming that the weight matrix P is known and applying the Least Squares principle given by Eq. (3.71), the following mathematical development can be applied.

$$v^T P v = (x^T A^T - l_0^T) P (Ax - l_0)$$
$$v^T P v = x^T A^T P A x - x^T A^T P l_0 - l_0^T P A x + l_0^T P l_0 \tag{3.81}$$

Since $x^T A^T P l_0 = l_0^T P A x$

$$v^T P v = x^T A^T P A x - 2 l_0^T P A x + l_0^T P l_0 \tag{3.82}$$

Equation (3.82) has extremes when its first derivative is zero. Thus

$$\frac{\partial (v^T P v)}{\partial x} = 2 x^T A^T P A - 2 l_0^T P A = 0 \tag{3.83}$$

Note that the second derivative is greater than zero, indicating that it is a minimum (not a maximum) point. Thus, the following equation is obtained by dividing Eq. (3.83) by two and applying the appropriate transposes.

$$A^T P A x - A^T P l_0 = 0 \tag{3.84}$$

Considering

$$N = A^T P A$$
$$n = A^T P l_0 \tag{3.85}$$

gives

$$Nx = n \tag{3.86}$$

Matrix Eq. (3.86) is called a matrix of *normal equations* whose unknowns can be calculated by

$$x = N^{-1} n \tag{3.87}$$

Equation (3.84) is general and can be used for unweighted and weighted adjustments. If all observations have the same weight, the matrix P will be an identity matrix with values equal to 1 for all diagonal elements.

In Eq. (3.87), the vector x contains the corrections to be added to the approximate values of the unknowns. Thus, by substituting (3.87) in (3.80), the adjusted unknowns are obtained, and consequently, the residual error vector v and the adjusted values by using Eqs. (3.79) and (3.73), respectively. Finally, the standard deviation of unit weight is determined using Eq. (3.88).

$$s_0^2 = \frac{v^{\mathrm{T}} P v}{n - u} \tag{3.88}$$

In addition to calculating the unknowns, which is the first objective of the adjustment computation, it is also possible to explore the results to obtain other important statistical indicators, such as the variance-covariance matrix for the unknowns, for the residual error vector and for the adjusted observation matrix. Such parameters can be obtained by using the *general law of propagation of variances* for the function vector f, which is simultaneously composed of the partial vectors:

l: observation vector
x: vector of the unknown
$\tilde{l}$: adjusted observation vector
v: residual error vector

Thus, the function vector f is given by Eq. (3.89).

$$f = \begin{pmatrix} l \\ x \\ \tilde{l} \\ v \end{pmatrix} = \begin{pmatrix} I \\ N^{-1} A^{\mathrm{T}} P \\ A N^{-1} A^{\mathrm{T}} P \\ A N^{-1} A^{\mathrm{T}} P - I \end{pmatrix} * l \tag{3.89}$$

Considering $\tilde{Q} = A N^{-1} A^{\mathrm{T}}$

$$f = \begin{pmatrix} I \\ N^{-1} A^{\mathrm{T}} P \\ \tilde{Q} P \\ \tilde{Q} P - I \end{pmatrix} * l = F^{\mathrm{T}} * l \tag{3.90}$$

And then, by the *general law of propagation of variances* and considering that $\tilde{Q} P \tilde{Q} = A N^{-1} A^{\mathrm{T}} = \tilde{Q}$

$$
Q_{ff} = \begin{pmatrix}
\overset{l}{P^{-1}} & \overset{x}{AN^{-1}} & \overset{\tilde{l}}{\tilde{Q}} & \overset{v}{\tilde{Q}-P^{-1}} \\
\cdot & N^{-1} & N^{-1}A^{\mathrm{T}} & 0 \\
\cdot & simmetrical & \tilde{Q} & 0 \\
\cdot & \cdot & \cdot & P^{-1}-\tilde{Q}
\end{pmatrix}
\begin{matrix}
l \\ x \\ \tilde{l} \\ v
\end{matrix}
\qquad (3.91)
$$

which means

$$
Q_{ll} = Q = P^{-1} \tag{3.92}
$$

$$
Q_{xx} = N^{-1} \tag{3.93}
$$

$$
Q_{\tilde{l}\tilde{l}} = \tilde{Q} = AN^{-1}A^{\mathrm{T}} \tag{3.94}
$$

$$
Q_{vv} = P^{-1} - AN^{-1}A^{\mathrm{T}} \tag{3.95}
$$

and so on.

Given the equations presented above, the following steps must be taken to perform an adjustment computation using the Parametric adjustment model:

1. Carefully formulate the functional and stochastic models.
2. Establish error equations.
3. Determine the vectors and matrices x_0, l_0, A, P.

   Calculate PA and then $N = (PA)^{\mathrm{T}}A$ e $n = (PA)^{\mathrm{T}}l_0$.

4. Calculate:

$$
Q_{xx} = N^{-1}
$$

$$
x = N^{-1}n = Q_{xx}n
$$

$$
\tilde{x} = x_0 + x
$$

$$
v = Ax - l_0
$$

$$
\tilde{l} = l + v
$$

5. Calculate:

$$
s_0^2 = \frac{v^{\mathrm{T}}Pv}{n-u}
$$

$$
\Sigma_{xx} = s_0^2 N^{-1} = s_0^2 Q_{xx}
$$

$$
\Sigma_{\tilde{l}} = s_0^2 AN^{-1}A^{\mathrm{T}}
$$

6. Carry out *à posteriori* checks.

Accept or reject the model.

Example 3.6

Given the plane triangle shown in Fig. 3.3 and the angle values in Table 3.8, calculate the adjusted values of the angles α_A, α_B and α_C using the Parametric adjustment model and assuming that all angles are of the same precision and uncorrelated.

Solution:

From Table 3.8.

Fig. 3.3 Triangle geometry

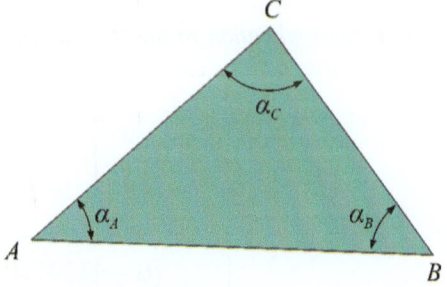

Table 3.8 Angle values

$\alpha_A = 43°28'20''$
$\alpha_B = 51°32'51''$
$\alpha_C = 84°58'52''$

$$l = \begin{bmatrix} 43°\,28'20'' \\ 51°\,32'51'' \\ 84°\,58'52'' \end{bmatrix}$$

By the triangle geometry: $\tilde{\alpha}_A + \tilde{\alpha}_B + \tilde{\alpha}_C = 180°$

Considering the approximate value α_{i0} and the correction value $\delta\alpha_i$, as given by Eq. (3.74), the observation equations can be written as follows.

$$\tilde{\alpha}_A = \alpha_{A_0} + \delta\alpha_A$$

$$\tilde{\alpha}_B = \alpha_{B_0} + \delta\alpha_B$$

$$\tilde{\alpha}_C = 180 - \alpha_{A_0} - \delta\alpha_A - \alpha_{B_0} - \delta\alpha_B$$

Thus, the following error equations can be written.

$$v_A = \alpha_{A_0} + \delta\alpha_A - \alpha_A$$

$$v_B = \alpha_{B_0} + \delta\alpha_B - \alpha_B$$

$$v_C = 180 - \alpha_{A_0} - \delta\alpha_A - \alpha_{B_0} - \delta\alpha_B - \alpha_C$$

Taking the approximate values α_{i0} equal to zero gives

$$v_A = \delta\alpha_A - \alpha_A$$

$$v_B = \delta\alpha_B - \alpha_B$$

$$v_C = 180 - \delta\alpha_A - \delta\alpha_B - \alpha_C$$

For the application of the Parametric adjustment model, using Eq. (3.79), the above error equations can be written in matrix form as $v = Ax - l_0$, where

$$A = \begin{bmatrix} 1 & 0 \\ 0 & 1 \\ -1 & -1 \end{bmatrix} \qquad v = \begin{bmatrix} v_A \\ v_B \\ v_C \end{bmatrix}$$

$$l_0 = \begin{bmatrix} \alpha_A = 43°\,28'20'' \\ \alpha_B = 51°\,32'51'' \\ \alpha_C - 180° = -95°\,01'08'' \end{bmatrix} \qquad x = \begin{bmatrix} \delta\alpha_A \\ \delta\alpha_B \end{bmatrix}$$

Using Eqs. (3.84) to (3.87) and assuming that observations are independent and of the same precision, i.e., $P = I$.

$$N = \begin{bmatrix} 2.00 & -1.00 \\ -1.00 & 2.00 \end{bmatrix} \qquad n = \begin{bmatrix} 138.4911 \\ 146.5664 \end{bmatrix} \qquad x = \begin{bmatrix} 43°\,28'19'' \\ 51°\,32'50'' \end{bmatrix}$$

Using Eqs. (3.79) and (3.73), the residual error vector and the vector of the adjusted values are given as follows.

$$v = \begin{bmatrix} -1'' \\ -1'' \\ -1'' \end{bmatrix} \qquad \tilde{l} = \begin{bmatrix} 43°\,28'19'' \\ 51°\,32'50'' \\ 84°\,58'51'' \end{bmatrix}$$

The standard deviations of the unit weight, considering that there are three observations (n) and two unknowns (u) is given as follows.

$$s_0^2 = \frac{v^{\mathrm{T}} P v}{3-2} = 3.0''^2$$

$$s_0 = \pm 1.7''$$

Using Eq. (3.93), the variance-covariance matrix for the unknowns (angles) is given as follows.

$$Q_{xx} = \begin{bmatrix} 0.666667 & -0.333333 \\ -0.333333 & 0.666667 \end{bmatrix}$$

$$\Sigma_{xx} = s_0^2 * Q_{xx} = \begin{bmatrix} 2.00 & -1.00 \\ -1.00 & 2.00 \end{bmatrix}$$

Using the diagonal values in the variance-covariance matrix for the parameters, their standard deviation is $s_{\alpha_A} = s_{\alpha_B} = \pm 1.4''$.

Finally, the standard deviations of the adjusted values can be calculated as follows.

$$Q_{\tilde{l}\tilde{l}} = A N^{-1} A^{T} = \begin{bmatrix} 0.6667 & -0.3333 & -0.3333 \\ -0.3333 & 0.6667 & -0.3333 \\ -0.3333 & -0.3333 & 0.6667 \end{bmatrix}$$

$$\Sigma_{\tilde{l}\tilde{l}} = s_0^2 * Q_{\tilde{l}\tilde{l}} = \begin{bmatrix} 2.0 & -1.0 & -1.0 \\ -1.0 & 2.0 & -1.0 \\ -1.0 & -1.0 & 2.0 \end{bmatrix} = \begin{bmatrix} s_{\tilde{\alpha}A}^2 & s_{\tilde{\alpha}A\tilde{\alpha}B} & s_{\tilde{\alpha}A\tilde{\alpha}C} \\ s_{\tilde{\alpha}B\tilde{\alpha}A} & s_{\tilde{\alpha}B}^2 & s_{\tilde{\alpha}B\tilde{\alpha}C} \\ s_{\tilde{\alpha}C\tilde{\alpha}A} & s_{\tilde{\alpha}C\tilde{\alpha}B} & s_{\tilde{\alpha}C}^2 \end{bmatrix}$$

And then,

$$\tilde{\alpha}_A = 43°28'19'' \pm 1.4'' \quad \tilde{\alpha}_B = 51°32'50'' \pm 1.4'' \quad \tilde{\alpha}_C = 84°58'51'' \pm 1.4''$$

3.9.1.2 Conditional Adjustment Model

In some adjustment computation problems, it is easier to express the mathematical relationships between observations than between observations and several unknowns. The equations formulated in this case are called *conditional equations*, which contain only observations and no parameters. In such a situation, the adjusted observation functions, where all conditions are satisfied, are given by Eq. (3.96).

$$f\left(\tilde{l}_1, \tilde{l}_2, \cdots, \tilde{l}_n\right) = 0 \tag{3.96}$$

On the other hand, the original observation functions are given by Eq. (3.97), where the conditions are not perfectly satisfied, and a closing error w has to be considered.

$$f(l_1, l_2, \cdots, l_n) = w \tag{3.97}$$

Then, consider the existence of n observations and the established conditional equations that satisfy the model. In this situation, if there are more conditional equations than necessary to solve the problem, it is possible to apply an adjustment computation model based on the number of redundant independent conditional equations r. Based on this assumption, consider the existence of r linearised conditional equations expressed as follows

$$b_{10} + b_{11}\tilde{l}_1 + b_{12}\tilde{l}_2 + \cdots + b_{in}\tilde{l}_n = 0$$
$$\ddots \tag{3.98}$$
$$b_{r0} + b_{r1}\tilde{l}_1 + b_{r2}\tilde{l}_2 + \cdots + b_{m}\tilde{l}_n = 0$$

Considering that $\tilde{l}_i = l_i + v_i$ and regrouping the numerical terms, the following residual equations are obtained:

$$b_{10} + b_{11}v_1 + b_{12}v_2 + \cdots + b_{in}v_n + w_1 = 0$$
$$\ddots \tag{3.99}$$
$$b_{r0} + b_{r1}v_1 + b_{r2}v_2 + \cdots + b_{m}v_n + w_r = 0$$

Which in matrix form is expressed as follows.

$$b_0 + B^\mathrm{T}v + w = 0 \tag{3.100}$$

Considering that in Eq. (3.100), the value of b_0 has no algebraic function, the final residual equation can be written as follows:

$$B^\mathrm{T}v + w = 0 \tag{3.101}$$

Next, to consider the stochastic model in the adjustment computation, it is necessary to formulate the weight matrix P so that

$$\Sigma_{ll} = s_0^2 * Q = s_0^2 * P^{-1} \tag{3.102}$$

Similar to the parametric adjustment model, the Least Squares criterion for this problem will be

$$v^{\mathrm{T}}Pv = v^{\mathrm{T}}Q^{-1}v = minimum \tag{3.103}$$

To comply with the least squares criterion and simultaneously have a solution to Eq. (3.101), the method of constrained minima by Lagrange multipliers can be used.

$$\Omega = v^{\mathrm{T}}Pv - 2k\left(B^{\mathrm{T}}v + w\right) = minimum \tag{3.104}$$

where k represents the unknown Lagrange multiplier.

The minimum value for Eq. (3.104) is reached when its partial derivatives are equal to zero.

$$\frac{\partial \Omega}{\partial v} = 2v^{\mathrm{T}}P - 2k^{\mathrm{T}}B^{\mathrm{T}} = 0 \tag{3.105}$$

Dividing the Eq. (3.105) by 2, transposing and rearranging gives

$$Pv = Bk \tag{3.106}$$

Multiplying the left side of Eq. (3.106) by $P^{-1} = Q$ gives

$$P^{-1}Pv = P^{-1}Bk \tag{3.107}$$

And then,

$$v = QBk \tag{3.108}$$

Equation (3.108) gives the residual error values as a function of k, which can be calculated by replacing Eq. (3.101) with Eq. (3.108).

$$B^{\mathrm{T}}QBk + w = 0 \tag{3.109}$$

Considering $N = B^{\mathrm{T}}QB$
Equation (3.109) can be expressed as follows:

$$Nk + w = 0 \tag{3.110}$$

$$k = -N^{-1}w \tag{3.111}$$

Finally, the values of v can be calculated using Eq. (3.108) and the standard deviation of unit weight using Eq. (3.112).

$$s_0^2 = \frac{v^{\mathrm{T}}Bk}{r} = \frac{-k^{\mathrm{T}}w}{r} \tag{3.112}$$

where r is the number of redundant independent conditional equations.

As with the parametric adjustment model, it is possible to explore the results of the conditional adjustment model to obtain the variance-covariance values involved in the adjustment computation. To do this, it is necessary to express the observations l, the correction values w, the constant k, the residual errors v and the adjusted observations $\tilde{l}$, as a function of the observations l. Thus, knowing that

$$l = I * l$$
$$w = b_0 + B^T l$$
$$k = -N^{-1}w = -N^{-1}b_0 - N^{-1}B^T l$$

The function vector f is given by Eq. (3.113).

$$
f = \begin{pmatrix} l \\ w \\ k \\ v \\ \tilde{l} \end{pmatrix} = \begin{pmatrix} 0 \\ b_0 \\ -N^{-1}b_0 \\ -QBN^{-1}b_0 \\ -QBN^{-1}b_0 \end{pmatrix} + \begin{pmatrix} I \\ B^T \\ -N^{-1}B^T \\ -QBN^{-1}B^T \\ I - QBN^{-1}B^T \end{pmatrix} * l \tag{3.113}
$$

Summarised as

$$y = f_0 + F^T l \tag{3.114}$$

And then,

$$Q_{yy} = F^T Q_{ll} F \tag{3.115}$$

Finally,

$$
Q_{ff} = \begin{pmatrix}
\overset{l}{Q} & \overset{w}{QB} & \overset{k}{-QBN^{-1}} & \overset{v}{-Q_{vv}} & \overset{\tilde{l}}{Q_{\tilde{l}l}} \\
B^T Q & N & -I & -B^T Q & 0 \\
-N^{-1}B^T Q & -I & N^{-1} & N^{-1}B^T Q & 0 \\
-Q_{vv} & -QB & QBN^{-1} & QBN^{-1}B^T Q & 0 \\
Q_{\tilde{l}l} & 0 & 0 & 0 & Q - Q_{vv}
\end{pmatrix}
\begin{matrix} l \\ w \\ k \\ v \\ \tilde{l} \end{matrix}
\tag{3.116}
$$

which means

$$Q_{ww} = N \tag{3.117}$$

$$Q_{kk} = N^{-1} \tag{3.118}$$

$$Q_{vv} = QBN^{-1}B^{\mathrm{T}}Q \tag{3.119}$$

$$Q_{\bar{l}\bar{l}} = Q - Q_{vv} \tag{3.120}$$

And so on.

Given the equations presented above, the following steps must be taken to perform an adjustment calculation using the conditional adjustment model:

1. Establish a detailed structure of observations.
2. Establish and eventually linearise the conditional equations. Then, form the vectors and matrices l, Q, B and eventually b_0.
3. Calculate w.
4. Calculate:

$$QB$$

$$N = B^{\mathrm{T}}QB$$

$$N^{-1}$$

$$k = -N^{-1}w$$

$$v = QBk$$

5. Calculate:

$$\bar{l} = l + v$$

6. Calculate:

$$s_0^2 = \frac{v^{\mathrm{T}}Pv}{r} = \frac{-k^{\mathrm{T}}w}{r}$$

$$Q_{vv} = QBN^{-1}B^{\mathrm{T}}Q$$

$$Q_{\bar{l}\bar{l}} = Q - Q_{vv}$$

And so on.

Accept or reject the model.

Example 3.7

Using the values in Example 3.6, calculate the adjusted values of the angles α_A, α_B and α_C by applying the Conditional adjustment model.

Solution:

As this is the solution of a triangle with three angle observations, there is one
  redundant observation (r). Therefore, a conditional equation must be written to
  apply the conditional adjustment model. Thus, using Eq. (3.97).

$$\alpha_A + \alpha_B + \alpha_C - 180° = w$$
$$43°\,28'2'' + 51°\,32'51'' + 84°\,58'52'' - 180° = 3''$$

which allows the residual equation to be written in matrix form as given by
Eq. (3.101).

$$B^{\mathrm{T}}v + w = 0$$

where

$$B^{\mathrm{T}} = \begin{bmatrix} 1 & 1 & 1 \end{bmatrix} \qquad v = \begin{bmatrix} v_A \\ v_B \\ v_C \end{bmatrix} \qquad w = 3'' \qquad l = \begin{bmatrix} 43°\,28'20'' \\ 51°\,32'51''' \\ 84°\,58'52'' \end{bmatrix}$$

Assuming that all observations are of equal precision and uncorrelated.

$$Q = \begin{bmatrix} 1 & 0 & 0 \\ 0 & 1 & 0 \\ 0 & 0 & 1 \end{bmatrix}$$

Using Eqs. (3.110) and (3.111).

$$N = B^{\mathrm{T}}QB = [3]$$
$$k = -N^{-1}w = -1''$$

Using Eqs. (3.108), the residual error vector and the vector of the adjusted values
are given as follows.

$$v = \begin{bmatrix} -1'' \\ -1'' \\ -1'' \end{bmatrix} \qquad \tilde{l} = l + v = \begin{bmatrix} 43°\,28'20'' \\ 51°\,32'51'' \\ 84°\,58'52'' \end{bmatrix} + \begin{bmatrix} -1'' \\ -1'' \\ -1'' \end{bmatrix} = \begin{bmatrix} 43°\,28'19'' \\ 51°\,32'50'' \\ 84°\,58'51'' \end{bmatrix}$$

The standard deviations of the unit weight, considering that there are three
observations (n) and two unknowns (u) is given as follows:

$$s_0^2 = \frac{v^T P v}{3 - 2} = 3.0''^2$$

$$s_0 = \pm 1.7''$$

Finally, the standard deviations of the adjusted values can be calculated using Eqs. (3.119) and (3.120).

$$Q_{vv} = QBN^{-1}B^T Q = \frac{1}{3} * \begin{bmatrix} 1 & 1 & 1 \\ 1 & 1 & 1 \\ 1 & 1 & 1 \end{bmatrix} \qquad Q_{\tilde{l}\tilde{l}} = Q - Q_{vv} = \frac{1}{3} * \begin{bmatrix} 2 & -1 & -1 \\ -1 & 2 & -1 \\ -1 & -1 & 2 \end{bmatrix}$$

$$\Sigma_{\tilde{l}\tilde{l}} = s_0^2 * Q_{\tilde{l}\tilde{l}} = \begin{bmatrix} 2.0 & -1.0 & -1.0 \\ -1.0 & 2.0 & -1.0 \\ -1.0 & -1.0 & 2.0 \end{bmatrix} = \begin{bmatrix} s_{\tilde{\alpha}A}^2 & s_{\tilde{\alpha}A\tilde{\alpha}B} & s_{\tilde{\alpha}A\tilde{\alpha}C} \\ s_{\tilde{\alpha}B\tilde{\alpha}A} & s_{\tilde{\alpha}B}^2 & s_{\tilde{\alpha}B\tilde{\alpha}C} \\ s_{\tilde{\alpha}C\tilde{\alpha}A} & s_{\tilde{\alpha}C\tilde{\alpha}B} & s_{\tilde{\alpha}C}^2 \end{bmatrix}$$

And then,

$$\tilde{\alpha}_A = 43°28'19'' \pm 1.4'' \quad \tilde{\alpha}_B = 51°32'50'' \pm 1.4'' \quad \tilde{\alpha}_C = 84°58'51'' \pm 1.4''$$

3.9.1.3 General Adjustment Model

Although common, the applications of the parametric and conditional adjustment models can be seen as special cases of a general adjustment computation model, called the Gauss-Helmert model, in which observations and parameters (unknowns) are intertwined. The general model equation, in this case, after linearisation, is given by Eq. (3.121).

$$b_1 v_1 + \ldots\ldots + b_n v_n + a_1 x_1 + \ldots\ldots + a_u x_u + w = 0 \tag{3.121}$$

with n observations, u unknowns and r conditional equations, resulting in $r - u$ redundant independent conditional equations. Then, if $n > r - u$, the general adjustment computation model can be applied.

For the mathematical development of the model, consider Eqs. (3.122) and (3.123) as already indicated for the Parametric adjustment model.

$$\tilde{x}_i = x_{i_0} + x_i \tag{3.122}$$

$$\tilde{l}_i = l_i + v_i \tag{3.123}$$

Under these conditions, for $i = 1, 2, \ldots., r$, it is possible to write r equations of the form

$$b_{i1}v_1 + \ldots\ldots + b_{in}v_n + a_{i1}x_1 + a_{iu}x_u + w_i = 0 \tag{3.124}$$

With

$$B^{\mathrm{T}} = \begin{bmatrix} b_{11} & \cdots & b_{1n} \\ b_{21} & \cdots & \vdots \\ \vdots & \ddots & \vdots \\ b_{r1} & \cdots & b_{rn} \end{bmatrix} \qquad A = \begin{bmatrix} a_{11} & \cdots & a_{1u} \\ a_{21} & \cdots & \vdots \\ \vdots & \ddots & \vdots \\ a_{r1} & \cdots & a_{ru} \end{bmatrix} \qquad v = \begin{bmatrix} v_1 \\ \cdot \\ \cdot \\ \cdot \\ v_n \end{bmatrix}$$

$$x = \begin{bmatrix} x_1 \\ \cdot \\ \cdot \\ \cdot \\ x_u \end{bmatrix} \qquad w = \begin{bmatrix} w_1 \\ \cdot \\ \cdot \\ \cdot \\ w_r \end{bmatrix}$$

Which in matrix form is given by Eq. (3.125).

$$B^{\mathrm{T}}v + Ax + w = 0 \tag{3.125}$$

Similarly to the conditional adjustment model, to comply with the least squares criterion and simultaneously have a solution to Eq. (3.125), the method of constrained minima by Lagrange multipliers can be used, leading to

$$\Omega = v^{\mathrm{T}}Pv - 2k^{\mathrm{T}}\left(B^{\mathrm{T}}v + Ax + w\right) \tag{3.126}$$

where k represents the vector with the unknown Lagrange multipliers.

The extreme values for Eq. (3.126) are reached when partial derivatives of the function with respect to v and x are set to zero.

$$\frac{\partial\Omega}{\partial v} = 2v^{\mathrm{T}}P - 2k^{\mathrm{T}}B^{\mathrm{T}} = 0 \tag{3.127}$$

$$\frac{\partial\Omega}{\partial x} = -2k^{\mathrm{T}}A = 0 \tag{3.128}$$

Then, considering Eqs. (3.106), (3.108), (3.125) and (3.128)

$$B^{\mathrm{T}}QBk + Ax + w = 0 \tag{3.129}$$

Finally, by combining Eqs. (3.128) and (3.129) into a system of matrix equations and assuming $B^T QB = N$, the following system of equations can be expressed:

$$\begin{bmatrix} N & A \\ A^T & 0 \end{bmatrix} * \begin{bmatrix} k \\ x \end{bmatrix} + \begin{bmatrix} w \\ 0 \end{bmatrix} = \begin{bmatrix} 0 \\ 0 \end{bmatrix} \tag{3.130}$$

The solution to this system of equations can be obtained by inverting the coefficient matrix, as given by Eq. (3.131).

$$\begin{bmatrix} k \\ x \end{bmatrix} = - \begin{bmatrix} Q_{kk} & N^{-1}AQ_{xx} \\ Q_{xx}A^T N^{-1} & -Q_{xx} \end{bmatrix} * \begin{bmatrix} w \\ 0 \end{bmatrix} \tag{3.131}$$

And then,

$$x = -\left(A^T N^{-1} A\right)^{-1} A^T N^{-1} w \tag{3.132}$$

$$k = \left[N^{-1}A\left(A^T N^{-1} A\right)^{-1} A^T N^{-1} - N^{-1}\right] w \tag{3.133}$$

Thus, knowing the values of k and x, it is possible to obtain the vector v and then.

$$v^T P v = v^T PQBk = v^T Bk = -k^T w \tag{3.134}$$

Finally

$$s_0^2 = \frac{v^T Bk}{r - u} = \frac{-k^T w}{r - u} \tag{3.135}$$

As with the Parametric and Conditional adjustment models, it is possible to explore the results of the general adjustment model to obtain the variance-covariance values involved in the adjustment computation, as given by the equations shown below.

$$Q_{\bar{l}\bar{l}} = Q_{ll} - Q_{vv} \tag{3.136}$$

$$Q_{ww} = N = B^T Q_{ll} B \tag{3.137}$$

$$Q_{kk} = N^{-1} - N^{-1} A Q_{xx} A^T N^{-1} \tag{3.138}$$

$$Q_{xx} = \left(A^T N^{-1} A\right)^{-1} \tag{3.139}$$

$$Q_{lv} = -Q_{vv} \tag{3.140}$$

$$Q_{\hat{l}l} = Q_{\bar{l}\bar{l}} \tag{3.141}$$

$$Q_{lx} = Q_{\bar{l}x} = -Q_{ll} B N^{-1} A Q_{xx} \tag{3.142}$$

$$Q_{xv} = 0 \qquad (3.143)$$

$$Q_{kx} = 0 \qquad (3.144)$$

$$Q_{k\bar{l}} = 0 \qquad (3.145)$$

$$Q_{v\bar{l}} = 0 \qquad (3.146)$$

An example of the use of this adjustment computation mode is given in Sect. 5.9.1.2.2.

3.10   Absolute Error Ellipses

Whenever the (X, Y) coordinates of a point are determined by an observation adjustment model, the values of their respective standard deviations, s_X and s_Y, are obtained in addition to the values of the adjusted coordinates. These values represent the positional accuracy of the point in the directions of the Cartesian axes (X, Y). However, to better evaluate the positional accuracy of the point, it would be interesting to know these values in any direction, or even to determine them independently of the axis system. From this point of view, the first positional accuracy parameter that can be determined is the standard deviation in the direction of the resultant vector s_P (*linear standard deviation*), given by Eq. (3.147).

$$S_P{}^2 = S_X{}^2 + S_Y^2 \qquad (3.147)$$

Similarly, it is shown that for any two orthogonal directions (ξ, η) with respect to the Cartesian axes (X, Y), the error propagation from the coordinate system (X, Y) into the orthogonal coordinate system (ξ, η) gives

$$\left(\xi^2 + \eta^2\right)^2 = p^2 * \xi^2 + q^2 * \eta^2 \qquad (3.148)$$

with

$$p^2 = \left(\sin^2\theta * q_{XX} + \cos^2\theta * q_{YY} + 2\sin\theta_1 * \cos\theta * q_{XY}\right) * s_0{}^2 \qquad (3.149)$$

$$q^2 = \left(\cos^2\theta * q_{XX} + \sin^2\theta * q_{YY} - 2\sin\theta * \cos\theta * q_{XY}\right) * s_0{}^2 \qquad (3.150)$$

$$\tan(2\theta) = \frac{2q_{XY}}{q_{YY} - q_{XX}} \quad \rightarrow \quad \theta = \frac{1}{2}\,\mathrm{atan}\left(\frac{2q_{XY}}{q_{YY} - q_{XX}}\right) \qquad (3.151)$$

where q_{XX}, q_{YY} and q_{XY} are the cofactors of (X) and (Y), the angle θ is the azimuth of the semi-major axis of the standard deviations along the axes (η, ξ) with respect to the ground coordinate system, and p (semi-major axis) and q (semi-minor axis) are the standard deviations along the axes (η, ξ), as shown in Fig. 3.4. To correctly calculate the angle θ it is necessary to analyse the quadrant in which the major semi-major axis lies. This can be done using the same rezoning explained in Sect. 11.2.

The values of the parameters p, q and θ can also be determined using the following equations:

$$p^2 = \frac{s_0^2}{2} * \left[(q_{XX} + q_{YY}) + \sqrt{(q_{XX} - q_{YY})^2 + 4q_{XY}^2} \right] \qquad (3.152)$$

$$q^2 = \frac{s_0^2}{2} * \left[(q_{XX} + q_{YY}) - \sqrt{(q_{XX} - q_{YY})^2 + 4q_{XY}^2} \right] \qquad (3.153)$$

or

$$p^2 = \frac{(s_X^2 + s_Y^2) + \sqrt{(s_X^2 - s_Y^2)^2 + 4s_{XY}^2}}{2} \qquad (3.154)$$

$$q^2 = \frac{(s_X^2 + s_Y^2) - \sqrt{(s_X^2 - s_Y^2)^2 + 4s_{XY}^2}}{2} \qquad (3.155)$$

$$\theta = \text{atan}\left(\frac{s_{XY}}{p^2 - s_X^2} \right) \qquad (3.156)$$

Equation (3.148) is a fourth-degree equation represented geometrically by the dotted line highlighted in Fig. 3.4. Because it is a fourth-degree curve, it is more complex to calculate or even represent. However, it can be replaced by an inscribed ellipse, which is much easier to construct and calculate. Thus, instead of the fourth-degree error curve, an *absolute error ellipse* is used for simplicity to indicate the distribution of standard deviations of the coordinates (X, Y) of the point (P) in any direction (ξ, η). Equation (3.157) is the mathematical representation of this ellipse.

Fig. 3.4 Error curve and error ellipse

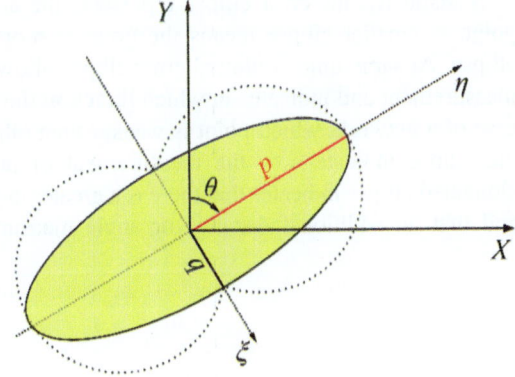

$$\frac{\eta^2}{p^2} + \frac{\xi^2}{q^2} - 1 = 0 \tag{3.157}$$

The parameters of this ellipse can be determined as follows:

(a) Its semi-axes are rotated about the axes of the coordinate system (X, Y) by an angle equal to θ, whose value can be determined by Eqs. (3.151) or (3.156).
(b) The lengths of its semi-axes p and q can be determined by Eqs. (3.152) and (3.153) or (3.154) and (3.155) depending on the availability of s_0^2.

The angle θ gives the direction of the maximum value of p. To calculate its value, it is necessary to verify the correct quadrant of 2θ by examining the sign of the numerator and denominator in Eq. (3.151) or of θ in Eq. (3.156), as given in Table 11.1.

From Fig. 3.5, it can be seen that the error ellipse is inscribed in the rectangle formed by the sides s_X and s_Y. The linear standard deviation s_P, defined by Eq. (3.147), is the diagonal of this rectangle, i.e., its value is always greater than the value of the semi-major axis of the error ellipse.

Fig. 3.5 Relationship between the absolute error ellipse and the linear standard deviation s_P

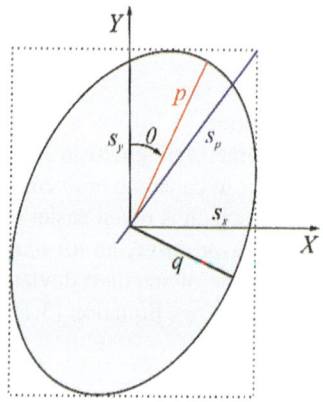

Statistically, the error ellipse represents the area of uncertainty around a given point. A smaller ellipse means the point is more reliable than one with a larger ellipse. At same time, a plotted error ellipse allows analysis of the geometry of the measurement and indicates in which direction the measurement is weaker or, in the case of a network, which point is weaker than others. The shape and orientation of the ellipse indicate how the measurement or network can be strengthened. An elongated ellipse indicates that there is a greater uncertainty in one of the coordinates and that an additional distance or angle measurement may be required for that station. A large but circular ellipse means that the point is balanced in (X, Y) but more accurate measurement techniques are required, or additional measurements from several other stations are needed.

Note that the error ellipse presented is a geometric entity in two dimensions, defined by two quantities (unknowns). The same principle can be generalised to any number of unknowns, that is, in a space $n > 2$, to obtain a hyper ellipsoid in n dimensions.

Example 3.8

Given the variance-covariance matrix for the adjusted coordinates (X, Y) of point (P_1) as shown below, calculate the parameters of the absolute standard error ellipse of point (P_1).

$$\Sigma_{XX} = \begin{bmatrix} 0.0000823 & -0.0000251 \\ -0.0000251 & 0.0001803 \end{bmatrix}_{m^2}$$

Solution:

Using Eqs. (3.154) and (3.155).

$$p^2 = \frac{(0.0000823 + 0.0001803) + \sqrt{(0.0000823 - 0.0001803)^2 + 4 * (-0.0000251)^2}}{2}$$

$$= 0.0001863 \text{ m}^2 \longrightarrow p = 13.6 \text{ mm}$$

$$q^2 = \frac{(0.0000823 + 0.0001803) - \sqrt{(0.0000823 - 0.0001803)^2 + 4 * (-0.0000251)^2}}{2}$$

$$= 0.0000762 \text{ m}^2 \longrightarrow q = 8.7 \text{ mm}$$

Using Eq. (3.156).
$$\theta = \text{atan}\left(\frac{-0.0000251}{0.000186 - 0.0000823}\right) = -0.2364379 rad = -13° 32' 49'' \text{ Quadrant 4, then}$$

$$\theta = 346° 27' 11''$$

The graphical representation of the corresponding absolute error ellipse is shown in Fig. 3.6.

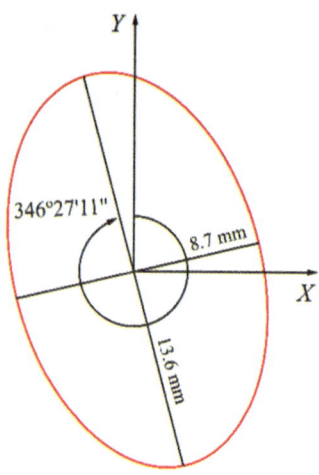

3.11   Relative Error Ellipses

As formulated in Eq. (3.67), the cofactor matrix $\boldsymbol{Q}_{XX}$ of (X, Y) is given by
Eq. (3.158).

$$\boldsymbol{Q}_{XX} = \begin{pmatrix} q_{XX} & q_{XY} \\ q_{YX} & q_{YY} \end{pmatrix} \tag{3.158}$$

If more than one point is considered, the matrix $\boldsymbol{Q}_{XX}$ has more than two parameters, as follows for the generalised case.

$$\boldsymbol{Q}_{XX} = \begin{bmatrix} q_{X_1 X_1} & q_{X_1 Y_1} & \cdots & \cdots & q_{X_1 Y_n} \\ q_{Y_1 X_1} & q_{Y_1 Y_1} & \cdots & \cdots & \vdots \\ \vdots & \vdots & \ddots & \ddots & \vdots \\ \vdots & \vdots & \cdots & q_{X_n X_n} & q_{X_n Y_n} \\ q_{Y_n X_1} & \cdots & \cdots & q_{Y_n X_n} & q_{Y_n Y_n} \end{bmatrix} \tag{3.159}$$

Using the law of propagation of cofactors, it is possible to calculate the variances
and covariances of any value related to the coordinates (X_i, Y_i). Consider, therefore,
the case of the difference in coordinates between points (P_1) and (P_2), given by
Eq. (3.160).

$$\Delta X = X_2 - X_1 \tag{3.160}$$

Given that the variance-covariance matrix for the adjusted coordinates of (P_1) and
(P_2) is known from a Least Squares adjustment, it is possible to determine the
variance-covariance matrix for the coordinate differences by using Eq. (3.28).

The variance-covariance matrix Σ_{XX} for the coordinates of (P_1) and (P_2) is given by Eq. (3.161).

$$\Sigma_{XX} = \begin{bmatrix} s^2_{X_1} & s_{X_1Y_1} & s_{X_1X_2} & s_{X_1Y_2} \\ s_{Y_1X_1} & s^2_{Y_1} & s_{Y_1X_2} & s_{Y_1Y_2} \\ s_{X_2X_1} & s_{X_2Y_1} & s^2_{X_2} & s_{X_2Y_2} \\ s_{Y_2X_1} & s_{Y_2Y_1} & s_{Y_2X_2} & s^2_{Y_2} \end{bmatrix} \tag{3.161}$$

The variance-covariance matrix $\Sigma_{\Delta X\Delta X}$ for the difference of the coordinates is given by Eq. (3.162).

$$\Sigma_{\Delta X\Delta X} = \begin{bmatrix} -1 & 0 & 1 & 0 \\ 0 & -1 & 0 & 1 \end{bmatrix} * \begin{bmatrix} s^2_{X_1} & s_{X_1Y_1} & s_{X_1X_2} & s_{X_1Y_2} \\ s_{Y_1X_1} & s^2_{Y_1} & s_{Y_1X_2} & s_{Y_1Y_2} \\ s_{X_2X_1} & s_{X_2Y_1} & s^2_{X_2} & s_{X_2Y_2} \\ s_{Y_2X_1} & s_{Y_2Y_1} & s_{Y_2X_2} & s^2_{Y_2} \end{bmatrix}$$

$$* \begin{bmatrix} -1 & 0 \\ 0 & -1 \\ 1 & 0 \\ 0 & 1 \end{bmatrix} = \begin{bmatrix} s^2_{\Delta X} & s_{\Delta X\Delta Y} \\ s_{\Delta Y\Delta X} & s^2_{\Delta Y} \end{bmatrix} \tag{3.162}$$

The $s^2_{\Delta X}$ and $s^2_{\Delta Y}$ values indicate the variances between the difference of the abscissa and ordinate coordinates of points (P_1) and (P_2) and $s_{\Delta X\Delta Y}$ their covariance. These values, therefore, allow to calculate the ellipse of errors of the relative precision of the position of these two points, called the *Relative Error Ellipse*.

The parameters of the relative error ellipse between points (P_1) and (P_2) can then be determined as follows.

$$p^2 = \frac{\left(s^2_{\Delta X} + s^2_{\Delta Y}\right) + \sqrt{\left(s^2_{\Delta X} - s^2_{\Delta Y}\right)^2 + 4s^2_{\Delta X\Delta Y}}}{2} \tag{3.163}$$

$$q^2 = \frac{\left(s^2_{\Delta X} + s^2_{\Delta Y}\right) - \sqrt{\left(s^2_{\Delta X} - s^2_{\Delta Y}\right)^2 + 4s^2_{\Delta XY}}}{2} \tag{3.164}$$

$$\theta = \mathrm{atan}\left(\frac{s_{\Delta X\Delta Y}}{p^2 - s^2_{\Delta X}}\right) \tag{3.165}$$

The graphical representation of the relative error ellipse is shown in Fig. 3.7. The error ellipse represents the area of uncertainty around a given point.

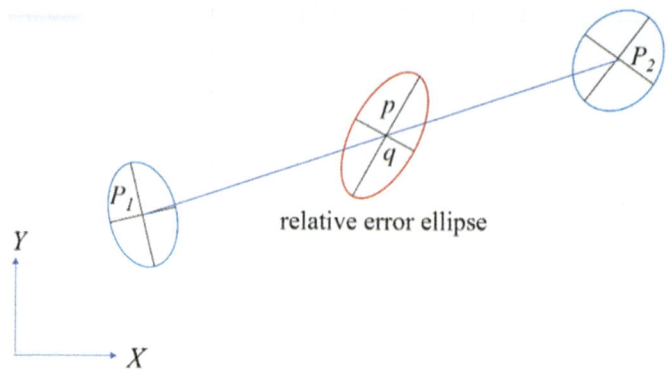

Fig. 3.7 Relative error ellipse

Example 3.9

The (X, Y) coordinates of points (P_1) and (P_2) have been determined by a traversing adjustment. Using the variance-covariance matrix for the adjusted points given below, determine the relative error ellipse parameters between (P_1) and (P_2).

$$\Sigma_{XX} = \begin{bmatrix} 8.23 * 10^{-5} & -2.51 * 10^{-5} & 2.87 * 10^{-5} & -9.91 * 10^{-6} \\ -2.51 * 10^{-5} & 1.80 * 10^{-4} & 2.33 * 10^{-6} & 1.04 * 10^{-4} \\ 2.87 * 10^{-5} & 2.33 * 10^{-6} & 4.42 * 10^{-5} & -3.57 * 10^{-6} \\ -9.91 * 10^{-6} & 1.04 * 10^{-4} & -3.57 * 10^{-6} & 9.72 * 10^{-5} \end{bmatrix}_{m^2}$$

Solution:

Using Eq. (3.162),

$$\Sigma_{\Delta X \Delta X} = \begin{bmatrix} s_{\Delta X}^2 & s_{\Delta X \Delta Y} \\ s_{\Delta Y \Delta X} & s_{\Delta Y}^2 \end{bmatrix} = \begin{bmatrix} 6.90 * 10^{-5} & -2.11 * 10^{-5} \\ -2.11 * 10^{-5} & 7.02 * 10^{-5} \end{bmatrix}_{m^2}$$

Using Eqs. (3.163) to (3.165),

$$p^2 = \frac{\left(6.90 * 10^{-5} + 7.02 * 10^{-5}\right) + \sqrt{\left(6.90 * 10^{-5} - 7.02 * 10^{-5}\right)^2 + 4 * 4.44 * 10^{-10}}}{2}$$

$$= 9.07 * 10^{-5} \, m^2$$

$$p = 0.0095 \, m = 9.5 \, mm$$

$$q^2 = \frac{(6.90*10^{-5}+7.02*10^{-5}) - \sqrt{(6.90*10^{-5}-7.02*10^{-5})^2 + 4*4.44*10^{-10}}}{2}$$

$$= 4.85*10^{-5}\,\text{m}^2$$
$$q = 0.0070\,\text{m} = 7.0\,\text{mm}$$

$$\theta = \text{atan}\left(\frac{-2.11*10^{-5}}{9.07*10^{-5} - 6.90*10^{-5}}\right) = -0.7707297\,rad = -45°\,50'26''$$

$$= 315°\,50'26''$$

The graphical representation of the corresponding relative error ellipse is shown in Fig. 3.8.

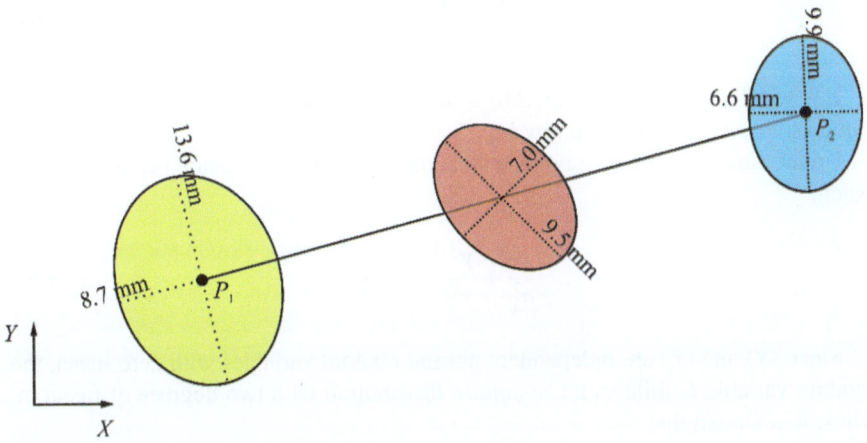

Fig. 3.8 Relative error ellipse between points (P_1) and (P_2)

3.12   Error Ellipse Confidence Level

The previous sections presented the equations that allow calculating the size and orientation of the geometric figure that represents the error ellipse of two random and independent variables (X, Y). The error ellipse thus determined delimits the surface of the possible geometric location of these two variables. However, it remains to determine the probability that any pair of values (X, Y) lies within the ellipse.

To determine this probability, it is necessary to consider the cumulative joint probability distribution function given by Eq. (3.166).

$$f(X,Y) = \frac{1}{2\pi\sigma_X\sigma_Y\sqrt{1-\rho_{XY}^2}} \, exp \left\{ \frac{-1}{2(1-\rho_{XY}^2)} \left[\frac{X^2}{\sigma_X^2} - 2\rho_{XY} * \left(\frac{X*Y}{\sigma_X*\sigma_Y} \right) + \frac{Y^2}{\sigma_Y^2} \right] \right\}$$

(3.166)

Which can be rewritten as

$$\frac{X^2}{\sigma_X^2} - 2\rho_{XY} * \left(\frac{X*Y}{\sigma_X*\sigma_Y} \right) + \frac{Y^2}{\sigma_Y^2} = \left(1 - \rho_{XY}^2 \right) * c^2$$

(3.167)

where c is a constant that depends on the values of the standard deviations σ_X, σ_Y and the correlation coefficient ρ_{XY}. Thus, knowing that the errors in (X) and (Y) are independent, $\rho_{XY} = 0$, Eq. (3.167) becomes

$$\frac{X^2}{\sigma_X^2} + \frac{Y^2}{\sigma_Y^2} = c^2$$

(3.168)

It is worth noting that Eq. (3.168) is an ellipse equation, which for $c^2 = 1$ is the same as the error ellipse given by Eq. (3.157).

Under these conditions, a point with coordinates (X, Y) is inside the error ellipse when,

$$\frac{X^2}{\sigma_X^2} + \frac{Y^2}{\sigma_Y^2} = U \le c^2$$

(3.169)

Since (X) and (Y) are independent normal random variables with zero mean, the random variable U follows a *Chi-square* distribution with two degrees of freedom. Thus, it is known that.

$$P\left\{ \frac{X^2}{\sigma_X^2} + \frac{Y^2}{\sigma_Y^2} \le \chi_{2,1-\alpha}^2 \right\} = 1 - \alpha$$

(3.170)

If $c^2 = 1$, then $\chi^2 = 1$. Equation (3.170) can then be rewritten as

$$P\left\{ \frac{X^2}{\sigma_X^2} + \frac{Y^2}{\sigma_Y^2} \le 1 \right\} = P\{\chi^2 \le 1\} = 1 - \alpha$$

(3.171)

The value of $(1 - \alpha)$ can be obtained as a function of the *Chi-square* probability distribution equation, where $(1 - \alpha = 0.3935)$, i.e., the probability of a point falling inside the standard error ellipse is 39.35%. In other words, there is a 39.35% probability that a point will be inside the error ellipse with dimensions equal to $1\sigma_0$.

On the other hand, for a significance level $\alpha = 0.05$ and two degrees of freedom.

$$P\left\{\frac{X^2}{\sigma_X^2} + \frac{Y^2}{\sigma_Y^2} \le \chi_{2,0.05}^2\right\} = P\{\chi^2 \le \chi_{2,0.05}\} = 0.95 \tag{3.172}$$

Then

$$c^2 = \chi_{2,0.05}^2 = 5.99 \tag{3.173}$$

In other words, there is a 95% probability that a given pair of variables (X, Y) lies within the error ellipse with dimensions equal to $\sqrt{5.99} = 2.447$ times the standard error ellipse, i.e., $2.447\sigma_0$.

Generalising and considering (A) and (B) as being, respectively, the major and minor semi-axes of the confidence ellipse.

$$A^2 = \left[\frac{1}{2}(\sigma_X^2 + \sigma_Y^2) + \sqrt{\frac{1}{4}(\sigma_X^2 - \sigma_Y^2)^2 + \sigma_{XY}^2}\right] * \chi_{2,1-\alpha}^2 \tag{3.174}$$

$$B^2 = \left[\frac{1}{2}(\sigma_X^2 + \sigma_Y^2) - \sqrt{\frac{1}{4}(\sigma_X^2 - \sigma_Y^2)^2 + \sigma_{XY}^2}\right] * \chi_{2,1-\alpha}^2 \tag{3.175}$$

The inclination of the axes remains equal to θ.

The mathematical developments presented so far have assumed that the true values of the unknowns and the values of σ_0^2 are known. In the most common cases, however, these values are unknown and σ_0^2 must be estimated as a function of the empirical value s_0^2. Thus, by the definition of χ^2, it is known that:

$$\frac{s_0^2}{\sigma_0^2} = \frac{1}{r} * \chi_r^2 \tag{3.176}$$

where

$r = n - u =$ degree of freedom
$n =$ number of observations
$u =$ number of unknowns

Thus,

$$\sigma_0^2 = \frac{s_0^2 * r}{\chi_r^2} \tag{3.177}$$

Also, by the definition of the *F-distribution*.

$$F_{f_1 f_2} = \frac{f_2}{f_1} * \frac{\chi_{f_1}^2}{\chi_{f_2}^2} \tag{3.178}$$

From where

$$\sigma_0^2 * \chi_2^2 = \frac{s_0^2 * r * \chi_2^2}{\chi_r^2} = 2s_0^2 * F_{2,r,1-\alpha} \tag{3.179}$$

And then,

$$P\left\{ \frac{X^2}{q_{XX}} + \frac{Y^2}{q_{YY}} \le 2s_0^2 * F_{2,r,1-\alpha} \right\} = 1 - \alpha \tag{3.180}$$

By analogy with Eqs. (3.152) and (3.153),

$$A^2 = s_0^2 * F_{2,r,1-\alpha} * (q_{XX} + q_{YY} + w) \tag{3.181}$$

$$B^2 = s_0^2 * F_{2,r,1-\alpha} * (q_{XX} + q_{YY} - w) \tag{3.182}$$

$$w = \sqrt{(q_{XX} + q_{YY})^2 + 4q_{XY}^2} \tag{3.183}$$

According to Eq. (3.173), note that

$$c^2 = 2F_{2,r,1-\alpha} \tag{3.184}$$

Thus, the relationship between the standard error ellipse and the confidence ellipse depends on the desired confidence level and the degrees of freedom, as shown in Table 3.9. For example, if the coordinates (X, Y) of a point (P) are determined with 3 degrees of freedom, the *error ellipse expansion factor* (or *scaling factor*) for a 95% confidence level is equal to 4.37, i.e., it will have dimensions equal to 4.37 times the dimensions of the standard error ellipse.

Table 3.9 Expansion factor to transform the standard error ellipse into the confidence ellipse

Degree of freedom	Confidence level		
	90%	95%	99%
1	$\sqrt{2 * 49.50} = 9.95$	$\sqrt{2 * 199.40} = 19.97$	$\sqrt{2 * 5,000} = 100.00$
2	$\sqrt{2 * 9.00} = 4.24$	$\sqrt{2 * 19.00} = 6.16$	$\sqrt{2 * 99.00} = 14.07$
3	3.30	4.37	7.85
5	2.75	3.40	5.15
10	2.42	2.86	3.89
20	2.28	2.64	3.42
120	2.35	2.48	3.10

Example 3.10
Suppose the coordinates of the point (P_1) in Example 3.8 were determined with a redundancy of 10. Calculate the parameters of the absolute 95% confidence error ellipse.

Solution:

Considering $r = 10$ and $1 - \alpha = 95\%$.

$$F_{2,10.95\%} = 4.10$$

$c = \sqrt{2 * 4.10} = 2.86$ *(Same as indicated in Table 3.9)*
Thus,

$$A = 2.86 * 13.6 = 38.9 \text{ mm}$$
$$B = 2.86 * 8.7 = 24.9 \text{ mm}$$

With the same orientation angle.

3.13   Review Questions

1. To determine the distance between points (A) and (B), three distances were measured, as shown in Fig. 3.9. Distance d_1 was measured n_1 times, d_2 was measured n_2 times, and d_3 was measured n_3 times. Write the equations for the distance AB and its standard deviation from the measured values.

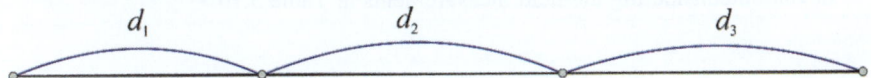

Fig. 3.9 Measurement geometry

2. Considering Fig. 3.10, write the equation for the standard deviation of δ.

Fig. 3.10 Measurement geometry

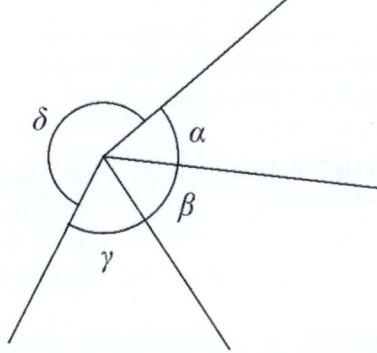

3. Suppose the elevation of a benchmark (A) is known with a standard deviation of ±3 mm. Given that the standard deviation of differential levelling over 1 km is ±2 mm, calculate the maximum distance that must be travelled from point (A) to determine the elevation of point (B) with an accuracy of ±10 mm.

4. An angle α was determined with three different theodolites, as shown below. Calculate the weighted mean and its standard deviation.

$$45° 27'20'' \pm 6''$$

$$45° 27'24'' \pm 9''$$

$$45° 27'23'' \pm 3''$$

5. A baseline was measured with four different total stations. The readings are given below. Calculate the weighted mean and its standard deviation.

$$1000.192 \text{ m} \pm 4 \text{ mm}$$

$$1000.403 \text{ m} \pm 6 \text{ mm}$$

$$1000.531 \text{ m} \pm 2 \text{ mm}$$

$$1000.285 \text{ m} \pm 5 \text{ mm}$$

6. Given the elevations of H_A, H_B and H_C (assumed free of errors) in a levelling network, as shown in Fig. 3.11, calculate the elevation H_D and its standard deviation considering the field measurements in Table 3.10.

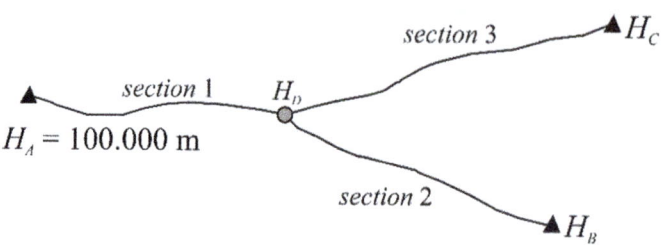

Fig. 3.11 Levelling network

Table 3.10 Field measurements

Elevation [m]	Section	d [km]	ΔH [m]
$H_A = 100.0000$	1	2.0	8.670
$H_B = 112.3411$	2	3.0	−3.668
$H_C = 99.1013$	3	3.5	9.574

7. Seven distances have been measured to determine the distance between points (A) and (B), as shown in Fig. 3.12. Given the measured values, write the error and conditional equations to apply the Parametric and Conditional adjustment models to determine the distance AB.

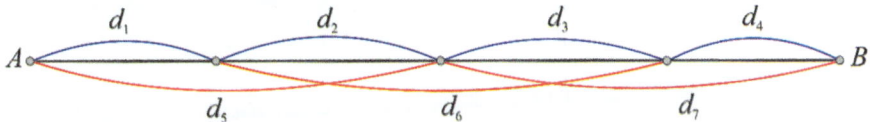

Fig. 3.12 Measurement geometry

8. In Fig. 3.13, the slope distance d' is 1000.000 ± 0.010 m and the angle β is $10° \, 00'00'' \pm 7''$. Calculate the values of the horizontal distance d and the vertical distance h and their precisions.

Fig. 3.13 Measurement geometry

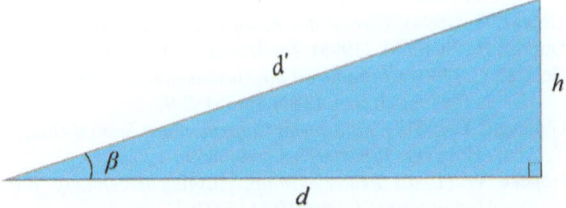

9. Three angles of a triangle were measured with three different theodolites. The measurement values are indicated below. Determine the values of the adjusted angle that satisfy the condition $\alpha + \beta + \gamma = 180°$.

$$\alpha = 30°\,00'05'' \pm 6''$$

$$\beta = 60°\,00'07'' \pm 9''$$

$$\gamma = 90°\,00'10'' \pm 3''$$

10. In Fig. 3.14, $d_{AB}=500.000$ m is the horizontal distance between points (A) and (B). Suppose a distance of 559.010 m ± 5 cm and an angle of $26°\,33'54'' \pm 7''$ were measured to locate a point (C) perpendicular to AB at (B). Assuming that points (A) and (B) are error-free, determine the value of the distance d_{BC} and its precision using the Parametric adjustment model.

Fig. 3.14 Measurement
geometry

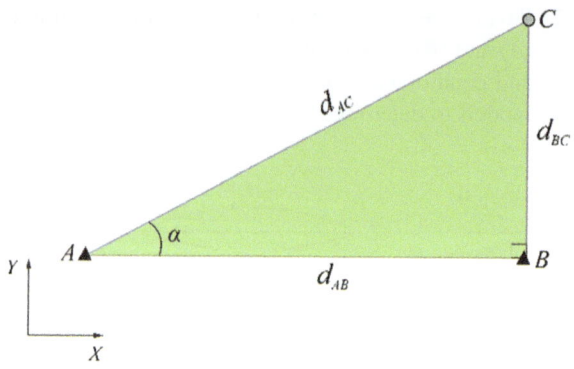

References

Bjerhammar, A. (1973). *Theory of Errors and Generalized Matrix Inverses*. Elsevier, Amsterdam, The Netherlands. ISBN 0-444-40981-5.

Dupraz, H. (1994). *Theorie des Erreurs 2*. EPFL, Lausanne, Switzerland.

Dupraz, H., Stahl, M. (1995). *Theorie des Erreurs 3*. EPFL, Lausanne, Switzerland.

Gemael, C. (2004). *Introdução ao Ajustamento de Observações—Aplicações geodésicas*. Editora UFPR. Curitiba, Brazil. ISBN 85-85132-92-2.

Ghilani, C.D. (2017). *Adjustment Computations: Spatial Data Analysis*. 6th Edition, John Wiley & Sons, Hoboken. ISBN: 978-1-119-38598-1

Höpcke, W. (1980). *Fehlerlehre und Ausgleichsrechnung*. Berlin, New York: Walter de Gruyter GmbH. doi: https://doi.org/10.1515/9783110838206.

Howald, P. (1994). *Theorie des Erreurs 1*. EPFL, Lausanne, Switzerland.

Mikhail, E. M. (1982). Observations and Least Squares. University Press of America. USA. ISBN 9780819123978, 0819123978.

Ogundare, John. (2018). Understanding Least Squares Estimation and Geomatics Data Analysis. https://doi.org/10.1002/9781119501459. 2019 John Wiley & Sons, Inc.

Tiestra, J. M. (1956). *Theory of the adjustment of normally distributed observations*. Nv. Uitgeverij Argus. Delft, Holland.

Chapter 4
Reference Systems

Irineu da Silva and Marcelo Monari

4.1 Introduction

When collecting field data for an engineering project, it needs to be mathematically modelled and the corresponding geographic features georeferenced to generate appropriate geospatial information for project design.

A mathematical model is a set of rules describing real-world situations, their interactions and dynamics through mathematical equations. In the case of Geomatics, it is about establishing geometric and algebraic relationships between data collected in the field to represent the physical situation in which the measurements were carried out, aiming to determine the value of geospatial data. To address this problem, it is first necessary to establish the geometry of the space in which the data are inserted. This means defining a spatial reference system that allows the precise location of points on the Earth's surface, defining a mathematical model composed of a *Coordinate Reference System* and a *terrestrial frame of control points*.

In Geomatics, a Coordinate Reference System is based on two geometric elements: the *datum* and the *coordinate system*. The *datum* defines how the Coordinate Reference System is related to the Earth, while the coordinate system describes how geospatial data is positioned in the *datum*. The reference frame, in turn, is the realisation of the Coordinate Reference System that provides coordinate values of points (control points) distributed over a small project area or even over a region as large as an entire continent, depending on the reference system being adopted. In

I. da Silva (✉)
São Carlos School of Engineering, University of São Paulo, São Carlos, SP, Brazil
e-mail: irineu@sc.usp.br

M. Monari
Civil Engineer, Assistant Professor in the Department of Civil Engineering at the Federal University of São Carlos (UFScar), São Paulo, Brazil
e-mail: marcelo.monari@ufscar.br

© The Author(s), under exclusive license to Springer Nature Switzerland AG 2025
I. da Silva, P. C. L. Segantine, *Geomatics Applied to Civil Engineering*,
https://doi.org/10.1007/978-3-031-75737-2_4

other words, the reference frame provides the physical means for using the Coordinate Reference System. In Geomatics, control points are materialised on the terrain in different ways, depending on their level of accuracy and whether they are classified as an engineering (local) or a geodetic control point. Figure 4.1 shows four examples of materialised control points currently used for Civil Engineering purposes.

a) Concrete pillar geodetic monument.

b) Continuous monitoring GNSS station on a concrete pillar.

c) Geodetic survey marker.

d) Engineering survey monument.

Fig. 4.1 Examples of control points available for Civil Engineering purposes. Courtesy: IBGE. (**a**) Concrete pillar geodetic monument. (**b**) Continuous monitoring of GNSS station on a concrete pillar. (**c**) Geodetic survey marker. (**d**) Engineering survey monument

In view of the above, and the fact that Civil Engineers are confronted with a combination of different surveying instruments and different surface dimensions in their projects, they need to have an adequate knowledge of the different reference systems used in Geomatics. To this end, this chapter discusses the main fundamentals of *datum* definition for Civil Engineering and its influence on Civil Engineering projects, while Chap. 5 presents the fundamentals of coordinate systems and their interrelationships.

4.2   Datum Definition

Given that the geospatial data dealt with in Geomatics are located on or near the Earth's surface, the geometric space in which they are embedded is related to the shape and dimensions of the Earth and, above all, to the geometric models used to represent its surface (*datum* definition). In this context, due to the roughness of the

Earth's relief, the irregularity of its gravitational field,[1] and the need for an adequate framework of control points for Civil Engineering projects, the definition of *datum* is divided into *horizontal* and *vertical*. The *horizontal datum* is the surface to which horizontal distances are referenced (see Chap. 7), and the *vertical datum* is the surface to which elevations are referenced, as presented in the following sections. On the other hand, depending on the size of the project, the *horizontal datum* may be either an *engineering (local) datum* or a *geodetic datum*, as described below.

4.2.1 Ground Surface

The primary surface of interest for Geomatics is the ground surface. This is the surface on which engineering and geodetic measurements are carried out and on which engineering projects are set out. However, its irregularity does not allow it to be used as a reference system, requiring the measured values to be reduced onto another surface where the corresponding calculations are carried out. Thus, the ground surface is where the field operations are carried out, and the reference surface is where the corresponding engineering or geodetic calculations are performed.

4.2.2 Vertical Datum: Geoid Model

Before studying the horizontal datum, it is important to understand the concepts of the *geoid model* of the Earth's shape, which defines the vertical component of the Coordinate Reference System and is, therefore, called the *vertical datum*. Conceptually, the geoid is the level surface[2] of the Earth's gravitational field, generated normal to the vertical line (plumb line) at any point on the Earth's surface, at the mean sea level elevation. It can be understood as the *equipotential surface* that best fits the undisturbed mean sea level surface, extending over the continents. The mean sea level surface is not equipotential since, in addition to the gravitational force, the action of the wind, salinity, temperature, etc., also changes its shape.

The geoid, in turn, is defined only by gravity. Thus, due to the variations in the distribution of mass and the effect of the Earth's rotation, it has an irregular geometric shape without a rigorous mathematical definition, as shown in Fig. 4.2. Due to the same effect, the Earth's gravitational field has a wavy shape so that the level surfaces at different elevations are not parallel.

---

[1] Region of the gravitational distribution that a body generates around itself, which, in the case of the Earth, is the vector field that represents the gravitational attraction that the Earth exerts on other bodies.

[2] The surface of constant gravitational potential, i.e., the intensity of the gravitational force at any point on this surface, is constant. Also known as the *equipotential surface*.

Fig. 4.2 Geoid model of
the Earth (vertical
exaggeration of 15,000).
(Adapted from NASA 1990)

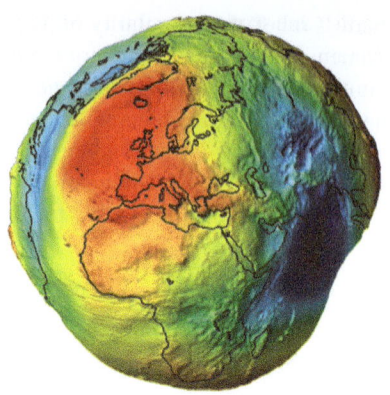

The mathematical determination of the geoid surface is complex, as it depends on the geoid undulations resulting from the variation in the direction of the vertical line along any point on the Earth's surface. For this reason, the geoid model is inadequate for the horizontal positioning of geospatial data and is primarily used for vertical positioning (elevation), i.e., defining the *vertical datum* of a Coordinate Reference System. Further details on the vertical datum are given in Chap. 13.

4.2.3   Engineering Datum

The simplest *horizontal datum* used for Civil Engineering projects is that in which a *local ground-based horizontal plane* is adopted as the reference surface within the boundaries of the survey area. This means that the Earth's curved surface is replaced by a horizontal plane tangent to the level surface at the tangent point, which defines the surveying concept known as *Plane Surveying*. The tangent point, for this purpose, is usually a point on the ground surface at the centre of the project area, with assumed spatial coordinates (x, y, H). In this case, the Earth's curved surface is completely neglected, and all distances and directions within the project area are reduced to the same horizontal plane. Regarding elevations, all level lines[3] are mathematically considered as straight lines perpendicular to the direction of the plumb line at the tangent point, i.e., the convergence of the plumb lines towards the Earth's centre of gravity within the boundaries of the project area is disregarded. On the other hand, it is important to point out that this is a hypothetical surface whose use in engineering projects must be carefully evaluated since geometric deformations will result from the flattening of the corresponding curved surface. Details of the values of these deformations are discussed in Chap. 7.

The widespread use of the *engineering datum* as a reference surface for Civil Engineering projects is because it is geometrically intuitive and easy to understand,

---

[3] A horizontal straight-line tangent to a level surface.

mainly due to the possibility of using the Two-Dimensional Cartesian Coordinate System. Furthermore, as described throughout the chapters of this book, it is compatible with the *geodetic datum* (presented below) through relatively easy-to-apply mathematical equations. Figure 4.3 illustrates the geometric details of an *engineering datum* compared to the Earth's curved surface.

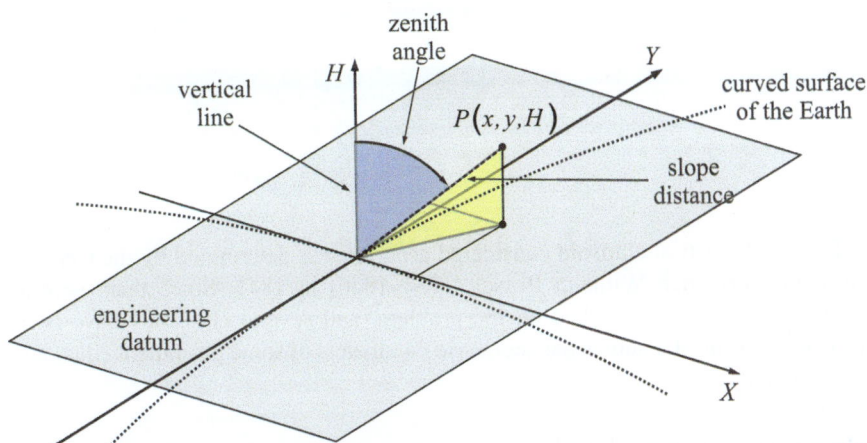

Fig. 4.3 Geometry of an engineering datum

4.2.4 Geodetic Datum

To avoid the problem of the rigorous mathematical indeterminacy of the Earth's shape and the difficulty of using the engineering datum as a *horizontal datum* over large areas, geodesists proposed the adoption of the *ellipsoid of revolution*, already known since the seventeenth century, as a reference surface. This surface is called the *reference ellipsoid*, from which the term *ellipsoidal surface* is derived. It is the mathematical surface that best fits the irregular shape of the geoid. Because it is a *datum* that considers the curvature of the Earth, its use defines the surveying concept known as *Geodetic Surveying*.

In Geodesy, a reference ellipsoid can be either global or regional. It is global if it covers the entire terrestrial surface and regional if it is intended to cover only a part of it. It is global and geocentric if its centre coincides with the Earth's centre of mass. In this case, the difference between its semi-axes (a, b) is about 25 km for the GRS80 ellipsoid. Figure 4.4 shows a 2D cross-section of a *geocentric global ellipsoid* fitted to the geoid.

In Fig. 4.4, the geoid undulation is highlighted by a dotted line for better graphical visualization. To understand the difference between the two surfaces, taking the GRS80 ellipsoid as a reference (Table 4.1), the geoid is 85 m above the ellipsoid in the west of Ireland and 106 m below the ellipsoid in the northern region of Sri Lanka.

Fig. 4.4 Cross-sectional comparison between the ellipsoid and the geoid (exaggerated for clarity)

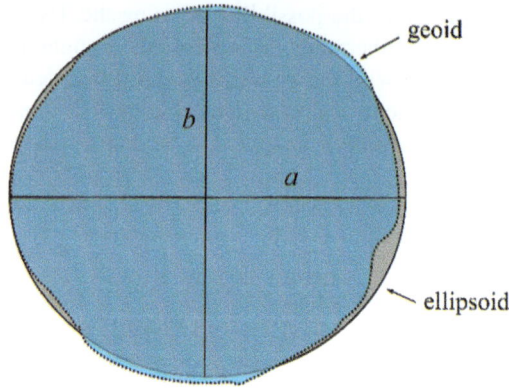

The first terrestrial ellipsoid considered accurate was determined by the German astronomer Friedrich Wilhelm Bessel (1784–1846) in 1841. Since then, several other researchers have worked on the problem, and several ellipsoids have been determined. Table 4.1 shows the geometric parameters of some prominent ellipsoids for Geomatics.

Table 4.1 Important ellipsoids for Geomatics

Name	Values of geometric parameters[*]	
Bessel 1841	$a = 6,377,397.155$ m	$f = 1/299.15281285$
International 1924 (Hayford)	$a = 6,378,388.000$ m	$f = 1/297$
North American 1927 (Clarke 1866)	$a = 6,378,206.400$ m	$f = 1/294.9786982$
North American 1983 (GRS80)	$a = 6,378,137.000$ m	$f = 1/298.257222101$
South American Datum 1969 (SAD69) International 1967	$a = 6,378,160.000$ m	$f = 1/298.25$
European 1987	$a = 6,378,388.000$ m	$f = 1/297$
Geodetic Reference System 1980 (GRS80)	$a = 6,378,137.000$ m	$f = 1/298.257222101$
World Geodetic System 1984 (WGS84)	$a = 6,378,137.000$ m	$f = 1/298.257223563$

*Note the non-compliance in the number of significant figures indicated for the geometric parameters of the ellipsoids. The values indicated are in accordance with those indicated in the official texts that define the geometric parameters of each ellipsoid
a semi-major axis of the ellipse, f flattening of the ellipse

As a reference for Civil Engineers, Table 4.2 lists some ellipsoid parameters that are commonly used in geodetic calculations and, for that reason, also included in several equations throughout this book.

Table 4.3 presents the values of some of the parameters for the GSR80 ellipsoid.

Although the use of geodetic datum in Civil Engineering projects is relatively limited, it is important to understand its principles. It is used to position highly accurate points, such as national control points to which the less precise local control points used in Civil Engineering projects are linked.

Another point to note is that the Global Navigation Satellite System (GNSS) measurements are referenced to the geodetic datum. Therefore, to use them efficiently in Civil Engineering projects, engineers must have sufficient geodetic knowledge to correctly configure measurement instruments, evaluate their measurement results, and perform coordinate transformations to make GNSS and conventional surveying measurements compatible.

Table 4.2 Geometric parameters of the ellipsoid

f = flattening of the ellipsoid: $$f = \frac{a-b}{a}$$	(4.1)	e' = second eccentricity: $$e' = \frac{\sqrt{a^2-b^2}}{b}$$	(4.5)
b = semi-minor axis of the ellipsoid: $$b = a*(1-f)$$	(4.2)	e'^2 = second eccentricity squared: $$e'^2 = \frac{a^2-b^2}{b^2} = \frac{e^2}{1-e^2}$$	(4.6)
e = first eccentricity: $$e = \frac{\sqrt{a^2-b^2}}{a}$$	(4.3)	M = radius of curvature in the meridian: $$M = \frac{a*(1-e^2)}{\left(1-e^2*\sin^2\phi_g\right)^{\frac{3}{2}}}$$	(4.7)
e^2 = first eccentricity squared: $$e^2 = \frac{a^2-b^2}{a^2} = f*(2-f)$$	(4.4)	N = radius of curvature in the prime vertical: $$N = \frac{a}{\sqrt{\left(1-e^2*\sin^2\phi_g\right)}}$$	(4.8)

a semi-major axis of the ellipsoid ϕ_g = geodetic latitude

Table 4.3 Parameter values for the GRS80 ellipsoid

Parameter	Value	Parameter	Value
b = semi-minor axis of the ellipsoid	6,356,752.3141 m	e' second eccentricity	0.082094438
e = first eccentricity	0.081819191	e'^2 second eccentricity squared	0.006739497
e^2 = first eccentricity squared	0.006694380		

4.2.5 Spherical Model

Geodetic calculations on the ellipsoid require the application of spherical geometry concepts, which are often excessive for the scale of measurements carried out for engineering projects. For this reason, in many cases, it is replaced by a sphere whose mathematical model is much simpler. Thus, considering that most engineering projects cover a limited area of the globe, the ellipsoidal surface is replaced by a spherical surface with a radius of curvature calculated as a function of the *radius of curvature of the ellipsoid* in the north-south direction, called the *radius of curvature in the meridian (M)* and the radius of curvature in the east-west direction, called the *radius of curvature in the prime vertical (N)*. The values of these radii depend on the latitude (ϕ_g) of the central point of the area under consideration. Consequently, two radii of curvature are used to define the dimensions of the sphere: the *local mean radius* of the Earth (in this book, denoted by R_0 and the *radius of curvature in azimuth* (in this book, denoted by R_α). The use of R_α is restricted to cases where there is a preferred direction for calculating the mean radius of the Earth. The equations for each of these calculations are given below.

$$R_0 = \sqrt{M * N} = \frac{a * \sqrt{1 - e^2}}{1 - e^2 * \sin^2(\phi_g)} \tag{4.9}$$

$$R_\alpha = \frac{M * N}{M * \sin^2(A_{Zg}) + N * \cos^2(A_{Zg})} \quad \text{or} \quad \frac{1}{R_\alpha} = \frac{\cos^2(A_{Zg})}{M} + \frac{\sin^2(A_{Zg})}{N} \tag{4.10}$$

where

a = semi-major axis of the ellipsoid
e^2 = first eccentricity squared
A_{Zg} = geodetic azimuth of the alignment of interest

In some situations, the value of the mean radius of the Earth is assumed to be the weighted average of the axes of the ellipsoid, as shown below.

$$R_0 = \frac{2a + b}{3} = a * \left[\frac{2}{3} + \frac{\sqrt{1 - e^2}}{3}\right] \tag{4.11}$$

Thus, in the case of the GRS80 ellipsoid, this value would be equal to approximately 6371 km, which is the value that can be adopted when it is difficult to calculate the local mean radius of the Earth with greater precision. In the literature, the reader will also find other definitions that are less interesting to Civil Engineering.

Example 4.1 Using the values given in Table 4.3, calculate the following geometric parameters for a place with a geodetic latitude equal to $-22°\ 00'\ 17.8160''$ and a geodetic azimuth[4] equal to $283°\ 33'\ 22.4158''$.

(a) radius of curvature in the meridian (M).
(b) radius of curvature in the prime vertical (N).
(c) local mean radius of the Earth.
(d) radius of curvature in azimuth (R_α).

Solution:
   Using Eq. (4.7),

$$M = \frac{6,378,137.000 * (1 - 0.006694380)}{\left[1 - 0.006694380 * \sin^2(-22°\ 00'\ 17.8160'')\right]^{3/2}} = 6,344,381.135 \text{m}$$

   Using Eq. (4.8),

---

[4]For more details on geodetic latitude and geodetic azimuth, please refer to Chap. 5.

$$N = \frac{6,378,137.000}{\sqrt{1 - 0.006694380 * \sin^2(-22°00'17.8160'')}} = 6,381,136.280\,m$$

Using Eq. (4.9),

$$R_0 = \sqrt{6,344,381.135 * 6,381,136.280} = 6,362,732.167m \cong 6,362,732m$$

Using Eq. (4.10),

$$R_\alpha =$$
$$\frac{6,344,381.135 * 6,381,136.280}{6,344,381.135 * \sin^2(283°33'22.4158'') + 6,381,136.280 * \cos^2(283°33'22.4158'')}$$
$$R_\alpha = 6,379,105.785\,m \cong 6,379,106\,m$$

It is important to note that for Civil Engineering calculations, it is optional to include the decimal digits of the local mean radius of the Earth. However, in this book, they are retained in some cases to help the reader check their calculations.

4.3   Geodetic Reference System (GRS)

The information that describes the geodetic datum, along with its respective coordinate system and reference framework, is called the *Geodetic Reference System (GRS)*. As mentioned, they are fundamental for georeferencing geospatial data from engineering projects that need to be tied to an official network of geodetic control points. Over the years, several geodetic reference systems have been developed for different regions of the world. The most important of these, for Geomatics applied to Civil Engineering, are described below.

The reader should be aware of the importance of using a single reference system for Civil Engineering projects. This means that all survey data is related to a single network of control points. All survey data (new and old) would perfectly combine to ensure the continuity of contiguous projects. Large projects can be subdivided into several independent surveys which, although physically separated, are computationally linked by sharing the same control point network. Likewise, different data can be shared between different projects as they are all in the same reference system. In addition to all these advantages, it is important to note that using a single reference system means that no point will be lost, as it can always be retrieved through its known coordinates.

4.3.1   *Horizontal Geodetic Reference System*

The most important *Horizontal Geodetic Reference Systems* for Geomatics applied
to Civil Engineering are presented below.

4.3.1.1   International Terrestrial Reference System (ITRS)

The ITRS is a global reference system based on a Three-Dimensional Cartesian
coordinate system.[5] Its origin is near the Earth's centre of mass, and its orientation is
equatorial. Its vertical Z-axis coincides with the Earth's axis of rotation, its X-axis
coincides with the direction of the Greenwich astronomical meridian, and its Y-axis
is perpendicular to the X-axis. The unit of length is the metre (SI). The reference
ellipsoid used for calculating latitudes and longitudes in this system is the GRS80
(see Table 4.3). The realisation and maintenance of this system is the responsibility
of the *International Earth Rotation and Reference Systems Service* (IERS), which
maintains a network of known coordinate points using modern geodetic techniques,
such as Very Long Baseline Interferometry (VLBI), Lunar Laser Ranging (LLR),
Satellite Laser Ranging (SLR), GNSS and Doppler Orbitography and
Radiopositioning Integrated by Satellite (DORIS).

The network of known coordinate points of the ITRS system is called the
International Terrestrial Reference Frame (ITRF). It is defined as the realisation
of the ITRS through the completion of its origin, orientation axes and scale, and their
temporal evolution. Physically, it consists of an international network of geodetic
points materialised over the entire terrestrial surface, with an accuracy of the order of
centimetres. The ITRF is the most accurate terrestrial reference frame to date. It is,
therefore, frequently used as a basis for other reference frames or as an intermediary
to describe relationships between the coordinate systems (Van der Marel 2014).

Successive ITRF realisations are referred to as ITRFyy, where (yy) specifies the
last year of the data validation for that specific data frame.

4.3.1.2   World Geodetic System 1984 (WGS84)

The WGS84 is a terrestrial reference system defined and initially realised by the US
Defense Mapping Agency (DMA). It is also a Global Geocentric Geodetic Reference System based on a Three-Dimensional Cartesian coordinate system oriented in
the same way as the ITRS. It uses the WGS84 ellipsoid as its geodetic reference (see
Table 4.1), and the determination of its network of geodetic control points is based
on the measurements made by the IERS.

The importance of WGS84 for Geomatics is because it is the Geodetic Reference
System used by the *Global Positioning System* (GPS). Therefore, as the GPS

---

[5]For more details, see Sect. 5.6.

undergoes periodic refinements, similar to the ITRS, the WGS84 has different realisations of its network of reference points, depending on the needs of the GPS.

The reader should note that the acronym WGS84 refers to the ellipsoid and the Geodetic Reference System.

4.3.1.3   Geodetic Reference System for the Americas (SIRGAS)

As described on the SIRGAS website,[6] SIRGAS is the Geodetic Reference System for the Americas. Its definition corresponds to the International Terrestrial Reference System (ITRS) and is realised by regional densification of the International Terrestrial Reference Frame (ITRF) in the Americas. This means that the reference ellipsoid adopted by SIRGAS is the GRS80. The current realisation of the SIRGAS is given by a network of continuously operating GNSS stations distributed over the Americas and the Caribbean called SIRGAS-CON (SIRGAS Continuously Operating Network). SIRGAS-CON is processed weekly to produce instantaneous weekly station positions aligned to the ITRF and multi-year (cumulative) reference frame solutions.

It is important to note that although the SIRGAS uses the GRS80 ellipsoid as the geometric figure of the Earth, in many Civil Engineering works, it is considered that there is a compatibility between the GRS80 and WGS84 ellipsoids at the centimetre level. Therefore, in many practical cases, where point positioning does not require accuracy better than the centimetre, engineers can use either of the two ellipsoids without requiring coordinate transformations between them.

4.3.2   Geodetic Altimetry Reference System

The Geodetic Altimetry Reference Systems have the exclusive function of providing the references for determining elevations from geospatial data. Unlike horizontal reference systems, they are physical in nature since they depend on the Earth's gravitational field and are consequently linked to the complex concept of the geoid.

In Geomatics, however, for the sake of simplicity, the definition of a geodetic altimetry reference system is based on the assumption that the local mean sea level is an adequate approximation of the original equipotential surface. A fundamental point or vertical datum is defined on this surface, the value of which corresponds to the mean sea level height calculated from tide gauge records over a given period. The realisation of this system is carried out through a differential levelling network,[7] complemented by gravimetric observations to account for the fact that equipotential surfaces are not parallel to each other in the levelled positions. The materialised

---

[6] https://sirgas.ipgh.org/en/organization/about-us/
[7] For more information on differential levelling, see Chap. 13.

objects bearing a marked point whose elevation is known are called *Benchmarks* (*BM*). Details of the components of the altimetric point network are given in Sect. 4.4.1.

4.4   Relationship between the Ellipsoid and the Geoid

As mentioned, the ellipsoid and the geoid are two specific references used for different purposes in Geomatics. Because the equipotential surface that defines the geoid is difficult to define mathematically, and the verticality of the structures and the flow of water are due to the action of the force of gravity, the determination of elevations on the ground is carried out on a mathematical surface, close to the geoid, called the *quasi-geoid*. On the other hand, as previously mentioned, the ellipsoid is used for the horizontal positioning of spatial data. Therefore, as they are not coincident surfaces and, in most cases, not even parallel, two geometric components must be considered to make surveying measurements and mathematical modelling compatible with the ellipsoid reference surface. These are the *geoid undulation* and the *deflection of the vertical*. The geometrical and algebraic details of these two components are presented below.

4.4.1   Geoid Undulation

Due to the variation of the masses of the terrestrial globe and the fact that some parts of the Earth are denser than others, the sum of the gravity vectors affects the shape of the equipotential surfaces, making them irregular. This irregularity causes the lines orthogonal to the ellipsoid surface and the geoid surface not to coincide. As shown in Fig. 4.5, the line perpendicular to the ellipsoid is called the *Normal line*. The line orthogonal to the geoid (and to the other true equipotential surfaces) is called the *Vertical line*. Many countries use this simplified elevation model. In cases where geopotential values are used, a different elevation model is adopted, as shown in Fig. 4.6.

From Figs. 4.5 and 4.6, the *ellipsoidal height* or *geometric height* (h_p) is the distance (on the Normal line) between the point (*P*) on the ground surface and the point (*Q*) on the ellipsoid. The *orthometric altitude* (H_p) is the distance (on the Vertical line) between the point (*P*) on the ground surface and the point (P_0) on the geoid. The *normal elevation* $\left(H_P^N\right)$ is the distance (on the Normal Vertical line) between the point (*P*) on the ground surface and point (Q'_0) on the quasi-geoid. The latter is linked to the ellipsoidal surface by the height anomaly ζ and is not an equipotential surface. This subject involves several concepts typical of Physical Geodesy, which are not discussed in this book. Interested readers are suggested to consult specialised references.

Fig. 4.5 Relationship between ellipsoidal height and orthometric altitude

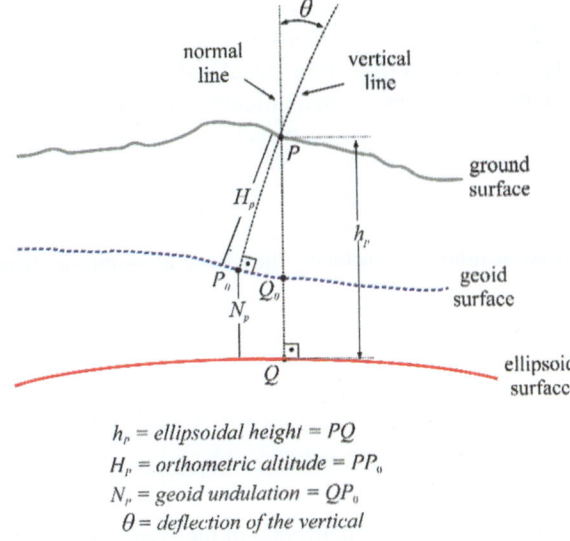

h_p = ellipsoidal height = PQ
H_p = orthometric altitude = PP_0
N_p = geoid undulation = QP_0
θ = deflection of the vertical

Fig. 4.6 Relationship between ellipsoidal height, orthometric altitude, and normal elevation. (curvature exaggerated for clarity)

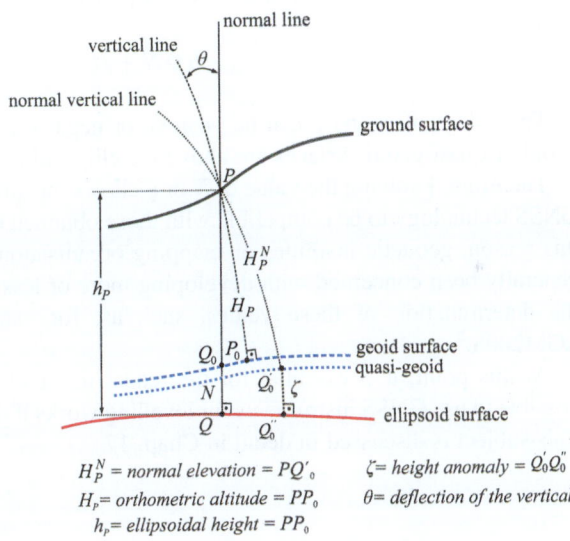

H_P^N = normal elevation = PQ'_0 ζ = height anomaly = Q_0Q_0''
H_p = orthometric altitude = PP_0 θ = deflection of the vertical
h_p = ellipsoidal height = PP_0

It is noteworthy that the curvatures shown in the *Vertical line* and the *Normal Vertical line*, illustrated in Figs. 4.5 and 4.6, are due to the non-parallelism of the equipotential surfaces and, in the case of the Vertical line, also due to the heterogeneous distribution of terrestrial masses. This subject is discussed in more detail in Chap. 13.

The ellipsoidal height has a purely geometric meaning, as it refers to the mathematical figure of the ellipsoid. Its value can be calculated by measurements using GNSS technology, which allows for determining the values of *geocentric Cartesian coordinates* from any point on or near the Earth's surface. Thus, as described in Sect. 5.9.2.3.2, if there is an ellipsoid related to the Cartesian system, the value of the ellipsoidal height (h) can be calculated.

Orthometric and normal altitudes, in turn, have a physical meaning, as they are related to the gravimetric potential of the point. As already mentioned, they are determined by differential levelling and gravimetric measurements. They are, therefore, the altitudes used as references in Civil Engineering works.

Because the ellipsoidal height and the orthometric and normal altitudes have different reference surfaces, they have a geometric difference. When related to the geoid, this difference is called *geoid undulation*, denoted by the letter (N). When related to the quasi-geoid, as already mentioned, it is called *height anomaly*. Thus, according to Figs. 4.5 and 4.6.

$$h = N + H * \cos \theta \qquad (4.12)$$

$$h = \zeta + H^N \qquad (4.13)$$

As the value of the deflection of the vertical θ is only a few seconds of arc, it can be ignored, and, in this case, Eq. (4.12) is replaced by Eq. (4.14).

$$h \cong N + H \qquad (4.14)$$

The values of N and ζ can be positive or negative. They are positive when the geoid or quasi-geoid surfaces are above the ellipsoid and negative otherwise.

Therefore, knowing the value of N or ζ allows the altimetric measurements using GNSS technology to be compatible with those obtained by differential levelling. For this reason, geodetic institutes or mapping organisations in several countries have generally been concerned with developing more or less accurate models that allow the determination of these values, such as, for example, the Geoidal Model EGM2008.[8]

At this point, it is essential for the reader to clearly understand that it is only possible to use GNSS instruments for levelling works if the value of N or ζ is known. This subject is discussed in detail in Chap. 17.

4.4.2 Deflection of the Vertical

Another crucial geometric effect resulting from the geometric difference between the geoid and the ellipsoid is the *deflection of the vertical*,[9] represented in this book by the Greek letter θ. It should be understood as the angular difference between the

---

[8] For more information, readers can use the following link: https://earth-info.nga.mil/index.php. Accessed in May 2023.

[9] Also called the deviation of the vertical by some authors.

direction of the gravity vector, or plumb line at a point, and the ellipsoidal normal through the same point for a particular ellipsoid, as shown in Fig. 4.7. This angular difference is a function of the slope variation of the geoid surface concerning the reference ellipsoid, i.e., as a function of the equipotential energy.

Fig. 4.7 Components of the deflection of the vertical

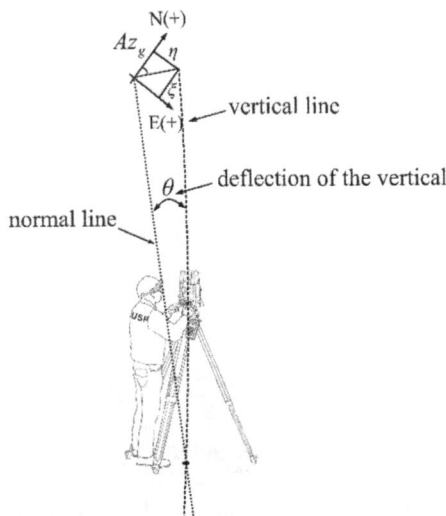

The deflection of the vertical usually does not exceed a few seconds of arc. It is generally around 5″ in flat or slightly undulating areas but can reach exceptionally 70″ in mountainous regions. As it is most of the time a relatively small value, the deflection of the vertical is often ignored in Civil Engineering projects.

For its calculation, the deflection of the vertical is decomposed into two mutually orthogonal components: a north–south or meridional component ξ, which is positive to the north (i.e., the plumb line intersects the celestial sphere north of the Normal line), and an east–west or prime vertical component η, which is positive to the east (i.e., the plumb line intersects the celestial sphere east of the Normal line). Both ξ and η are positive when the Vertical line is north-east of the Normal line, as shown in Fig. 4.7.

Although the deflection of the vertical is rarely used in Civil Engineering works, there are cases of high-precision measurements where it is necessary to know its value, as described below:

(a) In the conversion of azimuths determined by the gyrotheodolites[10] into geodetic azimuths;
(b) In the reduction of horizontal directions and vertical angles measured on Earth's surface to the ellipsoid;

---

[10] Surveying instrument composed of a gyroscope coupled to a theodolite and used to determine the direction of the true azimuth.

(c) In the reduction of slope distance (d') to ellipsoidal distance (d_0)[11];
(d) In the determination of elevation differences through vertical angles and slope distances, in cases of high-precision measurements;
(e) In the correction of elevation differences calculated through differential levelling.

According to Fig. 4.7,

$$\theta = \sqrt{\xi^2 + \eta^2} \tag{4.15}$$

or

$$\theta = \xi * \cos(A_{Zg}) + \eta * \sin(A_{Zg}) \tag{4.16}$$

where

$$Azg = arctg\left(\frac{\eta}{\xi}\right)$$

= geodetic azimuth as a function of the deflection of the vertical   (4.17)

Historically, determining the deflection of the vertical θ has been an astronomic or gravimetric geodetic problem, often beyond the capabilities of the typical surveyor. More recently, however, with the advent of GNSS, the deflection of the vertical components can be calculated by combining GNSS and differential levelling measurements. Since total station measurements are referenced to the Vertical line and GNSS measurements are referenced to the Normal line, it is possible to calculate the deflection from gravity anomalies, as shown in Fig. 4.8.

Fig. 4.8 Relationship between the geoid undulation and the deflection of the vertical. (Adapted from Heiskanen and Moritz 1967)

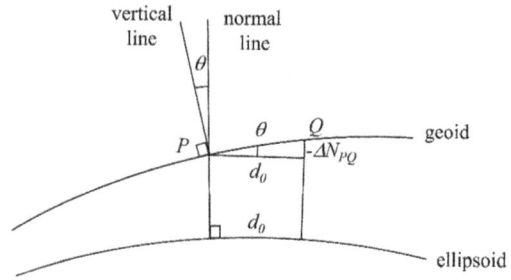

---

[11] For more details, see Chap. 7.

According to Fig. 4.8,

$$\theta = -\frac{\Delta N_{PQ}}{d_0} = \xi * \cos\left(A_{Zg_{PQ}}\right) + \eta * \sin\left(A_{Zg_{PQ}}\right) \qquad (4.18)$$

The minus sign is a convention.[12]

where

$$\Delta N_{PQ} \cong N_Q - N_P = (h_Q - H_Q) - (h_P - H_P) = (h_Q - h_P) - (H_Q - H_P)$$
$$= \Delta h_{PQ} - \Delta H_{PQ} \qquad (4.19)$$

ΔN_{PQ}= geoid undulation difference between points (Q) and (P)
d_0= ellipsoidal distance
h_P, h_Q= ellipsoidal heights
H_P, H_Q= orthometric altitudes

Equation (4.19) shows that the components of the deflection of the vertical can be calculated regardless of whether the geoid undulations are known, requiring only knowledge of the ellipsoidal heights, the orthometric altitudes and ellipsoidal distances, and the geodetic azimuth between the points considered in the calculation. In practice, the value of ΔH_{PQ} can be obtained by differential levelling and Δh_{PQ} and $A_{Zg_{PQ}}$ from GNSS measurements. Furthermore, assuming that the distance between points (P) and (Q) is small, the deflection of the vertical at point (P) is assumed to have a linear variation. Likewise, the ellipsoidal distance d_0 can be assumed equal to the slope distance between points (P) and (Q).

The solution to this problem is based on selecting a point (P) on the ground surface, from which side shots can be made to a series of points (Q_i) distributed around it. Using GNSS measurements, it is possible to determine the geodetic coordinates of the point (P) and all the measured points (Q_i), and thus the corresponding geodetic azimuths. A differential levelling must then be performed between the point (P) and each point (Q_i). Thus, with the values of the ellipsoidal heights and orthometric altitudes of each point, it is possible to establish a system of Eq. (4.18), which, once solved, gives the values of the components (ξ, η). Because there are two unknowns, at least two side shots are required. However, it is recommended to use more points, preferably distributed in the N-S and E-W directions. In this case, the problem can be solved by applying an adjustment computation model, such as the Parametric adjustment model.

To apply the Parametric adjustment model, the observation Eq. (4.18) can be rewritten as given by Eq. (4.20).

$$-\frac{\Delta N_{PQ}}{d_0} = \frac{\Delta H_{PQ} - \Delta h_{PQ}}{d_0} = \theta = \xi * \cos\left(A_{Zg_{PQ}}\right) + \eta * \sin\left(A_{Zg_{PQ}}\right) \qquad (4.20)$$

---

[12] Further information the reader can find in Heiskanen and Moritz (1967)—pages 111–114.

Equation (4.20) is a non-linear function involving the orthometric altitude, the ellipsoidal height, the geodetic azimuth, the slope distance between points (P) and (Q) and the unknowns (ξ, η). Thus, considering the observation θ_{obs}, the following equation can be written:

$$\Delta\theta_{Obs} + v_\theta = f\left(Az_g, \widetilde{\xi}, \widetilde{\eta}, \right) = \xi * \cos(Az_g) + \eta * \sin(Az_g) \tag{4.21}$$

The linearised form of Eq. (4.21) is:

$$f\left(Az_g, \widetilde{\xi}, \widetilde{\eta}, \right) = \theta_0 + \left(\frac{\partial\Delta N}{\partial\xi}\right)_0 \delta\xi + \left(\frac{\partial\Delta N}{\partial\eta}\right)_0 \delta\eta \tag{4.22}$$

The error equation is then given as follows.

$$v_\theta = \left(\frac{\partial\theta}{\partial\xi}\right)_0 \delta\xi + \left(\frac{\partial\theta}{\partial\eta}\right)_0 \delta\eta - (\theta_{Obs} - \theta_0) \tag{4.23}$$

where

$$\left(\frac{\partial\theta}{\partial\xi}\right)_0 = \cos(Az_g) \tag{4.24}$$

$$\left(\frac{\partial\theta}{\partial\eta}\right)_0 = \sin(Az_g) \tag{4.25}$$

$$l_{0_\theta} = \theta_{Obs} - \theta_0 \tag{4.26}$$

Each observation produces an error Eq. (4.23), whose unknown parameters are $\delta\xi$ and $\delta\eta$. For n observations, Eq. (4.23) can be written in matrix form as $v = Ax - l_0$ (Eq. 3.79), where

$$A = \begin{bmatrix} \left(\frac{\partial\theta_1}{\partial\xi}\right)_0 & \left(\frac{\partial\theta_1}{\partial\eta}\right)_0 \\ \left(\frac{\partial\theta_2}{\partial\xi}\right)_0 & \left(\frac{\partial\theta_2}{\partial\eta}\right)_0 \\ \vdots & \vdots \\ \left(\frac{\partial\theta_n}{\partial\xi}\right)_0 & \left(\frac{\partial\theta_n}{\partial\eta}\right)_0 \end{bmatrix} \quad v_\theta = \begin{bmatrix} v_{\theta 1} \\ v_{\theta 2} \\ \vdots \\ v_{\theta n} \end{bmatrix} \quad l_{0_\theta} = \begin{bmatrix} \theta_{Obs1} - \theta_{01} \\ \theta_{Obs2} - \theta_{02} \\ \vdots \\ \theta_{Obsn} - \theta_{0n} \end{bmatrix} \quad x = \begin{bmatrix} \delta\xi \\ \delta\eta \end{bmatrix}$$

If weights are to be considered, it is suggested to use the model $P = 1/d_i$, where d_i is the distance between points (P) and (Q_i).

Adjustment computation can be performed using Eqs. (3.84), (3.85), (3.86) and (3.87), from which $\delta\xi$ and $\delta\eta$ are obtained, followed by the adjusted values of (ξ, η).

Finally, as a total of n observations and two unknowns have been considered, the standard deviations of the unit weight can be calculated using Eq. (4.27). The remaining statistical parameters can be determined as indicated in Sect. 3.9.1.1.

$$s_0^2 = \frac{v^T P v}{n-2} \tag{4.27}$$

Example 4.2 To estimate the value of the deflection of the vertical at a point (P), a network of points was measured using GNSS technology. As a result, geodetic coordinates and ellipsoidal heights were obtained, as well as the geodetic azimuths between point (P) and each point in the network. A geometric levelling was also carried out to obtain the orthometric altitudes of each point. The calculated values are given in Table 4.4. Calculate the deflection of the vertical at point (P) using the Parametric adjustment model.

Table 4.4 Field measurements and calculated values

Point	Geodetic Azimuth	Ellipsoidal distance (d_o) [m]	$N = h-H$ [m]	ΔN_{PQi} [m]	$(\Delta N_{PQi}/d_o) * 10^{-6}$
P			−5.7310	0.0000	−
Q_1	53° 44′ 10.064728″	222.7800	−5.7160	0.0150	67.3310
Q_2	165° 40′ 34.744010″	494.2452	−5.7415	−0.0105	−21.2445
Q_3	211° 03′ 34.441320″	588.8063	−5.7230	0.0080	13.5868

Adapted from Souza and Garnés (2012)

Solution:
For the application of the Parametric adjustment model, using Eq. (3.79),
Eq. (4.23) *can be written in matrix form as* $v = Ax - l_0$, *where*

$$A = \begin{bmatrix} 0.591505 & 0.806301 \\ -0.968914 & 0.247400 \\ -0.856631 & -0.515929 \end{bmatrix} \quad v_\theta = \begin{bmatrix} v_{\theta 1} \\ v_{\theta 2} \\ v_{\theta 3} \end{bmatrix}$$

$$l_{0_\theta} = \begin{bmatrix} 6.733100 * 10^{-5} \\ -2.124452 * 10^{-5} \\ 1.358681 * 10^{-5} \end{bmatrix}_{rad} \quad x = \begin{bmatrix} \delta\xi \\ \delta\eta \end{bmatrix}$$

Using Eqs. (3.84), (3.85), (3.86) and (3.87) *and assuming that the weight matrix* **P** *is equal to the identity and the approximate values are equal to zero.*

$$N = \begin{bmatrix} 2.022489 & 0.679183 \\ 0.679183 & 0.977511 \end{bmatrix} \quad n = \begin{bmatrix} 4.877182^* \ 10^{-5} \\ 4.202337^* \ 10^{-5} \end{bmatrix}$$

$$x = \begin{bmatrix} \xi \\ \eta \end{bmatrix} = \begin{bmatrix} 1.262337 * 10^{-5} \\ 3.421934 * 10^{-5} \end{bmatrix}_{rad} = \begin{bmatrix} 2.6'' \\ 7.1'' \end{bmatrix}$$

And then,

$$\theta = \sqrt{2.6^2 + 7.1^2} = 7.5''$$

Using Eq. (3.79).

$$v^T = \begin{bmatrix} -3.227311 * 10^{-5} & 1.747941 * 10^{-5} & 4.205513 * 10^{-5} \end{bmatrix}_{rad}$$

Finally, the standard deviations of the unit weight and the variance-covariance matrix for the unknowns are given as follows.

$$s_0^2 = 3.115718 * 10^{-9} rad^2 \ (with \ n = 3, u = 2)$$
$$s_0 = \pm 11.5''$$

Using Eq. (3.93), the variance-covariance matrix for the unknowns (deflections of the vertical) is given as follows:

$$Q_{xx} = \begin{bmatrix} 0.644917 & -0.448094 \\ -0.448094 & 1.334346 \end{bmatrix}$$

$$\Sigma_{xx} = s_0^2 * Q_{xx} = \begin{bmatrix} 2.009381 * 10^{-9} & -1.394135 * 10^{-9} \\ -1.394135 * 10^{-9} & 4.157446 * 10^{-9} \end{bmatrix}_{rad^2}$$

Using the diagonal values in the variance-covariance matrix for the parameters, their standard deviations are

$$s_\xi = \pm 4.482611 * 10^{-5} rad = \pm 9.2''$$
$$s_\eta = \pm 6.447826 * 10^{-5} rad = \pm 13.3''$$

Finally,

$$\xi = 2.6'' \pm 9.2''$$
$$\eta = 7.1'' \pm 13.3''$$
$$\theta = 7.5'' \pm 14.0''$$

4.5 Review Questions

1. What is a control point, and why is it important for Civil Engineering projects?
2. Explain what a datum is.
3. Explain the differences between geoid and ellipsoid for Geomatics applied to Civil Engineering.
4. What are the different Data currently used to represent the shape of the Earth for Geomatics applications?
5. What is an equipotential surface, and why is it important for Geomatics applied to Civil Engineering?
6. What is a Geodetic Reference System?
7. What is the meaning of the acronym ITRS?
8. What is the meaning of the acronym ITRF?
9. What is the difference between WGS84 and GRS80?
10. Explain what geoid undulation is and why it is important for GNSS measurements.
11. Explain the difference between ellipsoidal height and orthometric altitude.
12. Visit the link https://www.unavco.org/software/geodetic-utilities/geoid-height-calculator/geoid-height-calculator.html and carry out some activities using the EGM2008.
13. Explain what deflection of the vertical is and why it is important in geodetic surveying.

References

Akkul, M. (2007). *Assessment of Deflection of The Vertical Components From GPS and Leveling Measurement*, Ms thesis, Selcuk University, Institute of the Natural and Applied Sciences, Konya.

Blitzkow, D.; Campos, I. O.; de Freitas, S. R., (2004). *Altitude: O que interessa e como equacionar?* Proceedings of the 1th Simpósio Brasileiro de Ciências Geodésicas e Tecnologias da Geoinformação, Recife-PE, Brasil.

Bomford, G. (1980). *Geodey*. 4 Ed. Oxford University Press. England.

Ceylan, A. (2009). *Determination of the deflection of vertical components via GPS and leveling measurement: A case study of a GPS test network in Konya, Turkey.* Scientific Research and Essay Vol.4 (12), pp. 1438–1444, December, 2009. Available online at http://www.academicjournals.org/SRE. ISSN 1992–2248. 2009 Academic Journals.

Featherstone, W. E. et Rueger, M. J. (2000). *The importance of Using Deviations of the Vertical for Reduction of Survey Data to a Geocentric Datum.* The Australian Surveyor, Vol 45, N. 2.

Heiskanen, W.A.; Moritz, H. (1967). *Physical geodesy.* W.H. Freeman and Company: San Francisco, CA, USA.

Hosmer, G.L. (1928). *Geodesy.* 2nd Edition. John Wiley & Sons, Inc. New York.

IERS Conventions (2010). *IERS Technical Note 36.* Frankfurt am Main: Verlag des Bundesamts für Kartographie und Geodäsie. Gérard Petit and Brian Luzum (eds.). 179 pp., ISBN 3-89888-989-6.

Jekeli, C. (2006). *Geometric Reference Systems in Geodesy.* Ohio State University. Columbus, Ohio, USA.

Souza, W. O., Garnés, S. J. A. (2012). *Determinação dos componentes do Desvio da Vertical pelo Método de Helmert: Relacionamento entre Nivelamento de Precisão com Altitude Elipsoidal e MAGEO2010*. IV Simposio Brasileiro de Ciencias Geodesicas e Tecnologias da Geoinformação. Recife, Brazil

VanderBerg, D. J. (1999). *Combining GPS and Terrestrial Observations to Determine Deflections of the Vertical*. Ms Thesis. Pardue University. USA.

Van der Marel, H. (2014). *Reference Systems for Surveying and Mapping (Lecture notes CTB3310)*. Faculty of Civil Engineering and Geosciences. Delft University of Technology. The Netherlands.

Van Sickle, J. (2004). *Basic GIS Coordinates*. CRC Press, Florida, USA.

Chapter 5
Coordinate Reference System

Irineu da Silva and Guilherme Poleszuk dos Santos Rosa

5.1 Introduction

As seen in the previous chapter, the mathematical modelling for defining geospatial data is based on adopting a reference system through which the data is georeferenced in space. In this case, georeferencing means determining the coordinates of points through mathematical relationships between values of quantities previously known or measured in the field concerning a terrestrial Coordinate Reference System. Thus, considering that the theories about reference systems were presented in Chap. 4, this chapter presents the theories about Coordinate Reference Systems regarding Geomatics applied to Civil Engineering.

The use of a coordinate system in Civil Engineering has several advantages. Firstly, it is a convenient method of recording position in space. Secondly, it facilitates and allows the standardisation of calculation methods so that each point is uniquely positioned. Finally, a coordinate system allows multiple individual systems to be unified into a global system, making it easier to identify and manage spatial data.

For Civil Engineering, there are seven coordinate systems to consider.

- Linear Referencing System (LSR)
- Two-Dimensional Cartesian Coordinate System
- Polar Coordinate System[1]

---

[1] Also known as the 2D Spherical Coordinate System.

I. da Silva (✉)
São Carlos School of Engineering, University of São Paulo, São Carlos, SP, Brazil
e-mail: irineu@sc.usp.br

G. Poleszuk dos Santos Rosa
Cartographic Engineer, Researcher at the Faculty of Science and Technology of the São Paulo State University (UNESP), São Paulo, Brazil

© The Author(s), under exclusive license to Springer Nature Switzerland AG 2025
I. da Silva, P. C. L. Segantine, *Geomatics Applied to Civil Engineering*,
https://doi.org/10.1007/978-3-031-75737-2_5

- Triangular Coordinate System
- Three-Dimensional Cartesian Coordinate System
- Geographic Coordinate System
- Geodetic Coordinate System

5.2   Linear Referencing System (LRS)

A *Linear Referencing System*, or LRS, stores and displays geospatial data through relative positioning along an axis rather than in classical Cartesian space. In this case, the axis can be a network of polylines (multiple axes) to which events (point locations) are referenced. As shown in Fig. 5.1, event locations along the axis are measured in terms of distance from a reference point.

In practice, the LRS is mainly used in GIS applications to store reference points for networks of linear features, such as roads, railways, pipelines, power lines and rivers. The great advantage of the LRS is that it allows attributes to be located using only one parameter, which facilitates data referencing in network analysis.

A detailed discussion of the characteristics of an LRS is beyond the scope of this book. Readers are suggested to consult specialised literature for further information.

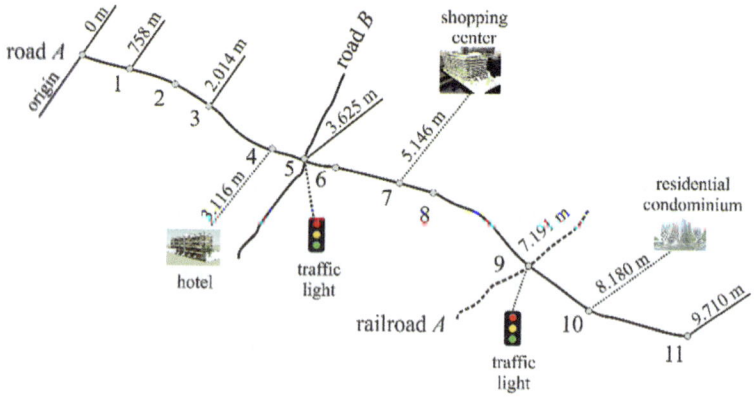

Fig. 5.1 Linear Referencing System

5.3   Two-Dimensional Cartesian Coordinate System

The most commonly used coordinate system in Civil Engineering is the *Two-Dimensional Cartesian Coordinate System* based on the Cartesian Coordinate System, named after its creator, the French philosopher René Descartes, also known as *Renatus Cartesius* (Latinised form) (1569–1650). It is widely used because it is a coordinate system in Euclidean space, allowing the use of plane geometry functions for point positioning.

This coordinate system consists of two geometric axes perpendicular to each other in the same plane, forming four quadrants, as shown in Fig. 5.2. The intersection of the two axes is the origin of the system. The primary axis, which is horizontal, is called the *abscissa axis* and is often represented by the letter *X*. The secondary axis, which is vertical and perpendicular to the abscissa axis, is called the *ordinate axis* and is often represented by the letter *Y*. Both axes are scaled equally according to the scale defined for the system. It is a right-handed system, which means that the *X*-axis is positive towards the right and the *Y*-axis is positive upwards in the plane view. The position of a point (*P*) in this system is described by two coordinate values, defined as the distances from the origin to the perpendicular projections of (*P*) on each axis. The pair of coordinate values (*X, Y*) is called a *2D Cartesian coordinate* (or *rectangular coordinate*).

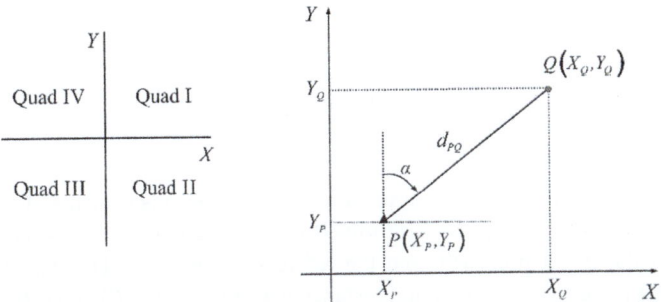

Fig. 5.2 Surveying 2D Cartesian coordinate system

In this coordinate system, a line segment can be defined by two points (as usual) or by a point and a direction. As shown in Fig. 5.2, unlike the *Mathematical Cartesian Coordinate System* (in which directions are reckoned counter-clockwise from the positive *X*-axis), the direction, in this case, is defined by the angle α, reckoned clockwise from the positive *Y*-axis. This convention is more intuitive in surveying, where directions like North (upward) and East (to the right) are commonly used as references. For this reason, some authors call it the *Surveying Two-Dimensional Cartesian Coordinate System* to distinguish it from the *Mathematical Two-Dimensional Cartesian Coordinate System*.

The reader will observe throughout this book that all geometric functions for determining the value of spatial data in an engineering reference system are established based on this coordinate system.

5.4 Polar Coordinate System

Another way of describing the position of a point (*P*) in 2D space is in polar coordinates, where the position is defined by a linear measurement and an angular measurement that defines the so-called polar coordinates (ρ_P, α_P). In this case, the pair of orthogonal axes of Cartesian coordinates are replaced by a single reference line passing through the origin (*O*) or pole of the system. The position of any point (*P*) in this system is defined with respect to this pole and the reference direction. Again, for Geomatics applications, angles are reckoned clockwise, as shown in Fig. 5.3.

Fig. 5.3 Surveying polar
coordinate system

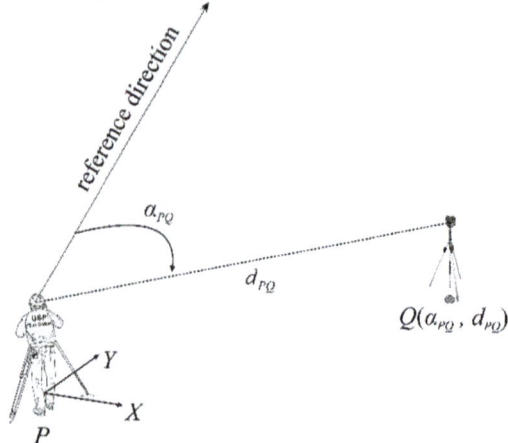

   The Two-Dimensional Polar Coordinate System is used in almost all field observations carried out with conventional surveying instruments. Considering Fig. 5.3, the polar angle α_{PQ} is determined by the angular difference between the direction of the target point (Q) and the reference direction.[2] The distance d_{PQ} is obtained by measuring the horizontal distance between points (P) and (Q). Although used in field observations, this coordinate system is rarely used in Civil Engineering projects. Generally, polar coordinates are converted into Cartesian coordinates before being used in Geomatics projects.

5.5   Triangular Coordinate System

When positioning points within a triangle, the coordinate system often used is the *Triangular Coordinate System.*[3] It describes the position of a point (P) within a triangle using the relative distance from the three vertices of the triangle, as shown in Fig. 5.4.

   This coordinate system has been used in several branches of engineering using finite elements. Concerning Geomatics applied to Civil Engineering, it is mainly used in digital surface modelling, since it allows for easy interpolation and manipulation of points inside triangles, which is a common scenario in digital surface modelling based on triangular meshes.

   The values of the triangular coordinates (L_1, L_2, L_3) can be calculated as a function of the areas of the corresponding triangles, as given by Eq. (5.1)

---

[2] See Chap. 6.

[3] Also known as the Barycentric Coordinate System.

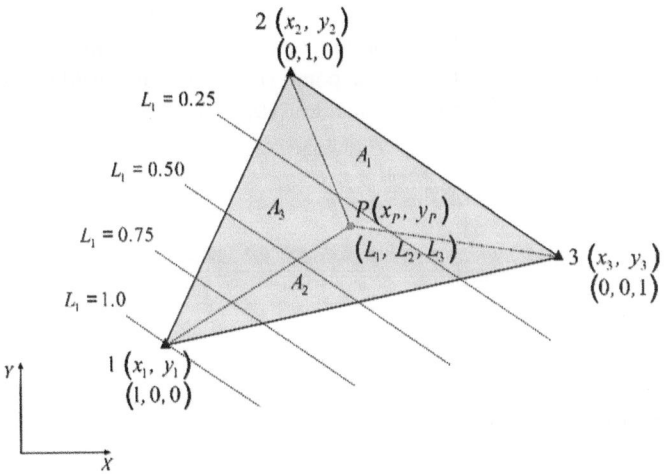

Fig. 5.4 Triangular coordinate system

$$L_1 = \frac{A_1}{A}; \quad L_2 = \frac{A_2}{A}; \quad L_3 = \frac{A_3}{A} \tag{5.1}$$

where

A_i = area of triangle P_{jk} opposite to vertex (i)
A = area of the triangle Δ_{123}

The relationships between the 2D Cartesian coordinates of the triangle's vertices and the 2D Cartesian and triangular coordinates of the point (P) can be written in matrix form, as given by Eq. (5.2).

$$
\begin{bmatrix} 1 \\ x_P \\ y_P \end{bmatrix}
=
\begin{bmatrix} 1 & 1 & 1 \\ x_1 & x_2 & x_3 \\ y_1 & y_2 & y_3 \end{bmatrix}
*
\begin{bmatrix} L_1 \\ L_2 \\ L_3 \end{bmatrix}
\tag{5.2}
$$

From where

$$
\begin{aligned}
1 &= L_1 + L_2 + L_3 \\
x_P &= L_1 * x_1 + L_2 * x_2 + L_3 * x_3 \\
y_P &= L_1 * y_1 + L_2 * y_2 + L_3 * y_3
\end{aligned}
\tag{5.3}
$$

The inverse matrix is given as follows.

$$
\begin{bmatrix} L_1 \\ L_2 \\ L_3 \end{bmatrix}
=
\frac{1}{2A}
\begin{bmatrix}
x_2 y_3 - x_3 y_2 & y_2 - y_3 & x_3 - x_2 \\
x_3 y_1 - x_1 y_3 & y_3 - y_1 & x_1 - x_3 \\
x_1 y_2 - x_2 y_1 & y_1 - y_2 & x_2 - x_1
\end{bmatrix}
*
\begin{bmatrix} 1 \\ x_P \\ y_P \end{bmatrix}
\tag{5.4}
$$

Example 5.1 Let the triangle Δ_{123} be defined by the 2D Cartesian coordinates in Table 5.1. Given the coordinates of the vertices of the triangle, calculate the triangular coordinates (L_1, L_2, L_3) of a point (P) positioned inside this triangle, which has 2D Cartesian coordinates equal to $(3.0, 5.0)$.

Table 5.1 Coordinates of the vertices

Vertex	X [m]	Y [m]
1	2.0	2.0
2	4.0	2.0
3	3.0	6.0

Solution:

Using Eq. (5.4).

$$
\begin{bmatrix} L_1 \\ L_2 \\ L_3 \end{bmatrix} = \frac{1}{2*4.0} \begin{bmatrix} 18 & -4 & -1 \\ -6 & 4 & -1 \\ -4 & 0 & 2 \end{bmatrix} * \begin{bmatrix} 1 \\ 3.0 \\ 5.0 \end{bmatrix} = \begin{bmatrix} 0.125 \\ 0.125 \\ 0.750 \end{bmatrix}
$$

5.6 Three-Dimensional Cartesian Coordinate Systems

The spatial position of a point can be determined by adding a third axis (Z-axis) to the Two-Dimensional Cartesian Coordinate System or by considering a second angle to the Two-Dimensional Polar Coordinate System, as shown in Figs. 5.5 and 5.6, respectively. In the first case, the Z-axis is added according to the normal to the plane formed by the (X, Y) axes, thus establishing the *Three-Dimensional Cartesian Coordinate System*. In the second case, the second angle is added perpendicularly to the plane of rotation of the first angle, creating what is known as the *Spherical Polar Coordinate System*.

Fig. 5.5 Three-dimensional Cartesian coordinate system

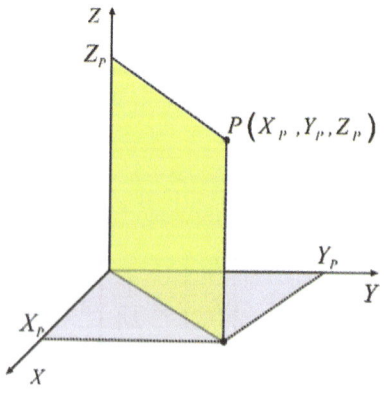

Fig. 5.6 Spherical polar
coordinate system

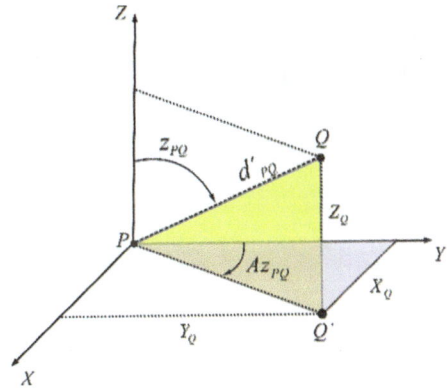

The Three-Dimensional Cartesian Coordinate System defines the position of a
point in space. It has always been of little use to Civil Engineering because
engineering projects are designed separately in horizontal and vertical planes.
More recently, however, new measurement technologies have allowed coordinates
to be determined directly in 3D space in some engineering applications, opening up
new perspectives for Civil Engineering projects. A typical example is the determination of 3D coordinates of points measured using laser scanning technology.
Another classic example is the positioning of GNSS antennas on the Earth's surface.

A Three-Dimensional Cartesian Coordinate System can be topocentric or geocentric. It is topocentric if its origin is defined somewhere on the Earth's surface and
geocentric if it coincides with the Earth's centre of mass.

5.6.1   Geocentric Three-Dimensional Cartesian Coordinate System

In a Geocentric Three-Dimensional Cartesian Coordinate System, the (X, Y) axes are
defined so that they belong to the plane of the equator, the Z-axis coincides with the
Earth's axis of rotation, and the X-axis is oriented to intersect a reference meridian
(usually the Greenwich Meridian). See Fig. 5.7. The system defined in this way is
called the *Global Geocentric Coordinate System*.

When working with Geodetic Reference Systems, the Global Geocentric Coordinate System is the basis for calculating all other coordinates of geospatial data. For
this reason, in many cases, its coordinates are the primary ones stored in a database.

As shown in Fig. 5.7, the reader should note the geometric difference between the
(Z_P) coordinate and the ellipsoidal height (h_P) in a Global Geocentric Coordinate
System. The (Z_P) coordinate is perpendicular to the equatorial plane, whereas the
ellipsoidal height (h_P) is normal to the reference surface. Therefore, an increase in
the value of (h_P) will not result in an equal increase in (Z_P) (except at the poles).

For example, Table 5.2 shows the geocentric Cartesian coordinates of the GNSS CORS and the EUNO pillar located at the São Carlos campus of the University of São Paulo, Brazil.

Fig. 5.7 Global geocentric coordinate system

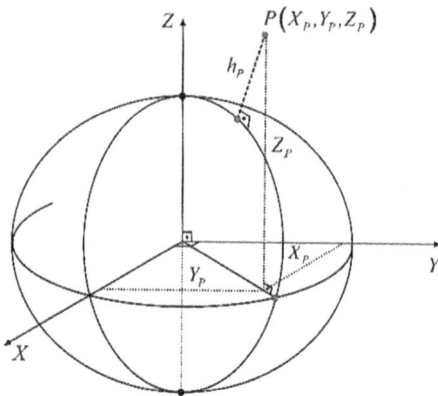

Table 5.2 Examples of global geocentric Cartesian coordinates (SIRGAS2000, 2000,4)

Point ID	X [m]	Y [m]	Z [m]
GNSS CORS	3,967,006.9729	−4,390,247.3691	−2,375,229.9370
EUNO pillar	3,965,000.9384	−4,392,462.9612	−2,374,512.8145

5.6.2   *Topocentric Three-Dimensional Cartesian Coordinate System*

Terrestrial measurements, however, generally refer to the observation point on the Earth's surface. Therefore, the coordinates derived from these observations are often expressed in a local (topocentric) coordinate system, which describes the position of objects relative to a given point on the Earth's surface. To distinguish it from the Global Geocentric Cartesian Coordinate System, its axes are denoted as (e, n, u), as shown in Fig. 5.8. The u-axis is usually aligned with the normal of the reference ellipsoid at the local origin point (P). It is positive in the zenith direction. The n-axis is orthogonal to the u-axis and directed towards the ellipsoidal north pole, defined by the geodetic meridian through (P). The e-axis is orthogonal to u and n, forming a right-handed Cartesian coordinate system. The system defined in this way is called the *Topocentric Three-Dimensional Cartesian Coordinate System*.

A Topocentric Coordinate System can be tied to the Earth or to a surveying instrument. In this case, the difference is the direction of the u-axis, which is referenced to the Normal line when tied to the figure of the Earth, and to the Vertical line (gravity vector) when tied to the instrument.

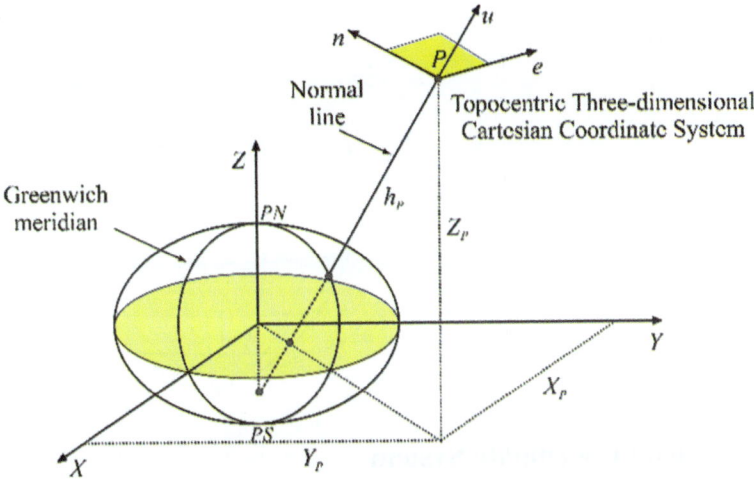

Fig. 5.8 Topocentric three-dimensional Cartesian coordinate system

Topocentric coordinates can be arbitrary local coordinates. In this case, they can be 2D assumed plane coordinates, 3D spatial object coordinates, or 3D Cartesian coordinates with assumed elevation and plane coordinates. Although useful in some applications, arbitrary local coordinates are usually incompatible with other local systems and are of limited value in georeferencing geospatial data.

5.7   Geographic Coordinate System (GCS)

Geographic coordinates are defined as spatial coordinates determined on a spherical reference surface, on which the points are positioned as a function of the angular values of arcs conveniently measured on it. Since the spherical reference surface is a model of the Earth's surface, the Geographic Coordinate System is defined along the Earth's axis of rotation, using the poles as reference points through which imaginary north-south lines define the *meridians of longitude*. Perpendicular to the meridians, running east-west and parallel to the equatorial plane are the parallel circles known as *parallels of latitude*, as shown in Fig. 5.9.

Locations on the Earth's surface in the GCS are measured in angular units from the centre of the Earth relative to the plane defined by the equator and the plane defined by the prime meridian (Meridian of Greenwich). Latitude and longitude values, therefore, define a location. Latitude is the angle from the equatorial plane to the location of interest on the Earth's surface. Longitude is the angle between the meridian of the Greenwich plane and the meridian plane intersecting the location of interest. See Fig. 5.10.

Fig. 5.9 Meridians and parallels on the sphere

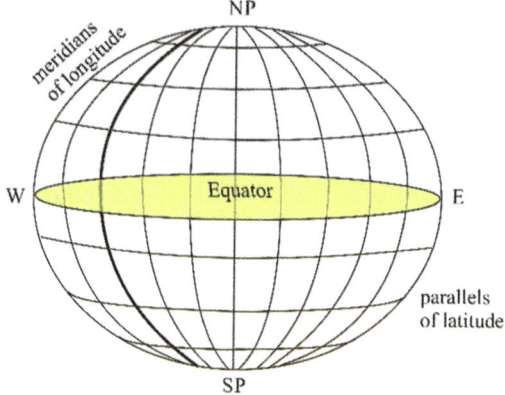

5.8 Geodetic Coordinate System

A *Geodetic Coordinate System*[4] is a 3D coordinate system in which position is given by *geodetic latitude* (ϕ_g), *geodetic longitude* (λ_g) and *ellipsoidal height* (h), taking the ellipsoid as the reference surface, as shown in Fig. 5.10.

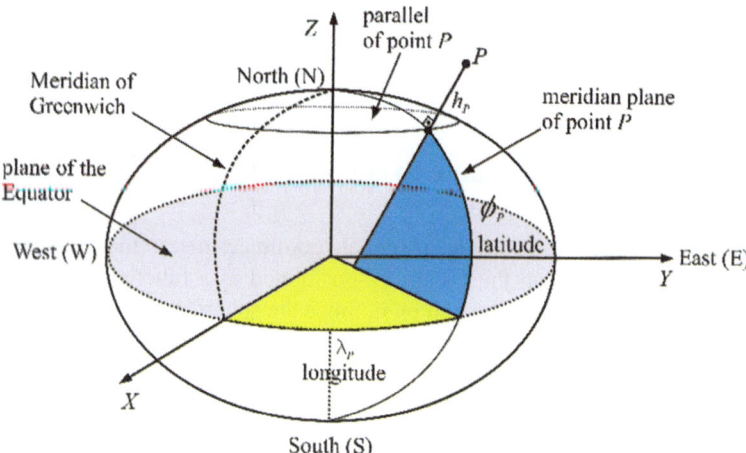

Fig. 5.10 Geodetic coordinate system

   Geodetic latitudes specify positions north and south in terms of the angle subtended at the plane of the equator and the parallel that intersects the point of interest. It ranges from 0° (equator) to 90° (north pole) in the northern hemisphere and from 0° to −90° in the southern hemisphere, or simply from 0° to 90°, followed

---

[4] Some authors prefer the term *Ellipsoidal coordinate system*.

by either North or South or the initials N and S, respectively. For example, a point in the southern hemisphere would have a latitude of - $22°00'18''$ or $22°00'18''$ S.

Geodetic longitudes are referenced to the Greenwich Meridian. They can be positioned in two ways: from 0° to 360°, positive in the East (E) direction, or from 0° to 180°, positive in the east (E) direction and negative in the west (W) direction. For example, a point west (W) of Greenwich would have a longitude of - $47°53'57''$ or $47°53'57''$ W.

For example, Table 5.3 shows the geodetic coordinates related to the SIRGAS2000 Geodetic Reference System of the GNSS CORS and the EUNO pillar, located at the São Carlos Campus of the University of São Paulo, Brazil.

It is important to note that a physical point on the Earth's surface can have different geodetic coordinates depending on which ellipsoid the coordinate systems refer to.

Table 5.3 Example of geodetic coordinates (SIRGAS2000, 2000,4)

Point ID	Latitude	Longitude	h [m]
GNSS-CORS	−22° 00'17.81599″	−47°53′ 57.04968″	824.587
EUNO pillar	−21° 59′ 52.55201″	−47°55′ 40.71034″	833.825

5.9   Coordinate Transformation

In many Geomatics applications, it is necessary to transform or convert the coordinates of points from one system to another or even to specify the coordinates of the same points in more than one system. Typically, if the change is made from one coordinate system to another on the same datum, it is referred to as a conversion of coordinates; otherwise, it is referred to as a coordinate transformation. In practice, however, the term coordinate transformation is used for both, as will be the case in this book.

5.9.1   *Transformations in Two-Dimensional Space*

The primary coordinate transformations in two-dimensional space are between Cartesian and Polar coordinate systems and from one Cartesian Coordinate System to another, as described in the following subsections.

5.9.1.1   Transformation between Cartesian and Polar Coordinate Systems

The most common type of two-dimensional coordinate transformation is between Cartesian and Polar Coordinate Systems. In addition to being simple, it is frequently used in Geomatics because the surveying measurements made in the field with total stations are based on a Polar Coordinate System, and the coordinates of the computed geospatial data are usually based on a Cartesian Coordinate System. Figure 5.11 shows the geometric relationships between the two systems.

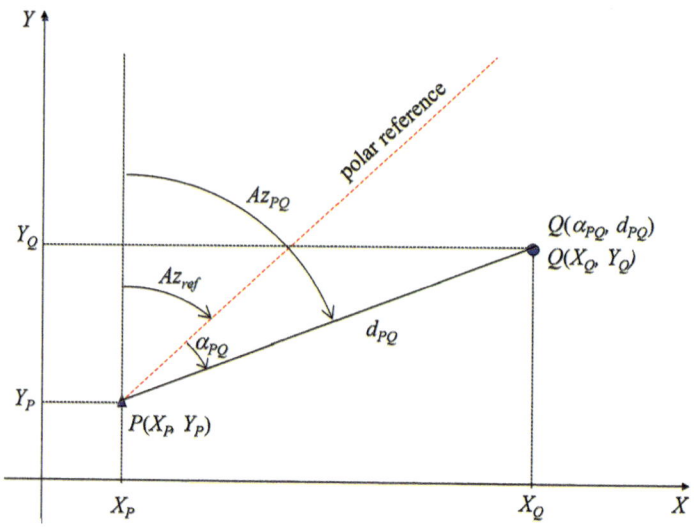

Az_{PQ} = horizontal direction (azimuth) of line PQ relative to the Y-axis
Az_{ref} = horizontal direction (azimuth) of the polar reference line
α_{PQ} = horizontal angle of line PQ regarding the reference line
d_{PQ} = horizontal distance between points (P) and (Q)

Fig. 5.11   Point positioning in a 2D Cartesian coordinate system and a 2D polar coordinate system

According to Fig. 5.11, the mathematical relationships for the aforementioned coordinate transformations can be written as follows.

$$Az_{PQ} = Az_{ref} + \alpha_{PQ} \tag{5.5}$$

$$X_Q = X_P + d_{PQ} * \sin(Az_{PQ}) \tag{5.6}$$

$$Y_Q = Y_P + d_{PQ} * \cos(Az_{PQ}) \tag{5.7}$$

Az_{PQ} = horizontal direction of line PQ regarding the Y-axis,
Az_{ref} = horizontal direction of the polar reference line,
α_{PQ} = polar angle of line PQ
d_{PQ} = horizontal distance between points (P) and (Q).

$$Az_{PQ} = \operatorname{atan}\left(\frac{X_Q - X_P}{Y_Q - Y_P}\right) + \alpha_{quad} \tag{5.8}$$

The α_{quad} value included in Eq. (5.8) takes into account the quadrant in which the line lies, as explained in more detail in Chap. 6.

$$d_{PQ} = \sqrt{\Delta X^2 + \Delta Y^2} \tag{5.9}$$

$$d_{PQ} = \Delta X * \mathrm{cossec}(Az_{PQ}) \tag{5.10}$$

$$d_{PQ} = \Delta Y * \sec(Az_{PQ}) \tag{5.11}$$

$$\sin(Az_{PQ}) = \frac{\Delta X}{d_{PQ}} \tag{5.12}$$

$$\cos(Az_{PQ}) = \frac{\Delta Y}{d_{PQ}} \tag{5.13}$$

Example 5.2 Using the values given in Table 5.4, calculate the Cartesian coordinates of the point (Q).

Table 5.4 Geometric elements

Geometric element	X [m]	Y [m]
P	5378.161	10,954.487
Az_{PQ}	34 ° 23'56"	
d_{PQ}	734.931 m	

Solution:
   Using Eqs. (5.6) and (5.7).

$$X_Q = 5,378.161 + 734.931 * \sin(34°23'56'')$$
$$X_Q = 5,793.361 \ m$$

$$Y_Q = 10,954.487 + 734.931 * \cos(34°23'56'')$$
$$Y_Q = 11,560.897 \ m$$

Example 5.3 Given the Cartesian coordinates of two points (A) and (B) in Table 5.5, calculate the horizontal distance between them and the reference angle Az of line AB.

Table 5.5 2D Cartesian coordinates

Point ID	X [m]	Y [m]
A	2389.762	14,178.269
B	1965.348	13,241.536
B-A	- 424.414	- 936.733

Solution:

   Using Eq. (5.9).

$$d_{AB} = \sqrt{(-424.414)^2 + (-936.733)^2} = 1,028.395 \ \ \mathrm{m}$$

Using Eq. (5.8), *it is possible to calculate the reference direction Az of the line AB. Thus, ignoring the quadrant of the line AB, an intermediate value Az'_{AB} is calculated as follows.*

$$Az'_{AB} = \mathrm{atan}\left(\frac{-424.414}{-936.733}\right) = 0.425411428 \ \ \mathrm{rad} = 24° \, 22'27''$$

Since it is a tangent, it is necessary to check the quadrant in which line AB lies. In this case, it is the third quadrant.

$$Az_{AB} = Az'_{AB} + 180° = 24° \, 22'27'' + 180° = 204° \, 22'27''$$

5.9.1.2 Two-Dimensional Cartesian Coordinate Transformation

Among the various methods available in the literature to transform Cartesian coordinates between Two-Dimensional systems, the linear ones are of particular interest for Geomatics applied to Civil Engineering. Mathematically, they are represented in matrix form as given by the system of Eqs. (5.14).

$$\begin{aligned} X &= a * \xi + b * \eta + c \\ Y &= d * \xi + e * \eta + f \end{aligned} \tag{5.14}$$

Where (X, Y) and (ξ, η) represent the coordinates of the same points in different coordinate systems, and the parameters a, b, c, d, e and f are the unknowns of the transformation, called the *transformation parameters*. In many cases, these parameters are already known so that the coordinates can be easily transformed between the two systems. In other cases, however, they must be determined using points with known coordinates in both systems, called *homologous points*.

Each homologous point establishes a pair of Eq. (5.14). Thus, whenever the number of homologous points (observations) exceeds the minimum amount necessary to obtain the unique solution, there will be more transformation equations than parameters, and the solution of the system of equations is, generally, performed using an adjustment computation model, as shown later in this chapter.

Two methods of coordinate transformation between two-dimensional Cartesian coordinate systems are mainly used in Geomatics applied to Civil Engineering.

- Two-Dimensional Conformal Coordinate Transformation
- Two-Dimensional Affine Coordinate Transformation

Both methods involve converting coordinates from one Two-Dimensional Cartesian Coordinate System to another as a function of common homologous points between the two systems, such as the points (P_1 to P_4) shown in Fig. 5.12.

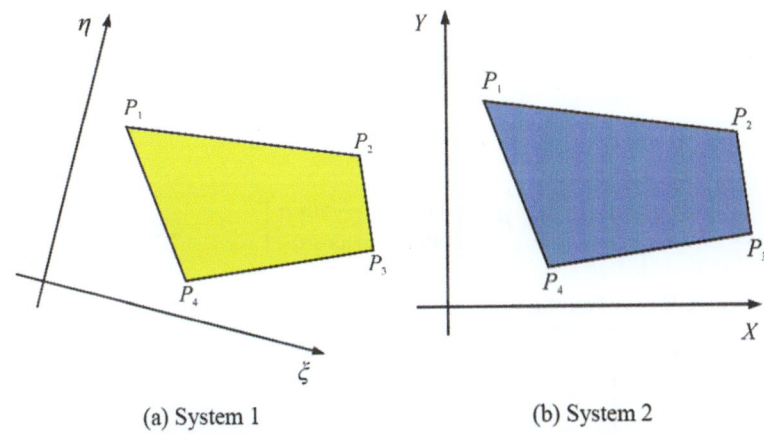

(a) System 1   (b) System 2

Fig. 5.12 Principle of homologous point configuration

5.9.1.2.1 Two-Dimensional Conformal Coordinate Transformation

The two-dimensional conformal coordinate transformation[5] consists of converting 2D Cartesian coordinates from a (ξ, η) system to a (X, Y) system using the coordinates of two or more homologous points, preserving the shapes of the existing geometric elements present in each system.

The mathematical concepts of this type of coordinate transformation are based on the rotation, translation and scaling of the axes of one of the coordinate systems to adapt it to the other. There are, therefore, three steps to consider, as shown in Fig. 5.13.

1. Rotate the (ξ, η) axes to make them parallel to the (X, Y) axes.
2. Change the scale of the (ξ, η) system to match the dimensions of the (X, Y) system.
3. Translate the origin of the (ξ, η) system to match the origin of the (X, Y) system.

Rotation
According to Fig. 5.13, to make the coordinate systems (ξ, η) and (X, Y) parallel, it is necessary to rotate the (ξ, η) system clockwise by an angle of rotation equal to α. This

---

[5] Also known as four-parameter transformation and similarity transformation.

creates a new coordinate system (ξ_a, η_a), parallel to the (X, Y) system, as indicated below.

$$\xi_a = \xi * \cos \alpha - \eta * \sin \alpha$$
$$\eta_a = \xi * \sin \alpha + \eta * \cos \alpha \qquad\qquad (5.15)$$

Which in matrix form can be written as follows.

$$\xi_a = R * \xi \qquad\qquad (5.16)$$

where

$$\xi_a = \begin{bmatrix} \xi_a \\ \eta_a \end{bmatrix} \qquad R = \begin{bmatrix} \cos \alpha & -\sin \alpha \\ \sin \alpha & \cos \alpha \end{bmatrix} \qquad \xi = \begin{bmatrix} \xi \\ \eta \end{bmatrix}$$

The matrix R is called the *rotation matrix.*

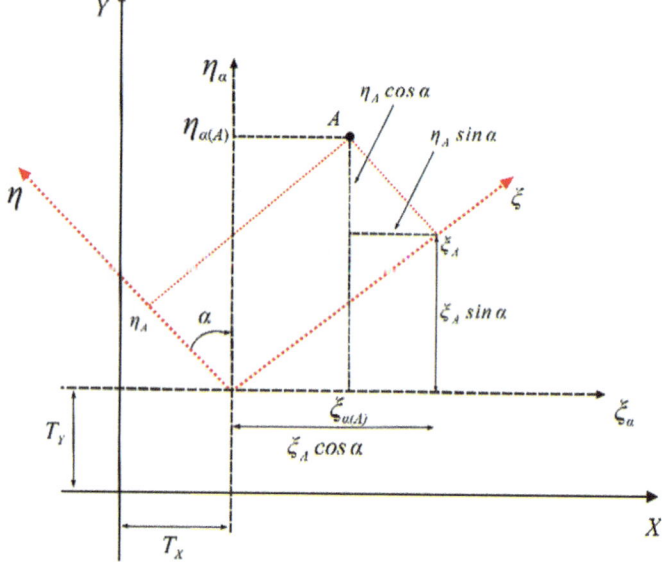

Fig. 5.13 Geometry of the coordinate transformation between two-dimensional Cartesian coordinate systems

Scale Factor

After rotating the (ξ, η) system, it is necessary to make it dimensionally equivalent to the (X, Y) system. This can be done by multiplying the coordinates of the (ξ_a, η_a) system by a scale factor k, from which a new coordinate system (ξ_k, η_k) is obtained. Taking into account Eq. (5.16), the following matrix equation can be written.

$$\boldsymbol{\xi}_k = k * \boldsymbol{R} * \boldsymbol{\xi} \tag{5.17}$$

Translation

Once the rotation and scaling steps have been completed, it is finally necessary to translate the origin of the (ξ, η) system so that it coincides with the origin of the (X, Y) system. Such an operation can be performed by applying the translation values (T_X, T_Y) to the coordinate system (ξ_k, η_k) to obtain the coordinates of the (ξ, η) system in the (X, Y) system. Therefore, using Eq. (5.17), the final matrix equation is given by

$$\boldsymbol{X} = k * \boldsymbol{R} * \boldsymbol{\xi} + \boldsymbol{T} \tag{5.18}$$

$$\begin{bmatrix} X \\ Y \end{bmatrix} = k * \begin{bmatrix} \cos\alpha & -\sin\alpha \\ \sin\alpha & \cos\alpha \end{bmatrix} * \begin{bmatrix} \xi \\ \eta \end{bmatrix} + \begin{bmatrix} T_x \\ T_y \end{bmatrix} \tag{5.19}$$

Such a transformation is orthogonal. The orthogonality condition occurs because the rotation matrix $\boldsymbol{R}$ is orthogonal, i.e., $\boldsymbol{R}^T = \boldsymbol{R}^{-1}$. This means that both axes in the two coordinate systems are orthogonal to each other and remain orthogonal after the transformation.

This coordinate transformation has four parameters and requires at least two homologous points. Each point allows a pair of transformation Eq. (5.20) to be written.

$$\begin{aligned} X &= k * (\xi * \cos\alpha - \eta * \sin\alpha) + T_X \\ Y &= k * (\xi * \sin\alpha + \eta * \cos\alpha) + T_Y \end{aligned} \tag{5.20}$$

Considering.

$$a = k * \cos\alpha \tag{5.21}$$

$$b = k * \sin\alpha \tag{5.22}$$

$$c = T_X \tag{5.23}$$

$$d = T_Y \tag{5.24}$$

Equations (5.20) can be rewritten as

$$\begin{aligned} X &= a * \xi - b * \eta + c \\ Y &= b * \xi + a * \eta + d \end{aligned} \tag{5.25}$$

or in matrix form, as given by Eq. (5.26).

$$\begin{bmatrix} X \\ Y \end{bmatrix} = \begin{bmatrix} a & -b \\ b & a \end{bmatrix} * \begin{bmatrix} \xi \\ \eta \end{bmatrix} + \begin{bmatrix} c \\ d \end{bmatrix} \tag{5.26}$$

As an example, consider two homologous points with coordinates (ξ_1, η_1), (ξ_2, η_2), and (x_1, y_1), (x_2, y_2). The four parameters to be determined are the scale factor k, the rotation angle α and the translations (T_x, T_y), described by the elements a, b, c and d. Thus, replacing the four parameters in Eq. (5.25) gives the following set of linear equations.

$$\begin{aligned} x_1 &= a * \xi_1 - b * \eta_1 + c \\ y_1 &= a * \eta_1 + b * \xi_1 + d \\ x_2 &= a * \xi_2 - b * \eta_2 + c \\ y_2 &= a * \eta_2 + b * \xi_2 + d \end{aligned} \tag{5.27}$$

Which can be expressed in matrix form as

$$\underbrace{\begin{bmatrix} x_1 \\ y_1 \\ x_2 \\ y_2 \end{bmatrix}}_{X} = \underbrace{\begin{bmatrix} \xi_1 & -\eta_1 & 1 & 0 \\ \eta_1 & \xi_1 & 0 & 1 \\ \xi_2 & -\eta_2 & 1 & 0 \\ \eta_2 & \xi_2 & 0 & 1 \end{bmatrix}}_{A} * \underbrace{\begin{bmatrix} a \\ b \\ c \\ d \end{bmatrix}}_{x} \tag{5.28}$$

Or,

$$X = Ax \tag{5.29}$$

Considering Eq. (5.29), the vector x can be calculated as given by Eq. (5.30).

$$x = \begin{bmatrix} a \\ b \\ c \\ d \end{bmatrix} = A^{-1}X \tag{5.30}$$

With the four parameters, Eq. (5.31) and (5.32) can be used to determine the scale factor k and the angle of rotation α.

$$k = \sqrt{a^2 + b^2} \tag{5.31}$$

$$\alpha = \operatorname{atan}\left(\frac{b}{a}\right) \tag{5.32}$$

If

$\alpha > 0 \Rightarrow$ Counter-clockwise (CCW) rotation
$\alpha < 0 \Rightarrow$ Clockwise (CW) rotation

The inverse of Eq. (5.26) is given as follows:

$$\begin{bmatrix} \xi \\ \eta \end{bmatrix} = \frac{1}{k^2} \begin{bmatrix} a & b \\ -b & a \end{bmatrix} * \begin{bmatrix} X - c \\ Y - d \end{bmatrix} \tag{5.33}$$

Because it is an orthogonal transformation, it has three important properties.

- The rotation matrix R is orthogonal and then $R^T = R^{-1}$.
- The lengths of the same geometric elements in the two coordinate systems, original and transformed, are related by a scale factor k.
- The transformation does not deform the angles, as it is conformal.[6]

Example 5.4 Station points (E010), (E011) and (E020) have known UTM easting and northing coordinates[7] determined by a GNSS RTK survey. The coordinates of points (E010) and (E020) were also determined using a total station based on a local (X, Y) coordinate system. It is required to determine the coordinates (X, Y) of point (E011). All coordinate values are given in Table 5.6.

Table 5.6 UTM and local coordinates

Point ID	E [m]	N [m]	X [m]	Y [m]
E010	204,729.018	7,560,947.116	152,596.092	254,100.825
E020	204,584.481	7,559,967.185	152,433.303	253,124.303
E011	204,543.592	7,560,971.516	?	?

Solution:
  To solve the problem, the values (200,000.000; 7,550,000.000) were subtracted from the coordinates (E, N) and (150,000.000; 250,000.000) from the coordinates (X, Y). This results in the following matrices.

$$A = \begin{bmatrix} 4,729.018 & -10,947.116 & 1 & 0 \\ 10,947.116 & 4,729.018 & 0 & 1 \\ 4,584.481 & -9,967.185 & 1 & 0 \\ 9,967.185 & 4,584.481 & 0 & 1 \end{bmatrix} \qquad X = \begin{bmatrix} 2,596.092 \\ 4,100.825 \\ 2,433.303 \\ 3,124.303 \end{bmatrix}_m$$

Using Eq. (5.30).

---

[6] Refer to Chap. 16 for more information.
[7] Refer to Chap. 16 for more information.

$$x = \begin{bmatrix} a \\ b \\ c \\ d \end{bmatrix} = A^{-1}X = \begin{bmatrix} 0.999284012 \\ -0.018731407 \\ -2,334.594970 \\ -6,749.871837 \end{bmatrix}$$

Using Eqs. (5.31) and (5.32).

$$k = \sqrt{a^2 + b^2} = 0.999459555$$

$$\alpha = \text{atan}\left(\frac{b}{a}\right) = -0.018742633 \ \text{rad} = -1°04'25.9'' \ (\text{CW rotation})$$

Finally, the coordinates (X, Y) of the point (E011) are obtained using Eq. (5.26).

$$\begin{bmatrix} a & -b \\ b & a \end{bmatrix} * \begin{bmatrix} E_{E011} \\ N_{E011} \end{bmatrix} + \begin{bmatrix} c \\ d \end{bmatrix} = \begin{bmatrix} 0.999284012 & 0.018731407 \\ -0.018731407 & 0.999284012 \end{bmatrix}$$

$$* \begin{bmatrix} 4,543.592 \\ 10,971.516 \end{bmatrix} + \begin{bmatrix} -2,334.5950 \\ -6,749.8718 \end{bmatrix} = \begin{bmatrix} 2,411.2558 \\ 4,128.6808 \end{bmatrix}_m$$

$$\begin{bmatrix} X_{E011} \\ Y_{E011} \end{bmatrix} = \begin{bmatrix} 2,411.2558 \\ 4,128.6808 \end{bmatrix} + \begin{bmatrix} 150,000.000 \\ 250,000.000 \end{bmatrix} = \begin{bmatrix} 152,411.256 \\ 254,128.681 \end{bmatrix}_m$$

5.9.1.2.2 Two-Dimensional Conformal Coordinate Transformation with Redundant Observations

There are few situations where a Two-Dimensional Coordinate Transformation occurs with only two homologous points. In addition to the poor quality of the results, this type of transformation does not allow the accuracy of the calculated values to be known. It is, therefore, recommended to use a transformation with more than two homologous points whenever possible. In this case, the solution to the problem can be obtained by using a Least Squares adjustment method. Two adjustment computation models can be used to solve the problem: the Parametric adjustment model and the Generalised Gauss-Helmert adjustment model, depending on whether the standard deviations of the coordinates of the homologous point are known or not.

Case 1: Homologous points with unkown coordinate standard deviation
Suppose the standard deviations of the coordinates of the homologous points in the two coordinate systems are not available in the data set. In this case, the solution can be obtained by the Parametric adjustment model, as shown below.

To better understand the adjustment computation modelling, consider the following:

$P : (X, \ Y) = $ point (P) in the coordinate system (X, Y).
$P' : (\xi, \eta) = $ point (P) in the coordinate system (ξ, η).
$\widetilde{P} : (\widetilde{X}, \widetilde{Y}) = $ point (P) in the coordinate system (X, Y) after adjustment computation.
$n = $ number of homologous points.

As already explained in Chap. 3, what differentiates the coordinates (X, Y) and the transformed coordinates $(\widetilde{X}, \widetilde{Y})$ are the residual errors v_X and v_Y, as given by Eq. (5.34).

$$v_X = \widetilde{X} - X$$
$$v_Y = \widetilde{Y} - Y \tag{5.34}$$

Substituting Eq. (5.20) into Eq. (5.34) gives

$$v_X = k * (\xi * \cos \alpha - \eta * \sin \alpha) + T_X - X$$
$$v_Y = k * (\xi * \sin \alpha + \eta * \cos \alpha) + T_Y - Y \tag{5.35}$$

Considering the parameters a, b, c, d, in Eqs. (5.21), (5.22), (5.23) and (5.24), Eq. (5.35) can be rewritten as given in Eq. (5.36).

$$v_X = a * \xi - b * \eta + c - X$$
$$v_Y = a * \eta + b * \xi + d - Y \tag{5.36}$$

Each homologous point generates 2 error equations. For n homologous points, there are, therefore, $2n$ error Eq. (5.36) containing the four unknowns $a, b, c,$ and d, which can be expressed as follows.

$$
\begin{bmatrix} v_{x1} \\ v_{y1} \\ v_{x2} \\ v_{y2} \\ \vdots \\ v_{xn} \\ v_{yn} \end{bmatrix}
=
\begin{bmatrix}
\xi_1 & -\eta_1 & 1 & 0 \\
\eta_1 & \xi_1 & 0 & 1 \\
\xi_2 & -\eta_2 & 1 & 0 \\
\eta_2 & \xi_2 & 0 & 1 \\
\vdots & \vdots & \vdots & \vdots \\
\xi_n & -\eta_n & 1 & 0 \\
\eta_n & \xi_n & 0 & 1
\end{bmatrix}
*
\begin{bmatrix} a \\ b \\ c \\ d \end{bmatrix}
-
\begin{bmatrix} X_1 \\ Y_1 \\ X_2 \\ Y_2 \\ \vdots \\ X_n \\ Y_n \end{bmatrix}
\tag{5.37}
$$

For the application of the Parametric adjustment model, the above equations can be written in matrix form as

$$v = A \, x - l_0 \tag{5.38}$$

where

$$
v = \begin{bmatrix} v_{x1} \\ v_{y1} \\ v_{x2} \\ v_{y2} \\ \vdots \\ v_{xn} \\ v_{yn} \end{bmatrix} \quad A = \begin{bmatrix} \xi_1 & -\eta_1 & 1 & 0 \\ \eta_1 & \xi_1 & 0 & 1 \\ \xi_2 & -\eta_2 & 1 & 0 \\ \eta_2 & \xi_2 & 0 & 1 \\ \vdots & \vdots & \vdots & \vdots \\ \xi_n & -\eta_n & 1 & 0 \\ \eta_n & \xi_n & 0 & 1 \end{bmatrix} \quad x = \begin{bmatrix} a \\ b \\ c \\ d \end{bmatrix} \quad l_0 = \begin{bmatrix} X_1 \\ Y_1 \\ X_2 \\ Y_2 \\ \vdots \\ X_n \\ Y_n \end{bmatrix}
$$

Using Eq. (3.84), (3.85), (8.86) and (3.87), the solution to Eq. (5.38) is given as follows.

$$
x = \left(A^{\mathrm{T}} P A\right)^{-1} A^{\mathrm{T}} P l_0 \tag{5.39}
$$

As before, knowing the values of the parameters a, b, c and d, it is possible to calculate the values of the scale factor k and the rotation angle α using Eqs. (5.31) and (5.32), respectively. It is also possible to calculate the values of the residual errors using Eq. (5.38), the values of the transformed coordinates using Eq. (5.25) and the standard deviation of unit weight using Eq. (5.40).

$$
s_0 = \pm \sqrt{\frac{v_X^{\mathrm{T}} v_X + v_Y^{\mathrm{T}} v_Y}{2n - 4}} \tag{5.40}
$$

where

n = number of homologous points

The variance-covariance matrix for the unknowns (four transformation parameters) is given by Eq. (5.41).

$$
\Sigma_{xx} = s_0^2 * \left(A^{\mathrm{T}} P A\right)^{-1} \tag{5.41}
$$

The standard deviation of the transformed coordinates can be calculated by applying the general law of propagation of variances in Eq. (5.25) using the variance-covariance matrix for the transformation parameters given in Eq. (5.41).

Case 2: Homologous points with known coordinate standard deviation

Suppose the standard deviations of the coordinates of the homologous points are known. In this case, the problem can be solved by substituting the true values of the observations (X, Y, ξ, η) into Eq. (5.25) to obtain Eq. (5.42).

$$\tilde{X} - a * \tilde{\xi} + b * \tilde{\eta} - c = 0$$
$$\tilde{Y} - b * \tilde{\xi} - a * \tilde{\eta} - d = 0 \tag{5.42}$$

Equation (5.42) relates, in terms of conditions, observations $l^T = [X \; Y \; \xi \; \eta]$ and parameters $x^T = [a \; b \; c \; d]$. These are, therefore, conditional equations with parameters that can be solved by using the Generalised Gauss-Helmert adjustment model.

Since the functional model is linear, it can be solved directly using the residual errors and the approximate values of the transformation parameters, as described in Sect. 3.9.1. Thus, considering

$$a = a_0 + \delta_a$$
$$b = b_0 + \delta_b$$
$$c = c_0 + \delta_c$$
$$d = d_0 + \delta_d \tag{5.43}$$
$$\tilde{X} = X + v_x$$
$$\tilde{Y} = Y + v_y$$

Equation (5.42) can be written as follows.

$$X + v_X - (a_0 + \delta a) * (\xi + v_\xi) + (b_0 + \delta_b) * (\eta + v_\eta) - (c_0 + \delta_c) = 0$$
$$Y + v_Y - (b_0 + \delta b) * (\xi + v_\xi) - (a_0 + \delta_a) * (\eta + v_\eta) - (d_0 + \delta_d) = 0 \tag{5.44}$$

where a_0, b_0, c_0, d_0 are the approximate values vector and $\delta_a, \delta_b, \delta_c, \delta_d$ the unknowns to be added to them to obtain the final parameter values. Then, by expanding and regrouping Eq. (5.44), in addition to neglecting the second-order terms $\delta_a V_\xi$, gives

$$v_X - a_0 v_\xi + b_0 v_\eta - \delta_a \xi + \delta_b \eta - \delta_c + X - (a_0\xi - b_0\eta + c_0) = 0$$
$$v_Y - b_0 v_\xi - a_0 v_\eta - \delta_b \xi - \delta_a \eta - \delta_d + Y - (b_0\xi + a_0\eta + d_0) = 0 \tag{5.45}$$

which in matrix form is

$$
\begin{bmatrix} 1 & 0 & -a_0 & b_0 \\ 0 & 1 & -b_0 & -a_0 \end{bmatrix}
\begin{bmatrix} v_X \\ v_Y \\ v_\xi \\ v_\eta \end{bmatrix}
+
\begin{bmatrix} -\xi & \eta & -1 & 0 \\ -\eta & -\xi & 0 & -1 \end{bmatrix}
\begin{bmatrix} \delta_a \\ \delta_b \\ \delta_c \\ \delta_d \end{bmatrix}
$$
$$
+ \begin{bmatrix} X - X_0 \\ Y - Y_0 \end{bmatrix} = \begin{bmatrix} 0 \\ 0 \end{bmatrix} \tag{5.46}
$$

Or

$$B^T v + A x + w = 0 \tag{5.47}$$

As presented in Sect. 3.9.1.3, the solution to Eq. (5.47) is as follows:

$$x = -\left(A^T N^{-1} A\right)^{-1} A^T N^{-1} w \tag{5.48}$$

$$k = -N^{-1}(A x + w) = \left[N^{-1} A \left(A^T N^{-1} A\right)^{-1} A^T N^{-1} - N^{-1}\right] w \tag{5.49}$$

$$v = Q B k \tag{5.50}$$

$$s_0^2 = \frac{v^T B k}{r - u} \tag{5.51}$$

where

$$N = B^T Q B \tag{5.52}$$

Q = cofactor matrix
r = conditional equations
u = four unknowns

As with the Parametric adjustment model, once the values of the unknowns a, b, c and d are obtained, the scale factor k and the rotation angle α can be calculated using Eqs. (5.31) and (5.32), respectively.

It is then possible to calculate the residual errors using Eq. (5.50), the adjusted values using Eq. (5.25), and the standard deviation of unit weight using Eq. (5.51). The variance-covariance matrix for the transformation parameters can be calculated using Eq. (5.53).

$$\Sigma_{xx} = s_0^2 * \left(A^T N^{-1} A\right)^{-1} \tag{5.53}$$

The standard deviation of the transformed coordinates can be calculated by applying the general law of propagation of variances in Eq. (5.25) using the standard deviation values of the transformation parameters given in Eq. (5.53).

Example 5.5 Using the values of the UTM and local coordinates given in Table 5.7, determine the (X, Y) local coordinates of points $(E011)$ and $(E014)$ using the two-dimensional conformal coordinate transformation adjusted by the Parametric adjustment model.

Solution:

To solve the problem, the values (200,000; 7,550,000) were subtracted from the coordinates (E, N) and (150,000.000; 250,000.000) from the coordinates (X, Y).

For the application of the Parametric adjustment model, using Eq. (5.38), the following matrices are obtained.

Table 5.7 UTM and local coordinates

Point	UTM coordinates		Local coordinates	
	E [m]	N [m]	X [m]	Y [m]
E007	204,031.218	7,560,476.948	151,889.987	253,644.066
E010	204,729.018	7,560,947.116	152,596.092	254,100.825
E020	204,584.481	7,559,967.185	152,433.303	253,124.303
E031	203,446.780	7,560,309.706	151,302.841	253,487.893
E011	204,543.592	7,560,971.516	?	?
E014	203,551.820	7,560,860.473	?	?

$$
A = \begin{bmatrix}
4{,}031.218 & -10{,}476.948 & 1 & 0 \\
10{,}476.948 & 4{,}031.218 & 0 & 1 \\
4{,}729.018 & -10{,}947.116 & 1 & 0 \\
10{,}947.116 & 4{,}729.018 & 0 & 1 \\
4{,}584.481 & -9{,}967.185 & 1 & 0 \\
9{,}967.185 & 4{,}584.481 & 0 & 1 \\
3{,}446.780 & -10{,}309.706 & 1 & 0 \\
10{,}309.706 & 3{,}446.780 & 0 & 1
\end{bmatrix}
\quad
l_0 = \begin{bmatrix}
1{,}889.987 \\
3{,}644.066 \\
2{,}596.092 \\
4{,}100.825 \\
2{,}433.303 \\
3{,}124.303 \\
1{,}302.841 \\
3{,}487.893
\end{bmatrix}
\quad
x = \begin{bmatrix}
a \\
b \\
c \\
d
\end{bmatrix}
\quad
v = \begin{bmatrix}
v_{X007} \\
v_{Y007} \\
v_{X010} \\
v_{Y010} \\
v_{X020} \\
v_{Y020} \\
v_{X031} \\
v_{Y031}
\end{bmatrix}
$$

Using Eq. (5.39) and $P = I$.

$$
x = \begin{bmatrix}
a \\
b \\
c \\
d
\end{bmatrix} = \begin{bmatrix}
0.999279344 \\
-0.018732740 \\
-2{,}334.5866 \\
-6{,}749.8163
\end{bmatrix}
$$

Residual errors are calculated by substituting the values of the transformation parameters into Eq. (5.38).

$$v^{T} = [0.0013 \quad -0.0003 \quad 0.0009 \quad -0.0019 \quad 0.0003 \quad 0.0029 \quad -0.0025 \quad -0.0007]_{m}$$

The values of the adjusted coordinates are calculated using Eq. (5.34).

$$\tilde{l}^T = [151,889.9883 \ \ 253,644.0657 \ \ 152,596.0929 \ \ 254,100.8231 \ \ 152,433.3033$$
$$253,124.3059 \ \ 151,302.8385 \ \ 253,487.8923]_m$$

The standard deviation of unit weight is given by Eq. (5.40).

$$v_x^T = [0.0013 \quad 0.0009 \quad 0.0003 \quad -0.0025]_m$$
$$v_y^T = [-0.0003 \quad -0.0019 \quad 0.0029 \quad -0.0007]_m$$

Using Eq. (5.40) *with* $2n = 8$, *gives*

$$s_0 = \pm 0.0023 \text{ m} = \pm 2.3 \text{ mm}$$

The variance-covariance matrix for the unknowns (four transformation parameters) can be calculated using Eq. (5.41).

$$\Sigma_{xx} = \begin{bmatrix} 3.45916 * 10^{-12} & 0.00000 & -1.45211 * 10^{-8} & -3.60625 * 10^{-8} \\ 0.00000 & 3.45916 * 10^{-12} & 3.60625 * 10^{-8} & -1.45211 * 10^{-8} \\ -1.45211 * 10^{-8} & 3.60625 * 10^{-8} & 4.38234 * 10^{-4} & 0.00000 \\ -3.60625 * 10^{-8} & -1.45211 * 10^{-8} & 0.00000 & 4.38234 * 10^{-4} \end{bmatrix}$$

Using the diagonal values in the variance-covariance matrix for the parameters, the following results are obtained:

$$a = 0.999279344 \pm 0.000001860$$
$$b = -0.018732740 \pm 0.000001860$$
$$c = -2,334.5866 \text{ m} \pm 0.0209 \text{ m}$$
$$d = -6,749.8163 \text{ m} \pm 0.0209 \text{ m}$$

Using Eqs. (5.31) and (5.32).

$$k = \sqrt{a^2 + b^2} = 0.999454913$$

$$\alpha = \text{atan}\left(\frac{b}{a}\right) = -0.018744054 \text{ rad} = -1°04'26.2''$$

Finally, using Eq. (5.26).

$$\begin{bmatrix} X_{E011} & X_{E014} \\ Y_{E011} & Y_{E014} \end{bmatrix} = \begin{bmatrix} 152,411.258 & 151,418.120 \\ 254,128.679 & 254,036.295 \end{bmatrix}_m$$

The standard deviation of the coordinates of the points (E011) and (E014) can be calculated by using the general law of propagation of variances in Eq. (5.25) *using the covariance matrix for the transformation parameters.*

$$s_{X_{E011}} = s_{Y_{E011}} = \pm 0.0304 \text{ m} = \pm 3.04 \text{ cm}$$
$$s_{X_{E014}} = s_{Y_{E014}} = \pm 0.0298 \text{ m} = \pm 2.98 \text{ cm}$$

Example 5.6 In this example, consider the same coordinate values as in Table 5.7. Note also that the coordinates determined using GNSS technology have precisions of ± 2.0 cm, and the coordinates determined using the total station have precisions of ± 1.0 cm. It is required to determine the (X, Y) local coordinates of points $(E011)$ and $(E014)$ using the two-dimensional conformal coordinate transformation adjusted by the Generalised Gauss-Helmert adjustment model.

Solution:
To solve the problem, it is first necessary to reduce the coordinates (E, N) and (X, Y) as in Example 5.5. *Then, since the precisions of the homologous points are known, the solution can be obtained using the Gauss-Helmert adjustment model, as follows.*
From Example 5.5, *the approximate values of the parameters can be taken as*

$$a_0 = -0.999279344$$
$$b_0 = -0.018732740$$
$$c_0 = -2,334.5866 \text{ m}$$
$$d_0 = -6,749.8163 \text{ m}$$

The standard deviation of the unit weight s_0 will be assumed as equal to 1.0. The following matrices can then be written.

$$B = \begin{bmatrix}
1 & 0 & 0 & 0 & 0 & 0 & 0 & 0 \\
0 & 1 & 0 & 0 & 0 & 0 & 0 & 0 \\
0 & 0 & 1 & 0 & 0 & 0 & 0 & 0 \\
0 & 0 & 0 & 1 & 0 & 0 & 0 & 0 \\
0 & 0 & 0 & 0 & 1 & 0 & 0 & 0 \\
0 & 0 & 0 & 0 & 0 & 1 & 0 & 0 \\
0 & 0 & 0 & 0 & 0 & 0 & 1 & 0 \\
0 & 0 & 0 & 0 & 0 & 0 & 0 & 1 \\
-0.999279344 & 0.018732740 & 0 & 0 & 0 & 0 & 0 & 0 \\
-0.018732740 & -0.999279344 & 0 & 0 & 0 & 0 & 0 & 0 \\
0 & 0 & -0.999279344 & 0.018732740 & 0 & 0 & 0 & 0 \\
0 & 0 & -0.018732740 & -0.999279344 & 0 & 0 & 0 & 0 \\
0 & 0 & 0 & 0 & -0.999279344 & 0.018732740 & 0 & 0 \\
0 & 0 & 0 & 0 & -0.018732740 & -0.999279344 & 0 & 0 \\
0 & 0 & 0 & 0 & 0 & 0 & -0.999279344 & 0.018732740 \\
0 & 0 & 0 & 0 & 0 & 0 & -0.018732740 & -0.999279344
\end{bmatrix}$$

$$A^{\mathrm{T}} = \begin{bmatrix} -4{,}031.218 & -10{,}476.948 & -4{,}729.018 & -10{,}947.116 & -4{,}584.481 & -9{,}967.185 & -3{,}446.780 & -10{,}309.706 \\ 10{,}476.948 & -4{,}031.218 & 10{,}947.116 & -4{,}729.018 & 9{,}967.185 & -4{,}584.481 & 10{,}309.706 & -3{,}446.780 \\ -1 & 0 & -1 & 0 & -1 & 0 & -1 & 0 \\ 0 & -1 & 0 & -1 & 0 & -1 & 0 & -1 \end{bmatrix}$$

$$Q_{ll} = \begin{bmatrix} 0.0001 & 0 & 0 & 0 & 0 & 0 & 0 & 0 & 0 & 0 & 0 & 0 & 0 & 0 & 0 & 0 \\ 0 & 0.0001 & 0 & 0 & 0 & 0 & 0 & 0 & 0 & 0 & 0 & 0 & 0 & 0 & 0 & 0 \\ 0 & 0 & 0.0001 & 0 & 0 & 0 & 0 & 0 & 0 & 0 & 0 & 0 & 0 & 0 & 0 & 0 \\ 0 & 0 & 0 & 0.0001 & 0 & 0 & 0 & 0 & 0 & 0 & 0 & 0 & 0 & 0 & 0 & 0 \\ 0 & 0 & 0 & 0 & 0.0001 & 0 & 0 & 0 & 0 & 0 & 0 & 0 & 0 & 0 & 0 & 0 \\ 0 & 0 & 0 & 0 & 0 & 0.0001 & 0 & 0 & 0 & 0 & 0 & 0 & 0 & 0 & 0 & 0 \\ 0 & 0 & 0 & 0 & 0 & 0 & 0.0001 & 0 & 0 & 0 & 0 & 0 & 0 & 0 & 0 & 0 \\ 0 & 0 & 0 & 0 & 0 & 0 & 0 & 0.0001 & 0 & 0 & 0 & 0 & 0 & 0 & 0 & 0 \\ 0 & 0 & 0 & 0 & 0 & 0 & 0 & 0 & 0.0004 & 0 & 0 & 0 & 0 & 0 & 0 & 0 \\ 0 & 0 & 0 & 0 & 0 & 0 & 0 & 0 & 0 & 0.0004 & 0 & 0 & 0 & 0 & 0 & 0 \\ 0 & 0 & 0 & 0 & 0 & 0 & 0 & 0 & 0 & 0 & 0.0004 & 0 & 0 & 0 & 0 & 0 \\ 0 & 0 & 0 & 0 & 0 & 0 & 0 & 0 & 0 & 0 & 0 & 0.0004 & 0 & 0 & 0 & 0 \\ 0 & 0 & 0 & 0 & 0 & 0 & 0 & 0 & 0 & 0 & 0 & 0 & 0.0004 & 0 & 0 & 0 \\ 0 & 0 & 0 & 0 & 0 & 0 & 0 & 0 & 0 & 0 & 0 & 0 & 0 & 0.0004 & 0 & 0 \\ 0 & 0 & 0 & 0 & 0 & 0 & 0 & 0 & 0 & 0 & 0 & 0 & 0 & 0 & 0.0004 & 0 \\ 0 & 0 & 0 & 0 & 0 & 0 & 0 & 0 & 0 & 0 & 0 & 0 & 0 & 0 & 0 & 0.0004 \end{bmatrix}$$

$$w^{\mathrm{T}} = \begin{bmatrix} -0.0013 & 0.0003 & -0.0009 & 0.0019 & -0.0003 & -0.0029 & 0.0025 & 0.0007 \end{bmatrix}_{\mathrm{m}}$$

$$x^{\mathrm{T}} = \begin{bmatrix} -4.1746 * 10^{-14} & -6.1407 * 10^{-15} & 5.3227 * 10^{-11} & 5.6735 * 10^{-10} \end{bmatrix}$$

Then,

$$a = 0.999279344 \quad b = -0.018732740$$

$$c = -2{,}334.5866 \text{ m} \quad d = -6{,}749.8163 \text{ m}$$

$$k^{\mathrm{T}} = \begin{aligned}[t] & [2.517634665 \quad -0.673315838 \quad 1.839711155 \quad -3.746497819 \\ & \ 0.576130381 \quad 5.795706448 \quad -4.933476201 \quad -1.375892791] \end{aligned}$$

$$v^{\mathrm{T}} = \begin{aligned}[t] & [0.000252 \quad -0.000067 \quad \ \ 0.000184 \quad -0.000375 \quad 0.000058 \\ & \ \ 0.000580 \quad -0.000493 \quad -0.000138 \quad -0.001011 \quad 0.000250 \\ & -0.000763 \quad \ \ 0.001484 \quad -0.000187 \quad -0.002321 \quad 0.001962 \\ & \ \ 0.000587]_{m} \end{aligned}$$

$$s_0 = \pm\sqrt{\frac{v^T Bk}{r-u}} = \pm 0.1026 \text{ m} = \pm 10.26 \text{ cm} \quad (\text{with } r=8 \text{ and } u=4)$$

$$\Sigma_{xx} = \begin{bmatrix} 3.45916*10^{-12} & -1.15137*10^{-26} & -1.45211*10^{-8} & -3.60625*10^{-8} \\ -1.15137*10^{-26} & 3.45916*10^{-12} & 3.60625*10^{-8} & -1.45211*10^{-8} \\ -1.45211*10^{-8} & 3.60625*10^{-8} & 0.000438234 & 0 \\ -3.60625*10^{-8} & -1.45211*10^{-8} & 0 & 0.000438234 \end{bmatrix}$$

Using the diagonal values in the variance-covariance matrix for the parameters given above, the following results are obtained:

$$a = 0.999279344 \pm 0.0000018598 \quad b = -0.0187327400 \pm 0.0000018598$$

$$c = -2,334.5866 \text{ m} \pm 0.0209 \text{ m} \quad d = -6,749.8163 \text{ m} \pm 0.0209 \text{ m}$$

$$k = \sqrt{a^2 + b^2} = 0.999454913$$

$$\alpha = \text{atan}\left(\frac{b}{a}\right) = -0.018744054 rad = -1°04'26.2''$$

Finally, using Eq. (5.42).

$$\begin{bmatrix} X_{E011} & X_{E014} \\ Y_{E011} & Y_{E014} \end{bmatrix} = \begin{bmatrix} 152,411.258 & 151,418.120 \\ 254,128.679 & 254,036295 \end{bmatrix}_m$$

The standard deviation of the coordinates of the points (E011) and (E014) can be calculated by applying the general law of propagation of variances in Eq. (5.25) using the covariance matrix for the transformation parameters.

$$S_{X_{E011}} = S_{Y_{E011}} = \pm 0.0304 \text{ m} = \pm 3.04 \text{ cm} \quad s_{X_{E014}} = s_{Y_{E014}} = \pm 0.0298 \text{ m} = \pm 2.98 \text{ cm}$$

5.9.1.2.3 Two-Dimensional Affine Coordinate Transformation

The *Two-Dimensional Affine Coordinate Transformation* is also a linear coordinate transformation. In this case, however, the rotation matrix does not satisfy the orthogonality condition, i.e., the Affine coordinate transformation does not preserve angles and, consequently, neither the areas.

The mathematical treatment for the Affine transformation is similar to that of the conformal transformation. However, it allows the inclusion of different scale factors and rotation angles for the (X, Y) axes. In these terms, the Affine transformation can be defined mathematically by Eq. (5.54).

$$X = k_X * \xi * \cos(\alpha_X) - k_Y * \eta * \sin(\alpha_Y) + T_X$$
$$Y = k_X * \xi * \sin(\alpha_X) + k_Y * \eta * \cos(\alpha_Y) + T_Y$$

(5.54)

where

k_X= scale factor for the X-axis
k_Y= scale factor for the Y-axis
α_X, α_Y= angle of rotation in the X-axis and in the Y-axis
T_X= translation in the X direction
T_Y= translation in the Y direction

The Affine transformation is, therefore, a transformation with six parameters, while the Conformal transformation is a transformation with four parameters. Figure 5.14 shows the geometric relationship of the Two-Dimensional Affine Coordinate Transformation. To determine the transformation parameters, the following simplifications are considered.

$$a = k_X * \cos(\alpha_X)$$

(5.55)

$$b = k_Y * \sin(\alpha_Y)$$

(5.56)

$$d = k_X * \sin(\alpha_X)$$

(5.57)

$$e = k_Y * \cos(\alpha_Y)$$

(5.58)

$$c = T_X$$

(5.59)

$$f = T_Y$$

(5.60)

Thus, Eq. (5.54) can be rewritten as follows.

$$X = a * \xi - b * \eta + c$$
$$Y = d * \xi + e * \eta + f$$

(5.61)

which in matrix form results in Eq. (5.62).

$$\begin{bmatrix} X \\ Y \end{bmatrix} = \begin{bmatrix} a & -b \\ d & e \end{bmatrix} * \begin{bmatrix} \xi \\ \eta \end{bmatrix} + \begin{bmatrix} c \\ f \end{bmatrix}$$

(5.62)

or

$$\begin{bmatrix} X \\ Y \end{bmatrix} = \begin{bmatrix} \xi & -\eta & 1 & 0 & 0 & 0 \\ 0 & 0 & 0 & \xi & \eta & 1 \end{bmatrix} * \begin{bmatrix} a \\ b \\ c \\ d \\ e \\ f \end{bmatrix} \tag{5.63}$$

The inverse of Eq. (5.62) is given as follows.

$$\begin{bmatrix} \xi \\ \eta \end{bmatrix} = \begin{bmatrix} a & -b \\ d & e \end{bmatrix}^{-1} * \begin{bmatrix} X - c \\ Y - f \end{bmatrix} \tag{5.64}$$

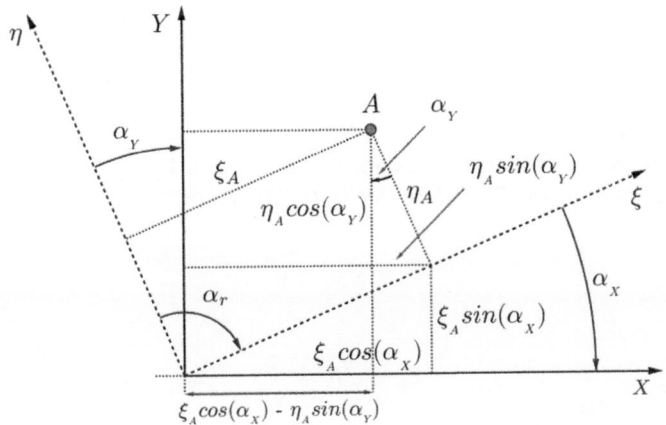

Fig. 5.14 Geometric relations of the two-dimensional affine coordinate transformation

As this is a coordinate transformation with six parameters, it is necessary to have at least three homologous points in both coordinate systems. Each point makes it possible to write two transformation Eq. (5.61). For example, considering the case of three homologous points (ξ_1, η_1), (ξ_2, η_2), (ξ_3, η_3), and (x_1, y_1), (x_2, y_2), (x_3, y_3), the calculation of the transformation parameters can be carried out using Eq. (5.65).

$$
\begin{bmatrix} x_1 \\ y_1 \\ x_2 \\ y_2 \\ x_3 \\ y_3 \end{bmatrix} = \begin{bmatrix} \xi_1 & -\eta_1 & 1 & 0 & 0 & 0 \\ 0 & 0 & 0 & \xi_1 & \eta_1 & 1 \\ \xi_2 & -\eta_2 & 1 & 0 & 0 & 0 \\ 0 & 0 & 0 & \xi_2 & \eta_2 & 1 \\ \xi_3 & -\eta_3 & 1 & 0 & 0 & 0 \\ 0 & 0 & 0 & \xi_3 & \eta_3 & 1 \end{bmatrix} * \begin{bmatrix} a \\ b \\ c \\ d \\ e \\ f \end{bmatrix} \qquad (5.65)
$$

$$
X \qquad\qquad A \qquad\qquad x
$$

Considering Eq. (5.65), the vector x can be calculated as given by Eq. (5.66).

$$
x = \begin{bmatrix} a \\ b \\ c \\ d \\ e \\ f \end{bmatrix} = A^{-1} X \qquad (5.66)
$$

Given the six transformation parameters, the values k_X, k_Y, α_X, α_Y, α_r can be calculated using Eqs. (5.67), (5.68), (5.69), (5.70) and (5.71).

$$
k_X = \sqrt{a^2 + d^2} \qquad (5.67)
$$

$$
k_Y = \sqrt{b^2 + e^2} \qquad (5.68)
$$

$$
\alpha_X = \operatorname{atan}\left(\frac{d}{a}\right) \qquad (5.69)
$$

$$
\alpha_Y = \operatorname{atan}\left(\frac{b}{e}\right) \qquad (5.70)
$$

$$
\alpha_r = 90° + \alpha_X - \alpha_Y \qquad (5.71)
$$

Whenever more than three homologous points are available, the solution can be obtained by applying an adjustment computation model similar to the conformal transformation. In the case of the Parametric adjustment computation model, the error equations are given by Eq. (5.72).

$$v_X = a * \xi - b * \eta + c - X$$
$$v_Y = d * \xi + e * \eta + f - Y \tag{5.72}$$

Each homologous point generates 2 error equations. For n homologous points, there are, therefore, $2n$ error Eq. (5.72) containing the six unknowns a, b, c, d, e, f, which can be expressed as follows.

$$
\begin{bmatrix} v_{X_1} \\ v_{Y_1} \\ v_{X_2} \\ v_{Y_2} \\ \vdots \\ \vdots \\ v_{X_n} \\ v_{Y_n} \end{bmatrix}
=
\begin{bmatrix}
\xi_1 & -\eta_1 & 1 & 0 & 0 & 0 \\
0 & 0 & 0 & \xi_1 & \eta_1 & 1 \\
\xi_2 & -\eta_2 & 1 & 0 & 0 & 0 \\
0 & 0 & 0 & \xi_2 & \eta_2 & 1 \\
\vdots & \vdots & \vdots & \vdots & \vdots & \vdots \\
\vdots & \vdots & \vdots & \vdots & \vdots & \vdots \\
\xi_n & -\eta_n & 1 & 0 & 0 & 0 \\
0 & 0 & 0 & \xi_n & \eta_n & 1
\end{bmatrix}
*
\begin{bmatrix} a \\ b \\ c \\ d \\ e \\ f \end{bmatrix}
-
\begin{bmatrix} X_1 \\ Y_1 \\ X_2 \\ Y_2 \\ \vdots \\ \vdots \\ X_n \\ Y_n \end{bmatrix}
\tag{5.73}
$$

The remaining calculations are performed similarly to the conformal transformation.

Example 5.7 Using the same values as in Example 5.5, calculate the coordinates of the points (E011) and (E014) in the coordinate system (X, Y) using the Two-Dimensional Affine Coordinate Transformation adjusted by the Parametric adjustment model.

Solution:
To solve the problem, the values (200,000.000; 7,550,000.000) were subtracted from the coordinates (E, N) and (150,000.000; 250,000.000) from the coordinates (X, Y).
For the application of the Parametric adjustment model, using Eq. (5.72), the following matrices are obtained.

$$
A =
\begin{bmatrix}
4,031.218 & -10,476.948 & 1 & 0 & 0 & 0 \\
0 & 0 & 0 & 4,031.218 & 10,476.948 & 1 \\
4,729.018 & -10,947.116 & 1 & 0 & 0 & 0 \\
0 & 0 & 0 & 4,729.018 & 10,947.116 & 1 \\
4,584.481 & -9,967.185 & 1 & 0 & 0 & 0 \\
0 & 0 & 0 & 4,584.481 & 9,967.185 & 1 \\
3,446.780 & -10,309.706 & 1 & 0 & 0 & 0 \\
0 & 0 & 0 & 3,446.780 & 10,309.706 & 1
\end{bmatrix}
$$

$$l_0 = \begin{bmatrix} 1{,}889.987 \\ 3{,}644.066 \\ 2{,}596.092 \\ 4{,}100.825 \\ 2{,}433.303 \\ 3{,}124.303 \\ 1{,}302.841 \\ 3{,}487.893 \end{bmatrix}_m \qquad x = \begin{bmatrix} a \\ b \\ c \\ d \\ e \\ f \end{bmatrix} \qquad v = \begin{bmatrix} v_{X007} \\ v_{Y007} \\ v_{X010} \\ v_{Y010} \\ v_{X020} \\ v_{Y020} \\ v_{X031} \\ v_{Y031} \end{bmatrix}$$

Using Eq. (5.39) *and* $P = I.$

$$x = \begin{bmatrix} a \\ b \\ c \\ d \\ e \\ f \end{bmatrix} = \begin{bmatrix} 0.999277269 \\ -0.018732082 \\ 2{,}334.5710 \\ -0.018734302 \\ 0.999284401 \\ -6{,}749.8625 \end{bmatrix}$$

The residual errors are calculated by substituting the values of the transformation parameters into the Eq. (5.38).

$$v^T = \begin{bmatrix} 0.001569510 & 0.000185523 & -0.000526309 & -0.000062212 \\ -0.000213087 & -0.000025188 & -0.000830116 & -0.000098123 \end{bmatrix}_m$$

As in Example 5.5, *the standard deviation of the unit weight is given by* Eq. (5.40) *with* $2n = 8$ *and* 6 *parameters (unknowns).*

$$s_0 = \pm 0.001327 \text{ m} = 1.3 \text{ mm}$$

The variance-covariance matrix for the unknowns (six transformation parameters) is given by Eq. (5.41).

$$\Sigma_{xx} = \begin{bmatrix} 1.835667*10^{-12} & 6.568417*10^{-13} & -8.581695*10^{-10} & 0 & 0 & 0 \\ 6.568417*10^{-13} & 3.771240*10^{-12} & 3.655874*10^{-8} & 0 & 0 & 0 \\ -8.581695*10^{-10} & 3.655874*10^{-8} & 3.851765*10^{-4} & 0 & 0 & 0 \\ 0 & 0 & 0 & 1.835667*10^{-12} & 6.568417*10^{-13} & -8.581695*10^{-10} \\ 0 & 0 & 0 & 6.568417*10^{-13} & 3.771240*10^{-12} & 3.655874*10^{-8} \\ 0 & 0 & 0 & -8.581695*10^{-10} & 3.655874*10^{-8} & 3.851765*10^{-4} \end{bmatrix}$$

Using the diagonal values in the variance-covariance matrix for the parameters given above, the following results are obtained:

$$a = 0.999277269 \pm 0.00000135 \quad b = -0.018732082 \pm 0.00000194$$
$$c = -2,334.5710 \text{ m} \pm 0.0196259 \text{ m} \quad d = -0.018734302 \pm 0.00000135$$
$$e = 0.99928401 \pm 0.00000194 \quad f = -6,749.8625 \text{ m} \pm 0.0196259 \text{ m}$$

Using Eqs. (5.67), (5.68), (5.69), (5.70) and (5.71).

$$k_X = \sqrt{a^2 + d^2} = 0.999452826 \quad k_Y = \sqrt{b^2 + e^2} = 0.999459957$$

$$\alpha_X = \text{atan}\left(\frac{d}{a}\right) = -1° 04'26.6'' \quad \alpha_X = \text{atan}\left(\frac{b}{e}\right) = -1° 04'26.1''$$

$$\alpha_r = 90° + \alpha_X - \alpha_Y = 89° 59'59.5''$$

Finally, using Eq. (5.62).

$$\begin{bmatrix} X_{E011} & X_{E014} \\ Y_{E011} & Y_{E014} \end{bmatrix} = \begin{bmatrix} 152{,}411.257 & 151{,}418.121 \\ 254{,}128.681 & 254{,}036.298 \end{bmatrix}_{\text{m}}$$

$$S_{X_{E011}} = S_{Y_{E011}} = \pm 2.96 \text{ cm} \quad S_{X_{E014}} = S_{Y_{E014}} = \pm 2.92 \text{ cm}$$

5.9.2   Transformation in Three-Dimensional Space

There are four Three-Dimensional Coordinate Transformations of interest for Geomatics applied to Civil Engineering.

- The transformation between Cartesian and Spherical Polar Coordinate Systems
- The transformation between Cartesian coordinate systems
- The transformation between Geocentric and Geodetic coordinate systems
- The transformation between Geocentric and Topocentric coordinate systems

5.9.2.1   Transformation between Cartesian and Spherical Polar Coordinate Systems

The transformation between Cartesian and Spherical Polar coordinate systems is similar to the 2D coordinate transformation presented in Sect. 5.9.1.1. For 3D space, see Fig. 5.15. In this case, the polar orientation refers to the Y-axis, and the point (P)

is at the origin of the coordinate system. Under these conditions, the following mathematical relationships can be written for the coordinate transformation.

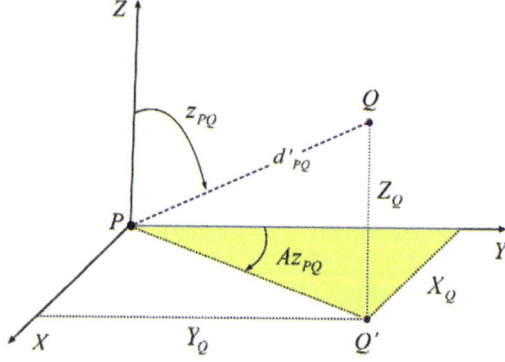

Fig. 5.15 Geometric relations between spatial Cartesian and Spherical polar coordinates

$$X_Q = X_P + d'_{PQ} * \sin(z_{PQ}) * \sin(Az_{PQ})$$
$$Y_Q = Y_P + d'_{PQ} * \sin(z_{PQ}) * \cos(Az_{PQ})$$
$$Z_Q = Z_P + d'_{PQ} * \cos(z_{PQ}) \qquad (5.74)$$

$$d'_{PQ} = \sqrt{\Delta X^2_{PQ} + \Delta Y^2_{PQ} + \Delta Z^2_{PQ}} \qquad (5.75)$$

$$Az_{PQ} = \mathrm{atan}\left(\frac{\Delta X_{PQ}}{\Delta Y_{PQ}}\right) + \alpha_{\mathrm{quad}} \qquad (5.76)$$

$$Z_{PQ} = \mathrm{acos}\left(\frac{\Delta Z_{PQ}}{d'_{PQ}}\right) \qquad (5.77)$$

Example 5.8 Using the 3D Cartesian Coordinates in Table 5.13, calculate the spherical polar coordinates of the point (Q).

Table 5.13 3D Cartesian Coordinates

Point	X [m]	Y [m]	Z [m]
P	1000.000	1000.000	1000.000
Q	1173.787	1034.268	1015.369
Q-P	173.787	34.268	15.369

Solution:

Using Eqs. (5.75), (5.76) and (5.77).

$$d'_{PQ} = \sqrt{(173.787^2 + 34.268^2 + 15.369^2)} = 177.799\,\text{m}$$

$$Az_{PQ} = \text{atan}\left(\frac{173.787}{34.268}\right) = 1.376110001\,\text{rad} = 78°\,50'43'' \quad \text{(First quadrant)}$$

$$z_{PQ} = \text{acos}\left(\frac{15.369}{177.799}\right) = 1.484247927\,\text{rad} = 85°\,02'28''$$

Example 5.9 Using the 3D Cartesian coordinates of the point (P) and the spherical polar coordinates calculated for the point (Q) in the previous example, calculate the 3D Cartesian coordinates of the point (Q).

Solution:

Using Eq. (5.74).

$$X_Q = 1,000.000 + 177.799 * \sin(85°\,02'28'') * \sin(78°\,50'43'') = 1,173.787 \text{ m}$$
$$Y_Q = 1,000.000 + 177.799 * \sin(85°\,02'28'') * \cos(78°\,50'43'') = 1,034.268 \text{ m}$$
$$Z_Q = 1,000.000 + 177.799 * \cos(85°\,02'28'') = 1,015.369 \text{ m}$$

5.9.2.2 Transformation between 3D Cartesian Coordinate Systems

The coordinate transformation between the (X', Y', Z') and (X, Y, Z) Three-Dimensional Cartesian coordinate systems, similar to the Two-Dimensional Cartesian transformation, is performed in three steps, as shown in Fig. 5.16.

1. Rotate the (X', Y', Z') axes to make them parallel to the (X, Y, Z) axes;
2. Change the scale of (X', Y', Z') system to match the dimensions of the (X, Y, Z) system.
3. Translate the origin of the (X', Y', Z') system to match the origin of the (X, Y, Z) system.

Rotation

The rotations occur counter-clockwise as seen from the origin of the coordinate frame and are performed sequentially, as shown in Fig. 5.16.

Angle ω is the angle of rotation about the X-axis, φ is the angle of rotation about the Y-axis and κ is the angle of rotation about the Z-axis. With these definitions, rotations about each coordinate axis are conveniently expressed by proper orthogonal matrices as given by Eq. (5.78). The final result is, therefore, non-commutative and depends on the specific sequence of the individual rotations applied.

Fig. 5.16 Principle of coordinate transformation between spatial Cartesian coordinate systems

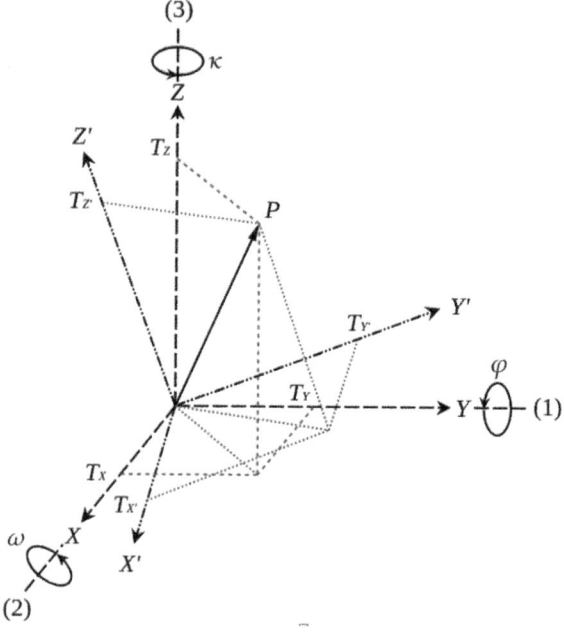

$$R_3(k) = \begin{bmatrix} \cos\kappa & \sin\kappa & 0 \\ -\sin\kappa & \cos\kappa & 0 \\ 0 & 0 & 1 \end{bmatrix} \quad R_2(\varphi) = \begin{bmatrix} \cos\varphi & 0 & -\sin\varphi \\ 0 & 1 & 0 \\ \sin\varphi & 0 & \cos\varphi \end{bmatrix} \quad R_1(\omega) = \begin{bmatrix} 1 & 0 & 0 \\ 0 & \cos\omega & \sin\omega \\ 0 & -\sin\omega & \cos\omega \end{bmatrix}$$

$$(5.78)$$

Considering that rotations are performed as $R(\kappa, \varphi, \omega) = R_3(\kappa) * R_2(\varphi) * R_1(\omega)$,[8] the final rotation matrix is given by Eq. (5.79).

$$R(\kappa,\varphi,\omega) = \begin{bmatrix} \cos\varphi * \cos\kappa & \sin\omega * \sin\varphi * \cos\kappa + \cos\omega * \sin\kappa & -\cos\omega * \sin\varphi * \cos\kappa + \sin\omega * \sin\kappa \\ -\cos\varphi * \sin\kappa & -\sin\omega * \sin\varphi * \sin\kappa + \cos\omega * \cos\kappa & \cos\omega * \sin\varphi * \sin\kappa + \sin\omega * \cos\kappa \\ \sin\varphi & -\sin\omega * \cos\varphi & \cos\omega * \cos\varphi \end{bmatrix}$$

$$(5.79)$$

The short form of Eq. (5.79) is given by Eq. (5.80).

---

[8] Note that, in some references, the authors prefer to use the transposed matrix $R(\omega,\varphi,\kappa)$.

$$\mathbf{R}(\kappa, \varphi, \omega) = \begin{bmatrix} r_{11} & r_{12} & r_{13} \\ r_{21} & r_{22} & r_{23} \\ r_{31} & r_{32} & r_{33} \end{bmatrix} \tag{5.80}$$

Scale Factor and Translations

The scale factor and translations occur similar to the 2D coordinate transformation. Furthermore, the 3D coordinate transformation can also be orthogonal or non-orthogonal, i.e., the axes (X', Y', Z') and (X, Y, Z) can be orthogonal or non-orthogonal.

The scale factor is often given in ppm rather than a multiplication constant. In these cases, it is necessary to transform it into the multiplication constant k for further application, as given by Eq. (5.81).

$$k = 1 + \frac{\text{ppm scale factor}}{1{,}000{,}000.000} \tag{5.81}$$

The reader unfamiliar with the ppm concept should understand it as mm/km, i.e., 1 mm in 1,000,000 mm, the same as 1 mm in 1 km.

The main purpose of coordinate transformation between spatial Cartesian coordinate systems is to convert geodetic coordinates between different data sets. In this case, it is necessary to perform three transformation steps. First, geodetic coordinates in datum A must be converted to Cartesian coordinates. The transformed Cartesian coordinates in datum A can then be converted to Cartesian coordinates in datum B. The transformed Cartesian coordinates in datum B can then be converted back to geodetic coordinates in datum B.

5.9.2.2.1 Transformation of Three-Dimensional Cartesian Coordinates with Three Parameters

The three-parameter transformation of spatial Cartesian coordinates assumes that the coordinate axes of the two systems are parallel and that there is no scale factor between them. Therefore, the transformation consists of applying only the translations (T_X, T_Y, T_Z) between the origins of the two systems, as given by Eq. (5.82).

$$\begin{bmatrix} X \\ Y \\ Z \end{bmatrix} = \begin{bmatrix} X' \\ Y' \\ Z' \end{bmatrix} + \begin{bmatrix} T_X \\ T_Y \\ T_Z \end{bmatrix} \tag{5.82}$$

5.9.2.2.2   Transformation of Three-Dimensional Cartesian Coordinates
            with Five Parameters—Molodensky Transformation

A variant of the Three-Dimensional Cartesian coordinates transformation of spatial
coordinates is the five-parameter transformation of spatial coordinates, called the
Molodensky transformation. In this case, the coordinate transformation is performed
directly from one Geodetic Coordinate System to another without any intermediary
Cartesian transformation. It allows directly transforming the coordinates (ϕ_g, λ_g, h)
from system (A) to system (B), provided that the (T_X, T_Y, T_Z) values and the differences $(\Delta a, \Delta f)$ between the geometric parameters (a) and (f) of both ellipsoids are
known. The equations for this coordinate transformation are as follows.

$$\Delta\phi = \frac{1}{M+h} * \left[\begin{array}{l} -T_X * \sin\phi * \cos\lambda - T_Y * \sin\phi * \sin\lambda + T_Z * \cos\phi + \dfrac{N * e^2 * \sin\phi * \cos\phi * \Delta a}{a} + \\[2mm] \sin\phi * \cos\phi * \left(M * \dfrac{a}{b} + N * \dfrac{b}{a} \right) * \Delta f \end{array} \right]$$

$$\tag{5.83}$$

$$\Delta\lambda = \frac{1}{(N+h) * \cos\phi} * (-T_X * \sin\lambda + T_Y * \cos\lambda) \tag{5.84}$$

$$\Delta h = T_X * \cos\phi * \cos\lambda + T_Y * \cos\phi * \sin\lambda + T_Z * \sin\phi - \frac{a}{N} * \Delta a$$
$$+ \frac{N * b * \sin^2\phi * \Delta f}{a} \tag{5.85}$$

where

$a=$ semi-major axis of the ellipsoid of the system (A)
$b=$ semi-minor axis of the ellipsoid of the system (A)
$h=$ ellipsoidal height of the points in the system (A)
$M=$ radius of curvature in the meridian of the ellipsoid of the system (A), as given by
     Eq. (4.7)
$N=$ radius of curvature in the prime vertical of the ellipsoid of the system (A), as
     given by Eq. (4.8)
$\phi, \lambda =$ latitude and longitude of the points in the system (A)
$T_X, T_Y, T_Z=$ translations in the respective directions (X, Y, Z) between systems (B)
     and (A), that is, (B-A)
$\Delta a=$ difference between the semi-major axes of the ellipsoid (B) and the ellipsoid (A)
$\Delta f=$ difference between the flattening of the ellipsoid (B) and the ellipsoid (A)

The values obtained by Eqs. (5.83), (5.84) and (5.85) must be added algebraically to the geodetic coordinates of the points in system (A) to give their geodetic coordinates in the system (B).

Example 5.10 Using the coordinates of point (*SPC1*) given in Table 5.14 and the transformation parameters given in Table 5.15, calculate the global geocentric Cartesian coordinates of the point (*SPC1*) in the SIRGAS2000 Geodetic Reference System, using the three-parameter transformation of spatial Cartesian coordinates. Then calculate the corresponding geodetic coordinates using the five-parameter transformation of spatial Cartesian coordinates.

Table 5.14 3D Cartesian and geodetic coordinates of the point (*SPC1*) in the SAD69 Geodetic Reference System

Point	X [m]	Y [m]	Z [m]
SPC1	4,007,282.4577	−4,306,654.0737	−2,458,182.3397
	Latitude	Longitude	h [m]
	−22° 48′56.88510″	−47° 03′44.05965″	630.156

Table 5.15 Coordinate transformation parameters from SAD69 to SIRGAS2000 geodetic reference systems[a]

Translation	Value [m]
T_X	−67.350
T_Y	3.880
T_Z	−38.220

[a]Costa et al. (2005)

Solution:

1. *Geocentric Cartesian coordinates using the three-parameter transformation of spatial Cartesian coordinates:*

Considering the transformation parameters indicated in Table 5.15 *and using* Eq. (5.82).

$$\begin{bmatrix} X \\ Y \\ Z \end{bmatrix}_{\text{SIRGAS2000}} = \begin{bmatrix} X' \\ Y' \\ Z' \end{bmatrix}_{\text{SAD69}} + \begin{bmatrix} T_X \\ T_Y \\ T_Z \end{bmatrix} = \begin{bmatrix} 4,007,282.4577 \\ -4,306,654.0737 \\ -2,458,182.3397 \end{bmatrix}$$

$$+ \begin{bmatrix} -67.350 \\ 3.880 \\ -38.220 \end{bmatrix} = \begin{bmatrix} 4,007,215.108 \\ -4,306,650.194 \\ -2,458,220.560 \end{bmatrix}_{\text{m}}$$

2. *Geocentric Cartesian coordinates using the five-parameter transformation of spatial Cartesian coordinates:*

To apply this coordinate transformation method, it is first necessary to calculate the differences between the semi-major axes Δa and between the flattening Δf of the ellipsoids GSR80 and International Ellipsoid of 1967 (South American Datum 1969—SAD69), which are the reference ellipsoids of SIRGAS200 and SAD69 Geodetic Reference systems, respectively. See Table 4.1. Thus,

$$\Delta a = a_{GRS80} - a_{\text{Internacional}1967} = -23.000 \text{ m}$$
$$\Delta f = f_{GRS80} - f_{\text{Internacional}1967} = -8.118805 * 10^{-8}$$

Then, it is necessary to calculate the radius of curvature in the meridian and the radius of curvature in the prime vertical of the point (SPC1) regarding the International Ellipsoid of 1967 (SAD69), as indicated below.

$$M = \frac{6{,}378{,}160.000 * (1 - 0.006694542)}{\left[1 - 0.006694542 * \sin^2(-22°48'56.88510'')\right]^{\frac{3}{2}}} = 6{,}345{,}039.3674 \text{ m}$$

$$N = \frac{6{,}378{,}160.000}{\sqrt{1 - 0.006694542 * \sin^2(-22°48'56.88510'')}} = 6{,}381{,}372.6417 \text{ m}$$

Then, using Eqs. (5.83), (5.84) and (5.85).

$$\Delta\phi = -8.462 * 10^{-6}\text{rad} = -1.74542''$$
$$\Delta\lambda = -7.932 * 10^{-6}\text{rad} = -1.63617''$$
$$\Delta h = -7.1762 \text{ m}$$

Finally, the geodetic coordinates of the point (SPC1) are obtained in the SIRGAS2000 Geodetic Reference System.

$$\phi_{SIRGAS2000} = \phi_{SAD69} + \Delta\phi$$
$$= -22°48'56.88510'' + (-1.74542'') = -22°48'58.63052''$$
$$\lambda_{SIRGAS2000} = \lambda_{SAD69} + \Delta\lambda$$
$$= -47°03'44.05965'' + (-1.63617'') = -47°03'45.69582''$$
$$h_{SIRGAS2000} = h_{SAD69} + \Delta h = 630.156 + (-7.1762) = 622.9798 \text{ m}$$

5.9.2.2.3   Transformation of Three-Dimensional Cartesian Coordinates with Seven Parameters—Helmert 3D or Bursa-Wolf Transformation

In this case, it is assumed that there are three rotations, three translations and a single scale factor for all axes. Therefore, the transformation has seven parameters, and the matrix equation is formulated as follows.

$$\begin{bmatrix} X \\ Y \\ Z \end{bmatrix} = k * R(\kappa, \varphi, \omega) * \begin{bmatrix} X' \\ Y' \\ Z' \end{bmatrix} + \begin{bmatrix} T_X \\ T_Y \\ T_Z \end{bmatrix} \qquad (5.86)$$

where

$k =$ scale factor
$R(\kappa, \varphi, \omega)=$ rotation matrix
$T_X, T_Y, T_Z=$ translations in the directions (X, Y, Z)

In this type of transformation, only the scale of the geometric elements is modified, and the shape is maintained. It is, therefore, a conformal transformation.

In addition to the coordinate transformations presented, there are others of less interest to Geomatics. These are the Spatial Coordinate Transformations with 9 and 12 parameters. In the first case, it is assumed that there is a different scale factor for each axis. In the second case, the axes are considered to be non-orthogonal, thus adding three more parameters of non-orthogonality to the system of equations. These two coordinate transformation methods are beyond the scope of this book.

Example 5.11 Using the values from the previous example and the transformation parameters in Table 5.16, calculate the geocentric Cartesian coordinates of point (*SPC*1) in the SIRGA2000 Geodetic Reference System using the 3D Helmert Transformation.

Table 5.16 Full coordinate transformation parameters from SAD69 to SIRGAS2000 geodetic reference systems[a]	Translation [m]		Rotation		Scale factor (k)
	T_X	−67.350	ω	−0.001″	0.005 ppm
	T_Y	3.880	φ	−0.001″	
	T_Z	−38.220	κ	−0.003″	

[a]Costa et al. (2005)

Solution:
   The first step in solving this problem is to calculate the elements of the rotation matrix.

$R(\kappa, \varphi, \omega)$

$$= \begin{bmatrix} 1 & 1.454441046 * 10^{-8} & 4.848136741 * 10^{-9} \\ -1.454441043 * 10^{-8} & 1 & -4.848136882 * 10^{-9} \\ -4.848136811 * 10^{-9} & 4.848136811 * 10^{-9} & 1 \end{bmatrix}$$

The value of the scale factor k given in ppm must be converted to decimal.

$$k = 1 + \left(\frac{0.005}{1,000,000.000} \right) = 1.000000005 = 1 + 0.5 * 10^{-8}$$

Using Eq. (5.86).

$$k * \boldsymbol{R}(\kappa, \varphi, \omega) * \begin{bmatrix} X \\ Y \\ Z \end{bmatrix} = \left(1 + 0.5 * 10^{-8}\right)$$

$$* \begin{bmatrix} 1 & 1.45444 * 10^{-8} & 4.84814 * 10^{-9} \\ -1.45444 * 10^{-8} & 1 & -4.84814 * 10^{-9} \\ -4.84814 * 10^{-9} & 4.84814 * 10^{-9} & 1 \end{bmatrix} * \begin{bmatrix} 4{,}007{,}282.457 \\ -4{,}306{,}654.074 \\ -2{,}458{,}182.340 \end{bmatrix}$$

$$k * \boldsymbol{R}(\kappa, \varphi, \omega) * \begin{bmatrix} X \\ Y \\ Z \end{bmatrix} = \begin{bmatrix} 4{,}007{,}282.402 \\ -4{,}306{,}654.142 \\ -2{,}458{,}182.393 \end{bmatrix}_{\mathrm{m}}$$

And then,

$$\begin{bmatrix} X \\ Y \\ Z \end{bmatrix}_{\text{SIRGAS 2000}} = \begin{bmatrix} 4{,}007{,}282.402 \\ -4{,}306{,}654.142 \\ -2{,}458{,}182.393 \end{bmatrix} + \begin{bmatrix} -67.35 \\ 3.88 \\ -38.22 \end{bmatrix} = \begin{bmatrix} 4{,}007{,}215.052 \\ -4{,}306{,}650.262 \\ -2{,}458{,}220.613 \end{bmatrix}_{\mathrm{m}}$$

5.9.2.3   Transformation between Geocentric and Geodetic Coordinates

With the advent of GNSS technology, Geomatics and Civil Engineering professionals have been compelled to quickly adapt to the constant need to perform transformations between global geocentric Cartesian and geodetic coordinates. This need is because GNSS determines the coordinates of points in space in the Global Geocentric Cartesian Coordinate System (X, Y, Z), and engineers prefer to work with grid[9] or local coordinates in their projects. Therefore, in most cases, geocentric Cartesian coordinates must be converted into geodetic coordinates (ϕ_g, λ_g, h) before determining grid coordinates.

5.9.2.3.1   Transformation from Geodetic to Global Geocentric Cartesian Coordinates

The conversion from geodetic (ϕ_g, λ_g, h) to Cartesian (X, Y, Z) coordinates is given by the well-known Eq. (5.87).

---

[9] Refer to Chap. 16.

$$X = (N + h) * \cos(\phi_g) * \cos(\lambda_g)$$
$$Y = (N + h) * \cos(\phi_g) * \sin(\lambda_g) \qquad (5.87)$$
$$Z = [N(1 - e^2) + h] * \sin(\phi_g)$$

where

$X, Y, Z =$ global geocentric Cartesian coordinates
$N =$ radius of curvature in the prime vertical, as shown by Eq. (4.8)
$h =$ ellipsoidal height
$\phi_g, \lambda_g =$ geodetic latitude and longitude
$e^2 =$ first eccentricity squared, as given by Eq. (4.4)

5.9.2.3.2 Transformation from Global Geocentric Cartesian to Geodetic Coordinates

Several mathematical models can be used to transform global geocentric Cartesian into geodetic coordinates. Usually, this problem is solved iteratively, although a closed-form solution is possible. The iterative solution can be obtained by applying the following steps[10]:

1. Compute the longitude.

$$\tan(\lambda_g) = \frac{Y}{X} \quad \rightarrow \quad \lambda_g = \text{atan}\left(\frac{Y}{X}\right) \qquad (5.88)$$

2. Compute the radius of a parallel.

$$p = \sqrt{X^2 + Y^2} \qquad (5.89)$$

3. Compute an approximate value for the latitude ϕ_0.

$$\phi_0 = \text{atan}\left[\frac{Z}{p * (1 - e^2)}\right] \qquad (5.90)$$

4. Compute an approximate value for N_0.

---

[10]There are several other models for this type of coordinate transformation in the literature. The reader interested in further information should consult specialised technical references.

$$N_0 = \frac{a^2}{\sqrt{a^2 * \cos^2(\phi_0) + b^2 * \sin^2(\phi_0)}} \qquad (5.91)$$

5. Compute an approximate value for the ellipsoidal height.

$$h_0 = \frac{p}{\cos \phi_0} - N_0 \qquad (5.92)$$

6. Compute an improved value for the latitude ϕ.

$$\phi = \mathrm{atan}\left\{\frac{Z}{p} * \frac{1}{1 - \left[e^2 * \left(\frac{N_0}{N_0 + h_0}\right)\right]}\right\} \qquad (5.93)$$

7. Check if another iteration is needed: if $\phi = \phi_0$ the iteration is completed; otherwise, set $\phi_0 = \phi$ and continue from step 4.

Example 5.12 Using the geodetic coordinates of the point (SPC1) in the SIRGAS2000 Geodetic Reference System given below, calculate the values of their corresponding global geocentric Cartesian coordinates.

$$\phi_g = -22° \, 48'58.63052'' \quad \lambda_g = -47° \, 03'45.69584'' \quad h = 622.980 \text{ m}$$

Solution:

To solve this problem, it is necessary to consider the value of the first eccentricity squared of the SIRGAS2000 Geodetic Reference System (see Table 4.3) and calculate the radius of curvature in the prime vertical.

$$N = \frac{6,378,137.000}{\sqrt{[1 - 0.006694380 * \sin^2(-22° \, 48'58.63052'')]}} = 6,381,349.6817 \text{ m}$$

Using Eq. (5.87).

$X = (6,381,349.6817 + 622.980) * \cos(-22°48'58.63052'') * \cos(-47°03'45.69584'')$

$\quad = 4,007,215.1075 \text{ m}$

$Y = (6,381,349.6817 + 622.980) * \cos(-22°48'58.63052'') * \sin(-47°03'45.69584'')$

$\quad = -4,306,650.1937 \text{ m}$

$Z = [6,381,349.6817 * (1 - 0.006694380) + 622.980] * \sin(-22°48'58.63052'')$

$\quad = -2,458,220.5598 \text{ m}$

Example 5.13 Given the values of the global geocentric Cartesian coordinates of the point (*SPC*1) calculated in the previous example, calculate the corresponding geodetic coordinates using the iterative solution.

Solution:
  Considering the iterative solution steps.

$$\tan(\lambda_g) = \frac{-4,306,650.1937}{4,007,215.1075} = -1.074723986 \rightarrow \lambda_g = -47°03'45.69584''$$

$$p = \sqrt{4,007,215.1075^2 + (-4,306,650.1937)^2} = 5,882,602.2140 \text{ m}$$

$$\phi_0 = \operatorname{atan}\left(\frac{-2,458,220.5598}{5,882,602.2140 * (1 - 0.006694380)}\right) = -22°48'58.67902''$$

$$N_0 = \frac{6,378,137.000^2}{\sqrt{6,378,137.0002 * \cos^2(-22°48'58.67902'') + 6,356,752.314 * \sin^2(-22°48'58.67902'')}}$$
$$= 6,381,349.6853 \text{ m}$$

$$h_0 = \frac{5,882,602.2202}{\cos(-22°48'58.67902'')} - 6,381,349.6853 = 623.6078 \text{ m}$$

$$\phi_0 = \operatorname{atan}\left[\frac{-2,458,220.5598}{5,882,602.2140} * \frac{1}{1 - \left(0.006694380 * \frac{6,381,349.6853}{6,381,349.6853+623.6078}\right)}\right] =$$

$$-22°48'58.63047''$$

  Since $\phi \neq \phi_0$, *another iteration must be performed from step 4. The results of successive iterations are shown in* Table 5.17.

Table 5.17 Results of successive iterations

Parameter	1st iteration	2nd iteration	3rd iteration
N_i [m]	6,381,349.6853	6,381,349.6817	6,381,349.6817
h_i [m]	623.6078	622.9794	622.9800
ϕ_g	$-22°48'58.63047''$	$-22°48'58.63052''$	$-22°48'58.63052''$

5.9.2.4 Transformation from Global Geocentric Cartesian to Topocentric Cartesian Coordinates

Terrestrial geodetic measurements generally refer to the observation point located on the surface of the Earth. Coordinates derived from these observations are, therefore, often expressed in a local (topocentric) reference coordinate system (*e*, *n*, *u*), as explained in Sect. 5.6.2 and shown in Fig. 5.8. In that figure, (*X*, *Y*, *Z*) are the global geocentric Cartesian coordinates, and $(e, n, u)^{11}$ are the topocentric Cartesian coordinates. The Topocentric Coordinate System has its origin at point (P), whose geodetic and Cartesian coordinates are (ϕ_P, λ_P, h_P) and (X_P, Y_P, Z_P) respectively.

The transformation from Geocentric to Topocentric coordinate systems can be built up from simple translations and rotations. The first step is to transform the origin from the centre of the Earth to the origin of the Topocentric system by subtracting the values of (X_P, Y_P, Z_P) from the (X, Y, Z) coordinates. A new coordinate system is then created at point (P), denoted as (X_1, Y_1, Z_1).

$$\begin{bmatrix} X_1 \\ Y_1 \\ Z_1 \end{bmatrix} = \begin{bmatrix} X \\ Y \\ Z \end{bmatrix} - \begin{bmatrix} X_P \\ Y_P \\ Z_P \end{bmatrix} \tag{5.94}$$

To make the two coordinate systems parallel, a rotation about the Z_1-axis by an amount of $\lambda + 90°$ must be performed to align the X_1-axis with the local e-axis. According to the rotation matrix $R\kappa$ presented in Sect. 5.9.2.2 and applying the corresponding trigonometric function, a new coordinate system (X_2, Y_2, Z_2) is obtained, as given in matrix form by Eq. (5.95).

$$\begin{bmatrix} X_2 \\ Y_2 \\ Z_2 \end{bmatrix} = \begin{bmatrix} -\sin\lambda_P & \cos\lambda_P & 0 \\ -\cos\lambda_P & -\sin\lambda_P & 0 \\ 0 & 0 & 1 \end{bmatrix} * \begin{bmatrix} X_1 \\ Y_1 \\ Z_1 \end{bmatrix} \tag{5.95}$$

The (X_2, Y_2, Z_2) system must then be rotated again, but this time around the X_2-axis, by an angle ω equal to $(90° - \phi)$, so that the Z_2-axis coincides with the u-axis. A new coordinate system (X_3, Y_3, Z_3) is thus obtained in matrix form by Eq. (5.96).

$$\begin{bmatrix} X_3 \\ Y_3 \\ Z_3 \end{bmatrix} = \begin{bmatrix} 1 & 0 & 0 \\ 0 & \sin\phi_P & \cos\phi_P \\ 0 & -\cos\phi_P & \sin\phi_P \end{bmatrix} * \begin{bmatrix} X_2 \\ Y_2 \\ Z_2 \end{bmatrix} \tag{5.96}$$

Substituting Eq. (5.95) into Eq. (5.96) gives

$$\begin{bmatrix} X_3 \\ Y_3 \\ Z_3 \end{bmatrix} = \begin{bmatrix} -\sin\lambda_P & \cos\lambda_P & 0 \\ -\cos\lambda_P * \sin\phi_P & -\sin\lambda_P * \sin\phi_P & \cos\phi_P \\ \cos\lambda_P * \cos\phi_P & \sin\lambda_P * \cos\phi_P & \sin\phi_P \end{bmatrix} * \begin{bmatrix} X_1 \\ Y_1 \\ Z_1 \end{bmatrix} \tag{5.97}$$

Finally, considering the translations (X_P, Y_P, Z_P)

---

[11] Some authors may use (n, e, u) or even (u, n, e), which affects the position of the elements of the rotation matrix.

$$
\begin{bmatrix} e \\ n \\ u \end{bmatrix} = \begin{bmatrix} -\sin\lambda_P & \cos\lambda_P & 0 \\ -\cos\lambda_P * \sin\phi_P & -\sin\lambda_P * \sin\phi_P & \cos\phi_P \\ \cos\lambda_P * \cos\phi_P & \sin\lambda_P * \cos\phi_P & \sin\phi_P \end{bmatrix} * \begin{bmatrix} X - X_P \\ Y - Y_P \\ Z - Y_P \end{bmatrix} \quad (5.98)
$$

or,

$$
\begin{bmatrix} e \\ n \\ u \end{bmatrix} = R_1(90° - \phi_P) * R_3(\lambda_P + 90°) * \begin{bmatrix} X - X_P \\ Y - Y_P \\ Z - Z_P \end{bmatrix} \quad (5.99)
$$

In Geomatics applied to Civil Engineering, the use of the transformation method described above is of little interest. In most cases, the engineer is concerned with transforming a set of points with known geocentric Cartesian coordinates (X_i, Y_i, Z_i) into a local ground coordinate system (X_L, Y_L, H_L). The transformation calculation is then performed for different points (Q_i) with respect to the origin point (P), as given by Eq. (5.100).

$$
\begin{bmatrix} X - X_P \\ Y - Y_P \\ Z - Z_P \end{bmatrix} = \begin{bmatrix} X_Q - X_P \\ Y_Q - Y_P \\ Z_Q - Z_P \end{bmatrix} = \begin{bmatrix} \Delta X \\ \Delta Y \\ \Delta Z \end{bmatrix} \quad (5.100)
$$

And then,

$$
\begin{bmatrix} \Delta e \\ \Delta n \\ \Delta u \end{bmatrix} = R_1(90° - \phi_P) * R_3(\lambda_P + 90°) * \begin{bmatrix} \Delta X \\ \Delta Y \\ \Delta Z \end{bmatrix} \quad (5.101)
$$

Applying the matrix calculations above, Eq. (5.101) can be written in linear form, as shown in Eq. (5.102).

$$
\begin{aligned}
\Delta e &= (-\Delta X * \sin\lambda_P) + (\Delta Y * \cos\lambda_P) \\
\Delta n &= (-\Delta X * \cos\lambda_P * \sin\phi_P) - (\Delta Y * \sin\lambda_P * \sin\phi_P) + (\Delta Z * \cos\phi_P) \quad (5.102) \\
\Delta u &= (\Delta X * \cos\lambda_P * \cos\phi_P) + (\Delta Y * \sin\lambda_P * \cos\phi_P) + (\Delta Z * \sin\phi_P)
\end{aligned}
$$

Assuming that point (P) has coordinates (X_{LP}, Y_{LP}, H_{LP}) with respect to a predefined local ground coordinate system (X_L, Y_L, H_L), the coordinate of point (Q), in this coordinate system, can be determined by adding the values of $(\Delta e, \Delta n, \Delta u)$ to the corresponding values of the local coordinates (X_{LP}, Y_{LP}, H_{LP}) of the origin point (P), as given by Eq. (5.103).

$$
\begin{bmatrix} X_{L_Q} \\ Y_{L_Q} \\ H_{L_Q} \end{bmatrix} = \begin{bmatrix} \Delta e \\ \Delta n \\ \Delta u \end{bmatrix} + \begin{bmatrix} X_{L_P} \\ Y_{L_P} \\ H_{L_P} \end{bmatrix} \tag{5.103}
$$

As will be seen in Chap. 16, the coordinate transformation presented above is useful for applications where the area to be mapped is sufficiently small that the curvature of the Earth can be ignored, allowing, for example, the results from GNSS and total stations to be combined into a local coordinate system. However, it should be noted that for high-precision measurements, the results based on terrestrial observations are referenced to the astronomical topocentric system, which is aligned with the local gravity vector (plumb line) through (P) and not the ellipsoid normal, and therefore it is necessary to consider the defection of the vertical before performing calculation on the ellipsoid. The reader must therefore bear in mind that, at the level of accuracy generally used in engineering surveying, this assumption is usually ignored and both coordinate systems are considered to be parallel.

More details on this type of coordinate transformation are presented in Sect. 16.12.

Example 5.14 Using the geocentric Cartesian and the geodetic coordinates in Tables 5.18 and 5.19, and taking point $(E010)$ as the reference point with local ground coordinates as given below, calculate the local ground coordinates of points $(E011)$ and $(E014)$.

Local ground coordinates of point $(E010)$.

$$
X_{L_{E010}} = 152{,}596.092 \text{ m}
$$
$$
Y_{L_{E010}} = 254{,}100.825 \text{ m}
$$
$$
H_{L_{E010}} = 866.869 \text{ m}
$$

Table 5.18 Global geocentric coordinates of points $(E010)$, $(E011)$ and $(E014)$— SIRGAS2000 GRS

Point	X [m]	Y [m]	Z [m]
E010	3,969,275.4490	−4,386,776.4898	−2,377,946.7280
E011	3,969,145.4939	−4,386,908.3527	−2,377,920.9610
E014	3,968,379.3952	−4,387,541.8840	−2,378,002.7270
E011-E010	−129.9551	−131.8629	25.7670
E014-E010	−896.0538	−765.3942	−55.9990

Table 5.19 Geodetic coordinates of points $(E010)$ and $(E011)$ in the SIRGAS2000 GRS

Point	Latitude (ϕ_g)	Longitude (λ_g)	h [m]
E010	−22° 01′52.52292″	−47° 51′37.23361″	866.869
E011	−22° 01′51.61736″	−47° 51′43.67800″	?
E014	−22° 01′54.61923″	−47° 52′18.30450″	?

Solution:
Using Eq. (5.101) *for the point (E011).*

$$
\begin{bmatrix} \Delta e \\ \Delta n \\ \Delta u \end{bmatrix} = \begin{bmatrix} 0.741511627 & 0.670940018 & 0.000000000 \\ 0.251677881 & -0.278150162 & 0.926979359 \\ 0.621947547 & -0.687365972 & -0.375112341 \end{bmatrix}
$$
$$
* \begin{bmatrix} -129.9551 \\ -131.8629 \\ 25.7670 \end{bmatrix} = \begin{bmatrix} -184.8353 \\ 27.8563 \\ 0.1473 \end{bmatrix}_m
$$

Using Eq. (5.103).

$$X_{L_{E011}} = -184.8353 + 152{,}596.092 = 152{,}411.257 \text{ m}$$

$$Y_{L_{E011}} = 27.8563 + 254{,}100.825 = 254{,}128.681 \text{ m}$$

$$H_{L_{E011}} = 0.1473 + 866.869 = 867.016 \text{ m}$$

By the same reasoning, the following local ground coordinates are obtained for the point (E014).

$$X_{L_{E014}} = -1{,}177.9679 + 152{,}596.092 = 151{,}418.124 \text{ m}$$

$$Y_{L_{E014}} = -64.5323 + 254{,}100.825 = 254{,}036.293 \text{ m}$$

$$H_{L_{E014}} = -10.1866 + 866.869 = 856.682 \text{ m}$$

5.9.2.5 Transformation from Topocentric to Global Geocentric Cartesian Coordinates

According to Eq. (5.99) and considering the rotation matrix $R_{1,3} = R(90° - \phi_P) * R(\lambda_P + 90°)$ as orthogonal, the following equation can be expressed in matrix form.

$$
\begin{bmatrix} X \\ Y \\ Z \end{bmatrix} - \begin{bmatrix} T_X \\ T_Y \\ T_Z \end{bmatrix} = R_{1,3}^T * \begin{bmatrix} e \\ n \\ u \end{bmatrix} \tag{5.104}
$$

Thus, after corresponding rotations and translations, the resulting transformation from topocentric to global geocentric Cartesian coordinates is given as follows.

$$
\begin{bmatrix} X \\ Y \\ Z \end{bmatrix} = \begin{bmatrix} -\sin\lambda_P & -\sin\phi_P * \cos\lambda_P & \cos\phi_P * \cos\lambda_P \\ \cos\lambda_P & -\sin\phi_P * \sin\lambda_P & \cos\phi_P * \sin\lambda_P \\ 0 & \cos\phi_P & \sin\phi_P \end{bmatrix} * \begin{bmatrix} e \\ n \\ u \end{bmatrix} + \begin{bmatrix} T_X \\ T_Y \\ T_Z \end{bmatrix}
$$

$$(5.105)$$

Or in the linear form.

$$
\begin{aligned}
\Delta X &= (-\Delta e * \sin\lambda_P) - (\Delta n * \sin\phi_P * \cos\lambda_P) + (\Delta u * \cos\phi_P * \cos\lambda_P) \\
\Delta Y &= (\Delta e * \cos\lambda_P) - (\Delta n * \sin\phi_P * \sin\lambda_P) + (\Delta u * \cos\phi_P * \sin\lambda_P) \quad (5.106) \\
\Delta Z &= (\Delta n * \cos\phi_P) + (\Delta u * \sin\phi_P)
\end{aligned}
$$

Example 5.15 Given the results of the previous example, calculate the global geocentric Cartesian coordinates of the point $(E011)$ from the topocentric Cartesian coordinates of the point $(E010)$.

Solution:
  Using Eq. (5.105).

$$
\begin{bmatrix} X \\ Y \\ Z \end{bmatrix}_{E011} = \begin{bmatrix} 0.741511627 & 0.251677881 & 0.621947547 \\ 0.670940018 & -0.278150162 & -0.687365972 \\ 0.000000000 & 0.926979359 & -0.375112341 \end{bmatrix}
$$

$$
* \begin{bmatrix} -184.835314 \\ 27.8563400 \\ 0.14729508 \end{bmatrix} + \begin{bmatrix} 3,969,275.4490 \\ -4,386,776.4898 \\ -2,377,946.7280 \end{bmatrix} = \begin{bmatrix} 3,969,145.4939 \\ -4,386,908.3527 \\ -2,377,920.9610 \end{bmatrix}_m
$$

5.9.2.6 Transformation from Geodetic to Topocentric Cartesian Coordinates

Local ground coordinates can also be obtained directly from geodetic coordinates using Eq. (5.107).

$$
\begin{bmatrix} \Delta n \\ \Delta e \\ \Delta u \end{bmatrix} = \begin{bmatrix} M + h_P & 0 & 0 \\ 0 & (N + h_P) * \cos\phi_P & 0 \\ 0 & 0 & 1 \end{bmatrix} * \begin{bmatrix} \Delta\phi \\ \Delta\lambda \\ \Delta h \end{bmatrix}
$$

$$(5.107)$$

where

M = radius of curvature in the meridian, as given by Eq. (4.7)
N = radius of curvature in the prime vertical, as shown by Eq. (4.8)
Δh = difference between the ellipsoidal heights of the considered points, taking $(h_Q - h_P)$

$\Delta\phi$ = difference between geodetic latitudes of the considered points, taking $(\phi_Q - \phi_P)$

$\Delta\lambda$ = difference between geodetic longitudes of the considered points, taking $(\lambda_Q - \lambda_P)$

The local ground coordinates of the point (Q) are obtained by adding the calculated values from Eq. (5.107) to the local ground coordinates of the point (P).

Example 5.16 Using the geodetic coordinates in Table 5.19 and taking point ($E010$) as the reference point, calculate the local ground coordinates of the point ($E011$).

Solution:

Using Eq. (5.107) *and the corresponding geodetic parameters.*

$$M = \frac{6{,}378{,}137.000 * (1 - 0.006694380)}{\left(1 - 0.006694380 * \sin^2(-22°\,01'52.52292'')\right)^{\frac{3}{2}}} = 6{,}344{,}401.4869 \text{ m}$$

$$N = \frac{6{,}378{,}137.000}{\sqrt{\left(1 - 0.006694380 * \sin^2(-22°\,01'52.52292'')\right)}} = 6{,}381{,}143.1031 \text{ m}$$

$$\Delta H = H_{E011} - H_{E010} = 867.016 - 866.869 = 0.1470 \text{ m}$$

$$\begin{bmatrix} \Delta n \\ \Delta e \\ \Delta u \end{bmatrix} = \begin{bmatrix} 6{,}345{,}268.355908 & 0 & 0 \\ 0 & 5{,}915{,}991.511648 & 0 \\ 0 & 0 & 1 \end{bmatrix}$$
$$* \begin{bmatrix} 4.390278771 * 10^{-6} \\ -3.124328438 * 10^{-5} \\ 0.1470 \end{bmatrix} = \begin{bmatrix} 27.8557 \\ -184.8350 \\ 0.1470 \end{bmatrix}_m$$

Thus,

$$X_{L_{E011}} = -184.8353 + 152{,}596.092 = 152{,}411.257 \text{ m}$$

$$Y_{L_{E011}} = 27.8575 + 254{,}100.825 = 254{,}128.682 \text{ m}$$

$$H_{L_{E011}} = 0.1470 + 866.869 = 867.016 \text{ m}$$

5.10 Problems

1. To help the reader practice the transformation between the 2D Cartesian and 2D polar coordinate systems, a series of problems are presented below, relating to the geometric elements in Fig. 5.17 and the values in Tables 5.20 and 5.21.

 (a) Calculate the angle $Az_{P_1 Ref}$.
 (b) Calculate the angles $Az_{P_1 P_2}$ and $Az_{P_1 P_3}$.
 (c) Calculate the coordinates (X_{P_2}, Y_{P_2}).
 (d) Calculate the angle α_2.
 (e) Calculate the angles $Az_{P_2 P_3}$ and $Az_{P_3 P_1}$.
 (f) Calculate the coordinates (X_{P_3}, Y_{P_3}).
 (g) Calculate the distance $d_{P_1 P_3}$.
 (h) Calculate the distance $d_{P_2 P_4}$.
 (i) Calculate the angle $Az_{P_2 P_4}$.
 (j) Calculate the coordinates (X_{P_4}, Y_{P_4}).

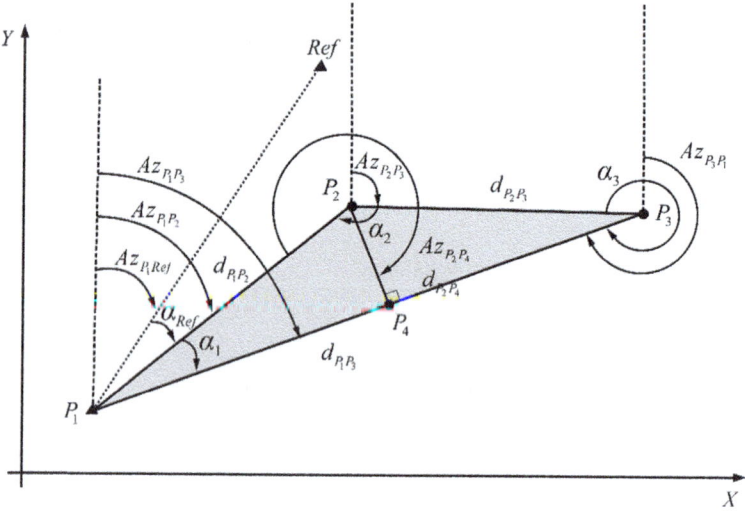

Fig. 5.17 2D Cartesian and Polar coordinates

Table 5.20 2D Cartesian coordinates

Point	X [m]	Y [m]
P_1	5444.739	12,535.120
Ref	5677.300	13,002.741

Table 5.21 2D Polar coordinates

α_{Ref}	28°10′53″
α_1	17°02′32″
α_3	339°14′8″
$d_{P_1 P_2}$	388.026 m
$d_{P_2 P_3}$	320.934 m

2. Given the 3D Cartesian coordinates of two points (P) and (Q) in Table 5.22, calculate the spatial distance between them.

Table 5.22 3D Cartesian coordinates

Point	X [m]	Y [m]	Z [m]
P	3,967,006.9741	−4,390,247.3702	−2,375,229.9392
Q	3,968,000.9381	−4,392,462.9610	−2,374,512.8141

3. Given the coordinates of the homologous points in Table 5.23, determine the coordinates (X, Y) of the point (A501) using the 2D conformal coordinate transformation method.

Table 5.23 Homologous points

	Coordinate system 1		Coordinate system 2	
Point	X' [m]	Y' [m]	X [m]	Y [m]
A343	580,852.326	134,705.632	580,852.310	134,705.740
A346	579,637.790	135,310.018	579,637.740	135,310.010
A501	580,477.281	135,342.065	?	?

4. Given the coordinates of the homologous points in Table 5.24, calculate the coordinates (X, Y) of point (A501) using the 2D Conformal and Affine coordinate transformation methods adjusted by the Parametric adjustment model.

Table 5.24 Homologous points

	Coordinate system 1		Coordinate system 2	
Point	X' [m]	Y' [m]	X [m]	Y [m]
A343	580,852.326	134,705.632	580,852.310	134,705.740
A344	579,885.880	134,583.440	579,885.880	134,583.440
A346	579,637.790	135,310.018	579,637.740	135,310.010
A348	580,244.849	135,349.318	580,244.840	135,349.310
A501	580,477.281	135,342.065	?	?

5. Given the values of the UTM and local coordinates in Table 5.25, determine the local coordinates (X, Y) of point (E011) using the Affine coordinate transformation method.

Table 5.25 UTM and local coordinates

	UTM coordinates		Local coordinates	
Point	E [m]	N [m]	X [m]	Y [m]
E007	204,031.218	7,560,476.948	151,889.987	253,644.066
E010	204,729.018	7,560,947.116	152,596.092	254,100.825
E020	204,584.481	7,559,967.185	152,433.303	253,124.303
E011	204,543.592	7,560,971.516	?	?

6. Given the rotation angles in Table 5.26, calculate the elements of the corresponding 3D rotation matrix $R(\kappa, \varphi, \omega)$.

Table 5.26 Spatial angle of rotation

Rotation	Angle
κ	$-0°45'47''$
φ	$0°05'37''$
ω	$-1°15'19''$

7. Given the global geocentric coordinates of the point (P) in the SIRGAS2000 Geodetic Reference System in Table 5.27 and the transformation parameters between the SIRGAS2000 and SAD69 reference systems shown in Table 5.28, calculate the coordinates of the point (P) in the SAD69 reference system.

Table 5.27 Coordinate of point (P) in the SIRGAS2000 Geodetic Reference System

Cartesian coordinates [m]
$X_P = 3{,}967{,}006.974$
$Y_P = -4{,}390{,}247.372$
$Z_P = -2{,}375{,}229.939$

Table 5.28 Coordinate transformation parameters between the SIRGAS2000 and SAD69 Geodetic Reference System

Translation	Value [m]	Rotation	Value	Scale factor	Value [ppm]
T_X	67.41	ω	$-0.001''$	k	0.005
T_Y	-3.79	φ	$-0.001''$		
T_Z	38.27	κ	$0.003''$		

8. Given the geodetic coordinates of the point (P) in Table 5.29, calculate its geocentric Cartesian coordinates in the SIRGAS200 Geodetic Reference System.

Table 5.29 Geodesic coordinates SIRGAS2000 of point (P)

Latitude (ϕ_g)	Longitude (λ_g)	h [m]
$-23° 00'17.81602''$	$-47° 53'57.04971''$	824.590

9. Given the geocentric Cartesian coordinates of the point (P) in the SIRGAS200 geodetic reference system in Table 5.30, calculate its geodetic coordinates.

Table 5.30 3D Cartesian coordinates of the point (P) in the SIRGAS200 Geodetic Reference System

X [m]	Y [m]	Z [m]
3,967,006.974	$-4{,}390{,}247.372$	$-2{,}375{,}229.939$

10. Given the geocentric Cartesian coordinates of points (P) and (Q) in Table 5.31 and the geodetic coordinates of point (P) in Table 5.32, calculate the topocentric Cartesian coordinates of point (Q).

Table 5.31 Geocentric Cartesian coordinates of points (P) and (Q)

Point	X [m]	Y [m]	Z [m]
P	4,014,842.014	−4,253,716.991	−2,536,508.161
Q	4,014,214.435	−4,255,117.986	−2,535,104.567

Table 5.32 Geodetic coordinates SIRGAS2000 of point (P)

Point	Latitude (ϕ_g)	Longitude (λ_g)
P	−23° 35′05.2445″	−46 39′17.2412″

References

Anderson, J. M., Mikhail, E. M. (1998). *Surveying, Theory and Practice.* 7th Edition. WCB/ McGraw-Hill, Boston, USA.

Burkholder, E. F. (1997a). *Definition and Description of a Global Spatial Data Model (GSDM).* Circleville, Ohio, EUA. 30p.

Burkholder, E. F. (1997b). *Spatial Data, Coordinate Systems, and the Science of Measurement.* Journal of Surveying Engineering, Vol. 127, No. 4, November, 2001. ISSN 0733-9453/01/ 0004-0143–0156.

Cattin, P-H. (2015). *Referentiels Geodesiques.* heig-vd – Haute Ecole d'Ingénierie et de Gestion du Canton de Vaud. Yverdon-les-Bains. Suisse.

Costa, Sonia M.A., Lima, Marco A. (2005). *Parâmetros de transformação entre o SAD69 e SIRGAS2000.* XXII Congresso Brasileiro de Cartografia. Macaé, Rio de Janeiro.

Deakin, R. E. (2004). *The standard and Abridge Molodensky coordinate transformation formulae.* Department of Mathematical and Geospatial Sciences, RMIT University, Melbourne, Australia.

Direction Fédérale des Mensurations Cadastrales. (2008). *Guide pour l'application des transformations géométriques en mensuration officielle.* Office fédéral de topographie. Direction fédérale des mensuration cadastrales. Switzerland.

Gerdan, P., Deaki, R. E. (1999). *Transforming Cartesian coordinates X, Y, Z to Geographical coordinates φ λ h.* The Australian Surveyor, Vol. 44, No. 1, pp. 55–63.

Greenfeld, J.S. (1997). *Least square weighted coordinate transformation formulas and their applications.* Journal of Surveying Engineering, USA.

Huguenin, L., Gillieron, P-Y. (2005). *Conversion des coordonnées.* EPFL, Switzerland.

Van Sickle, J. (2004). *Basic GIS Coordinates.* CRC Press, Florida, USA.

Janssen, V. (2009). *Understanding Coordinate Systems, Datums and Transformations in Australia.* In: Ostendorf, B., Baldock, P., Bruce, D., Burdett, M. and P. Corcoran (eds.), Proceedings of the Surveying & Spatial Sciences Institute Biennial International Conference, Adelaide 2009, Surveying & Spatial Sciences Institute, pp. 697–715. ISBN: 978-0-9581366-8-6.

Lu, Z., Qu, Y. Qiao. S. (2014). *Geodesy – Introduction to Geodetic Datum and Geodetic Systems.* Springer. ISBN 978-3-642-41244-8.

Ministério da Guerra, Diretoria do Serviço Geográfico (1960). *Manual Técnico, Cálculos geodésicos.* 3° fascículo. Rio de Janeiro.

Morais Junior, J. T. (2019). *Estudo de comparação dos métodos de transformação de coordenadas geodésicas para coordenadas topocêntricas para fins de implantação de obras na engenharia civil*. Master Dissertation, EESC-USP, São Carlos, SP.

Soler, T. (1998). *A compendium of transformation formulas useful in GPS work*. Journal of Geodesy, vol 72, p. 482–490.

Wolf, P. R., Bon A. D., Benjamin, E. W. (2014). *Elements of photogrammetry with applications in GIS*, 4th ed. New York: McGraw-Hill Education, 676p.

Chapter 6
Directions and Angles

Irineu da Silva and Guilherme Poleszuk dos Santos Rosa

6.1 Definitions

As discussed in previous chapters, angles and distances are the two fundamental quantities for Geomatics. Positioning points on the ground surface, determining elevation differences, and setting out designed feature elements are examples of surveying operations based on angle and distance measurements. Therefore, this chapter discusses the concepts related to angular measurements, while the concepts related to distance measurements are discussed in Chaps. 7 and 8.

There are several definitions of an angle in the literature. In Geometry, an angle is a figure where two lines meet at a common vertex point. In Trigonometry, an angle is the measure of the rotation required to bring two lines into coincidence. In Geomatics, an angle is defined as the difference in direction between two converging lines, where the direction is the angular relationship between a given line and an arbitrary or predefined reference line (also called the *orientation line*). It is quantified in units of angular measurement, as presented in Chap. 2.

For Geomatics, spatial angles are decomposed into *horizontal* and *vertical*, as described in the following sections.

I. da Silva (✉)
São Carlos School of Engineering, University of São Paulo, São Carlos, SP, Brazil
e-mail: irineu@sc.usp.br

G. Poleszuk dos Santos Rosa
Cartographic Engineer, Researcher at the Faculty of Science and Technology of the São Paulo State University (UNESP), São Paulo, Brazil

© The Author(s), under exclusive license to Springer Nature Switzerland AG 2025   177
I. da Silva, P. C. L. Segantine, *Geomatics Applied to Civil Engineering*,
https://doi.org/10.1007/978-3-031-75737-2_6

6.1.1 Horizontal Angle

> **A horizontal angle is the dihedral angle between two intersecting vertical planes containing two arbitrary spatial directions.**

Figure 6.1 shows the geometry of a horizontal angle between two spatial directions, PQ and PR, contained in the vertical planes π and π', respectively. Points (Q') and (R') are the respective projections of (Q) and (R) onto the horizontal plane passing through the point (P), defining the horizontal directions PQ' and PR'. The horizontal angle is the dihedral angle α between the two vertical planes.

Fig. 6.1 Horizontal angle definition

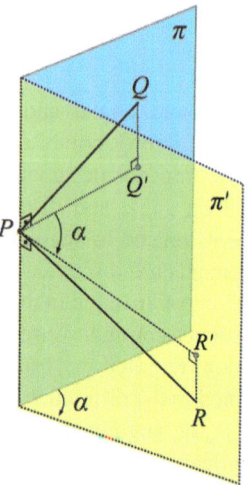

The value of a horizontal angle can be either positive or negative, depending on the difference between the values of the horizontal directions PR' and PQ'. However, it is usually counted clockwise for practical reasons and ranges from 0° to 360°. Therefore, if the difference between the horizontal directions is negative, 360° must be added to make it positive.

6.1.2 Vertical Angle

> **A vertical angle is the direction difference between two lines intersecting in a vertical plane.**

For Geomatics, as shown in Fig. 6.2, there are two types of vertical angles:

- *Altitude angle*[1]: angle β referenced to the horizontal plane passing through the observation point (P), with values ranging from 0° to +90° when measured above the horizontal plane and from 0° to −90° when measured below the horizontal plane, as shown in Fig. 6.3. In practice, it is observed with a vertical circle zeroed on the horizontal line.

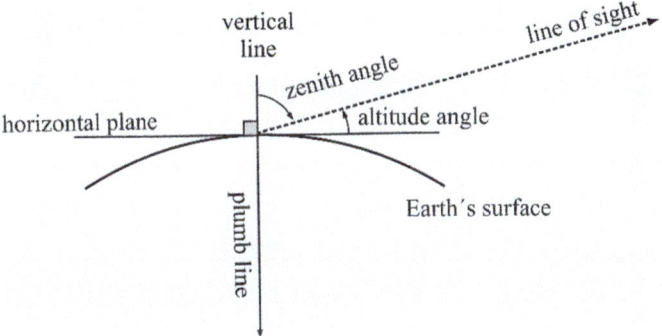

Fig. 6.2 Vertical angles related to the Earth's surface

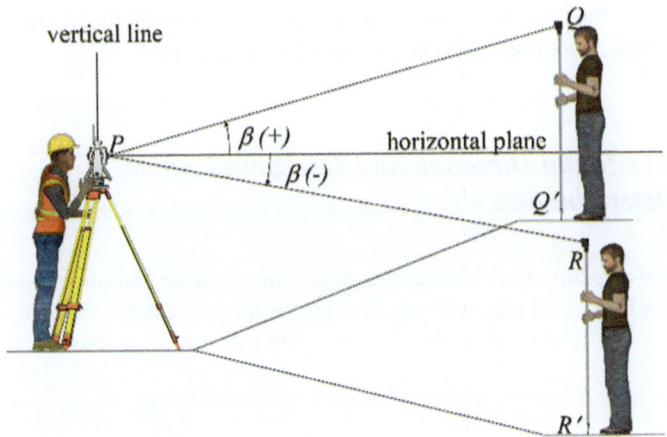

Fig. 6.3 Altitude angle

- *Zenith angle*: angle z referenced to the vertical line passing through the observation point (P), with values ranging from 0° to 360° and counted clockwise, as shown in Fig. 6.4. In practice, it is observed with a vertical circle zeroed at the zenith.

For more information on horizontal and vertical instrument circles, readers are suggested to refer to Chap. 9. Readers should also note that in the error-free condition, the sum of the altitude angle β and the zenith angle z must equal 90° or 270°, depending on the value of z, i.e., $\beta + z = 90°$ or $\beta + z = 270°$ (Figs. 6.3 and 6.4).

---

[1] Or simply vertical angle.

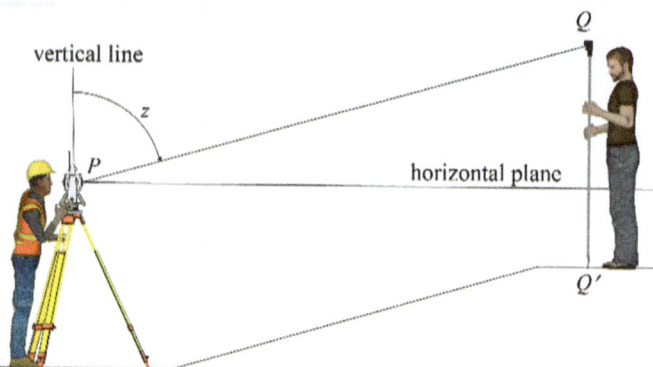

Fig. 6.4 Zenith angle

As explained in Chap. 9, angular measurements are performed using surveying instruments called theodolites, which can be mechanical or electronic, and total stations, which are exclusively electronic. The vertical circles of mechanical theodolites are graduated according to the zenith angle to avoid the occurrence of negative vertical angles. On the other hand, total stations and electronic theodolites allow the user to configure the most convenient type of vertical angle.

6.1.3 Horizontal Direction and Horizontal Angle Determination

A *horizontal direction* is defined as the projection of the spatial direction of a given line onto the horizontal plane. In practice, in Geomatics, it is the angular value read on the horizontal graduated circle of a surveying instrument, as shown in Fig. 6.5.

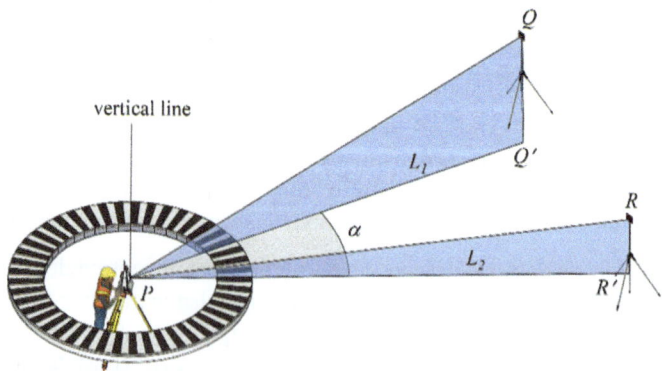

Fig. 6.5 Horizontal directions on a horizontal circle

Figure 6.5 shows the situation of an operator performing two direction readings, L_1 and L_2, to define the horizontal angle α between them. To do this, it is first necessary to determine which of these directions is the reference direction. For example, assuming that observation L_1 is the reference direction, the value of α counted clockwise is calculated according to Eq. (6.1).

$$\alpha = L_2 - L_1 \qquad (6.1)$$

In Geomatics applied to Civil Engineering, it is advisable to work with clockwise angles. Thus, by generalising Eq. (6.1), the following relationship can be written for the determination of a horizontal angle:

Horizontal clockwise angle = foresight horizontal direction reading—backsight horizontal direction reading

$$\alpha = L_F - L_B \qquad (6.2)$$

In Geomatics, the term used to describe the observation of a direction is *sighting*, i.e. directing the line of sight of the instrument's telescope to a specific target point. In practice, the first direction observed is called the *backsight* direction, and all subsequent directions are called the *foresight* direction.

In some cases, instead of determining an angle by reading the forward and backward directions, it is possible to read the deflection angle, i.e., the horizontal angle determined from the direction of the extension of the previous line (backward direction), right or left, to the direction of the following line, as shown in Fig. 6.6. However, as almost all engineering projects are now based on coordinates, the use of deflection angle measurements is declining.

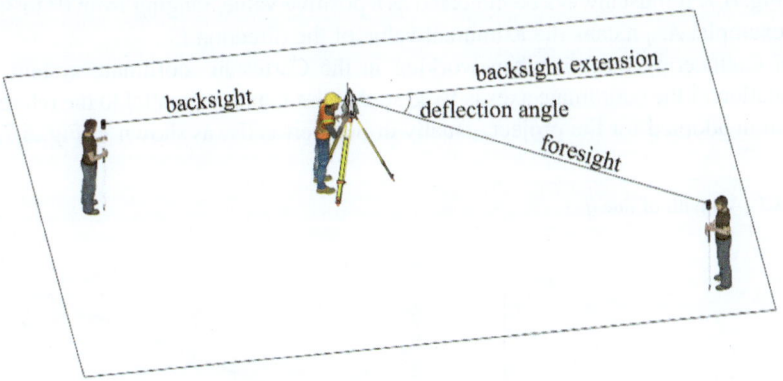

Fig. 6.6 Principle of deflection angle reading

Example 6.1 To determine a horizontal clockwise angle α between two lines, two horizontal directions $L_1 = 321° \ 10' \ 54''$ and $L_2 = 24° \ 40' \ 04''$ have been observed with a total station. Assuming that L_1 is the backsight direction, calculate the value of α.

Solution:

   Using Eq. (6.1).

$$\alpha = 24°\,40'\,04'' - 321°\,10'\,54'' = -296°\,30'\,50'' = 63°\,29'10''$$

Readers should note that horizontal angles are calculated rather than measured, as determining a horizontal angle with a surveying instrument uses two directions. On the other hand, the determination of vertical angles is based on a fixed horizontal or vertical reference line materialised by the mechanical or electronic compensator of the surveying instrument. Therefore, horizontal angles are calculated, and vertical angles are measured.

6.1.4   Azimuth and Bearing

As explained earlier, the value of a horizontal angle is the difference between two horizontal directions. In this context, whenever the first direction is a reference line, i.e., a line whose direction is known, commonly called the *meridian*, the angle determined by the second direction is called the *Azimuth* or *Bearing*, as described below.

6.1.4.1   Azimuth

Azimuth Az_{ij} can be understood as a general term referring to the horizontal angle reckoned clockwise from a chosen reference direction, usually the North direction. See Fig. 6.7. It must always be indicated as a positive value, ranging from 0° to 360°. For example, Az_{ij} means the azimuthal value of the direction i,j.

   In engineering design, when working in the Cartesian coordinate system, the orientation of the coordinate axes is fixed so that the Y-axis is parallel to the reference direction adopted for the project, usually drawn vertically, as shown in Fig. 6.7.

Fig. 6.7 Azimuth of line *ij*

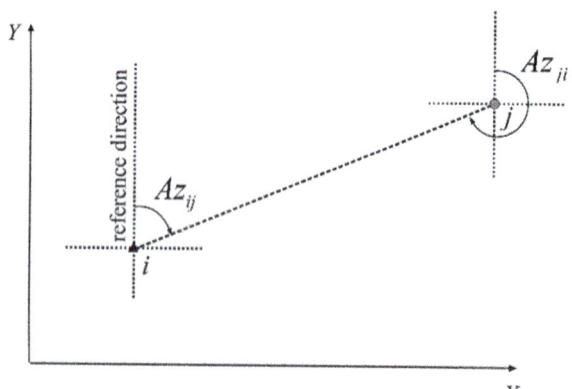

   As presented in Chap. 11, azimuths and distances are essential information for determining the coordinates of points in a Civil Engineering project.

   The reference direction (meridian) may be an arbitrary line with the directional value assigned by the surveyor, characterising an *assumed azimuth*. This is sometimes used in engineering surveying work (Plane Surveying). The *assumed reference line* (*assumed meridian*) can be any easily retrieved line in the project area or the direction between two points defined for this purpose. The critical point to remember here is that such a line is specific to the project in which it is defined and is, therefore, not related to any other reference line, not allowing data sharing between contiguous surveys. Furthermore, the direction may not be reproducible if the original reference points are removed or lost. For large projects, however, the reference line is determined according to the official geodetic framework of control points, which is referenced to the meridian passing through the rotation axis of the reference ellipsoid. The azimuth derived from this meridian is called a *Geodetic Azimuth*. Another azimuth to consider is the *Grid Azimuth*, discussed later in Chap. 16.

   As will be seen later in this book, azimuths can be calculated from Cartesian coordinates. They can also be called forward azimuths or backward azimuths, depending on the point of origin of the line being considered. As shown in Fig. 6.7, if Az_{ij} is considered the forward azimuth, its opposite direction Az_{ji} is the backward azimuth, and the difference between them is equal to 180°, as given by Eq. (6.3).

$$Az_{ji} = Az_{ij} \pm 180° \qquad (6.3)$$

   If $Az_{ij} < 180°$, add 180° to get Az_{ji}.
   If $Az_{ij} > 180°$, subtract 180° to get Az_{ji}.

Example 6.2 Assuming that the forward azimuth of line PQ is equal to $Az_{PQ} = 321° \, 10' \, 54''$, calculate the backward Azimuth Az_{QP}.

Solution:
   As the forward Azimuth Az_{PQ} is greater than 180°, it is necessary to subtract 180° from it to obtain the backward Azimuth Az_{QP}.

$$Az_{QP} = 321° \, 10' \, 54'' - 180° = 141° \, 10' \, 54''$$

   Figure 6.8 *illustrates the geometry of the forward and backward azimuths.*

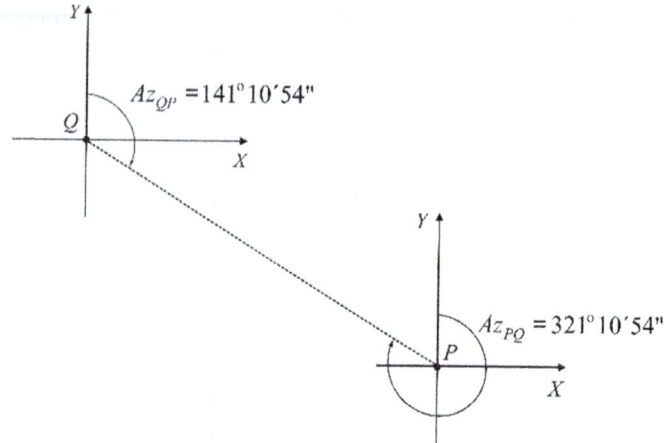

Fig. 6.8 Forward and backward azimuths for the line PQ

6.1.4.2 Bearing

Instead of azimuth, directions are sometimes referred to as *Bearing R_{ij}*, which is obtained from the acute angle reckoned clockwise or counterclockwise from a reference line with North or South as the reference direction, as shown in Fig. 6.9. Its values range from 0° to 90° and should always be given positive values. The bearing of a line is indicated by the quadrant in which it falls. Therefore, bearings are indicated as *NE, NW, SE* or *SW*. Taking Fig. 6.9 as an example, bearings would be indicated as

$R_{PA} = 45°\ 32'15''NE$ or $R_{PA} = N45°\ 32'15''E$
$R_{PB} = 45°\ 32'15''SE$ or $R_{PB} = S45°\ 32'\ 15''\ E$
$R_{PC} = 45°\ 32'15''SW$ or $R_{PC} = S45°\ 32'15''\ W$
$R_{PD} = 45°\ 32'15''NW$ or $R_{PD} = N45°\ 32'\ 15''\ W$

Fig. 6.9 Representation of bearings on a trigonometric circle

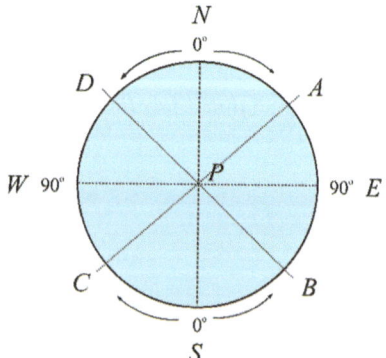

Analogously to the azimuth, the bearing can be classified as *assumed, geodetic* and *grid*.

Azimuths and bearings can be related when referenced to the same line, as given in Table 6.1.

Table 6.1 Mathematical relationship between azimuth and bearing

Quadrant	Mathematical relationship
I	$R(NE) = Az$
II	$R(SE) = 180° - Az$
III	$R(SW) = Az - 180°$
IV	$R(NW) = 360° - Az$

Example 6.3 Using the azimuth values shown in the first column of Table 6.2, convert them to bearing.

Table 6.2 Azimuth to bearing conversion

Azimuth	Bearing		
$Az = 35° \, 14' \, 56''$	$R(NE) = Az$	→	$R = 35° \, 14' \, 56'' NE$
$Az = 118° \, 26' \, 31''$	$R(SE) = 180° - Az$	→	$R = 61° \, 33' \, 29'' SE$
$Az = 245° \, 44' \, 09''$	$R(SW) = Az - 180°$	→	$R = 65° \, 44' \, 09'' SW$
$Az = 311° \, 52' \, 27''$	$R(NW) = 360° - Az$	→	$R = 48° \, 07' \, 33'' NW$

Solution:
   The converted values are shown in the last column of Table 6.2.

Example 6.4 Using the bearing values shown in the first column of Table 6.3, convert them to azimuth.

Solution:
   The converted values are shown in the last column of Table 6.3.

Table 6.3 Bearing to azimuth conversion

Bearing	Azimuth		
$R = 45° \, 04' \, 16'' NE$	$Az = R(NE)$	→	$Az = 45° \, 04' \, 16''$
$R = 18° \, 46' \, 51'' SE$	$Az = 180° - R(SE)$	→	$Az = 161° \, 13' \, 09''$
$R = 45° \, 24' \, 59'' SW$	$Az = 180° - R(SW)$	→	$Az = 225° \, 24' \, 59''$
$R = 11° \, 42' \, 37'' NW$	$Az = 360° - R(NW)$	→	$Az = 348° \, 17' \, 23''$

6.2 Angular Measurement

The measurement of angular values in Geomatics is carried out using surveying instruments equipped with two graduated circles installed in the horizontal and the vertical positions concerning the vertical axis of the instrument, as described in Sect. 9.3.1.

Briefly, the angular measurements with a surveying instrument occur through an index attached to the rotation axes of the instrument. By referring to this index, the operator can read the horizontal and vertical directions of a given line on the horizontal and vertical circles. The start and end points of this line are the centre of the graduated circle (P) and the sighting point (Q), respectively, as shown in Fig. 6.10.

As for the angular reading on the horizontal circle, it corresponds to the horizontal direction of the line PQ. For the angular reading on the vertical circle, it is generally assumed that the "zero" index of the circle coincides with the vertical line passing through the point (P). Thus, the value read corresponds to the vertical angle of the line PQ.

Fig. 6.10 The relative position of a total station's horizontal and vertical graduated circles

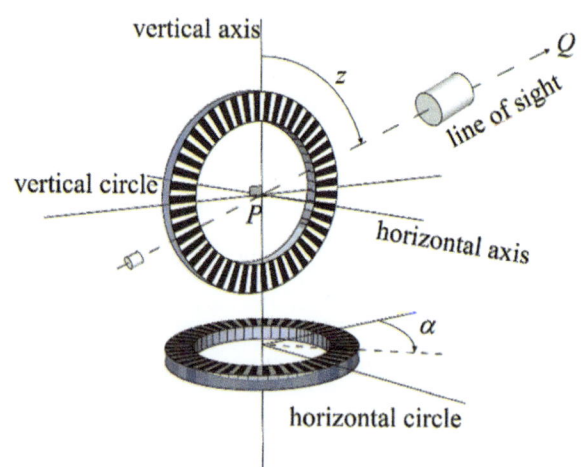

Observations of horizontal directions and vertical angles are performed by setting up a surveying instrument at a station mark (P) on the ground and aiming at the target point (Q) at the other end of the line, as shown in Fig. 6.11. The point (Q) may be a mark on a target, a well-defined feature of an object, or a reflector prism attached to a centred survey pole held vertically over a survey mark that defines the vertical to the point (Q) on the ground (point Q'). Typically, the surveyor triggers a measurement after sighting the target. Readings are displayed on the instrument screen and, if configured, also recorded to a data logger.

Fig. 6.11 Principle of direction and vertical angle observations with a surveying instrument

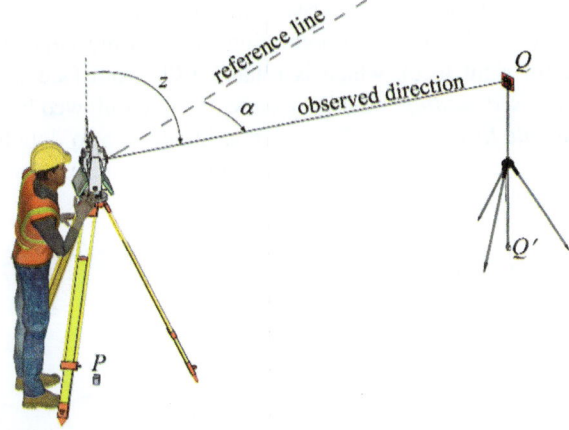

The instrument orientation value (reference direction) can be set to zero or any value computed by the instrument or manually entered by the surveyor. It is essential to understand that all subsequent measurements and coordinate computations are referenced to the orientation value.

6.2.1 Angular Observation Methods for Calculating the Horizontal Angle

The procedures to be followed for angular readings of directions to determine horizontal angles with a survey instrument depend on the number of angles to be determined and the sequence of observations from the same station. Therefore, assuming that angle observations are performed using an electronic survey instrument, the following methods of angle observations can be considered.

- Direct and reversed observations method
- Direction method
- Combined directions method
- Closing the horizon

6.2.1.1 Direct and Reversed Observations Method

Direct and reversed observations, *Ld* and *Li*, is a method of angular observation based on readings from two symmetrical positions of a graduated circle (horizontal or vertical) of a surveying instrument to define the horizontal direction or the vertical angle of a line. See Fig. 6.12. The field procedure consists of performing an observation of the angular direction of a given target with the instrument on face

1 (or left face), characterising the so-called *direct reading* or *single-face reading*, followed by another observation of the same target with the instrument on face 2 (or right face), which is rotated 180° from face 1, characterising the so-called *reversed reading*. Usually, a direct reading followed by a reversed reading is called a *double-face reading*. In this case, in addition to doubling the number of readings, this method of angular observation minimises several instrumental errors, as described in Sect. 10.2.1.2.

Fig. 6.12 Principle of direct and reversed observations

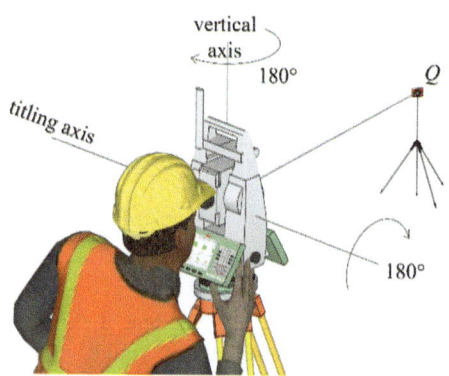

In practice, the field procedure is carried out by the following steps:

1. Aim at point (Q) and take the first reading L_d on face 1. Let $L_d = a_0$;
2. Rotate the instrument's telescope around the *Tilting axis* until the line of sight is opposite to the point (Q). Then, rotate the instrument about its *Vertical axis* and re-sight the point (Q) with a horizontal direction value approximately equal to the initial value plus 180°. This gives the value of the reversed reading, which is equal to $L_i \approx a_0 + 180°$;
3. Repeat the procedure as many times as necessary.

The mean (adjusted value) of the horizontal direction, in the direct position, is given by Eq. (6.4).

$$\bar{L} = \frac{\sum L_d + \sum L_i}{2n} \pm 90° \qquad (6.4)$$

Subtract 90° when the direct reading is less than 180°; otherwise, add 90°. where

$\bar{L} =$ mean of the horizontal direction in the direct position
$L_d =$ direct reading
$L_i =$ reversed reading
$n =$ number of double-face readings

For vertical angles, see Sect. 6.2.2.

The circumstances in which a double-face reading is preferable to a single-face reading depend mainly on the measurement situation and accuracy requirements. It is always recommended when:

- The highest degree of accuracy is required.
- Sighting within $\pm 20°$ of the zenith.
- High-temperature differences occur within a short period.
- Calibration values are not determined or, for some reason, do not seem correct.

The number of repetitions depends on the type of work to be performed. For current engineering work, one double-face reading is sufficient.

Regarding error propagation for the direct and reversed observations method, it is essential to consider the nominal angular accuracy of the surveying instrument being used. When using a total station, the standard deviation for observing a direction (horizontal and vertical) is specified by ISO 17123-3 and given as $\pm s_L$ seconds of arc. This value corresponds to the standard deviation of a direction measured in double-face readings and is usually considered to be the combined pointing and reading errors, as described in Sect. 10.2.2. Therefore, if a direction is observed on single-face readings, considering Eq. (6.4), the standard deviation s_{Li} of the single-face observation can be calculated as follows.

$$s_L^2 = \frac{s_{Li}^2 + s_{Ld}^2}{4} \tag{6.5}$$

Assuming that $s_{Li} = s_{Ld}$ and that Li and Ld are independent (a reasonable assumption).

$$s_{L_1} = \pm \sqrt{2} s_L \tag{6.6}$$

If a set of n single-face readings have been taken with a total station with an assumed standard deviation of $\pm s_L$, the precision of the adjusted direction is given by Eq. (6.7) and in the case of double-face readings by Eq. (6.8).

$$\text{Single face}: s_{\overline{L}} = \pm \frac{\sqrt{2} s_L}{\sqrt{n}} \tag{6.7}$$

$$\text{Double-face}: s_{\overline{L}} = \pm \frac{s_L}{\sqrt{n}} \tag{6.8}$$

Example 6.5 Five double-face readings have been taken using a total station to determine the direction of a line PQ, as shown in Table 6.4. Given the readings and considering that the standard deviation of the total station specified by the manufacturer is $\pm 2''$, calculate the adjusted value of the horizontal direction PQ in the direct position and its standard deviation.

Table 6.4 Direct and reversed readings values

Station	Sighting point	Direct reading (Face I)	Reversed reading (Face II)
P	Q	46° 14′08″	226° 14′10″
	Q	46° 14′07″	226° 14′04″
	Q	46° 14′13″	226° 14′17″
	Q	46° 14′17″	226° 14′19″
	Q	46° 14′14″	226° 14′16″
Sum		231° 10′59″	1,131° 11′06″

Solution:

The values in Table 6.4 *show that direct readings are less than* $180°$. *Thus, using* Eq. (6.4),

$$\overline{L} = \frac{231°\ 10'\ 59'' + 1,131°\ 11'\ 06''}{2*5} - 90° = 46°\ 14'\ 12.5''$$

Using Eq. (6.8).

$$s_{\overline{L}} = \frac{s_L}{\sqrt{n}} = \pm \frac{2}{\sqrt{5}} = \pm 0.9''$$

6.2.1.2   Determination of the Horizontal Angle by the Direction Method

Determining a horizontal angle by the direction method is a surveying measurement procedure that consists of observing directions by horizontal circle readings taken from the instrument station to successive target points, as described in the previous sections. It is a prevalent method as almost all modern instruments are built in direction mode. The instrument's horizontal circle is fixed, and once set, each reading on the horizontal circle is referenced to the same orientation. Therefore, using Eq. (6.2), the difference between any two circle readings L_F and L_B gives an angle α. If the readings are taken individually on *single-face*, then the standard deviation of the angle α is calculated using the law of propagation of variances, as given in Eq. (6.9).

$$s\alpha = \pm \sqrt{s_{LF}^2 + s_{LB}^2} = \pm \sqrt{2s_L^2 + 2s_L^2} = \pm 2s_L \tag{6.9}$$

If the readings are taken in *double-face*, the standard deviation of the angle α is calculated using the law of propagation of variances, as given in Eq. (6.10).

$$s\alpha = \pm \sqrt{s_{LF}^2 + s_{LB}^2} = \pm \sqrt{s_L^2 + s_L^2} = \pm \sqrt{2}s_L \qquad (6.10)$$

If *multiple single-face readings* are taken, the standard deviation of the resulting angle is given by Eq. (6.11).

$$s_\alpha = \pm \frac{2s_L}{\sqrt{n}} \qquad (6.11)$$

where n is the number of repetitions forwards and backwards for the determination of the angle.

If *multiple double-face readings* are taken, the standard deviation of the resulting angle is given by Eq. (6.12).

$$s_\alpha = \pm \frac{\sqrt{2}s_L}{\sqrt{n}} \qquad (6.12)$$

where n is the number of repetitions forwards and backwards for the determination of the angle.

Equations (6.11) and (6.12) lead to the conclusion that the use of multiple observations is a good way to increase precision, as it is expected that with more measurements there will be fewer errors, as random errors are more likely to cancel out. However, it is important to be sure that all outliers and systematic errors have been detected and corrected. It is also important to consider that, depending on the task, the repetition process may be time-consuming and, in these cases, the engineer may consider using a higher-quality instrument.

When several angles are determined from the same station, it is a common practice to perform sequential direction observations, usually in direct and reversed modes, as shown in Fig. 6.13. In this case, the field procedure consists of installing, centring and levelling the instrument over the station mark and then taking a backsight reading to the first point (P_1), followed by foresight readings on the sequential points ($P_2, \ldots, P_4$). Generally, for accurate measurements, multiple sets of direct and reversed readings are taken by repeating the same procedure for each set.

Note that the $0°00'00''$ direction in Fig. 6.13 corresponds to the zero-graduation direction of the total station.

As discussed in Chap. 10, in addition to reading and pointing errors, the standard deviation of an angle determined by the direction method is also affected by instrument and target centring errors.

Fig. 6.13 Determination of horizontal angles by the direction method

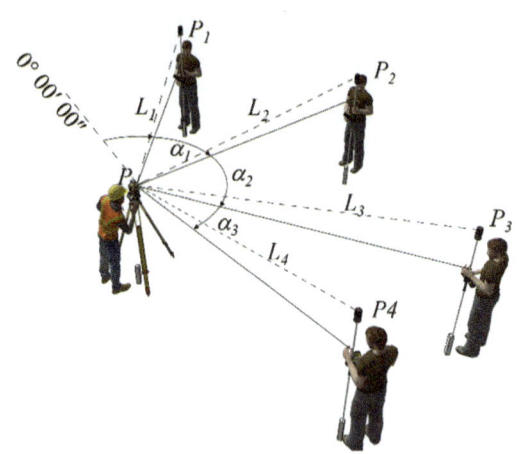

Example 6.6 Using Fig. 6.13 and the field measurements in Table 6.5, calculate the horizontal angles between the sighting pairs and the corresponding standard deviations. Assume that the measurements were carried out using a total station with a nominal angular accuracy of $\pm 2''$.

Table 6.5 Field measurements and resulting angle

Station	Sighting point	Direct horizontal direction (L_d)	Reversed horizontal direction (Li)	Adjusted horizontal direction ($\overline{L}_d$)	Angle (α)
P	P_1	26° 33′ 50″	206° 33′ 53″	26° 33′ 51.5″	
	P_2	63° 26′ 11″	243° 26′ 07″	63° 26′ 09.0″	36° 52′ 17.5″
	P_3	95° 42′ 31″	275° 42′ 30″	95° 42′ 30.5″	32° 16′ 21.5″
	P_4	122° 54′ 35″	302° 54′ 31″	122° 54′ 33.0″	27° 12′ 02.5″

Solution:

The resulting angles are indicated in the last highlighted column of Table 6.5. The standard deviation of each angle can be calculated using Eq. (6.10).

$$s_{\alpha i} = \pm 2'' \sqrt{2} = \pm 2.8''$$

If observations were taken on a single-face, the standard deviation of each angle should be calculated using Eq. (6.9) and the following result would be obtained.

$$s_{\alpha i} = \pm 2'' * 2 = \pm 4.0''$$

6.2.1.3   Determination of Horizontal Angles by the Method of the Combined Direction

Determining a horizontal angle by direction is easy to apply and requires little effort in the field. However, it allows little quality control of the calculated angular values. For this reason, whenever more than one angle needs to be determined in a high-precision survey, it is recommended to use the method of the combined direction, as shown in Fig. 6.14.

In this case, several independent direction values are observed from the point (P) with an estimated propagation of the standard deviation, as indicated in the previous sections. Note that all angles from α_1 to α_6 are calculated using Eq. (6.2). The mathematical model of this problem comprises six observations and three unknowns (x_1, x_2, x_3), resulting in three redundant observations.

The problem can be solved by applying an adjustment computation model, such as the Parametric adjustment model. Thus, from Fig. 6.14, the following error equations can be written:

Fig. 6.14 Principle of the method of the combined direction

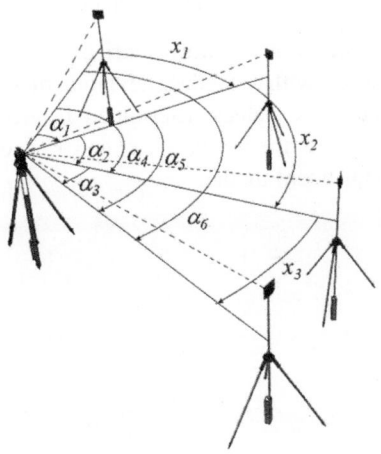

$$v_1 = x_1 - \alpha_1$$
$$v_2 = x_2 - \alpha_2$$
$$v_3 = x_3 - \alpha_3$$
$$v_4 = x_1 + x_2 - \alpha_4 \tag{6.13}$$
$$v_5 = x_2 + x_3 - \alpha_5$$
$$v_6 = x_1 + x_2 + x_3 - \alpha_6$$

Equation (6.13) can be written in matrix form as $v = Ax - l_0$ (Eq. 3.79), where

$$A = \begin{bmatrix} 1 & 0 & 0 \\ 0 & 1 & 0 \\ 0 & 0 & 1 \\ 1 & 1 & 0 \\ 0 & 1 & 1 \\ 1 & 1 & 1 \end{bmatrix} \quad v = \begin{bmatrix} v_1 \\ v_2 \\ v_3 \\ v_4 \\ v_5 \\ v_6 \end{bmatrix} \quad l_0 = \begin{bmatrix} \alpha_1 \\ \alpha_2 \\ \alpha_3 \\ \alpha_4 \\ \alpha_5 \\ \alpha_6 \end{bmatrix} \quad x = \begin{bmatrix} x_1 \\ x_2 \\ x_3 \end{bmatrix}$$

The adjustment computation can then be performed using Eqs. (3.84), (3.85), (3.86) and (3.87), from which x_i is obtained, followed by the observation residuals using Eq. (3.79) and the adjusted observations $\tilde{\alpha}_i$ using Eq. (6.14).

$$\tilde{\alpha}_i = \alpha_i + v_{\alpha i} \tag{6.14}$$

The standard deviation of the unit weight is determined using Eq. (3.88), and the variance-covariance matrix of the parameters (angles) using Eq. (3.93) multiplied by the variance of the unit weight value.

Finally, the standard deviations of the adjusted angles are given using the diagonal values in the variance-covariance matrix.

Example 6.7 Twelve horizontal directions on face 1 were observed using a total station with a standard deviation of $\pm 2''$ (ISO 17123-3), as shown in Fig. 6.15. Given the angle values determined from the observed directions, shown in Table 6.6, calculate the angles x_1, x_2 and x_3 and their standard deviation using the parametric adjustment model.

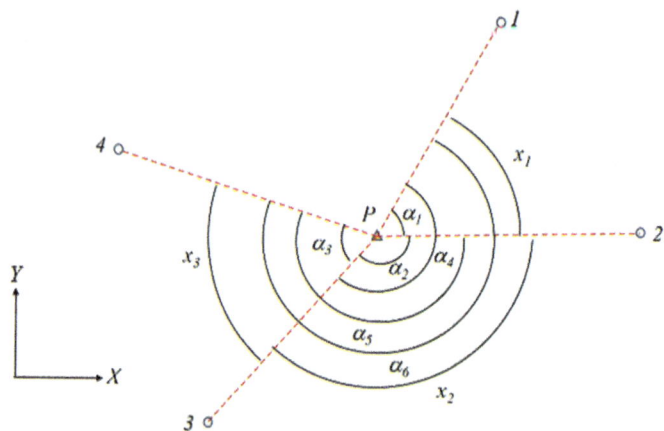

Fig. 6.15 Field observation geometry

Table 6.6 Angle values

Observation set	Angle	
$L_2 - L_1$	α_1	58° 52′ 09″
$L_3 - L_2$	α_2	133° 48′ 49″
$L_4 - L_3$	α_3	65° 17′ 28″
$L_3 - L_1$	α_4	192° 40′ 53″
$L_4 - L_2$	α_5	199° 06′ 18″
$L_4 - L_1$	α_6	257° 58′ 29″

Solution:

Considering that the total station has a nominal standard deviation of $\pm 2''$, using Eq. (6.9) the precision $s_{\alpha i}$ of the angles determined from observations is given as follows.

$$s_{\alpha_i} = \pm 2'' * \sqrt{2} * \sqrt{2} = \pm 4''.$$

Using Eq. (6.13), the following error equations can be written as

$$v_1 = x_1 - 58° 52' \ 09''$$
$$v_2 = x_2 - 133° 48' \ 49''$$
$$v_3 = x_3 - 65° 17' \ 28''$$
$$v_4 = x_1 + x_2 - 192° 40' \ 53''$$
$$v_5 = x_2 + x_3 - 199° 06' \ 18''$$
$$v_6 = x_1 + x_2 + x_3 - 257° 58' \ 29''$$

For the application of the Parametric adjustment model, using Eq. (3.79), the above equations can be written in matrix form as $v = Ax - l_0$, where

$$A = \begin{bmatrix} 1 & 0 & 0 \\ 0 & 1 & 0 \\ 0 & 0 & 1 \\ 1 & 1 & 0 \\ 0 & 1 & 1 \\ 1 & 1 & 1 \end{bmatrix} \quad v = \begin{bmatrix} v_1 \\ v_2 \\ v_3 \\ v_4 \\ v_5 \\ v_6 \end{bmatrix} \quad l_0 = \begin{bmatrix} 58° 52' \ 09'' \\ 133° 48' \ 49'' \\ 65° 17' \ 28'' \\ 192° 40' \ 53'' \\ 199° 06' \ 18'' \\ 257° 58' \ 29'' \end{bmatrix} \quad x = \begin{bmatrix} x_1 \\ x_2 \\ x_3 \end{bmatrix}$$

Using Eqs. (3.84), (3.85), (3.86) and (3.87) and assuming that the weight matrix P is equal the identity.

$$N = \begin{bmatrix} 3.0 & 2.0 & 1.0 \\ 2.0 & 4.0 & 2.0 \\ 1.0 & 2.0 & 3.0 \end{bmatrix} \quad n = \begin{bmatrix} 509° 31'31.0'' \\ 783° 34'29.0'' \\ 522° 22'15.0'' \end{bmatrix} \quad x = \begin{bmatrix} 58° 52'08.2'' \\ 133° 48'48.0'' \\ 65° 17'30.2'' \end{bmatrix}$$

Using Eq. (3.79).

$$\mathbf{v}^{\mathrm{T}} = [-0.8'' \quad 1.0'' \quad 2.2'' \quad 3.3'' \quad 0.3'' \quad -2.5'']$$

In this example, since the approximate values are equal to zero, the adjusted values of the unknowns are equal to the correction values given in the x vector. Thus,

$$x_1 = 58° 52' 08.2''$$
$$x_2 = 133° 48' 48.0''$$
$$x_3 = 65° 17' 30.2''$$

Finally, for $n = 6$ and $u = 3$, the standard deviation of the unit weight and the variance-covariance matrix for the unknowns (angles) are given as follows.

$$s_0^2 = 7.833333''^2$$

$$s_0 = \pm 2.8''$$

$$\Sigma_{xx} = s_0^2 * N^{-1} = \begin{bmatrix} 3.9167 & -1.9583 & 0 \\ -1.9583 & 3.9167 & -1.9583 \\ 0 & -1.9583 & 3.9167 \end{bmatrix}$$

Using the diagonal values in the variance-covariance matrix for the parameters, the standard deviations of the adjusted angles are given as follows.

$$x_1 = 58° 52'8.2'' \pm 2.0''$$
$$x_2 = 133° 48'48.0'' \pm 2.0''$$
$$x_3 = 65° 17'30.2'' \pm 2.0''$$

6.2.1.4 Closing the Horizon

The method of angular measurement called *closing the horizon* consists of observing all the necessary directions of a measurement section until the full clockwise rotation of the graduated circle is completed and the first direction (reference) is observed again, as shown in Fig. 6.16. As all angular directions have been observed, the sum of the calculated angles must be 360°.

The difference between the sum of the calculated angles and 360° is the error of closure e of the observation, which must be less than a predefined tolerance value.

To improve the quality of the calculated angles, it is recommended to observe each horizontal direction in the direct and reversed positions of the telescope.

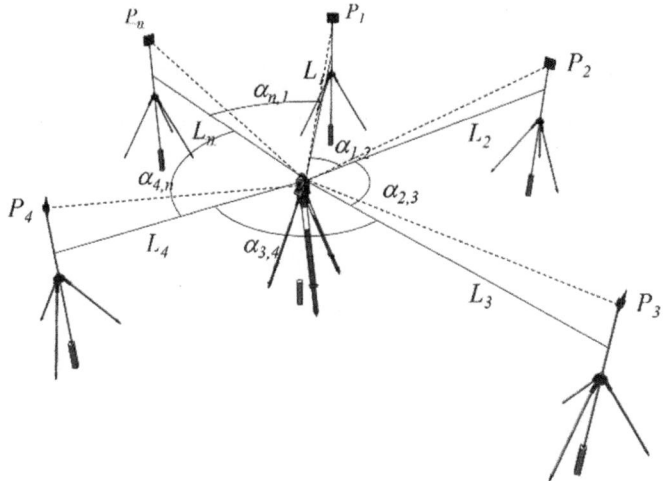

Fig. 6.16 Angular measurement by closing the horizon

As two readings are taken in the initial direction L_1, the error of closure is given by Eq. (6.15).

$$e = L_{1_2} - L_{1_1} \tag{6.15}$$

where

L_{1_2} = second reading at L_1
L_{1_1} = first reading at L_1

The calculation sequence for obtaining the other angles is as follows:

1. Calculate the error of closure e according to Eq. (6.15)
   Calculate the new reference observation value according to Eq. (6.16).

$$\overline{L}_1 = \frac{L_{1_2} + L_{1_1}}{2} \tag{6.16}$$

2. Calculate the new value of each observation according to Eq. (6.17)

$$\overline{L}_i = L_i - \overline{L}_1 \tag{6.17}$$

The observation $\overline{L}_1$ then becomes zero, and the others are corrected according to the error of closure. It is, therefore, advisable to choose the longest distance with the best visibility as the initial direction L_1.

Another way of solving this problem is to apply the same observations adjustment reasoning used in the previous section. Thus, according to Fig. 6.16 and considering that the angles to be calculated are $\alpha_{1,2}$ to $a_{4,n}$, the observation $a_{n,1}$ is redundant. The values of the adjusted angles can then be calculated by the Parametric adjustment model, as presented in the previous section. See Example 6.8 for further details.

Example 6.8 Table 6.7 shows the results of a field survey using the angle measurement technique by closing the horizon. Using the values given for the horizontal directions, calculate the corresponding adjusted angles.

Table 6.7 Field measurement

Station	Sighting point	Horizontal direction
P	P_1	28° 57′ 50″
	P_2	87° 49′ 50″
	P_3	221° 38′ 32″
	P_4	286° 56′ 08″
	P_1	28° 57′ 43″

Solution:

Since the closing the horizon method was applied, the error of closure can be calculated using Eq. (6.15).

$$e = 28°\ 57'\ 43'' - 28°\ 57'\ 50'' = -7''$$

According to Eq. (6.16), the average value of the reference reading is equal to:

$$\bar{L}_1 = \frac{28°\ 57'\ 43'' + 28°\ 57'\ 50''}{2} = 28°\ 57'\ 46.5''$$

The values of the individual observations can be calculated according to Eq. (6.17). The results are shown in Table 6.8.

Table 6.8 Corrected values

Station	Sighting point	Final horizontal direction
P	P_1	0° 00′ 00″
	P_2	58° 52′ 3.5″
	P_3	192° 40′ 45.5″
	P_4	257° 58′ 21.5″

Finally, the horizontal angles can be calculated using Eq. (6.1).

$$\alpha_{1,2} = \bar{L}_2 - \bar{L}_1 = 58°\ 52'03.5''$$
$$\alpha_{2,3} = \bar{L}_3 - \bar{L}_2 = 133°\ 48''42.0''$$
$$\alpha_{3,4} = \bar{L}_4 - \bar{L}_3 = 65°\ 17'36.0''$$
$$\alpha_{4,1} = \bar{L}_1 - \bar{L}_4 = 102°\ 01'38.5''$$
$$\alpha_{1,2} + \alpha_{2,3} + \alpha_{3,4} + \alpha_{4,1} = 360°\ 00'00''$$

6.2.2   Vertical Angle Determination

The measurement of vertical angles in Geomatics is carried out using a graduated vertical circle installed in a plane parallel to the plane containing the vertical axis of the instrument, as shown in Fig. 6.17. In this case, the vertical circle allows vertical angular measurements between the direction of a spatial line and the horizontal plane intersecting the observation point, as shown in Figs. 6.2 and 6.3.

For the angular reading, the vertical circle is held fixed in the instrument's alidade with a reading index that is considered to be coincident with the vertical line due to corrections of the instrument's electronic compensator. The instrument displays the value of the vertical angle observed as the telescope rotates around its Tilting-axis, as shown in Fig. 6.17.

Fig. 6.17 Illustration of measuring a vertical angle

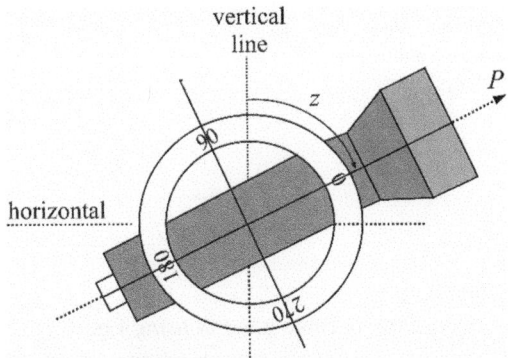

As mentioned in Sect. 6.2.1.1, whenever possible, it is advisable to carry out double-face-readings when measuring a vertical angle. In this case, considering the instrument configured for measurements of zenith vertical angles, the adjusted direct reading of the zenith angle can be calculated using Eq. (6.18).

$$\bar{z}_D = \frac{n360° + \sum z_D - \sum z_I}{2n} \tag{6.18}$$

where

$\bar{z}_D =$ adjusted direct reading of the zenith angle
$z_D =$ direct reading
$z_I =$ reversed reading
$n =$ number of direct and reversed readings

For the determination of the altitude angle β from the zenith angle, the following equations can be applied:

$$\text{In direct mode}: \quad \beta = 90° - z \tag{6.19}$$

$$\text{In reversed mode}: \quad \beta = z - 270° \tag{6.20}$$

As for the horizontal directions, assuming that a set of n double-face reading have been made with a total station with a standard deviation of $\pm s_L$ (ISO 17123-3) and that the observations are independent, the precision of the adjusted vertical angle is given by Eq. (6.8).

Example 6.9 A series of zenith angle readings have been taken for the same target in the direct and reverse positions of the telescope. Given the values in Table 6.9 and knowing that the standard deviation of the total station specified by the manufacturer is $\pm 2''$, calculate the adjusted value of the zenith angle in the direct position of the telescope and its standard deviation.

Table 6.9 Vertical angular readings in direct and reverse telescope positions

Station	Sighted point	Direct reading	Reversed reading
A	B	88° 16′ 42″	271° 43′ 04″
	B	88° 16′ 51″	271° 43′ 03″
	B	88° 16′ 42″	271° 43′ 10″
	B	88° 16′ 38″	271° 43′ 08″
	B	88° 16′ 40″	271° 43′ 14″
Sum		441° 23′ 33″	1,358° 35′ 39″

Solution:
   According to Table 6.9 *and using* Eq. (6.18),

$$\bar{z}_D = \frac{5 * 360° + 441° 23′ 33″ - 1,358° 35′ 39″}{2 * 5} = 88° 16′ 47″$$

The standard deviation of the adjusted value of the zenith angle is calculated using Eq. (6.8).

$$s_{\bar{z}} = \frac{s_L}{\sqrt{n}} = \pm \frac{2''}{\sqrt{5}} = \pm 0.9''$$

6.3 Angular Error Corrections

As mentioned, angular measurements in Geomatics are always carried out using surveying instruments. For this reason, most of the errors in angular measurements are related to the use of these instruments. For Geomatics applied to Civil Engineering, the following sources of angular errors must be considered: *instrumental, operational,* and *errors due to atmospheric refraction.* Instrumental and operational errors are discussed in Chap. 10, and errors due to vertical atmospheric refraction are discussed in Chaps. 7 and 13. The angular error due to lateral atmospheric refraction and its correction are discussed in the next section.

6.3.1   Angular Correction Due to Lateral Atmospheric Refraction

Because angular measurements are performed with optical instruments, it is important to consider that as the light beam reflected from the target passes through the layers of the atmosphere, its path is affected by the difference in air density. The temperature gradient causes it to bend towards the warmer air, where the speed of light is higher. As shown in Fig. 6.18, this effect causes the electromagnetic wave to follow a curvilinear path, and the direction of the sight AB observed in the instrument becomes AB', causing an angular variation equal to δ_α.

In the case of lateral displacement, the angular variation occurs, for example, when an alignment is measured very close (about 1 m) to a long linear structure, with lateral temperature variation along it, such as a building wall, a tunnel, etc.

According to the recommendations of the US Army Corps of Engineers—USACE (2018), the approximate angular correction δ_α, in arc seconds of the observed direction, to be applied in this case may be approximated by Eq. (6.21).

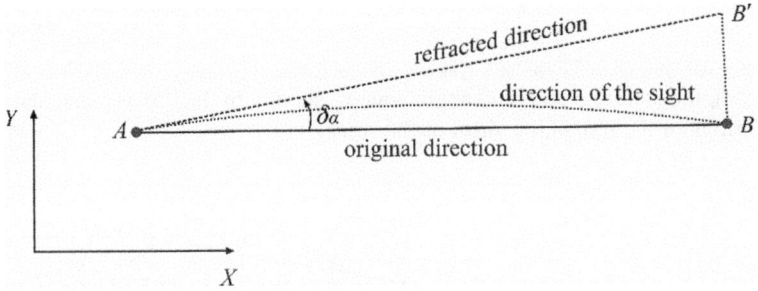

Fig. 6.18 Angular variation due to lateral refraction

$$\delta\alpha = \frac{8'' * P * d}{T^2} * \frac{\partial t}{\partial y} \tag{6.21}$$

where

d = measured distance, in metres
P = atmospheric pressure, in mb
T = temperature, in Kelvin (273.17 K+ $T°$C)

$\frac{\partial t}{\partial y}$ = temperature gradient in the direction perpendicular to the propagation of the electromagnetic wave (°C/m)

Note that using Eq. (6.21) requires the measurement of the lateral temperature gradient along the line of sight, which is a difficult procedure. For this reason, this type of correction is rarely used. However, as the angular variation in these cases is not negligible, it is recommended to avoid carrying out measurements under these conditions whenever possible and to look for alternative methods.

Using the example given in the USACE Manual N. 1110-2-1009, for $\partial t/ \partial y = 0.1°$ C/m, $d = 500$ m, $P = 1000$ mb and $t = 27°C$, the angular correction $\delta_\alpha = 4.4''$.

6.4   Problems

1. Given the definitions of altitude and zenith vertical angle (β and z), complete the equivalences in Table 6.10.

Table 6.10 Vertical angle

Altitude angle (β)	Zenith angle (z)
$-2°$ 45' 56"	–
$2°$ 45' 56"	–
–	92° 27' 44"
–	88° 15' 59"

2. Given the bearing and the angle values shown in Fig. 6.19, calculate the bearings of the remaining sides of the route.

Fig. 6.19 Geometry of the route

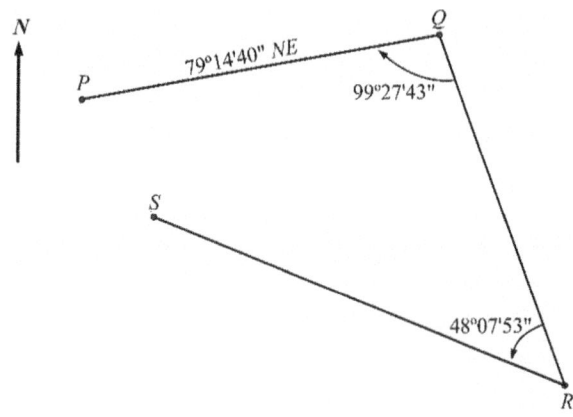

3. Given the azimuth values shown in Fig. 6.20, calculate the internal angles α_A to α_E, of the polygon.

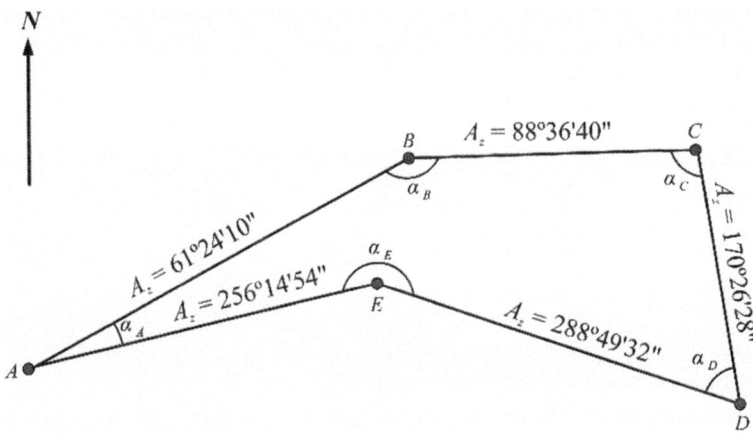

Fig. 6.20 Azimuth values

4. Given the horizontal direction observations in Table 6.11, calculate the clockwise horizontal angles α_{PQ}, α_{PR} and α_{QR}.

Table 6.11 Horizontal direction observations	Station	Sighting point	Observation (L_i)
	E	P	125° 45′ 56″
		Q	322° 12′ 32″
		R	32° 27′ 44″

5. Given the linear accuracies below, calculate the corresponding expected angular error.

   (a) 1:1,000
   (b) 1: 5,000
   (c) 1:10,000
   (d) 1:25,000
   (e) 1:50,000
   (f) 1:100,000
   (g) 1:1,000,000

6. Assuming that all the numbers to the right of the decimal point are significant, what is the resolution in millimetres of the coordinates $Lat = -22°\ 55′\ 34.2215″$; $Long = -47°\ 49′\ 22.0145″$ for a mean radius of the Earth equal to 6371 km?

7. To define a horizontal direction to be used as a reference for the setting-out work on a construction site using a total station, multiple double-face readings were carried out between points (P) and (Q). Using the observation given in Table 6.12, calculate the adjusted horizontal direction at the direct position of the instrument.

Table 6.12 Direct and reversed readings

Station	Sighting point	Direct reading (L_d)	Reversed reading (L_i)
P	Q	22° 32′ 08	202° 32′ 10″
		22° 32′ 07	202° 32′ 04″
		22° 32′ 13	202° 32′ 17″
		22° 32′ 17	202° 32′ 19″
		22° 32′ 14	202° 32′ 16″

8. To compare the quality of horizontal direction observations using a manual and an automatic total station, two series of horizontal directions were performed using the closing the horizon method. Given the readings in Table 6.13, calculate the error of closure and adjusted values for each instrument.

Table 6.13 Observed horizontal directions

Station	Sighting point	Automatic total station	Manual total station
P	1	168° 22′ 43.1″	168° 22′ 42.0″
	2	195° 45′ 35.6″	195° 45′ 32.2″
	3	262° 03′ 38.5″	262° 03′ 38.2″
	4	297° 20′ 52.1″	297° 20′ 47.9
	5	19° 30′ 51.1″	19° 30′ 49.6″
	1	168° 22′ 44.6″	168° 22′ 41.1″

9. The same points as in the previous problem were observed in the direct and reversed method with both total stations. Using the readings in Table 6.14, repeat the calculations and compare the results of the sets of measurements.

Table 6.14 Direct and reversed readings

Station	Sighting point	Automatic total station		Manual total station	
		Direct reading	Reversed reading	Direct reading	Reversed reading
P	1	168° 22′ 43.5″	348° 22′ 43.7″	168° 22′ 39.6″	348° 22′ 43.3″
	2	195° 45′ 32.4″	15° 45′ 36.0″	195° 45′ 30.1″	15° 45′ 31.3″
	3	262° 03′ 38.8″	82° 03′ 40.4″	262° 03′ 34.2″	82° 03′ 38.0″
	4	297° 20′ 50.7″	117° 20′ 52.9″	297° 20′ 47.5″	117° 20′ 51.2″
	5	19° 30′ 49.7″	199° 30′ 54.0″	19° 30′ 44.1″	199° 30′ 50.7″
	1	168° 22′ 42.0″	348° 22′ 41.8″	168° 22′ 34.7″	348° 22′ 38.2″

10. The angular difference between the vertical zenith angles of two total stations installed at points (P) and (Q), as shown in Fig. 6.21, is to be checked. With the first instrument positioned at point (P), the reticules of the telescope of the second instrument were aimed at point (Q) and vice versa. Using the readings in Table 6.15, calculate the deviation between the observed vertical zenith angles.

Fig. 6.21 Measurement geometry

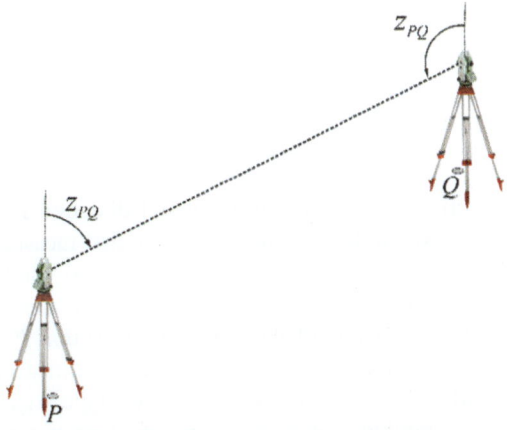

Table 6.15 Field measurements

Station	Sighting point	Zenith vertical angle
P	Q	88° 21′ 03″
Q	P	91° 39′ 23″

11. The same procedure as in the previous problem was repeated in direct and reversed reading. Using the readings in Table 6.16, calculate the deviation between the observed vertical zenith angles.

Table 6.16 Direct and reversed readings

Station	Sighting point	Zenith angle direct reading	Zenith angle reversed reading
P	Q	88° 21′ 07″	271° 38′ 29″
		88° 21′ 05″	271° 38′ 28″
		88° 21′ 09″	271° 38′ 29″
		88° 21′ 08″	271° 38′ 23″
Q	P	91° 39′ 21″	268° 20′ 29″
		91° 39′ 23″	268° 20′ 28″
		91° 39′ 20″	268° 20′ 29″
		91° 39′ 21″	268° 20′ 32″

12. Using the field measurements given in Table 6.17 and assuming that the total station used for the measurement has a nominal angular standard deviation of $\pm 3''$, calculate the following propagated standard deviations:

Table 6.17 Field measurements

Station	Series	Sighting point	Horizontal directions	
			Direct reading	Reversed reading
P	1	1	168° 22′ 43.5″	348° 22′ 43.7″
		2	195° 45′ 32.4″	15° 45′ 36.0″
	2	1	168° 22′ 42.6″	348° 22′ 43.4″
		2	195° 45′ 31.1″	15° 45′ 34.3″

(a) The value of the horizontal direction P-1 and its standard deviation for a single-face reading using the first measurement series.

(b) The value of the horizontal angle $\alpha_{1,2}$ and its standard deviation for a single-face reading using the first measurement series.

(c) The value of the horizontal direction P-1 and its standard deviation for a double-face reading using the first measurement series.

(d) The value of the horizontal angle $\alpha_{1,2}$ and its standard deviation for a double-face reading using the first measurement series.

(e) The value of the horizontal direction P-1 and its standard deviation for a single-face reading using the two series of measurements.

(f) The value of the horizontal angle $\alpha_{1,2}$ and its standard deviation for a single-face reading using the two series of measurements.

(g) The value of the horizontal direction P-1 and its standard deviation for a double-face reading using the two series of measurements.

(h) The value of the horizontal angle $\alpha_{1,2}$ and its standard deviation for a double-face reading using the two series of measurements.

References

Brabant, M. (2012). *Topographie opérationnelle.* Eyrolles Paris, France.

Ghilani, C. D., Wolf, P. R. (2007). *Elementary Surveying – An Introduction to Geomatics.* 12 Edition. Pearson Prentice Hall, EUA.

ISO 17.123-3:2001. *Optics and optical instruments – Field procedures for testing geodetic and surveying* instruments – Part 3: Theodolites.

ISO 9.849/2017. *Optics and optical instruments – Geodetic and surveying instruments – Vocabulary.*

US Army Corps of Engineers – USACE (2018). Engineering and Design. Structural Deformation Surveying. EM 1110-2-1009.

Chapter 7
Distances

Irineu da Silva

7.1 Introduction

As already explained, distance and angular measurements are fundamental in Geomatics for determining the position of a point in space and for setting-out Civil Engineering works. In addition, distance measurements, in particular, are essential for scaling control point networks, as presented in Chap. 11.

Although fundamental to Geomatics, measuring a distance to acceptable accuracies was a laborious and time-consuming operation in most surveying works for a long time. Simple instruments with good angular measurement accuracies already existed in the seventeenth century, while the problem of accurate distance measurement was only solved in the twentieth century. In addition, there has always been the problem of the influence of the Earth's curvature, which imposes restrictions on the types of distances used in engineering projects. For this reason, it is essential that, besides a broad knowledge of measurement techniques, Civil Engineers understand the different types of distances used in Geomatics to be able to apply them conveniently in their projects. Therefore, this Chapter aims to study mathematical concepts related to the types of distances used in Civil Engineering.

7.2 Types of Distances

Five types of distances must be considered for Geomatics applied to Civil Engineering.

I. da Silva (✉)
São Carlos School of Engineering, University of São Paulo, São Carlos, SP, Brazil
e-mail: irineu@sc.usp.br

© The Author(s), under exclusive license to Springer Nature Switzerland AG 2025   207
I. da Silva, P. C. L. Segantine, *Geomatics Applied to Civil Engineering*,
https://doi.org/10.1007/978-3-031-75737-2_7

- Slope distance
- Horizontal distance
- Vertical distance
- Ellipsoidal distance
- Grid distance

The geometric details and mathematical concepts regarding each of them are presented in the following subsections.

7.2.1 Plane Surveying: Slope and Horizontal Distance

As already described in Chap. 4. Plane Surveying is a subset of Geomatics Reference Systems that ignore the curvature of the Earth for horizontal point positioning. Under this condition, all planimetric measurements (distances and directions) carried out on the ground surface are reduced to a horizontal reference plane considered a local ground-based plane, as shown in Fig. 7.1.

In Fig. 7.1, the horizontal plane contains the point (P) and is normal to the vertical line that passes through it. If a distance is measured from (P) to any point (Q) above or below the reference plane, it is called the *slope distance* d'_{PQ}. Note that this is the distance generally measured in the field with a surveying instrument. Its projection on the horizontal reference plane defines the *horizontal distance* d_{PQ}.

Fig. 7.1 Distances in Plane Surveying

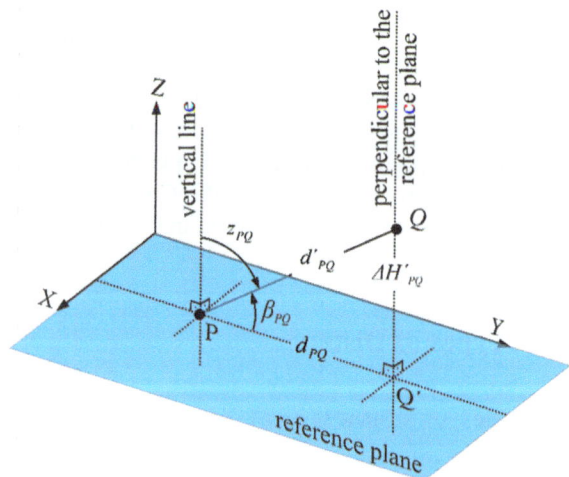

Horizontal distance is of great importance in Civil Engineering because it is used in calculations to determine the planimetric coordinates of points, to indicate the dimensions of objects in a project, and to set out geometric elements on a construction site. Thus, according to Fig. 7.1, the following geometric elements are of interest in Plane Surveying:

d'_{PQ} = slope distance between points (P) and (Q)
d_{PQ} = horizontal distance between points (P) and (Q)
$\Delta H'_{PQ}$ = vertical distance between points (P) and (Q)
β_{PQ} = altitude angle of alignment PQ
z_{PQ} = zenith angle of alignment PQ

Note that if points (P) and (Q) are located at the same elevation relative to the horizontal reference plane, that is, if the distance $\Delta H'_{PQ} = 0$, the slope, and the horizontal distances have the same value. Also, note that point (Q') is located on the horizontal reference plane, on the same level as point (P).

Since a local ground-based plane is adopted as the reference surface, the verticals at points (P) and (Q) are considered parallel. In this case, the convergence of plumb lines to the Earth's centre of mass is disregarded. For this reason, the horizontal plane can be moved vertically to the elevation of the vertex of the slope distance measurement without affecting the calculated distance for short lines. The horizontal distance thus generated is the base of the right triangle, whose hypotenuse is the slope distance. Such geometry allows for establishing the mathematical relationships given by Eq. (7.1). Note that for calculating the horizontal distance, besides the slope distance value, it is necessary to know the value of the zenith angle or the altitude angle, which is also measured in the field using a surveying instrument.

$$d_{PQ} = d'_{PQ} * \cos(\beta_{PQ}) \quad \text{or} \quad d_{PQ} = d'_{PQ} * \sin(z_{PQ}) \tag{7.1}$$

Adopting the local ground-based plane as the reference surface generates systematic errors that can be significant as the plane expands about the point of tangency with the Earth's surface or as the differences in elevation among points in the project area increase. For more details on these matters, refer to Sect. 7.2.6.

Although affected by systematic errors, depending on the size of the project area, their effects can be neglected. Therefore, many Civil Engineering projects consider the local ground-based plane as a surface of reference. It is essential, however, to emphasise that this simplification of Geodetic Surveying to Plane Surveying must be carefully evaluated and should only be adopted for projects of reduced dimensions and local context, as is the case with most Civil Engineering projects. Otherwise, it is necessary to consider the Earth's curvature and vertical atmospheric refraction, in addition to adequately applying the concepts of cartographic projections. The decision on which reference surface to use must be taken jointly by all professionals involved in the project. Particular attention should be given to cases where measurements are carried out by combining total stations and GNSS instruments. This subject is described in detail in Chap. 16.

7.2.1.1   Precision of the Horizontal Distance Reduced to Local Ground-Based Plane

According to Eq. (7.1), the calculated horizontal distance is a function of the slope distance and the measured vertical angle. Therefore, the precision of the calculated horizontal distance depends on the precision $s_{d'}$ with which the slope distance is measured, the precision (s_z or s_β) with which the vertical angle is measured and the values of the slope distance d' and the vertical angle (z or β). Thus, by the general law of propagation of variances, the precision of the horizontal distance is given as follows.

$$s_d = \pm\sqrt{\sin^2(z) * s_{d'}^2 + [d' * \cos(z)]^2 * s_z^2} = \pm\sqrt{\cos^2(\beta) * s_{d'}^2 + [-d' * \sin(\beta)]^2 * s_\beta^2}$$

$$(7.2)$$

Example 7.1

Figure 7.2 shows the slope distance d'_{PQ} between points (P) and (Q) and the vertical angles of line PQ. Considering that the measured slope distance should be reduced to the local ground-based plane, tangent to the vertical line at point (P), calculate the corresponding horizontal distance d_{PQ} and its precision.

Fig. 7.2 Field measurement geometry

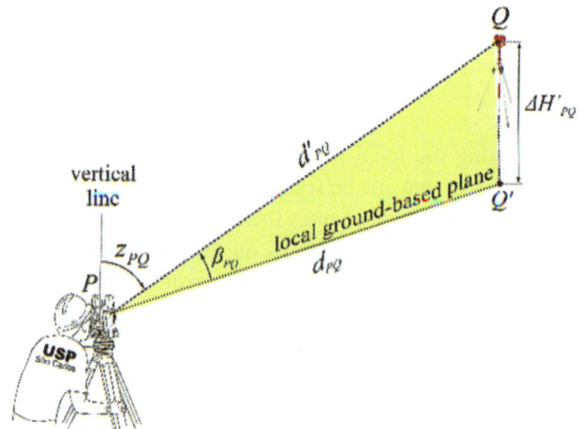

Measured values:

$$d'_{PQ} = 1045.022 \text{ m} \pm 4.1 \text{ mm}$$

$$\beta_{PQ} = 5°27'01'' \pm 5''$$

$$z_{PQ} = 84°32'59'' \pm 5''$$

Solution:

By Eqs. (7.1) and (7.2),

$$d_{PQ} = 1.045,022 * \cos(5°\,27'\,01'') = 1.040,297\,m$$
$$= 1.045,022 * \sin(84°32'\,59'') = 1.040,297\,m$$

$$s_{d_{PQ}} = \pm \sqrt{\sin^2(84°32'59'') * \left(\frac{4.1}{1000}\right)^2 + [-1045.022 * \cos(84°32'59'')]^2 * \left(\frac{5''}{206,264.8062}\right)^2}$$

$$s_{d_{PQ}} = \pm0.0047\,m = \pm4.7\,mm \approx \pm4.6\,ppm$$

7.2.2   Vertical Distance Considering the Local Ground-Based Plane

Generically, the term vertical distance refers to the length measured or reduced to the vertical line at the place of the target point. When considering the local ground-based plane as a reference surface, as shown in Fig. 7.1, the vertical distance between points (P) and (Q) is the projection of the slope distance d'_{PQ} on the opposite side of the right triangle contained in the vertical plane passing through (P) and (Q). Mathematically, it is defined as given by Eq. (7.3).

$$\Delta H'_{PQ} = d'_{PQ} * \sin(\beta_{PQ}) \quad \text{or} \quad \Delta H'_{PQ} = d'_{PQ} * \cos(z_{PQ}) \tag{7.3}$$

where $\Delta H'_{PQ}$ is the vertical distance between points (P) and (Q).

7.2.2.1   Precision of the Calculated Vertical Distance

Same as for the calculated horizontal distance, the precision of the calculated vertical distance is given by Eq. (7.4).

$$s_{\Delta H'} = \pm \sqrt{\cos^2(z) * s_{d'}^2 + [-d' * \sin(z)]^2 * s_z^2} =$$
$$\pm \sqrt{\sin^2(\beta) * s_{d'}^2 + [d' * \cos(\beta)]^2 * s_\beta^2} \tag{7.4}$$

Example 7.2

Using the values from Example 7.1, calculate the vertical distance $\Delta H'_{PQ}$ between points (P) and (Q) and its precision, considering the horizontal reference plane tangent to the vertical line at point (P).

Solution:

   Using Eqs. (7.3) and (7.4),

$$\Delta H'_{PQ} = 1045.022 * \sin(5°\,27'01'') = 1045.022 * \cos(84°\,32'59'') = 99.258\,m$$

$$s_{\Delta H'_{PQ}} = \pm \sqrt{\cos^2\left(84°32'59''\right) * \left(\frac{4.1}{1000}\right)^2 + \left[1045.022 * \sin\left(84°32'59''\right)\right]^2 * \left(\frac{5''}{206,264.8062}\right)^2}$$

$$s_{\Delta H'_{PQ}} = \pm 0.0252\,\mathrm{m} = \pm 25.2\,\mathrm{mm}$$

Note the more significant influence of the vertical angle on the precision of the calculated vertical distance.

7.2.3 Geometric Elements of Distance Measurement with a Total Station Considering the Local Ground-Based Plane

To easily understand the geometry of distance measurements using a total station, it is better to move the horizontal reference plane vertically to the level of the optical centre of the instrument, point (I), as shown in Fig. 7.3. The slope distance ds measured, in this case, is the spatial distance between points (I) and (R), situated at the centre of the reflector prism. Points (P) and (Q) are located on the ground and vertically to points (I) and (R). In general, what is sought is to determine the values of the horizontal and vertical distances between them.

As presented in Chap. 8, when using a total station, the distance ds measurement is carried out using electromagnetic signals, which are influenced by the atmosphere, causing their path to be curved instead of a straight line, as shown in Fig. 7.3. However, the difference between the two distances is minimal, and, in practice, the chord d'_{IR} is adopted as the slope distance. The difference between them can be calculated by Eq. (7.5).

Fig. 7.3 Geometric elements of distance measurement with a total station

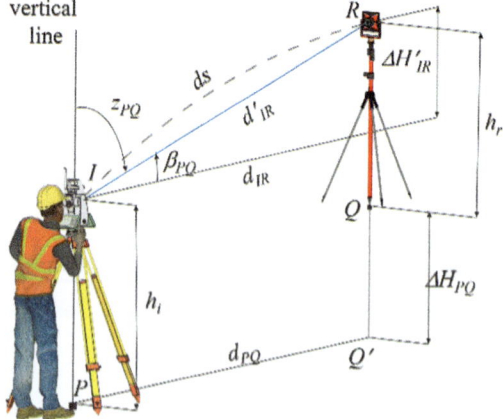

$$ds - d'_{IR} = \frac{k^2 * ds^3}{24 R_0{}^2} \tag{7.5}$$

where k is the coefficient of refraction, as described in Sect. 7.2.7 and R_0 is the local mean radius of the Earth.

As shown in Fig. 7.3, the vertical distance between points (P) and (I) is equal to h_i (height of the instrument), and between points (Q) and (R) is equal to h_r (height of the reflector prism). Because these two heights may differ, there are two vertical distances to consider when measuring the slope distance using a total station. The first is the vertical distance $\Delta H'_{IR}$ between points (I) and (R). It is purely geometric and has no geographic significance. For that reason, it is sometimes called the *trigonometric component of measure*. The second is the vertical distance ΔH_{PQ} between points (P) and (Q), which represents the vertical distance between the points on the ground and on which the total station and the prism pole are installed. This is the vertical distance usually sought to determine in a surveying measurement.

Note that, as all distances are orthogonally projected in the horizontal reference plane, the horizontal distance d_{PQ} between points (P) and (Q) is equal to the horizontal distance d_{IR} between points (I) and (R). The same is not valid for the vertical distance, i.e., the values of $\Delta H'_{IR}$ and ΔH_{PQ} are different. They are only equal if $h_i = h_r$.

In addition to the aforementioned geometric elements, the reader should note the need to know the zenith angle z_{PQ} or the vertical altitude angle β_{PQ}, also measured by the total station.

The values of the horizontal distance d_{IR} and the vertical distance $\Delta H'_{IR}$ can be calculated using Eqs. (7.1) and (7.3). The value of the vertical distance ΔH_{PQ} can be calculated using the equations indicated in Sect. 13.8.

7.2.4   Relationship between Horizontal and Spherical Distance on the Ground Surface

To evaluate the influence of the Earth's curvature on Civil Engineering projects, it is essential to relate the value of the horizontal distance to its respective arc on the Earth's curved surface, as shown in Fig. 7.4. For simplification, the curved surface, in this case, is associated with a sphere whose radius is equal to the local mean radius of the Earth at the latitude of the point (P).

According to Fig. 7.4, d_{PQ} is the horizontal distance obtained by reducing the slope distance to the local ground-based plane. Its projection onto the terrestrial sphere at the altitude of the point (P) generates the spherical distance $ds_{PQ'}$ and the chord $c_{PQ'}$. Under these circumstances, the geometric values in Fig. 7.4 are as follows:

Fig. 7.4 Relationships
between horizontal and
spherical distances on the
Earth's surface

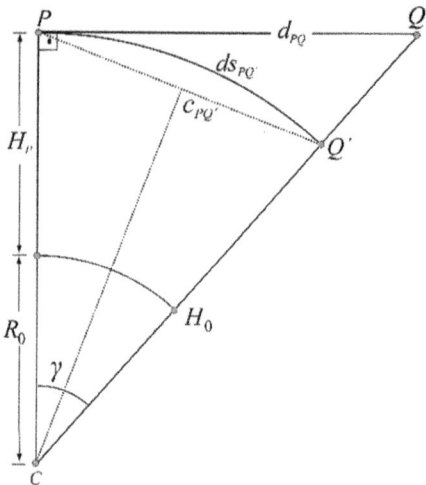

$Q'=$ projection of (Q) onto the spherical surface of the Earth at the elevation of
    point (P)
$d_{PQ}=$ horizontal distance (tangent) at (P)
$ds_{PQ'}=$ spherical distance at the level of point (P)
$c_{PQ'}=$ chord of the arc PQ'
$H_P=$ altitude of point (P)
$R_0=$ local mean radius of the Earth
$\gamma=$ angle at the centre of the sphere
$C=$ centre of the sphere

The geometric relationships between the elements of Fig. 7.4 allow for determining the following equations:

$$\text{Arc } PQ' : ds_{PQ'} = (R_0 + H_P) * \gamma \tag{7.6}$$

$$\text{Chord } PQ' : c_{PQ'} = 2(R_0 + H_P) * \sin\left(\frac{\gamma}{2}\right) \tag{7.7}$$

$$\text{Tangent } PQ : d_{PQ} = (R_0 + H_P) * \tan(\gamma) \tag{7.8}$$

Table 7.1 lists the differences between the arc PQ' and the chord PQ'; between
the tangent (horizontal distance) PQ and the arc PQ'; and between the tangent PQ
and the chord PQ'. The values were calculated considering $R_0 = 6{,}371{,}000.000$ m
and $H_P = 800$ m.

Table 7.1 shows that for distances less than 10 km, the difference between the arc
and the chord on the surface of the Earth is of the order of 1 mm, i.e., 1:10,000,000.
The same applies to the difference between the tangent (horizontal distance) and the

arc for distances of less than 5 km. Under these conditions, for the practical works of Geomatics applied to Civil Engineering, the horizontal and the spherical terrestrial distances are considered equal for distances up to 10 km (1:1,250,000). Therefore, when a slope distance of up to 10 km is projected on the horizontal reference plane, tangent to point (P), the spherical distance at the level of (P) is obtained. It should also be noted that in Civil Engineering, the measured distances are often less than 2 km, which further emphasises the lack of concern about the difference between the measured horizontal distance and spherical distances and between the horizontal distance and the chord. In summary, horizontal, spherical and chord are considered to be the same for current Civil Engineering applications.

Table 7.1 Differences in distances between the arc, the chord, and the tangent on the Earth's surface

Tangent [m] d_{PQ}	Difference between arc and chord [mm] $ds_{PQ'} - c_{PQ'}$	Difference between tangent and arc [mm] $d_{PQ} - ds_{PQ'}$	Difference between tangent and chord [mm] $d_{PQ} - c_{PQ'}$
1000	0.00	0.01	0.01
2000	0.01	0.07	0.07
3000	0.03	0.22	0.25
5000	0.13	1.03	1.15
10,000	1.03	8.21	9.24

7.2.5 Ellipsoidal Distance

As discussed in previous chapters, when using the local ground-based plane is not recommended, the Earth's curvature must be considered by adopting the ellipsoid of revolution as the reference surface. In this situation, as distance measurements are mainly carried out on the ground surface at altitudes different from that of the ellipsoid, they must first be reduced to the curved surface of the ellipsoid for most geodetic computations. The equivalent distance on the ellipsoid is an arc known as the *ellipsoidal distance*.

Considering that the magnitudes of the distances measured in Geomatics applied to Civil Engineering are small compared to the radius of the Earth, the ellipsoidal surface can be replaced by a sphere of radius equal to the local mean radius R_0 of the Earth. See Fig. 7.5. It is shown that this simplification does not change the results for distances up to 50 km.

Fig. 7.5 Geometric
relationships between
spherical surfaces

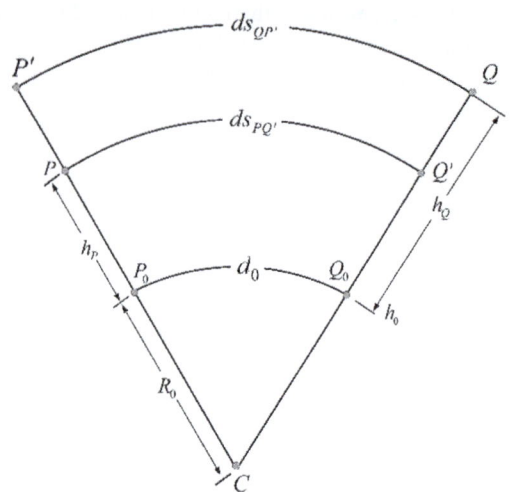

Fig. 7.5 Geometric
relationships between
spherical surfaces

From Fig. 7.5, the following geometric elements are of interest:

h_P= ellipsoidal height of point (P)
h_Q= ellipsoidal height of point (Q)
R_0= local mean radius of the Earth at the latitude of (P)
$ds_{PQ'}$ = spherical distance at the elevation of (P)
$ds_{QP'}$ = spherical distance at the elevation of (Q)
d_0= ellipsoidal distance
C = centre of the sphere

The surfaces indicated in Fig. 7.5 are concentric spheres, which allows for writing the following equations:

$$\frac{d_0}{R_0} = \frac{ds_{PQ'}}{R_0 + h_P} = \frac{ds_{QP'}}{R_0 + h_Q} \rightarrow d_0 = \frac{ds_{PQ'} * R_0}{R_0 + h_P} = \frac{ds_{QP'} * R_0}{R_0 + h_Q} \quad (7.9)$$

The projection of the ground spherical distance $ds_{PQ'}$ on the ellipsoid surface generates the ellipsoidal distance d_0, whose value can be calculated using a *reduction factor (Red)* in ppm (parts per million) or by an *elevation scale factor* k_{alt}, given by Eqs. (7.10) and (7.12).

$$Red = -\frac{h_P}{R_0 + h_P} * 10^6 \text{ ppm} \quad (7.10)$$

The ellipsoidal distance is then calculated by Eq. (7.11).

$$d_0 = ds_{PQ'} + \left(ds_{PQ'} * Red\right) \quad (7.11)$$

Note the need for dimensional compatibility between the distance $ds_{PQ'}$ and the reduction factor Red in Eq. (7.11).

$$k_{alt} = \frac{R_0}{R_0 + h_P} \tag{7.12}$$

In this case, the ellipsoidal distance is calculated by Eq. (7.13).

$$d_0 = ds_{PQ'} * k_{alt} \tag{7.13}$$

Equations (7.11) and (7.13) allow calculating the value of the ellipsoidal distance as a function of the elevation of the reference point (P) and the local mean radius R_0 of the Earth. The reduction is performed on the geoid surface if the orthometric altitude (H) is used. If the ellipsoidal height (h) is used, the reduction is performed on the ellipsoidal reference surface. Strictly speaking, the ellipsoidal height and instrument height h_i must be used. However, for many cases of Geomatics applied to Civil Engineering, it is indifferent to use any of them, mainly when the geoid undulation at the considered point is small. On that subject, Eq. (7.14) shows the difference in the ellipsoidal distance as a function of the difference in the orthometric altitude and the local mean radius of Earth.

$$\delta d_0 = \sqrt{\left[\frac{H}{(R_0 + H)^2}\right]^2 * \Delta R_0^2 + \left[\frac{-R_0}{(R_0 + H)^2}\right]^2 * \Delta H^2} \tag{7.14}$$

where,

δd_0 = difference in ellipsoidal distance d_0
H = local orthometric altitude
R_0 = local means the radius of the Earth
ΔR_0 = difference in the local mean radius of the Earth
ΔH = difference in the orthometric altitude

It becomes evident that the orthometric altitude difference has a much bigger effect on the ellipsoidal distance than the local mean radius difference. Table 7.2 shows the influence of these differences on the ellipsoidal distance for different values of R_0 and H.

Table 7.2 Difference in the ellipsoidal distance as a function of R_0 and H differences for a place with $R_0 = 6, 371, 000.000$ m and $H = 850.000$ m

ΔR_0	50 km	0 m	10 km	10 km	10 km
ΔH	0 m	10 m	6 m	10 m	20 m
δd_0	1.05 ppm	1.57 ppm	0.96 ppm	1.58 ppm	3.15 ppm

It should be noted from Table 7.2 that for the difference in the local mean radius of the Earth, the difference in the ellipsoidal distance is of the order of 1 ppm for $\Delta R_0 = 50$ km. In most calculations, this means that one can use the value of the local mean radius of the Earth as equal to 6,371,000 metres without causing a significant variation in the ellipsoidal distance. As for the difference in altitude, the difference in the ellipsoidal distance is of the order of 1 ppm for $\Delta H = 6$ metres, indicating that, in many regions of the Earth, it is necessary to consider the geoidal undulation and the height of the instrument for the calculation of the elevation scale factor k_{alt}.

Example 7.3

If the point (P) in Example 7.1 is located at the latitude $22°01'49''$ S and orthometric altitude of 700.456 metres, calculate the ellipsoidal distance d_{0PQ}, considering the parameters of the SIRGAS2000 Geodetic Reference System.

Solution:

The first step in solving this problem is to calculate the value of the local mean radius of the Earth. Then, using Eqs. (3.7), (3.8), and (3.9).

$$M = 6,344,400.729 \text{ m} \quad N = 6,381,142.849 \text{ m} \quad R_0 = 6,362,745 \text{ m}$$

Using Eq. (7.9),

$$Red = -\frac{700.456}{6,362,745 + 700.456} * 10^6 \text{ ppm} = -110.1 \text{ ppm}$$

The horizontal distance calculated in Example 7.1 is equal to 1040.297 m. As presented in the previous section, this distance can be considered equivalent to the spherical distance at (P). Thus, the value of the ellipsoidal distance d_{0PQ} can be calculated using Eq. (7.10).

$$d_{0PQ} = 1040.297 - (1040.297 * 0.000110075) = 1040.182 \text{ m}$$

Another solution is to use the concept of k_{alt}. Thus, using Eq. (7.12),

$$k_{alt} = \left(\frac{6,362,745}{6,362,745 + 700.456}\right) = 0.999889925$$

Using Eq. (7.13),

$$d_{0PQ} = 1040.297 * 0.999889925 = 1040.182 \text{ m}$$

Considering that the geoid undulation at point (P) is equal to -6.29 m and instrument height is equal to 1.295 m, the value of ellipsoidal distance, taking these values into account, is equal to $d_{0PQ} = 1040.297 * 0.999890710 = 1040.183$ m, i.e., a difference of 1 mm from the previously calculated value.

As an indication of the difference in length when reducing the ground distance to the ellipsoid, Table 7.3 shows computed values for different ellipsoidal heights, considering the local mean radius of Earth equal to $R_0 = 6,371,000$ metres and ellipsoid GRS80.

Table 7.3 Relationship between topographical spherical distance and ellipsoidal distance for different values of ellipsoidal height

Ellipsoidal height [m]	Ground spherical distances [m]				
	1000	2000	3000	5000	10,000
500	0.078	0.157	0.235	0.392	0.785
1000	0.157	0.314	0.471	0.785	1.569
2000	0.314	0.628	0.941	1.569	3.138
5000	0.784	1.568	2.353	3.921	7.842

According to Table 7.3, a spherical ground distance of 2000.000 m at an ellipsoidal height of 1000.000 m undergoes a reduction of 0.314 m when reduced to the ellipsoid, i.e., 314 ppm. Thus, the value of the ellipsoidal distance, in this case, equals 1999.686 m.

At this point in studies, one can be confused about the usefulness of the ellipsoidal distance since it, in addition to being purely mathematical, is located at an altitude, in most cases, different from the altitude of the engineering project site. Clarifications on this matter are presented in Chap. 16.

7.2.6 Effect of the Earth's Curvature in Reducing the Slope Distance to Horizontal and Ellipsoidal Distances

In Sect. 7.2.1, it was established that Eq. (7.1) expresses the reduction of the slope distance to the horizontal reference plane. In Sect. 7.2.4, it was shown that the spherical distance and the chord could be considered equal to the horizontal distance reduced on the horizontal plane, at the level of the ground surface, for distances less than 10 km. The question now is to assess the effects on horizontal and ellipsoidal distances when considering the curvature of the Earth. To analyse these effects, consider the geometric elements of the sphere section representing the shape of the Earth shown in Fig. 7.6.

R_0 = local mean radius of the Earth at the latitude of (P)
γ = angle at the centre of the sphere PQ
d'_{PQ} = slope distance of the alignment PQ
β'_{PQ} = vertical altitude angle of alignment PQ
β'_{QP} = vertical altitude angle of alignment QP
z'_{PQ} = vertical zenith angle of alignment PQ
z'_{QP} = vertical zenith angle of alignment QP
$ds_{PQ'} = \text{arc } PQ' = \text{chord } PQ' \ (c_{PQ'})$

$ds_{QP'} = \text{arc } QP' = \text{chord } QP' \ (c_{QP'})$
$h_P=$ ellipsoidal height of point (P)
$h_Q=$ ellipsoidal height of point (Q)
$\Delta h=$ difference in ellipsoidal height between (P) *and* (Q)
$d_0=$ ellipsoidal distance between points (P) and (Q)
$C =$ centre of the sphere with local mean radius R_0

Fig. 7.6 Geometric relationship between spherical distance and slope distance, considering the curvature of the Earth

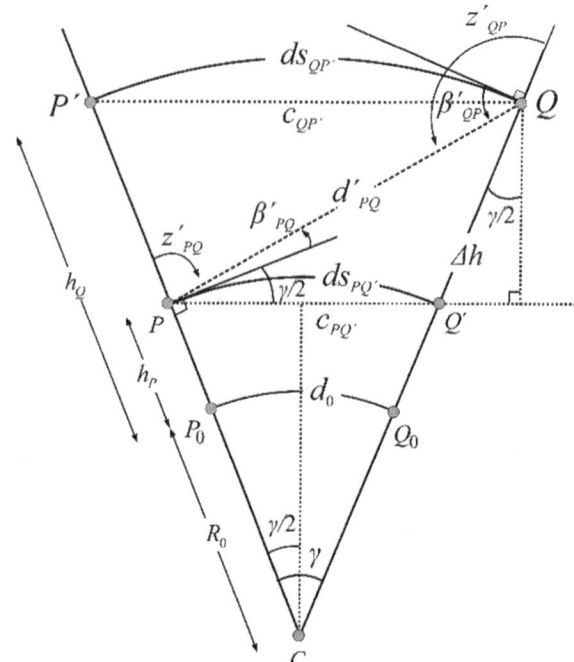

To understand the geometry in Fig. 7.6, consider the enlargement of the quadrilateral $P'QQ'P$ shown in Fig. 7.7, where horizontal distances are replaced by their respective chords.

Fig. 7.7 Enlarged view of the quadrilateral $P'QQ'P$

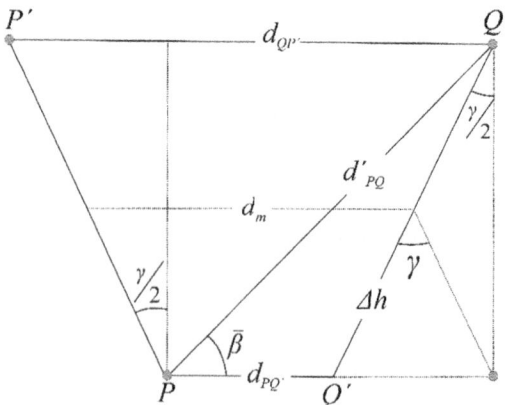

Figure 7.7 shows that due to the Earth's curvature, the horizontal distance $d_{QP'}$ at the level of (P) and the horizontal distance $d_{PQ'}$ at the level of (Q) are different. Figures 7.6 and 7.7 show that

$$\overline{\beta} = \beta'_{PQ} + \left(\frac{\gamma}{2}\right) \tag{7.15}$$

$$\overline{\beta} = \beta'_{QP} - \left(\frac{\gamma}{2}\right) \tag{7.16}$$

$$\gamma = \text{atan}\left(\frac{d'_{PQ}}{R_0 + h_P}\right) \tag{7.17}$$

Note that the negative sign of β'_{QP} was not considered in the equations above. Thus, at the level of (P):

$$d_{PQ'} = d'_{PQ} * \cos\left(\overline{\beta}\right) - \Delta h * \sin\left(\frac{\gamma}{2}\right) \tag{7.18}$$

At the level of (Q),

$$d_{QP'} = d'_{PQ} * \cos\left(\overline{\beta}\right) + \Delta h * \sin\left(\frac{\gamma}{2}\right) \tag{7.19}$$

At the average level between (P) and (Q),

$$dm = d'_{PQ} * \cos\left(\overline{\beta}\right) \tag{7.20}$$

In this case, the value of Δh must be calculated using Eq. (7.21).

$$\Delta h = d'_{PQ} * \sin\left(\overline{\beta}\right) * \sec\left(\frac{\gamma}{2}\right) \tag{7.21}$$

However, for a distance of 5 km, for example, the value of $\sec\left(\frac{\gamma}{2}\right) = \sec\left(\frac{5000}{2*6,371,000}\right) = 1.0000001$, indicates that the value of Δh can be calculated by Eq. (7.22) without affecting the precision of the values calculated later.

$$\Delta h = d'_{PQ} * \sin\left(\overline{\beta}\right) \tag{7.22}$$

Under these conditions and considering the Earth's curvature, the ellipsoidal distance can be calculated according to the equations indicated below:

$$d_0 = \left(\frac{R_0}{R_0 + H_P}\right) * \left[d'_{PQ} * \cos\left(\bar{\beta}\right) - \Delta h * \sin\left(\frac{\gamma}{2}\right)\right] \qquad (7.23)$$

$$d_0 = \left(\frac{R_0}{R_0 + H_Q}\right) * \left[d'_{PQ} * \cos\left(\bar{\beta}\right) + \Delta h * \sin\left(\frac{\gamma}{2}\right)\right] \qquad (7.24)$$

$$d_0 = \left(\frac{R_0}{R_0 + H_m}\right) * d'_{PQ} * \cos\left(\bar{\beta}\right) \qquad (7.25)$$

where

$$h_m = h_P + \frac{\Delta h}{2} \quad \text{or} \quad h_m = h_Q - \frac{\Delta h}{2} \qquad (7.26)$$

Example 7.4

Considering the data in Table 7.4, calculate the horizontal and the ellipsoidal distance reduced from (P), (Q) and the mean altitudes. Check that the value of the ellipsoidal distance is the same, regardless of the altitude considered. Consider that the height of the instrument at (P) is equal to 1.295 metres and at (Q) is equal to 1.273 metres.

Table 7.4   Measured values

Position	Target point	Slope distance [m]	Altitude angle	H [m]
P	Q	1045.022	5° 26′54.8″	700.456
Q	P		5° 27′28.7″	

Solution:

The geoid undulation at the surveying site, obtained using the geoid model EGM2008, equals -6.290 metres. Thus, the ellipsoidal height for the point (P) is

$$h_P = 700.456 - 6.290 + 1.295 = 695.461 \text{ m}$$

The value of γ can be calculated using Eq. (7.17).

$$\gamma = \text{atan}\left(\frac{1045.022}{6,363,745 + 696.461}\right) = 0°00′33.87″$$

Considering the absolute values of β_{PQ} and β_{QP}, the value of $\bar{\beta}$ at the altitudes of points (P) and (Q) are obtained using Eqs. (7.14) and (7.16).

$$\bar{\beta}_{PQ} = 5°26′54.8″ + 0°00′16.94″ = 5°27′11.74″$$

$$\bar{\beta}_{QP} = 5°27′28.7″ - 0°00′16.94″ = 5°27′11.76″$$

$$\bar{\beta} = \frac{5°27'11.74'' + 5°27'11.76''}{2} = 5°27'11.75''$$

Then

$$\Delta h = 1045.022 * \sin(5°27'11.75'') = 99.3124 \text{ m}$$

which, considering the difference between the heights of the instruments at (P) and (Q), becomes

$$\Delta h = 99.3124 - 0.022 = 99.2904 \text{ m}$$

Then, using Eqs. (7.18), (7.19), and (7.20).

$$d_{PQ'} = 1045.022 * \cos(5°27'11.75'') - 99.2904 * \sin(0°00'16.94'') = 1040.284 \text{ m}$$

$$d_{QP'} = 1045.022 * \cos(5°27'11.75'') + 99.2904 * \sin(0°00'16.94'') = 1040.300 \text{ m}$$

$$d_m = 1045.022 * \cos(5°27'11.75'') = 1040.292 \text{ m}$$

Considering that $h_P = 695.461$ m, $h_Q = 695.461 + 99.290 = 794.751$ m, *and* $h_m = 695.461 + \left(\frac{99.290}{2}\right) = 745.106$ m, *using Eqs. (7.23), (7.24) and (7.25).*

$$d_0 = 1,040.170 \text{ m } \textit{for the three reductions.}$$

Note that the horizontal distances at (P) and (Q) disregarding the curvature of the Earth (Eq. 7.1), that is, considering the local ground-based plane, are equal to

$$d_{PQ} = 1045.022 * \cos(5°26'54.8'') = 1040.300 \text{ m}$$

$$d_{QP} = 1045.022 * \cos(5°27'28.7'') = 1040.284 \text{ m}$$

Thus, there is a difference of approximately 16 ppm *between horizontal distances, considering and disregarding the curvature of the Earth. The same difference will be found for the ellipsoidal distance.*

7.2.7 Effect of the Curvature of the Earth and Vertical Atmospheric Refraction in Reducing the Slope Distance to Horizontal and Ellipsoidal Distances

When measuring a vertical angle with a surveying instrument, it is neither the angle $\bar{\beta}$ nor the angle β'_{PQ} shown in Figs. 7.6 and 7.7. The effectively measured angle is the angle β_{PQ}, that undergoes the effect of vertical atmospheric refraction. The atmosphere affects the geometry of the trajectory of a line of sight as the temperature gradient varies. Light rays passing through the Earth's atmosphere are bent toward the Earth's surface, as shown in Fig. 7.8. Due to this variation, the straight line of sight (dashed line) becomes a curved line in the vertical plane (dotted), producing a vertical displacement of the target image.

Fig. 7.8 Slope sight and vertical atmospheric refraction

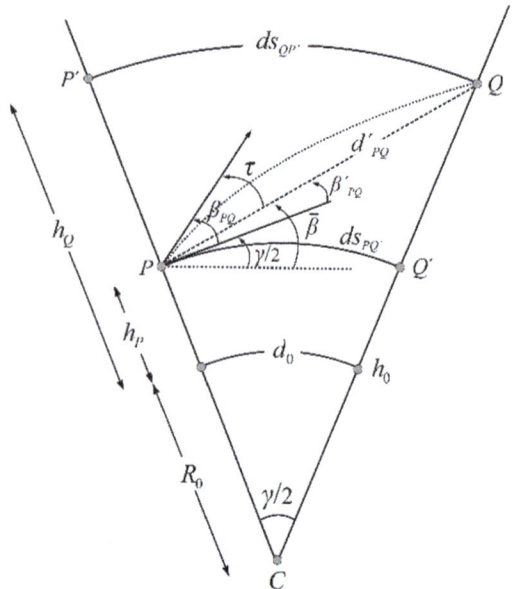

According to Fig. 7.8, the effectively measured angle β_{PQ} with the surveying instrument installed at point (P) differs from the geometric angle β'_{PQ}, as given by Eq. (7.27).

$$\beta_{PQ} = \beta'_{PQ} + \tau \qquad (7.27)$$

The angle τ is the *angle of refraction,* and its value varies with the pressure, humidity and temperature, making its computation difficult. Empirical observations have shown, however, that it is proportional to the curvature of the Earth and can be calculated by Eq. (7.28).

$$\tau = k * \frac{\gamma}{2} = k * \frac{d'_{PQ}}{2R_0} \qquad (7.28)$$

The coefficient of refraction k in Eq. (7.28) is defined as the ratio between the local mean radius R_0 of the Earth and the radius of the line-of-sight r, as given by Eq. (7.29).

$$k = \frac{R_0}{r} \qquad (7.29)$$

Note that a positive sign of the coefficient of refraction indicates a convex shape of the ray of light, i.e., it follows the Earth's curvature.

The value of k varies considerably depending on the atmospheric conditions. It can assume different values depending on factors such as the height at which the line of sight is from the ground surface, the type of vegetation, and atmospheric variation along the day, among others. Field research suggests values ranging from -4 to $+16$ for different atmospheric conditions, different distances and different heights from the ground level. In practice, however, in the absence of a precise value, it is recommended to adopt the value $k = 0.13$, determined by Gauss and often used as a default value by surveyors. However, reducing the uncertainty about the value of k is recommended using reciprocal observations, as presented in Sect. 7.2.11, especially in measurements close to the ground surface.

Due to the effect of vertical atmospheric refraction, the equations given in the previous section for calculating the horizontal and ellipsoidal distances must be modified to account for the measured vertical angle. In this respect, there is a series of equations proposed by different authors. In this book, three of them are presented as follows.

7.2.8 Calculation of the Ellipsoidal Distance Considering the Difference in Altitude between the Endpoints

When the value of the difference in altitude between the endpoints, measured by differential levelling, is known, the value of the ellipsoidal distance can be calculated by Eq. (7.30), which is considered a rigorous mathematical equation for this calculation.

$$d_0 = \sqrt{\frac{\left(d'_{PQ}\right)^2 - \Delta H^2}{\left(1 + \frac{H_P}{R_0}\right) * \left(1 + \frac{H_Q}{R_0}\right)}} \tag{7.30}$$

7.2.9 Calculation of the Ellipsoidal Distance Considering the Curvature of the Earth, Vertical Atmospheric Refraction, and Deviation of the Vertical

For the accurate calculation of the ellipsoidal distance, it is also necessary to consider the effect of the deflection of the vertical θ at the station point, since the slope distance is measured using surveying instruments levelled concerning the vertical line, and the ellipsoidal distance is calculated by reducing the slope distance, according to the normal line. Under these conditions, the ellipsoidal distance can be calculated using Eq. (7.31).

$$d_0 = R_0 * \mathrm{atan}\left(\frac{d'_{PQ} * \sin\left(z_{PQ} + \theta + \frac{k*d'_{PQ}}{2R_0}\right)}{R_0 + h_P + d'_{PQ} * \cos\left(z_{PQ} + \theta + \frac{k*d'_{PQ}}{2R_0}\right)}\right) \tag{7.31}$$

If the deflection of the vertical is unknown, the value of θ in Eq. (7.31) must be equal to zero.

7.2.10   Calculation of the Ellipsoidal Distance through Reciprocal and Simultaneous Observations

The value of the coefficient of refraction k used in the equations presented in the previous sections is difficult to determine. Therefore, whenever possible, its use is avoided in Civil Engineering projects. For this reason, to eliminate the effect of the Earth's curvature and atmospheric refraction, it is recommended to use the reciprocal and simultaneous observations method, as shown in Fig. 7.9.

This measurement method requires the joint work of two operators with two surveying instruments stationed at the endpoints of the distance to be measured. At a given moment, both operators perform vertical measurements on the objective lens of each measuring instrument, configuring the geometric situation indicated in Fig. 7.9.

Fig. 7.9 Geometric relationship between reciprocal views

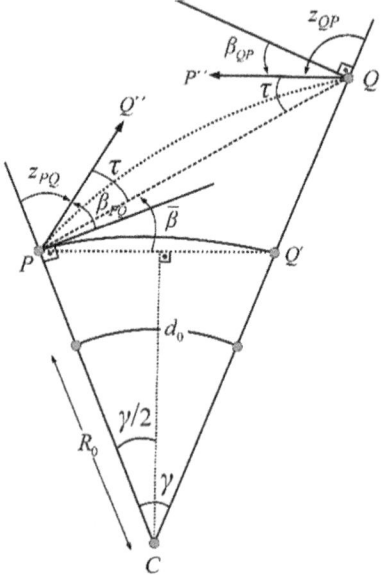

In Fig. 7.9, the dotted line represents the refracted line of sight between points (P) and (Q). For the operator stationed at (P), the point (Q) will appear in the direction PQ'', and for the operator placed at (Q), the point (P) will appear in the direction QP''. As they are observed simultaneously, the angles $Q''PQ = PQP'' = \tau$ are considered equal, where τ is the angle of refraction at (P) and (Q).

According to Fig. 7.9 and disregarding the negative sign of β_{PQ}, the following mathematical relationship can be expressed as

$$90° + \beta_{PQ} - \tau + \gamma + 90° - \beta_{QP} - \tau = 180° \qquad (7.32)$$

Hence,

$$\tau = \frac{\beta_{PQ} - \beta_{QP} + \gamma}{2} \qquad (7.33)$$

From Fig. 7.8,

$$\bar{\beta} = \beta_{PQ} + \left(\frac{\gamma}{2}\right) - \tau \qquad (7.34)$$

And then,

$$\bar{\beta} = \left(\frac{\beta_{PQ} + \beta_{QP}}{2}\right) = \left(\frac{z_{QP} - z_{PQ}}{2}\right) \qquad (7.35)$$

Similarly to the previous cases, considering Eqs. (7.18) to (7.20) and (7.23) to (7.25), the values of horizontal and ellipsoidal distances are obtained without the intervention of the vertical refraction coefficient, as indicated below.

$$d_{PQ'} = d'_{PQ} * \cos\left(\frac{z_{QP} - z_{PQ}}{2}\right) - \Delta h * \sin\left(\frac{\gamma}{2}\right) \qquad (7.36)$$

$$d_{QP'} = d'_{PQ} * \cos\left(\frac{z_{QP} - z_{PQ}}{2}\right) + \Delta h * \sin\left(\frac{\gamma}{2}\right) \qquad (7.37)$$

$$d_m = d'_{PQ} * \cos\left(\frac{z_{QP} - z_{PQ}}{2}\right) \qquad (7.38)$$

$$d_0 = \left(\frac{R_0}{R_0 + h_P}\right) * \left[d'_{PQ} * \cos\left(\frac{z_{QP} - z_{PQ}}{2}\right) - \Delta h * \sin\left(\frac{\gamma}{2}\right)\right] \qquad (7.39)$$

$$d_0 = \left(\frac{R_0}{R_0 + h_Q}\right) * \left[d'_{PQ} * \cos\left(\frac{z_{QP} - z_{PQ}}{2}\right) + \Delta h * \sin\left(\frac{\gamma}{2}\right)\right] \qquad (7.40)$$

$$d_0 = \left(\frac{R_0}{R_0 + h_m}\right) * d'_{PQ} * \cos\left(\frac{z_{QP} - z_{PQ}}{2}\right) \qquad (7.41)$$

$$\text{where} \quad \Delta h = d'_{PQ} * \sin\left(\frac{z_{QP} - z_{PQ}}{2}\right) \tag{7.42}$$

Note It is important to point out that the ellipsoidal distance calculated considering the curvature of the Earth and the vertical atmospheric refraction is always close to the altitude of the target point. In this case, for an approximate calculation of the ellipsoidal distance, one can use Eq. (7.1) for horizontal distance calculation and the height of the target point for distance reduction. See the result of Example 7.4.

7.2.11 Calculation of the Value of the Vertical Refraction Coefficient k

Calculating the value of the coefficient of vertical refraction is a topic of constant discussion. Different authors propose several methods in the literature, ranging from meteorological measurements to geometric measurements with total stations. For this book, considering Geomatics applied to Civil Engineering, two methods are presented: the first is based on the difference in altitude between the measured points, and the second is based on the reciprocal and simultaneous observations described in the previous section.

7.2.11.1 Calculation of k Knowing the Difference in Altitude Between the Measured Points

To understand this calculation procedure, the reader should consult (Chap. 13). As described in Sect. 13.8.2, the value of the vertical refraction coefficient k can be determined using Eq. (7.43), derived from Eq. (13.49).

$$k = 1 - \frac{2R_0 * \left(\Delta H_{PQ} - \Delta H'_{PQ} - h_i + h_r\right)}{d^2_{PQ}} \tag{7.43}$$

where

R_0= local mean radius of the Earth
ΔH_{PQ}= altitude difference between points (P) and (Q) calculated by differential levelling[1]
h_i= instrument height
h_r= reflector prism height

---

[1] For more details, see Sect. 13.3.

$\Delta H'_{PQ} =$ vertical distance between points (P) and (Q), given by Eq. (7.3)

$d_{PQ} =$ horizontal distance between points (P) and (Q)

In principle, the value of k, using Eq. (7.43), can be determined by a single measurement from (P) to (Q) or the other way around. However, due to the different atmospheric conditions at the two ends, the values of k calculated from point (P) and point (Q) are different. The solution, in this case, is to perform the reciprocal and simultaneous observations method and calculate the value of k as the average of the values obtained from (P) and from (Q). See Example 7.5 for more details.

7.2.11.2 Direct Calculation of the Value of k Through Reciprocal and Simultaneous Observations

When reciprocal and simultaneous observations are carried out between the two endpoints, as shown in Fig. 7.9, the following mathematical relationship can be expressed as

$$k * \frac{\gamma}{2} = \frac{\beta_{PQ} - \beta_{QP} + \gamma}{2} \tag{7.44}$$

From which the value of the coefficient of refraction k is given as follows.

$$k = \frac{\beta_{PQ} - \beta_{QP} + \gamma}{\gamma} = 1 + \frac{R_0}{d'_{PQ}} * (\beta_{PQ} - \beta_{QP}) = 1 + \frac{R_0}{d'_{PQ}}$$

$$* (\pi - z_{PQ} - z_{QP}) = 1 - \frac{R_0}{d'_{PQ}} * (z_{PQ} + z_{QP} - \pi) \tag{7.45}$$

where d'_{PQ} is the slope distance measured between points (P) and (Q).

It is important to note that the value of k derived from Eq. (7.45) is a mean value along the line of sight between (P) and (Q).

Considering Eq. (7.45) and applying the general law of error propagation, the precision s_k of the coefficient of refraction can be calculated as a function of the precision of the zenith angle s_z.

$$s_k = \frac{\sqrt{2} R_0 * s_z}{\rho'' * d'} \tag{7.46}$$

where $\rho'' = 206{,}264.8062''$ and s_z is given in seconds of arc.

It is interesting to note the inverse relationship between the precision s_k and the distance d', which means that the larger the distance, the more precise the determination of k.

Note As seen in earlier sections of this chapter, the effect of Earth's curvature and vertical atmospheric refraction on the determination of horizontal and ellipsoidal distance depends on the length of the slope distance measured and the altitude difference between the measured points. For this reason, in Civil Engineering works where the measured distance values are only a few hundred metres long, and the terrain relief is not very rough, the adoption of the local ground-based plane has little influence on the values of the horizontal distances. However, adopting more accurate calculation strategies for control point networks is recommended. Even so, some professionals prefer to disregard these effects, accepting them as accidental errors of observation in network adjustments. In addition, it is important to note that most total stations already have standard corrections for refraction and Earth's curvature built into their operating system. The reader should be aware of this fact and consult the Instruction Manual of his instrument to verify how these reductions are applied to their measurements. In addition, it is essential to ensure which coefficient of refraction is used in the equations and whether the instrument's operating system allows changes to be made. The use of unrealistic coefficients of refraction tends to impair the calculated value of the horizontal distance and, consequently, the ellipsoidal distance.

7.2.12   Grid Distance

In cases where the use of the local ground-based plane is not recommended or for implementation of control point networks connected to pre-existing geodetic networks, it is necessary to reduce the slope distance to the ellipsoid, as presented in the previous sections and then to a projection plane, as shown in Chap. 16. The new reduced distance on the cartographic projection plane is called the *grid distance*. Because it is a distance affected by a scale factor, which varies from one point to another on the ellipsoid's surface, its calculation requires applying equations related to the geometry of the ellipsoid, also presented in detail in Chap. 16.

Example 7.5

A field survey was carried out using the reciprocal and simultaneous observations method to apply the mathematical formulations presented in this chapter. Two total stations with 1 arc second angular precision were installed at points (P) and (Q), as indicated in Table 7.5. For vertical angle measurements, observations were made aiming at the objective lens of the opposite total station. The slope distance was measured using electronic distance measurement, as presented in Sect. 8.3.2. Field-measured values are shown in Table 7.5. A differential levelling between points (P) and (Q) was also performed, through which the orthometric altitudes $H_P = 700.456$ m and $H_Q = 799.736$ m were obtained. It is known that the local mean radius of the Earth is equal to 6,362,745 m. Considering the measured values, calculate the value of the coefficient of refraction and the horizontal and ellipsoidal distances by applying each of the equations presented in this chapter. The geoid undulation at the measurement site is −6.27 m.

Table 7.5 Values measured in the field

Station	Height of instrument [m]	Target point	Altitude angle	Zenith angle
P	1.269	Q	5°27′01.0″	84° 32′ 59.0″
Q	1.273	P	5°27′2.0″	95°27′2.0″
Slope distance $d'_{PQ} = 1{,}045.0220$ $d'_{QP} = 1{,}045.0233$				

Solution:

1. *Calculation of the coefficient of refraction k*

    1.1 *Considering the difference in altitude between the measured points*

    Using the values indicated in Table 7.5.

$\Delta H_{PQ} = 799.736 - 700.456 = 99.280\,\text{m}$ $\Delta H_{QP} = 700.456 - 799.736 = -99.280\,\text{m}$

$\Delta H'_{PQ} = 1{,}045.020 * \cos(84°32'59.0'') = 99.2582\,\text{m}$

$\Delta H'_{QP} = 1{,}045.0233 * \cos(95°27'22.0'') = -99.3642\,\text{m}$

Using Eq. (7.43),

$$k_P = 1 - \left[\frac{2 * 6{,}362{,}745.000 * (99.280 - 99.2582 - 1.269 + 1.273)}{1040.2975^2}\right] = 0.696$$

$$k_Q = 1 - \left[\frac{2 * 6{,}362{,}745.000 * (-99.280 - 99.3642 - 1.273 + 1.269)}{1040.2886^2}\right] = -0.057$$

$$k = \left[\frac{0.696 + (0.057)}{2}\right] = 0.377$$

1.2 *Through reciprocal and simultaneous observations method*

   Using Eq. (7.45).

$$k = 1 - \left[\frac{6{,}362{,}745.000}{1045.0220} * (1.475671034 + 1.666023430 - \pi)\right] = 0.380$$

Considering that the angular precision of the measuring instrument is equal to 1″, the precision of k can be calculated using Eq. (7.46).

$$s_k = \frac{\sqrt{2}}{206{,}264.8062} * \frac{6{,}362{,}745 * 1''}{1{,}045.022} = \pm 0.042$$

2. *Calculation of the ellipsoidal distance*

   2.1 *Considering the difference in altitude between the measured points*

    Using Eq. (7.30).

$$d_0 = \sqrt{\frac{1045.022^2 - 99.280^2}{\left(1 + \frac{700.456}{6,362,745.000}\right) * \left(1 + \frac{799.736}{6,362,745.000}\right)}} = 1040.173 \text{ m}$$

2.2 *Using* Eq. (7.31)

Considering the deflection of the vertical equal to zero.
From point (P):

$$d_0 = 6,362,745.000 * \operatorname{atan}\left[\frac{1045.022 * \sin\left(1.475671034 + \dfrac{0.380 * 1,045.022}{2 * 6,362,745.000}\right)}{6,362,745.000 + 695.443 + 1045.022 * \cos\left(1.475671034 + \dfrac{0.380 * 1,045.022}{2 * 6,362,745.000}\right)}\right]$$

$d_0 = 1040.171 \text{ m}$

   From point (Q):

$$d_0 = 6,362,745.000 * \operatorname{atan}\left[\frac{1045.023 * \sin\left(1.666023430 + \dfrac{0.380 * 1045.023}{2 * 6,362,745.000}\right)}{6,362,745.000 + 794.719 + 1,045.023 * \cos\left(1.666023430 + \dfrac{0.380 * 1045.023}{2 * 6,362,745.000}\right)}\right]$$

$d_0 = 1040.171 \text{ m}$

2.3 *Through reciprocal and simultaneous observations*

   Using Eqs. (7.39) to (7.42).

$$\Delta h = 1045.022 * \sin\left(\frac{95°27'22.0'' - 84°32'59.0''}{2}\right) = 99.3111 \text{ m}$$

$$d_0 = \left(\frac{6,362,745.000}{6,362,745.000 + 695.435}\right) * \left[1045.022 * \cos\left(\frac{95°27'22.0'' - 84°32'59.0''}{2}\right)\right.$$

$$\left. - 99.3111 * \sin(0°00'16.9'')\right]$$

$$d_0 = 1040.171 \, \text{m}$$

$$d_0 = \left(\frac{6,362,745.000}{6,362,745.000 + 794.719}\right) * \left[1045.023 * \cos\left(\frac{95°27'22.0'' - 84°32'59.0''}{2}\right)\right.$$

$$\left. + 99.3111 * \sin(0°00'16.9'')\right]$$

$$d_0 = 1040.171 \, \text{m}$$

$$d_0 = \left(\frac{6,362,745.000}{6,362,745.000 + 745.077}\right) * 1045.022 * \cos\left(\frac{95°27'22.0'' - 84°32'59.0''}{2}\right)$$

$$d_0 = 1040.171 \, \text{m}$$

The horizontal distances at (P) and at (Q), considering the curvature of the Earth and the vertical atmospheric refraction, can be calculated using the ellipsoidal distance.

$$d_{PQ} = 1040.171 * \left(\frac{6,362,745.000 + 695.435}{6,362,745.000}\right) = 1040.284 \, \text{m}$$

$$d_{QP} = 1040.171 * \left(\frac{6,362,745.000 + 794.719}{6,362,745.000}\right) = 1040.301 \, \text{m}$$

If the curvature of the Earth and the atmospheric refraction are disregarded, the horizontal distances calculated by Eq. (7.1) are equal to

$$d_{PQ} = 1045.022 * \sin(84°32'59.0'') = 1040.297 \, \text{m}$$

$$d_{QP} = 1045.023 * \sin(95°27'22.0'') = 1040.289 \, \text{m}$$

The above values indicate that not considering the Earth's curvature and atmospheric refraction would cause an error of 12.3 ppm in the horizontal distance calculated by Eq. (7.1).

Note that as the slope distance and the difference in altitudes between endpoints decrease, the error by not considering the curvature of the Earth and the vertical atmospheric refraction also decreases. For the hypothetical case that the measurements in this example are $d'_{PQ} = 500.000$ m and $\beta = 2°00'00''$, the error would be 3 ppm.

7.3 Problems

1. Point (Q) is to be stakeout at a horizontal distance of 512.394 m with a total station set up at point (P). Assuming that the vertical zenith angle read in the total station is $93°22'56''$, determine the slope distance to be measured in the field.

2. A slope distance of 340.725 m was measured from point (P) to (Q) using a total station. Knowing that the elevations and the heights of the instrument and the reflector prism at (P) and (Q) were $H_A = 875.127$ m, $H_B = 870.693$ m, $h_i = 1.580$ m and $h_r = 2.150$ m, respectively, determine the horizontal distance.

3. In Problem 1, assume that the slope distance can be measured with a precision of ±(2 mm + 2 ppm). Determine the allowable tolerance in the zenith angle to obtain a horizontal distance within ±12 mm.

4. In Problem 1, assume that the zenith angle can be measured with a precision of ±7''. Determine the allowable tolerance in the slope distance to have a horizontal distance within ±10 mm.

5. Considering the geometric elements shown in Fig. 7.10 and the observation values in Table 7.6, calculate the distance ΔH_{PQ}.

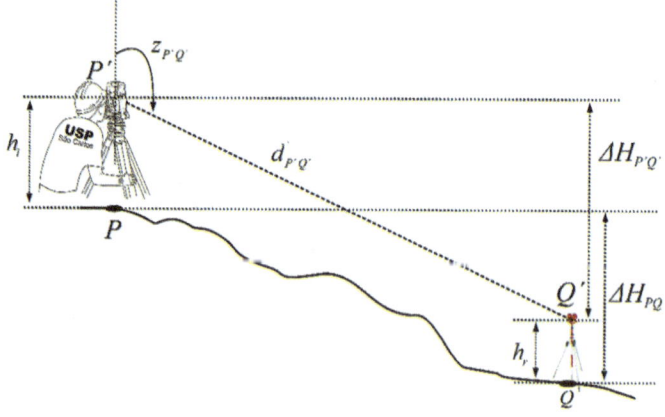

Fig. 7.10 Measurement geometry

Table 7.6 Measured values

Geometric element	Observation
Zenith angle $Z_{P'Q'}$	$93°33'51''$
Slope distance $d'_{P'Q'}$	428.177 m
Instrument height h_i	1.650 m
Reflector prism height h_r	2.150 m

6. Knowing that the spherical distance between points (P) and (Q) on the surface of the Earth is 2645.289 metres, calculate the ellipsoidal distance between them, assuming that they are at an orthometric altitude of 995.531 metres and that point (P) is at a latitude of $27° 19'09''$S.

7. Given that the geoidal height N at points (P) and (Q) in the previous problem is 5.65 m, calculate the ellipsoidal distance using the ellipsoidal height instead of the orthometric altitude. Compare the results.

8. Knowing that the distance between two points at sea level is equal to 1242.536 m, calculate what would be the distance between them at an elevation of 815.497 m. Assume that the local mean radius of Earth at the latitude of measurement is 6371 km.

9. Calculate the difference between the spherical distance and the chord, and between the spherical distance and the horizontal distance, between two points separated by a distance of 5478.259 metres at an altitude of 818.154 metres. Assume that the local mean radius of the Earth at the latitude of measurement is 6371 km.

10. In Problem 9, assuming that the elevation of the measurement location is 675.000 m, determine the corresponding ellipsoidal distance.

11. A slope distance of 1000 m and a zenith angle of $93°45'12''$ has been measured using a total station. Assuming that the elevation of the measuring point is $H = 725.000$ m and the local mean radius of the Earth at the latitude of the measurement is 6371 km, calculate the difference in the horizontal distance at the instrument level and the target point level.

12. Supposing that a slope distance of 2000.000 m has been measured with a vertical altitude angle of $\beta = 8°44'13''$, calculate the horizontal distance considering and ignoring the curvature of the Earth and the vertical atmospheric refraction, assuming a refraction coefficient $k = 0.13$ and $R_0 = 6371$ km. Compare results.

13. Explain why horizontal and spherical distances at ground levels are considered equivalent quantities for Geomatics applied to Civil Engineering.

14. Explain the calculation steps to determine ellipsoidal distance from slope distance.

15. Discuss the difference between horizontal distance, ellipsoidal distance and grid distance and when they are used in Civil Engineering projects.

16. Discuss when the curvature of the Earth and the vertical atmospheric refraction can be ignored in Civil Engineering projects.

17. Explain what can be done to avoid the effects of the curvature of the Earth and vertical atmospheric refraction in precise distance measurements.

References

Anderson, J. M., Mikhail, E. M. (1998). *Surveying, Theory and Practice.* 7th Edition. WCB/ McGraw-Hill, Boston, USA.

Burkholder, E. F. (1991). *Computation of Horizontal/Level Distances.* Journal of Surveying Engineering—ASCE, Vol 117, N 3, pages 104-116.

Burkholder, E. F. (2004). *Accuracy of Elevation Reduction Factor.* Journal of Surveying Engineering—ASCE, Vol 130, N. 3, Pages 134-137.

Burkholder, E. F. (2016). *Using the Global Spatial Model to Compute Combined Factors*. Journal of Surveying Engineering, ASCE, ISSN 0733-9453.

Milles, S., Lagofun, J. (2011). *Topographie et topométrie modernes*. Eyrolles, Paris.

Rüeger, J. M., (1996). *Electronic Distance Measurement—An Introduction*. Fourth Edition. Springer-Verlag Berlin Heidelberg, Germany.

Scherrer, R. (1995). *Reduction of distance measured with infra-red EDM instruments*. Wild Heerbrugg—White paper, Switzerland.

Schofield, W. (2001). *Engineering Surveying*. 5th Edition. Butterworth Heinemann. Oxford. England.

Vicenty, T. (1986). *Geometric Reduction of Measured Lines*, Surveying and Mapping, Vol. 46, No. 3.

Chapter 8
Distance Measurement

Irineu da Silva and Paulo C. L. Segantine

8.1 Introduction

Measuring a distance consists of performing operations to determine the linear dimension of an object or the length of a line between two points on the ground surface. For this purpose, engineers have several measuring instruments at their disposal, the choice of which depends on several factors, such as the desired accuracy, the roughness of the terrain, the type of object to be measured, the distance length, and the operator's skill. In this context, distance measurement methods available for Geomatics applied to Civil Engineering can be divided into two classes, according to the surveying instruments used for the measurement.

- Direct distance measurement methods
- Indirect distance measurement methods

8.2 Direct Distance Measurement Methods

Direct distance measurement refers to any method that directly compares the distance to be measured with a predefined measurement standard. Among these methods, the most commonly used in Civil Engineering works are presented in the following subsections.

I. da Silva (✉) · P. C. L. Segantine
São Carlos School of Engineering, University of São Paulo, São Carlos, SP, Brazil
e-mail: irineu@sc.usp.br; pclsegantine@usp.br

© The Author(s), under exclusive license to Springer Nature Switzerland AG 2025   237
I. da Silva, P. C. L. Segantine, *Geomatics Applied to Civil Engineering*,
https://doi.org/10.1007/978-3-031-75737-2_8

8.2.1   Pacing

Pacing is the process of walking the distance to be measured and counting the number of paces taken to cover that distance. It is considered practical and satisfactory on reconnaissance surveys where an estimated value is acceptable.

Pace length varies according to the individual's height, walking speed, and physical endurance, as well as the slope of the terrain. In general, it is estimated that the pace length of a man 1.70 meters tall is about 78 centimetres and that for every 5 cm difference in height, the pace length increases or decreases by one centimetre. Studies show that this measurement method has a relative accuracy between 1/50 and 1/100, depending on the roughness of the terrain, the amount of vegetation, and the surveyor's experience.

Paces are usually counted using a digital *pedometer* equipped with a gear and an electronic circuit that records the operator's movement and, therefore, the number of paces.

8.2.2   Wheel Odometer

The *wheel odometer*[1] is an instrument that counts the number of revolutions of a wheel to measure a distance along a surface, as shown in Fig. 8.1. The distance can be determined by multiplying the number of revolutions of the wheel by its circumference. This equipment gives an accuracy of about 1/200 when the measuring surface is smooth, as in the case of paved surfaces.

Fig. 8.1 Example of wheel odometer

---

[1] Also known as the surveyor's wheel.

8.2.3   Tape Measure

Taping is a classic method used daily by professional in Geomatics. Although the use of handheld laser distance meters and total stations has become more common, in many cases, Civil Engineers still prefer the traditional tape measure due to its practical use.

In the case of land surveying, most distance measurements are carried out using a total station. However, using a traditional tape measure or even a handheld laser distancemeter can be easier in setting-out works or small distance measurements. In this context, some relevant technical and operational details about these measuring instruments are presented below.

8.2.3.1   Traditional Fibreglass Measuring Tapes

Fibreglass measuring tapes, as the name suggests, are made of fibreglass bonded to polyvinyl chloride, which ensures durability, flexibility, and thermal and chemical resistance. They consist of a graduated tape (in metres, centimetres, and millimetres) rolled inside a protective container, as shown in Fig. 8.2.

Fig. 8.2 Examples of fibreglass measuring tapes

Commercial fibreglass tapes are available in lengths of 15, 30 or 50 m, with the 30-m length most commonly used in Civil Engineering applications. The accuracy achieved when measuring a distance with this type of tape depends on the distance, the shape of the object and the experience of the operator. On average, it is between 1/1000 and 1/5000.

8.2.3.2   Steel Measuring Tapes

Like fibreglass tape, steel tape is graduated in metres, centimetres, and millimetres. As the name suggests, they are made from steel strips and are sometimes referred to as engineer's or surveyor's tape. Typically, pocket steel tapes are manufactured in lengths of 3–5 m, and long tapes are manufactured in lengths of 20–30 m. See Fig. 8.3. Although more accurate than fibreglass tapes, they are also more difficult to handle and are subject to significant changes in temperature. On average, the measuring accuracy of steel tapes ranges from 1/1000 to 1/10,000.

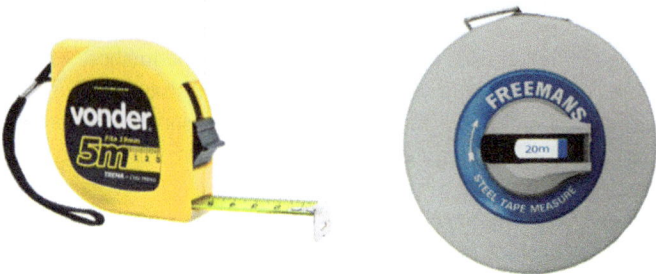

Fig. 8.3 Examples of steel measuring tapes

Measuring distance with a tape in the field is not a simple task. Depending on the distance and the slope of the terrain, the measurement method to be used varies, as shown below.

8.2.4 Taping on Smooth Level Ground

The tape measure procedure for determining the length of lines on smooth or level ground is relatively simple. It consists of stretching the tape between two measuring points on the surface and reading the horizontal distance directly from the zero end of the tape, as shown in Fig. 8.4. The accuracy of this measurement method is approximately ±5 mm for a distance of 50 m.

Fig. 8.4 Taping on smooth, level ground

8.2.5   *Taping on Sloping Ground*

Taping on the sloping ground for horizontal distance measurement can be done in the same way as described for level ground, provided that the slope of the terrain is constant and the slope rate i is known. The horizontal distance d_{PQ} can then be determined by measuring the slope distance d'_{PQ} and using Eq. (8.1).

$$d_{PQ} = \frac{d'_{PQ}}{\sqrt{1 + i^2}} \tag{8.1}$$

It is estimated that the accuracy of this measurement method is approximately ± 10 mm at a distance of 50 m.

Example 8.1

A slope distance of 254.785 m was measured on a site with an average slope rate of 2%, as shown in Fig. 8.5. Calculate the corresponding horizontal distance.

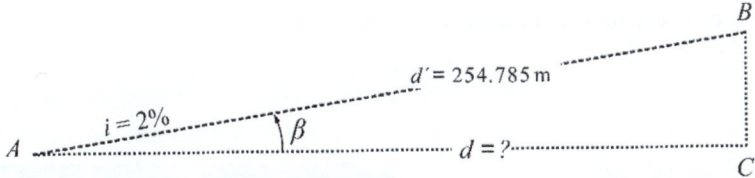

Fig. 8.5 Slope correction

Solution:

   Using Eq. (8.1)

$$d = \frac{254.785}{\sqrt{1 + 0.02^2}} = 254.734 \text{ m}$$

The problem can also be solved by considering that the tangent of a line represents its slope.

$$\tan(\beta) = 2\% = 0.02 \quad \rightarrow \quad \text{atan}(0.02) = 1°\,08'44.7''$$

$$d = d' * \cos(\beta) = 254.785 * \cos(1°\,08'44.7'') = 254.734 \text{ m}$$

8.2.6 *Horizontal Taping on Irregular Ground*

If the terrain is uneven, using any of the methods described in the previous sections may be difficult. In this case, in addition to the tape, the operator should use a range pole and a staff bubble, as shown in Fig. 8.6, and follow the measurement procedures shown in Figs. 8.7 and 8.8.

Depending on the operator's skill, a plumb line can replace the range pole at the lowest end of the terrain (left side of Fig. 8.7). However, depending on the slope of the terrain, this procedure can make the measurement much slower and less accurate due to the difficulty of keeping the plumb bob over the measurement point.

To measure the distance, the operator at the highest end of the terrain must place the zero end of the tape directly over the measuring point on the ground. In turn, the operator at the lower end must place a range pole over the measurement point on the ground and plumb it using the staff bubble. Then, by moving the measuring end of the tape in the vertical plane over the range pole, it is possible to find the horizontal distance, which is the shortest length reading. See Fig. 8.8.

The quality of this measurement procedure depends on the skill of the operator and the precautions taken in the field. In the best case, an accuracy of the order of ±10 mm can be achieved for distances up to 50 m.

Fig. 8.6 Range pole and staff bubble

Fig. 8.7 Taping on uneven ground

Fig. 8.8 Shortest distance determination

8.2.7 Blunders, Systematic and Random Errors in Taping

The primary sources of error affecting the quality of taping linear distances are presented below:

1. *Error counting the number of intermediate measurements made with the tape.*
2. *Tape reading error:* This is a blunder resulting from the operator's lack of attention. It is recommended that the operator gets used to observing the adjacent figures on the tape during reading to avoid this type of error.
3. *Reading uncertainty due to the thickness of the range pole:* This is a random error due to the operator's difficulty in using the range pole axis as a reference line.
4. *Error due to incorrect tape length:* Although measuring tapes are meticulously manufactured, their nominal length rarely corresponds exactly to the values specified by the manufacturer, mainly when used over a long period of time. It is recommended to verify the tape length periodically to avoid this type of systematic error.
5. *Error due to temperature variation:* Measuring tapes are manufactured and verified at a standard temperature specified by the manufacturer, usually between 15°C and 20°C. Therefore, a variation in their lengths (systematic error) will occur when they are submitted to a different temperature. The value of the length variation can be calculated using Eq. (8.2). A higher temperature means that the measured distance is shorter than indicated and vice versa.

$$\Delta d = d_0 * \delta * \Delta t \qquad (8.2)$$

where

Δd = tape length correction at the time of measurement
d_0 = nominal length of the tape
δ = coefficient of thermal expansion. For steel measuring tapes, this coefficient is
    equal to $0.0000116°C^{-1}$
Δt = temperature difference between the standard temperature specified by the
    manufacturer and the temperature at the time of measurement

6. *Error due to tape sagging:* this is a systematic error due to the difficulty of keeping the tape stretched during the measurement. Due to its weight, the tape will sag and take the form of a catenary, as shown in Fig. 8.9. A sag correction must, therefore, be applied. Specific equations for this type of correction can be found in the literature.
7. *Error due to variations in tension:* When a tape is verified, it is subjected to a certain tension at its ends (usually specified by the manufacturer). Thus, each time a greater or lesser tension is applied, the tape is lengthened or shortened accordingly (systematic error). Specific equations for this type of correction can be found in the literature.

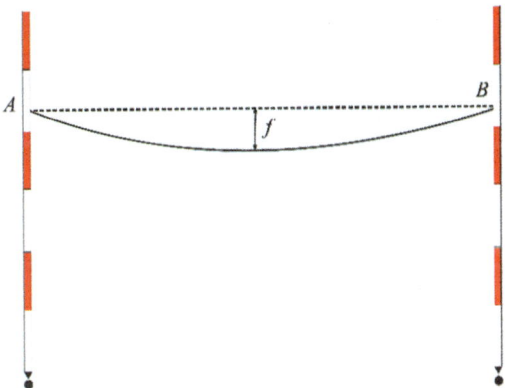

Fig. 8.9 Distance measurement error due to tape sagging

8. *Error due to the range pole not being held vertically*: This is one of the most common types of random error that occurs when measuring distances with a tape, as shown in Fig. 8.10. To minimise this, the operator must use a staff bubble, as shown in Fig. 8.6 and, wherever possible, make the measurement using the lowest part of the range pole. Note that even with a staff bubble, it is not always possible to keep the range pole vertical, especially on steep terrain or windy days.

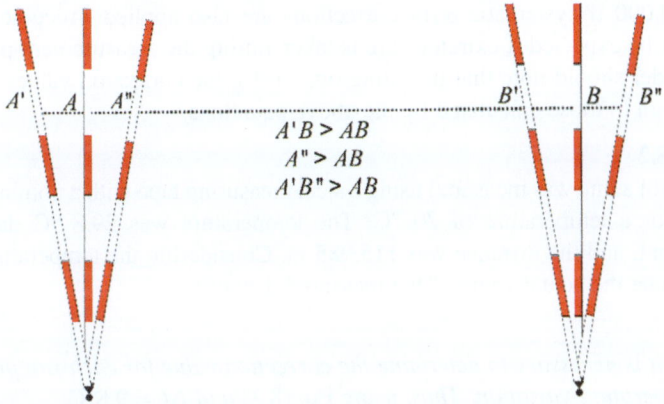

Fig. 8.10 Distance measurement error due to the range pole not being held vertically

9. *Error due to imperfect taping alignment*: this is a random error resulting from the horizontal misalignment of the intermediate range poles with respect to the two end points of the measurement, as shown in Fig. 8.11.

Typically, this type of error has little effect on the total length of the distance. The value of the variation in the measured distance is given by Eq. (8.3), where e is the misalignment error and d is the measured distance,

$$\Delta d = \frac{e^2}{2d} \tag{8.3}$$

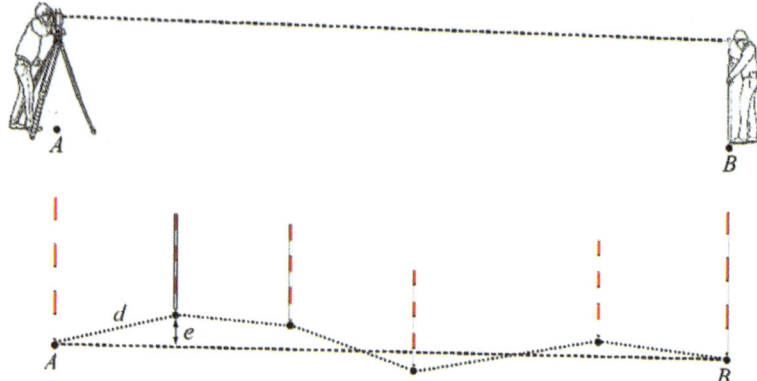

Fig. 8.11 Distance measurement error due to imperfect taping alignment

For example, for a horizontal misalignment of 32 cm at a distance of 50 metres, the variation in the measured distance is only 1.0 mm. The distance to be considered is, therefore, equal to 49.999 metres. Note that a misalignment error of the order of 30 cm should be considered a blunder.

In summary, considering all the possible errors that can occur, traditional tape measures achieve accuracies of about 1:1000 if only the blunders are corrected and about 1:10,000 if systematic error corrections are also applied. Accuracies of 1: 50,000 can be expected if extreme care is taken during the measurement process.

The reader should note that in setting-out works, the correction values have the opposite sign to those calculated by the above equations.

Example 8.2

The length of a line was measured using a steel measuring tape with a nominal length of 20 m for a temperature of 20 °C. The temperature was 29.8 °C during the measurement, and the distance was 115.985 m. Considering the temperature variation, calculate the actual value of the measured distance.

Solution:

Firstly, it is necessary to determine the correction value for each run performed due to temperature variation. Thus, using Eq. (8.2) *and* $\Delta t = 9.8°C$.

$$\Delta d = 20 * 0.0000116 * 9.8 = 0.00227 \text{ m} = 2.3 \text{ mm}$$

It took 5.8 runs to measure the line length. Therefore, the total correction for temperature variation is -13.34 mm. *The actual line length is* 115.972 m

8.2.8   Handheld Laser Distance Meters

Handheld laser distance meters were introduced to the world market in the 1990s and were quickly adopted by professionals who perform distance measurements on a daily basis. As the name suggests, these instruments emit a visible laser beam reflected by the target at the end of the distance to be measured, as shown in Fig. 8.12.

A handheld laser distance meter cannot be classified as a direct distance measurement method. Nevertheless, it is included in this section for the convenience of the reading as it is a direct replacement for the traditional tape measure.

The quality of measurements taken with a handheld laser distance meter varies depending on the manufacturer and can reach an accuracy of around ±(2 mm + 2 ppm). The range varies from 30 m (standard instruments) to 500 metres (high-performance instruments). Because they are electronic instruments, handheld laser distance meters are similar in size to a mobile phone and consist of a VGA display and a series of keys for various instrument functions.

The primary function of a handheld laser distance meter is to measure slope distances without physical contact with the target. However, some manufacturers incorporate various features to facilitate measurements of horizontal and vertical distances, as well as trigonometric functions that allow the operator to perform quick mathematical operations during the measurement process, such as area and volume calculations.

Fig. 8.12 Distance measurement with a handheld laser distance meter. (Courtesy: Bosch Co. *Location*: (https://www. bosch-professional.com/br/ pt/))

8.3   Indirect Distance Measurement Methods

Indirect distance measurement methods allow a distance to be calculated by measuring indirect quantities without using physical standards to compare measurements. Such a distance measurement method is also called *Tacheometry*. For

Geomatics applied to Civil Engineering, there are three commonly used indirect measurement methods: optical, electronic and GNSS distance measurements.

8.3.1   Optical Distance Measurement Method

This measurement method is based on the similarity of triangles, as shown in Fig. 8.13. In this case, from the similar triangles $\triangle ACB$ and $\triangle DCE$:

$$\frac{CG}{CF} = \frac{AB}{DE} \quad \rightarrow \quad CG = AB * \frac{CF}{DE} \tag{8.4}$$

Thus, knowing the distances AB, CF and DE, it is possible to determine the value of CG.

Two measurement methods were developed using the above principle. The first one was based on the use of a horizontal bar with a fixed AB distance (2 metres), which made it possible to achieve accuracies of between 1/5000 and 1/10,000 in distance measurements, depending on the accuracy of the angular measurement γ and the value of the CG distance. The second measurement method was based on the use of a vertical graduated rod (instead of the horizontal bar) and on the stadia principle developed by the mining engineer William Green in 1778. Later, the German mechanical engineer Georg von Reichenbach (1771–1826) built a telescope called *Stadia*. It consisted of a tube with three horizontal wires called *stadia hairs* or stadia *marks*, positioned at the end of the tube, corresponding to points (D), (F) and (E) in Fig. 8.13. See Fig. 8.14. Therefore, the vertical distance AB can be determined by placing a stadia rod at the end AB and projecting onto it the images of points (D) and (E) in Fig. 8.13. Then, since the distances CF and DE are fixed, substituting Eq. (8.4) in (8.5) gives Eq. (8.6), which allows the distance CG to be calculated knowing the vertical distance AB.

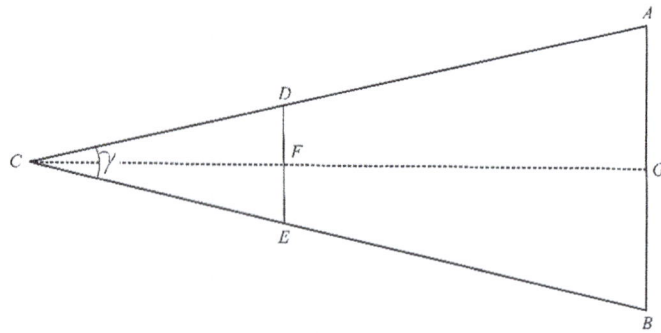

Fig. 8.13 Principle of the optical distance measurement method

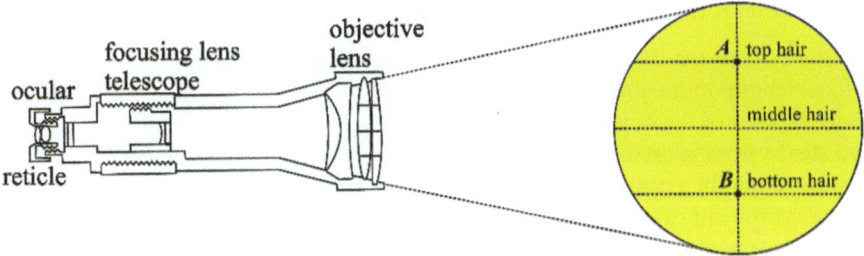

Fig. 8.14 Stadia hairs

$$\frac{CF}{DE} = \text{constant} = w \qquad (8.5)$$

$$CG = AB * w \qquad (8.6)$$

To facilitate calculations, telescopes are manufactured so that the value of w equals 100.

Over the years, surveying instrument manufacturers have perfected the principle of the Stadia method and built telescopes with high-precision lens arrangements, improving the quality and ease of measurement. This type of instrument is called an *Optical Tacheometer*. The mathematical principle of measuring a horizontal distance using such an instrument is as follows.

Consider the case of measuring a distance between two points (*P*) and (*Q*), as shown in Fig. 8.15. The measuring telescope, positioned over point (*P*), is the telescope of a theodolite; the stadia rod, positioned over point (*Q*), is a levelling staff similar to the one used for differential levelling, as described in Sect. 9.6.

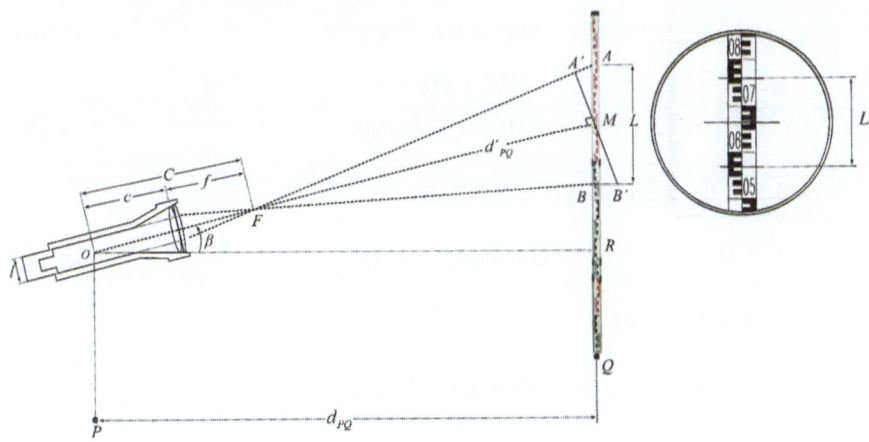

Fig. 8.15 Principle of stadia distance measurement and levelling rod readings

From Fig. 8.15, the following values are of interest:

l: distance between the two extreme stadia hairs on the reticule ring
f: focal length of the objective lens
F: focal point of the objective lens
c: distance from instrument centre (vertical axis) to objective lens centre
$C = c + f$: additive constant of the tacheometer, whose value is always indicated on the instrument cases (usually equal to zero)
$\beta =$ altitude angle
d'_{PQ} slope distance between the instrument centre and the point (M) on the stadia rod
d_{PQ} horizontal distance between the instrument centre and the stadia rod
$AB = L$: difference between upper and lower stadia hairs readings in the rod intercept
M: central reading (projected onto the stadia rod)

The optical beam is focused obliquely on the stadia rod, intersecting it at points (A), (M), and (B). Drawing the segment $A'B'$, perpendicular to OM at point (M), so that (A') is on the extension of FA and (B') is on the segment FB, the triangles $\Delta AA'M$, and $\Delta BB'M$ are obtained. The angles with the point (M) as a vertex are equal to (β) in these two triangles because they have perpendicular sides. To simplify the calculation and because the distances MA' and MB' are very small compared to the distances OA' and OB', the angles at (A') and at (B') are considered to be right angles. Thus, the sides MB' and MA' are considered as cathetus and MB and MA as hypotenuses.

$$MA' = MA * \cos(\beta) \tag{8.7}$$

$$MB' = MB * \cos(\beta) \tag{8.8}$$

$$MA' + MB' = (MA + MB) * \cos(\beta) \tag{8.9}$$

$$MA' + MB' = A'B' \tag{8.10}$$

$$MA + MB = L \tag{8.11}$$

$$A'B' = L * \cos(\beta) \tag{8.12}$$

Considering the right triangle OMR in (R).

$$OR = OM * \cos(\beta) \tag{8.13}$$

According to Eq. (8.6).

$$OM = g * A'B' + C$$

Or,

$$OM = 100 * A'B' + C \tag{8.14}$$

Substituting (8.12) in (8.14) gives

$$OM = 100L * \cos(\beta) + C \tag{8.15}$$

$$OR = [100L * \cos(\beta) + C] * \cos(\beta) \tag{8.16}$$

Since $OR = d_{PQ}$.

$$d_{PQ} = 100L * \cos^2(\beta) + C * \cos(\beta) \tag{8.17}$$

If the telescope is manufactured so that $C = 0$, Eq. (8.17) becomes

$$d_{PQ} = 100L * \cos^2(\beta) = 100L * \sin^2(z) \tag{8.18}$$

where

$z = $ zenith angle

The accuracy of this measurement method depends fundamentally on the vertical angle reading value and the quality of the readings on the stadia rod. A reading error of just 1 mm in the stadia rod will result in a 10 cm error in the horizontal distance, regardless of the length of the measured distance and the accuracy of the angle reading. This distance measurement method was widely used before the advent of electronic distance measurement. It is now rarely used and should only be considered as an option if the operator has no other more accurate measuring instrument.

Example 8.3

A theodolite with a telescope equipped with stadia hairs was installed over point (P) and a stadia rod over point (Q) to determine the horizontal distance between the two points. The readings of the stadia hairs were $FS = 3286$ mm, $FM = 2940$ mm and $FI = 2594$ mm. The measured altitude angle β was $5°32'$. Using the measured values, calculate the horizontal distance between points (P) and (Q).

Solution:

Assuming that the measuring instrument has an additive constant $g = 100$ and $C = 0$, using Eq. (8.18).

$$d_{PQ} = 100 * (3.286 - 2.594) * \cos^2(5°\,32') = 68.557 \text{ m}$$

8.3.2   Electronic Distance Measurement (EDM)

Electronic distance measurement (EDM) was introduced to the market in the 1960s. The measurement principle is based on the direct or indirect observation of the propagation time of a signal carried by an electromagnetic wave. To determine the distance between points (A) and (B), the EDM instrument and a target are positioned over them, as shown in Fig. 8.16. The target can be a reflector prism or a natural surface such as a wall or natural terrain feature with sufficient reflection coefficients. The first case is referred to as a *prism measurement*, and the second is a *reflectorless measurement*.

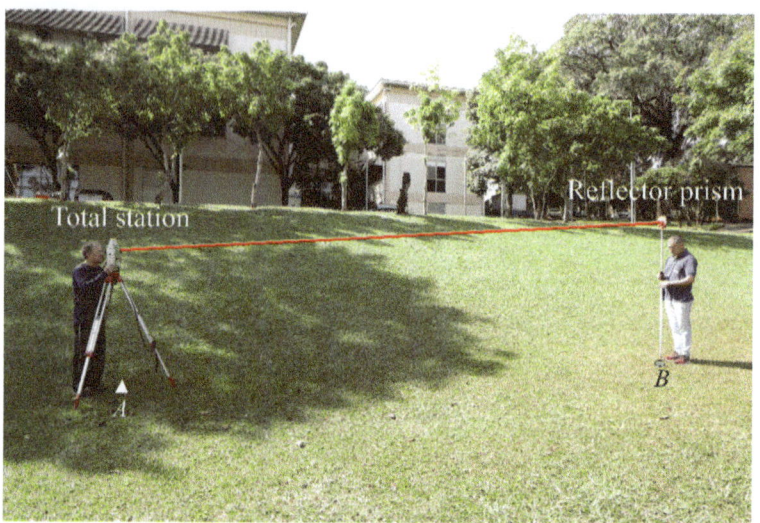

Fig. 8.16 Principle of distance measurement with a total station

EDM instruments are easy to operate and achieve high accuracy at a relatively low cost. Almost all surveying instruments on the market use three EDM methods: *Time-of-flight*, *Phase-shift* and *Wave Form Digitiser* (WFD).

8.3.2.1   Time-*of-Flight* Method

The time-of-flight (or pulse) method calculates the distance based on the time it takes for an electromagnetic pulse to travel from the EDM instrument to the target and back to the instrument, as shown in Fig. 8.17.

In brief, an emitter is positioned at the starting point of the distance d to be measured. A pulse of high-intensity laser radiation is emitted, followed by a timer start. Simultaneously, in the receiver, a photosensor of the EDM instrument receives the signal reflected from the target. The time of flight of the pulse Δt is then used to calculate the distance, as given by Eqs. (8.19) and (8.20).

$$2d = c * \Delta t \tag{8.19}$$

$$d = c * \frac{\Delta t}{2} \tag{8.20}$$

where c is the speed of electromagnetic waves in the medium through which they are transmitted.

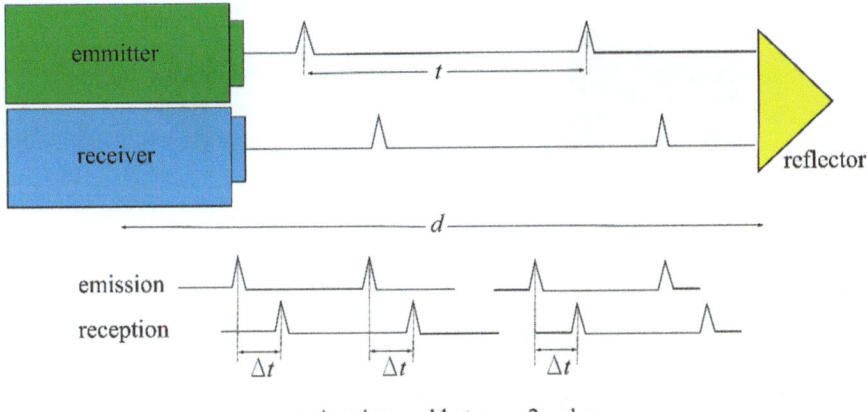

t time interval between 2 pulses

Fig. 8.17 Principle of time-of-flight distance measurement

The propagation speed of electromagnetic waves in a vacuum is a physical constant considered equal to $c = 299{,}792{,}458$ m/s. However, in the atmosphere, propagation speed varies with temperature, humidity, and water vapour pressure, as described later in this chapter. For the sake of simplicity, taking a speed value of 300,000 km/s and a stopwatch accurate to 1 ps $(ps = 10^{-12}s)$, it is possible to measure distances by electronic means with an accuracy of about 1.0 millimetre.

The range and speed of time-of-flight distance measurement depend on the power of the emitted pulse and the type of reflector technology used. When reflector prisms are used, the range can be up to 10 km with a measurement time of approximately 2 to 3 s. For reflectorless measurements, the range can be up to 2.0 km with a measurement time of 10 to 12 s.[2]

8.3.2.2 *Phase-Shift* Method

This method calculates the distance based on the phase shift $\Delta\lambda$ between the emitted and reflected signals and the number of integer wavelengths in the medium, as shown in Fig. 8.18.

In the measurement process, the emitter positioned at point (A) emits an electromagnetic signal with a wavelength equal to λ, divided into two beams. One of the beams is directed to the target at point (B), and the other, called the reference beam, is directed to an internal phase counter. The first beam is then reflected back to the receiver.

---

[2]Different instrument manufacturers may quote different values. As this technology is developing rapidly, the reader is advised to seek updated information whenever necessary.

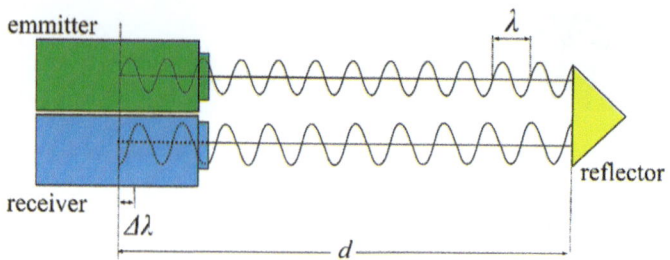

Fig. 8.18 Distance measurement by the phase-shift method

As the signal travels in space, there will be a phase shift $\Delta\lambda$ between the reference and the reflected beams. The phase-shift measurement itself is only unambiguous within one wavelength λ of the signal. Therefore, the number of integer wavelengths N between the two measurement endpoints must be determined to obtain an absolute distance measurement.

Referring to Fig. 8.18, it can be observed that the modulated beam travels twice the distance $2d$ between the endpoints and that the received beam is out of phase $\Delta\lambda$ with the transmitted beam. See Eq. (8.21).

$$2d = N * \lambda + \Delta\lambda \tag{8.21}$$

$$d = \frac{N * \lambda}{2} + \frac{\Delta\lambda}{2} \tag{8.22}$$

The phase-shift $\Delta\lambda$ measurement method uses a digital modulation method called *quadrature amplitude modulation*. The phase shift is measured based on a quartz crystal E' oscillation, as shown in Fig. 8.19.

Current equipment can measure the phase shift of an electromagnetic wave with a resolution of about 1:10,000. Thus, using wavelengths between 5 and 10 metres, it is possible to measure the phase shift with an accuracy of 0.5 mm to 1 mm. However, to transmit a wave of this size, it is necessary to use a disproportionately large transmitter compared to surveying instruments. To solve this problem, the measurement signal is modulated[3] with a much higher frequency wave called the carrier wave. The carrier wave is thus the transmitted wave, but the phase shift measurement is performed on the measurement wave (now called the modulated wave) as if it had been transmitted directly by the system.

After measuring the phase shift, it is still necessary to resolve the ambiguity of the system, i.e., to calculate the number of integer wavelengths N between the emitter and the target. By combining the results, the measured distance value is obtained.

---

[3]Modulating a wave means combining some characteristic of its signal with a carrier wave, modifying the carrier's wave signal to carry the desired information from the modulated wave when it is transmitted.

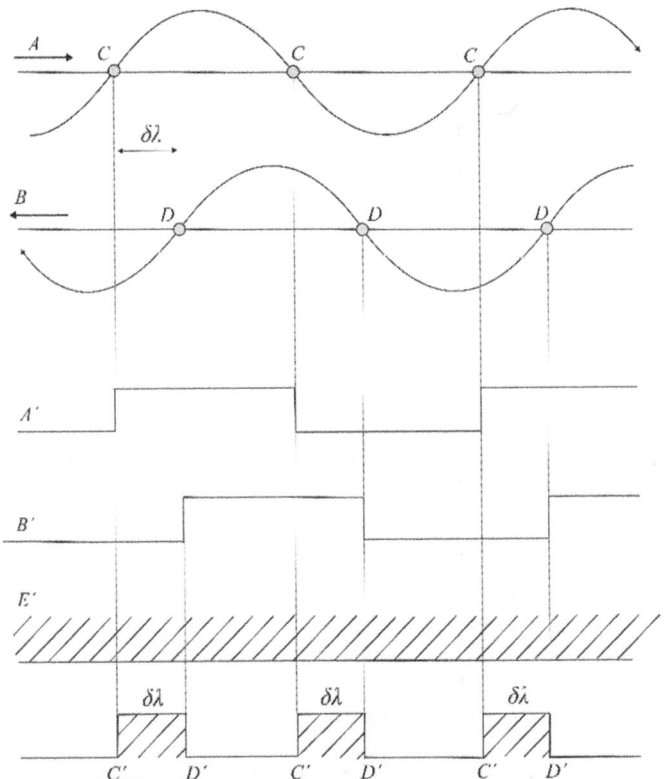

Fig. 8.19 Principle of quadrature amplitude modulation

The ambiguity N in Eq. (8.22) is resolved by the continuous emission of several signals at different wavelengths, as shown in Fig. 8.20.

Modulation is typically performed with two or three waves of different lengths. The first one, with the highest frequency, determines the accuracy of the measurement, and the others are used to resolve the ambiguity (value of N). In the case of Fig. 8.20, suppose that a first wave was emitted at a frequency $f = 15$ MHz ($\lambda = 20$ m) and that the measured phase shift was equal to 0.6821 λ, i.e., 6.821 metres for half a wave. A second wave was then emitted at a frequency $f = 1.5$ MHz ($\lambda = 200$ m), and a phase-shift of 0.2672 λ was obtained, i.e., equivalent to 26.72 metres for half a wave, which means two integer waves of 10 metres each. Finally, a third wave was emitted at a frequency $f = 0.15$ MHz ($\lambda = 2000$ m), and a phase-shift equal to 0.627λ was obtained, i.e., equivalent to 627 metres for half a wave, which means six integer waves of 100 metres. Under these conditions, the measured distance is calculated according to the values given in Table 8.1.

The range and speed of a phase-shift distance measurement depend on the quality of the EDM instrument used and the type of surface reflecting the signal. When reflector prisms are used, the range can be up to 12 km, with a measurement time of 0.5 s. When using natural terrain surfaces (reflectorless measurement), the range can be up to 2 km, with a measurement time of 2 s.

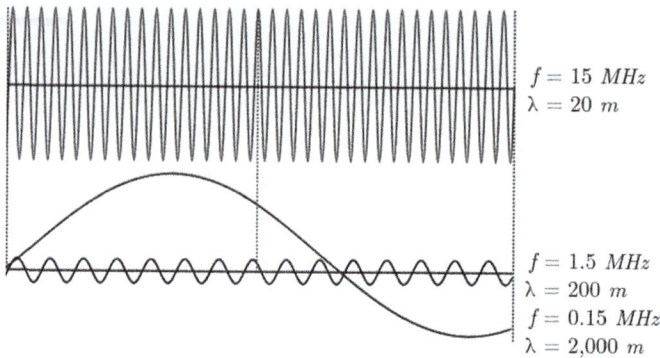

$f = 15\ MHz$
$\lambda = 20\ m$

$f = 1.5\ MHz$
$\lambda = 200\ m$
$f = 0.15\ MHz$
$\lambda = 2,000\ m$

Fig. 8.20 Principle of EDM

Table 8.1 Partial values for calculating the distance measured with the EDM instrument

Frequency	Half wavelength [m]	Phase-shift converted to length [m]
15 MHz	10	6.821
1.5 MHz	100	$2*10 = 20.00$
0.15 MHz	1000	$6*10 = 600.00$
Total distance = 626.821 meters		

8.3.2.3 *Wave Form Digitizer* (WFD) Method

More recently, a new distance measurement technique, the WFD method, has been incorporated into EDMs, combining the time-of-flight and phase-shift measurement techniques described above. In this case, the measurement principle is based on calculating the propagation time of a given pulse detected in the reference beam with its counterpart in the reflected beam by correlating the sinusoidal shape of the emitted and received pulses. The distance is obtained by accumulating several measurements with different pulses. See Fig. 8.21 for details.

The WFD measuring principle, as described by the Leica Geosystems White Paper WFD (2021), is based on the emission of short pulses with a frequency of up to 2 MHz by the EDM sensor. Similar to the phase-shift method, a small portion of each pulse is directed to a waveform digitiser inside the instrument. Most of the pulse is directed at the target, reflected and detected by the photosensor in the instrument. Both pulses are digitised as full waveforms and accumulated from multiple signals. The time difference between the accumulated pulses is then used to calculate the distance, similar to the time-of-flight method.

As with the phase-shift method, the ambiguity of the detected signal must be resolved if a single frequency repetition rate is used. This is achieved by using different repetition rates during each distance measurement.

This type of distance measurement is faster, more accurate and allows reaching greater distances when compared to its predecessors.

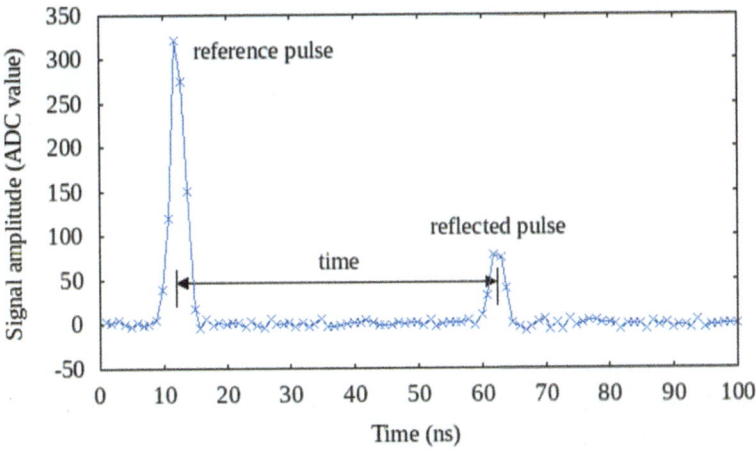

Fig. 8.21 Principle of WFD distance measurement

8.3.2.4   Correction of Distances Measured with an EDM Instrument

The distance measured with an EDM instrument has to be corrected, and there are two types of correction to consider:

- Geometric corrections
- Atmospheric corrections

Both corrections are considered to be systematic errors of the instrument. The discussion of the geometric corrections is presented in Sect. 10.2.3. The effects of atmospheric corrections are discussed in the next section.

8.3.2.4.1   Atmospheric corrections in distance measurement

As previously mentioned, the basic principle of distance measurement with an EDM instrument is the propagation of an electromagnetic wave between the endpoints points of the distance to be measured. Thus, since the power of the signal is reduced as it propagates through the atmosphere, it is necessary to consider this effect when calculating the measured distance.

It is well known that the propagation speed of an electromagnetic wave in the atmosphere varies with the medium's temperature, humidity and water vapour pressure. Typically, the instruments are calibrated for an atmospheric pressure of 760 mmHg (1013.25 mbar), air temperature of 12 °C and relative humidity of 60%; therefore, under these conditions, the atmospheric correction for distance measurements is zero. However, it is necessary to correct the measured distance value for measurements in other atmospheric conditions. Each manufacturer of EDM instruments provides a mathematical formulation for this purpose. The empirical formulas

(8.23) and (8.24), suggested by Barrel & Sears, are the closest to most of those given by the manufacturers:

$$\Delta d = 281.5 - \left(\frac{0.29035\,p}{1 + 0.00366\,t}\right) + \left(\frac{11.27\,rh}{100(273.16 + t)} * 10^x\right) \tag{8.23}$$

$$x = \left(\frac{7.5t}{237.3 + t}\right) + 0.7857 \tag{8.24}$$

where.

Δd = atmospheric correction, in ppm
p = atmospheric pressure, in bar
t = temperature, in °C
rh = relative humidity (%)

The atmospheric correction, which is almost always positive, is proportional to the distance and is added to the slope distance value measured by the EDM instrument. Therefore, once the distance value is displayed on the instrument, it has already been corrected.

Example 8.4

An EDM measurement was carried out under the following conditions: atmospheric pressure of p = 745 mmHg = 993 mbar and ambient temperature and relative humidity of 32 °C and 80%, respectively. Calculate the atmospheric correction value for the measured distance.

Solution:
Using Eq. (8.23).

$$\Delta d = 281.5 - \left[\frac{0.29035 * 993}{1 + 0.00366 * 32}\right] + \left[\frac{11.27 * 80}{100 * (273.16 + 32)} * 10^x\right]$$

The parameter (x) is calculated according to Eq. (8.24).

$$x = \frac{7.5 * 32}{237.3 + 32} + 0.7857 = 1.68$$

Replacing the value of (x) in the expression of Δd.

$$\Delta d = 281.5 - \frac{(0.29035 * 993)}{(1 + 0.00366 * 32)} + \frac{11.27 * 80}{100 * (273.16 + 32)} * 10^{1.68} = 281.5$$

$$- 258.09 + 1.40 = 24.8\,ppm$$

Most EDMs already have the atmospheric correction equation stored in their memory and only require the atmospheric pressure and ambient temperature of the measurement location to be specified to calculate the correction value in ppm. As a general rule, it is usually assumed that

- A 1 °C change in temperature causes a 1.4 ppm change in the measured distance.
- A 1.0 mbar change in atmospheric pressure causes a 0.3 ppm change in the measured distance.
- A relative humidity value of 60% is sufficient to maintain the atmospheric correction variation in the order of ±2 ppm for temperatures up to 40 °C.

8.3.2.5   Precision of Measurements Performed with EDM Instruments

Typically, the precision of an EDM instrument is specified by the manufacturer as an absolute value plus a variable value depending on the measured distance in the format ±(a [mm] + b [ppm]). In other words, the absolute value a is independent of the measured distance and occurs due to internal factors of the instrument, while the value of b is proportional to the distance to be measured. In the latter case, it is a consequence of the uncertainty in the measurement of the electromagnetic signal in a vacuum, the uncertainty in determining the refractive index and the error due to the inconsistency of the oscillator.

Instrument manufacturers usually specify the linear precision of an EDM as an additive calculation, as given by Eq. (8.25).

$$s_d = \pm (a + b * d[\text{ km}]) \text{ mm} \qquad (8.25)$$

However, if the coefficients a and b are considered non-correlated random errors, the linear precision should be calculated using the general law of propagation of variances, as given by Eq. (8.26).

$$s_d = \pm \sqrt{a^2 + (b * d[\text{km}])^2} \text{ mm} \qquad (8.26)$$

To verify the difference between using Eq. (8.25) or (8.26), suppose a distance of 1.3 km is measured using an EDM with a linear precision specified by the manufacturer as ±(1 mm + 1.5 ppm). Using the additive calculation, the precision of the measured distance would be equal to ±(1.0 + 1.5 * 1.3) = ± 3.0 mm. Using the propagation method, the precision of the measured distance would be equal to $\pm \sqrt{1^2 + (1.5 * 1.3)^2} = \pm 2.2$ mm. It is up to the user to decide which calculation method to use.

Table 8.2 gives examples of precisions quoted in manufacturers' technical catalogues for the three EDM methods.

Table 8.2 Maximum precision achieved in EDM with reflector prism and reflectorless

EDM method	Precision	
	With prism	Reflectorless
Phase-shift	±(1 mm + 2 ppm)	±(2 mm + 2 ppm)
Time-of-flight	±(5 mm + 2 ppm)	±(4 mm + 2 ppm)
WFD	±(1 mm + 1.5 ppm)	±(2 mm + 2 ppm)

Standard deviation ISO 17123-4

8.3.2.6 Reflectorless Distance Measurement

After EDM was introduced in the 1960s, another important innovation in distance measurement occurred in the 1990s, including the reflectorless distance measurement method. From then on, engineers could rely on instruments capable of measuring inaccessible remote points and, as the technology improved, also laser scanning, as described in Chap. 18.

The paradigm of reflectorless distance measurement has always been the range of the laser beam used to perform the measurement. To standardise range comparisons between different technologies, standardised reflective plates known as *Kodak Gray Cards* are used, which have different reflectance on each side. The grey side (*Kodak Gray*) has a reflectance of 18%, while the white side (*Kodak White*) has a reflectance of 90%. In theory, the reflectorless measurement could reach distances of up to 2000 metres for Kodak White and 600 metres for Kodak Grey. In practice, however, each surface (target) has a different reflectivity, in addition to being highly influenced by its roughness, humidity and the angle of incidence of the laser beam, making it difficult to establish an exact rule for the range of reflectorless distance measurement.

In addition to the laser beam range, another factor to consider in reflectorless distance measurement is the size of the spot where the beam hits the target. This varies considerably depending on the type of laser, the distance to the target and the angle of incidence on the surface. With current technology, EDM instruments available for reflectorless distance measurement produce laser spots with different dimensions, depending on the manufacturer and type of instrument. A typical case is a laser spot with dimensions of approximately 8 mm × 20 mm at 50 m.

The size of the laser spot is an important factor to be considered when the operator needs to measure distances where the reflecting surfaces are corners, edges or small elements of a structure. The surface covered by the laser beam defines the measured distance value. If the beam covers several different surfaces, the resulting measurement will be a combination of the measured distances. Therefore, the smaller the diameter of the laser spot, the more accurate the measured distance.

Another critical factor to consider in the reflectorless distance measurement is the laser beam classification. In this context, the lasers used in surveying instruments are classified as Class 1, Class 2 and Class 3R, as described below:

- *Laser Class* 1 is classified as a product that is safe for the human eye, even if incident directly on the eye through an optical device.
- *Laser Class* 1 M is classified as safe under all conditions except when passing through an optical magnifying device, such as a telescope.
- *Laser Class* 2 emits radiation in the visible spectrum and is classified as a hazardous product if it directly hits the human eye. Most total stations and laser plummets[4] use this type of laser beam.
- *Laser Class* 2 M has the same characteristics as *Laser Class* 1 M.
- *Laser Class* 3R is classified as a high-power laser and, therefore, has a long range but is also very unsafe for the human eye. This type of laser beam is governed by safety regulations that do not allow it to be used indiscriminately.

EDM instruments capable of reflectorless distance measurement generally also measure with a prism. The additive constants of the two measurement methods, as described in Sect. 10.2.3.1 are different, and the operator must configure them correctly in the instrument, depending on the method used. The target reflecting the laser beam can be any structure with a dimension greater than the dimension of the laser spot.

8.3.3   Distance Measurement with GNSS Technology

The basic operating principle of satellite navigation systems is to generate geocentric Three-Dimensional Cartesian coordinates of points in space. These coordinates can be used to calculate the distance between any two surveyed points, i.e., the relative vector between the corresponding receiving antennas, as shown in Fig. 8.22.

The vector distance calculated from the GNSS coordinates corresponds to the spatial distance between the surveyed points, as given in Eq. (5.75). Therefore, this technology can be used to measure distances, provided due care is taken in its use, as described in Chap. 17.

The accuracy achieved with this distance measurement method depends on several factors, such as the location and interval of data collection, the type of instrument and positioning method used, the effects of satellite geometry, data processing techniques and adjustment methods. Typically, the accuracy of the order of ±(3 mm + 0.1 ppm) is achieved when measurements are performed using relative positioning in static mode (without considering operational errors).

---

[4]For more details, see Sect. 9.2.5.

Fig. 8.22 GNSS vector line

Example 8.5

The geocentric Cartesian coordinates of points (P) and (Q) in Example 7.4 were obtained from a GNSS survey, as shown in Table 8.3. Calculate the slope distance between points (P) and (Q).

Table 8.3 Coordinates determined from GNSS observations

Station	X [m]	Y [m]	Z [m]
P	3,964,001.2561	-4,390,918.1127	−2,378,632.3622
Q	3,964,666.6981	−4,390,900.8861	−2,377,826.7893
Difference (Q-P)	665.4420	17.2266	805.5729

Solution:

To calculate the slope distance from geocentric Cartesian coordinates, Eq. (5.75) *must be used. Thus,*

$$d'_{PQ} = \sqrt{665.4420^2 + 17.2266^2 + 805.5729^2} = 1045.0156 \text{ m}$$

The same distance was measured with a total station, whose result was equal to 1045.0220, i.e., a difference of 6.4 mm.

8.4 Review Questions

1. What is the expected difference in error when measuring a distance of 100 m using fibreglass and steel tape measures?

2. Explain the differences between the time-of-flight, phase-shift and waveform digitiser methods of electronic distance measurement.
3. Explain the corrections that should be applied to EDM measurements and how they affect the measurement result.
4. If a total station has been configured for distance measurement at an atmospheric pressure of 760 mmHg (1013.25 mb), air temperature of 20 °C and relative humidity of 60%, what is the expected error when a distance of 1000 m is measured at the same location at a temperature of 30 °C?
5. What causes a greater error in an EDM distance measurement: a neglected 20 °C temperature variation or a neglected 10 mm atmospheric pressure difference from the standard?
6. Explain how manufacturers specify the standard deviation (precision) of a total station and what error can be expected from an instrument specified by the manufacturer as having a precision of ±(4 mm + 2 ppm) when measuring a distance of 500.342 m, neglecting instrument and target centring errors.
7. What is the expected error in ppm when a total station is used to measure a distance of 1500.000 metres with an accuracy of ±(2 mm + 2 ppm)?
8. What is more precise, a ±(2 mm + 2 ppm) or ±(3 mm + 1 ppm)?
9. Explain what a reflectorless distance measurement using a total station is and its advantages and disadvantages compared to reflector prism measurement.
10. Discuss how to determine the horizontal distance between two ground stations using GNSS instruments.

References

Alojzy Dzierzega, René Scherrer (2003). *Measuring with Electronic Total Stations*. White paper, Switzerland, 12p.

Grimm, H.; Pache, F.; Giger, K. (1986). *Timed-pulse distance measurement with geodetic accuracy*. Wild Heerbrugg - White paper. Switzerland.

Hannes Maar, Hans-Martin Zogg. WFD (2014). *Wave Form Digitizer Technology*. White Paper. Leica Geosystems AG, Heerbrugg, Switzerland.

Shih, P. T.-Y. (2013). *On accuracy specification of electronic distance meter*. Survey Review. Vol. 45, No 331. London.

Uren, J., Price, B. (2010). *Surveying for Engineers*. 5th Edition. Palgrave Macmillan. Hampshire, England.

Chapter 9
Surveying Instruments

Irineu da Silva

9.1 Introduction

The term "*surveying instrument*" refers to any equipment used to perform a surveying measurement. They range from simple measuring devices to sophisticated measuring instruments. Understanding the characteristics of each is, therefore, essential for the engineer to carefully select the appropriate instrument to meet the needs of an engineering project. Significant financial and human resources can be saved if engineers properly integrate equipment and field crews into their project schedules and technical requirements. Therefore, the engineer must understand the technical aspects of the surveying instrument that need to be considered when deciding on which instrument to choose. To assist them, the following guidelines are suggested:

- The measuring instrument must ensure that the accuracy of the measured values meets the requirements of the project. The engineer must always be sure that the results obtained are correct.
- The measuring instrument must guarantee the stability of the measured results without the need for recurring maintenance or calibration.
- The measuring instrument must be easy to use and, at the same time, have all the features to ensure effectiveness in the field. In addition, it must be customisable to allow the development of user applications according to the needs of the project, avoiding the "black boxes" of restricted functionalities and ensuring that operators do not depend on exclusive solutions of some models specified by the manufacturer.
- The measuring instrument must be robust enough to be used in adverse situations of environmental conditions of fieldwork.

I. da Silva (✉)
São Carlos School of Engineering, University of São Paulo, São Carlos, SP, Brazil
e-mail: irineu@sc.usp.br

- Users should always verify the existence of maintenance workshops based in their country, availability of parts, reasonable costs and speed in providing services.
- The measuring instrument must ensure reliable storage of the collected data and easy transfer to other media and different formats without corrupting the original data structure. When selecting a surveying instrument, the engineer must be cautious to avoid incompatibility between data formats from different manufacturers, which would require extensive editing in the office or, in the worst case, loss of data collected in the field. Office editing should be kept to a minimum.
- The measuring instrument must ensure speed and efficiency in all surveying operations. In addition, it must provide facilities for collecting structured data in the field and transferring data to third-party software in the office.
- Users must always remember that a surveying measurement is a set of operations that includes the instruments, their accessories, the working methodology, the measurement techniques, the operator and the assistants. The correct combination of all these elements is the key to the success of any surveying work.

Choosing the most appropriate measuring instrument can be a challenging task. It depends on the type and amount of data to be measured, the measurement purposes, the required accuracy, the time available for the work, the site conditions and the measurement technologies available. To help the engineer in this choice, Chart 9.1 shows a classification of the available measurement technologies according to the type of data to be measured and the most commonly used instruments in each class.

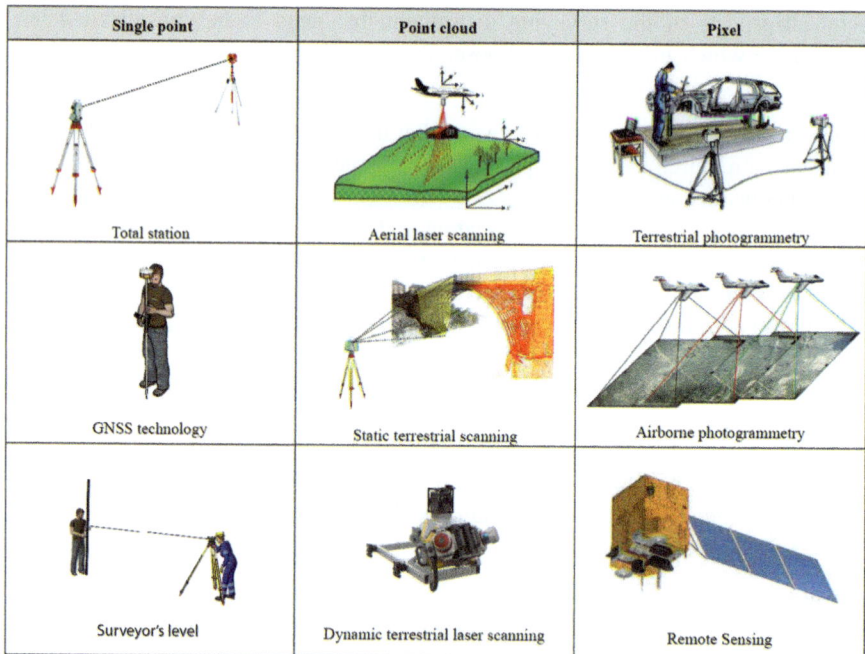

Single point	Point cloud	Pixel
Total station	Aerial laser scanning	Terrestrial photogrammetry
GNSS technology	Static terrestrial scanning	Airborne photogrammetry
Surveyor's level	Dynamic terrestrial laser scanning	Remote Sensing

Chart 9.1 Surveying measurement technologies currently available depending on the type of data to be measured

The *Single-point* class refers to the methods and instruments used to collect field data based on single-point measurement. Measurements performed with a total station, GNSS-based measuring instruments and surveyor's level are included in this classification. These instruments require more time to collect data in the field but provide better-quality measurements than other measurement techniques. In that case, the accuracy can be in the order of the millimetre.

The *point cloud* class is associated with methods and instruments that collect many points from a single setup. In this case, it is said that a point cloud has been measured. The measured points carry the spatial position (X, Y, H) and information about the quality of the signal reflected from the target. Measurements in this class are based on instruments using *Light Detection and Ranging* (LiDAR) technology, as described in *Chap. 18*, which can achieve accuracy in the order of centimetres in horizontal and vertical components.

As for the *pixel* class, it is related to the methods and instruments that perform measurements on digital images. Through the methods and instruments of this class, it is possible to obtain geometric and radiometric information about the objects measured. Digital images can be obtained from short distances, on the ground, or even through aircraft or satellites, thus defining Terrestrial Photogrammetry, Airborne Photogrammetry and Remote Sensing.

The following sections of this chapter describe the most relevant characteristics of the instruments and accessories associated with the single-point class. Technical details of the instruments of other classes are presented in the chapters corresponding to the description of their technologies. The details of instrument handling will not be discussed as they vary between models and manufacturers. In this regard, it is suggested that users consult the technical manuals of their respective instruments.

To facilitate the reading of this chapter, however, attention should be paid to the meaning of the following terms related to the operation of surveying instruments routinely used by professional in Geomatics:

- *Station*: point materialised on the ground and over which a surveying instrument is installed.
- *Sighting point*: point, materialised or not, at which a sighting is carried out for surveying measurements.
- *Instrument levelling*: field operation consisting of positioning the vertical axis of the surveying instrument parallel to the Vertical line (plumb line) passing through the station point.
- *Instrument centring*: field procedure consisting of positioning the vertical axis of the surveying instrument to coincide with the station point.
- *Instrument setup*: field procedure consisting of levelling and centring the surveying instrument over the station point.
- *Line of sight*: imaginary line coinciding with the line-of-sight axis of the telescope of a surveying instrument.

Given that modern surveying instruments are almost all electronic, it is essential to classify them according to the degree of protection provided by the enclosures of electrical equipment. In this sense, they are classified according to the International Electrotechnical Commission (IEC) 60529:2005 standard, based on the Ingress protection (IP) rating, which indicates the degree of protection provided by an enclosure against access to hazardous parts, ingress of solid foreign objects, and ingress of water. The IP code consists of the letter IP followed by two digits and an optional letter. The first digit indicates the level of protection provided by the enclosure against access to hazardous parts and the ingress of solid foreign objects, as shown in Table 9.1. The second digit indicates the level of protection against harmful ingress of water, as shown in Table 9.2. The optional letter indicates additional product standard information (not applicable to surveying instruments).

Table 9.1 IP ratings: solid objects

| Level | IP code—first digit | |
	Protection against	Effective against
0–4	Not applicable for surveying instruments	
5	Dust protected	Ingress of dust is not entirely prevented, but it must not enter in sufficient quantity to interfere with the satisfactory operation of the equipment
6	Dust-tight	No ingress of dust

Table 9.2 IP ratings: moisture

| Level | IP code—second digit | |
	Protection against	Effective against
0–2	Not applicable for surveying instruments	
3	Spraying water	Protected against sprays of water up to 60° from the vertical. Limited ingress permitted
4	Splashing water	Protected against water splashes from all directions. Limited ingress permitted
5	Water jets	Protected against jets of water. Limited ingress permitted
6	Powerful water jets	Protected against powerful jets of water. Limited ingress permitted
7	Immersion, up to a 1-metre depth	Watertight against the effects of temporary immersion in water between 15 cm and 1 m for 30 min
8	Continuous immersion	Watertight against the effects of immersion in water under pressure for long periods
9	High-pressure and temperature water jets	Water projected at high pressure and high temperature against the enclosure from any direction shall not have harmful effects

Before discussing the surveying instruments themselves, the following section presents the technical details of the main components that are usually part of single-point class instruments. These are:

- Alidade
- Tripod
- Level bubble[1]
- Electronic level
- Tribrach
- Telescope

9.2   Main Parts of a Surveying Instrument

9.2.1   Alidade

According to the ISO 9849:2017 standard, an *alidade is the upper part of a surveying instrument consisting of a sighting device with an index and accessories for reading or recording data*. In this case, accessories may include graduated circles and other angular measurement elements. It contains three reference axes for surveying measurements, as shown in Fig. 9.1.

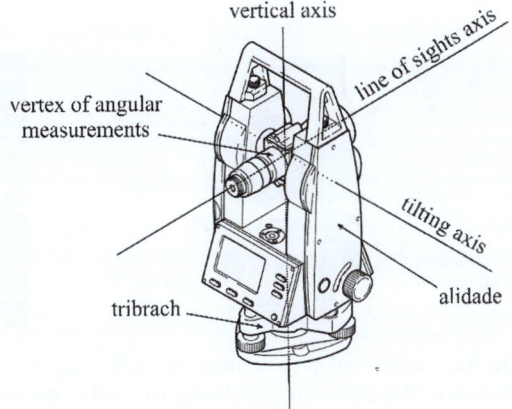

Fig. 9.1 Alidade and instrument axes. (Adapted from Leica Geosystems–T100 user manual)

[1]Referred by some authors as level vial.

The axis that attaches the alidade to the bottom part of the instrument (tribrach) is called the *vertical axis*.[2] It is the axis around which the alidade of a surveying instrument rotates horizontally. It is, therefore, made of a material with high structural stability and balanced to ensure that the instrument rotates around itself without angular deformation. To allow the instrument's line of sight to point in any direction in space, the alidade has two uprights that support a second axis called the *tilting axis*,[3] which is perpendicular to the vertical axis and around which the telescope rotates. The telescope, in turn, defines a third axis perpendicular to the second axis, called the *line-of-sight axis*, as shown in Fig. 9.1.

The instrument's three axes converge at a single point, called the centre of the instrument, which is the vertex of angular measurements.

Inside the body of the instrument are several electronic components that control its functions and interact with the operator through an LCD/VGA[4] screen or another form of interface, such as data loggers, remote control API and notebooks. As mentioned, depending on the measuring instrument, its body contains a telescope (combined or not with an EDM), graduated circles for angular measurements, and tangent screws, as shown in Fig. 9.2a.

If the instrument's telescope is assembled to rotate around the tilting and vertical axes, it has two tangent screws. These two movements allow the operator to position the telescope's line of sight precisely in line with the target point. See Fig. 9.2b.

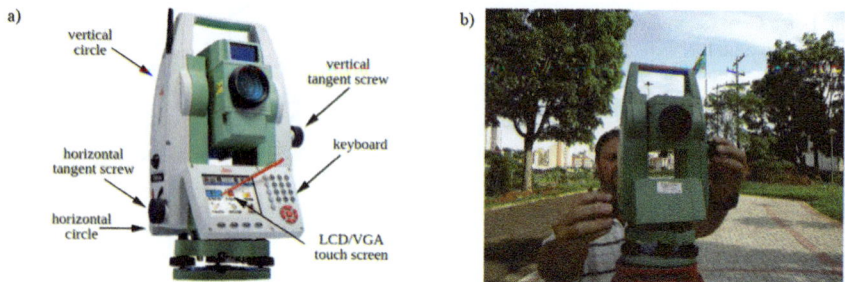

Fig. 9.2 Alidade and its components. (**a**) Main components contained in the alidade of a surveying instrument (Adapted from Leica Geosystems). (**b**) Operator handling the tangent screws of a total station

---

[2]Referred to by some authors as the standing axis.

[3]Referred to by some authors as the horizontal axis.

[4]Some manufacturers provide instruments with different VGA qualities for the user interface display, such as QVGA, HVGA, and full-VGA.

As will be seen in the following chapters, there are variations in the types of components that go into the alidade of surveying instruments. The general design is the same; but the number of axes, angle circles and tangent screws, as well as the type of telescope, can vary between different classes of instrument.

9.2.2   Tripod

The surveying tripod is a basic device for installing surveying instruments over the station point. It usually consists of the following parts: three adjustable legs (each with a metal cone called a "shoe" which can be driven into the ground by applying pressure to the leg), a tripod head plate, quick-release clamps and thumbscrews to lock the legs in the desired position. See Fig. 9.3.

Fig. 9.3 Surveying tripod

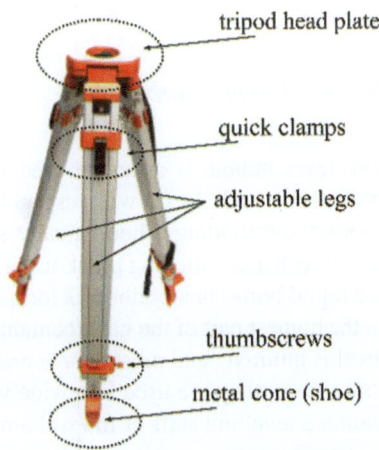

tripod head plate

quick clamps

adjustable legs

thumbscrews

metal cone (shoe)

The tripods used in Geomatics are made of wood, fibreglass or aluminium. The ISO 12858-2:2020 standard defines the requirements for tripods in terms of vertical stability under load and torsional rigidity. According to ISO standards, tripods can be classified as heavy or lightweight. A heavy tripod must have a mass of more than 5.5 kg. This type of tripod can support instruments weighing up to 15 kg.

Lightweight tripods, on the other hand, are only suitable for instruments weighing less than 5 kg. In terms of vertical stability, the ISO standard defines that the position of the tripod head must not shift vertically by more than 0.05 mm when subjected to twice the maximum weight of the instrument. The defined vertical deformation of 0.05 mm is so small that the effect on angular accuracy is negligible. However, for precision levelling applications, vertical stability must be considered.

When an instrument rotates, rotational forces affect the horizontal movements of the tripod head plate. The resistance of a stationary tripod to such rotational forces (torque) is called torsional rigidity, and the precision with which the tripod's orientation returns to its original position is called *hysteresis*.[5] According to ISO standards, when the tripod plate is rotated 180″, the maximum allowable hysteresis for heavy and lightweight tripods is 3″ and 1″, respectively.

Another requirement to consider is the horizontal drift of the tripod head. Such a drift is a measure of how its orientation changes over time. This is not an ISO requirement so there are no guideline values. However, for precision applications, users should also consider this requirement.

Operators often use lightweight tripods for ease of transport and handling. However, it is important to remember that the stability of the instrument/tripod combination is essential for precision work, and therefore, the robustness of the tripod is an important factor to consider.

Another problem associated with survey tripods is the loss of rigidity of their components due to loosening connections. It is, therefore, recommended that the operator regularly checks all the tripod connections. Under no circumstances should a tripod be used that does not have all its joints securely fastened.

9.2.3   Level Bubble

The level bubble is a device used to verify the horizontal inclination of various instruments and objects with respect to the direction of gravity. It consists of a metal support surrounding a hermetically sealed cylindrical glass bulb containing a thin liquid with a low melting point, usually ether. A space is left in the bulb so that when the liquid boils, an air bubble is formed, which, due to the Earth's gravity, will lodge in the highest part of the glass container, as shown in Fig. 9.4. The use of this type of level is intuitive, and most people understand the principles of use immediately. For this reason, they are used in a wide variety of surveying situations, for example, to plumb a levelling staff or to level a measuring instrument.

For Geomatics, there are two types of level bubbles to consider:

- circular bubble or bull's-eye level
- Tubular or plate bubble.

A *circular bubble* is manufactured using circular glass vials assembled in a round glass container encased in metal. The upper side of the glass container has a spherical surface engraved with centred circular marks. An object is considered level when the bubble is placed in the centre of the circular marks, as shown in Fig. 9.4.

---

[5]Hysteresis is the ability of the tripod to absorb the horizontal rotation of the instrument and return to its original position when the instrument is stationary.

Fig. 9.4 Circular level
bubble

Due to the radius of curvature of the spherical surface (usually 7–8 metres), the circular level bubbles are not very sensitive for surveying instruments and are only used for low-precision levelling. In the case of a theodolite or a total station, it is attached to the tribrach[6] and used for the initial levelling of the instrument. This type of level bubble is also used for the plumbing of levelling staffs and prism poles, as described in *Sect. 9.4.2* and *Sect. 9.6*, respectively.

A tubular level bubble consists of a metal-encased glass tube filled with liquid and an air bubble. The air bubble always moves to the highest point of the level. The tubular glass has graduation lines, separated every 2 mm, perpendicular to its axis, to allow verifying the bubble position, as shown in Fig. 9.5. As the bubble can only move in one direction, to level the equipment, it is necessary to check the position of the bubble in two perpendicular directions, as also shown in Fig. 9.5.

Fig. 9.5 Tubular level
bubble

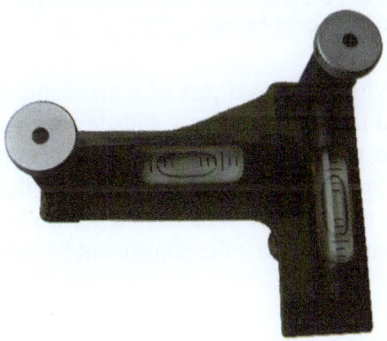

---

[6]See Sect. 9.2.5.

The sensitivity σ of a tubular level bubble is a function of the radius of curvature of the cylinder r and the interval between graduation lines a, as shown in Eq. (9.1). The greater the r-value, the more sensitive the level bubble and the more difficult to level the instrument attached to it.

$$\sigma = \frac{a}{r} = \frac{2[\,\mathrm{mm}\,]}{r[\,\mathrm{mm}\,]} \tag{9.1}$$

Typically, the level bubbles used in surveying instruments have radii of curvature equal to 21 m ($\sigma = 20''$), 42 m ($\sigma = 10''$) and 84 m ($\sigma = 5''$), depending on the classification of the instrument.

On some levelling instruments, the level bubble and the index levels are made as coincidence bubbles using a prism system arranged so that half the ends of the bubbles are side by side, as shown in Fig. 9.6. When the bubble is levelled, both ends match. This levelling method is 3 to 4 times more accurate than graduated lines.

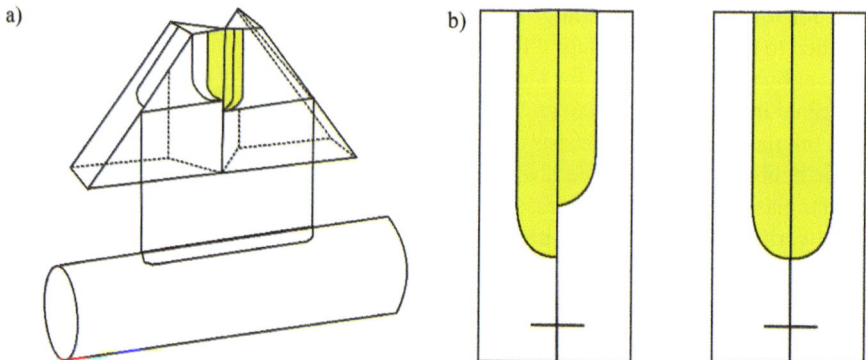

Fig. 9.6 Level bubbles with prism system. (**a**) Principle of the level bubble with prism system. (**b**) Matching ends of bubbles

When used as a part of a theodolite or total station, the tubular level bubble is fixed in the instrument's alidade and is generally used for final levelling. Furthermore, to maintain its quality, the level bubble must be periodically adjusted to ensure it is horizontal concerning the axes of the instrument. To do this, it has screws at its ends that allow it to be moved vertically. Although the user can easily do this, it is recommended that the adjustment be carried out in a certified laboratory.

9.2.4   Electronic Level

In the 1990s, surveying instrument manufacturers began replacing the tubular level bubble with an electronic level. In this case, instead of having a tubular level attached to the alidade, the instrument uses an electronic compensator built into its body, as shown in Fig. 9.7a. In this case, the levelling procedure is similar to that of the tubular bubble level but simpler, since the operator does not have to rotate the instrument's alidade to check the level. The whole procedure is carried out using an image of the electronic level displayed on the instrument's screen, as shown in Fig. 9.7b.

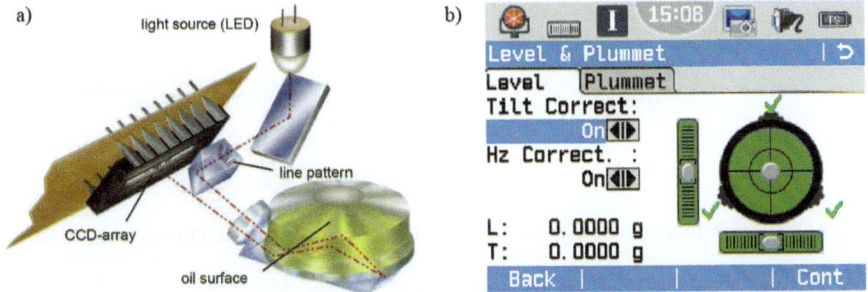

Fig. 9.7 Electronic compensator and electronic level bubble display. (**a**) Example of the electronic compensator (Courtesy Leica Geosystems AG). (**b**) Example of the electronic level display

The electronic compensator shown in Fig. 9.7a is a dual-axis compensator, which consists of a line of patterns engraved on a prism, which, when illuminated, is reflected twice by the horizontal surface of a reflecting liquid. Each image reflected from the line of patterns is read by a linear array of *Charged Coupled Device* (CCD[7]) or *Complementary Metal Oxide Semiconductor* (CMOS[8]) sensors, through which the two components of the instrument's inclination (longitudinal and transversal) are mathematically determined. The values obtained are then used to correct the measured angular values affected by the non-verticality of the instrument, as described in *Sect. 10.2.1.2.3*. The electronic compensator is assembled in the centre of the instrument to minimise sensitivity to vibration and instrument rotation. Typically, this type of compensator has a working range of between $2'$ and $6'$ and an accuracy ranging from $0.5''$ to $1.5''$.

9.2.5   Tribrach

The tribrach is the lower part of a surveying instrument, by means of which it is attached to the tripod and from which the measuring instrument is centred and levelled over the station point. It usually consists of a base plate and an upper plate connected by three footscrews 120° apart, as shown in Fig. 9.8a. By turning the footscrews differently, the upper tribrach plate can be positioned approximately parallel to the horizontal plane using the circular level bubble attached to the

---

[7]Image sensor consisting of a network of light-sensitive electronic elements. See *Chap. 19* for more details.

[8]Image sensors similar to CCD sensors but with different data transfer characteristics.

tribrach. See Fig. 9.8b. Accurate final levelling of the instrument is achieved using the footscrews and the electronic level displayed on the instrument's screen.

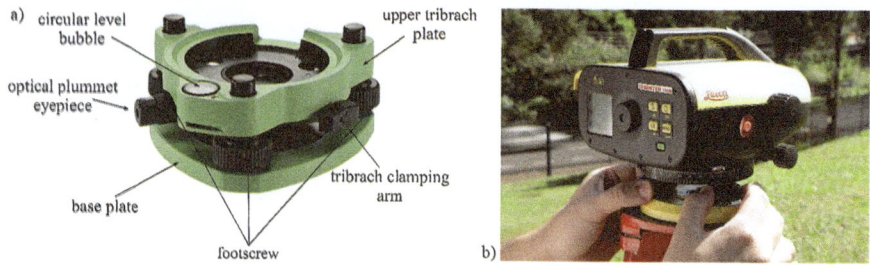

Fig. 9.8 Tribrach. (**a**) Parts of a tribrach. (**b**) Principle of footscrew turning for instrument levelling

The stability between the tripod and the tribrach is the first factor to be considered to ensure reliable measurements. The instrument is fixed to the tripod by attaching it to the tribrach and securing the set (instrument + tribrach) to the tripod using a 5/8″ threaded screw, as shown in Fig. 9.9a, b. Once positioned on the tripod, the instrument can be centred over the station point using an *optical plummet* fixed in the tribrach (Fig. 9.9c) or a *laser or built-in video plummet* incorporated in the instrument's alidade, as shown in Fig. 9.9d. Positioning over the station point is achieved by moving the tribrach over the tripod head. The centring accuracy of the instrument is assumed to be of the order of:

- ±0.5 mm at 1.50 m height, with a good quality optical plummet
- ±1.5 mm at 1.50 m height, with a laser plummet
- ±0.5 mm at 1.55 m height, with a built-in video plummet

In surveying measurements, measuring the height of the instrument above the ground is often necessary, as described in *Sect. 13.8.1*. Surveyors have always used conventional measuring tapes for this purpose. However, a new technology is now available that replaces the traditional laser plummet with an EDM sensor that can be used as a laser pointer, centring and measuring the height of the instrument. The EDM is based on a biaxial optical design centred on the instrument's vertical axis and the time-of-flight distance measurement principle.

The verticality of any of the three types of plummet must be adjusted periodically and preferably in a certified laboratory. The procedures for this operation are also provided in the instrument's user manual.

To ensure the quality of the tribraches, they are manufactured in accordance with the ISO 12858-3:2005 standard. Among the recommendations of this standard, it is essential to highlight the torsional rigidity that a tribrach must have to guarantee the angular accuracy of the instrument, especially when used in conjunction with a robotic total station.[9] The hysteresis effect on a tribrach occurs when the upper plate drifts around the lower base, and this effect must be as small as possible. On high-quality tribraches, it should not exceed one arc second.

---

[9] See Sect. 9.3.

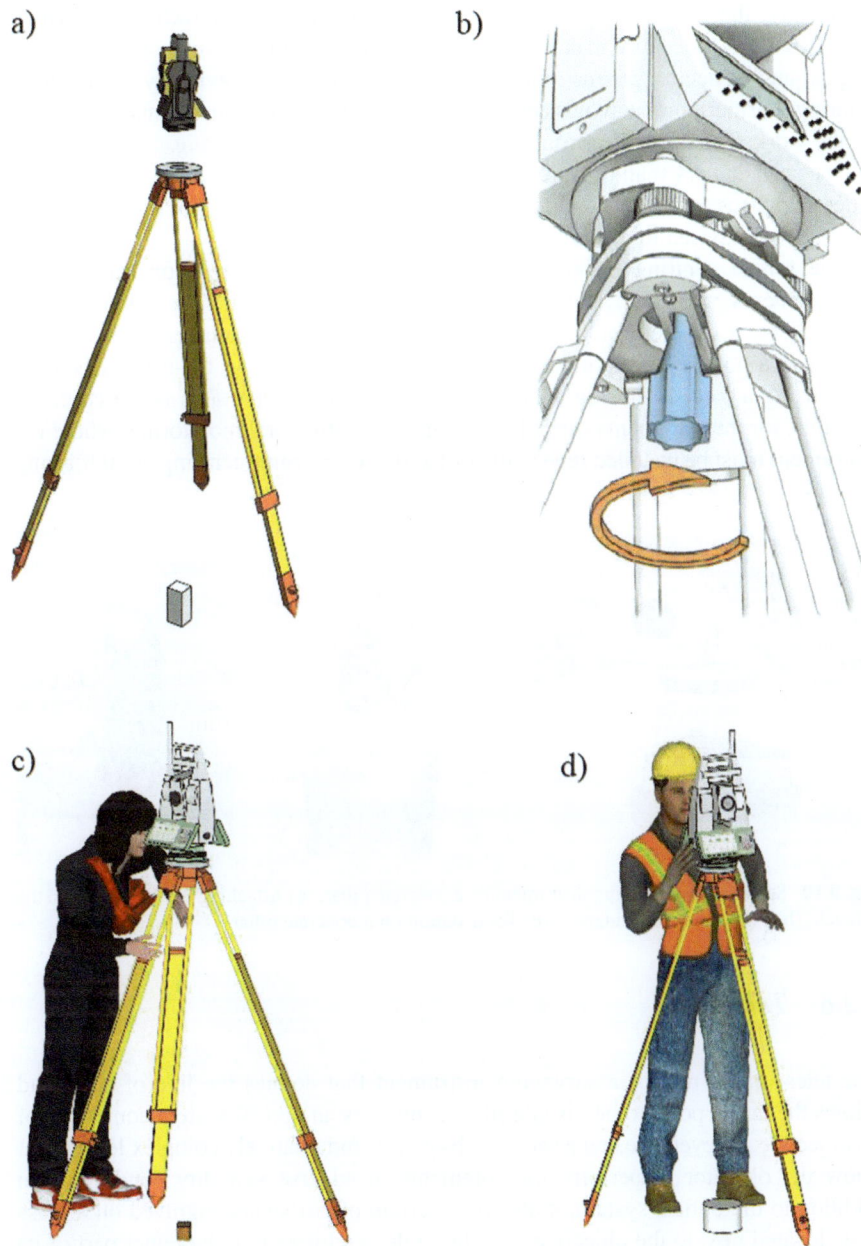

Fig. 9.9 Installing a surveying instrument. (**a**) Attaching the instrument to the tripod. (**b**) Fixing the tribrach to the tripod. (**c**) Instrument centring using optical plummet. (**d**) Instrument centring using laser or built-in video plummet

Usually, the instrument's body can be removed from the tribrach, as shown in
Fig. 9.10. This procedure makes it possible to exchange the instrument for sighting
targets and reflector prisms without compromising the previously established
centring. Furthermore, it avoids centring errors in repeated measurements and speeds
up the work. This type of instrument setup is called *forced centring*.

Special adapters manufactured exclusively for this purpose are used to mount
reflector prisms and GNSS antennas on a tripod using a tribrach. For further details,
the reader is advised to consult manufacturers' catalogues.

The forced centring technique is mainly used on high-precision traversing[10]
measurements with total stations or for measurements where the instrument is
mounted on a concrete pillar, as shown in Fig. 9.10. In the latter case, a metal
plate with a hole is placed at the top of the pillar, which allows the tribraches to be
fixed using a forced centring adapter, as shown in Fig. 9.10. This is the case, for
example, for measurements carried out for geodetic structural monitoring, where the
instrument must be installed repeatedly or fixed over the same centring point for long
periods.

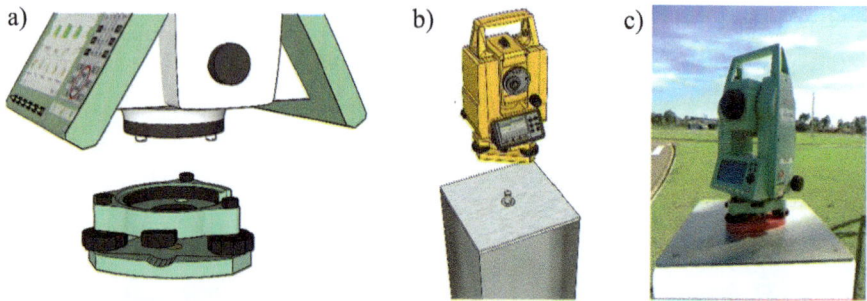

Fig. 9.10   Setting up a surveying instrument on a concrete pillar. (**a**) Attaching the instrument to the
tribrach. (**b**) Forced centring adapter. (**c**) Total station on a concrete pillar

9.2.6   Telescope

The telescope is part of a surveying instrument that defines the line of sight and
allows the target point to be visualised. It comprises an optical system consisting of
an objective, an eyepiece and a set of high-quality and relatively complex lenses that
allow the operator to perform measurements in adverse visibility conditions. In
addition to the optical system, it also has a group of crosshairs engraved on a glass
plate located next to the objective; a sighting device located on the upper part of its
body, and two focus controls, one of them to focus on the target and the other to
focus on the crosshairs. The correct focusing of the two lens systems is essential to

---

[10] See Chap. 12.

avoid the *parallax* effect, i.e., the occurrence of an apparent movement of the target concerning the crosshairs due to the movement of the position of the eye of the observer. Such an effect must be eliminated before each measurement. Fig. 9.11a shows a schematic drawing of a conventional telescope of a theodolite with its main components. The straight line passing through the centre of the objective and crossing the crosshairs defines the line of sight, as shown in Fig. 19.11b.

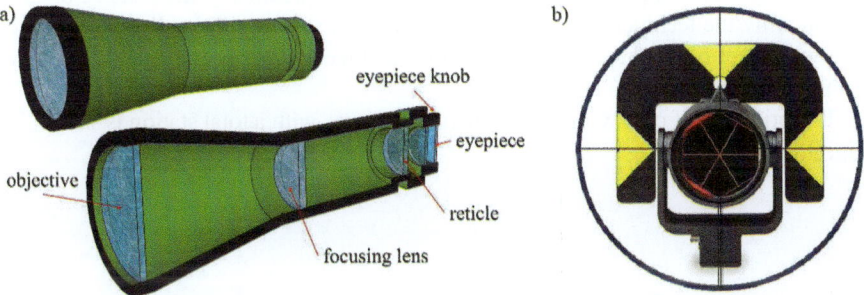

Fig. 9.11 Telescope cross-section sketch and crosshairs over a prism. (**a**) Telescope sketch. (**b**) Crosshairs over a prism

In addition to the telescope's mechanical components, two other factors related to the optical system interfere with its quality. These are the *telescope magnification power* and the *field of view*, as described below.

Telescope Magnification Power
The magnification power of a telescope indicates how large a focused object appears when viewed through the telescope rather than with the naked eye. In practice, it is the ratio between the objective's focal length and the eyepiece's focal length. Typically, the telescope magnification power of a conventional surveying instrument is in the order of 30 times.

Field of View
The field of view of a telescope is the angle at which the entrance window of an optical system subtends the centre of the entrance pupil. It is typically of the order of 1.5°, or 27 metres at a distance of 1 km.

Although the term telescope refers to a device for optical sighting, it should be understood as the device through which the measuring reference axis is defined in many surveying instruments. Therefore, in many instruments, it is through this axis that distance measurements are made, images are captured and collimation and scanning laser beams are emitted. Some instruments of the single-point class do not even allow any optical visualisation, as shown in Fig. 9.23.

Having presented the main parts of the surveying instruments in general, the technical details of the most important instruments of the single-point class, which are the total stations and the surveyor's levels, are discussed next.

9.3 Electronic Theodolite and Total Station

The *electronic theodolite* and the *total station* are the most well-known and used surveying instruments in Geomatics. The electronic theodolite is primarily used to measure horizontal directions and vertical angles, while the total station is used to measure distances as well as directions and angles.

The physical composition of both instruments is practically the same. For this reason, only the technical characteristics of the total station will be discussed in the following sections, which can be extended to theodolites, as these are evidently simpler instruments.

The principle of observing direction and distances with a total station is shown in Fig. 9.12.

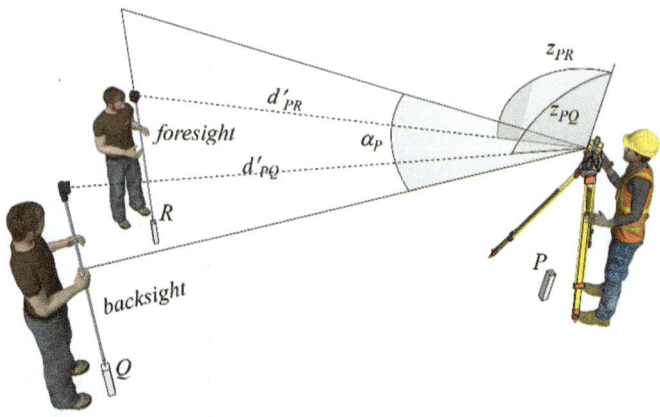

Fig. 9.12 Principle of measuring angles and distances with a total station

Generally, the total station is the most commonly used surveying instrument for engineering applications. As shown in Fig. 9.13, it has all the basic components of a surveying instrument: the alidade, the telescope, the EDM, the vertical and horizontal graduated circles, the LCD or VGA display, the tangent screws, the centring plummet and rotation axes. Some total stations also have automatic target recognition systems and digital sensors for image capture, as described later in this chapter.

9.3.1 Graduated Circles

Graduated circles and the EDM instrument are the basic parts of a total station. They are protected inside the alidade and positioned in such a way concerning the rotating axes to allow the observation of horizontal directions, vertical angles and distances, as described in previous chapters.

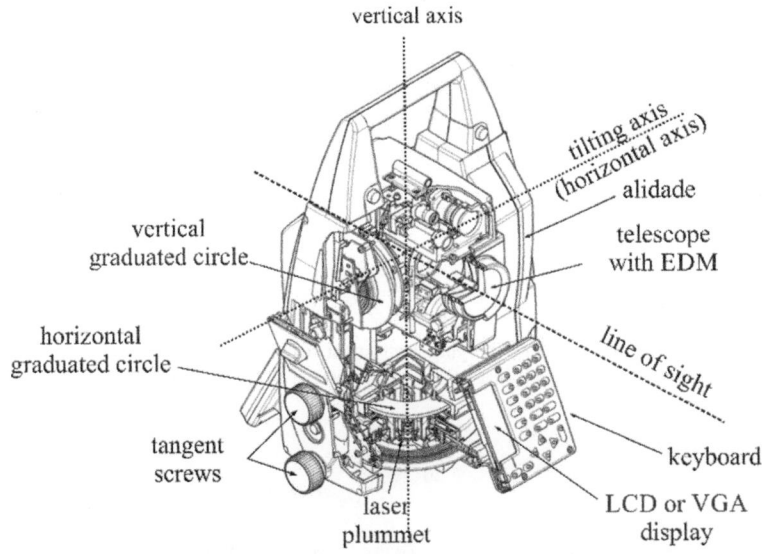

vertical axis

tilting axis
(horizontal axis)
alidade

telescope
with EDM

vertical
graduated circle

line of sight

horizontal
graduated circle

tangent
screws

keyboard

laser
plummet

LCD or VGA
display

Fig. 9.13 Basic components of a total station. (Source: Adapted from Leica Geosystems)

The horizontal circle is assembled at the bottom of the alidade, centred on the vertical axis of the instrument, while the vertical circle is assembled vertically at the side of the alidade and centred on the tilting axis of the instrument. This means that when the instrument is levelled, the horizontal circle is parallel to the horizontal plane passing through the centre of the instrument, and the vertical circle is parallel to the vertical plane perpendicular to that horizontal plane. See Fig. 9.14a.

For angular readings, most instruments use a measuring system consisting of a circular glass plate of variable diameter from 70 to 100 mm, with coded marks engraved on sectors of the plate and capable of resolving values to tenths of angular seconds. See Fig. 9.14b. Reading is performed by illuminating the coded circle using a combination of LEDs, which project the illuminated portion onto an array of CMOS sensors. The image projected onto the sensors is encoded and converted into angular information. The angular value is first approximated to an accuracy of 16 arc seconds. The precise angular value is obtained from the position of the centroid of the code lines hit by the light beam. Typically, 30 lines and four sensor arrays are used to enable accurate interpolation.

There are two types of angle encoders: absolute and incremental. The main difference between them is that, with absolute encoders, the limb's position always corresponds to a fixed zero position. With an incremental encoder, this position varies, and the encoder always displays the difference between the current and previous positions. Through these measurement processes, it has been possible to achieve accuracies of the order 0.5″ or better. The observed values are then displayed on the instrument's screen. Figure 9.15a illustrates the principle of the angular measurement system, and Fig. 9.15b shows a 3D view of the complete assembly of such a system.

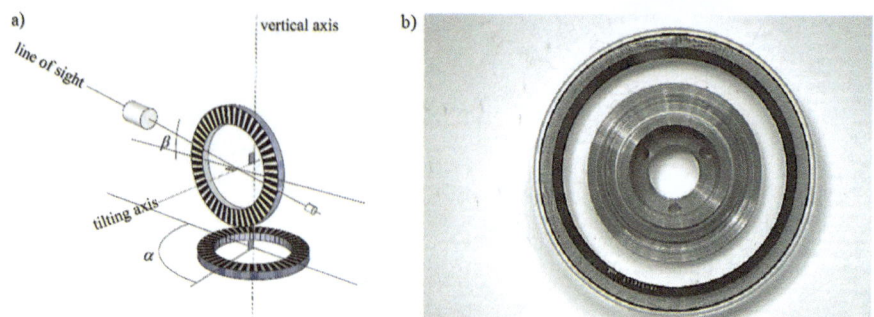

Fig. 9.14 Graduated circles of a surveying instrument. (**a**) Positioning of graduated circles in relation to the axes of a surveying instrument. (**b**) Example of a graduated circle of a surveying instrument

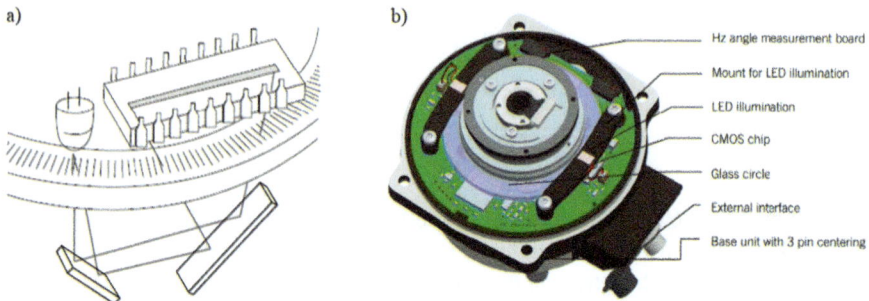

Fig. 9.15 Angular measuring system of a total station. (**a**) Principle of the angular measurement system. (**b**) Main parts of an angle measuring device. (Courtesy Leica Geosystems AG)

9.3.2 EDM Instrument

For many years, the most commonly used surveying instruments in Geomatics were the theodolite for measuring directions and, in some special situations, also for stadia distance measurement, and the measuring tape for distance measurements. Field notes were initially recorded in field books, and the corresponding field data were then transferred manually to spreadsheets or computer software. The first significant evolution in surveying measurements occurred with the advent of EDM instruments, as described in *Sect. 8.3.2*. As they are autonomous measuring devices, they were initially attached to the telescope of the theodolite for use in surveying measurements, allowing angular and distance measurements to be made almost simultaneously. See details in Fig. 9.16.

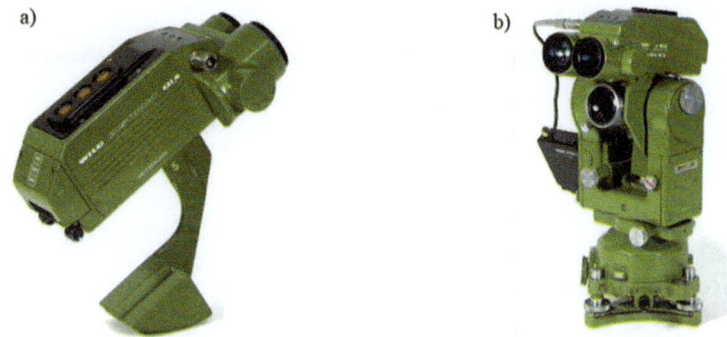

Fig. 9.16 EDM instrument. (**a**) Example of an autonomous EDM instrument. (**b**) EDM instrument attached to the theodolite's telescope

In the 1980s, the EDM instrument began to be inserted into the theodolite's telescope, coaxial with the line of sight, forming a single assembly for angular and distance observations. At about the same time, with the development of micropro-cessors, this new instrument also began to store field data and attributes on an internal storage device or an external electronic field book. This new instrument was called a *total station*. As a result of this new technology, field measurements became more structured and much faster than previous methods. See Fig. 9.17.

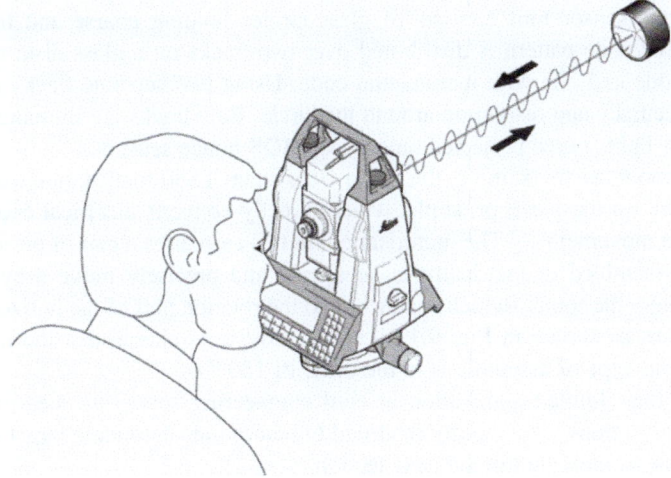

Fig. 9.17 Principle of total station distance measurement

Current total stations have different modes for distance measurements depending on the manufacturer. The difference between them is the time taken to measure the distance and the accuracy of the measured value. Table 9.3 shows an example of the relationship between these values.

Table 9.3 Distance measurement modes with a total station

Instrument	Distance measurement mode	Accuracy	Measuring time
Leica Flexline	Precise	1 mm + 1.5 ppm	2.4 s
	Fast	2 mm + 1.5 ppm	2.0 s
	Tracking	3 mm + 1.5 ppm	0.15 s
	Averaging	1 mm + 1.5 mm	--
	Long range	5 mm + 2 ppm	2.5 s

9.3.3 Robotic Total Station (RTS)

A *Robotic Total Station* (RTS) is a surveying instrument with built-in servo drivers for the automatic 360° rotation of the alidade and the telescope around their axes. Due to this capability, a designed position can be accurately reached without any iterative correction by the operator. Therefore, given these characteristics, it has been considered a powerful tool, especially when used for setting-out works.

Each instrument manufacturer uses a different servo drive. They are mounted directly on the horizontal and vertical axes of the instrument to eliminate additional mechanical gears.

For example, the Trimble S6 total station uses integrated servo and angle technology based on a frictionless electromagnetic drive and direct drive technology called MagDrive®. The servo drive consists of a holder containing magnetic and soft iron areas that are distributed in two concentric cylindrical structures separated by an air gap, as shown in Fig. 9.18.

The angle sensor unit consists of glass circles holding coarse and fine code patterns. The code pattern is distributed over two tracks on a glass disk: one with absolute code and one with incremental code. Using two separate tracks provides uniform accuracy and resolution around the circle. Both tracks are illuminated by a single laser light source projected onto two CMOS image sensors.

Furthermore, as an example, the Leica Geosystem TS30 total station uses direct drives based on the piezo principle, which directly converts electrical energy into mechanical movements.[11] The motorisation, in this case, uses a pair of piezoelectric ceramics assembled diametrically to accelerate and precisely move a cylindrical ceramic ring—the rotor—which is attached to the rotation part of the horizontal and vertical axes, as shown in Fig. 9.19. Leica Geosystems claims that the rotational speed for this type of assembly is in the order of 180 °/s.

Due to their limited application in civil engineering surveying measurements, robotic total stations were rapidly modified to incorporate automatic target recognition devices, as presented in the next section.

---

[11] For more information, readers are suggested to consult Uchino, K. (2019): MicroMechatronics.

Fig. 9.18 Integrated angle and servo system used by Trimble. (Source: Lemmon and Jung 2005)

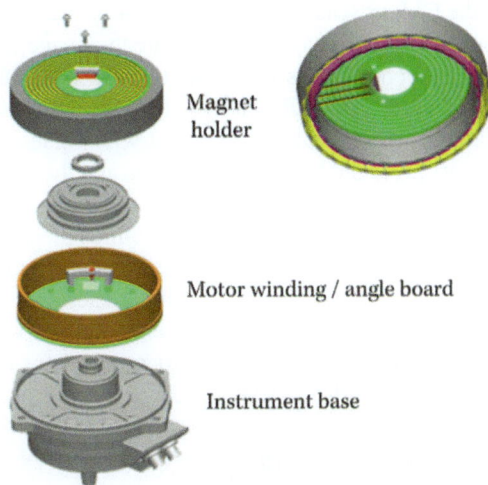

Magnet holder

Motor winding / angle board

Instrument base

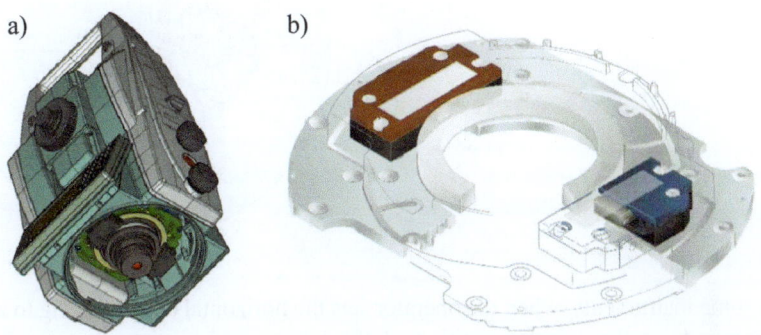

a)

b)

Fig. 9.19 Principle of piezo motors system from Leica Geosystems AG. (**a**) Piezo motor system installed in a Leica Geosystems total station. (**b**) Direct drive of the Leica Geosystem TS30 total station coupled with coded glass circle. (Source: Zogg et al. 2009)

9.3.4 Automatic Target Recognition (ATR)

The ATR device is a target recognition system based on a CMOS sensor array built into the total stations, which allows the instrument to actively find a stationary prism or track a moving reflector prism and determine its angular position with high accuracy. Today, it is an indispensable part of every RTS.

For automatic target recognition, the instrument emits an infrared laser beam coaxial with the telescope to locate the reflector prism. When the laser beam hits a prism, the beam reflects back to the telescope. A beam splitter then decouples the beam from the optical path and guides the light through a laser beam bandpass filter to the CMOS sensor. The reflected laser beam appears as a light spot on the CMOS sensor. Various algorithms evaluate the image data, identify the prism and calculate the pixel coordinates of the spot centre with subpixel accuracy. Using these pixel coordinates, the ATR sensor calculates the offset of the spot centre regarding the optical axis. Combining the offset with values from the instrument's angle and tilt sensors gives the horizontal direction and the final vertical angle of the point relative to the prism centre.

Note that the ATR correction requires measuring the distance to the prism. Each time the distance is measured, the instrument corrects the offset values. This means that when operators look through the telescope, they will see that the crosshairs are not positioned precisely in the centre of the prism. There will always be an ATR offset as shown in Fig. 9.20 (exaggerated for better visualisation), which is automatically corrected as explained.

Fig. 9.20 Principle of ATR offset

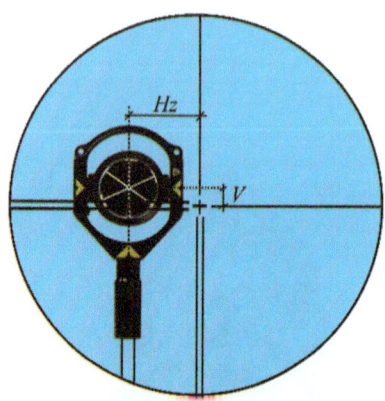

On some instruments, when the operator sets the horizontal circle reading to zero, it corresponds to the angular position of the telescope, however, the instrument internally corrects for the offset and records the zero-value corresponding to the centre of the target.

The prism search process is carried out by automatically moving the telescope within a search window defined by the user. To speed up the search process, some total stations have an additional sensor mounted on top of the telescope, which emits a laser beam in a 20° by 0.5° window, as shown in Fig. 9.21. As soon as this laser beam finds the prism, the instrument triggers the ATR module to search along the vertical axis. If the instrument does not find the prism, it repeats the search algorithm until it finds it or displays a prism not found warning.

The ATR module has algorithms to evaluate the characteristics of the image and the reflected energy to confirm whether they come from a reflector prism of a total station to avoid the detection of non-relevant reflectors. At the current stage of this technology, the performance of the detection algorithms is not error-free, and the user of an RTS must pay attention to this occurrence.

Fig. 9.21 Laser beam for automatic target searching. (Courtesy Leica Geosystems AG)

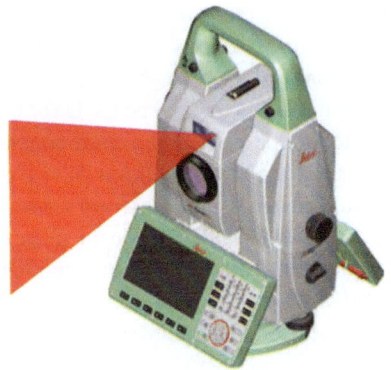

The detection range of a reflector prism by the ATR module of a total station depends on the type of prism. It varies from 55 m for an adhesive prism to 1500 m for a circular prism. Readers are advised to consult the technical instructions of their instruments to check the recommended distances.

To ensure that ATR measurements are correct, the operator can perform a manual angular measurement operation to a prism approximately 100 m away from the instrument, write down the value, then move the telescope and perform an automatic measurement. Both measured values must be in the accuracy range specified by the manufacturer.

RTS can also have a prism tracking module associated with the prism detection capability. This means the telescope moves as the prism moves, detecting its position almost instantly. Some manufacturers claim their instruments can track the prism movement at a speed of up to 50 km/h, depending on the type of prism and the distance to the instrument. For a field assistant (prism holder) moving across the terrain, a speed of around 18 km/h is sufficient.

In terms of precision, manufacturers claim that robotic instruments achieve the same values as manual measurements. The operator must confirm these values, especially for different types of prisms and distances.

9.3.5   Image-Assisted Total Station (IATS)

Another major advance in acquiring spatial data through surveying instruments has been the integration of digital imagery, resulting in the so-called *Image-assisted Total Stations* (IATS). The basic operating principle of the image capture system consists of placing an array of CCD or CMOS sensors coaxial with the optical axis of the instrument's telescope. The camera streams the image onto the instrument's display or to an external field computer at a magnification of about 30 times, in addition to orienting and georeferencing the line of sight using the angular sensor units. When standing at the instrument, the operator has the option of looking at the live video stream or through the telescope. This means that the adjustment and calibration processes are critical factors in achieving the highest quality of image functionality.

Some IATS have three CCD (or CMOS) cameras operating together, two outside the telescope body and a third coaxial with the line of sight. Figure 9.22 shows an example of an IATS setup with two cameras installed. The first of them, installed in the upper part of the telescope's body, captures a panoramic image of the scene. The second one, installed inside the telescope, captures the magnified image used for data acquisition. Note that the image capture system and the resolution of the CCD or CMOS sensors vary between manufacturers. The example in Fig. 9.22 is a schematic cross-section of the Leica Nova MS50 total station telescope.

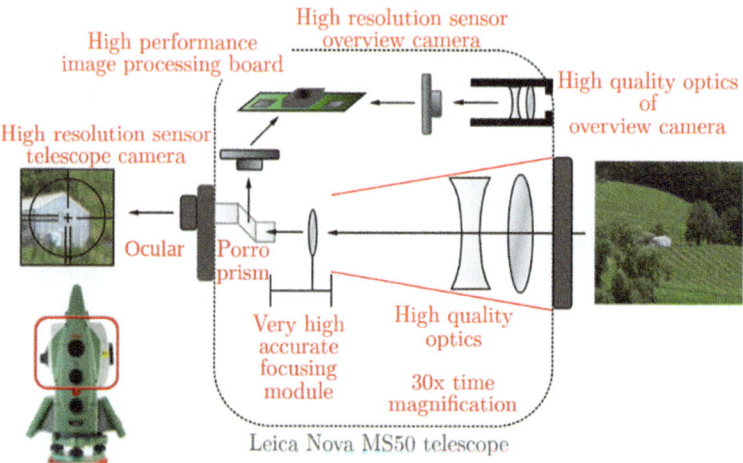

Fig. 9.22 Schematic cross-section of the Leica Nova MS50 total station telescope. (Source: Grimm and Zogg 2013)

The imaging system of an IATS can help the operator quickly search and detect the object to be measured or even record images for documentation or later photogrammetric processing, as described in *Chap. 19*. For this, an image processing application program must accompany IATS, or the images must be imported into appropriate programs to allow the instrument to be used as a photogrammetric sensor.

The development of autofocus, motorisation and reflectorless capabilities allows a feature point to be selected in the displayed image. By tapping on the selected point on the screen, the instrument rotates to where it is located in the field, and a surveying measurement can be taken. Instruments operating entirely on imagery and without needing a visual observation telescope are already available on the market, such as the Trimble SX10 total station, shown in Fig. 9.23.

Fig. 9.23 Trimble S10 total
station, operated by image
sensors. (Courtesy Trimble
Navigation Limited)

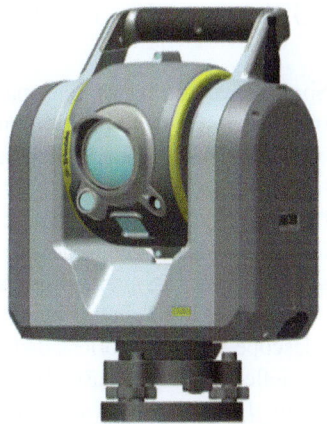

In an IATS, the measurement points and connecting lines graphically overlay the
image data on the screen. To work correctly, the image data and graphic data are
displayed as a central projection using collinearity equations so that both are
correctly superimposed on the image. More information on the collinearity equation
is presented in *Chap. 18*. To help the reader understand image-based surveying,
Fig. 9.24 shows a promotional image of the Leica Geosystems TS15 total station.

Fig. 9.24 Leica TS15 total
station as advertised by
Leica Geosystems. Courtesy
Leica Geosystems AG

9.3.6   Scanning Total Station

Some surveying instrument manufacturers have recently introduced scanning capabilities into their total station hardware. This innovation has been made possible by combining instrument motorisation, reflectorless distance measurement and IATS technologies. The result is a total station that combines the conventional surveying capabilities of geospatial data collection with laser scanning capabilities in a single instrument. Some manufacturers refer to this new type of instrument as a *scanning total station* (STS), and others as a *multi-station*.

Laser scanning, in this case, is performed by robotic movement of the total station telescope and reflectorless distance measurements on a predefined grid of points or by deflection of the laser beam based on a rotating polygon mirror, as shown in *Chap. 18*. In the latter case, the area to be scanned can be defined by selecting a viewing window corresponding to a rectangle defined by two points or a polygon for a more complex shape.

All laser scanning total stations are image-assisted. This allows the operator to define the scan area on the instrument screen, as shown in Fig. 9.25a. Other scan parameters are also specified on the same screen, such as point spacing, maximum scan distance and others. The scanning rate for a piezo motor-based instrument is in the order of 1000 points per second for a surface located at 300 m. For instruments based on laser beam deflection, the same rate is of the order of 26,000 points per second. The grid spacing varies between manufacturers and scanning methods (always in the order of a few millimetres at a distance of 10 m). For these and other details, readers should consult the manufacturers' catalogues. The Trimble SX10 total station shown in Fig. 9.25 is an example of a laser scanning total station.

The procedure for laser scanning measurements is shown in Fig. 9.25b. As detailed in *Chap. 18*, the result is a point cloud with 3D coordinates referenced to the total station's current coordinate system.

Fig. 9.25 Surveying measurements with scanning total station. (**a**) Definition of the laser scanning area. (Courtesy Trimble Navigation Limited). (**b**) Principle of surveying measurement with scanning total station. (Courtesy Leica Geosystems AG)

9.3.7   *Operational Characteristics of a Total Station*

As an electronic instrument, the total station can store all information related to its operational configuration and the values of spatial data measured in the field, including images and 3D point clouds. As mentioned, for ease of operation during data collection, the instrument is equipped with an LCD/VGA screen and generally accepts the connection of an external field controller, which allows the user to graphically control the type of spatial data collected and the geometry of the field measurement.

Structuring the data collection can be done using code tables (code lists), which, if well structured, will allow the survey software to understand the type of geometric elements being measured and automatically display them on the computer screen. This allows the experienced operator to collect and encode the data to be downloaded into the software without needing field annotation. Two types of coding are used for this purpose, depending on the application program and the type of instrument.

- *Free code*: the code is recorded as a line of text related to the measurand. It is entered into the instrument before point measurement. It can be entered manually or by commands associated with a code list. The surveying software understands the code as a command line to be executed, associated with software commands previously defined by the user in the software database.
- *Thematic code (or point code)*: the code is recorded as an attribute of a point, line or polygon. In this case, the recorded code is associated with the measurand. It can also be entered into the instrument manually or by commands associated with a code list. The surveying software then reads the point attribute and executes a command previously defined by the user in the software database.

The best practice is to predefine a list of codes suitable for the type of work. This makes data processing in the office a straightforward task. The engineer must simply attach the code table to the project management software and import the data. If a CAD 3D view is available, the points are automatically plotted with the blocks and connected by the lines that define the features that belong to the assigned layers.

In addition to measuring and coding configuration, total stations make it possible to carry out mathematical calculations using the values measured or entered into the instrument. Depending on the manufacturer, the operator can obtain the values of the following geometric elements in the field:

- Horizontal distance
- Vertical distance
- Local ground-based or grid coordinates
- Grid distance
- Azimuths
- Others

In addition, total stations also have internal application programs that allow for in-field processing, such as:

- Coordinate geometry calculations, as described in *Chap. 11*
- Coordinate transformation
- Height transfer
- Point stakeout functions
- Stakeout of 2D and 3D road geometric elements, as described in *Chap. 23*
- Tunnelling stakeout
- Free station coordinates determination
- Tie distance
- Traversing computation
- Area and volume computation
- Remote height
- Hidden point
- Reference Plane
- As-Built checking
- Area subdivision
- Instrument checking and adjustment
- Many others

All total stations have resources for indicating parameters for the atmospheric corrections of the distance meter, and some of them also have resources for indicating parameters for geodetic corrections, such as the map projection scale factor.

The raw data format recorded by total stations and how data is imported and exported varies between manufacturers. A standard still needs to specify how the data should be made available. Likewise, wireless mode operations and external media connections vary between classes of instruments.

Operational details on using a total station in the field are described in *Chap. 20*.

9.3.8 Classification of Total Stations According to Their Practical Use

There is no official classification of total stations according to their practical use. In general, they are classified according to their accuracy and automation capabilities. However, to help the reader to select the best instrument for different applications, the following classification is recommended by the authors:

- Total stations for building construction in Civil Engineering (*construction total station*)
- Total stations for large Civil Engineering construction sites, such as transportation, water dam, subdivisions, etc.
- Total stations for underground works
- Total stations for mapping survey

- Total stations for high precision surveying and geodetic surveying
- Total stations for industrial measurement or 3D metrology
- Total stations for machine control
- Total stations for surface modelling

The main technical features of each listed total station are presented below. Note that the accuracies and distance ranges suggested for each class are based on the values currently available from the instrument manufacturers.

Total stations for building construction (construction total stations) have their characteristics related to the needs of a construction site. Typically, they are designed to be operated by users with little experience in Geomatics but with knowledge of site work. They have menus and configuration parameters related to determining alignments, offsets, curves and stakeout works, graphically displayed. Instruments in this class do not need to measure distances exceeding 1000 m and may have angular and linear precisions of approximately ±7 arc seconds and ±(3 mm + 3 ppm), respectively. They must have a graphical interface to display the geometric relationships of the measurements, make reflectorless measurements, and store large amounts of information about the job in progress.

Total stations for large Civil Engineering construction sites are designed to be operated by users familiar with Geomatics. Such instruments must have high data storage capacity, the ability to perform coordinate geometry calculations in the field and processors to apply atmospheric and geodetic corrections to angular and linear measurements. To be used in civil construction works, they do not need to measure distances longer than 2000 m with a prism and 500 m reflectorless. Angular precision for this class of total station can vary from ±2 to ±5 arc seconds. Linear precision can be in the order of ±(2 mm + 2 ppm).

In most cases, fully manual total stations are preferred, with or without the ability to connect external data collectors. Although optional, the inclusion of a graphical interface (to display the georeferencing of the measurement) and the ability to store large amounts of information about the job can improve the instrument's performance. It is also important that the instrument be equipped with an electronic compensator to correct the verticality of the instrument, as shown in *Sect. 10.2.1.2.3.*

Total stations for underground works generally have the same basic characteristics as those used on large Civil Engineering construction sites. However, there are two fundamental differences. The first is the instrument's precision, which, for underground work, must generally have an angular precision of ±0.5 or ±1 arc second and a linear precision of ±(0.6 mm + 1 ppm) or ±(1 mm + 1 ppm). The second difference is that, depending on the type of work, total stations for underground work must be robotic, operate via Bluetooth, and have laser scanning capability.

Total stations for mapping surveys, in general, need to be able to work with code lists and have resources for including attributes of the measured spatial data. Graphic displays or image capture capabilities are recommended. They must be able to measure distance with and without prisms, in the order of 3000 and 1000 metres, respectively. They do not necessarily have to be robotic. For cadastral surveys, the instrument should be able to work with an external data collector. Angular precision for this class of total station can range from ±2 to ±5 arc seconds, and linear precision can be on the order of ±(2 mm + 2 ppm).

High-precision and geodetic surveying total stations are manufactured for long-distance measurements, reaching 5 km with a prism. They do not necessarily require resources for describing spatial data since they are preferably indicated for precision measurements in geodetic surveys. Typically, they have angular precision ranging from ±0.5 to ±1 arc second and linear precision ranging from ±(0.6 mm + 1 ppm) to ±(1 mm + 1 ppm). For geodetic monitoring of structures, they must be robotic and have an ATR module. Reflectorless distance measurement capability is optional. On the other hand, they must have processors for atmospheric and geodetic corrections and an electronic compensator for correcting the instrument's verticality.

Total stations for industrial measurement or 3D metrology are part of a specific type of surveying instrument designed specifically for this application. They do not require operators specialised in Geomatics, so they have menus and operating systems for industrial dimensioning. They are usually supplied with special application programs for specific industrial applications. They also do not need a long-range or geodetic correction capability, but they must have reflectorless distance measurement capability and allow connection to an external data collector. Like total stations for high-precision surveying and geodetic surveying, total stations for industrial measurement or 3D metrology typically have angular precision ranging from ±0.5 to ±1 arc second and linear precision ranging from ±(0.6 mm + 1 ppm) to ±(1 mm + 1 ppm). This instrument always has an electronic compensator to correct the instrument's verticality.

The total stations used for machine control must primarily be high-precision instruments. Using instruments with angular precision equal to ±0.5 arc seconds and linear precision in the order of ±(0.6 mm + 1 ppm) is recommended. They do not need to measure distances longer than 1 km with a prism, nor longer than 500 m reflectorless. However, they must be robotic and have ATR capability. They must also have geodetic correction capability and be able to connect to external devices. Bluetooth connectivity is essential. They do not necessarily need to have resources for describing spatial data. However, it is recommended that they store large amounts of design and measurement data even if, most of the time, the machine positioning calculations are performed in the machine. It is also essential that the instrument is equipped with an electronic compensator to correct its verticality.

The eighth class of total stations includes instruments capable of laser scanning, as described in *Sect. 9.3.6*. They are often used in surface modelling surveying, especially in the case of complex three-dimensional structures or for Digital Terrain Modelling. The main parameters to consider for this class of instruments are accuracy, scanning rate and range, point spacing and resolution of the digital image.

9.4   Reflector Prism and Prism Pole

The quality of the accessories is of fundamental importance to achieve high perfor-mance and accuracy in surveying works. Professionals often worry about the instrument itself but sometimes neglect the accessories. A faulty or poorly maintained accessory can compromise the quality of the entire work. Using an instrument of the highest quality and precision is only possible if the accessories are up to the job. Rigorous care with the instrument, accessories and field procedures is the key to any professional's success. In this context, in addition to the parts of the surveying instrument presented in *Sect. 9.2*, the reflector prism and the prism pole are two accessories to be considered when measuring with a total station. Technical details about them are given in the following sections.

9.4.1   Reflector Prism

A reflector prism is an accessory used together with a total station to reflect the electromagnetic signal emitted by the EDM instrument when measuring a distance. It is usually attached to the end of a surveying pole (prism pole) so that the offset between the centre of reflection of the electromagnetic signal and the surveying pole axis is known. Manufacturers currently offer different types of reflector prism, as shown in Fig. 9.26.

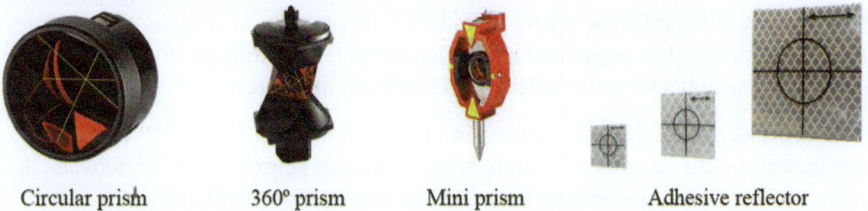

Circular prism   360° prism   Mini prism   Adhesive reflector

Fig. 9.26 Reflector prisms used for distance measurement with a total station. (**a**) Circular prism. (**b**) 360° prism. (**c**) Mini prism. (**d**) Adhesive reflector (Courtesy Leica Geosystems AG)

Except for the adhesive prism, reflector prisms consist of a set of glass prisms made by cutting the corner of a solid glass cube to form mutually parallel reflecting surfaces that reflect any incident electromagnetic beam parallel to the angle of incidence, as shown in Fig. 9.27.

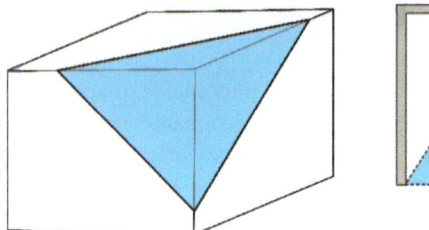 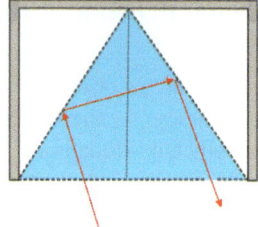

Fig. 9.27 Reflector surfaces of a reflector prism. (**a**) Principle of cutting glass cubes for the production of prisms. (**b**) Parallel signals incident and reflected on a prism

The measurement quality using a reflector prism depends on some characteristics related to its construction and use, as indicated below:

- Reflector constants
- Signal path
- Beam deviation
- Reflective coating
- Anti-reflective coating
- Glass quality

The details of each of these characteristics are presented below.

Reflector Constants

Whenever measurements are made from an EDM instrument to a reflector prism, they are referenced between two points: the vertical axis of the instrument and the vertical axis of the reflector prism. It is, therefore, essential to consider the mechanical design of the glass prism body, housing and pole mounting pin to ensure a well-defined distance between the reflector and instrument centres.

The offset between the electronic and mechanical centres in the EDM instrument is adjusted during instrument manufacture. As shown below, the offset between the refraction centre and its vertical axis depends on some geometric elements.

Since the speed of an electromagnetic wave is influenced by the reflection index of the propagation medium, all electromagnetic beams entering the prism are decelerated, and the optical path inside the prism body is given by Eq. (9.2). See Fig. 9.28.

$$w = n * s \tag{9.2}$$

where

$w=$ distance from the front surface of the prism to the theoretical reflection point (s_0)
$n=$ glass body refractive index
$s=$ distance from the front surface of the prism to its corner point

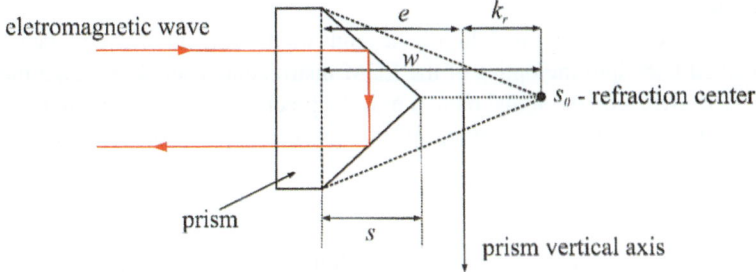

Fig. 9.28 Cross-section of a circular prism showing its geometric elements

To relate the measured distance to the vertical axis of the prism, the prism constant or addition constant k_r is applied, as given in Eq. (9.3).

$$k_r = e - (n * s) \tag{9.3}$$

where

k_r prism constant
$e=$ distance from the vertical axis to the front surface of the prism

The value of k_r, set by the manufacturer, is a function of the instrument and the prism type. It is usually between -30 and -40 mm, but can also be zero. Therefore, the operator must be concerned with the value configured in the instrument when replacing the prism.

The position of the prism's vertical axis varies with the prism's inclination concerning the line of sight. Therefore, if the prism or the line of sight is tilted by an angle α, an error Δd will occur in the measured distance, which can be calculated using Eq. (9.4).

$$\Delta d = e * (1 - \cos \alpha) - s * \left(n - \sqrt{n^2 - \sin^2 \alpha}\right) \tag{9.4}$$

where

$\Delta d=$ EDM distance correction
$\alpha=$ tilting angle of the front face of the prism concerning the line of sight

For example, a prism with geometric values of $e = 40$ mm, $s = 60$ mm, and $n = 1.5$ would have an addition constant $k_r = -50$ mm. In this condition, if the line of sight is tilted by an angle $\alpha = 30°$, a distance error of $\Delta d = 0.2$ mm would be observed. Note that this value could reach 3.5 mm for a tilting angle of $\alpha = 60°$.

In most cases, the tilting angle is small and can be neglected. However, the operator must always ensure that the front face of the prism is as perpendicular as possible to the instrument's line of sight axis. The prism itself has a close-up crosshair, which must be used to align the prism face with the line of sight.

Signal Path

The distance measurement range highly depends on the amount of energy generated and received back into the optics of the EDM instrument. Therefore, in addition to atmospheric effects on the signal path, it must be precisely collimated with the centre of the prism to maximise the amount of energy reaching it. For example, a given prism of 5 cm in diameter provides a surface of $0.002 \, m^2$, representing 1/20,000 of the emitted signal. This is considered the ideal reflection surface for the EDM signal. A larger prism would not increase the distance measurement range as the receiver optics can only process the reflected signal within a specific diameter. A larger prism would undoubtedly contribute to the amount of returned signal, but most of the returned signal would pass through the optics of the EDM instrument. In this case, the solution is to use multiple reflectors to increase the amount of reflected energy and, consequently, the distance measurement range (Mao and Nindl 2009).

Beam Deviation

The angle between the incident and reflected electromagnetic beam from the prisms is called the *beam deviation*. If the angle value is significant, there will be a reduction in signal strength and, therefore, a reduction in the distance that can be measured with the instrument. Good quality prisms do not deviate by more than a few seconds of arc, typically around $1''$.

Reflective Coating

The prism's reflective index represents its ability to reflect electromagnetic signals used in EDM instruments. This phenomenon depends on the material from which the prism is made and the quality of the reflective coating. Good quality prisms usually have copper-coated faces that achieve a reflective index of about 75%. This index, therefore, also influences the distance measurement range.

Anti-reflective Coating

During distance measurement, most of the signal emitted by the EDM instrument passes through the prism body to be reflected. However, there is a portion of the energy reflected by the prism's front face, disturbing the quality of the return signal, as it has a shorter return time than the original beam because it does not pass through the prism body.

Good quality prisms have an anti-reflective coating on their front surface to avoid this phenomenon. Measurements with prisms that do not have this coating or whose coating is not suitable for the type of electromagnetic wave emitted may have errors of up to 3 mm in the measured distance. This phenomenon is, therefore, more pronounced for short distances.

Glass Quality

To ensure the quality of the measurement, the glass with which the prism is manufactured must have a homogeneous refractive index, contain as few air bubbles as possible and must be robust enough to withstand changes in climatic conditions without altering its physical and geometrical characteristics.

9.4.1.1   Circular Prism

The circular prism consists of a triple-prism glass assembly conveniently fitted into a circular housing (Fig. 9.26a) by grinding down its three corners. It is the most commonly used reflector prism in Geomatics surveying. For use in the field, it is protected by a resin housing mounted on a support that allows it to rotate vertically about its horizontal axis and horizontally about the pole axis, as shown in Fig. 9.29. To avoid misalignment errors, circular prisms require line of sight alignment within a tolerance of ±10° on the vertical and horizontal axes (Mao and Nindl 2009).

Fig. 9.29 Pole with prism

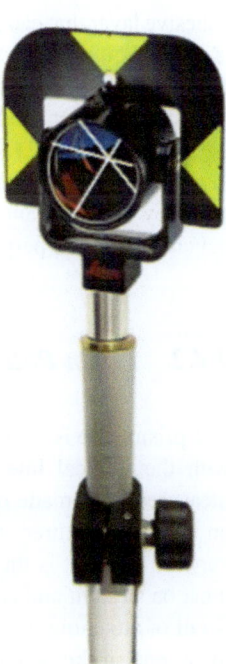

9.4.1.2   360° Prism

The 360° prism is a particular type of reflector that consists of an assembly of six individual prisms forming an arrangement that allows the incident electromagnetic beam to be reflected from any incidence position without the need to direct it to the line of sight. Because of this feature, such a prism is indicated for use in conjunction with the RTS. In this case, the field assistant does not need to worry about aligning the prism face with the line of sight.

9.4.1.3   Mini-prism

For short distances and special applications, it is also possible to use a circular mini-prism or a 360° mini-prism. These prisms are suitable for situations where the size of the prism may be a limiting factor or where the use of a smaller reflector will facilitate fieldwork, for example, for stakeout work. All recommendations and characteristics indicated for circular, and 360° prisms are also applied to mini prisms.

9.4.1.4   Adhesive Reflector

As an alternative to reflector prisms, the engineer can choose to use a particular type of reflector called an adhesive reflector. This type of flat reflector is made of plastic and available in various sizes, shapes and colours. Because it is a flat prism, it has an adhesive layer that can be stuck to a flat surface to which a distance is to be measured. It is generally used when the reflector needs to remain in a fixed position for long periods.

The range of such a reflector varies according to its size, but typically, a square adhesive reflector of about 30 mm will have a range of about 100 m. In addition to size, another critical factor to consider when using adhesive reflectors is the angle of incidence of the electromagnetic beam. As a flat reflective surface, the range and accuracy of a distance measurement can deteriorate rapidly if the line of sight deviates from the perpendicular to the reflector plane.

9.4.2   Prism Pole

The prism pole is an accessory that supports the reflector prism and keeps it aligned with the vertical line that passes through the point where the distance is to be measured. It is made of metal or carbon fibre and has a variable length, depending on the manufacturer. A circular level bubble attached to its body ensures its axis's verticality. As it is the extension of the prism's axis to the ground station, its axis must be straight and coincide with the centre of its two ends. Figure 9.30 shows the detail of attaching the reflector prism to the pole assembling pin and positioning the lower end of the pole over the ground station mark.

Fig. 9.30   Upper and lower parts of a prism pole. (**a**) Upper end of a prism pole and assembling pin. (**b**) Lower end of a prism pole over the ground station mark

9.5 Surveyor's Level

In Geomatics, the generic name *"surveyor's level"*[12] is given to the surveying instrument used to establish horizontal reference planes on the Earth's surface to determine the difference in elevation between points on the terrestrial surface. It is recommended that readers less familiar with levelling read *Sect. 13.3*, for more details on this subject. As described in this section, the horizontal plane is determined by rotating the surveyor's level telescope around a vertical axis attached to the tribrach mounted on a tripod. The telescope's crosshairs can be used to make vertical measurements on a levelling staff, as shown in Fig. 9.31.

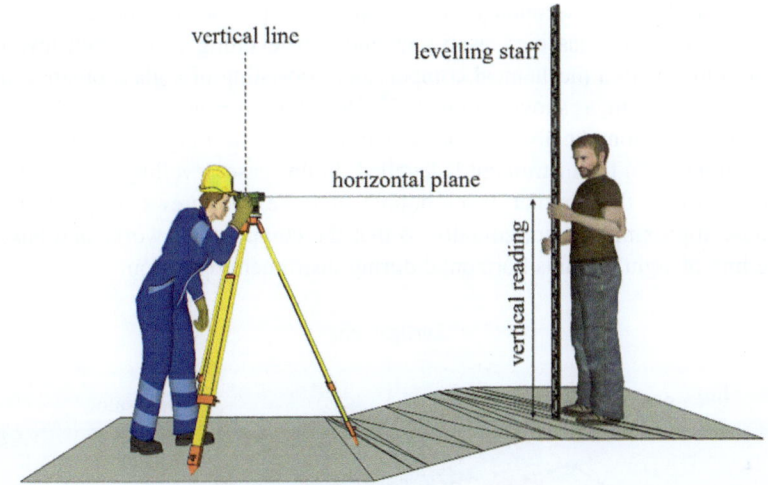

Fig. 9.31 Principle of differential levelling

Details of the levelling staff are given in *Sect. 9.6*. Details of the use of the surveyor's level in conjunction with the levelling staff for vertical measurements are given in *Chap. 13*. Only the technical details of the surveyor's level are described in the present chapter.

There are three types of surveyor's levels for Geomatics Applied to Civil Engineering.

- Optical level
- Digital level
- Laser level

---

[12] Also referred by some authors as *engineering level*, *dumpy level* or simply *level*.

Although the operating mode varies, the working principle of all surveyor's levels is quite similar. The main technical details of each of them are presented below.

9.5.1   Optical Level

The optical level is a surveying instrument used for direct levelling, i.e. the direct measurement of vertical distances using a levelling staff, as described in *Sect. 13.3*. It consists of the alidade, a telescope that can rotate about the vertical axis of the instrument, a single tangent screw for horizontal movements, a circular level bubble and a tribrach. It has no secondary axis, so the telescope does not move vertically.

Optical levels are classified as tilting and self-levelling levels. Self-levelling levels are those with a mechanical compensator consisting of a glass prism, mirror and pendulum system, as shown in Fig. 9.32. With this compensator, the telescope's line of sight is automatically horizontal within a specific range (specified by the manufacturer) when the instrument is levelled. In this case, levelling the instrument consists of using the circular level bubble and the footscrews to position the instrument approximately horizontally so that the compensator works and ensures that the line of sight remains horizontal during instrument operation.

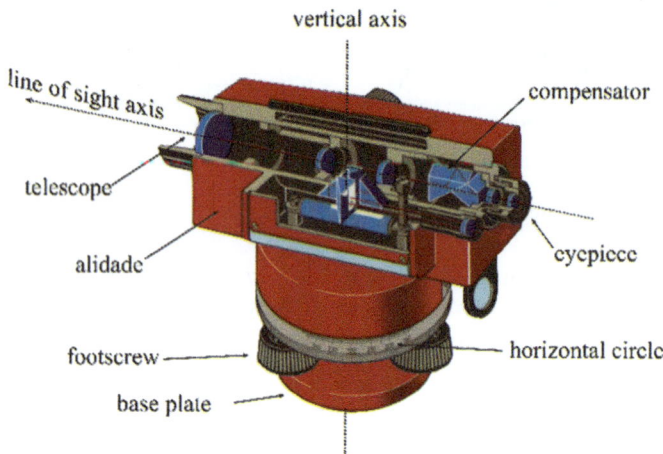

Fig. 9.32 Main components of a self-levelling optical level

Tilting levels, on the other hand, do not have an automatic compensator. In this case, the line of sight is levelled using a tubular level bubble that must be centred in each sight to keep the line-of-sight horizontal. The level bubble is centred by slightly tilting the vertical plane of the telescope by turning a micrometre knob attached to it without changing the height of the instrument. The footscrews should not be moved during this procedure.

In modern self-levelling optical levels, the compensator is factory-set at a controlled temperature to ensure a horizontal line of sight. However, this can change if the temperature changes more than 10 °C or the instrument vibrates excessively. The operator should inspect the line of sight whenever this occurs to ensure it remains horizontal. *Section 10.4.1.2* describes the field procedure for this inspection.

Some optical levels also have a horizontal circle, with a resolution of 1°, installed below the telescope and attached to it. See Fig. 9.32. Although the resolution is low, it can be used to observe directions in low-accuracy Civil Engineering works.

The crosshairs built into the telescope define the line-of-sight axis and the horizontal line of the instrument. They also allow the operator to measure distances using the principle of stadia distance measurement, as described in *Sect. 8.3.1*.

As for the theodolite or total station, the telescope of an optical level has two focus controls, one focusing on the target and the other on the crosshairs. This means that the operator must also be concerned with the occurrence of parallax when focusing on the target.

In terms of precision, optical levels are classified according to the standard deviation of a 1-km double-run levelling, as described in *Sect. 13.6*. Table 9.4 shows the nominal values of some surveyors' levels available on the market.

Table 9.4 Nominal precision of self-levelling optical levels

Class	Standard deviation of a 1-km double-run levelling (ISO 17123-2)	Application
Building construction level	2.5 mm	Construction sites, profile levelling
Engineering level	1.5–2.0 mm	Public works, road construction
High-precision level	0.7–1.5 mm	Industrial construction, high-precision civil works
Geodetic level	0.3–0.5 mm	High-precision geodetic levelling and structure monitoring

9.5.2 Digital Level

A particular type of self-levelling instrument was launched on the market in the 1990s, whose main characteristic is the ability to read the elevation values on a barcode levelling staff digitally. Hence the name *"digital level"*. It works the same way as self-levelling optical levels, but with the advantage that the operator only needs to aim and focus on the levelling staff (if the instrument is not already equipped with autofocus). The instrument does the rest: reads and stores the readings. To make this possible, the instrument is equipped with a linear array of CCD sensors that capture the image of the displayed portion of the barcode levelling staff and compare it with a standard image of the crosshair recorded in the instrument's memory.

The instrument's internal processor finds the best correlation using an image correlation method and displays the reading value on the instrument's screen, as shown in Fig. 9.33. The digital levels can also be operated as conventional optical levels in adverse environmental conditions or in the event of a power failure.

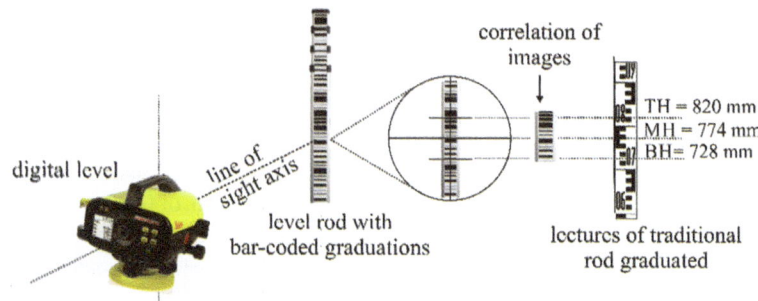

Fig. 9.33 Reading on the bar-coded levelling staff and lectures of traditional rod graduated

As briefly shown in Fig. 9.34, in addition to the vertical and line-of-sight axes common to all surveyors' levels, the digital level consists of the following components: a set of lenses with horizontal and vertical crosshairs, an automatic levelling compensator, a damping compensator, a vibration compensator, a beam splitter, a spectral radiation filter for CCD sensor array, a CCD sensor array, an eyepiece, an electronic mainboard and a temperature sensor.

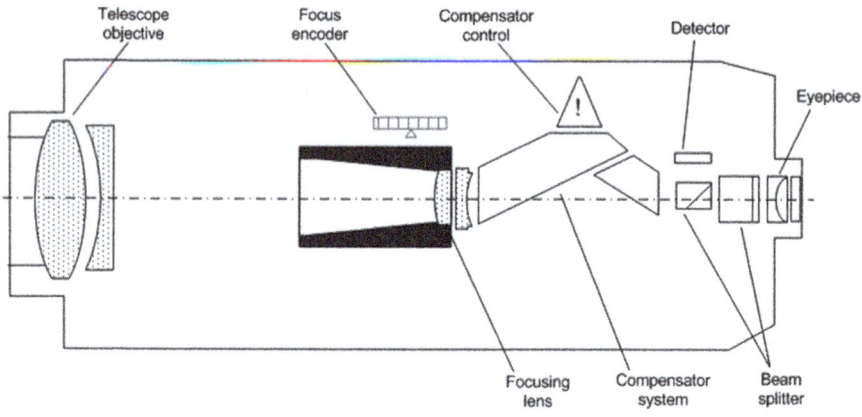

Fig. 9.34 Basic optomechanical design of a digital level. (From Schofield 2001)

The portion of the barcode levelling staff captured by the CCD camera depends on the distance of the levelling staff from the instrument. For this reason, the digital level measures the distance between the instrument and the levelling staff, which facilitates the compensation process for the levelling error of closure, as indicated in *Sect. 13.4.*

The maximum distance for a digital level to read the levelling staff is 110 metres, with a distance measurement precision of approximately 3 to 5 mm/10 m. The maximum reading resolution for the height difference is in the order of 0.01 mm.

As electronic instruments, digital levels have internal application programs that automatically record measurements and perform mathematical operations according to the recorded data. Some instruments also have routines to assist the user in field readings, such as line levelling and point stakeouts.

Regarding coding the observed points, the digital levels use the same types of codes as total stations, as described in *Sect. 9.3*.

The method of measurement varies from manufacturer to manufacturer. However, most instruments use the following procedure:

- The operator sights the levelling staff, focuses the image and starts the measurement.
- The instrument controls whether the compensator is active. Some instruments have devices to check the tilting of the instrument in the longitudinal and transverse directions to the line of sight.
- The instrument captures an infrared image of the displayed portion of the levelling staff and calculates an approximate distance from the position of the internal focusing lens.
- The infrared image is temporarily transformed into a binary image for an initial comparison between the image obtained and the image recorded in the instrument's memory (rough correlation).
- With the image roughly positioned, the system uses the complete 8-bit infrared image and carries out a refined search for the most probable position in relation to the image recorded in the instrument's memory (fine correlation), displaying the reading on the instrument's screen.

Using a digital level has advantages and inconveniences, as mentioned below.

Advantages
- Elimination of blunders such as misreading on the levelling staff and annotation errors.
- Automatic distance measurement, which can help to compensate for levelling error of closure.
- Faster operating speed in the field and the office. Operators have reported up to a 50% reduction in time.
- Consistency in the quality of measurements.

Inconveniences
- Power requirements for operation.
- Need for sufficient infrared light for the instrument to perform the measurement.
- Need to capture an infrared image of a portion of the levelling staff to perform the measurement. The portion of the crosshairs alone is not enough.

In terms of precision, digital levels, like optical levels, are classified according to the standard deviation of a 1 km double run. Table 9.5 shows the nominal values of some digital levels currently available on the market.

Table 9.5 Nominal precision of some digital levels currently available on the market

Manufacturer/ model	Standard deviation of a 1-km double-run levelling (ISO 17123-2)	Other characteristics
Leica LS15	0.2 mm	Measuring distance: 1.8 m to 110 m; Magnification power of the telescope up to 32 times; Internal storage of up to 30,000 levelling lines; Digital compass; Digital level bubble; Autofocus; International Protection Rating: IP55
Sokkia SDL1X	0.2 mm	Measuring distance: 1.6 m to 100 m; Magnification power of the telescope up to 32 times; Internal storage of up to 10,000 levelling lines; Autofocus; International Protection Rating: IP54
Trimble DiNi 0.3	0.3 mm	Measuring distance: 1.5 m to 100 m; Ability to take measurements with just a 30 cm portion of the levelling staff barcode; Magnification power of the telescope up to 32 times; Internal storage of up to 30,000 levelling lines; International Protection Rating: IP55
Topcon DL-502	0.6 mm	Measuring distance: 1.6 m to 100 m; Magnification power of the telescope up to 32 times; Internal storage of up to 2000 levelling lines; International Protection Rating: IP54

9.5.3 Laser Level

The laser level is the name given to the surveyor's level capable of generating a beam of amplified, monochromatic, directional and highly collimated electromagnetic radiation of the light spectrum, called LASER (*Light Amplification by Stimulated Emission of Radiation*). It can be in the visible or infrared spectrum and has a very small angle of divergence and a high degree of similarity in phase, direction and amplitude. These properties make it an excellent tool for determining planes in space or reflectorless distance measurements. Based on these characteristics, since the 1990s, levelling instruments have been developed based on the principle of emitting a laser beam in a horizontal or vertical direction to determine points, alignments or planes in space and have been given the name laser level.

A laser level is a laser beam emitter built into the body of a measuring instrument with characteristics similar to those of a self-levelling optical level to produce controlled lines of sight. Although physically different from an optical level, it can be understood as a surveyor's level in which the telescope has been replaced by a laser beam emitting diode working in conjunction with a pentaprism, as shown in Fig. 9.35.

Fig. 9.35 Laser beam generation on a laser level

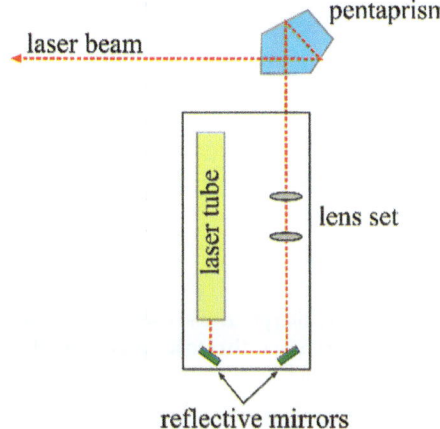

Due to their physical characteristics, laser levels are most commonly used in civil construction applications, such as alignment and levelling on construction sites (indoor and outdoor), earthworks, drainage works and others. They are highly automated and reliable electronic instruments that can be levelled manually or self-levelled and can detect levelling losses during field operation.

The simplest laser level is the one in which the laser beam is emitted in a single direction, producing alignments in the horizontal, vertical or inclined directions, as shown in Fig. 9.36a. A derivate of this type of laser level is the one in which an optical prism capable of splitting the laser beam into horizontal and vertical lines is inserted to produce right angles on the beam projection surface, as shown in Fig. 9.36b.

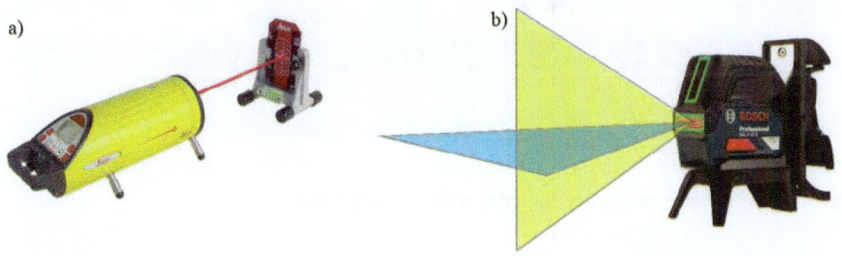

Fig. 9.36 Laser level with fixed diodes. (**a**) Laser level for liner alignment (Courtesy Leica Geosystems AG). (**b**) Laser level projecting horizontal and vertical lines for orthogonal orientation. (Courtesy Bosch Co. Adapted by authors)

A second type of laser level used in Civil Engineering works is one in which the pentaprism can rotate around the vertical axis of the instrument, producing a plane in space that can be detected by a laser detector. The generated plane can be horizontal, vertical or tilted (in one or two directions), as shown in Fig. 9.37a.

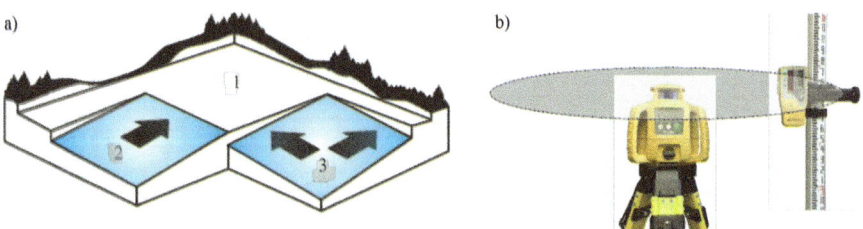

Fig. 9.37 Levelling planes and laser detector installed in a levelling staff. (**a**) Levelling planes from a rotating laser level. (**b**) Rotating laser level and laser detector installed in a levelling staff. (Courtesy: Bosch Co)

The dimensions of the laser detector vary depending on the application. For typical Civil Engineering levelling applications, they are similar in size to a mobile phone. They can be used to mark points directly on surfaces or mounted on a levelling staff for engineering levelling, as shown in Fig. 9.37b.

In the case of machine automation in Civil Engineering, the laser detectors are larger and more robust to allow them to be installed in mast or machine implements as shown in Fig 9.38.

In terms of precision, rotating laser levels are classified according to the deflection of the laser beam concerning its line of collimation. Table 9.6 shows the nominal values of some rotating laser levels currently available on the market.

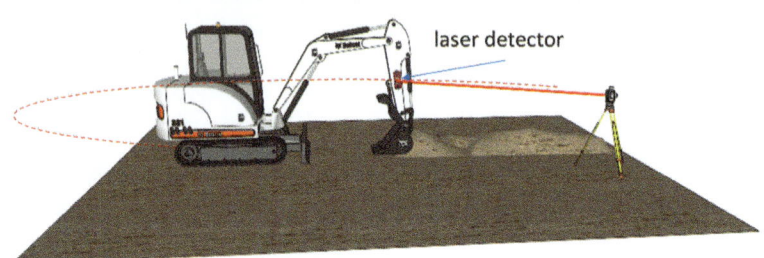

Fig. 9.38 Earthworks machinery equipped with a laser detector

Table 9.6 Nominal precision of some rotating laser levels currently available on the market

Model	Laser beam deflection	Other characteristics
LEICA Rugby 680 laser level with digital Rod Eye 160 detector	Precision: ±1.5 mm/30 m	Laser class: Class 1, Rotation: 600 rpm
		International Protection Rating: IP67
		Maximum range: 1350 m
		Detection precision: 0.5 mm
TOPCON H4CRL-H4C laser level with LS-80L detector	Precision: ±1.5 mm/30 m	Laser class: Class 3R, Rotation: 600 rpm
		International Protection Rating: IP66
		Maximum range: 800 m
		Detection precision: 1.0 mm
SPECTRA PRECISION LL500 laser level with HL700 detector	Precision: ±1.5 mm/30 m	Rotation: 600 rpm
		Maximum range: 500 m
		Detection precision: 0.5 mm

9.6 Levelling Staff

The levelling staff is a graduated rod, usually made of wood or aluminium and 3 or 4 metres long. It must be held vertically over the point at which a vertical distance measurement is to be taken, using a circular level bubble attached to it. The measurement is then carried out by reading the value corresponding to the intersection of the telescope's horizontal crosshair on the levelling staff, as shown in Fig. 9.39a. The graduation is in centimetres and mirrored every 5 cm to facilitate reading. Usually, levelling staffs are manufactured in two folding parts or several parts that fit together (sliding joints), as shown in Fig. 9.39b. The latter should be avoided for precision work.

For digital levels, the levelling staff is usually made of aluminium and graduated on both sides, one with a barcode and the other with conventional centimetre graduations. See Fig. 9.39c.

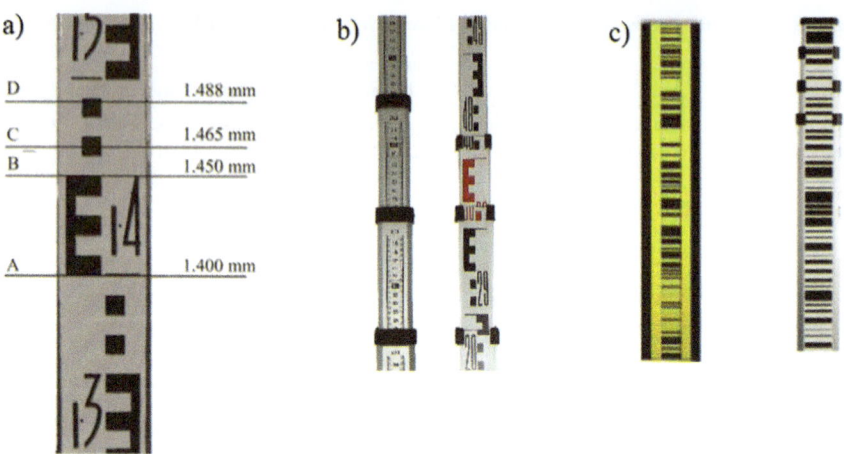

Fig. 9.39 Types of levelling staff. (**a**) Levelling staff with examples of readings. (**b**) Folding levelling staff and levelling staff with sliding connections. (**c**) Barcode levelling staff (invar and aluminium)

For example, a barcode levelling staff manufactured by Leica Geosystems is manufactured in bands 2.025 mm wide, with 15 bands (30.375 mm) forming a code. The 134 system codes (approximately 4.05 m) are all different. The instrument must observe at least two complete codes (approximately a 70 mm portion of the staff) to perform a measurement. For both the conventional staff and the barcode staff, the graduation precision is estimated to be equal to ±0.5 mm/m, and the trace position precision is equal to ±0.15 mm.

The reading precision on a conventional levelling staff is of the order of ±1 mm, which considers that the operator can distinguish the position of the stadia hairs on the staff with an accuracy of 1 mm. Knowing that the visual separation power of the human eye is equal to $1 = 0.0003$ rad, the maximum distance to guarantee a 1 mm distinction on a levelling staff with the naked eye is given by the following calculation:

$$d = \frac{0.001 \text{ m}}{0.0003 \text{ rad}} = 3.3 \text{ m}$$

In this way, the maximum reading distance to guarantee a 1 mm distinction with a telescope depends on the magnification of the telescope. So:

$$d[\text{m}] = 3.3 * G \tag{9.5}$$

Table 9.7 shows the relationship between the magnification of the telescope and the maximum distance at which the value of 1 mm can be distinguished using a conventional staff.

Table 9.7 Relationship between telescope magnification power and distance from the levelling staff

Magnification power	Maximum distance [m]	Recommended distance [m]
20×	67	30
25×	83	40
30×	100	50

Special invar[13] levelling staff is available for high-precision levelling. In this case, the graduation is made on the invar alloy protected by a wooden or aluminium support. Due to the high precision required, this levelling staff is manufactured in a single piece, with double graduation, vertically displaced, as shown in Fig. 9.40a, b. In this case, the reading procedure consists of using a micrometre attached to the telescope of the instrument, through which the line of sight is displaced so that it coincides with an integer value of the graduation of the invar tape. To coincide with the integer value, the reticule movement is measured by the micrometre (resolution in hundredths of a millimetre) and added to the integer value read, thus allowing the operator to have a greater resolution in the measurement carried out.

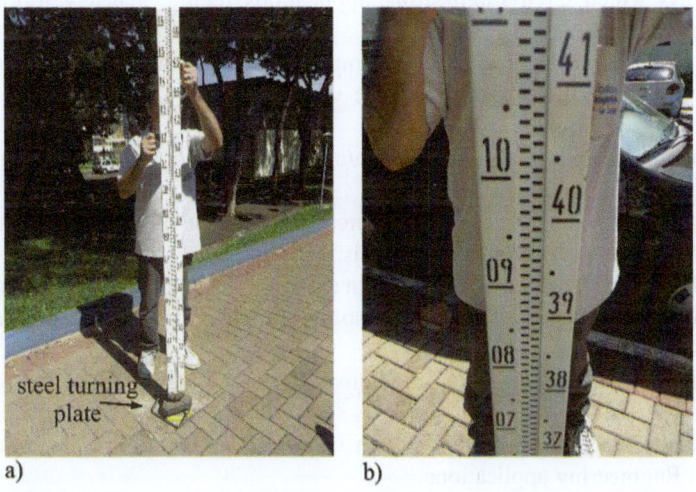

a)   b)

Fig. 9.40 Invar levelling staff. (**a**) Invar levelling staff on a steel turning plate. (**b**) Invar levelling staff graduation detail

The choice of levelling staff depends directly on the precision of the work to be carried out. The engineer must always remember that it is a measuring instrument and must be handled and cared for. The sensitive points of a levelling staff are:

---

[13] Invar is an alloy from a mixture of nickel and iron with an extremely low thermal expansion coefficient.

- Expansion due to atmospheric conditions.
- Ensuring its straightness and the perpendicularity of the graduations to its axis.
- Protection of its base against wear and tear.

Levelling staffs that have been dropped, have faulty sliding joints, or poor graduation must be checked and calibrated before any field work. In some cases, using a steel turning plate (Fig. 9.40) should also be considered to prevent the staff from sinking into the ground during levelling operations.

9.7   Review Questions

1. What is the difference between a theodolite, a total station and a surveyor's level?
2. When measuring with a total station, where is the reference point of the measurement located on the instrument?
3. Discuss what the axes of a total station are and their purpose in surveying measurement.
4. Discuss which Civil Engineering applications require a heavy tripod when surveying with a total station.
5. Briefly explain how an electronic compensator works with a total station.
6. Briefly discuss the importance of a well-calibrated optical (laser or digital) plummet when using a total station.
7. What are the advantages (or disadvantages) of a laser or built-in video plummet over an optical plummet?
8. Briefly discuss the importance of forced centring in surveying measurements and the field procedures for doing this.
9. Explain how graduated circles work in a total station and what they are used for.
10. Explain the difference between absolute and incremental encoders and the advantages of using either.
11. Explain the differences between manual, robotic, image-assisted and scanning total stations.
12. Explain what automatic total station target recognition is and its benefits for Civil Engineering applications.
13. Explain the purpose of using a code list in a surveying measurement, the types of coding currently available and their advantages for Civil Engineering applications.
14. What type of total station would you recommend for the following surveying applications: (a) road construction, (b) volume calculations, (c) tunnelling, (d) building set out, (e) data collection and (f) point setting out for land parcelling and (g) structural monitoring?
15. Discuss the different types of reflector prisms currently available for total station surveying and the Civil Engineering applications in which you would use each of them.
16. Briefly explain the difference between an optical and a digital level.

17. Briefly explain how a self-levelling (automatic level) works.
18. Explain what a laser level is and give some examples of Civil Engineering work where its use would be appropriate.

References

Bayoud, F. (2006). *Leica's Pinpoint EDM Technology with Modified Signal Processing and Novel Optomechanical Features*. Shaping the Change. XXIII FIG Congress. Munich, Germany.

Bayoud, F. (2007). *Leica Geosystems Total Station Series TPS1200*. White Paper. Leica Geosystems AG, Heerbrugg, Switzerland. 12p.

Ehrhart, M. (2012). *Automated, Total Station-based Verification of Reflector Pole Heights*. Masterarbeit zur Erlangung des akademischen Grades Diplom-Ingenieur/in Masterstudium Geomatics Science. Technische Universität Graz. Institut für Ingenieurgeodäsie und Messsysteme

Garget, D. (2005). *Testing of Robotic Total Stations for Dynamic Tracking*. Ms. Thesis. University of Southern Queensland. Faculty of Engineering and Surveying.

Grimm, D. E., Zogg H-M. (2013). *Leica Nova MS50 – The new dimension in measuring technology*. Leica Geosystems AG, Heerbrugg, Switzerland.

Grimm, D., Kleemaier, G., Zogg, H. (2015). *ATRplus*. White paper. Leica Geosystems AG. Heerbrugg, Switzerland.

ISO 17.123-2:2001. *Optics and optical instruments – Field procedures for testing geodetic and surveying instruments – Part 2: Levels.*

ISO 12.858-2:2020. *Optics and optical instruments – Ancillary devices for geodetic instruments – Part 2: Tripods.*

ISO 12858-3:2005. *Optics and optical instruments – Ancillary devices for geodetic instruments – Part 3: Tribrachs.*

Kahmen, H., Reiterer A., (2006). *Video theodolite measurement systems – State of the art*. ISPRS Commission V. Symposium 'Image Engineering and Vision Metrology', ISPRS Volume XXXVI, Part 5, Dresden 25–27 September, 5p.

Lachat, E., Landes, T., Grussenmeyer, P. (2017). *Investigation of a Combined Surveying and Scanning Device: The Trimble SX10 Scanning Total Station*. Sensors, 17, 730. 23p.

Lemmon, T, & Jung, R, (2005a). *Trimble S6 with Magdrive Servo Technology*. Trimble Navigation White Paper. 9p.

Levin, E., Nadolinets, L., Akhmedov, D. (2017). *Surveying Instruments and Technology*. CRC Press. ISBN 13: 978-1-4987-6238-0.

Mao, J., Nindl D. (2009). *Surveying reflectors*. With paper. Leica Geosystems AG. Heerbrugg, Switzerland.

Schneider, F.; Dixon, D. (2002). *The new Leica digital levels DNA03 and DNA10*. FIG XXII International Congress, Washington-DC.

Schofield, W. (2001). *Engineering Surveying*. 5th Edition. Butterworth Heinemann. Oxford. England.

Sokol, Š., Bajtala, M. and Ježko, J. (2014). *Verification of Selected Precision Parameters of the Trimble S8 DR Plus Robotic Total Station*. INGEO 2014 – 6th International Conference on Engineering Surveying, Prague, Czech Republic, April pages 3–4.

T. Lemmon, R. Jung (2005b). *TRIMBLE S6 with magdrive servo technology*. White paper. TRIMBLE SURVEY, Westminster, Colorado, USA.

Uchino, K. (2019). *MicroMechatronics*. CRC Press, Boca Raton, USA. https://doi.org/10.1201/9780429260308. 584 pages. 2nd Edition.

Zogg, H-M., Lienhart, W., Nindl, D. (2009). *Leica TS30*. With paper. Leica Geosystems AG. Heerbrugg, Switzerland.

Chapter 10
Instrumental and Operational Errors

Irineu da Silva and Guilherme Poleszuk dos Santos Rosa

10.1 Introduction

Whenever a surveying instrument is used to make measurements, it is necessary to consider the quality with which it can perform such an operation. As mentioned, this quality depends on several factors, the most important of which are instrument imperfections, natural environmental influences and operational errors. While most of these errors are irrelevant to typical Civil Engineering works, some need to be investigated and corrected to ensure that the quality of the measurement results is within the limits specified by the manufacturer or by calibration procedures.

Although the automation of measurement procedures has made fieldwork increasingly easier and reliable, Civil Engineers must be aware of the measuring errors that can occur when performing surveying works. Therefore, they must have sufficient knowledge to answer the following questions related to surveying instruments:

- What systematic and random errors are inherent in surveying measurement?
- What is the precision specification of a surveying instrument, how is it determined, and how can it be used for surveying specifications?
- What are the advantages of performing double-face readings when using a total station?
- What is the effect of instrument error on surveying measurements, and how can it be determined?
- What is the effect of operational errors on surveying measurements, and how they can be determined?

I. da Silva (✉)
São Carlos School of Engineering, University of São Paulo, São Carlos, SP, Brazil
e-mail: irineu@sc.usp.br

G. Poleszuk dos Santos Rosa
Cartographic Engineer, Researcher at the Faculty of Science and Technology of the São Paulo State University (UNESP), São Paulo, Brazil

- How do environmental errors affect surveying measurements, and how they can be reduced?

As discussed earlier in this book, surveying measurements are subject to systematic and random errors. Most of the time, systematic errors occur due to instrumental errors and environmental conditions and random errors due to operational conditions. Systematic instrumental errors are verified through instrument calibrations and must be corrected whenever detected. Systematic environmental errors are determined using mathematical models that allow their values to be quantified and corrected.

Calibrating a surveying instrument means comparing the results of its measurements to a standard. This operation allows determining how much the measured values deviate from the standard to certify that the instrument works according to the manufacturer's technical specifications or specific technical standards.

Usually, the instrument is adjusted after calibration to reduce the effects of systematic instrumental errors on subsequent measurements. Therefore, it is recommended that instrument users carry out frequent field verifications to verify the quality of their instruments. Guidelines for performing these verifications and the respective adjustments are usually specified by the manufacturers or technical standards, such as the International Organization for Standardization (ISO) specifications.

No specific standard indicates how often a surveying instrument should be verified and/or calibrated. However, it is recommended to verify it whenever working procedures require it to operate close to the limits of its accuracy. As for calibration, common sense indicates that it should be done at least once a year in a laboratory officially designated for this purpose.

As mentioned, random errors do not follow any specific mathematical law and, therefore, cannot be corrected. It is through them that the instrumental precision (standard deviation) is determined, as presented in Chap. 3. Currently, the standard deviation for a surveying instrument used with its ancillary equipment (tripod, tribrach, reflectors, surveying pole, staff, etc.) is specified in accordance with ISO 17123 *Optics and optical instruments—Field procedures for testing geodetic and surveying instruments*, which consists of the following parts:

- ISO 17123-1:2014 Part 1: Theory
- ISO 17123-2:2001 Part 2: Levels
- ISO 17123-3:2001 Part 3: Theodolites
- ISO 17123-4:2012 Part 4: Electro-optical distance metres (EDM measurements to reflectors)
- ISO 17123-5:2018 Part 5: Total stations
- ISO 17123-6:2022 Part 6: Rotating lasers
- ISO 17123-7:2005 Part 7: Optical plumbing instruments
- ISO 17123-8:2015 Part 8: GNSS field measurement systems in real-time kinematic (RTK)
- ISO 17123-9:2018 Part 9: Terrestrial laser scanners
- ISO/WD 17123-11.2 Part 11 GNSS Instruments (under development)

Manufacturers specify the standard deviation values of their instruments according to measurement procedures following the specifications of these standards. Specified values may, of course, differ from those observed in the field, which are subject to systematic instrumental and environmental errors and operator skill. Knowing the influence of these errors on surveying measurements is, therefore, essential to obtain a high-quality result.

It is important to emphasise that systematic errors, as already explained, can be corrected, whereas random errors are uncontrollable. They can only be treated statistically and compared with predetermined values which determine whether the job should be accepted or rejected.

It should also be noted that when choosing a surveying instrument, in addition to precision, it is also necessary to consider its reading resolution, as presented in Chap. 3. Generally, resolution values are indicated in the instrument's technical specifications.

This chapter briefly discusses instrumental and operational errors related to the total station and surveyor's level. The influence of environmental errors is presented in the chapters concerning angular observations, distance measurement, and levelling methods. Systematic and random errors related to other surveying instruments presented in this book are discussed in their respective chapters.

10.2   Total Station Instrumental and Operational Errors

As discussed in Chap. 9, total stations are primarily used to measure horizontal directions, vertical angles and slope distances. Their instrumental and operational errors are, therefore, related to these quantities. These are unavoidable errors that occur during the manufacture of the instrument and during field surveying operations. Their influence on the measurement results can be minimised by selecting suitable measurement methods and by correcting systematic errors, as described in the following sections.

10.2.1   Systematic Angular Errors

The systematic angular errors of a total station occur whenever horizontal direction and vertical angle observations are carried out during a surveying measurement. They are mainly related to the instrument components, as indicated below:

- Graduated circle errors
- Axial errors
- Plummet verticality error
- Automatic Target Recognition (ATR) error
- Laser pointer collimation error

For a total station to reach the nominal precisions indicated by the manufacturer, it must guarantee the following geometric conditions:

(a) The plane containing the horizontal circle must be perpendicular to the vertical axis of the instrument.
(b) The tilting axis must be perpendicular to the vertical axis of the instrument.
(c) The plane containing the vertical circle must be perpendicular to the tilting axis of the instrument.
(d) The line of sight must be perpendicular to the tilting axis of the instrument.
(e) The vertical axis (often referred to as the Standing axis or Trunnion axis) must be perfectly vertical.
(f) The centres of the horizontal and vertical graduated circles must coincide with the vertical and tilting axes, respectively.
(g) The line of sight must be horizontal when the zenith angle is equal to 90° or 270°.
(h) The axis of the EDM must coincide with the line of sight.
(i) In image-assisted total stations, the centre of the image must coincide with the line of sight.
(j) The three axes of the instrument—vertical, tilting and line of sight axes—must coincide at the same point, which is the vertex from which the angles are determined with the instrument.

The geometric conditions (a), (c) and (j) depend exclusively on the manufacture of the instrument and do not allow any adjustment by the user. Condition (h) can only be adjusted in specialised laboratories. All others can be verified and adjusted in the field.

The geometric conditions (f) and (g) are *graduated circle errors*. In turn, the geometric conditions (c), (d) and (e) are referred to as *axial errors*. Details on each of these errors are presented in the following subsections.

10.2.1.1   Graduated Circle Errors

Four graduated circle errors must be considered when measuring a horizontal direction or a vertical angle with a total station: the *scale graduation error, circle eccentricity error, vertical index error,* and *compensator index error*. All of them are considered systematic errors of the instrument.

10.2.1.1.1   Scale graduation and circle eccentricity errors

For the instrument to correctly indicate the observed direction values, the centre of the horizontal circle must coincide with the centre of the vertical axis, and the centre of the vertical circle must coincide with the centre of the tilting axis. When these conditions are not fulfilled, there are errors due to the circle eccentricities. Likewise, angular reading errors may occur due to the imperfect graduation of the circles.

The influence of the eccentricity error on the angular readings is directly proportional to the magnitude of the eccentricity value and inversely proportional to the graduation radius of the circle. These magnitudes, however, are unknown, making it impossible to estimate the influence of the eccentricity error on the angular readings. The same difficulty exists for the determination of the scale graduation error. In both cases, the solution is to take readings at two opposite indexes of the circle and consider the average of these readings as the final read value.

This procedure, however, only compensates for errors in reading the horizontal circle. For vertical circles, laboratory tests are performed, which indicate a (sinusoidal) error curve. Based on this curve, a correction factor is estimated and stored in the instrument's memory for each vertical angle reading.

Most modern surveying instruments already measure in two opposite indexes of horizontal circles. Simultaneously, the vertical circle correction factor is stored in the instrument's memory, requiring no operator action to eliminate the corresponding errors.

In addition to the above, a reading error can also occur due to the circle tilting concerning its axis. However, the influence of this error is minimal and is not considered in angular observations.

10.2.1.1.2   Compensator index error

A compensator index error occurs when the zero point of the compensator does not match the plumb line. With a dual-axis compensator, this error is split into two components, one transverse t and one *lonfitudinal* l to the line of sight. To simplify the user interface, most instruments determine the mechanical vertical and compensator index errors in a single step. The compensator index error is then stored in temporary memory to correct the measured vertical axis components.

For the compensator to indicate the tilting values correctly, it needs to be adjusted periodically. The adjustment can be carried out in specialised laboratories or by the users themselves, following the guidelines in the technical manual of the instrument. Note, however, that if the tilting values are set incorrectly, the compensator reading will be wrong, affecting measurements similarly to the vertical axis error.

Another error that can occur during field measurements is due to the temperature difference between the instrument and the environment around it, which can affect the compensator's functioning. It can be partially corrected by double-face readings. Even so, letting the instrument acclimate to the environment before starting angular observations is recommended. For most instruments, the adjustment time to ambient temperature is approximately 2 minutes/°C of difference between the instrument[1] and ambient temperatures.

---

[1]Most surveying instruments have an internal thermometer to indicate the instrument"s temperature.

10.2.1.1.3   Vertical index error

Vertical angles are measured by rotating the vertical circle and reading the angular values relative to an index line parallel to the vertical axis of surveying instruments. It is, therefore, expected that when the instrument is level, the vertical index line will coincide with the plumb line, and the telescope will be horizontal whenever the instrument gives a reading of the zenith angle equal to 90° or 270°. Suppose, however, that the zero point of the vertical scale reading is not parallel to the vertical axis of the instrument. In this case, a *mechanical vertical index error iv* occurs, as shown in Fig. 10.1.

Fig. 10.1 Vertical index error

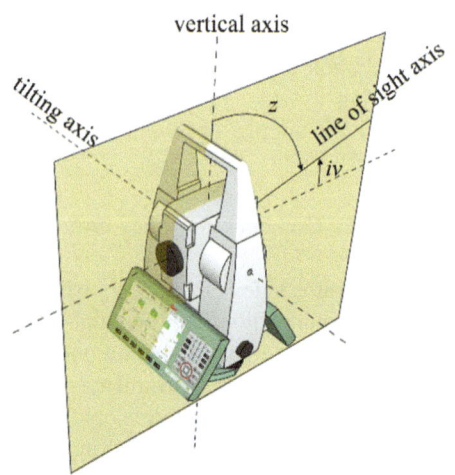

The value of *iv* can be determined by sighting accurately at a target about 100 m from the instrument with an altitude angle within ±9°. Next, double-face readings of zenith angles (z_d and z_r) should be taken. The value of the vertical index error is then given by Eq. (10.1).

$$iv = \frac{z_d + z_r}{2} - 180° \tag{10.1}$$

where,

z_d, z_r= direct and reversed zenith angle readings

The calculated *iv* value accounts for the mechanical and the compensator-generated zero-point errors.

When the zenith angle is not measured in the direct and reversed mode, but the value of the vertical index error is known, the measured value of the zenith angle can be corrected using Eq. (10.2).

$$\bar{z} = z' - iv \qquad\qquad (10.2)$$

where

$\bar{z} =$ corrected value of the zenith angle
$z' =$ measured value of the zenith angle

To be retained:

> The vertical index error is eliminated by performing double-face readings and using the average of these readings as the observation value.

Example 10.1
To verify the vertical index error of the vertical circle of a total station, double-face readings of the zenith angle were carried out as shown in Table 10.1. Calculate the value of the vertical index error.

Table 10.1 Zenith angle readings and the calculated vertical index error

Observation	Reading	Zenith angle	iv
1	Direct	89°25′23″	−1.5″
	Reversed	270°34′ 34″	

Solution:
Using Eq. (10.1),

$$iv = \frac{89°25'23'' + 270°34'34''}{2} - 180° = -1.5''$$

Thus, the corrected value for the direct reading of the zenith angle is given by Eq. (10.2):

$$z_d = 89°25'23'' - (-1.5'') = 89°25'24.5''$$

Similarly, using Eq. (6.18),

$$\bar{z}_D = \frac{360° + 89°25'23'' - 270°34'34''}{2} = 89°25'24.5''$$

10.2.1.2   Axial Errors

As the name implies, axial errors are related to the instrument axes. They are caused by the instrument axes not being perpendicular to each other. When this is the case, their influences change the values of angular observations, which means that surveyors cannot disregard them without the risk of generating systematic errors in angular measurements. They are:

- Horizontal collimation or line of sight error
- Tilting axis error
- Instrument levelling error or vertical axis error

The first two are residual instrument calibration errors and can vary over time. They can be eliminated by double-face readings. Vertical axis error, on the other hand, is not a genuine axis error but an instrument setup error, and therefore, it cannot be eliminated by double-face readings.

10.2.1.2.1 Horizontal collimation error

Horizontal collimation error c occurs when the line of sight is not perpendicular to the tilting axis, as shown in Fig. 10.2a. Under this condition, when the telescope rotates around the tilting axis, the line of sight follows a conical path that influences the observations in the horizontal direction, depending on the value of the altitude angle, as shown in Fig. 10.2b.

where

ε_c = horizontal reading error due to the horizontal collimation error
c = horizontal collimation error
OZ = vertical axis
HH' = tilting axis
OP = line of sight
β= altitude angle
OZI = vertical plane perpendicular to HH'
OZE = vertical plane of the direction OP

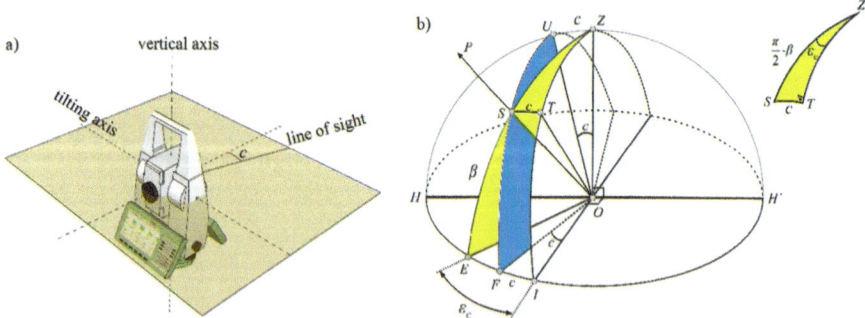

Fig. 10.2 Principle of horizontal collimation error. (**a**) Horizontal collimation error. (**b**) Influence of the horizontal collimation error

If the line of sight is horizontal, it should be perpendicular to the line *HH'* and pass through point (*I*). However, due to the horizontal collimation error, it passes through point (*F*), and as the telescope rotates, it describes a cone whose trace will be *FSU*. From the *OP* line of sight, the vertical plane *OZS* intersects the horizontal circle at (*E*), producing the horizontal reading error $IE = \varepsilon_c$.

According to the spherical triangle *ZST*,

$$\frac{\sin(\varepsilon_c)}{\sin(c)} = \frac{\sin\left(\frac{\pi}{2}\right)}{\sin\left(\frac{\pi}{2} - \beta\right)} \quad \rightarrow \quad \sin(\varepsilon_c) = \frac{\sin(c)}{\cos(\beta)} \tag{10.3}$$

Since *c* and ε_c are small,

$$\varepsilon_c = \frac{c}{\cos(\beta)} = \frac{c}{\sin(z)} \tag{10.4}$$

where

$z =$ zenith angle

When determining the horizontal angle by direction mode, the influence of the single-face reading $\Delta\varepsilon_c$ is given by Eq. (10.5). Therefore, the influence of the horizontal collimation error on the calculated horizontal angle decreases as the difference between the zenith angles decreases. It is equal to zero if the zenith angles are equal.

$$\Delta\varepsilon_c = \frac{c}{\sin(z_F) - \sin(z_B)} \tag{10.5}$$

To be retained:

The influence of the horizontal collimation error on the horizontal direction reading increases with the slope of the line of sight. It has an opposite sign depending on whether the telescope is in the direct or reversed position, which means it is eliminated by measuring in both telescope positions. It is equal to *c* for horizontal sightings. It does not influence vertical angle reading.

The horizontal collimation error *c* can be determined by sighting accurately a target about 100 m from the instrument, in both telescope positions, with an altitude angle of $0° \pm 9°$. Under these conditions, the interferences of the other axis errors are minimised. The value of the horizontal collimation error can be determined using Eq. (10.6). The adjusted value can then be calculated by adding the additive inverse of the collimation error to the direct reading. As a check, the adjusted horizontal direct reading can also be calculated using Eq. (6.4).

$$c = \frac{L_d - (L_r \pm 180\,°)}{2} \tag{10.6}$$

where

L_d, L_r= direct and reversed readings.

Add 180° if L_r is greater than 180° and subtract if it is less.

In this case, the reader should be aware that the direct and reversed sights must be carried out on well-defined targets so that the pointing error does not influence the determined horizontal collimation error. In addition, the readings must be repeated as many times as the operator considers necessary to obtain more accurate results.

Example 10.2

To determine the horizontal collimation error of a total station, the double-face readings shown in Table 10.2 were carried out in the field. Calculate the horizontal collimation error.

Table 10.2 Field measurements and the calculated value of c

Observation	Reading	Horizontal direction	c
1	Direct	297°45'46"	−1.5"
	Reverse	117°45'49"	

Solution:

Using Eq. (10.6),

$$c = \frac{297°45'46" - (117°45'49" + 180°)}{2} = -1.5"$$

$$\bar{L} = 297°45'\,46" + 0°00'1.5" = 297°45'47.5"$$

Using Eq. (6.4)

$$\bar{L} = \frac{297°45'46" + 117°45'49"}{2} + 90° = 297°45'47.5"$$

10.2.1.2.2   Tilting axis error

Tilting axis error i occurs when the tilting axis of the instrument is not perpendicular to its vertical axis, as illustrated in Fig. 10.3a. Under this condition, when the telescope rotates around the tilting axis, the line of sight follows an inclined path, which influences the observations in the horizontal direction, depending on the value of the zenith angle, as shown in Fig. 10.3b.

where

ε_i = horizontal reading error due to the tilting axis error
i = tilting axis error

OZ = vertical axis
HH' = instrument tilting axis inclined due to the error (i)
OP = line of sight
β = altitude angle
OZI = vertical plane of the direction OI
OZE = vertical plane of the direction OP
OTI = inclined plane of the line of sight

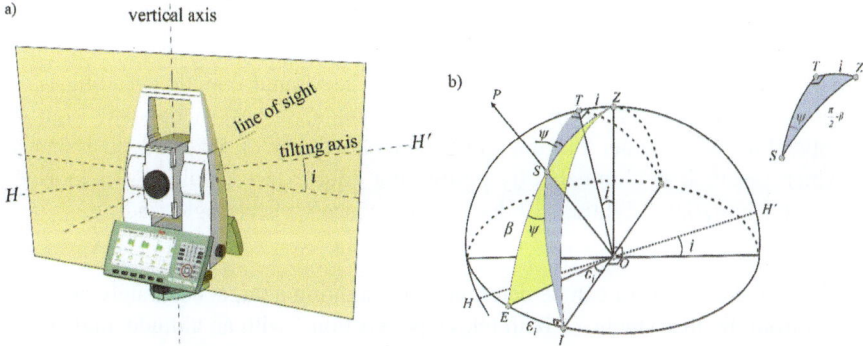

Fig. 10.3 Principle of tilting axis error. (**a**) Tilting axis error. (**b**) Influence of the tilting axis error

According to the spherical triangle OIE,

$$\frac{\sin{(\varepsilon_i)}}{\sin{(\psi)}} = \frac{\sin{(\beta)}}{\sin{\left(\frac{\pi}{2}\right)}} \quad \rightarrow \quad \frac{\sin{(\varepsilon_i)}}{\sin{(\psi)}} = \sin{(\beta)} \tag{10.7}$$

According to the spherical triangles OTZ *and* STZ,

$$\frac{\sin{(i)}}{\sin{(\psi)}} = \frac{\sin{\left(\frac{\pi}{2} - \beta\right)}}{\sin{\left(\frac{\pi}{2}\right)}} \quad \rightarrow \quad \frac{\sin{(i)}}{\sin{(\psi)}} = \cos{(\beta)} \tag{10.8}$$

Then, substituting Eq. (10.8) into Eq. (10.7) gives

$$\sin{(\varepsilon_i)} = \sin{(i)} * \tan{(\beta)} \tag{10.9}$$

Since i and ε_i are small,

$$\varepsilon_i = i * \tan{(\beta)} = \frac{i}{\tan{(z)}} \tag{10.10}$$

where
z = zenith angle.

When determining the horizontal angle by direction mode, the influence of the single-face reading $\Delta\varepsilon_i$ is given by Eq. (10.11). Therefore, the influence of the tilting axis error on the calculated horizontal angle decreases as the difference between the zenith angles decreases. It is equal to zero if the zenith angles are equal.

$$\Delta\varepsilon_i = i * \frac{\sin\left(z_F - z_B\right)}{\sin\left(z_B\right) * \sin\left(z_F\right)} \tag{10.11}$$

To be retained:

The influence of the tilting axis error on the horizontal direction reading is proportional to the tangent of the altitude angle. It has an opposite sign depending on whether the telescope is in the direct or reversed position, which means it is eliminated by performing double-face readings. It equals zero for horizontal sightings. It does not influence vertical angle reading.

The tilting axis error i can be determined by sighting a target accurately at about 100 m from the instrument, in both telescope positions, with an altitude angle of at least ±27°. Under these conditions, there is an influence of the horizontal collimation and tilting axis errors, as given by Eq. (10.12).

$$\frac{i}{\tan\left(z\right)} + \frac{c}{\sin\left(z\right)} = \frac{L_d - \left(L_r \pm 180°\right)}{2} \tag{10.12}$$

where

L_d, L_r = direct and reversed readings
i = tilting axis error
c = horizontal collimation error
z = zenith angle

On the other hand, when the horizontal collimation error c is already known, the tilting axis error i can be determined using Eq. (10.13).

$$i = \left(\frac{L_d - \left(L_r \pm 180°\right)}{2} - \frac{c}{\sin\left(z\right)}\right) * \tan\left(z\right) \tag{10.13}$$

Note that when the horizontal collimation error has already been corrected, Eq. (10.13) becomes:

$$i = \frac{L_d - \left(L_r \pm 180°\right)}{2} * \tan\left(z\right) \tag{10.14}$$

The influence of horizontal collimation and tilting axis errors on the horizontal direction reading is given by Eq. (10.15).

$$\bar{L} = L' - \frac{c}{\sin (z)} - \frac{i}{\tan (z)} \tag{10.15}$$

where

$\bar{L} =$ horizontal direction adjusted reading value
$L' =$ horizontal direction reading value

Example 10.3

To determine the tilting axis error of a total station, double-face readings and zenith angle were carried out as shown in Table 10.3. Assuming that this total station has the horizontal collimation error calculated in Example 10.2, calculate the horizontal direction reading error.

Table 10.3 Field measurements

Observation	Reading	Horizontal direction	Zenith angle
1	Direct	79°40′45″	62°12′29″
	Reverse	259°40′37″	

Solution:
    Using Eq. (10.13)

$$i = \left[\frac{79°40'45'' - (259°40'37'' - 180°)}{2} - \frac{-1.5''}{\sin(62°12'29'')} \right] * \tan(62°12'29'')$$

$$= 10.8''$$

Using Eq. (10.15).

$$\bar{L} = 79°40'45'' - \frac{-1.5''}{\sin (62°12'29'')} - \frac{10.8''}{\tan (62°12'29'')} = 79°40'41''$$

10.2.1.2.3 Vertical axis error

Vertical axis error occurs whenever the instrument is not levelled correctly, causing the vertical axis not to align with the plumb line. This situation influences both the horizontal direction and the vertical angle readings. Its influence can only be mathematically determined by measuring the tilt of the vertical axis in the direction of the telescope l and crosswise t using a dual-axis compensator, as described in Sect.

10.2.1.1.2. Therefore, as it cannot be eliminated by measuring in both telescope positions, special attention must be paid to levelling the instrument.

As shown in Fig. 10.4, the two vertical axis error components, longitudinal-*l* and transversal-*t*, influence the horizontal direction and the vertical angle readings similarly to the tilting axis and vertical index errors. However, if the instrument is equipped with a dual-axis compensator, it will measure the small tilts of the instrument and apply them to calculate the inclined plane.

Fig. 10.4 Principle of the vertical axis error

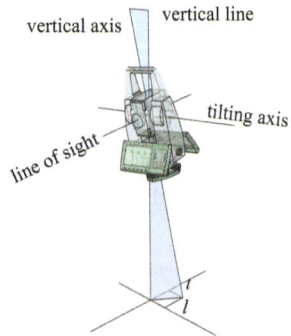

Once the inclined plan components are determined, they are used to determine the inclination angles for each new pointing direction and the corrected horizontal direction and vertical angle values.

It is worth noting that almost all total stations and electronic theodolites are currently equipped with dual-axis compensators that can be switched on and off if required.

10.2.1.3   Plummet Verticality Error

Levelling the instrument and correcting its vertical axis errors (*l, t*) does not guarantee that the instrument's vertical axis coincides with the vertical line passing through the station point. As described below, controlling the verticality of the optical (laser or video) plummet is also necessary.

In the case of an instrument with a laser or built-in video plummet installed on the alidade, the verticality verification can be performed according to the following procedure:

> Install and level the instrument on a flat surface; fix a sheet of paper below it and mark the centre of the point of incidence of the laser beam on the paper. Then, rotate the instrument around its vertical axis and verify whether the laser beam signature on the sheet of paper remains fixed. If the position of the laser beam incidence point moves more than specified for the instrument, it must be sent to a specialised laboratory for the necessary adjustments. The same procedure can be performed for the digital plummet case, replacing the laser beam with the centre of the digital plummet image.

In the case of an instrument equipped with an optical plummet installed on the tribrach, the verticality verification can be carried out according to the following procedure:

> Lay the instrument on a table so the tribrach's bottom face is positioned in front of a smooth wall. Mark the point of intersection of the optical plummet sight on the wall. Then, rotate the tribrach and verify the signature of the intersection point on the wall. If the position displaces more than specified by the manufacturer, the instrument must be sent to a specialised laboratory for proper adjustments.

10.2.1.4   ATR Collimation Error

The ATR collimation error is the angular divergence between the line of sight and the axis of the ATR digital camera. It consists of two components. The horizontal component affects the horizontal direction, whereas the vertical component affects the vertical angle. This type of error is significant in cases where manual measurements are mixed with automatic measurements. For correction, it is recommended to follow the manufacturer's recommendations, generally described in the instrument's User Manual. As mentioned, when measuring with ATR, the crosshairs may not be exactly coincident with the centre of the prism, even if the ATR is newly calibrated. As explained, the ATR measures the divergence from the centre, and the angular measurements are automatically corrected.

10.2.1.5   Laser Pointer Collimation Error

For most reflectorless total stations with a visible laser pointer, the laser beam is both the pointer and the measuring beam. Therefore, the laser beam must be collimated with the line of sight, ensuring that measurements are taken at the same point the crosshairs are pointing. In this case, calibration can be performed using a well-defined target, centring the crosshairs in the centre of the target and verifying the position of the laser pointer on the crosshairs. If they match, the instrument is in adjustment. Otherwise, a mechanical adjustment must be made to align them. The instrument's User Manual generally describes how to perform such a verification.

10.2.2   Random Angular Errors

Assuming that all systematic instrumental errors have been eliminated, as described in the previous section, it is still necessary to consider random instrumental errors, among which, depending on the application, the most important for Geomatics applied to Civil Engineering are as follows.

- Instrument and target centring errors
- Pointing error
- Instrument levelling error
- Horizontal tripod drift

10.2.2.1 Instrument and Target Centring Errors

Instrument and target centring errors are random errors caused when the vertical axis of the instrument or target does not coincide with the reference mark on the ground. It can significantly influence the observation of the horizontal direction, especially for short distances. However, despite their importance, there still needs to be specific standardised procedures for evaluating them. In practice, they can be minimised using high-quality tripods, calibrated optical (laser or video) plummets, and accurate instrument and target centring over the station and the sighting point. When these requirements are not met, centring errors will influence the horizontal direction observations.

The instrument centring error depends on factors such as plummet calibration, tripod quality, site lighting, operator visual acuity, optical plummet zoom level, and instrument height above the station mark. Installing an instrument at a height of 1.5 m results in an accuracy of 0.5 mm using optical plummet and between 1 mm and 1.5 mm using laser plummet. With built-in video plummet, this accuracy is about 0.5 mm for an instrument installed at a height of 1.55 m. In extreme cases of forced-centring, it can reach accuracies ranging from 0.03 mm to 0.1 mm.

The instrument centring error ε_i influences the quality of all the horizontal directions observed from that instrument installation and, consequently, the quality of the horizontal angles α determined from these directions. The maximum angular error occurs when the centring error vector is located on the bisector of the angle to be determined, as shown in Fig. 10.5.

The effect of the instrument centring error on the horizontal angle calculated using two directions observed from the same station is given by Eq. (10.16).

Fig. 10.5 Principle of instrument centring error

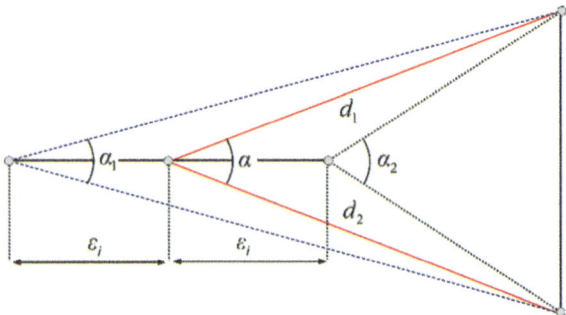

$$s_{\alpha_i} = \pm \frac{\varepsilon_i * \rho''}{d_1 * d_2} * \sqrt{d_1^2 + d_2^2 - 2 * d_1 * d_2 * \cos(\alpha)} \qquad (10.16)$$

where

s_{α_i} = error in angle due to instrument centring error
d_i = horizontal distance
α = horizontal angle
ρ'' = 206,264.806″ (radians to arc second conversion factor)

Target centring error ε_r occurs as illustrated in Fig. 10.6. Its influence on the quality of the calculated horizontal angle for two directions observed from the same station is given by Eq. (10.17).

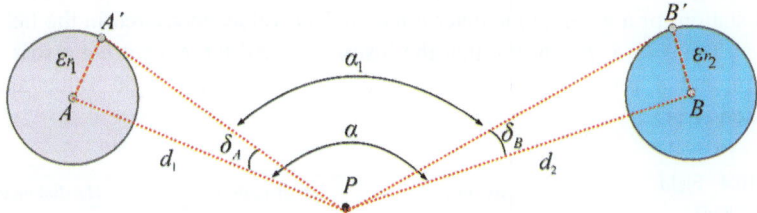

Fig. 10.6 Principle of target centring error

$$s_{\alpha_r} = \pm \rho'' \sqrt{\left(\frac{\varepsilon_{r_1}}{d_1}\right)^2 + \left(\frac{\varepsilon_{r_2}}{d_2}\right)^2} \qquad (10.17)$$

where

s_{α_r} = error in angle due to target centring error
$\varepsilon_{r_1}, \varepsilon_{r_2}$ = target centring errors
ρ'' = 206,264.806″ (radians to arc second conversion factor)

Supposing $\varepsilon_{r1} = \varepsilon_{r2} = \varepsilon_r$, Eq. (10.17) can be rewritten as follows:

$$s_{\alpha_r} = \pm \frac{\varepsilon_r * \rho'' \sqrt{d_1^2 + d_2^2}}{d_1 * d_2} \qquad (10.18)$$

where

ε_r = target centring error
d_1 = distance to target A
d_2 = distance to target B

Usually, centring errors (instrument and target) are difficult to measure. Therefore, for simplicity, the combined error in angle $s_{\alpha max}$ is calculated assuming that both errors are equal to ε and the distances between the instrument and the targets are both equal to d. Under these conditions, the maximum error in angle due to the simultaneous centring errors occurs for an angle of 180°, as given in Eq. (10.19).

$$s_{\alpha \max_{i,r}} = \pm \rho'' \frac{\sqrt{6}}{d} * \varepsilon \tag{10.19}$$

If measuring eccentricities and distances is not an option, taking repeated sets of angular measurements by re-centring the instrument and target between sets may be a recommended solution. Another solution may be to increase the sighting distances to the targets.

Example 10.4
The backsight and foresight horizontal directions and distances were measured with a total station for a given angle determination. The values measured in the field are given in Table 10.4. Assuming that the instrument and target centring errors were 1.5 mm and 4 mm, respectively, calculate the horizontal angle value and its estimated error.

Table 10.4 Field measurements

Direction	Hz direction	Hz distance [m]
L_1	46°47′ 23″	37.322
L_2	151°11′ 33″	41.564

Solution:
   Using Eq. (6.1)

$$\alpha = L_2 - L_1 = 104°24'10''$$

The error in angle due to instrument centring error can be calculated using Eq. (10.16).

$$s_{\alpha_i} = \pm \frac{0.0015 * 206,264.8062}{37.322 * 41.564} * \sqrt{37.322^2 + 41.564^2 - 2 * 37.322 * 41.564 * \cos(104°24'10'')}$$
$$= \pm 12.4''$$

The error in angle due to the target centring error can be calculated using Eq. (10.18).

$$s_{\alpha_r} = \pm \frac{0.004 * 206,264.8062 \sqrt{37.322^2 + 41.564^2}}{37.322 * 41.564} = \pm 29.7''$$

The combined error in angle due to instrument and target centring errors is then calculated by error propagation as follows:

$$s\alpha_{i,r} = \pm \sqrt{12.4^2 + 29.7^2} = \pm 32.2''$$

Example 10.5
Assuming that the instrument and target centring errors in Example 10.4 were both approximately 2.5 mm and the measured horizontal distances were approximately 40 m, calculate the maximum error in angle of this measurement.

Solution:
  Using Eq. (10.19)

$$s_{\alpha \max_{i,r}} = \pm 206,264.8062 * \frac{\sqrt{6}}{40} * 0.0025 = \pm 32''$$

10.2.2.2   Pointing Error

The pointing error when using a total station corresponds to the misalignment between the telescope crosshairs and the centre of the sighting point. It is a random error that varies depending on the optical quality of the telescope, the operator's visual limitations, atmospheric conditions, the size and shape of the reflector or target point, the backlighting of the target point and the thickness of the horizontal and vertical crosshairs of the telescope. This error can be minimised by increasing the number of readings and by carefully pointing to the reflector. For a well-defined point and good visibility conditions, the pointing error for a single-face observation can be estimated using Eq. (10.20).

$$\varepsilon_p = \pm \frac{k}{M} \tag{10.20}$$

where

ε_p= pointing error, in arc seconds
k = constant value ranging from 30″ to 60″ (usually 45″)
M = telescope magnification power

To minimise the pointing error in high-precision measurements, it is recommended to use special targets designed especially for this purpose or, if possible, to use additional lens systems that allow increasing the value of the telescope magnification power, which in some cases can reach 59 times. When none of these devices is available, it is recommended to repeat the measurements as many times as is considered appropriate. Twenty repetitions may be an adequate number. In this case, the standard deviation of the measurement includes the pointing and reading errors. Its value for a set of n repetitions of a single direction can be calculated as described in Sect. 3.6.3.2.

Air turbulence is one of the main factors interfering with pointing error, especially when targets are exposed to sunlight. Very little can be done to minimise this effect apart from carrying out measurements in more favourable conditions, such as the early morning or late afternoon, when the effect of air turbulence is minimal.

It is important to note that the angular standard deviation s_L of a total station, specified by the manufacturer or determined by instrument calibration, includes both the pointing error and the reading error. This is the value to be used for error propagation. The error in angle s_a, in this case, is given by Eqs. (6.6) or (6.7), as described in Sect. 6.2.1.2.

Example 10.6
Calculate the expected single-face observation pointing error for a total station with a telescope magnification power of 30x.

Solution:
   Using Eq. (10.20) *and assuming* $k = 45''$

$$\varepsilon_P = \frac{45''}{30} = \pm 1.5''$$

If a lens system for the telescope's magnification power of 59 times is used, the expected pointing error would be

$$\varepsilon_P = \frac{45''}{59} = \pm 0.8''$$

10.2.2.3   Instrument Levelling Error

If the instrument is not equipped with an electronic compensator, as described in Sect. 9.2.4, even if the instrument is levelled carefully, there will be a residual levelling error, which depends on the sensitivity of the tubular bubble attached to the instrument, according to Eq. (10.21).

$$\varepsilon_{inc} = \pm 0.2\gamma'' \tag{10.21}$$

where

ε_{inc}= levelling error, in arc seconds
γ''= level bubble sensitivity, in arc seconds

   The error in the horizontal direction due to the levelling error is given by Eq. (10.22).

$$s_{inc} = \pm \varepsilon_{inc} * \cot g(z) \tag{10.22}$$

where

s_{inc}= error in the horizontal direction due to the levelling error
z = zenith angle

Levelling error affects the accuracy of horizontal direction readings, especially for lines of sight with large altitude angles. This is the case, for example, in geodetic monitoring of dams, where the reference and measurement points have significant differences in level between them.

However, it is worth emphasising that if the instrument is equipped with a dual-axis compensator, the sensitivity to inclination is of the order of 0.5", which will not affect the observation of the horizontal direction even in lines of sight with large altitude angles.

10.2.2.4 Horizontal Tripod Drift

As stated in Sect. 9.2.5, the horizontal tripod drift can influence the horizontal direction readings. In general, wooden and aluminium tripods have little mass. Therefore, they are susceptible to torsion due to recurrent handling and differential dilatation of the legs due to temperature variation caused by the sun. It is an error that cannot be corrected, but that can be avoided by taking some precautions during measurements, such as:

- Keeping the instrument at the same station as little as possible.
- Using a sunshade or umbrella to protect the instrument and the tripod during measurements.
- Use solid, stable tripods or, where possible, make use of forced-centring.
- Regularly verifying the backsight orientation in high-precision work.

10.2.2.5 Combined Error Propagation

Considering the random angular errors discussed in the previous sections, the standard deviation of a single horizontal angle determination, using a total station or an electronic theodolite with a dual-axis compensator, can be calculated by combined error propagation, as given by Eq. (10.23).

$$s_{\alpha_T} = \pm \sqrt{s^2_{\alpha_i} + s^2_{\alpha_r} + S^2_{\alpha}} \qquad (10.23)$$

where

s_{α_T} = total combined error in angle
s_{α_i} = error in angle due to the instrument centring error
s_{α_r} = error in angle due to the target centring error
s_{α} = error in angle due to the pointing and reading errors

Example 10.7
Using the combined angle centring error calculated in Example 10.5 and assuming that the total station used for the measurement has an angular standard deviation

specified by the manufacturer of 5″, calculated the total combined error in angle for that measurement.

Solution:

Since a single back and foresight horizontal direction readings were carried out for the angle determination, the error in angle due to the instrument error can be calculated using Eq. (6.9).

$$s_a = \pm 2\, s_L = \pm 10''$$

Using Eq. (10.23)

$$s_{a_T} = \pm \sqrt{32^2 + 10^2} = 33.5''$$

10.2.3 Electronic Distance Measurement Errors

Analogously to the procedures for determining the angular precision of a surveying instrument, there is a specific standard for determining the precision of an EDM. In this case, ISO 17123-4:2012 is used, which specifies the procedures that must be followed to determine the standard deviation of a series of measurements performed on calibration bases with known distances. Thus, the systematic instrumental errors and the instrument's precision are determined by adjustment computation models. Some laboratories perform simplified procedures based on fixed distances determined by collimation reflectors distributed in a closed environment.

The calibration bases are built on concrete pillars with forced-centring when performed outdoors, as shown in Sect. 9.2.5. It is assumed that the instrument centring error, in this case, does not exceed 0.1 mm. The measurements of the distances between the pillars must be carried out with instruments considered to be of high precision or laser interferometers. It is also recommended that the stability of the concrete pillars be verified regularly to ensure calibration quality.

The systematic instrumental errors in measurement with an EDM are classified into geometric and atmospheric errors, as mentioned in Sect. 8.3.2.4. The atmospheric ones have already been presented in Sect. 8.3.2.4.1. In turn, the geometric ones are presented in the subsequent sections.

Three types of geometric errors can influence distances measured with an EDM.

- Zero error
- EDM scale factor
- EDM cyclic error

10.2.3.1   Zero Error (Additive Constant or Index Error)

Zero error is a systematic error which occurs due to electrical delay, geometric detours, eccentricities in the mechanical and electronic centre of the EDM relative to the plumb line, and also due to physical and geometrical properties of the reflector prism. It is an algebraic value independent of the length of the measured distance, which must be verified periodically and refers to a given pairing of instrument and reflector.

The major change arises from the use of different reflectors, especially when using cheap reflectors. The corresponding additive constant k_0 has to be added to the displayed value (measured value) in order to obtain the corrected distance between the physical zero-points of the EDM instrument and reflector (ISO definition). An incorrect value will affect all measurements by the same amount and sign. This is the most important error source, especially for short-range measurements.

The simplest method that can be used to verify the EDM additive constant k_0 for a given instrument/reflector is to take three distance measurements along three station points, as shown in Fig. 10.7.

Fig. 10.7 Verification of the EDM additive constant

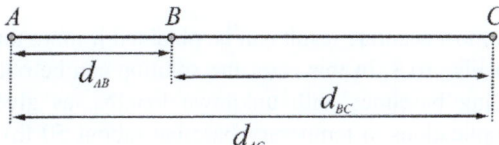

From Fig. 10.7, the following observation equations can be written as

$$d_{AB} = l_{AB} + k_0 \tag{10.24}$$

$$d_{BC} = l_{BC} + k_0 \tag{10.25}$$

$$d_{AB} + d_{BC} = l_{AC} + k_0 \tag{10.26}$$

where

d_{ij} = correct value of the distance d_{ij}
l_{ij} = measured distance between point (i) and (j)

Substituting Eqs. (10.24) and (10.25) into Eq. (10.26) gives.

$$l_{AC} + k_0 = l_{AB} + k_0 + l_{BC} + k_0 \tag{10.27}$$

And then,

$$k_0 = l_{AC} - (l_{AB} + l_{BC}) \tag{10.28}$$

For a better result, it is recommended to repeat the measurements several times and take the average value as the final result. Note that it is not necessary to know the baseline lengths to use this method. However, it is recommended to use three tripods with forced centring setups or concrete pillars.

Example 10.8

Three distance measurements were carried out to verify the additive constant value k_0 of a total station combined with a circular prism, as given in Table 10.5. Calculate the value of k_0.

Table 10.5 Distance measurements

Length	Distance [m]
d_{AB}	150.0175
d_{BC}	280.0327
d_{AC}	430.0302

Solution:

   Using Eq. (10.28)

$$k_0 = 430.0302 - (150.0175 + 280.0327) = -20.0 \text{ mm}$$

A most accurate result can be obtained if more than three points are used, as shown in Fig. 10.8. In this case, the solution can be obtained by adjustment computation using baselines with unknown lengths, as given by Eq. (10.29). For practical applications, a temporary baseline (about 50 m) consisting of at least four points (A, B, C and D, aligned in the same horizontal plane and using forced centring tripods must be set out. The distances between the tripods shall be an integer of the unit length ($U = \lambda/2$) of the EDM.

$$l_i + v_i = d_i + k_0 \tag{10.29}$$

where

l_i=measured length
v_i=residual error
d_i=adjusted distance
k_0= additive constant

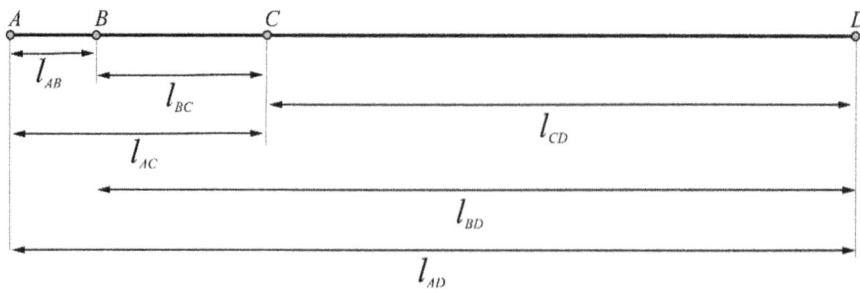

Fig. 10.8 Baseline configuration for the determination of the additive constant

Considering the four baselines and the observation configuration shown in Fig. 10.8, the following error equations can be written as

$$
\begin{aligned}
v_1 &= -k_0 + d_{AB} - l_{AB} \\
v_2 &= -k_0 + d_{BC} - l_{BC} \\
v_3 &= -k_0 + d_{CD} - l_{CD} \\
v_4 &= -k_0 + d_{AB} + d_{BC} - l_{AC} \\
v_5 &= -k_0 + d_{BC} + d_{CD} - l_{BD} \\
v_6 &= -k_0 + d_{AB} + d_{BC} + d_{CD} - l_{AD}
\end{aligned}
\tag{10.30}
$$

Equation (10.30) can be written in matrix form as $v = Ax - l_0$ (Eq. 3.79), where

$$
A = \begin{bmatrix}
-1 & 1 & 0 & 0 \\
-1 & 0 & 1 & 0 \\
-1 & 0 & 0 & 1 \\
-1 & 1 & 1 & 0 \\
-1 & 0 & 1 & 1 \\
-1 & 1 & 1 & 1
\end{bmatrix}
\quad
v = \begin{bmatrix}
v_1 \\ v_2 \\ v_3 \\ v_4 \\ v_5 \\ v_6
\end{bmatrix}
\quad
x = \begin{bmatrix}
k_0 \\ d_{AB} \\ d_{BC} \\ d_{CD}
\end{bmatrix}
\quad
l_0 = \begin{bmatrix}
l_{AB} \\ l_{BC} \\ l_{CD} \\ l_{AC} \\ l_{BD} \\ l_{AD}
\end{bmatrix}
$$

Adjustment computation can be carried out using Eqs. (3.84) to (3.87). If all measurements are considered to have the same precision, the weight matrix P is the identity matrix. If their variances are considered, the weight matrix P is a diagonal matrix with the inverse of the variances calculated for each distance measurement. The result is the unknown values of k_0, d_{AB}, d_{BC} and d_{CD}. Finally, as a total of six observations and four unknowns have been considered, the standard deviations of the unit weight can be calculated using Eq. (10.31). The remaining statistical parameters can be determined as indicated in Sect. 3.9.1.1.

$$
s_0^2 = \frac{v^T P v}{2}
\tag{10.31}
$$

Example 10.9
Suppose that an additional station set up has been included in the baseline of Example 10.8, so that six distances have been measured to verify the additive constant k_0, as shown in Fig. 10.8. The measured values are given in Table 10.6. Calculate the new value of k_0.

Table 10.6 Distance measurements

Length	Distance [m]
d_{AB}	150.0175
d_{BC}	280.0327
d_{CD}	970.0119
d_{AC}	430.0302
d_{BD}	1250.0248
d_{AD}	1400.0223

Solution:

Using the error Eqs. (10.30) and assuming that all measurements have the same precision, the following matrices can be written for applying the Parametric adjustment model.

$$A = \begin{bmatrix} -1 & 1 & 0 & 0 \\ -1 & 0 & 1 & 0 \\ -1 & 0 & 0 & 1 \\ -1 & 1 & 1 & 0 \\ -1 & 0 & 1 & 1 \\ -1 & 1 & 1 & 1 \end{bmatrix} \quad v = \begin{bmatrix} v_1 \\ v_2 \\ v_3 \\ v_4 \\ v_5 \\ v_6 \end{bmatrix} \quad l_0 = \begin{bmatrix} 150.0175 \\ 280.0327 \\ 970.0119 \\ 430.0302 \\ 1250.0248 \\ 1400.0223 \end{bmatrix} \quad x = \begin{bmatrix} k_0 \\ d_{AB} \\ d_{BC} \\ d_{CD} \end{bmatrix}$$

Using Eqs. (3.84) to (3.87) and assuming that the weight matrix P *is equal the identity.*

$$N = \begin{bmatrix} 6.0 & -3.0 & -4.0 & -3.0 \\ -3.0 & 3.0 & 2.0 & 1.0 \\ -4.0 & 2.0 & 4.0 & 2.0 \\ -3.0 & 1.0 & 2.0 & 3.0 \end{bmatrix} \quad n = \begin{bmatrix} -4480.13940 \\ 1980.07000 \\ 3360.11000 \\ 3620.05900 \end{bmatrix}$$

$$x = \begin{bmatrix} -0.0199 \\ 149.9976 \\ 280.0128 \\ 969.9921 \end{bmatrix}_m$$

Using Eq. (3.79)

$$v^T = \begin{bmatrix} -0.05 & 0.00 & 0.05 & 0.05 & -0.05 & 0.00 \end{bmatrix}_{mm}$$

From the vector $x, k_0 = -19.9$ *mm*

Finally, using Eq. (10.31), the standard deviations of the unit weight and the variance-covariance matrix for the unknowns are given as follows.

$$s_0^2 = 5.00 * 10^{-3} \ mm^2$$

$$\Sigma_{xx} = s_0^2 N^{-1} = \begin{bmatrix} 5.000 * 10^{-3} & 2.500 * 10^{-3} & 2.500 * 10^{-3} & 2.500 * 10^{-3} \\ 2.500 * 10^{-3} & 3.750 * 10^{-3} & 0.000 & 1.250 * 10^{-3} \\ 2.500 * 10^{-3} & 0.000 & 3.750 * 10^{-3} & 0.000 \\ 2.500 * 10^{-3} & 1.250 * 10^{-3} & 0.000 & 3.750 * 10^{-3} \end{bmatrix}$$

Using the diagonal values in the variance-covariance matrix for the unknowns, the standard deviations of k_0 is ±0.07 mm, and then

$$k_0 = -19.9 \text{ mm} \pm 0.07 \text{ mm}$$

10.2.3.2   EDM Scale Error

The EDM scale error s (ideally equal to 1.0000) is a systematic error that occurs usually due to a variation in the frequency of the modulation wave used by the instrument, from its standard value. This variation may be due to ageing or drift in frequency of the quartz crystal oscillator in the instrument, non-homogeneous emission/reception patterns from the emitting and receiving diodes (phase inhomo-geneities) and the difficulty of correctly modelling atmospheric variations, which affect the propagation speed of the electromagnetic beam over the distance between the source generating the signal (total station) and the reflector prism. It is an error that varies linearly proportional to the length of the measured distance (station-prism) and for this reason, given in ppm.

The method commonly used to determine the EDM scale factor is from measure-ments taken over a certified calibration baseline, which means that the baseline lengths have been measured to a better quality than the instrument being calibrated. As with the EDM additive constant, the scale factor is determined by comparing a series of measured distances along a linear array of pillars with known distances and applying a Parametric adjustment model. Once the constant scale factor s has been determined by the adjustment computation, it is stored in the instrument's memory as a multiplication factor for all subsequent distance measurements. As the calibra-tion method requires adjusted baselines, it is strongly recommended to perform it only in certified calibration laboratories.

10.2.3.3   EDM Cyclic Error

The EDM cyclic error is a systematic error that occurs due to the difficulty of correctly measuring the phase shift between the waves emitted and received by the instrument, which depends on its electronic system. It is an error that varies inversely proportional to the power of the return signal. It increases as the measured distance also increases. In a well-calibrated instrument, it is minimal, not exceeding ±2 mm. Although most of the time, it is practically zero, it should not be neglected and should be subjected to periodic calibrations to control its magnitude. As with the additive constant, it can be detected through repeated measurements over calibration baselines. However, it is a procedure that requires a specialised calibration baseline to detect the presence of cyclic error from the spacing of measurement intervals.

In Sect. 8.3.2.5, the distance measurement accuracy for a total station was specified as $\pm(a \ [mm] + b \ [ppm])$. The constant a is a function of zero and cyclic errors, and b is a function of the scale factor.

10.3 When to Calibrate a Surveying Instrument

Instrumental errors are determined and entered into the instrument during the manufacturing process. Even so, they may vary due to shocks, temperature changes and other factors. For this reason, it is recommended to calibrate the instrument periodically (at least once a year) or whenever one of the following situations occurs:

- Before using it for the first time.
- After long periods without use.
- After sudden temperature changes.
- After intensive use or long transport distances.
- Periodically, in cases of high precision work.

Once calibrated, most electronic theodolites and total stations can automatically consider the effects of instrumental errors on angular observations and, specifically for the latter case, also on linear ones. Thus, the values indicated on the instrument's display already consider instrumental errors, and the operator does not need to worry about them in subsequent observations. Nevertheless, if the operator wants to guarantee the quality of the observations or if the work is of high precision, it is always recommended to carry out direct and reversed readings. It should be noted, however, that instrumental errors are not as worrisome as the theory suggests. They are easily avoided, provided good quality equipment is used, and correct measurement procedures are applied. Particular attention should be given to rented instruments.

A simple two-point baseline check can be used for periodic quality verifications for most Civil Engineering projects. In this case, the procedure is based on establishing a baseline with high-quality tape and comparing the distance measured with the total station with the baseline value. The instrument must be sent to a calibration centre for maintenance and calibration if an unacceptable difference is found. If necessary, zero error can be determined on-site following the procedures described above. However, the cyclic and scale factors are small and must be determined in the laboratory or can be ignored for the short distances typically used in Civil Engineering projects.

10.4 Instrumental and Operational Errors for the Surveyor's Level

The surveyor's level is a high-precision instrument that must be verified and calibrated regularly. The operators themselves can conduct some verifications in the field; however, more specific ones must be carried out in specialised laboratories. Calibrations are performed using invar levelling staff and autocollimator instruments. In the case of field tests, specific operating conditions indicated by technical standards must be respected to ensure that any error found is really a result of the instrument's conditions and not an operational error. On the procedures for field testing of the surveyor's level, the reader should refer to ISO 17123-2:2001.

As it is a relatively simple instrument to manufacture, there is a wide variety of surveyor's levels on the market. However, it is advisable not to choose an instrument on the basis of price alone, and never neglect regular verifications and calibrations.

The most common systematic and random errors that may occur during measurements with a surveyor's level are presented in the subsequent sections.

10.4.1 Systematic Errors

The systematic errors that may occur during measurements using a surveyor's level can be categorised as internal and external. The external ones are those related to the environmental and geometric conditions of the surveying work and are presented in detail in Sect. 13.5. In turn, the internal ones are related to instrumental errors, such as:

- Crosshair horizontality error
- Vertical collimation error
- Levelling staff scale and index errors

10.4.1.1 Crosshair Horizontality Error

Generally, operators take their readings on the levelling staff by centring the vertical crosshair on its axis. However, there are cases where this procedure is impossible; therefore, if the horizontal reticule is not perfectly horizontal, there will be a reading error. To verify the existence of this error, the operator must sight a clearly visible point in the field or laboratory, making one end of the horizontal crosshair coincide with it, and then move the telescope horizontally over that point to the other end of the horizontal crosshair. If the horizontal crosshair does not keep its trace over the target point during the telescope movement, there is a misalignment of the crosshair that must be corrected. In this case, it is recommended to send the instrument to a specialised laboratory for proper correction.

10.4.1.2   Vertical Collimation Error

The vertical collimation error e_c of a surveyor's level occurs if the line of sight is not horizontal when the instrument is considered level. In practice, the vertical angle between the real line of sight and the ideal horizontal line is given by a collimation factor C, in mm/m. For standard instruments used in civil construction, this factor is of the order of $C = 0.1$ mm/m, and for precise instruments, it is of the order of 0.03 mm/m. This type of error is eliminated when the backsight and foresight observations are equidistant. However, as it is not always possible to keep them equidistant, as in profile levelling and stakeout works, an error may occur in the levelling staff reading value, which varies depending on the collimation factor and the difference between the backsight and foresight distances, as given by Eq. (10.32).

$$e_c = C * \Delta d \tag{10.32}$$

where

e_c= vertical collimation error
$C =$ collimation factor
Δd= difference between the backsight and foresight distances

For example, an instrument with a collimation factor equal to 0.05 mm/m and a difference between backsight and foresight distances of 3 m would result in a staff reading error of 0.15 mm.

There are several methods available in the literature for surveyor's level verification. More information can be found in the instrument manual. However, the most common procedure for Geomatics applied to Civil Engineering is the two-peg test, which should be carried out as described below.

Fig. 10.9 Principle of two-peg test

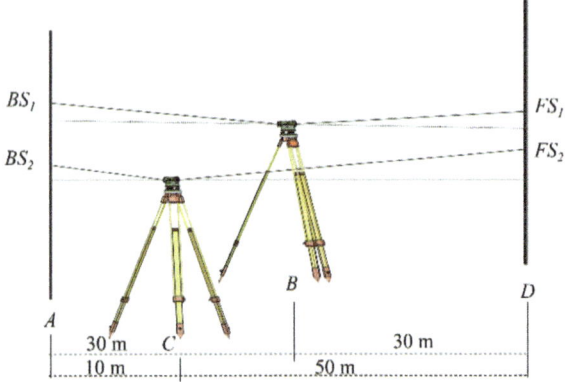

As shown in Fig. 10.9, on a flat ground surface, place a levelling staff on point (A) and another one on point (D). Set up the surveyor's level on point (B), about 30 m from the levelling staffs (preferably protected from the sun). Take a backsight reading BS_1 on the levelling staff (A) and a foresight reading FS_1 on the levelling staff (D). Then move the instrument to point (C), located at a distance of about 10 m from point (A), and perform a backsight reading BS_2 on the levelling staff (A) and a foresight reading FS_2 on the levelling staff (D). With the measured values, it is possible to calculate the vertical collimation error using Eq. (10.33).

$$e_c = (BS_1 - FS_1) - (BS_2 - FS_2) \tag{10.33}$$

For practical reasons, an auxiliary operator should take notes on the levelling work to calculate the value of FS_2 before it is read. Thus,

$$(BS_1 - FS_1) - (BS_2 - FS_{2(\text{calc})}) = 0 \tag{10.34}$$

$$FS_{2(\text{calc})} = (FS_1 - BS_1) + BS_2 \tag{10.35}$$

Consequently, the vertical collimation error can also be calculated using Eq. (10.36).

$$e_c = FS_{2(\text{calc})} - FS_{2(\text{read})} \tag{10.36}$$

Once the value of the vertical collimation error has been calculated, the value of the collimation factor can be calculated using Eq. (10.32). It is essential to point out that subsequent corrections, when necessary, must be performed by reversing the algebraic sign of the collimation error.

For example, to achieve a collimation factor in the order of 0.05 mm/m, the vertical collimation error calculated using the distances indicated in Fig. 10.9 must be less than 2.0 mm.

After the field verification, the instrument must be adjusted in the laboratory if the error exceeds the pre-established tolerance value. To obtain redundancy of values in the measurements, it is recommended to repeat them by changing the height of the instrument by a few centimetres between pairs of measurements. ISO 17123-2:2001 recommends measuring ten pairs of backsight and foresight readings.

Example 10.10

The observations shown in Table 10.7 were performed with an automatic level, following the distance specifications indicated in Fig. 10.9. Calculate the instrument's collimation error and the vertical collimation factor.

Table 10.7 Field measurements and calculated values

Observation	BS_1[mm]	FS_1[mm]	BS_2 [mm]	FS_2 [mm]	BS_1 - FS_1 [mm]	BS_2 - FS_2[mm]	e_c[mm]
1	1639	1622	1577	1562	17	15	2
2	1657	1640	1581	1566	17	15	2
3	1697	1680	1605	1590	17	15	2
$e_{c_{\text{mean}}}$	–	–	–	–	–	–	2

Solution:

    The last column of Table 10.7 *shows the values of the vertical collimation error for each pair of backsight and foresight readings, according to the geometric configuration of* Fig. 10.9.

    For the first pair of readings, using Eq. (10.33),

$$e_c = (1639 - 1622) - (1577 - 1562) = 2 \text{ mm}$$

    The calculations for the remaining pairs of measurements are given in Table 10.7. *The final vertical collimation error is given by the average of the partial errors, whose value is* $e_{c_{mean}} = 2$ mm.

    Collimation factor using Eq. (10.32)*:* $C = \frac{2}{50} = 0.04$ mm/m

Example 10.11

The same measurement procedure as in Example 10.10 was performed for a digital level. The measured and calculated values are shown in Table 10.8. Based on the measured values, calculate the collimation error and the vertical collimation factor.

Table 10.8 Field measurements and calculated values

Observation	BS_1[mm]	FS_1[mm]	BS_2 [mm]	FS_2 [mm]	BS_1 - FS_1 [mm]	BS_2 - FS_2[mm]	e_c[mm]
1	1603.3	1587.1	1531.3	1514.9	16.2	16.4	−0.2
2	1631.8	1615.6	1557.2	1540.7	16.2	16.5	−0.3
3	1655.6	1639.5	1567.7	1551.5	16.1	16.2	−0.1
$e_{c_{mean}}$		−	−	−	−	−	−0.2

Solution:

    Repeating the same calculation sequence from Example 10.10:
    Collimation error of the first pair of measurements:

$$e_c = (1603.3 - 1587.1) - (1531.3 - 1514.9) = -0.2 \text{ mm}$$

Average collimation error: $e_{c_{mean}} = -0.2$ mm
Collimation factor: $C = \frac{-0.2}{50} = -0.004$ mm/m

10.4.1.3   Levelling Staff Scale and Index Errors

Levelling Staff scale and index ("the zero mark") errors are systematic errors resulting from a defect in the graduation or wear at the base of the levelling staff. Scale error can be eliminated by calibrating the levelling staff using a steel tape measure. The index error is compensated for if the same levelling staff is used for backsight and foresight observations.

10.4.1.4 Unstable Ground

A systematic error occurs if the staff or tripod sinks into the ground during levelling procedures. In this situation, a misclosure error can be found in the foresight and backsight measurements (the absolute value of the rise is greater than that of the falls). This error can be minimised by firmly fixing the tripod legs on the ground and using a steel turning plate as a temporarily stable object where the levelling staff can be held during levelling observations. See Fig. 10.10. The essential point to be considered in this case is that the tripod or levelling staff heights stay the same during measurements. The operator should also avoid lifting and lowering the staff unnecessarily.

Fig. 10.10 Steel turning plate for precise levelling

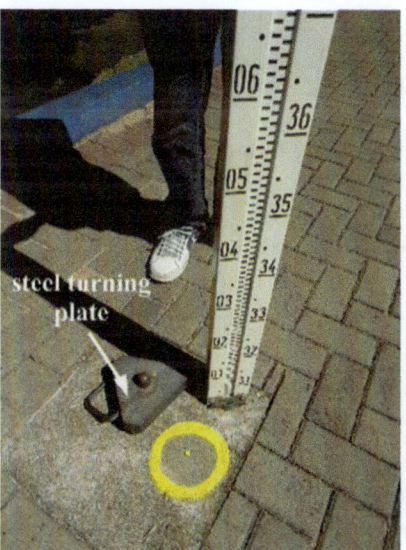

10.4.2 Random Errors

The random errors that can occur in a measurement with a surveyor's level are the following:

- Compensator error
- Levelling staff not held vertically
- Pointing error
- Parallax

10.4.2.1 Compensator Error

When levelling an automatic surveyor's level, there will be a levelling error that varies depending on the precision e_n of its compensator. Typical precision values for

compensators used in automatic levels vary from 0.1" to 0.5" and may be higher for
basic levels used in Civil Engineering construction. In this case, the levelling error is
given by Eq. (10.37).

$$e_n = \pm \frac{\gamma_C''}{\rho''} * d \qquad (10.37)$$

where

e_n = compensator error
γ_C'' = compensator accuracy
d = sighting distance
ρ'' = 206264.8062" (conversion factor from radian to arc second)

10.4.2.2   Levelling Staff Not Held Vertically

The reading error resulting from the levelling staff not being held vertically depends
on the quality of the circular bubble attached to the staff. As shown in Fig. 10.11, if
the staff is not vertical, the reading value will be higher than the correct one and can
be determined by Eq. (10.38).

$$e_m = \pm \frac{l}{2} * \left(\frac{\gamma''}{\rho''} \right)^2 \qquad (10.38)$$

where

e_m = reading error
l = reading value
γ'' = sensibility of the bubble level attached to the levelling staff, in arc second
ρ'' = 206,264.8062" (conversion factor from radian to arc second)

Fig. 10.11 Reading error
due to staff inclination

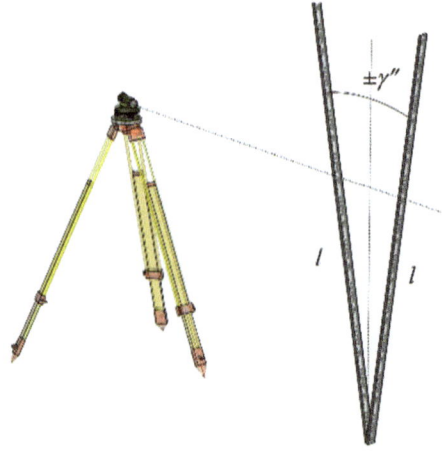

Example 10.12

Differential levelling was performed with a levelling staff whose circular bubble level had a sensitivity of $20'$. Considering the geometric configuration of Fig. 10.11 and that a reading of 2025 mm was taken on the levelling staff, calculate the value of the reading error.

Solution:

Using Eq. (10.38)

$$e_m = \pm \frac{2,025}{2} * \left(\frac{20'}{3,437.7468'}\right)^2 = \pm 0.03 \text{ mm}$$

10.4.2.3   Pointing Error

The pointing error e_p in a measurement with a surveyor's level corresponds to the precision with which it is possible to read values on the levelling staff. This error depends on atmospheric conditions, the value of the telescope magnification power and the distance between the level and the staff. The mathematical formulation used to calculate this error is given by Eq. (10.39).

$$e_p = \pm \frac{k * d}{M * \rho''} \tag{10.39}$$

where

ep = pointing error
k = constant value ranging from $30''$ to $60''$ (usually $45''$)
d = sighting distance
M = telescope magnification power
ρ'' = $206,264.8062''$ (conversion factor from radian to arc second)

Example 10.13

Calculate the pointing error for a sighting distance of 40 m when using an automatic level with a telescope magnification power of $20\times$.

Solution:

Using Eq. (10.39).

$$e_p = \pm \frac{45'' * 40,000}{20 * 206,264.8062''} = \pm 0.4 \text{ mm}$$

10.4.2.4   Parallax

When working with any surveying instrument equipped with a telescope and crosshairs for sighting, an observer simultaneously views the object focused on by the telescope and the crosshairs. Both images must be focused on a single plane to ensure a correct sighting. If this condition is not met, parallax is said to exist, and the observer will feel as if the crosshairs are moving over the object when he or she slightly moves their eyes. To eliminate parallax, the telescope focusing knob and eyepiece focusing ring must be rotated in a repeated procedure until all parallax are eliminated.

10.4.2.5   Digital Level Errors

In addition to the random and systematic errors for optical levels presented in the previous sections, depending on the manufacturer, digital levels also have a systematic error associated with particular sections of the staff or some specific sighting distances. Investigations have shown that this type of error occurs whenever the size of the levelling staff code line projected onto the CCD array is equal to the size of the photodiode. Taking the NA3000 level (Leica Geosystems) as an example, an error of 0.8 mm may occur when the sighting is performed at a distance of about 15 m from the staff. For different instrument manufacturers and models, this critical distance varies. Readers are advised to consult the manufacturer's technical manuals for more information on the occurrence of this type of error.

In addition to critical distance and collimation error, measurements with digital levels can produce errors related to levelling staff reading. Thus, as the determination of the measured value depends on the visualisation of a specific portion of the levelling staff, incorrect reading values may occur if only a part is visualised. Therefore, it is recommended to avoid measurements at the ends of the levelling staff.

Measurements should also be avoided in places with poor lighting, such as tunnels. In these cases, it is recommended to use special artificial lighting or a levelling staff with its own lighting. Likewise, operators should avoid readings with very dark shading on the levelling staff, as this can cause the same effect as interference on the line of sight. It is also important to point out that the digital levels do not perform measurements on blurred images and may have problems with readings on levelling staff with dirt or erasures in the barcode painting. Finally, as with optical levelling, the distance to the levelling staff should be approximately 25–30 m.

Bar code levelling staff used in conjunction with digital levels are calibrated in the laboratory through a system in which the staff is levelled on a device that can move vertically. A laser interferometer monitors this movement. At the same time, a digital level installed at an appropriate distance takes measurements of the levelling staff, which are compared with the values indicated by the interferometer. Based on the measured values, the reading quality of the staff is determined.

10.5 Comments on the Surveyor's Level Calibration

Calibrating a surveyor's level is a simple procedure performed by almost all laboratories that calibrate surveying instruments. Likewise, the verification procedure for this type of instrument is a simple task that can be performed by the operator in the field, following the manufacturer's recommendations. For these reasons, performing routine verifications and sending the instrument for calibration is recommended whenever there are signs of malfunction.

In the case of digital levels, verifications can also be carried out according to the manufacturer's recommendations. However, calibration should only be carried out in a certified laboratory with a proven ability to calibrate this type of instrument.

10.6 Problems

1. Direct and reversed zenith angle observations were taken to determine the vertical index error of a total station, as shown in Table 10.9. Determine the value of the vertical index error. Suppose a single-face reading of a zenith angle of $85°47'32''$ was measured. Calculate the corrected zenith angle and the corresponding altitude angle.

Table 10.9 Direct and reversed readings

Station	Sighting point	Zenith angle direct reading	Zenith angle reversed reading
P	Q	88° 21′ 07″	271° 38′ 29″
		88° 21′ 05″	271° 38′ 28″
		88° 21′ 09″	271° 38′ 29″
		88° 21′ 08″	271° 38′ 23″

2. Suppose the instrument in problem 1 is used to calculate the vertical distance between points (A) and (B). Assume that point (A) is at the same elevation as the centre of the instrument. If the slope distance and the zenith angle to point (B) are 250.3452 m and $75°12'05''$, respectively, calculate the error in the height of (B) if the vertical index error is not considered.

3. Direct and reversed horizontal direction observations of $19°30'49.7''$ and $199°30'54.0''$ were taken to determine the horizontal collimation error of a total station.

   (a) Determine the value of the horizontal collimation error.
   (b) Calculate the error that would be introduced in the determination of a horizontal clockwise angle under the following conditions: (i) from single-face readings at points at the same elevation of the instrument, (ii) from single-face reading at a point at a zenith angle of $85° 12'37''$ and (iii) from single-face readings at a backsight point at a zenith angle of $85° 12'37''$ and a foresight point at a zenith angle of $95° 55'26''$.

4. Suppose that the same instrument as in problem 3 is to be verified for tilting axis error. To do this, the horizontal directions of 19°30′ 44.1″ and 199°30′ 50.7″ were observed at a zenith angle of 61°45′35″. Under these conditions, calculate the tilting error, considering the horizontal collimation error has already been corrected. Calculate the error introduced in the horizontal angle determined by a backsight horizontal direction reading of 85°12′37″ and a foresight reading of 95°55′26″ if the collimation and the tilting errors are not corrected.

5. A total station measurement was requested to check the verticality of the abutment of a bridge approximately 50 m high. The measurement was made by positioning the instrument about 150 m from the base (A) of the abutment, with the centre of the instrument level with point (A), as shown in Fig. 10.12. The readings are given in Table 10.10.

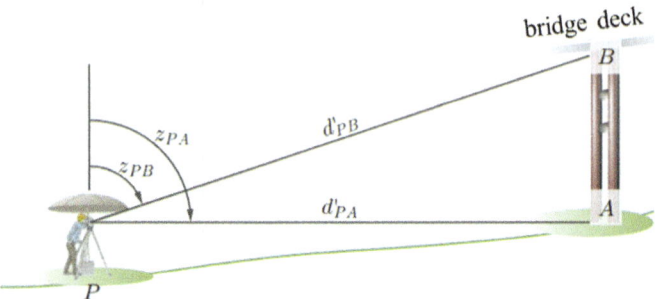

Fig. 10.12 Geometry of the measurement

Table 10.10 Field measurements

Station	Sighting point	Horizontal direction	Zenith angle	Slope distance [m]
P	A	110°25′13″	90°00′00″	150.0000
	B	110°28′40″	71°34′44″	158.2277

(a) Assuming a Cartesian coordinate system with the Y-axis aligned with the line of sight from (P) to (A), calculate if there is a displacement of point (B) relative to point (A), the amount of this displacement in the direction of the line of sight and perpendicular to it, the absolute displacement and its orientation relative to the Y-axis.

(b) After the measurement, it was verified that the total station had a horizontal collimation error. Double-readings were carried out to calculate this error, as shown in Table 10.11. Under this condition, calculate the horizontal collimation error, the corrected coordinates of the point (B) and the angle of inclination of the pilar.

Table 10.11 Direct and reversed readings

Direct reading	Reversed reading
90°12′15″	270°12′01″
90°12′16″	270°12′01″
90°12′15″	270°12′02″
90°12′14″	270°12′03″
90°12′16″	270°12′01″

6. To determine the horizontal clockwise angle between two lines, single-face readings of 32°27′35″ (backsight) and 147°30′54″ (foresight) were carried out using a total station. Assuming that the observations were taken at an instrument height of 1.61 m and a target centring error of 4 mm, calculate the corrected horizontal angle value.

7. Given the instrument and the target centring errors from problem 6 and assuming that the total station has a standard deviation specified by the manufacturer of ±5″, calculate the correction value to be used for an angle determination using this instrument.

8. A vertical collimation error should be checked on a digital level using the procedures described in Sect. 10.4.1.2. The reading values are given in Table 10.12. Calculate the collimation error and the corrected reading of 1175 mm taken on a levelling staff 30 m from the level.

Table 10.12 Reading values

BS_1[mm]	FS_1[mm]	BS_2[mm]	FS_2[mm]
1639	1622	1577	1562

9. Calculate the corrected value of a levelling staff reading of 1938 mm using an automatic level with a telescope magnification power of 30× and a levelling staff with a bubble level with a sensitivity of 20′.

10.7 Review Questions

1. Discuss the instrumental errors that can occur when using a total station and how they can be mitigated in field measurements.
2. Discuss which types of angular observation can be affected by vertical index error.
3. Briefly describe what horizontal collimation and tilting errors are.
4. Explain why direct and reversed observations with a total station cannot correct the vertical axis error of the instrument and what must be done to correct it.
5. Have a look at various instrument specification sheets from different manufacturers to find out the angular and linear standard deviations they specify for their total stations. Compare what effect they would have on a 1 km distance measurement.

6. Briefly describe the errors in electronic distance measurement that a total station is subject to, the procedures for checking them and how often they should be checked.
7. Briefly describe the instrumental errors that a surveyor's level is subject to, the procedures for checking them and how often they should be checked.

References

Deumlich, F. & Staiger, R., (2002). *Instrumentenkunde der Vermessungstechnik.* 9., völlig neu bearbeitete und erweiterte Auflage, Herbert Wichmann Verlag, Heidelberg.

DIN 18.723-1:1990-07 – *Field Procedure for Precision Testing of Surveying Instruments - General Information.* German Institute for Standardisation (Deutsches Institut für Normung).

FIG N⁰ 9 1994. *A technical Monograph (2007). Recommended Procedures for Routine Checks of Electro-Optical Distance Meters.* Edited by Rudolf STAIGER.

García-Balboa, J. L., Ruiz-Armenteros, A. M., Rodríguez-Avi, J., Antonio M., José, Reinoso-Gordo, J. F., and Robledillo-Román, R. (2018). *A Field Procedure for the Assessment of the Centring Uncertainty of Geodetic and Surveying Instruments.* Sensors, 26p.

ISO 17.123-1. *Optics and optical instruments — Field procedures for testing geodetic and surveying instruments — Part 1: Theory.*

ISO 17.123-2. *Optics and optical instruments — Field procedures for testing geodetic and surveying instruments — Part 2: Levels.*

ISO 17.123-3. *Optics and optical instruments — Field procedures for testing geodetic and surveying instruments — Part 2: Theodolites.*

ISO 17.123-4. *Optics and optical instruments — Field procedures for testing geodetic and surveying instruments — Part 4: Electro-optical distance meters (EDM measurements to reflectors).*

ISO 17.123-5: *Optic and optical instruments – Field procedures for testing geodetic and surveying instruments: Part 5: Total stations.*

ISO 17.123-6: *Optic and optical instruments – Field procedures for testing geodetic and surveying instruments: Part 5: Rotating lasers.*

Lambrou E., Nikolitsas, K. (2017). *Detecting the centring error of geodetic instruments over a ground mark through a tribrach-based optical plummet.* Appl Geomat 9:237–245. Access: https://doi.org/10.1007/s12518-017-0197-8.

Levin, E., Nadolinets, L., Akhmedov, D. (2017). *Surveying Instruments and Technology.* CRC Press. ISBN 13: 978-1-4987-6238-0.

Martin, D. et Gatta, G. (2006). *Calibration of Total Stations Instruments at the ESRF.* Shaping the Change, XXIII FIG Congress, Munich, Germany, October, pages 8–13.

Šiaudinytė, L. (2014). *Research and development of methods and instrumentation for the calibration of vertical angle measuring systems of geodetic instruments.* Doctoral dissertation. Vilnius Gediminas Technical University. 122p.

Staiger, R. (2004). *Le Contrôle des Instruments Géodésiques.* XYZ Revue de l'Association Français de Topographie, 99(2), pages 39–46.

Walser, H. P. (2004). *Development and calibration of an image assisted total station.* Doctoral dissertation. Swiss Federal Institute of Technology, Zurich, 168p.

Woschitz, H. Brunner, F. K. (2002). *System Calibration of Digital Levels – Experimental Results of Systematic Effects.* INGEO2002, end Conference of Engineering Surveying, November, pp 165–172, Bratislava, Slovakia.

Zeiske, K. (2001). *Current status of the ISO standardization of accuracy determination process for surveying instruments.* In Proceedings of FIG international conference Seoul, Korea, pages 1–9.

Chapter 11
Coordinate Geometry for Point Positioning

Irineu da Silva and Guilherme Poleszuk dos Santos Rosa

11.1 Introduction

Nowadays, most Civil Engineering projects are based on locating geospatial data, specified by their coordinates (X, Y, H), which provide point positioning for many purposes, such as establishing a network of control points, defining a cartographic database for project development and preparing a dataset for the setting out of designed features. Although working with coordinates in this context is fairly straightforward, especially when using CAD or GIS platforms, there may be instances where calculations are needed, especially when new control points or feature points need to be added or retrieved from a database. To this end, this chapter covers the main concepts of *coordinate geometry*[1] *for point positioning*.

As mentioned in previous chapters, the (X, Y) coordinates of points for most Civil Engineering projects are based on a Two-Dimensional Cartesian coordinate system, taking the local ground-based horizontal plane as the horizontal datum (engineering datum). Elevations (H), on the other hand, are based on a network of vertical benchmarks linked to a Geodetic Altimetry Reference System or an arbitrary local control point established exclusively for the project. The determination of the coordinates (X, Y) is the objective of this chapter, while the determinations of the elevations (H) are presented in Chap. 13. However, as the simultaneous determination of

---

[1] A branch of geometry where the positions of points and lines are described using numerical coordinates.

I. da Silva (✉)
São Carlos School of Engineering, University of São Paulo, São Carlos, SP, Brazil
e-mail: irineu@sc.usp.br

G. Poleszuk dos Santos Rosa
Cartographic Engineer, Researcher at the Faculty of Science and Technology of the São Paulo State University (UNESP), São Paulo, Brazil

three-dimensional coordinates is increasingly used in Civil Engineering, this chapter also deals with some methods for determining three-dimensional coordinates.

Coordinate geometry is often used to solve classic problems of Geomatics applied to Civil Engineering, such as determining distances, azimuths, angles, elevations and areas defined by point coordinates. However, more complex problems can also benefit from coordinate geometry for point positioning, such as determining coordinates in a network of control points, GNSS navigation, 3D modelling, building design, GIS management and many others. In the context of this book, only the basic concepts of coordinate geometry related to point positioning are covered in this chapter with the objective of applying algebraic methods to accurately determine and manage the positions of points within a coordinate system.

Finally, it is worth noting that although the point determination methods presented here are based on Plane Surveying assumptions, they can also be applied to grid coordinates, as long as the appropriate geodetic reductions are considered, as shown in Chap. 16.

11.2   Calculation of Azimuth and Distance by Coordinates (Inverse Problem)

Given the Two-Dimensional Cartesian coordinates of two points in the local ground-based horizontal plane, finding the forward and backward azimuths and the horizontal distance between these points is one of the fundamental problems in Geomatics, often referred to as the *inverse problem*. The solution to this problem is based on the coordinate transformation between Cartesian and Polar coordinate systems, as described in Sect. 5.9.1.1. The geometric details of such a coordinate transformation are shown in Fig. 11.1.

Fig. 11.1 Geometry of azimuth and distance calculations

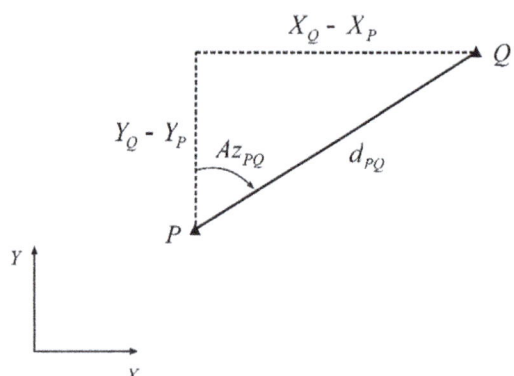

Given:

Cartesian coordinates of points (P) and (Q):

$$P : (X_P, Y_P)$$

$$Q : (X_Q, Y_Q)$$

To calculate:

$Az_{PQ}=$ azimuth of line PQ

$d_{PQ}=$ horizontal distance between points (P) and (Q)

According to Fig. 11.1, the following equations can be written to calculate azimuth and distance, which are general and valid for all four trigonometric quadrants.

$$\tan(Az_{PQ}) = \frac{\Delta X}{\Delta Y} = \frac{X_Q - X_P}{Y_Q - Y_P} \tag{11.1}$$

$$d_{PQ} = \sqrt{\Delta X^2 + \Delta Y^2} = \frac{\Delta X}{\sin(Az_{PQ})} = \frac{\Delta Y}{\cos(Az_{PQ})} \tag{11.2}$$

To calculate the azimuth of the line PQ, it is first necessary to calculate the coordinate differences $\Delta X = X_Q - X_P$ and $\Delta Y = Y_Q - Y_P$ and then apply Eq. (11.3) using the algebraic signs of ΔX and ΔY.

$$Az = \operatorname{atan}\left(\frac{\Delta X}{\Delta Y}\right) \tag{11.3}$$

The reader should note that the value of Az can be positive or negative depending on the algebraic signs of ΔX and ΔY. The azimuth of the line PQ is therefore given by Eq. (11.4).

$$Az_{PQ} = Az + \alpha_{\text{quad}} = \operatorname{atan}\left(\frac{X_Q - X_P}{Y_Q - Y_P}\right) + \alpha_{\text{quad}} \tag{11.4}$$

The value of α_{quad} depends on which quadrant the line lies. It can be $0°$, $180°$ or $360°$. The decision is taken considering the algebraic signs of ΔX and ΔY, according to Table 11.1.

Table 11.1 Quadrant analysis for azimuth determination

Quadrant	ΔX	ΔY	Az_{PQ}
I	+	+	$=Az + 0°$
II	+	−	$=Az + 180°$
III	−	−	$=Az + 180°$
IV	−	+	$=Az + 360°$

11.2.1   Error Propagation in Distance Calculation

Assuming that the coordinates (X_P, Y_P) and (X_Q, Y_Q) are known, the horizontal distance d_{PQ} can be calculated using Eq. (11.5).

$$d_{PQ} = \sqrt{(X_Q - X_P)^2 + (Y_Q - Y_P)^2} \tag{11.5}$$

In this case, if the standard deviations s_{XP}, s_{YP}, s_{XQ}, and s_{YQ} are also known, the standard deviation s_{dPQ} can be calculated by applying the general law of propagation of variances, as given by Eq. (11.6).

$$s_{d_{PQ}}^2 = J^T \Sigma_{ll} J \tag{11.6}$$

where

$$\Sigma_{ll} = \begin{bmatrix} s_{X_Q}^2 & s_{X_Q Y_Q} & 0 & 0 \\ s_{Y_Q X_Q} & s_{Y_Q}^2 & 0 & 0 \\ 0 & 0 & s_{X_P}^2 & s_{X_P Y_P} \\ 0 & 0 & s_{Y_P X_P} & s_{Y_P}^2 \end{bmatrix} \tag{11.7}$$

$$J^T = \left[\left(\frac{\partial d_{PQ}}{\partial X_Q} \right)_0 \left(\frac{\partial d_{PQ}}{\partial Y_Q} \right)_0 \left(\frac{\partial d_{PQ}}{\partial X_P} \right)_0 \left(\frac{\partial d_{PQ}}{\partial Y_P} \right)_0 \right] \tag{11.8}$$

$$\frac{\partial d_{PQ}}{\partial X_Q} = \frac{X_Q - X_P}{d_{PQ}} \tag{11.9}$$

$$\frac{\partial d_{PQ}}{\partial Y_Q} = \frac{Y_Q - Y_P}{d_{PQ}} \tag{11.10}$$

$$\frac{\partial d_{PQ}}{\partial X_P} = \frac{X_P - X_Q}{d_{PQ}} \tag{11.11}$$

$$\frac{\partial d_{PQ}}{\partial Y_P} = \frac{Y_P - Y_Q}{d_{PQ}} \tag{11.12}$$

It can be easily verified that if the standard deviations of the coordinates are uncorrelated and all equal to s, the standard deviation sd_{PQ} is given by Eq. (11.13).

$$sd_{PQ} = s\sqrt{2} \tag{11.13}$$

11.2.2   Error Propagation in Azimuth Calculation

Similarly to the distance, the standard deviation of the azimuth s_{AzPQ} can be calculated by applying the general law of propagation of variances to Eq. (11.3), as follows:

$$s_{Az_{PQ}}^2 = J^T \Sigma_{ll} J \tag{11.14}$$

where Σ_{ll} is given by Eq. (11.7) and

$$J^{\mathrm{T}} = \left[\left(\frac{\partial Az_{PQ}}{\partial X_Q} \right)_0 \left(\frac{\partial Az_{PQ}}{\partial Y_Q} \right)_0 \left(\frac{\partial Az_{PQ}}{\partial X_P} \right)_0 \left(\frac{\partial Az_{PQ}}{\partial Y_P} \right)_0 \right] \tag{11.15}$$

$$\frac{\partial Az_{PQ}}{\partial X_Q} = \frac{Y_Q - Y_P}{d_{PQ}^2} \tag{11.16}$$

$$\frac{\partial Az_{PQ}}{\partial Y_Q} = \frac{X_P - X_Q}{d_{PQ}^2} \tag{11.17}$$

$$\frac{\partial Az_{PQ}}{\partial X_P} = \frac{Y_P - Y_Q}{d_{PQ}^2} \tag{11.18}$$

$$\frac{\partial Az_{PQ}}{\partial Y_P} = \frac{X_Q - X_P}{d_{PQ}^2} \tag{11.19}$$

In this case, it can also be easily verified that if the standard deviations of the coordinates are uncorrelated and all equal to s, the standard deviation s_{AzPQ} is given by Eq. (11.20).

$$s_{AzPQ} = \frac{s\sqrt{2}}{d_{PQ}} \tag{11.20}$$

Example 11.1
Given the Cartesian coordinates of points (1) and (4) and their standard deviations shown in Table 11.2, calculate the azimuth $Az_{1,4}$, the horizontal distance $d_{1,4}$ and their standard deviations.

Table 11.2 Coordinates of points (1) and (4)

Point	X [m]	Y [m]
1	5035.880 ± 0.0024	10,064.838 ± 0.0014
4	4904.603 ± 0.0011	10,029.047 ± 0.0033
(4–1)	−131.277	−35.791

Solution:

Distance calculation:

Using Eq. (11.2) or (11.5),

$$d_{1,4} = 136.069 \text{ m}$$

Assuming that the standard deviations of the coordinates are uncorrelated, the matrix of variance-covariance for the observations is given as follows.

$$\Sigma_{ll} = \begin{bmatrix} 1.2100 * 10^{-6} & 0 & 0 & 0 \\ 0 & 1.0890 * 10^{-5} & 0 & 0 \\ 0 & 0 & 5.7600 * 10^{-6} & 0 \\ 0 & 0 & 0 & 1.9600 * 10^{-6} \end{bmatrix}$$

Using Eq. (11.8),

$$J^{\mathrm{T}} = [-0.964786 \quad -0.263037 \quad 0.964786 \quad 0.263037]$$

Using Eq. (11.6),

$$s_{d_{1,4}}^2 = 7.376827 * 10^{-6} \, \mathrm{m}^2$$

$$s_{d_{1,4}} = \pm 0.0027 \, \mathrm{m} = \pm 2.7 \, \mathrm{mm}$$

And then,

$$d_{1,4} = 136.069 \, \mathrm{m} \pm 2.7 \, \mathrm{mm}$$

Azimuth calculation:

Using Eq. (11.3),

$$Az = \operatorname{atan}\left(\frac{-131.277}{-35.791}\right) = 74°\, 44'58.9''$$

Considering that the line 1–4 is in the third quadrant.

$$Az_{1,4} = 74°\, 44'58.9'' + 180° = 254°\, 44'58.9''$$

Using Eq. (11.15),

$$J^{\mathrm{T}} = [-1.933118 * 10^{-3} \quad 7.090441 * 10^{-3} \quad 1.933118 * 10^{-3} \quad -7.090441 * 10^{-3}]$$

Using Eq. (11.14),

$$s_{Az_{1,4}}^2 = 6.720719 * 10^{-10} \mathrm{rad}^2$$

$$s_{Az_{1,4}} = \pm 2.592435 * 10^{-5} \mathrm{rad} = \pm 5.3''$$

And then,

$$Az_{1,4} = 254°\, 44'58.9'' \pm 5.3''$$

11.3 Point Positioning by Azimuth and Distance (Direct Problem)

Another fundamental problem for point positioning in Geomatics, known as the *direct problem*, is to determine the coordinates of a point (Q) from the coordinates of a control point (P), considering the horizontal distance and the azimuth of the line PQ. The solution to this problem is based on the Two-Dimensional coordinate transformation from the Polar to the Cartesian coordinate system, as described in Sect. 5.9.1.1. The geometric details of this coordinate transformation are shown in Fig. 11.2.

Fig. 11.2 Geometry of point positioning by azimuth and distance

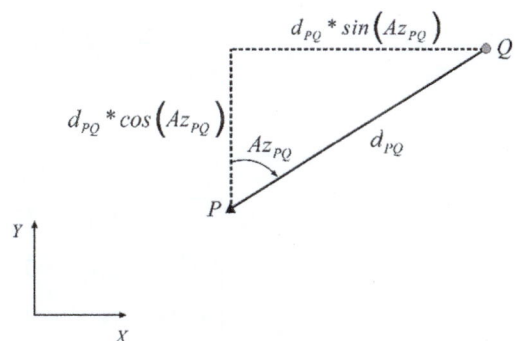

Given:

Cartesian coordinates of the control point (P): (X_P, Y_P)
Azimuth of line PQ: Az_{PQ}
Horizontal distance PQ: d_{PQ}

To calculate:

Coordinates of the point (Q): (X_Q, Y_Q)

According to Fig. 11.2, the coordinates of point (Q) can be calculated using Eq. (11.21).

$$X_Q = X_P + d_{PQ} * \sin(Az_{PQ})$$
$$Y_Q = Y_P + d_{PQ} * \cos(Az_{PQ})$$

(11.21)

11.3.1   Error Propagation in Point Positioning by Azimuth and Distance

For error propagation analysis, Eq. (11.21) can be rewritten as

$$
\begin{aligned}
X_Q &= X_P + 0Y_P + d_{PQ} * \sin(Az_{PQ}) \\
Y_Q &= 0X_P + Y_P + d_{PQ} * \cos(Az_{PQ})
\end{aligned}
\tag{11.22}
$$

In this case, assuming that the standard deviations s_{XP}, s_{YQ}, s_{dPQ} and sAz_{PQ} are known, the variance-covariance matrix for the function $\boldsymbol{F}$ can be calculated by applying the law of propagation of variances.

$$
\Sigma_{ff} = \boldsymbol{J}^{\mathrm{T}} \Sigma_{ll} \boldsymbol{J}
\tag{11.23}
$$

where

$$
\Sigma_{ff} =
\begin{bmatrix}
s_{X_Q}^2 & s_{X_Q Y_Q} \\
s_{Y_Q X_Q} & s_{Y_Q}^2
\end{bmatrix}
\tag{11.24}
$$

$$
\Sigma_{ll} =
\begin{bmatrix}
s_{X_P}^2 & s_{X_P Y_P} & 0 & 0 \\
s_{Y_P X_P} & s_{Y_P}^2 & 0 & 0 \\
0 & 0 & s_{d_{PQ}}^2 & 0 \\
0 & 0 & 0 & s_{Az_{PQ[\mathrm{rad}]}}^2
\end{bmatrix}
\tag{11.25}
$$

$$
\boldsymbol{J}^{\mathrm{T}} =
\begin{bmatrix}
\left(\dfrac{\partial X_Q}{\partial X_P}\right)_0 & \left(\dfrac{\partial X_Q}{\partial Y_P}\right)_0 & \left(\dfrac{\partial X_Q}{\partial d_{PQ}}\right)_0 & \left(\dfrac{\partial X_Q}{\partial Az_{PQ}}\right)_0 \\
\left(\dfrac{\partial Y_Q}{\partial X_P}\right)_0 & \left(\dfrac{\partial Y_Q}{\partial Y_P}\right)_0 & \left(\dfrac{\partial Y_Q}{\partial d_{PQ}}\right)_0 & \left(\dfrac{\partial Y_Q}{\partial Az_{PQ}}\right)_0
\end{bmatrix}
\tag{11.26}
$$

$$
\frac{\partial X_Q}{\partial X_P} = \frac{\partial Y_Q}{\partial Y_P} = 1
\tag{11.27}
$$

$$
\frac{\partial X_Q}{\partial Y_P} = \frac{\partial Y_Q}{\partial X_P} = 0
\tag{11.28}
$$

$$
\frac{\partial X_Q}{\partial d_{PQ}} = \sin(Az_{PQ})
\tag{11.29}
$$

$$
\frac{\partial X_Q}{\partial Az_{PQ}} = d_{PQ} * \cos(Az_{PQ})
\tag{11.30}
$$

$$
\frac{\partial Y_Q}{\partial d_{PQ}} = \cos(Az_{PQ})
\tag{11.31}
$$

$$\frac{\partial Y_Q}{\partial Az_{PQ}} = -d_{PQ} * \sin(Az_{PQ}) \tag{11.32}$$

In a more general case, Eq. (11.22) can be modified to consider the reference azimuth Az_{RefP} and the horizontal clockwise angle α_P instead of the azimuth Az_{PQ}, as given by Eq. (11.33).

$$X_Q = X_P + 0Y_P + d_{PQ} * \sin(Az_{RefP} + \alpha_P \pm 180)$$
$$Y_Q = 0X_P + Y_P + d_{PQ} * \cos(Az_{RefP} + \alpha_P \pm 180) \tag{11.33}$$

For the application of the law of propagation of variances, Eq. (11.25) then becomes

$$\Sigma_{ll} = \begin{bmatrix} s_{X_P}^2 & s_{X_P Y_P} & 0 & 0 & 0 \\ s_{Y_P X_P} & s_{Y_P}^2 & 0 & 0 & 0 \\ 0 & 0 & s_{d_{PQ}}^2 & 0 & 0 \\ 0 & 0 & 0 & s_{Az_{RefP}[rad]}^2 & 0 \\ 0 & 0 & 0 & 0 & s_{\alpha_P[rad]}^2 \end{bmatrix} \tag{11.34}$$

And the partial derivates of the function F are as follows.

$$\frac{\partial X_Q}{\partial X_P} = \frac{\partial Y_Q}{\partial Y_P} = 1 \tag{11.35}$$

$$\frac{\partial X_Q}{\partial Y_P} = \frac{\partial Y_Q}{\partial X_P} = 0 \tag{11.36}$$

$$\frac{\partial X_Q}{\partial d_{PQ}} = \sin(Az_{RefP} + \alpha_P \pm 180°) \tag{11.37}$$

$$\frac{\partial X_Q}{\partial Az_{RefP}} = d_{PQ} * \cos(Az_{RefP} + \alpha_P \pm 180°) \tag{11.38}$$

$$\frac{\partial X_Q}{\partial \alpha_P} = d_{PQ} * \cos(Az_{RefP} + \alpha_P \pm 180°) \tag{11.39}$$

$$\frac{\partial Y_Q}{\partial d_{PQ}} = \cos(Az_{RefP} + \alpha_P \pm 180°) \tag{11.40}$$

$$\frac{\partial Y_Q}{\partial Az_{RefP}} = -d_{PQ} * \sin(Az_{RefP} + \alpha_P \pm 180°) \tag{11.41}$$

$$\frac{\partial Y_Q}{\partial \alpha_P} = -d_{PQ} * \sin(Az_{RefP} + \alpha_P \pm 180°) \tag{11.42}$$

Example 11.2

Given the coordinates and standard deviations of point (*1*) in Table 11.2 and considering that a horizontal distance of 166.493 m ± 3.0 mm to point (*3*) has been measured using a total station, calculate the coordinates of point (*3*) and their standard deviations. From previous measurements, it is assumed that the azimuth $Az_{1,3}$ is equal to 216°02′04″ ± 10″.

Solution:

Using Eq. (11.21), *the observation equations are given as follows:*

$$X_3 = 5035.880 + 166.493 * \sin(216°02′04″) = 4937.937 \text{ m}$$

$$Y_3 = 10,064.838 + 166.493 * \cos(216°02′04″) = 9930.201 \text{ m}$$

Assuming that the standard deviations of the coordinates are uncorrelated, the matrix of variance-covariance for the observations is given as follows:

$$\Sigma_{ll} = \begin{bmatrix} 5.760000 * 10^{-6} & 0 & 0 & 0 \\ 0 & 1.960000 * 10^{-6} & 0 & 0 \\ 0 & 0 & 9.000000 * 10^{-6} & 0 \\ 0 & 0 & 0 & 2.350443 * 10^{-9} \end{bmatrix}$$

Using Eq. (11.26),

$$J^T = \begin{bmatrix} 1 & 0 & -0.588272 & -134.636810 \\ 0 & 1 & -0.808663 & 97.943087 \end{bmatrix}$$

Using Eq. (11.23),

$$\Sigma_{ff} = \begin{bmatrix} 5.148122 * 10^{-5} & -2.671327 * 10^{-5} \\ -2.671327 * 10^{-5} & 3.039287 * 10^{-5} \end{bmatrix}$$

Using the diagonal values in the variance-covariance matrix of functions Σ_{ff}, the standard deviations of the coordinates of point (3) are as follows:

$$X_3 = 4937.937 \text{ m} \pm 7.2 \text{ mm}$$

$$Y_3 = 9930.201 \text{ m} \pm 5.5 \text{ mm}$$

The correlation coefficient can be calculated using Eq. (3.14) *and the covariance value in the variance-covariance matrix* Σ_{ff} *as follows:*

$$r_{X_3Y_3} = \frac{-2.671327 * 10^{-5}}{0.0072 * 0.0055} = 67.5\%$$

11.4   Calculation of Consecutive Azimuths

Rather than being calculated from coordinates, the azimuth of a given line can also be determined by adding a horizontal clockwise angle α_i to a back azimuth, as shown in Fig. 11.3.

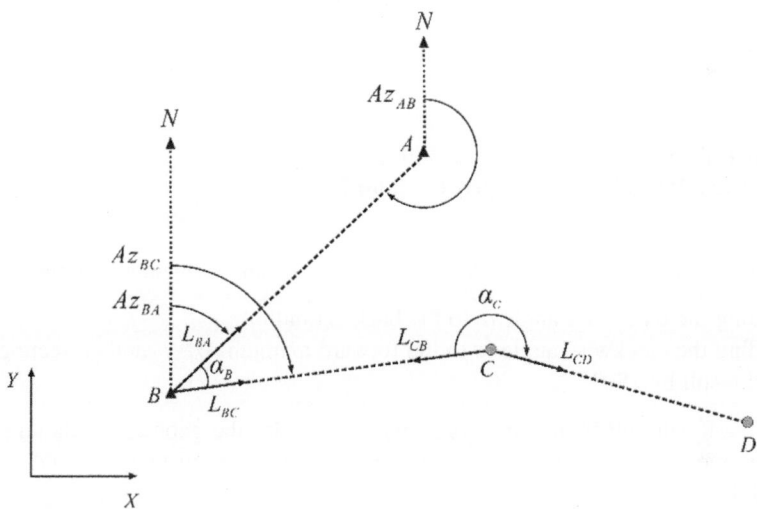

Fig. 11.3 Geometry of the calculation of consecutive azimuths

Given:

   Azimuth Az_{AB} or Az_{BA}

Observation:

$L_{BA}, L_{BC}, L_{CB}, L_{CD}$

To calculate:

   Horizontal clockwise angle α_i
   Azimuth $Az_{i,i+1}$

In this case, the problem is to determine the azimuth $Az_{i,i+1}$ of successive lines from the azimuth $Az_{i,i-1}$ or $Az_{i-1,i}$ of the previous line and the forward $L_{i,i+1}$ and backward $L_{i,i-1}$ horizontal directions observed around point (i). The calculation procedure requires knowledge of an initial back azimuth, usually determined from the coordinates of control points, such as points (A) and (B) in Fig. 11.3. Therefore, given the horizontal clockwise angles determined as a function of the foresight and backsight readings, the following equation can be written to express the calculation of successive azimuths:

$$\alpha_B = L_{\text{fore}} - L_{\text{back}} = L_{BC} - L_{BA} \qquad (11.43)$$

$$Az_{BC} = Az_{BA} + \alpha_B \qquad (11.44)$$

which, comprehensively, can be rewritten as given by the following equations:

$$Az_{1,i+1} = Az_{i,i-1} + \alpha_i \qquad (11.45)$$

$$Az_{i,i+1} = Az_{i-1,i} + \alpha_i \pm 180° \qquad (11.46)$$

Add 180° if $(Az_{i-1,i})$ is less than 180°.
Subtract 180° if $(Az_{i-1,i})$ is greater than 180°.

Thus, the forward azimuth $Az_{i,i+1}$ of a single line can be calculated in two ways:

1. Adding the clockwise angle α_i to the back azimuth $Az_{i,i-1}$
2. Adding the clockwise angle α_i to the forward azimuth $Az_{i-1,i}$ and correcting the final result by 180°

To clarify the algebraic equations above, consider the geometric relationships between the azimuths and the horizontal clockwise angles of lines BC and CD in Fig. 11.4.

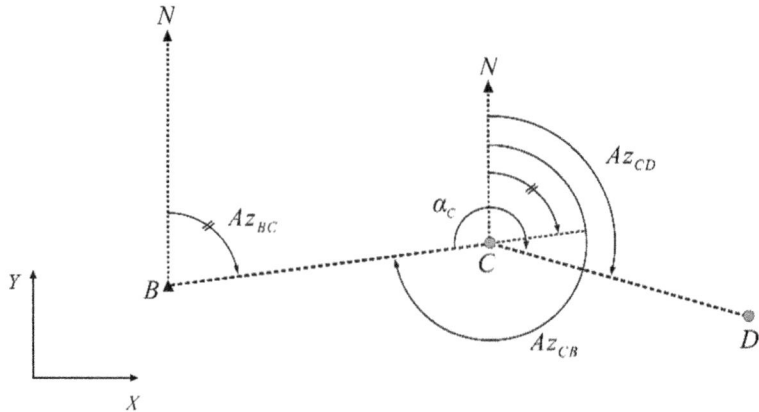

Fig. 11.4 Detail of angular relationships for calculating consecutive azimuths

From Fig. 11.4,

$Az_{CD} = Az_{CB} + \alpha_C - 360°$ (360° was subtracted because the sum is greater than 360°)

$Az_{CD} = Az_{BC} + \alpha_C + 180°$ (180° was added because the forward azimuth Az_{BC} is less than 180°)

When determining successive azimuths, the reader should note that the process is easier if the azimuths are calculated sequentially using Eq. (11.46), taking care to check the addition or subtraction of 180°.

Regarding error propagation, the reader should note that the calculation of successive azimuths is based on linear functions. Therefore, to calculate the standard deviation of a given forward azimuth, it is necessary to know the standard deviations of both the back azimuth and the horizontal clockwise angle. The error propagation must be calculated using Eq. (11.14) for the back azimuth. For the horizontal angle, the error propagation is calculated according to the angular observation method, adding the instrument and target centring errors, as described in Sect. 10.2.2.1.

Example 11.3
Given the coordinates of points (*1*) and (*4*) from Example 11.1 and the field measurements in Table 11.3, calculate the azimuth of the line *4–3*.

Table 11.3 Field measurements

Station	Sighting point	Horizontal direction
4	*1*	25°25′01″
	3	112°01′53″

Solution:
  Using Eq. (11.43),

$$\alpha_4 = 112°\,01'53'' - 25°\,25'01'' = 86°\,36'52''$$

From Example 11.1,

$$Az_{1,4} = 254°\,44'58.9''$$

Using Eq. (11.46),

$$Az_{4,3} = 254°\,44'58.9'' + 86°\,36'52'' - 180° = 161°\,21'51''$$

If the back azimuth $Az_{4,1} = 74°\,44'58.9''$ is used, the forward azimuth $Az_{4,3}$ can be calculated using Eq. (11.45).

$$Az_{4,3} = 74°\,44'59'' + 86°\,36'52'' = 161°\,21'51''$$

11.5   Calculation of Successive Coordinates

As seen in Sect. 11.3, it is possible to determine the coordinates of a foresight point by knowing the coordinates of the station point, the forward azimuth, and the distance between the station and the foresight point. Thus, considering the sequence of calculations for the successive azimuth determination presented in the previous section, it is also possible to calculate the successive coordinates by measuring the distances between pairs of sequential points, as shown in Fig. 11.5. In this case, point (B) is the station point, and point (A) is the azimuthal reference point with known coordinates. The coordinates of the following consecutive points of the course can be determined as a function of the forward and backward directions and the horizontal distances between them, all measured in the field. The geometric details of the successive coordinate calculation are shown in Fig. 11.5.

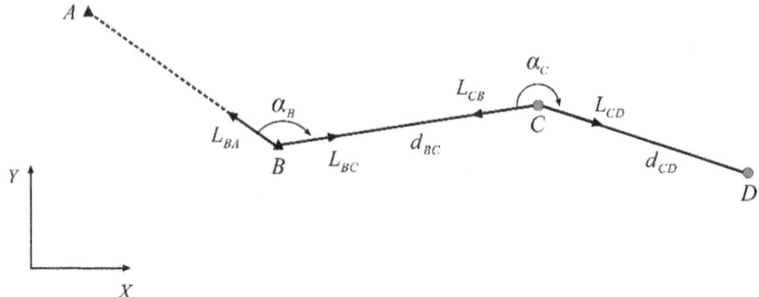

Fig. 11.5 Geometry of consecutive coordinate calculation

Given:

Point (A): (X_A, Y_A)
Point (B): (X_B, Y_B)

Observation:

$L_{BA}, L_{BC}, L_{CB}, L_{CD}, d_{BC}, d_{CD}$

To calculate:

Point $(i+1)$: (X_{i+1}, Y_{i+1})

According to Fig. 11.5, the steps for calculating successive coordinates are:

1. Calculate the first azimuth from the known coordinates of points (A) and (B) using Eq. (11.4)
2. Calculate the horizontal clockwise angles α_i, using Eq. (11.43)
3. Calculate the coordinates of the foresight point $(i+1)$ using Eq. (11.47)

$$X_{i+1} = X_i + d_{i,i+1} * \sin(Az_{i-1,i} + \alpha_i \pm 180°)$$
$$Y_{i+1} = Y_i + d_{i,i+1} * \cos(Az_{i-1,i} + \alpha_i \pm 180°)$$

(11.47)

11.5.1 Error Propagation in the Successive Coordinate Calculation

As in the calculation of the variance-covariance matrix of the coordinates of a single point (n) along the successive points of the course, the solution to the problem can be obtained by successively applying the law of propagation of variances to Eq. (11.47). For this, as explained above, it is necessary to know the standard deviations of the first azimuth, of the coordinates of the first point and of each horizontal angle, and of each distance. For further details on the application of successive propagation of coordinate variances, see Example 12.1.

Example 11.4
Using the results of Example 11.3 and the field measurements in Table 11.4, calculate the coordinates of points (5) and (2).

Solution:
  From Example 11.1,

Table 11.4 Field measurements

Station	Sighting point	Horizontal direction	Horizontal distance [m]
4	1	25°25′01″	–
	5	81°26′27″	142.476
5	4	0°00′00″	–
	2	87°14′26″	84.330

$$Az_{1,4} = 254°\,44'58.9''$$

Using Eq. (11.43),

$$\alpha_4 = 56°\,01'26''$$

$$\alpha_5 = 87°\,14'16''$$

Azimuths $Az_{4,5}$ and $Az_{5,2}$ can then be determined as follows:

$$Az_{4,5} = 254°\,44'58.9'' + 56°\,01'26'' - 180° = 130°\,46'24.9''$$

$$Az_{5,2} = 130°\,46'24.9'' + 87°\,14'26'' - 180° = 38°\,00'50.9''$$

Using Eq. (11.47),

$$X_5 = 4904.603 + 142.476 * \sin(130°\,46'24.9'') = 5012.500 \text{ m}$$

$$Y_5 = 10,029.047 + 142.476 * \cos(130°\,46'24.9'') = 9936.000 \text{ m}$$

And,

$$X_2 = 5012.500 + 84.330 * \sin(38°\,00'50.9'') = 5064.435 \text{ m}$$

$$Y_2 = 9936.000 + 84.330 * \cos(38°\,00'50.9'') = 10,002.440 \text{ m}$$

11.6   Azimuth Determination by Multiple Back Sightings

Due to the current ease of determining the coordinates of points in the field, it is advisable to carry out azimuth determinations by observing several reference points, as shown in Fig. 11.6, to increase the reliability of the results.

Fig. 11.6 Geometry of azimuth determination by multiple back sightings

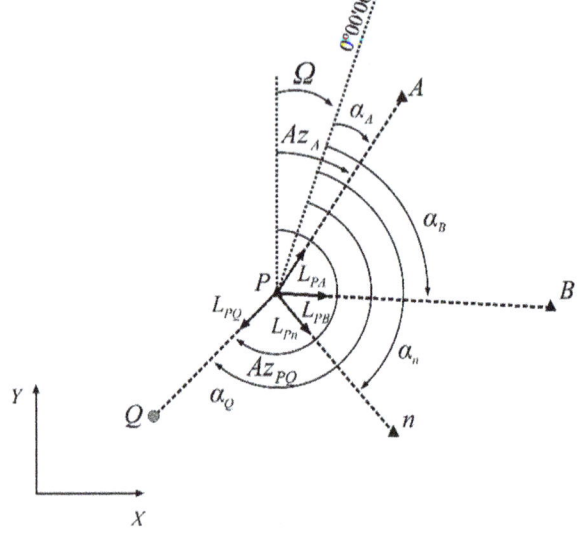

Given:

Point (P): (X_P, Y_P)
Point (A): (X_A, Y_A)
Point (B): (X_B, Y_B)
Point (n): (X_n, Y_n)

Observation:

$L_{PA}, L_{PB}, L_{Pn}, L_{PQ}$

To calculate:

Azimuth Az_{PQ}

The calculation sequence for determining the azimuth Az_{PQ} by multiple back sightings is shown in Table 11.5.

Table 11.5 Calculation sequence for determining the azimuth Az_{PQ}

Sighting point	Calculated azimuth (Eq. 11.4)	Horizontal angle	Unknown orientation (ω_i)	Preliminary azimuth	Residual error
A	Az_{PA}	$\alpha_A = L_{PA}$	$\omega_A = Az_{PA} - \alpha_A$	$Az'_{PA} = \alpha_A + \Omega$	$v_A = Az_{PA} - Az'_{PA}$
B	Az_{PB}	$\alpha_B = L_{PB}$	$\omega_B = Az_{PB} - \alpha_B$	$Az'_{PB} = \alpha_B + \Omega$	$v_B = Az_{PB} - Az'_{PB}$
n	Az_{P_n}	$\alpha_n = L_{P_n}$	$\omega_n = Az_{P_n} - \alpha_n$	$Az'_{P_n} = \alpha_n + \Omega$	$v_n = Az_{P_n} - Az'_{P_n}$
Q	–	$\alpha_Q = L_{PQ}$	–	–	–
			$\Omega = \dfrac{\sum \omega}{n}$	$Az_{PQ} = \alpha_Q + \Omega$	

For the reference direction, note that the orientation unknown Ω is calculated by averaging the individual observation unknowns ω_i. It should only be calculated if the values of residual errors v_i are less than an allowable tolerance. Under this condition, once calculated, it can be used to determine the azimuths of the remaining unknown points.

The reader should also note that the horizontal clockwise angles α_i are shown in Fig. 11.6 only for ease of understanding, as they correspond exactly to the values of the observed horizontal directions.

Example 11.5
Using the azimuth determination sketch shown in Fig. 11.6 and the field measurements and coordinates given in Table 11.6, calculate the azimuth of the line PQ.

Table 11.6 Field measurements and known coordinates

Station	Sighted point	Horizontal direction	X [m]	Y [m]
P	A	14° 34′04″	5254.368	10,514.587
$X_P = 5000.000$ m	B	102° 12′07″	5777.513	9654.853
$Y_P = 10,000.000$ m	C	145° 25′22″	5356.893	9152.742
	Q	190° 18′37″	–	–

Solution:

   Following the calculation sequence given in Table 11.5, *the calculated results are shown in* Table 11.7. *Note the residual error of 3″, which was considered acceptable for this particular azimuth determination.*

Table 11.7 Calculation sequence and results

Sighting point	Calculated azimuth (Eq. 11.4)	Horizontal angle	Unknown orientation (ω_i)	Preliminary azimuth ($\alpha i + \Omega$)	Residual error
A	26° 18′14″	$\alpha_A = 14° 34′04″$	$\omega_A = 11° 44′10″$	26° 18′11″	$v_A = 3″$
B	113° 56′13″	$\alpha_B = 102° 12′07″$	$\omega_B = 11° 44′06″$	113° 56′14″	$v_B = -1″$
C	157° 09′27″	$\alpha_C = 145° 25′22″$	$\omega_C = 11° 44′05″$	157° 09′29″	$v_C = -2″$
Q	–	$\alpha_Q = 190° 18′37″$	–	–	–
			$\Omega = 11° 44′07″$	$Az_{PQ} = 202° 02′44″$	

11.7   Point Setting-out by Coordinates

Point setting-out by coordinates is a standard procedure in Civil Engineering. The objective is to physically mark or "set out" the exact location of a point on the ground, using its coordinates (X, Y) as defined in the project plan. To do this, it is necessary to have at least two control points already stakeout on the construction site in a coordinate system common to all points. These control points are the references for setting out new points retrieved from the project database.

   A typical example of setting out points by coordinates is the case of an engineering work for which a framework of control points is available. Based on the location of these points, graphically represented in a CAD application software, designers can develop their projects and then generate a list of points to be set out in the field, whose coordinates can be retrieved from the graphical representation or even analytically (in the latter case, through coordinate geometry calculations). In general, the points to be set out are subdivision marks, road axes or centrelines of building walls, column foundation pile centres, etc. Figure 11.7 illustrates details of a generic geometric situation for setting out designed points.

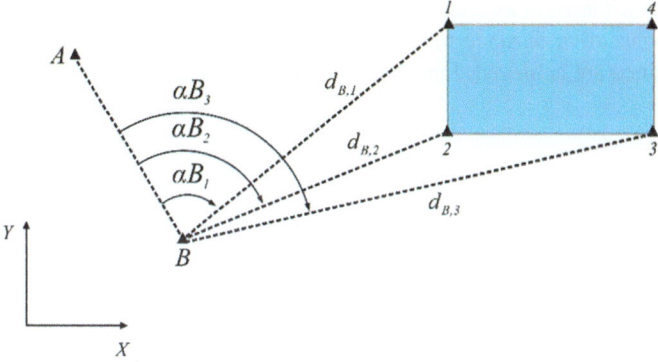

Fig. 11.7 Geometry of point setting-out by coordinates

Given:

Point (A): (X_A, Y_A)
Point (B): (X_B, Y_B)
Point (i): (X_i, Y_i)

To calculate:

Horizontal clockwise angle around (B): αB_i
Horizontal distance: $d_{B,i}$

The setting out parameters for point (i) are the horizontal clockwise setting out angle αB_i about the station point (B) and the horizontal distance $d_{B,i}$ between the station point (B) and the designed setting out point (i). According to the diagram in Fig. 11.7, these parameters are calculated using the following equations:

$$\alpha B_i = Az_{\text{for}} - Az_{\text{back}} = \operatorname{atan}\left(\frac{X_i - X_B}{Y_i - Y_B}\right) + \alpha_{\text{quad 1}} - \operatorname{atan}\left(\frac{X_A - X_B}{Y_A - Y_B}\right) + \alpha_{\text{quad 2}} \quad (11.48)$$

$$d_{B_i} = \sqrt{(X_i - X_B)^2 + (Y_i - Y_B)^2} \quad (11.49)$$

Once the setting out parameters have been calculated, the procedures for point setting-out using a total station can be carried out as follows:

1. Install the instrument over the most convenient control point to carry out the setting out procedures.
2. Aim at the backsight control point and set the horizontal direction of the instrument to zero.
3. Turn the instrument clockwise until it displays the value of the horizontal setting-out angle for the desired point.
4. Guide the field assistant operator in the direction of the line of sight.

5. Ensure that the field assistant operator remains in the setting-out direction and takes consecutive distance measurements until the exact setting-out orientation and distance are achieved. Finally, stake out the point.

Example 11.6
Given the coordinates in Table 11.8 and knowing that a total station will be installed over point (2) and backsighted on point (1), as shown in Fig. 11.8, calculate the setting out elements of points (4) and (5).

Table 11.8 Point coordinates

Point	X [m]	Y [m]
1	5035.880	10,064.838
2	5064.435	10,002.440
4	4904.603	10,029.047
5	5012.500	9936.000

Fig. 11.8 Measurement geometry

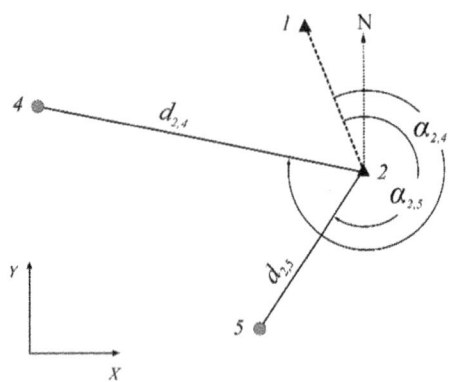

Solution:
Using Eqs. (11.48) and (11.49), the calculated results are given in Table 11.9.

Table 11.9 Calculated results

Line	ΔX [m]	ΔY [m]	Azimuth	Setting out angle	Setting out distance [m]
2–1	−28.555	62.398	335° 24′ 35.6″	–	
2–4	−159.832	26.607	279° 27′04.6″	279°27′04.6″ − 335°24′35.6″ = 304°02′29.0″	162.031
2–5	−51.935	−66.440	218° 00′50.9″	218°00′50.9″ − 335°24′35.6″ = 242°36′15.2″	84.330

11.8   Point Positioning by Angle and Distance

As shown in Sect. 11.3, if the coordinates of a control point are given, and the azimuth to a waypoint is known, it is possible to determine the coordinates of the waypoint by measuring the distance between the control point and the waypoint. Similarly, it is also possible to determine the coordinates of a waypoint by observing several horizontal directions and distances. If only horizontal directions from a waypoint to several control points are observed, the measurement method is called *point positioning by resection*. On the other hand, if horizontal directions from several control points to the waypoint are observed, the measurement method is called *point positioning by intersection*. If only distances between the waypoint and several control points are measured, the measurement method is called *point positioning by distance measurement,* also called a *bilateration* or *multilateration,* depending on the number of distance measurements.

Combinations of the above methods are also possible, as described later in this chapter. The computational methods for these point positioning procedures depend on the number of observations available. If redundant observations are taken, the solution is obtained by applying an Adjustment Computation model; otherwise, a single solution can be obtained using direct equations, as presented in the following subsections.

11.8.1   Resection

Determining the unknown coordinates (X, Y) of an occupied station point (P) using a theodolite or a total station can be accomplished by observing the horizontal directions to at least three control points, (A), (B) and (C), whose coordinates are known, as shown in Fig. 11.9. No distance is measured, which makes this method of point positioning extremely convenient when distance measurement is impossible or inaccurate. As a result, the coordinates and orientation of the station point (P) are calculated in the same coordinate system as the control points.

Fig. 11.9 Geometry of point positioning by resection

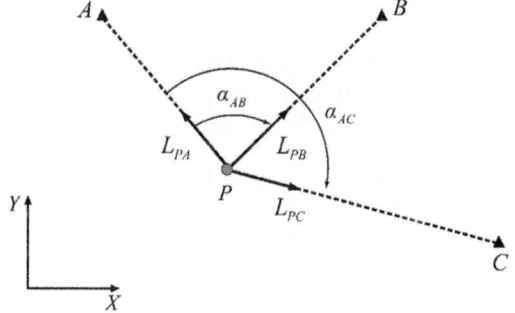

Given:

Point (A): (X_A, Y_A)
Point (B): (X_B, Y_B)
Point (C): (X_C, Y_C)

Observation:

L_{PA}, L_{PB}, L_{PC}

To calculate:

Coordinates of (P): (X_P, Y_P)

This point positioning method is very useful in topographic surveys where distance measurement is impractical, such as large construction sites where access to reflector prism control points may be limited.

The problem is solved by considering it as the intersection of three lines whose azimuths are known. Thus, using Eq. (11.43) for the calculation of the angles α_{AB} and α_{AC}, the following equations can be used to determine the coordinates of point (P).

$$\alpha_{AB} = L_{PB} - L_{PA} \tag{11.50}$$

$$\alpha_{AC} = L_{PC} - L_{PA} \tag{11.51}$$

$$\tan \varphi = \frac{(X_A - X_B) * \cot(\alpha_{AB}) - (X_A - X_C) * \cot(\alpha_{AC}) + (Y_B - Y_C)}{(Y_A - Y_B) * \cot(\alpha_{AB}) - (Y_A - Y_C) * \cot(\alpha_{AC}) - (X_B - X_C)} \tag{11.52}$$

$$Y_P = Y_A + \frac{(X_A - X_B) * [\cot(\alpha_{AB}) - \tan(\varphi)] - (Y_A - Y_B) * [1 + \cot(\alpha_{AB}) * \tan(\varphi)]}{1 + \tan^2(\varphi)}$$

$$\tag{11.53}$$

$$X_P = X_A + (Y_P - Y_A) * \tan(\varphi) \tag{11.54}$$

Note that replacing point (B) with (C) in Eq. (11.53) gives another similar equation for calculating (Y_P).

Once the coordinates are calculated, the accuracy can be verified by comparing the calculated and the measured angle to additional control points, if available.

Note: As this is a calculation method without data redundancy, the accuracy is highly dependent on the relative positions of the control points. If they are collinear or nearly collinear, the method becomes unreliable.

Example 11.7

According to Fig. 11.10 and using the field observations in Table 11.10, calculate the coordinates (X, Y) point (P) using the resection method. The coordinates of control points (2), (3), and (4) are given in Table 11.11.

Fig. 11.10 Observation geometry

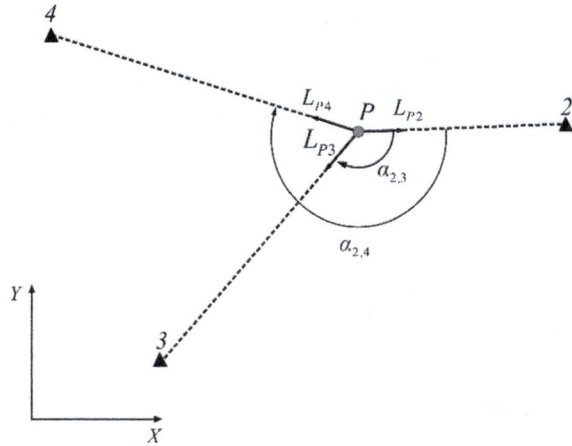

Table 11.10 Field observations

Station	Sighting point	Horizontal direction
P	2	87° 49′ 50″
	3	22° 38′ 36″
	4	286° 56′ 06″

Table 11.11 Control point coordinates

Control point	X [m]	Y [m]
2	5064.435	10,002.440
3	4937.937	9930.201
4	4904.603	10,029.047

Solution:

   Using Eqs. (11.50) to (11.54), the calculation results are given in Table 11.12.

Table 11.12 Calculation results for determining the coordinates of the point (*P*) by the resection method

$\alpha_{2,3} = 133°\,48′46.0″$	$\alpha_{2,4} = 199°\,06′16.0″$	$\cot_{2,3} = -0.959393715$	$\cot_{2,4} = 2.887103390$
$X_2 - X_3 = 126.4980$ m	$X_2 - X_4 = 159.8320$ m	$Y_3 - Y_4 = -98.8460$ m	$Y_2 - Y_3 = 72.2390$ m
$Y_2 - Y_4 = -26.6070$ m	$X_3 - X_4 = 33.3340$ m	$\tan(\varphi) = 26.3978837249$	
$X_P = 5000.001$ m	$Y_p = 9999.999$ m		

11.8.2 Intersection

Determining the coordinates of a point (P) by the intersection method consists of setting up a theodolite or a total station successively over two points with known coordinates, (A) and (B), and observing the horizontal directions from them to point (P), as shown in Fig. 11.11. Again, this is a method of determining coordinates without measuring the distances between the station and the sighting points. As with the resection method, the coordinates and orientation of point (P) are calculated in the same coordinate system as the control points.

Fig. 11.11 Geometry of point positioning by intersection

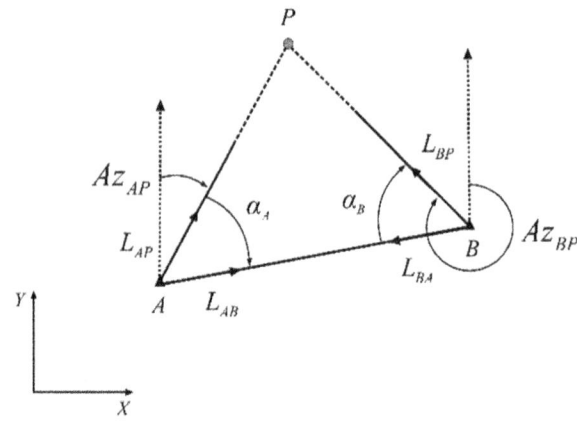

Given:

Point (A): (X_A, Y_A)
Point (B): (X_B, Y_B)

Observation:

$L_{AP}, L_{AB}, L_{BA}, L_{BP}$

To calculate:

Coordinates of (P): (X_P, Y_P)

This method is the reverse of resection and is often used when the unknown point is distant or difficult to reach directly.

According to Fig. 11.11, the following calculation sequence is proposed to determine the coordinates of point (P):

1. From the known coordinates of points (A) and (B), calculate the azimuths Az_{AB} and Az_{BA} using Eq. (11.4).
2. Calculate the horizontal clockwise angles α_A and α_B, using Eq. (11.43).
3. Calculate the azimuth $Az_{AP} = Az_{AB} - \alpha_A$.
4. Calculate the azimuth $Az_{BP} = Az_{BA} + \alpha_B$.

The coordinates of point (P) can then be calculated using the following equations:

$$Y_P = Y_A + \frac{(X_A - X_B) - (Y_A - Y_B) * \tan(Az_{BP})}{\tan(Az_{BP}) - \tan(Az_{AP})} \tag{11.55}$$

Or,

$$Y_P = Y_B + \frac{(X_B - X_A) - (Y_B - Y_A) * \tan(Az_{AP})}{\tan(Az_{AP}) - tg(Az_{BP})} \tag{11.56}$$

And then,

$$X_P = X_A + (Y_P - Y_A) * \tan(Az_{AP}) \tag{11.57}$$

Or,

$$X_P = X_B + (Y_P - Y_B) * \tan(Az_{BP}) \tag{11.58}$$

with Az_{AP} or Az_{BP} necessarily different from 90° or 270° due to undefined tangent values.

Note: As with the point positioning by resection, the intersection method is a calculation method without data redundancy. Therefore, for the measurement to give reliable results, it is necessary to avoid observing acute angles to point (P). It is also important to note that the accuracy of the angle measurement is critical. Any error in angle measurement can significantly affect the accuracy of the position determined.

Example 11.8

An intersection observation from two control points (1) and (2) was carried out to a point (P), as shown in Fig. 11.12. The coordinates of the control points are given in Table 11.13, and the field observations are given in Table 11.14. Calculate the coordinates of the point (P).

Fig. 11.12 Observation geometry

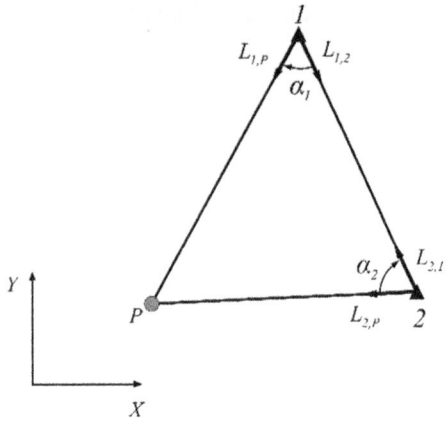

Table 11.13 Control point coordinates

Control point	X [m]	Y [m]
1	5035.880	10,064.838
2	5064.435	10,002.440

Table 11.14 Field observations

Station	Sighting point	Horizontal direction
1	2	260° 39′ 37″
	P	314° 12′ 45″
2	1	5° 25′ 19″
	P	297° 50′ 06″

Solution:

Using Eq. (11.43),

$$\alpha_1 = 314°\,12'45'' - 260°\,39'37'' = 53°\,33'08''$$

$$\alpha_2 = 360°\,00'00'' - (297°\,50'06'' - 5°\,25'19'') = 67°\,35'13''$$

Using Eq. (11.4),

$$Az_{1,2} = 155°\,24'35.6''$$

$$Az_{2,1} = 335°\,24'35.6''$$

From Fig. 11.12,

$$Az_{1,P} = 155°\,24'35.6'' + 53°\,33'08'' = 208°\,57'43.6''$$

$$Az_{2,P} = 355°\,24'35.6'' + 67°\,35'13'' = 267°\,49'22.6''$$

Finally, using for example Eqs. (11.55) and (11.57),

$$Y_P = 10,064.838 + \frac{-28.555 - 62.398 * \tan(267°\,49'22.6'')}{\tan(267°\,49'22.6'') - \tan(208°\,57'43.6'')}$$

$$Y_P = 9999.990 \text{ m}$$

$$X_P = 5035.880 + (9999.990 - 10,064.838) * \tan(208°\,57'43.6'')$$

$$X_P = 4999.990 \text{ m}$$

11.8.3   Bilateration

The bilateration method is a point positioning method based on distance measurements from two points with known coordinates, (A) and (B), to a point (P), as shown in Fig. 11.13. It is a mathematical problem of intersecting two circles with known centres and radii. Unlike the previous two methods, no direction is measured in this case.

Fig. 11.13 Geometry of point positioning by bilateration

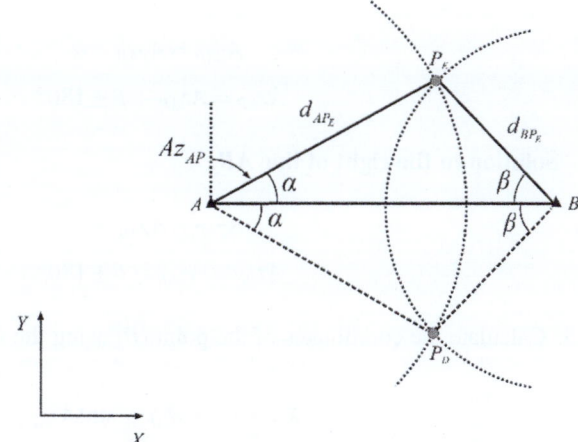

Given:

Point (A): (X_A, Y_A)
Point (B): (X_B, Y_B)

Observation:

d_{AP}, d_{BP}

To calculate:

Coordinates of (P): (X_P, Y_P)

This method can be very useful on construction sites for locating objects close to control points.

As shown in Fig. 11.13, there are two possible solutions for determining the coordinates of the point (P): one to the right of line AB and one to the left. Considering the latter, the following calculation sequence is proposed:

1. Calculate the azimuth Az_{AB} from the known coordinates of points (A) and (B) using Eq. (11.4).
2. Calculate the horizontal distance d_{AB} between points (A) and (B) using Eq. (11.2) or (11.5).
3. Calculate the values of α and β using the law of cosines to the triangle ABP:

$$\cos(\alpha) = \frac{d_{AB}^2 + d_{AP}^2 - d_{BP}^2}{2d_{AB} * d_{AP}} \quad \rightarrow \alpha \tag{11.59}$$

$$\cos(\beta) = \frac{d_{AB}^2 + d_{BP}^2 - d_{AP}^2}{2d_{AB} * d_{BP}} \quad \rightarrow \beta \tag{11.60}$$

4. Calculate the azimuths Az_{AP} and Az_{BP} using the following equations:

Solution to the left of line AB:

$$A_{ZAP} = A_{ZAB} - \alpha \tag{11.61}$$

$$A_{ZBP} = A_{ZAB} + \beta \pm 180° \tag{11.62}$$

Solution to the right of line AB:

$$A_{ZAP} = A_{ZAB} + \alpha \tag{11.63}$$

$$A_{ZBP} = A_{ZAB} - \beta \pm 180° \tag{11.64}$$

5. Calculate the coordinates of the point (P) using the following equations:

$$X_P = X_A + d_{AP} * \sin(Az_{AP})$$
$$Y_P = Y_A + d_{AP} * \cos(Az_{AP}) \tag{11.65}$$

or,

$$X_P = X_B + d_{BP} * \sin(Az_{BP})$$
$$Y_P = Y_B + d_{BP} * \cos(Az_{BP}) \tag{11.66}$$

Example 11.9

The bilateration method was applied to points (*3*) and (*4*) from Example 11.7 to determine the coordinates of the point (*P*). The horizontal distances measured in the field are given in Table 11.15, and the measurement sketch is shown in Fig. 11.14. Calculate the coordinates of point (*P*).

Table 11.15 Field measurements

Station	Sighting point	Horizontal distance [m]
3	P	93.399
4		99.713

Fig. 11.14 Observation geometry

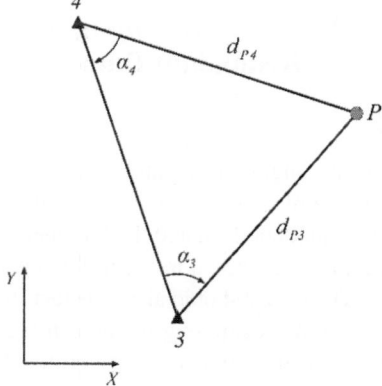

Solution:

From Table 11.11 *and using Eqs. (11.2) and (11.4),*

$$d_{3,4} = 104.3153 \text{ m}$$

$$Az_{3,4} = 341°\,21'51.4''$$

Using Eqs. (11.59) and (11.60),

$$\alpha_3 = \text{acos}\left(\frac{9662.3781}{19,485.8943}\right) = 60°\,16'23.4''$$

$$\alpha_4 = \text{acos}\left(\frac{12,100.9964}{20,803.1882}\right) = 54°\,25'49.9''$$

Using Eqs. (11.61) and (11.62),

$$Az_{3P} = 341° 21'51.4'' + 60° 16'23.4'' - 360° 00'00'' = 41° 38'14.8''$$

$$Az_{4P} = 341° 21'51.4'' - 54° 25'49.9'' - 180° = 106° 56'01.4''$$

Finally, the point (P) coordinates are determined using for example Eq. (11.65).

$$X_P = 4937.937 + 93.399 * \sin(41° 38'14.8'') = 4999.993 \text{ m}$$

$$Y_P = 9930.201 + 93.399 * \cos(41° 38'14.8'') = 10,000.004 \text{ m}$$

11.9   Point Positioning by Angles and Distances Using Redundant Data

In many cases, the number of control points available in the field for point positioning by angle and distance is greater than strictly necessary for a single solution. Therefore, if observations are made from or to all available control points, there will be redundant data, and the Parametric Adjustment model can be applied to solve the problem, as shown in the following subsections.

The reader should also consider that to apply the Parametric Adjustment model to any point positioning method; it is necessary to calculate the approximate coordinates of the point to be determined. Since there are redundant observations, the approximate coordinate values can be calculated by using the equations related to the unambiguous solution. Thus, as shown below, the approximate coordinates (X_0, Y_0) of the point (P) will be used in the adjustment computation.

11.9.1   Adjustment of Intersections

In the case of a point positioning by intersection with redundant observations, horizontal directions from n control points to point (P) are observed, as shown in Fig. 11.15. An observed azimuth Az_{Obs} must be calculated for each direction using Eq. (11.4). To do this, however, it is first necessary to determine the approximate coordinates of the point (P), which is then called (P_0). These coordinates can be determined by taking any two control points and using for example Eqs. (11.55) and (11.57).

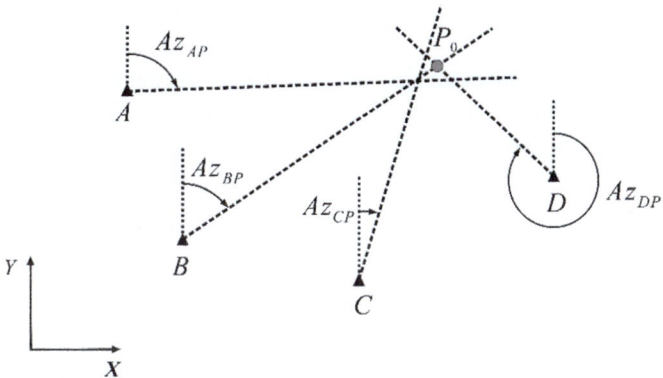

Fig. 11.15 Principle of the intersection method with redundant observations

Considering that only azimuths are observed from n control points to point (P), the azimuth observable Az_{iP} from (i) to (P) is given by Eq. (11.67).

$$Az_{iP} = \text{atan}\left(\frac{X_P - X_i}{Y_P - Y_i}\right) + \alpha_{\text{quad}} \tag{11.67}$$

Equation (11.67) is a non-linear function involving the fixed coordinates of the point (i) and the unknowns (X_P, Y_P). Thus, considering the observation values of Az_{iP}, the following equation can be written as

$$Az_{\text{obs}_{iP}} + vAz_{iP} = f_i(X_i, Y_i, \tilde{X}_P, \tilde{Y}_P) = \text{atan}\left(\frac{X_P - X_i}{Y_P - Y_i}\right) \tag{11.68}$$

The linearised form of Eq. (11.68) is

$$f_i(X_i, Y_i, \tilde{X}_P, \tilde{Y}_P) = Az_{iP_0} + \left(\frac{\partial Az_{iP}}{\partial X_P}\right)_0 \delta X_P + \left(\frac{\partial Az_{iP}}{\partial Y_P}\right)_0 \delta Y_P \tag{11.69}$$

And then,

$$vAz_{iP} = \left(\frac{\partial Az_{iP}}{\partial X_P}\right)_0 \delta X_P + \left(\frac{\partial Az_{iP}}{\partial Y_P}\right)_0 \delta Y_P - (Az_{\text{obs}_{iP}} - Az_{iP_0}) \tag{11.70}$$

where

$$\begin{aligned}
\left(\frac{\partial Az_{iP}}{\partial X_P}\right)_0 &= \frac{Y_{P_0} - Y_i}{d_{iP_0}^2} = \frac{\cos(Az_{iP_0})}{d_{iP_0}} \\
\left(\frac{\partial Az_{iP}}{\partial Y_P}\right)_0 &= -\frac{X_{P_0} - X_i}{d_{iP_0}^2} = -\frac{\sin(Az_{iP_0})}{d_{iP_0}}
\end{aligned} \tag{11.71}$$

$$l_{0Az_{iP}} = Az_{obs_{iP}} - Az_{iP_0} \tag{11.72}$$

Each observation produces an error Eq. (11.70), whose unknown parameters are the δX and δY_P coordinate variations. For n observations, Eq. (11.70) can be written in matrix form as $v = Ax - l_0$ (Eq. 3.79), where

$$
A = \begin{bmatrix}
\left(\frac{\partial Az_{AP}}{\partial X_P}\right)_0 & \left(\frac{\partial Az_{AP}}{\partial Y_P}\right)_0 \\
\left(\frac{\partial Az_{BP}}{\partial X_P}\right)_0 & \left(\frac{\partial Az_{BP}}{\partial Y_P}\right)_0 \\
\vdots & \vdots \\
\left(\frac{\partial Az_{nP}}{\partial X_P}\right)_0 & \left(\frac{\partial Az_{nP}}{\partial Y_P}\right)_0
\end{bmatrix}
\quad
v = \begin{bmatrix}
vAz_{AP} \\
vAz_{BP} \\
\vdots \\
vAz_{nP}
\end{bmatrix}
\quad
l_0 = \begin{bmatrix}
Az_{Obs_{AP}} - Az_{AP_0} \\
Az_{Obs_{BP}} - Az_{BP_0} \\
\vdots \\
Az_{Obs_{nP}} - Az_{nP_0}
\end{bmatrix}
\quad
x = \begin{bmatrix}
\delta X_P \\
\delta Y_P
\end{bmatrix}
$$

If all observations are assumed to have the same precision, no weight is assigned. However, as they are taken from different positions, they must be weighted to consider this geometric aspect. Thus, as the azimuths are the observables, the assignment of weights in the Least Squares adjustment is based on the azimuth value of each observation. In this case, the weight can be, for example, the inverse of the standard deviation of the azimuth Az_{iP}, which must account for the standard deviations of the azimuth Az_{ij} and the direction L_{iP} obtained, followed by the adjusted coordinates of the point (P). Finally, as a total of n observations and two unknowns have been considered, the standard deviations of the unit weight can be calculated using Eq. (11.73) and the standard deviation of the unknowns using the variance-covariance matrix given by Eq. (11.74). The remaining statistical parameters can be determined as indicated in Sect. 3.9.1.1.

$$s_0^2 = \frac{v^T P v}{n\text{-}2} \tag{11.73}$$

$$\Sigma_{xx} = s_0^2 N^{-1} \tag{11.74}$$

Example 11.10
For the same point (P) as in example 11.8, an intersection observation from four control points (1), (2), (3) and (4) was carried out, as shown in Fig. 11.16. The field observations are given in Table 11.16, and the coordinates of the control points are given in Table 11.17. Calculate the coordinates of point (P) using the Parametric adjustment model.

Fig. 11.16 Observation geometry

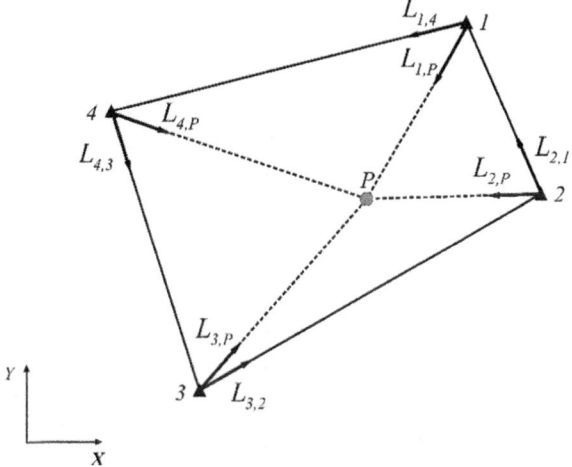

Table 11.16 Field observations

Station	Sighting point	Horizontal direction
1	4	278°38′25″
	P	232°50′58″
2	1	5°25′19″
	P	297°50′35″
3	2	356°24′38″
	P	337°47′11″
4	3	336°06′59″
	P	281°41′22″

Table 11.17 Control point coordinates

Control point	X [m]	Y [m]
1	5035.880	10,064.838
2	5064.435	10,002.440
3	4937.937	9930.201
4	4904.603	10,029.047

Solution:

From example 11.8, *the approximate coordinates of point (P) can be*

$$X_{P_0} = 4999.990 \text{ m}$$
$$Y_{P_0} = 9999.990 \text{ m}$$

From the observation in Table 11.16, *the horizontal clockwise angles are calculated using* Eq. (11.43).

$$\alpha_1 = 314° 12'33''$$

$$\alpha_2 = 292° 25'16''$$

$$\alpha_3 = 341° 22'33''$$

$$\alpha_4 = 305° 34'23''$$

Preliminary distance between points (i) and (P) can be calculated using Eqs. (11.2) or (11.5).

$$d_{1,P_0} = 74.1172 \text{ m}$$

$$d_{2,P_0} = 64.4916 \text{ m}$$

$$d_{3,P_0} = 93.3867 \text{ m}$$

$$d_{4,P_0} = 99.7145 \text{ m}$$

Using the observation horizontal clockwise angles calculated above and the coordinates of control point given in Table 11.17, the observation azimuth Az_{ObsiP} can be calculated using Eqs. (11.45) or (11.46). Preliminary azimuth of the line P_0-i can be calculated using Eq. (11.4). Observation and preliminary azimuths are given in Table 11.18.

Table 11.18 Observation and preliminary azimuth

Line	Azimuth	
	Observation	Preliminary
1-P	208°57′31.9″	208°57′44.1″
2-P	267°49′52.7″	267°49′22.5″
3-P	41°38′47.6″	41°38′31.0″
4-P	106°56′14.4″	106°56′30.8″

Considering the error Eq. (11.71) and Eq. (3.79), the following matrices can be written for applying the Parametric adjustment model.

$$A = \begin{bmatrix} -1.180481 * 10^{-2} & 6.533348 * 10^{-3} \\ -5.890612 * 10^{-4} & 1.549471 * 10^{-2} \\ 8.002335 * 10^{-3} & -7.115289 * 10^{-3} \\ -2.922361 * 10^{-3} & -9.593393 * 10^{-3} \end{bmatrix}$$

$$l_{0_{ip}} = \begin{bmatrix} -5.935840 * 10^{-5} \\ 1.425419 * 10^{-4} \\ 7.973282 * 10^{-5} \\ -7.993817 * 10^{-5} \end{bmatrix} \quad x = \begin{bmatrix} \delta X_P \\ \delta Y_P \end{bmatrix}$$

Assuming that all observations have the same weight and using Eqs. (3.84) *to* (3.87).

$$N = \begin{bmatrix} 2.122780 * 10^{-4} & -1.151558 * 10^{-4} \\ -1.151558 * 10^{-4} & 4.254313 * 10^{-4} \end{bmatrix}$$

$$n = \begin{bmatrix} 1.488406 * 10^{-6} \\ 2.020392 * 10^{-6} \end{bmatrix} \qquad x = \begin{bmatrix} 0.01124 \\ 0.00779 \end{bmatrix}_{m}$$

The adjusted coordinates of point (P) are then given as follows.

$$X_P = 4999.990 + 0.01124 = 5000.001 \text{ m}$$
$$Y_P = 9999.990 + 0.00779 = 9999.998 \text{ m}$$

The residual error vector can be calculated using Eq. (3.79).

$$v = \begin{bmatrix} -4.6'' \\ -5.9'' \\ -9.3'' \\ -5.7'' \end{bmatrix}$$

The standard deviations of the unit weight can be calculated using Eq. (11.73).

$$s_0^2 = 2.060790 * 10^{-9} \text{rad}^2 \text{ (with } n = 4)$$
$$s_0 = \pm 4.539593 * 10^{-5} \text{rad} = \pm 9.4''$$

Finally, using Eq. (11.74), *the variance-covariance matrix of the parameters is given as follows:*

$$\Sigma_{xx} = s_0^2 N^{-1} = \begin{bmatrix} 1.137881 * 10^{-5} & 3.080020 * 10^{-6} \\ 3.080020 * 10^{-6} & 5.677703 * 10^{-6} \end{bmatrix}$$

Using the diagonal values in the variance-covariance matrix of the parameters, the standard deviations of the adjusted coordinates are given as follows:

$$X_P = 5000.001 \text{ m} \pm 3.4 \text{ mm}$$
$$Y_P = 9999.998 \text{ m} \pm 2.4 \text{ mm}$$

with a correlation coefficient of 38.3%.
The absolute error ellipse for point (P) is given as shown in Fig. 11.17.

Fig. 11.17 Absolute error ellipse for point (P) determined by intersection

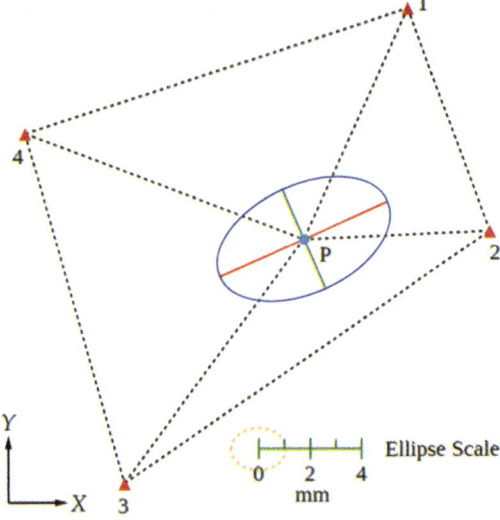

11.9.2 Adjustment of Resections

In the case of point positioning by resection with redundant observations, horizontal directions L_{Pi} from point (P) to n control points are observed, as shown in Fig. 11.18. Each pair of direction observations generates a horizontal angle α_{ij}. Analogous to the intersection method, the approximate coordinates (P_0) of the point to be determined must be calculated by Eqs. (11.50) to (11.54), taking any three control points. Furthermore, considering Fig. 11.18, point (P) is the adjusted point, and Az_{Pi} is the azimuth of the line Pi.

Fig. 11.18 Principle of the resection method with redundant observations

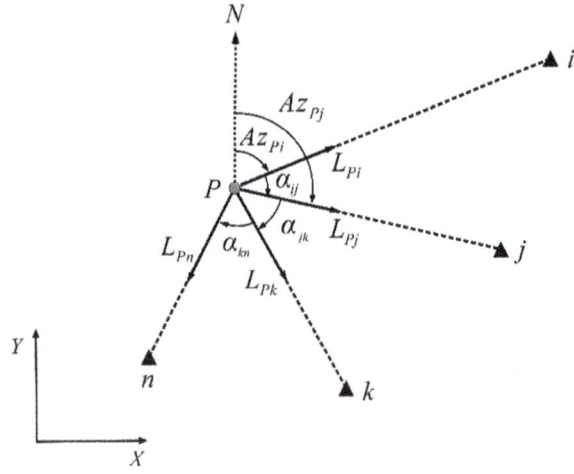

Considering that only horizontal directions are observed from point (P) to n control points, the observation equation for angle α_{ij} from (P) to (i) and (j) is given as follows:

$$\alpha_{ij} = \text{atan}\left(\frac{X_j - X_P}{Y_j - Y_P}\right) + \alpha_{\text{quad1}} - \text{atan}\left(\frac{X_i - X_P}{Y_i - Y_P}\right) + \alpha_{\text{quad2}} \tag{11.75}$$

Note that the quadrants must be checked for angle calculation.

Equation (11.75) is a non-linear function involving the fixed coordinates of points (i) and (j) and the unknowns (X_P, Y_P). Thus, considering the observation α_{ij}, the following equation can be written as

$$\alpha_{\text{Obs}_{ij}} + v\alpha_{ij} = f_i\left(X_i, Y_i, X_j, Y_j, \tilde{X}_P, \tilde{Y}_P\right) = \text{atan}\left(\frac{X_j - X_P}{Y_j - Y_P}\right) + \alpha_{\text{quad1}}$$

$$- \text{atan}\left(\frac{X_i - X_P}{Y_i - Y_P}\right) + \alpha_{\text{quad2}} \tag{11.76}$$

The linearised form of Eq. (11.76) is

$$f_i\left(X_i, Y_i, X_j, Y_j, \tilde{X}_P, \tilde{Y}_P\right) = \alpha_{ij_0} + \left(\frac{\partial \alpha_{ij}}{\partial X_P}\right)_0 \delta X_P + \left(\frac{\partial \alpha_{ij}}{\partial Y_P}\right)_0 \delta Y_P \tag{11.77}$$

The error equation is then given as follows:

$$v\alpha_{ij} = \left(\frac{\partial \alpha_{ij}}{\partial X_P}\right)_0 \delta X_P + \left(\frac{\partial \alpha_{ij}}{\partial Y_P}\right)_0 \delta Y_P - \left(\alpha_{\text{Obs}_{ij}} - \alpha_{ij0}\right) \tag{11.78}$$

where

$$\left(\frac{\partial \alpha_{ij}}{\partial X_P}\right)_0 = \frac{Y_i - Y_{P_0}}{d_{iP_0}^2} - \frac{Y_j - Y_{P_0}}{d_{jP_0}^2}$$

$$\left(\frac{\partial \alpha_{ij}}{\partial Y_P}\right)_0 = \frac{X_{P_0} - X_i}{d_{iP_0}^2} - \frac{X_{P_0} - X_j}{d_{jP_0}^2} \tag{11.79}$$

$$l_{0\alpha_{ij}} = \alpha_{\text{Obs}_{ij}} - \alpha_{ij_0} \tag{11.80}$$

Each observation produces an error Eq. (11.78), whose unknown parameters are the δX_P and δY_P coordinate variations. For n observations, Eq. (11.78) can be written in matrix form as $v = Ax - l_0$ (Eq. 3.79), where

$$A = \begin{bmatrix} \left(\dfrac{\partial \alpha_{1,2}}{\partial X_P}\right)_0 & \left(\dfrac{\partial \alpha_{1,2}}{\partial Y_P}\right)_0 \\[2ex] \left(\dfrac{\partial \alpha_{2,3}}{\partial X_P}\right)_0 & \left(\dfrac{\partial \alpha_{2,3}}{\partial Y_P}\right)_0 \\[1ex] \vdots & \vdots \\[1ex] \left(\dfrac{\partial \alpha_{n-1,n}}{\partial X_P}\right)_0 & \left(\dfrac{\partial \alpha_{n-1,n}}{\partial Y_P}\right)_0 \end{bmatrix} \qquad v = \begin{bmatrix} v\alpha_{1,2} \\[1ex] v\alpha_{2,3} \\[1ex] \vdots \\[1ex] v\alpha_{n-1,n} \end{bmatrix}$$

$$l_{0\alpha_{ij}} = \begin{bmatrix} \alpha_{\text{Obs}_{1,2}} - \alpha_{1,2_0} \\[1ex] \alpha_{\text{Obs}_{2,3}} - \alpha_{2,3_0} \\[1ex] \vdots \\[1ex] \alpha_{\text{Obs}_{n-1,n}} - \alpha_{n-1,n_0} \end{bmatrix} \qquad x = \begin{bmatrix} \delta X_P \\[1ex] \delta Y_P \end{bmatrix}$$

As with the intersection method, no weight is assigned if all observations are assumed to have the same accuracy. However, as they are taken from different positions, they can be weighted to account for this geometric aspect. In this case, the directions are the observables that can be used to weigh the observations. In this case, the weight can be the inverse of the standard deviation of the angle α_{ij}, which must account for the standard deviations of the azimuth Az_{Pi} and the direction L_{Pi}.

Adjustment computation can be performed using Eqs. (3.84) to (3.87), from which δX_P, δY_P are obtained, followed by the adjusted coordinates of the point (P). Finally, as a total of 4 observations and two unknowns have been considered, the standard deviations of the unit weight can be calculated using Eq. (11.73). The remaining statistical parameters can be determined as indicated in Sect. 3.9.1.1.

Example 11.11

The same point (P) from the previous examples is to be determined using the resection method. The horizontal directions from points (P) to control points (1), (2), (3) and (4) have been observed, as shown in Fig. 11.19. The observed values converted to azimuths are given in Table 11.19. The coordinates of the control points are given in Table 11.20. Calculate the point (P) coordinates and their standard deviation using the Parametric adjustment model.

Solution:

As in the previous example, the approximate coordinates of point (P) are assumed to be

$$X_{P_0} = 4999.990 \text{ m}$$
$$Y_{P_0} = 9999.990 \text{ m}$$

Using Eqs. (11.2) and (11.4), the preliminary distance and azimuth of the line $P_0\text{-}i$ are given in Table 11.21.

Fig. 11.19 Observation geometry

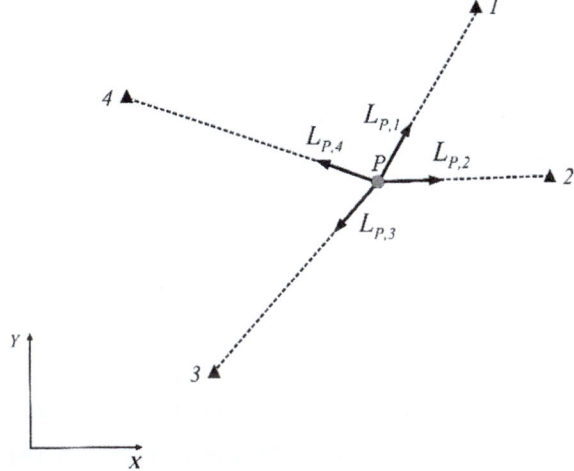

Table 11.19 Field observations

Station	Sighting point	Azimuth
P	1	28°57′36″
	2	87°49′50″
	3	221°38′36″
	4	286°56′06″

Table 11.20 Control point coordinates

Control point	X [m]	Y [m]
1	5035.880	10,064.838
2	5064.435	10,002.440
3	4937.937	9930.201
4	4904.603	10,029.047

From the observations in Table 11.19 *and the preliminary azimuths in* Table 11.21, *the preliminary and the observation horizontal clockwise angles are given in* Table 11.22.

Table 11.21 Preliminary distance and azimuth

Station	Sighting point	Preliminary horizontal distance [m]	Preliminary azimuth
P_0	1	74.1172	28°57′44.1″
	2	64.4916	87°49′22.2″
	3	93.3867	221°38′31.0″
	4	99.7145	286°56′30.8″

Table 11.22 Preliminary and observation angles

Angle from (P)	Preliminary angle (Eq. 11.76)	Observation angle (Eq. 11.43)
1,2	58°51'38.1"	58°52'14.0"
2,3	133°49'08.8"	133°48'46.0"
3,4	65°17'59.8"	65°17'30.0"
4,1	102°01'13.3"	102°01'30.0"

Considering the error Eq. (11.78) and Eq. (3.79), the following matrices can be written for applying the Parametric adjustment model.

$$A = \begin{bmatrix} 1.121575 * 10^{-2} & 8.961366 * 10^{-3} \\ 8.591396 * 10^{-3} & -2.261000 * 10^{-2} \\ -1.092470 * 10^{-2} & -2.478104 * 10^{-3} \\ -8.882447 * 10^{-3} & 1.612674 * 10^{-2} \end{bmatrix}$$

$$l_0 \alpha_{ij} = \begin{bmatrix} 1.739359 * 10^{-4} \\ -1.104570 * 10^{-4} \\ -1.446530 * 10^{-4} \\ 8.117418 * 10^{-5} \end{bmatrix}_{rad} \quad x = \begin{bmatrix} \delta X_P \\ \delta Y_P \end{bmatrix}$$

Assuming that all observations have the same weight and using Eqs. (3.84) to (3.87).

$$N = \begin{bmatrix} 3.978519 * 10^{-4} & -2.099155 * 10^{-4} \\ -2.099155 * 10^{-4} & 8.577310 * 10^{-4} \end{bmatrix} \quad n = \begin{bmatrix} 1.861106 * 10^{-6} \\ 5.723677 * 10^{-6} \end{bmatrix} \quad x = \begin{bmatrix} 0.0094 \\ 0.0090 \end{bmatrix}_m$$

The adjusted coordinates of the point (P) are then given as follows:

$$X_P = 4999.990 + 0.0094 = 4999.999 \text{ m}$$
$$Y_P = 9999.990 + 0.0090 = 9999.999 \text{ m}$$

The residual error vector can be calculated using Eq. (3.79).

$$v = \begin{bmatrix} 2.5" \\ -2.4" \\ 4.0" \\ -4.1" \end{bmatrix}$$

The standard deviations of the unit weight can be calculated using Eq. (11.73).

$$s_0^2 = 5.326298 * 10^{-10} \text{rad}^2 \ (with \ n = 4)$$

$$s_0 = \pm 3.263832 * 10^{-5} \text{rad} = \pm 4.8''$$

Finally, using Eq. (11.4), the variance-covariance matrix of the parameters is given as follows:

$$\Sigma_{xx} = s_0^2 N^{-1} = \begin{bmatrix} 3.074533 * 10^{-6} & 7.524410 * 10^{-7} \\ 7.524410 * 10^{-7} & 1.426098 * 10^{-6} \end{bmatrix}$$

Using the diagonal values in the variance-covariance matrix of the parameters, the standard deviations of the adjusted coordinates are given as follows.

$$X_P = 4999.999 \ m \pm 1.2 \ mm$$
$$Y_P = 9999.999 \ m \pm 0.8 \ mm$$

The absolute error ellipse for point (P) is given as shown in Fig. 11.20.

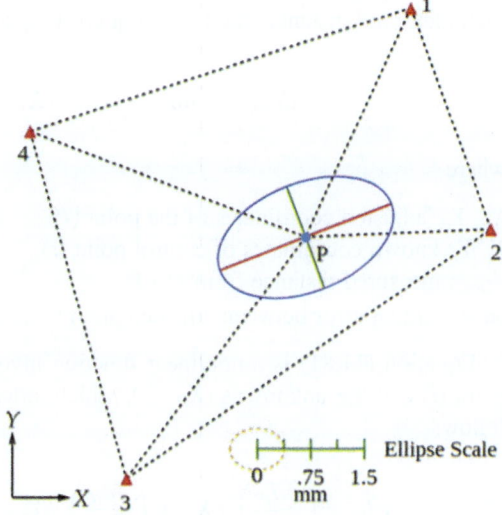

Fig. 11.20 Absolute error ellipse for point (P) determined by resection

11.9.3 Adjustment of Multilateration

In the case of point positioning by redundant distance measurements, the horizontal distances d_{ij} of n control points to (P) are observed, as shown in Fig. 11.21. Since more than two control points are used for point positioning, this measurement method is called *multilateration*. As with the previous methods, the coordinates of the approximate point (P_0) must be calculated using Eqs. (11.59) to (11.65), taking any two control points as the baseline.

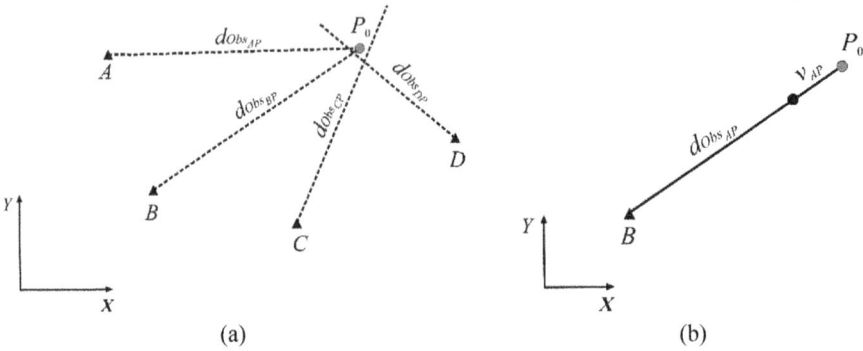

Fig. 11.21 Principle of the multilateration method

The adjustment computation, in this case, considers that there is a residual error between the adjusted and measured distances, as shown in Fig. 11.21. Therefore, for each measured distance, the following error equation can be written for the line iP:

$$d_{\mathrm{Obs}_{iP}} + vd_{iP} = \sqrt{(X_P - X_i)^2 + (Y_P - Y_i)^2} \tag{11.81}$$

where

X_P, Y_P: adjusted coordinates of the point (P)
X_i, Y_i: known coordinates of control point (i)
$d_{\mathrm{Obs}_{ij}}$: measured distance from i to P
vd_{iP}: residual error between adjusted and measured distances

Equation (11.81) is a nonlinear function involving the fixed coordinates of the point (i) and the unknowns (X_P, Y_P), which, after linearisation, can be rewritten as follows:

$$vd_{iP} = \left(\frac{\partial d_{iP}}{\partial X_P}\right)_0 \delta X_P + \left(\frac{\partial d_{iP}}{\partial Y_P}\right)_0 \delta Y_P - \left(d_{\mathrm{Obs}_{ip}} - d_{iP_0}\right) \tag{11.82}$$

where

$$\left(\frac{\partial d_{iP}}{\partial X_P}\right)_0 = \frac{X_{P0} - X_i}{d_{iP0}} \tag{11.83}$$

$$\left(\frac{\partial d_{iP}}{\partial Y_P}\right)_0 = \frac{Y_{P0} - Y_i}{d_{iP0}} \tag{11.84}$$

And

$$l_{0d_{iP}} = d_{\text{Obs}_{iP}} - d_{ip_0} \tag{11.85}$$

Each observation will produce an error Eq. (11.82), whose unknown parameters are the coordinate variations δX_P and δY_P. For n observations, Eq. (11.82) can be written in matrix form as $v = Ax - l_0$ (Eq. 3.79), where

$$A = \begin{bmatrix} \left(\frac{\partial d_{AP}}{\partial X_P}\right)_0 & \left(\frac{\partial d_{AP}}{\partial Y_P}\right)_0 \\ \left(\frac{\partial d_{BP}}{\partial X_P}\right)_0 & \left(\frac{\partial d_{BP}}{\partial Y_P}\right)_0 \\ \vdots & \vdots \\ \left(\frac{\partial d_{nP}}{\partial X_P}\right)_0 & \left(\frac{\partial d_{nP}}{\partial Y_P}\right)_0 \end{bmatrix} \quad v = \begin{bmatrix} vd_{AP} \\ vd_{BP} \\ \vdots \\ vd_{nP} \end{bmatrix} \quad l_0 = \begin{bmatrix} d_{\text{obs}_{1P} - d_{1P_0}} \\ d_{\text{obs}_{2P} - d_{2P_0}} \\ \vdots \\ d_{\text{obs}_{nP} - d_{nP_0}} \end{bmatrix} \quad x = \begin{bmatrix} \delta X_P \\ \delta Y_P \end{bmatrix}$$

As with the previous methods, no weight is assigned if all observations are assumed to have the same accuracy. However, as they differ in length, they can be weighted to account for this geometric aspect. Thus, as distances are the observables, they can be used to weight observations.

Adjustment computation can be performed using Eqs. (3.84) to (3.87), from which δX_P and δY_P are obtained, followed by the adjusted coordinates of the point (P). Finally, as a total of 4 observations and two unknowns have been considered, the standard deviations of the unit weight can be calculated using Eq. (11.73). The remaining statistical parameters can be determined as indicated in Sect. 3.9.1.1.

Example 11.12

The same point (P) from the previous examples is to be determined using the multilateration method. The horizontal distances from points (P) to control points (1), $(2$, (3) and (4) have been observed, as shown in Fig. 11.22. The observed values are given in Table 11.23. The coordinates of the known points are given in Table 11.24. Calculate the coordinates of the point (P) and their standard deviations using the Parametric adjustment model.

Fig. 11.22 Observation
geometry

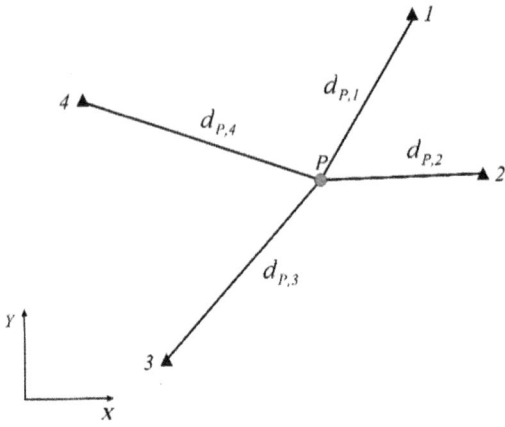

Table 11.23 Field
observations

Station	Sighting point	Horizontal distance [m]
P	1	74.105
	2	64.482
	3	93.391
	4	99.718

Table 11.24 Control point
coordinates

Control point	X [m]	Y [m]
1	5035.880	10,064.838
2	5064.435	10,002.440
3	4937.937	9930.201
4	4904.603	10,029.047

Solution:

As in the previous example, the approximate coordinates of point (P) are assumed to be

$$X_{P_0} = 4999.990 \text{ m}$$
$$Y_{P_0} = 9999.990 \text{ m}$$

Using Eqs. (11.2), the preliminary distances of the line iP_0 are as follows:

$$d_{P01} = 74.117 \text{ m}$$
$$d_{P02} = 64.492 \text{ m}$$
$$d_{P03} = 93.387 \text{ m}$$

$$d_{P04} = 99.715 \text{ m}$$

Considering the error Eq. (11.82) *and* Eq. (3.79), *the following matrices can be written for applying the Parametric adjustment model:*

$$A = \begin{bmatrix} -0.484233 & -0.874939 \\ -0.999278 & -0.037989 \\ 0.664473 & 0.747312 \\ 0.956601 & -0.291402 \end{bmatrix} \quad l_0 d_{iP} = \begin{bmatrix} -0.012172 \\ -0.009554 \\ 0.004281 \\ 0.003462 \end{bmatrix}_m \quad x = \begin{bmatrix} \delta X_P \\ \delta Y_P \end{bmatrix}$$

Assuming that all observations have the same weight and using Eqs. (3.84) *to* (3.87),

$$N = \begin{bmatrix} 2.589649 & 0.679450 \\ 0.679450 & 1.410351 \end{bmatrix} \quad n = \begin{bmatrix} 0.021598 \\ 0.013203 \end{bmatrix} \quad x = \begin{bmatrix} 0.006735 \\ 0.006117 \end{bmatrix}_m$$

The adjusted coordinates of the point (P) are then given as follows.

$$X_P = 4999.990 + 0.0067 = 4999.997 \text{ m}$$

$$Y_P = 9999.990 + 0.0061 = 9999.996 \text{ m}$$

The residual error vector can be calculated using Eq. (3.79).

$$v = \begin{bmatrix} 3.6 \\ 2.6 \\ 4.8 \\ 1.2 \end{bmatrix}_{mm}$$

The standard deviations of the unit weight can be calculated using Eq. (11.73).

$$s_0^2 = 2.176393 * 10^{-5} \text{ m}^2 (\text{with } n = 4)$$

$$s_0 = \pm 4.665183 * 10^{-3} \text{ m} = \pm 4.7 \text{ mm}$$

Finally, using Eq. (11.74), *the variance-covariance matrix of the parameters is given as follows:*

$$\Sigma_{xx} = s_0^2 N^{-1} = \begin{bmatrix} 9.620195 * 10^{-6} & -4.634621 * 10^{-6} \\ -4.634621 * 10^{-6} & 1.766434 * 10^{-5} \end{bmatrix}$$

Using the diagonal values in the variance-covariance matrix of the parameters, the standard deviations of the adjusted coordinates are given as follows:

$$X_P = 4999.997 \text{ m} \pm 3.1 \text{ mm}$$

$$Y_P = 9999.996 \text{ m} \pm 4.2 \text{ mm}$$

The absolute error ellipse for point (P) is given as shown in Fig. 11.23.

Fig. 11.23 Absolute error ellipse for point (P) determined by multilateration

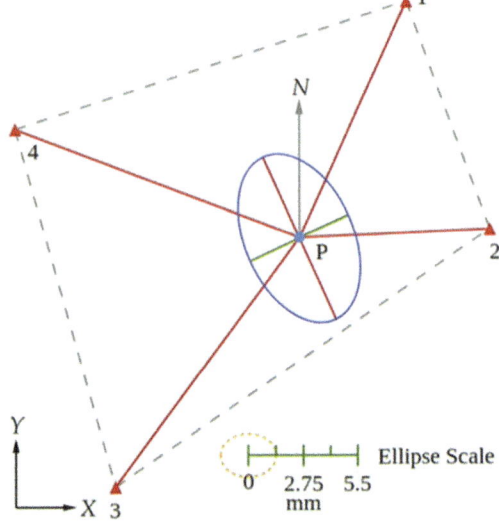

11.9.4   Adjustment Combining Intersection, Resection and Multilateration

If a point (*P*) is determined simultaneously by intersection, resection and multilateration, the approximate value of the point (P_0) can be calculated by any of the three methods. The adjustment computation can then be performed by applying the following set of error equations:

$$vAz_{iP} = \left(\frac{\partial Az_{iP}}{\partial X_P}\right)_0 * \delta X_P + \left(\frac{\partial Az_{iP}}{\partial Y_P}\right)_0 * \delta Y_P - (Az_{\text{Obs}_{iP}} - Az_{iP0})$$

$$v\alpha_{ij} = \left(\frac{\partial \alpha_{ij}}{\partial X_P}\right)_0 * \delta X_P + \left(\frac{\partial \alpha_{ij}}{\partial Y_P}\right)_0 * \delta Y_P - (\alpha_{\text{Obs}_{ij}} - \alpha_{ij_0}) \qquad (11.86)$$

$$vd_{iP} = \left(\frac{\partial d_{iP}}{\partial X_P}\right)_0 * \delta X_P + \left(\frac{\partial d_{iP}}{\partial Y_P}\right)_0 * \delta Y_P - (d_{\text{Obs}_{iP}} - d_{iP_0})$$

The adjustment computation is then carried out as for the previous adjustment methods. However, in this case, as a total of $n = n_{\text{int}} + n_{\text{res}} + n_{\text{mult}}$ observations and two unknowns are involved in the calculation, the standard deviations of the unit weight can be calculated using Eq. (11.87).

$$s_0^2 = \frac{v^T P v}{n_{int} + n_{res} + n_{mult} - 2} \qquad (11.87)$$

Usually, the combination of resection and multilateration is the most common case, which some authors have called the *Two-dimensional Free Station* method.

11.10   Three-Dimensional Point Positioning

In a Three-Dimensional point positioning, each point is represented by its (X, Y, H) coordinates, which locate the point in the Three-Dimensional coordinate system, as described in Sect. 5.6. In this case, the coordinates can be determined by measuring horizontal directions, vertical angles and distances, or only horizontal directions and vertical angles, as described in the following subsections.

11.10.1   Three-Dimensional Point Positioning by Horizontal Directions, Vertical Angle and Distance

The principle of Three-Dimensional point positioning by horizontal directions, vertical angle and distance between the station and the target points is illustrated in Fig. 11.24. In this case, as long as the coordinates of the station (P) and orientation point (Ref) are known, the coordinates of a given target point (Q) can be determined using Eqs. (11.88) to (11.91). Note that coordinates are related to the instrument and target centres.

Fig. 11.24 Principle of 3D point positioning by horizontal directions, vertical angle and distance

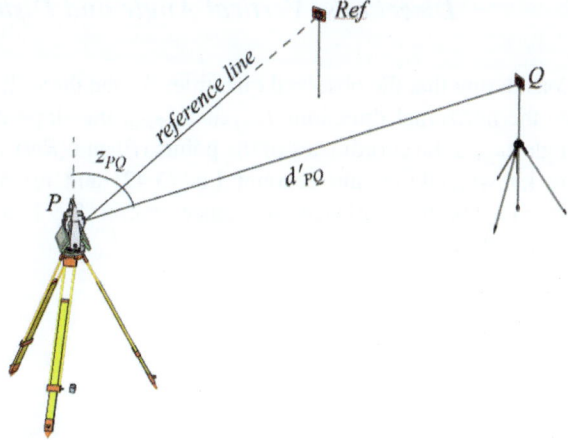

Given:

Points (P): (X_P, Y_P, H_P)
Point (Ref): $(X_{Ref}, Y_{Ref}, H_{Ref})$

Observation:

Horizontal directions: L_{PRef}, L_{PQ}
Slope distance: d'_{PQ}
Zenith angle: z_{PQ}

To calculate:

Coordinates of the point (Q): (X_Q, Y_Q, H_Q)

Equations to be used for the Three-Dimensional point positioning are

$$\alpha_P = L_{PQ} - L_{PRef} \tag{11.88}$$

$$Az_{PRef} = \text{atan}\left(\frac{X_{Ref} - X_P}{Y_{Ref} - Y_P}\right) + \alpha_{quad} \tag{11.89}$$

$$Az_{PQ} = Az_{PRef} + \alpha_P \tag{11.90}$$

$$
\begin{aligned}
X_Q &= X_P + d'_{PQ} * \sin(z_{PQ}) * \sin(Az_{PQ}) \\
Y_Q &= Y_P + d'_{PQ} * \sin(z_{PQ}) * \cos(Az_{PQ}) \\
H_Q &= H_P + d'_{PQ} * \cos(z_{PQ})
\end{aligned}
\tag{11.91}
$$

Note that, depending on the required accuracy and the distance measured, the curvature of the Earth and the vertical atmospheric refraction must be considered when calculating the elevation (H) of the target point, as described in Chap. 13.

11.10.2   Error Propagation in 3D Point Positioning by Directions, Vertical Angle and Distance

Considering that the observed quantities for the three-dimensional point positioning are the horizontal directions L_{PQ} and L_{PRef}, the slope distance d'_{PQ} and the zenith angle z_{PQ}, if the coordinates of the points (P) and (Ref) and their standard deviations are known, taking into account Eq. (3.42) and rewriting Eq. (11.91) to obtain Eq. (11.92), the variance-covariance matrix for function f is calculated using Eq. (11.93).

$$X_Q = X_P + 0Y_P + 0H_P + d'_{PQ} * \sin(z_{PQ}) * \sin(Az_{PQ})$$
$$Y_Q = 0X_P + Y_P + 0H_P + d'_{PQ} * \sin(z_{PQ}) * \cos(Az_{PQ}) \quad (11.92)$$
$$H_Q = 0X_P + 0Y_P + H_P + d'_{PQ} * \cos(z_{PQ})$$

$$\Sigma_{ff} = J^{\mathrm{T}} \Sigma_{ll} J \quad (11.93)$$

where

$$\Sigma_{ff} = \begin{bmatrix} s^2_{X_Q} & s_{X_Q Y_Q} & s_{X_Q H_Q} \\ s_{Y_Q X_Q} & s^2_{Y_Q} & s_{Y_Q H_Q} \\ s_{H_Q X_Q} & s_{H_Q Y_Q} & s^2_{H_Q} \end{bmatrix}$$

$$J^{\mathrm{T}} = \begin{bmatrix} \left(\frac{\partial X_Q}{\partial X_P}\right)_0 & 0 & 0 & \left(\frac{\partial X_Q}{\partial d'_{PQ}}\right)_0 & \left(\frac{\partial X_Q}{\partial z_{PQ}}\right)_0 & \left(\frac{\partial X_Q}{\partial Az_{PQ}}\right)_0 \\ 0 & \left(\frac{\partial Y_Q}{\partial Y_P}\right)_0 & 0 & \left(\frac{\partial Y_Q}{\partial d'_{PQ}}\right)_0 & \left(\frac{\partial Y_Q}{\partial z_{PQ}}\right)_0 & \left(\frac{\partial Y_Q}{\partial Az_{PQ}}\right)_0 \\ 0 & 0 & \left(\frac{\partial H_Q}{\partial H_P}\right)_0 & \left(\frac{\partial H_Q}{\partial d'_{PQ}}\right)_0 & \left(\frac{\partial H_Q}{\partial z_{PQ}}\right)_0 & 0 \end{bmatrix}$$

$$\left(\frac{\partial X_Q}{\partial X_P}\right)_0 = \left(\frac{\partial Y_Q}{\partial Y_P}\right)_0 = \left(\frac{\partial H_Q}{\partial H_P}\right)_0 = 1 \quad (11.94)$$

$$\left(\frac{\partial X_Q}{\partial d'_{PQ}}\right)_0 = \sin(z_{PQ}) * \sin(Az_{PQ}) \quad (11.95)$$

$$\left(\frac{\partial Y_Q}{\partial d'_{PQ}}\right)_0 = \sin(z_{PQ}) * \cos(Az_{PQ}) \quad (11.96)$$

$$\left(\frac{\partial H_Q}{\partial d'_{PQ}}\right)_0 = \cos(z_{PQ}) \quad (11.97)$$

$$\left(\frac{\partial X_Q}{\partial z_{PQ}}\right)_0 = d'_{PQ} * \cos(z_{PQ}) * \sin(Az_{PQ}) \quad (11.98)$$

$$\left(\frac{\partial Y_Q}{\partial z_{PQ}}\right)_0 = d'_{PQ} * \cos(z_{PQ}) * \cos(Az_{PQ}) \quad (11.99)$$

$$\left(\frac{\partial H_Q}{\partial z_{PQ}}\right)_0 = -d'_{PQ} * \sin(z_{PQ}) \quad (11.100)$$

$$\left(\frac{\partial X_Q}{\partial Az_{PQ}}\right)_0 = d'_{PQ} * \sin(z_{PQ}) * \cos(Az_{PQ}) \quad (11.101)$$

$$\left(\frac{\partial Y_Q}{\partial Az_{PQ}}\right)_0 = -d'_{PQ} * \sin(z_{PQ}) * \sin(Az_{PQ}) \quad (11.102)$$

$$\Sigma_{ll} = \begin{bmatrix} s^2_{X_P} & s_{X_PY_P} & s_{X_PH_P} & 0 & 0 & 0 \\ s_{Y_PX_P} & s^2_{Y_P} & s_{Y_PH_P} & 0 & 0 & 0 \\ s_{H_PX_P} & s_{H_PY_P} & s^2_{H_P} & 0 & 0 & 0 \\ 0 & 0 & 0 & s^2_{d'_{PQ}} & 0 & 0 \\ 0 & 0 & 0 & 0 & s^2_{z_{PQ}[\text{rad}]} & 0 \\ 0 & 0 & 0 & 0 & 0 & s^2_{Az_{PQ}[\text{rad}]} \end{bmatrix} \qquad (11.103)$$

Example 11.13

To determine the Three-Dimensional coordinates of point (3), a total station was set up at a control point (P) with an azimuthal orientation at point (2), as shown in Fig. 11.24. The field measurements are given in Table 11.25, and the coordinates of control points (P) and (2) in Table 11.26. Assuming that $s\alpha_P = s_{z_{PQ}} = \pm 10''$, calculate the spatial coordinates of the point (3) and their corresponding standard deviations.

Table 11.25 Field measurements

Station	Sighting point	Horizontal direction	Slope distance [m]	Zenith angle
P	2	87° 49′50″	–	–
	3	221° 38′36″	93.650 ± 0.004	85° 44′14″

Table 11.26 Control point coordinates

Point	X [m]	Y [m]	H [m]
P	5000.000 ± 0.003	10, 000.000 ± 0.003	100.000 ± 0.004
2	5064.435 ± 0.003	10, 002.440 ± 0.003	

Solution:

Using Eq. (11.43),

$$\alpha_P = 221° 38'36'' - 87° 49'50'' = 133° 48'46''$$

Using Eq. (11.4),

$$Az_{P,2} = \text{atan} \left(\frac{64.435}{2.440} \right) = 87° 49'53''$$

Using Eq. (11.2) or (11.5),

$$dp_{,2} = 64.4812 \text{ m}$$

Using Eq. (11.45),

$$Az_{P,3} = 87°49'53'' + 133°48'46'' = 221°38'39''$$

Using Eq. (11.20),

$$sAz_{P,2} = 0.003 * \frac{\sqrt{2}}{64.4812} = \pm 6.57966 * 10^{-5}\text{rad} = \pm 13.6''$$

Applying the law of propagation of variances,

$$sAz_{P,3} = \sqrt{13.6^2 + 10^2} = \pm 16.9'' \left(8.172901 * 10^{-5}\text{rad}\right)$$

Using Eq. (11.91),

$$X_3 = 5000.000 + 93.3650 * \sin(85°44'14'') * \sin(221°38'36'') = 4937.942 \text{ m}$$

$$Y_3 = 10,000.000 + 93.650 * \sin(85°44'14'') * \cos(221°38'36'') = 9930.210 \text{ m}$$

$$H_3 = 100.000 + 93.650 * \cos(85°44'14'') = 106.961 \text{ m}$$

Using Eq. (11.93) to (11.103) and assuming that the coordinates of point (P) are uncorrelated, the following matrices can be written as

$$J^T = \begin{bmatrix} 1 & 0 & 0 & -0.662664 & -4.625659 & -69.789750 \\ 0 & 1 & 0 & -0.745219 & -5.201925 & 62.058493 \\ 0 & 0 & 1 & 0.074331 & -93.390930 & 0 \end{bmatrix}$$

$$\Sigma_{ll} = \begin{bmatrix} 9.00*10^{-6} & 0 & 0 & 0 & 0 & 0 \\ 0 & 9.00*10^{-6} & 0 & 0 & 0 & 0 \\ 0 & 0 & 1.60*10^{-5} & 0 & 0 & 0 \\ 0 & 0 & 0 & 1.60*10^{-5} & 0 & 0 \\ 0 & 0 & 0 & 0 & 2.350443*10^{-9} & 0 \\ 0 & 0 & 0 & 0 & 0 & 6.679632*10^{-9} \end{bmatrix}$$

$$\Sigma_{ff} = \begin{bmatrix} 4.861015*10^{-5} & -2.097196*10^{-5} & 2.272762*10^{-7} \\ -2.097196*10^{-5} & 4.367420*10^{-5} & 2.555903*10^{-7} \\ 2.272762*10^{-7} & 2.555903*10^{-7} & 3.658865*10^{-5} \end{bmatrix}$$

Using the diagonal values in the variance-covariance matrix of functions, the standard deviations of point (3) coordinates are given as follows:

$$X_3 = 4937.942 \text{ m} \pm 7.0 \text{ mm}$$
$$Y_3 = 9930.210 \text{ m} \pm 6.6 \text{ mm}$$
$$H_3 = 106.961 \text{ m} \pm 6.0 \text{ mm}$$

11.10.3   Three-Dimensional Point Positioning by Horizontal Directions and Vertical Angles

The principle of three-dimensional point positioning by horizontal directions and vertical angles is shown in Fig. 11.25. As no distance is measured, at least two surveying instruments are required, both sighting the target point from the station points. In this case, the observables are horizontal directions and vertical angles, establishing a measurement model with redundant data. Thus, as long as the coordinates of the station points are known, the coordinates of the target point can be determined by applying the Parametric Adjustment model, as shown below.

Fig. 11.25 Principle of 3D point positioning by horizontal directions and vertical angles

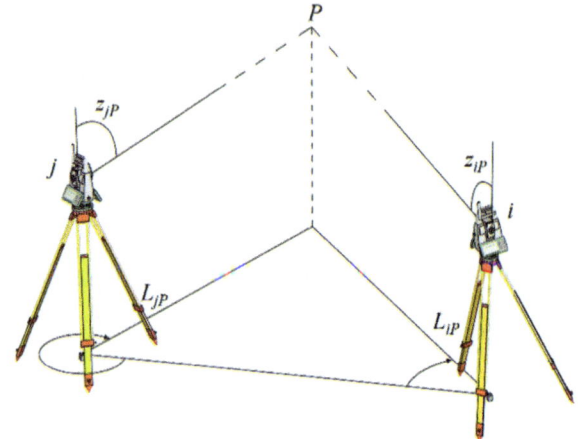

Given:

Points (i) and (j): (X_i, Y_i, H_i)

$$(X_j, Y_j, H_j)$$

Observation:

Horizontal directions: L_{iP}, L_{jP}
Zenith angles: z_{iP}, z_{jP}

To calculate:

Coordinates of the point (P): (X_P, Y_P, H_P)

As mentioned, to apply the Parametric adjustment model, it is necessary to determine the approximate (X_0, Y_0, H_0) coordinates of (P). This can be done using for example Eqs. (11.55) and (11.57) for horizontal positioning and (7.3) for vertical positioning.

Considering that the horizontal directions and vertical angles are observed from n control points to the point (P), as shown in Fig. 11.25, the error equation for the azimuth observations is obtained from Eq. (11.70). The vertical angle error equation is obtained from the vertical angle observable, as given by Eq. (11.104). Note that coordinates are related to the instrument and target centres.

$$z_{iP} = \text{atan} \left[\frac{\sqrt{(X_P - X_i)^2 + (Y_P - Y_i)^2}}{H_P - H_i} \right] \tag{11.104}$$

Thus, considering Eqs. (11.70) and (11.104), the error Eqs. (11.105) can be written for the application of the Parametric adjustment model.

$$vAz_{iP} = \left(\frac{\partial Az_{iP}}{\partial X_P} \right)_0 * \delta X_P + \left(\frac{\partial Az_{iP}}{\partial Y_P} \right)_0 * \delta Y_P + 0 \delta H_P - (Az_{Obs_{iP}} - Az_{iP0})$$

$$vz_{iP} = \left(\frac{\partial z_{iP}}{\partial X_P} \right)_0 * \delta X_P + \left(\frac{\partial z_{iP}}{\partial Y_P} \right)_0 * \delta Y_P + \left(\frac{\partial z_{iP}}{\partial H_P} \right)_0 * \delta H_P - (z_{Obs_{iP}} - z_{iP0}) \tag{11.105}$$

where

$$\left(\frac{\partial z_{iP}}{\partial X_P} \right)_0 = \frac{(X_{P0} - X_i) * (H_{P0} - H_i)}{d_{iP0} * d'^2_{iP0}} \tag{11.106}$$

$$\left(\frac{\partial z_{iP}}{\partial Y_P} \right)_0 = \frac{(Y_{P0} - Y_i) * (H_{P0} - H_i)}{d_{iP0} * d'^2_{iP0}} \tag{11.107}$$

$$\left(\frac{\partial z_{iP}}{\partial Z_P} \right)_0 = -\frac{\sqrt{(X_{P0} - X_i)^2 + (Y_{P0} - Y_i)^2}}{(X_{P0} - X_i)^2 + (Y_{P0} - Y)^2_i + (H_{P0} - H_i)^2} = -\frac{d_{iP0}}{d'^2_{iP0}} \tag{11.108}$$

d_{iP0}: horizontal distance between points (i) and (P_0)
d'_{iP0}: slope distance between points (i) and (P_0)

Each observation will produce a pair of error Eqs. (11.105), whose unknown parameters are the coordinate variations δX_P, δY_P and δH_P. For n observations, Eqs. (11.105) can be written in matrix form as $v = Ax - l_0$ (Eq. 3.79), where

$$A = \begin{bmatrix} \left(\frac{\partial Az_{iP}}{\partial X_P}\right)_0 & \left(\frac{\partial Az_{iP}}{\partial Y_P}\right)_0 & 0 \\ \left(\frac{\partial z_{iP}}{\partial X_P}\right)_0 & \left(\frac{\partial z_{iP}}{\partial Y_P}\right)_0 & \left(\frac{\partial z_{iP}}{\partial H_P}\right)_0 \\ \vdots & \ddots & \vdots \\ \left(\frac{\partial Az_{nP}}{\partial X_P}\right)_0 & \left(\frac{\partial Az_{nP}}{\partial Y_P}\right)_0 & 0 \\ \left(\frac{\partial z_{nP}}{\partial X_P}\right)_0 & \left(\frac{\partial z_{nP}}{\partial Y_P}\right)_0 & \left(\frac{\partial z_{nP}}{\partial H_P}\right)_0 \end{bmatrix} \quad v = \begin{pmatrix} vAz_{iP} \\ vz_{iP} \\ \vdots \\ vAz_{nP} \\ vz_{nP} \end{pmatrix}$$

$$l_0 = \begin{pmatrix} Az_{Obs_{iP}} - Az_{iP0} \\ z_{Obs_{iP}} - z_{iP0} \\ \vdots \\ Az_{Obs_{nP}} - Az_{nP0} \\ z_{Obs_{nP}} - z_{iP0} \end{pmatrix} \quad x = \begin{bmatrix} \delta X_P \\ \delta Y_P \\ \delta H_P \end{bmatrix}$$

The preliminary azimuth Az_{iP0} can be determined using Eq. (11.4).

The adjustment computation is then carried out as for the previous adjustment methods, from which δX_P, δY_P and δH_P are obtained. As a total of n observations and three unknowns have been considered, the standard deviations of the unit weight can be calculated using Eq. (11.109). The remaining statistical parameters can be determined as indicated in Sect. 3.9.1.1.

$$s_0^2 = \frac{v^T P v}{n-3} \tag{11.109}$$

Example 11.14

To determine the Three-Dimensional coordinates of point (P) from control points (3) and (4), horizontal direction and vertical angle observations were carried out using a total station, as shown in the sketch in Fig. 11.25. The azimuths obtained from the field observations are given in Table 11.27 to facilitate the resolution of the example. The coordinates of control points (3) and (4) are given in Table 11.28. Assuming that the azimuth observations have a standard deviation of $\pm 17''$ and the zenith angles have a standard deviation of $\pm 10''$, calculate the spatial coordinates of point (P) and their corresponding standard deviations using the Parametric adjustment model.

Table 11.27 Azimuth observation

Line	Azimuth	Zenith angle
3 - P_0	41°38′47.5″	94°15′58.7″
4 - P_0	106°56′14.4″	90°48′55.3″

Table 11.28 Control point coordinates

Control point	X [m]	Y [m]	H [m]
3	4937.937	9930.201	106.967
4	4904.603	10,029.047	101.419

Solution:

Using observations given in Table 11.27 *and coordinates in* Table 11.28, *the approximate coordinates of point (P) can be*

$$X_{P_0} = 4999.990 \text{ m}$$

$$Y_{P_0} = 9999.990 \text{ m}$$

$$H_{P_0} = 100.003 \text{ m}$$

Using Eqs. (11.2) and (5.75), the preliminary horizontal and slope distances of the line P_0 - i are given in Table 11.29.

Table 11.29 Preliminary distances

Line	Preliminary horizontal distance [m]	Preliminary slope distance [m]
3 - P_0	93.387	93.646
4 - P_0	99.715	99.725

Using Eqs. (11.4) and (5.77), the preliminary azimuths and zenith angles are given in Table 11.30.

Table 11.30 Preliminary azimuths and zenith angles

Line	Preliminary azimuth	Preliminary zenith angle
3 - P_0	41°38′31.0″	94°15″53.1″
4 - P_0	106°56′30.8″	90°48″55.3″

The diagonal values of the weight matrix P of observations using the inverse of the variances of the observations are given as follows:

Diagonal of $P = [1.0 \quad 2.890 \quad 1.0 \quad 2.890]$

Considering the error Eq. (11.105) and using Eq. (3.79), the following matrices can be written for the application of the Parametric adjustment model.

$$A = \begin{bmatrix} 8.002335 * 10^{-3} & -7.115289 * 10^{-3} & 0 \\ -5.277393 * 10^{-4} & -5.935313 * 10^{-4} & -1.064893 * 10^{-2} \\ -2.922361 * 10^{-3} & -9.593393 * 10^{-3} & 0 \\ -1.362039 * 10^{-4} & 4.149072 * 10^{-5} & -1.002661 * 10^{-2} \end{bmatrix}$$

$$l_0 = \begin{bmatrix} 7.996245 * 10^{-5} \\ 2.721727 * 10^{-5} \\ -7.971980 * 10^{-5} \\ 3.115322 * 10^{-5} \end{bmatrix} \qquad x = \begin{bmatrix} \delta X_P \\ \delta Y_P \\ \delta H_P \end{bmatrix}$$

Considering the weight matrix given above and using Eqs. (3.84) to (3.87).

$$N = \begin{bmatrix} 7.343584 * 10^{-5} & -2.801493 * 10^{-5} & 2.018587 * 10^{-5} \\ -2.801493 * 10^{-5} & 1.436833 * 10^{-4} & 1.706133 * 10^{-5} \\ 2.018587 * 10^{-5} & 1.706133 * 10^{-5} & 6.182658 * 10^{-4} \end{bmatrix}$$

$$n = \begin{bmatrix} 8.190885 * 10^{-7} \\ 1.528837 * 10^{-7} \\ -1.740347 * 10^{-6} \end{bmatrix}_m \qquad x = \begin{bmatrix} 0.013656 \\ 0.004127 \\ -0.003375 \end{bmatrix}$$

The adjusted coordinates of the point (P) are then given as follows:

$$X_P = 4999.990 + 0.0134 = 5000.004 \text{ m}$$
$$Y_P = 9999.990 + 0.0041 = 9999.994 \text{ m}$$
$$H_P = 100.003 - 0.0034 = 100.000 \text{ m}$$

The residual error vector can be calculated using Eq. (3.79).

$$v = \begin{bmatrix} 0.01'' \\ -0.19'' \\ 0.04'' \\ 0.21'' \end{bmatrix}$$

The standard deviations of the unit weight can be calculated using Eq. (11.109).

$$s_0^2 = 5.438423 * 10^{-12} \text{rad}^2 (with \ n = 4)$$

$$s_0 = \pm 2.332043 * 10^{-6} \text{rad} = \pm 0.5''$$

Finally, using Eq. (11.74), *the variance-covariance matrix of the parameters is given as follows:*

$$\Sigma_{xx} = s_0^2 N^{-1} = \begin{bmatrix} 8.107796 * 10^{-8} & 1.617566 * 10^{-8} & -3.093504 * 10^{-9} \\ 1.617566 * 10^{-8} & 4.120167 * 10^{-8} & -1.665101 * 10^{-9} \\ -3.093504 * 10^{-9} & -1.665101 * 10^{-9} & 8.943203 * 10^{-9} \end{bmatrix}$$

Using the diagonal values in the variance-covariance matrix of the parameters, the standard deviations of the adjusted coordinates are given as follows:

$$X_P = 5000.004 \text{ m} \pm 0.28 \text{ mm}$$
$$Y_P = 9999.994 \text{ m} \pm 0.20 \text{ mm}$$
$$H_P = 100.000 \text{ m} \pm 0.09 \text{ mm}$$

The absolute error ellipse for point (P) is given as shown in Fig. 11.26.

Fig. 11.26 Absolute error ellipse for point (*P*)

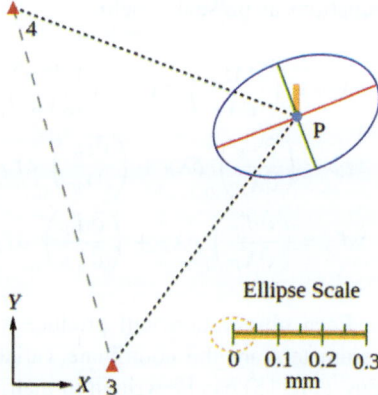

11.10.4   *Three-Dimensional Point Positioning by Horizontal Directions, Vertical Angles and Distances*

If slope distances are added to the Three-Dimensional point positioning by horizontal directions and vertical angles, the adjustment computation is performed by adding the spatial distance observation Eq. (11.110).

$$d'_{iP} = \sqrt{(X_P - X_i)^2 + (Y_P - Y_i)^2 + (H_P - H_i)^2} \qquad (11.110)$$

Developing Eq. (11.110) to its linear form gives the following error equation:

$$v'd_{iP} = \left(\frac{\partial d'_{iP}}{\partial X_P}\right)_0 \delta X_P + \left(\frac{\partial d'_{iP}}{\partial Y_P}\right)_0 \delta Y_P + \left(\frac{\partial d'_{iP}}{\partial H_P}\right)_0 \delta H_P - \left(d'_{Obs_{iP}} - d'_{iP_0}\right) \quad (11.111)$$

where

$$\left(\frac{\partial d'_{iP}}{\partial X_P}\right)_0 = \frac{X_{P_0} - X_i}{d'_{iP_0}} \qquad (11.112)$$

$$\left(\frac{\partial d'_{iP}}{\partial Y_P}\right)_0 = \frac{Y_{P_0} - Y_i}{d'_{iP_0}} \qquad (11.113)$$

$$\left(\frac{\partial d'_{iP}}{\partial H_P}\right)_0 = \frac{H_{P_0} - H_i}{d'_{iP_0}} \qquad (11.114)$$

The Parametric adjustment model is then obtained by combining all the error equations as presented below:

$$vAz_{iP} = \left(\frac{\partial Az_{iP}}{\partial X_P}\right)_0 \delta X_P + \left(\frac{\partial Az_{iP}}{\partial Y_P}\right)_0 \delta Y_P + 0\delta H_P - \left(AZ_{Obs_{ip}} - Az_{iP0}\right)$$

$$vz_{iP} = \left(\frac{\partial z_{iP}}{\partial X_P}\right)_0 \delta X_P + \left(\frac{\partial z_{ip}}{\partial Y_P}\right)_0 \delta Y_P \left(+\frac{\partial z_{iP}}{\partial H_P}\right)_0 \delta H_P - \left(Z_{Obs_{iP}} - z_{iP_0}\right) \qquad (11.115)$$

$$vd'_{iP} = \left(\frac{\partial d'_{iP}}{\partial X_P}\right)_0 \delta X_P + \left(\frac{\partial d'_{iP}}{\partial Y_P}\right)_0 \delta Y_P + \left(\frac{\partial d'_{iP}}{\partial H_P}\right)_0 \delta H_P - \left(d'_{Obs_{iP}} - d'_{iP_0}\right)$$

Each observation will produce a set of error Eqs. (11.115), whose unknown parameters are the coordinate variations δX_P, δY_P and δH_P. For n observations, Eqs. (11.115) can be written in matrix form as $v = Ax - l_0$ (Eq. 3.79), where

$$A = \begin{bmatrix}
\left(\dfrac{\partial Az_{iP}}{\partial X_P}\right)_0 & \left(\dfrac{\partial Az_{iP}}{\partial Y_P}\right)_0 & 0 \\[2ex]
\left(\dfrac{\partial z_{iP}}{\partial X_P}\right)_0 & \left(\dfrac{\partial z_{iP}}{\partial Y_P}\right)_0 & \left(\dfrac{\partial z_{iP}}{\partial H_P}\right)_0 \\[2ex]
\left(\dfrac{\partial d'_{iP}}{\partial X_P}\right)_0 & \left(\dfrac{\partial d'_{iP}}{\partial Y_P}\right)_0 & \left(\dfrac{\partial d'_{iP}}{\partial H_P}\right)_0 \\[2ex]
 & \ddots & \\[2ex]
\left(\dfrac{\partial Az_{nP}}{\partial X_P}\right)_0 & \left(\dfrac{\partial Az_{nP}}{\partial Y_P}\right)_0 & 0 \\[2ex]
\left(\dfrac{\partial z_{nP}}{\partial X_P}\right)_0 & \left(\dfrac{\partial z_{nP}}{\partial Y_P}\right)_0 & \left(\dfrac{\partial z_{nP}}{\partial H_P}\right)_0 \\[2ex]
\left(\dfrac{\partial d'_{nP}}{\partial X_P}\right)_0 & \left(\dfrac{\partial d'_{nP}}{\partial Y_P}\right)_0 & \left(\dfrac{\partial d'_{nP}}{\partial H_P}\right)_0
\end{bmatrix}
\qquad
v = \begin{bmatrix}
v Az_{iP} \\
v z_{iP} \\
v d'_{iP} \\
\vdots \\
v Az_{nP} \\
v z_{nP} \\
v d'_{nP}
\end{bmatrix}$$

$$l_0 = \begin{bmatrix}
Az_{Obs_{iP}} - Az_{iP_0} \\
z_{Obs_{iP}} - z_{iP_0} \\
d'_{Obs_{iP}} - d'_{iP_0} \\
\vdots \\
Az_{Obs_{nP}} - Az_{nP_0} \\
z_{Obs_{nP}} - z_{nP_0} \\
d'_{Obs_{nP}} - d'_{nP_0}
\end{bmatrix}
\qquad
x = \begin{bmatrix}
\delta X_P \\
\delta Y_P \\
\delta H_P
\end{bmatrix}$$

The preliminary azimuth Az_{iP0} and preliminary slope distance d'_{iP0} can be determined using Eqs. (11.4) and (11.110), respectively.

The adjustment computation is then carried out as for the previous adjustment methods. As a total of n observations and three unknowns have been considered, the standard deviations of the unit weight can be calculated using Eq. (11.109). The remaining statistical parameters can be determined as indicated in Sect. 3.9.1.1.

11.10.5   Three-Dimensional Free Station

Three-Dimensional Free Station or Spatial Resection point positioning is a method for determining an unknown station point's position (X_P, Y_P, H_P) by measuring horizontal directions, vertical angles, and slope distances to at least two control points, as shown in Fig. 11.27. The procedure consists of installing a total station over a point from which the operator has the best view to all the control points and computing the position and orientation of the instrument using the observation data regarding the control points. It is particularly useful in scenarios such as setting out construction features, performing quick checks, or during preliminary survey work.

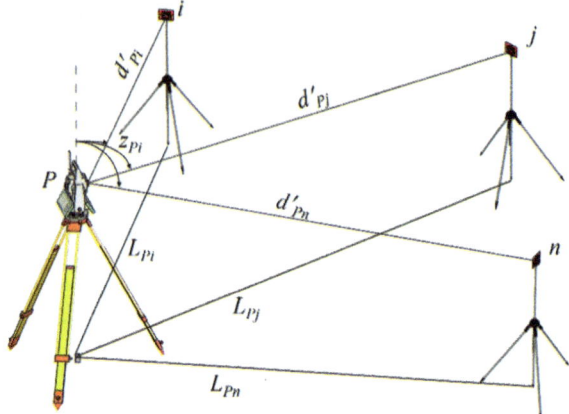

Since horizontal directions, vertical angles and slope distances are measured to several points, the Free-Station positioning can be computed by applying The Parametric Adjustment model, considering the observation equations for horizontal angle and distance (11.75) and (11.81), respectively. For the vertical angle, the observation Eq. (11.116) must be considered.

$$z_{Pi} = \operatorname{atan}\left[\frac{\sqrt{(X_i - X_P)^2 + (Y_i - Y_P)^2}}{H_i - H_P}\right] \tag{11.116}$$

$$va_{ij} = \left(\frac{\partial \alpha_{ij}}{\partial X_P}\right)_0 \delta X_P + \left(\frac{\partial \alpha_{ij}}{\partial Y_P}\right)_0 \delta Y_P - \left(\alpha_{\mathrm{Obs}_{ij}} - \alpha_{ij_0}\right)$$

$$vz_{Pi} = \left(\frac{\partial z_{Pi}}{\partial X_P}\right)_0 \delta X_P + \left(\frac{\partial z_{Pi}}{\partial Y_P}\right)_0 \delta Y_P + \left(\frac{\partial z_{Pi}}{\partial H_P}\right)_0 \delta H_P - \left(z_{\mathrm{obs}_{Pi}} - z_{P_{0i}}\right) \tag{11.117}$$

$$vd'_{Pi} = \left(\frac{\partial d'_{Pi}}{\partial X_P}\right)_0 \delta X_P + \left(\frac{\partial d'_{Pi}}{\partial Y_P}\right)_0 \delta Y_P + \left(\frac{\partial d'_{Pi}}{\partial H_P}\right)_0 \delta H_P - \left(d'_{\mathrm{Obs}_{Pi}} - d'_{P_{0i}}\right)$$

The partial derivatives for angles and distances are the same as given by Eqs. (11.79) and (11.83) and (11.84). The partial derivatives for the zenith angle are given below:

$$\left(\frac{\partial z_{Pi}}{\partial X_P}\right)_0 = \frac{(X_{P_0} - X_i) * (H_i - H_{P_0})}{d_{P_0i} * d'^2_{P_0i}} \tag{11.118}$$

$$\left(\frac{\partial z_{iP}}{\partial Y_P}\right)_0 = \frac{(Y_{P_0} - Y_i) * (H_i - H_{P_0})}{d_{P_0i} * d'^2_{P_0i}} \tag{11.119}$$

$$\left(\frac{\partial z_{ip}}{\partial H_P}\right)_0 = \frac{d_{P_0i}}{d'^2_{P_0i}} \tag{11.120}$$

Each observation will produce a set of error Eqs. (11.117), whose unknown parameters are the coordinate variations δX_P, δY_P and δH_P. For n observations, Eqs. (11.117) can be written in matrix form as $v = Ax - l_0$ (Eq. 3.79), where

$$A = \begin{bmatrix} \left(\frac{\partial a_{ij}}{\partial X_P}\right)_0 & \left(\frac{\partial a_{ij}}{\partial Y_P}\right)_0 & 0 \\ \left(\frac{\partial z_{Pi}}{\partial X_P}\right)_0 & \left(\frac{\partial z_{Pi}}{\partial Y_P}\right)_0 & \left(\frac{\partial z_{Pi}}{\partial H_P}\right)_0 \\ \left(\frac{\partial d'_{Pi}}{\partial X_P}\right)_0 & \left(\frac{\partial d'_{Pi}}{\partial Y_P}\right)_0 & \left(\frac{\partial d'_{Pi}}{\partial H_P}\right)_0 \\ \vdots & \vdots & \vdots \\ \left(\frac{\partial a_{n-1,n}}{\partial X_P}\right)_0 & \left(\frac{\partial a_{n-1,n}}{\partial Y_P}\right)_0 & 0 \\ \left(\frac{\partial z_{Pn}}{\partial X_P}\right)_0 & \left(\frac{\partial z_{Pn}}{\partial Y_P}\right)_0 & \left(\frac{\partial z_{Pn}}{\partial H_P}\right)_0 \\ \left(\frac{\partial z_{Pn}}{\partial X_P}\right)_0 & \left(\frac{\partial d'_{Pn}}{\partial Y_P}\right)_0 & \left(\frac{\partial d'_{Pn}}{\partial H_P}\right)_0 \end{bmatrix} \quad v = \begin{bmatrix} v_{a_{ij}} \\ v_{z_{Pi}} \\ v_{d'_{Pi}} \\ \vdots \\ v_{a_{n-1,n}} \\ v_{z_{Pn}} \\ v_{d'_{Pn}} \end{bmatrix} \quad l_0 = \begin{bmatrix} a_{Obs_{ij}} - a_{ij_0} \\ z_{Obs_{Pi}} - z_{P0i} \\ d'_{Obs_{Pi}} - d'_{PO_i} \\ \vdots \\ a_{Obs_{n-1,n}} - a_{n-1,n_0} \\ z_{Obs_{Pn}} - z_{P0n} \\ d'_{Obs_{Pn}} - d'_{POn} \end{bmatrix} \quad x = \begin{bmatrix} \delta X_P \\ \delta Y_P \\ \delta H_P \end{bmatrix}$$

The adjustment computation is then carried out as for the previous adjustment methods. As a total of n observations and three unknowns have been considered, the standard deviations of the unit weight can be calculated using Eq. (11.109). The remaining statistical parameters can be determined as indicated in Sect. 3.9.1.1.

Example 11.15

To determine the three-dimensional coordinates of point (P) using control points (3) and (4), horizontal direction, slope distances and vertical angle observations from point (P) were carried out using a total station, as the Free-station sketch in Fig. 11.27. Field observations are given in Table 11.31. Coordinates of control points are shown in Table 11.32. Calculate the spatial coordinates of point (P) and their corresponding standard deviations using the Parametric adjustment model. Assume that the horizontal and vertical angles were determined with a standard deviation of $\pm 10''$ and that the slope distances were measured with a standard deviation of ± 4 mm.

Table 11.31 contains observations for points (1) and (2), in case the reader wishes to extend the free station calculation to other points. The coordinates of these points are given in Table 11.32.

Table 11.31 Field observations

Station	Sighting point	Horizontal direction	Slope distance [m]	Zenith angle
P	1	28°57′36″	74.109	90°36′40″
	2	87°49′50″	64.482	90°11′04″
	3	221°38′36″	93.650	85°44′14″
	4	286°56′06″	99.728	89°11′08″

Table 11.32 Coordinates of control points

Control point	X [m]	Y [m]	H [m]
1	5035.880	10,064.838	99.213
2	5064.435	10,002.440	99.794
3	4937.937	9930.201	106.967
4	4904.603	10,029.047	101.419

Solution:

As in the previous examples, the approximate coordinates of point (P) are

$$X_{P_0} = 4999.990 \text{ m}$$

$$Y_{P_0} = 9999.990 \text{ m}$$

$$H_{P_0} = 100.003 \text{ m}$$

The preliminary horizontal and slope distances of the line i - P_0 are given in Table 11.29. Using Eqs. (11.43) and (5.77), the preliminary horizontal and zenith angles are given in Table 11.33.

Table 11.33 Preliminary horizontal and zenith angles

Angle at P	Preliminary horizontal angle	Line	Preliminary zenith angle
3,4	65°17′59.8″	P_0 - 3	85° 44′ 06.9″
		P_0 - 4	89° 11′ 11.1″

Considering the error Eq. (11.117) and using Eq. (3.79), the following matrices can be written for the application of the Parametric adjustment model.

$$A = \begin{bmatrix} -1.092470 * 10^{-2} & -2.478104 * 10^{-3} & 0 \\ 5.276644 * 10^{-4} & 5.934471 * 10^{-4} & 1.064894 * 10^{-2} \\ 6.626336 * 10^{-1} & 7.452426 * 10^{-1} & -7.436515 * 10^{-2} \\ 1.362039 * 10^{-4} & -4.149072 * 10^{-5} & 1.002661 * 10^{-2} \\ 9.565043 * 10^{-1} & -2.913725 * 10^{-1} & -1.419911 * 10^{-2} \end{bmatrix}$$

$$l_0 = \begin{bmatrix} -1.44653025 * 10^{-4} \\ 3.43540698 * 10^{-5} \\ 3.98232706 * 10^{-3} \\ -1.51543646 * 10^{-5} \\ 3.40888026 * 10^{-3} \end{bmatrix} \quad x = \begin{bmatrix} \delta X_P \\ \delta Y_P \\ \delta H_P \end{bmatrix}$$

$$\Sigma_{ll} = \begin{bmatrix} 2.35044305 * 10^{-9} & 0 & 0 & 0 & 0 \\ 0 & 2.35044305 * 10^{-9} & 0 & 0 & 0 \\ 0 & 0 & 1.60000000 * 10^{-5} & 0 & 0 \\ 0 & 0 & 0 & 2.35044305 * 10^{-9} & 0 \\ 0 & 0 & 0 & 0 & 1.60000000 * 10^{-5} \end{bmatrix}$$

*Assuming the weighting model equal to the inverse of the standard deviations, the weight matrix **P** is given as follows:*

$$P = s_0^2 * \Sigma_{ll}^{-1}$$

Assuming $s_O = 0.0011$ m,

$$P = \begin{bmatrix} 5.14796561 * 10^{+2} & 0 & 0 & 0 & 0 \\ 0 & 5.14796561 * 10^{+2} & 0 & 0 & 0 \\ 0 & 0 & 7.56250000 * 10^{-2} & 0 & 0 \\ 0 & 0 & 0 & 5.14796561 * 10^{+2} & 0 \\ 0 & 0 & 0 & 0 & 7.56250000 * 10^{-2} \end{bmatrix}$$

Considering the weight matrix given above and using Eqs. (3.84) to (3.87),

$$N = \begin{bmatrix} 1.63988353 * 10^{-1} & 3.03638760 * 10^{-2} & 1.15794770 * 10^{-3} \\ 3.03638760 * 10^{-2} & 5.17650612 * 10^{-2} & 8.39126344 * 10^{-4} \\ 1.15794770 * 10^{-3} & 8.39126344 * 10^{-4} & 1.10565341 * 10^{-1} \end{bmatrix}$$

$$n = \begin{bmatrix} 1.26794194 * 10^{-3} \\ 3.44680701 * 10^{-4} \\ 8.40520933 * 10^{-5} \end{bmatrix} \quad x = \begin{bmatrix} 7.29475626 * 10^{-3} \\ 2.39352336 * 10^{-3} \\ 8.54766122 * 10^{-4} \end{bmatrix}_{m}$$

The adjusted coordinates of the point (P) can then be calculated using Eq. (3.80):

$$X_P = 4999.9973 \text{ m}$$

$$Y_P = 9999.9924 \text{ m}$$

$$H_P = 100.0039 \text{ m}$$

The residual error vector can be calculated using Eq. (3.79):

$$v = \begin{bmatrix} 12.2'' \\ -4.1'' \\ 2.57 \text{ mm} \\ 5.1'' \\ 2.86 \text{ mm} \end{bmatrix}$$

The standard deviations of the unit weight can be calculated using Eq. (11.109) with n - u = 2.

$$s_0^2 = 1.71480219 * 10^{-6} \text{ (with } n = 5)$$

Finally, using Eq. (11.74), the variance-covariance matrix of the parameters is given as follows:

$$\Sigma_{xx} = s_0^2 N^{-1} = \begin{bmatrix} 1.17312571 * 10^{-5} & -6.88006830 * 10^{-6} & 7.06454265 * 10^{-8} \\ -6.88006830 * 10^{-6} & 3.71656850 * 10^{-5} & 2.10011074 * 10^{-7} \\ 7.06454265 * 10^{-8} & 2.10011074 * 10^{-7} & 1.55117345 * 10^{-5} \end{bmatrix}$$

Using the diagonal values in the variance-covariance matrix of the parameters, the standard deviations of the adjusted coordinates are given as follows:

$$X_P = 4999.9973 \text{ m} \pm 3.4 \text{ mm}$$

$$Y_P = 9999.9924 \text{ m} \pm 6.1 \text{ mm}$$

$$H_P = 100.0039 \text{ m} \pm 3.9 \text{ mm}$$

With the following correlation coefficients:

$$r_{XY} = -32.95\% \qquad r_{XZ} = 0.52\% \qquad r_{YZ} = 0.87\%$$

Using Eq. (3.94), the matrix of variance-covariance for the adjusted observations is given as follows.

$$\Sigma_{\bar{ll}} = \begin{bmatrix} 1.25582627*10^{-9} & -8.24415953*10^{-11} & -8.61524679*10^{-8} & -2.73874436*10^{-11} & -1.01324497*10^{-7} \\ -8.24415953*10^{-11} & 1.77452515*10^{-9} & 4.99699051*10^{-9} & 1.65738619*10^{-9} & -5.63293228*10^{-9} \\ -8.61524679*10^{-8} & 4.99699051*10^{-9} & 1.905279270*10^{-5} & -1.01270702*10^{-8} & -4.19779326*10^{-6} \\ -2.73874436*10^{-11} & 1.65738619*10^{-9} & -1.01270702*10^{-8} & 1.55981614*10^{-9} & 3.79315631*10^{-10} \\ -1.013244497*10^{-7} & -5.63293228*10^{-9} & -4.19779326*10^{-6} & 3.79315631*10^{-10} & 1.77261045*10^{-5} \end{bmatrix}$$

Using Eq. (3.73) and the diagonal values in the above variance-covariance matrix, the adjusted observations and their standard deviations are given as follows:

$$\alpha_{3,4} = 65° 17'42.18'' \pm 7.3''$$

$$z_{P,3} = 85° 44'09.88'' \pm 8.7''$$

$$d_{P,3} = 93.6526 \pm 4.4 \text{ mm}$$

$$z_{P,4} = 89° 11'13.08'' \pm 8.1''$$

$$d_{P,4} = 99.7309 \pm 4.2 \text{ mm}$$

The absolute error ellipse for point (P) using two control points is given as shown in Fig. 11.28.

Fig. 11.28 Absolute error
ellipse for point
(P) determined by free-
station using two control
points

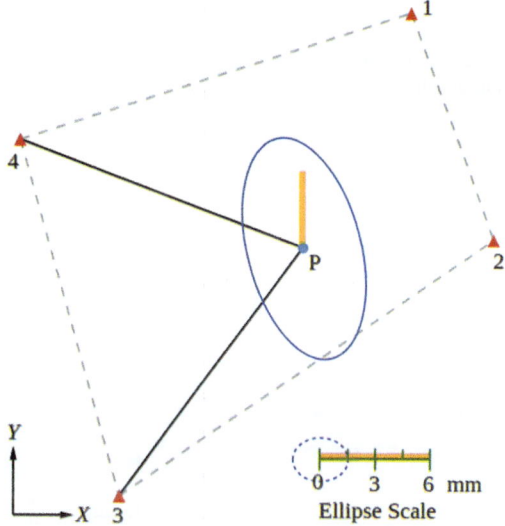

The absolute error ellipse in Fig. 11.28 and the coordinate values obtained using
the control points (3) and (4) indicate that it is advisable to include more control
points in the adjustment computation. Table 11.31 provides observations for points
(1) and (2), in case the reader wishes to extend the free station calculation to other
points. The coordinates of these points are given in Table 11.32. However, as the
matrices are too large, this solution is not shown here. Therefore, if the reader
includes the points (1) and (2), he/she will obtain the results for the first iteration as
shown below:

$$s_0^2 = 7.64520238 * 10^{-7} \text{ (with } n - u = 8)$$

$$X_P = 4999.9979 \text{ m} \pm 1.5 \text{ mm}$$

$$Y_p = 9999.9976 \text{ m} \pm 1.4 \text{ mm}$$

$$H_P = 100.0029 \text{ m} \pm 1.5 \text{ mm}$$

With the following correlation coefficients:

$$r_{XY} = -4.40\% \qquad r_{XZ} = 0.35\% \qquad r_{YZ} = 0.34\%$$

The absolute error ellipse for point (P) using four control points is given as
shown in Fig. 11.29.

Fig. 11.29 Absolute error ellipse for point (P) determined by free-station using four control points

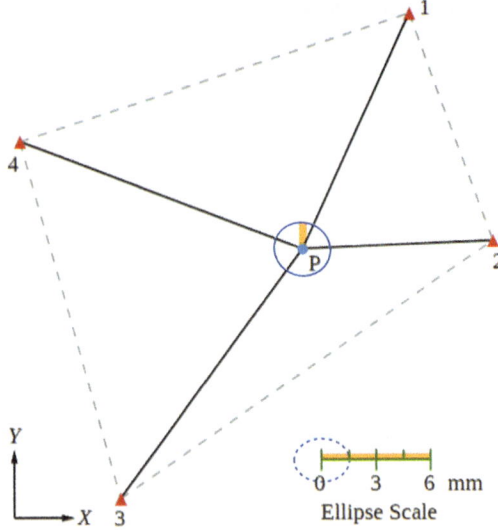

11.11   Eccentric Station

In some fieldwork, the terrain conditions may prevent the surveying instrument from being directly set up over a point with known coordinates (A). Therefore, if the measurements can be performed by occupying another point close to it, the *eccentric station solution* can be adopted, as shown in Fig. 11.30. In this case, the field operations consist of installing the instrument over the eccentric point (P), reading the horizontal directions at points (A) and (B) and measuring the distance d_{PA}.

Fig. 11.30 Geometry of an eccentric station

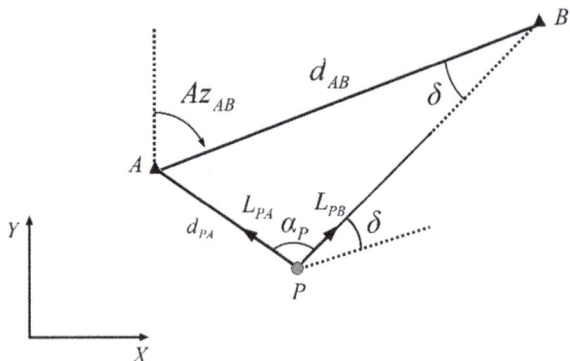

Given:

Point (A): (X_A, Y_A)
Point (B): (X_B, Y_B)

Measured:

d_{PA}, L_{PA}, L_{PB}

To calculate:

$\alpha_P = L_{PB} - L_{PA}$

Coordinates of point (P): (X_P, Y_P)

According to Fig. 11.30, the following steps must be taken to determine the coordinates of the point (P):

1. From the known coordinates of points (A) and (B), calculate the azimuth Az_{AB} according to Eq. (11.4).
2. Calculate the clockwise horizontal angle α_P using Eq. (11.43).
3. Calculate the horizontal distance d_{AB} between the known points (A) and (B), according to Eq. (11.2).
4. Calculate the angle (δ) using Eq. (11.121).

$$\sin(\delta) = \frac{d_{PA}}{d_{AB}} * \sin(\alpha_P) \qquad (11.121)$$

5. Calculate the azimuth $Az_{PB} = Az_{AB} - \delta$.
6. Calculate the azimuth $Az_{AP} = Az_{PB} - \alpha_P \pm 180°$.
7. Calculate the coordinates of the point (P) using Eq. (11.21), considering the azimuth Az_{AP} and the known coordinates of the point (A).

Example 11.16

On a construction site, it is necessary to carry out a series of measurements with a total station installed over point (A), using point (B) as an orientation point. However, point (A) cannot be occupied by the instrument. To solve the problem, it was decided to occupy an eccentric point (P) instead of the point (A), as shown in Fig. 11.31. The coordinates of points (A) and (B) are given in Table 11.34, and the field measurements are given in Table 11.35. Calculate the coordinates of the point (P).

Fig. 11.31 Field survey
geometry

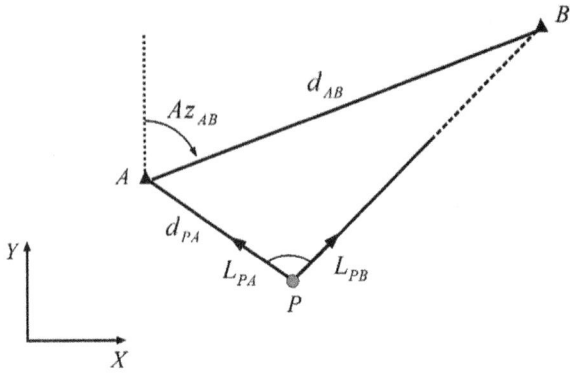

Table 11.34 Coordinates of
points (A) and (B)

Point	X [m]	Y [m]
A	6676.216	11,633.531
B	7453.743	12,743.125

Table 11.35 Field measurements

Station	Sighting point	Horizontal direction	Horizontal distance [m]
P	A	0°00′00″	87.943
	B	95°33′55″	–

Solution:
  Using Eq. (11.4),

$$Az_{AB} = 35°01'12''$$

Using Eq. (11.43),

$$\alpha_P = L_{PB} - L_{PA} = 95°33'55''$$

Using Eq. (11.2) or (11.5),

$$d_{AB} = 1.354{,}897 \text{ m}$$

Using Eq. (11.121),

$$\sin(\delta) = \frac{87.943}{1354.897} * \sin(95°33'55'') \rightarrow \delta = 3°42'14.3''$$

Using Eq. (11.45),

$$Az_{PB} = 35°01'12.4'' - 3°42'14.3'' = 31°18'58.1''$$

And then,

$$Az_{AP} = 31° 18'58.1'' - 95° 33'55.0'' + 180° = 115° 45'3.1''$$

Using Eqs. (11.21),

$$X_P = 6676.216 + 87.943 * \sin(115° 45'3.1'') = 6755.426\,m$$
$$Y_P = 11{,}633.531 + 87.943 * \cos(115° 45'3.1'') = 11{,}595.323\,m$$

11.12   Distance from a Point to a Line Segment

As shown in Fig. 11.32, calculating the distance from a point (P) with known coordinates to a line segment defined by two points, (A) and (B), also with known coordinates, is a simple problem of analytical geometry that can be solved by using Eq. (11.122).

Fig. 11.32 Distance from a
point to a line segment

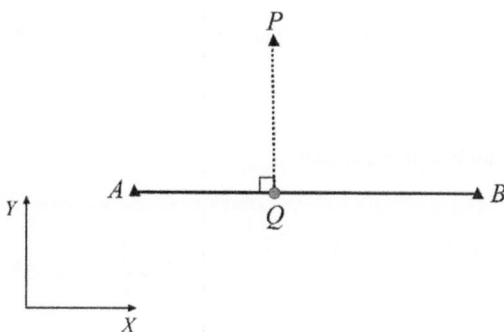

Given:

Point (A): (X_A, Y_A)
Point (B): (X_B, Y_B)
Point (P): (X_P, Y_P)

To calculate:

Horizontal distance: d_{PQ}

$$d_{PQ} = \frac{|(Y_B - Y_A) * X_P + (X_A - X_B) * Y_P - X_A * Y_B + X_B * Y_A|}{d_{AB}} \qquad (11.122)$$

Example 11.17
Using the coordinates given in Table 11.36, calculate the distance from the point (P) to the line segment defined by points (A) and (B).

Point	X [m]	Y [m]
A	7453.743	12,743.125
B	6676.216	11,633.531
P	7072.450	12,635.369

Table 11.36 Known coordinates

Solution:

Using Eq. (11.122);

$$d_{PQ} = \frac{\left| \begin{matrix} (11{,}633.531 - 12{,}743.125) * 7072.450 + (7453.743 - 6676.216) * 12{,}635.369 - \\ 7453.743 * 11{,}633.531 + 6676.216 * 12{,}743.125 \end{matrix} \right|}{1354.897}$$

$$= 250.423 \, m$$

11.13   Intersection of Two Oriented Line Segments

Determining the coordinates of the point of intersection (P) between two oriented line segments, as shown in Fig. 11.33, is a common analytical geometry problem in engineering projects. The solution to this problem is given as follows:

Fig. 11.33 Intersection between two line segments with known orientations

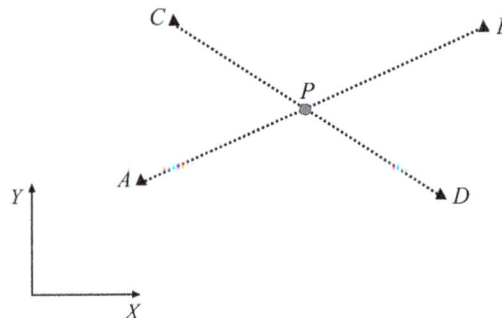

Given:

Point (A): (X_A, Y_A)
Point (B): (X_B, Y_B)
Point (C): (X_C, Y_C)
Point (D): (X_D, Y_D)

To calculate:

Point (P): (X_P, Y_P)

Calculation sequence:

1. Calculate the azimuths Az_{AC}, Az_{AB} and Az_{CD} and the distance d_{AC} from the known coordinates of the points (A), (B), (C) and (D) according to the Eqs. (11.4) and (11.2) or (11.5), respectively.

2. Calculate the clockwise horizontal angles around (A), (B) and (P), as shown below:

$$\alpha_A = Az_{AB} - Az_{AC} \tag{11.123}$$

$$\alpha_C = Az_{CA} - Az_{CD} \tag{11.124}$$

$$\alpha_P = 180° - \alpha_A - \alpha_C \tag{11.125}$$

3. Calculate the distance d_{AP} using Eq. (11.126).

$$d_{AP} = d_{AC} * \frac{\sin(\alpha_C)}{\sin(\alpha_P)} \tag{11.126}$$

4. Calculate the point (P) coordinates using Eq. (11.127).

$$X_P = X_A + d_{AP} * \sin(Az_{AB})$$
$$Y_P = Y_A + d_{AP} * \cos(Az_{AB}) \tag{11.127}$$

The same solution is obtained by exchanging point (A) for point (C).

Example 11.18

Given the coordinates of points (A), (B), (C) and (D) in Table 11.37, calculate the coordinates of the point (P) of the intersection between the lines AB and CD.

Table 11.37 Coordinates of points (A), (B), (C) and (D)

Point	X [m]	Y [m]
A	6359.487	11,744.626
B	6675.479	12,127.951
C	6421.483	12,254.247
D	6676.216	11,633.531

Solution:

Following the calculation sequence above and using Eqs. (11.4) and (11.2).

$$Az_{AC} = \text{atan}\left(\frac{61.996}{509.621}\right) = 6°56'09.7''$$

$$Az_{AB} = \text{atan}\left(\frac{315.992}{383.325}\right) = 39°30'01.0''$$

$$Az_{CD} = \text{atan}\left(\frac{254.733}{-620.716}\right) + 180° = 157°41'14.8''$$

$$d_{AC} = 513.3781\text{m}$$

Calculation of the clockwise horizontal angles around (A), (B) and (P):

$$\alpha_A = 39° 30'01.0'' - 6° 56'09.7'' = 32° 33'51.4''$$

$$\alpha_C = 6° 56'09.7'' - 157° 41'14.8'' + 180° = 29° 14'54.8''$$

$$\alpha_P = 180° - 32° 33'51.4'' - 29° 14'54.8'' = 118° 11'13.8''$$

Calculation of the distance d_{AP}, according to Eq. (11.126):

$$d_{AP} = 513.3781 * \frac{\sin(29° 14'54.8'')}{\sin(118° 11'13.8'')} = 284.5854 \text{ m}$$

Calculate the coordinates of the point (P) according to Eqs. (11.127).

$$X_P = 6359.487 + 284.5854 * \sin(39° 30'01.0'') = 6540.507 \text{ m}$$

$$Y_P = 11,744.626 + 284.5854 * \cos(39° 30'01.0'') = 11,964.218 \text{ m}$$

11.14   Centre and Radius of a Circle Defined by Three Known Points

As shown in Fig. 11.34, the determination of the radius R and the coordinates of the centre (P) of a circle defined by three points with known coordinates, (A), (B) and (C), can be determined using to the following calculation sequence:

Fig. 11.34 Circle defined by three points of known coordinates

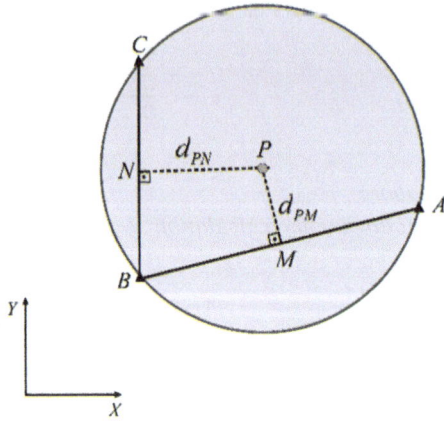

Given:

Point (A): (X_A, Y_A)
Point (B): (X_B, Y_B)
Point (C): (X_C, Y_C)

To calculate:

Coordinates of the centre (P): (X_P, Y_P)
Radius R of the circle

Calculation sequence:

1. Calculate the azimuths Az_{AB} and Az_{BC} from the known coordinates of points (A), (B) and (C) using Eq. (11.4).
2. Calculate the coordinates of points (M) and (N) using Eqs. (11.128) and (11.129):

$$X_M = \frac{X_A + X_B}{2} \quad Y_M = \frac{Y_A + Y_B}{2} \tag{11.128}$$

$$X_N = \frac{X_B + X_C}{2} \quad Y_N = \frac{Y_B + Y_C}{2} \tag{11.129}$$

3. Calculate the coordinates of the point (P) using the MP and NP bilateration calculation sequence, as described in Sect. 11.8.3.
4. Calculate the radius R of the circle, using the coordinates of its centre (P) and one of the points with known coordinates, (A), (B) or (C), according to Eq. (11.2) or (11.5).
5. Repeat the calculation using another point to control the results obtained.

11.15 Intersection of a Line Segment and a Circle

The intersection of a line segment AB with a circle defines two points, (M) and (N), on the circle, as shown in Fig. 11.35. Therefore, if the coordinates of two points, (A) and (B), along the line segment, are known, as well as the coordinates of the centre of the circle (C) and the length of its radius (R), it is possible to determine the coordinates of the points (M) and (N) according to the following the calculation sequence.

Fig. 11.35 Geometric relations of the intersection of a line segment and a circle

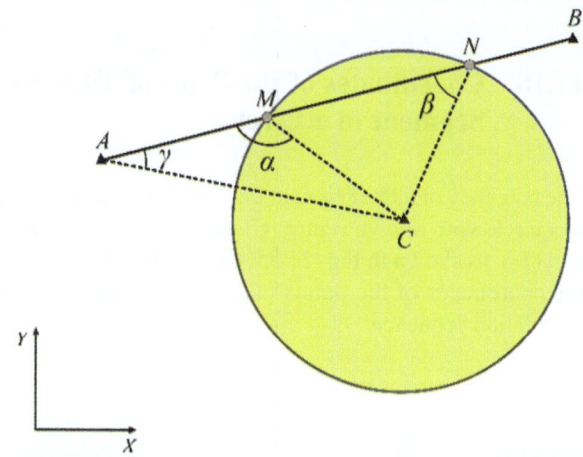

Given:

Point (A): (X_A, Y_A)
Point (B): (X_B, Y_B)
Point (C): (X_C, Y_C)
Radius of the circle: R

To calculate:

Coordinates of point M: (X_M, Y_M)
Coordinates of point N: (X_N, Y_N)

Calculation sequence:

1. Calculate the azimuths Az_{AB} and Az_{AC} from the known coordinates of points (A), (B) and (C) using Eq. (11.4).
2. Calculate the distance d_{AC} between the known points (A) and (C) using Eq. (11.2) or (11.5).
3. Calculate the angle $\gamma = Az_{AC} - Az_{AB}$.
4. Calculate the angle (α), using Eq. (11.130).

$$\alpha = \operatorname{asin}\left(\frac{d_{AC} * \sin \gamma}{R}\right) \qquad (11.130)$$

5. Calculate the azimuth Az_{CM} using Eq. (11.131).

$$Az_{CM} = Az_{AC} + (180° - \gamma - \alpha) \pm 180° \qquad (11.131)$$

6. Calculate the point (M) coordinates using Eq. (11.132).

$$X_M = X_C + R * \sin(Az_{CM})$$
$$Y_M = Y_C + R * \cos(Az_{CM}) \qquad (11.132)$$

7. Repeat Steps 4 to 6 to calculate the coordinates of the point (N).

11.16   Coordinates of the Point of Tangency of a Line Segment to a Circle

The tangency of a line segment that starts from any point (A) with known coordinates to a circle with known centre (C) and radius (R) defines two points of tangency (T_1) and (T_2), as shown in Fig. 11.35. Considering the known values given in Fig. 11.36, the coordinates of the points (T_1) and (T_2) can be determined using the following calculation sequence:

Fig. 11.36 Tangency of a
line segment to a circle

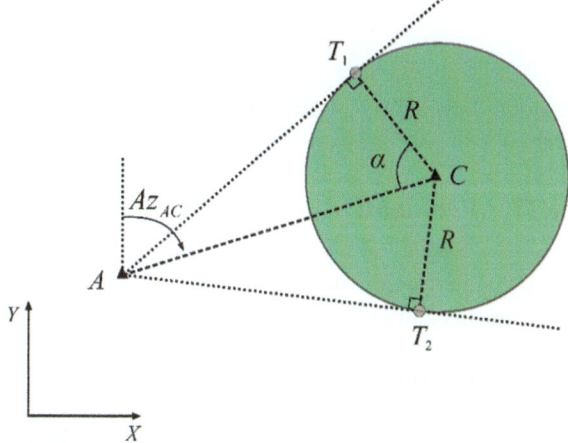

Given:

  Point (A): (X_A, Y_A)
  Point (C): (X_C, Y_C)
  Radius of the circle: R

To calculate:

  Coordinates of (T_1) : (X_{T1}, Y_{T1})
  Coordinates of (T_2) : (X_{T2}, Y_{T2})

Calculation sequence:

1. Calculate the distance d_{CA} and the azimuth Az_{CA} from the known coordinates of points (A) and (C) using Eqs. (11.2) and (11.4), respectively.
2. Calculate the angle α using Eq. (11.133).

$$\alpha = \mathrm{acos}\left(\frac{R}{d_{CA}}\right) \qquad\qquad (11.133)$$

3. Calculate the azimuths Az_{CT_1} and Az_{CT_2} using Eqs. (11.134) and (11.135).

$$Az_{CT_1} = Az_{CA} + \alpha \qquad\qquad (11.134)$$

$$Az_{CT_2} = Az_{CA} - \alpha \qquad\qquad (11.135)$$

4. Calculate the coordinates of the point (T_i) using Eqs. (11.136).

$$X_{T_i} = X_C + R * \sin(Az_{CT_i})$$
$$Y_{T_i} = Y_C + R * \cos(Az_{CT_i})$$

(11.136)

5. Control the results by checking if $d_{CT_1} = d_{CT_2} = R$.

11.17 Problems

1. Using the coordinates of points (1), (2), (3), and (4) in Table 11.38, calculate the azimuths and distances of the lines 1–2, 1–3, 1–4, 3–2 and 3–4.

Table 11.38 Coordinates of points (1), (2), (3) and (4)

Ponto	X [m]	Y [m]
1	5000.000	10,000.000
2	5845.000	10,175.000
3	6,211,167	9169.844
4	4773.746	9339.945

2. Using the coordinates of point (1) in the previous problem and the field measurements given in Table 11.39, calculate the coordinates of point (5).

Table 11.39 Field measurements

Line	Horizontal distance [m]	Azimuth
1–5	912.229	304° 25′39″

3. Using the coordinates of point (5) determined in the previous problem, calculate the coordinates of point (6), which is 500 metres perpendicular to line 1–5, as shown in Fig. 11.37. Point (M) is the midpoint of line 1–5.

Fig. 11.37 Geometric position of point (6)

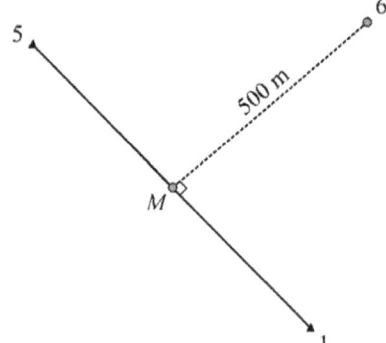

4. Using the coordinates of points (M) and (6) from the previous problem and the field measurements given in Table 11.40, calculate the coordinates of points (7) and (8).

Table 11.40 Field measurements

Station	Sighting point	Horizontal direction	Horizontal distance [m]
6	M	0° 00'00"	–
	7	219° 03'05"	534.586
	8	155° 52'18"	480.348
7	6	155° 11'25"	----
	8	168° 38'03"	533.663

5. Using the coordinates of points (1), (2), (6) and (7) from the previous problems, calculate the angle and distance for setting-out the points (1) and (2) with a total station set up at point (7) and a backsighted to point (6).

6. Suppose that points (1), (3), and (4) from the previous problems are marked in the ground. Suppose also that the coordinates of points (E1) and (E2) are to be determined. As points (1), (3) and (4) are not mutually visible, the engineer has decided to solve the problem by using an auxiliary point (A) from which points (E1) and (E2) can be sighted as shown in Fig. 11.38. The coordinates of point (A) were determined using the resection method. Using the field measurements given in Table 11.41, calculate the coordinates of points (E1) and (E2).

Fig. 11.38 Measurement geometry

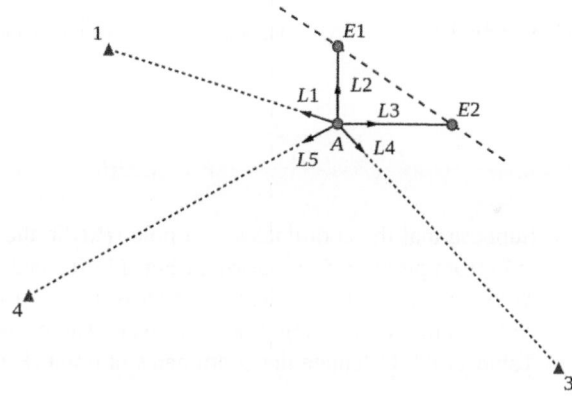

Table 11.41 Field measurements

Station	Sighting point	Horizontal direction	Horizontal distance [m]
A	1	L1 = 288° 21'35"	–
	E1	L2 = 2° 49'01"	169.691
	E2	L3 = 85° 33'13"	263.715
	3	L4 = 136° 24'11"	–
	4	L5 = 241° 32'46"	–

7. Suppose that the Two-Dimensional coordinates of an inaccessible point ($K1$) are to be determined from points (2) and (7) from the previous problem, as shown in Fig. 11.39. As no target or reflectorless measurements can be made to point ($K1$), its coordinates must be calculated using the intersection method. The field measurements are given in Table 11.42.

Fig. 11.39 Measurement geometry

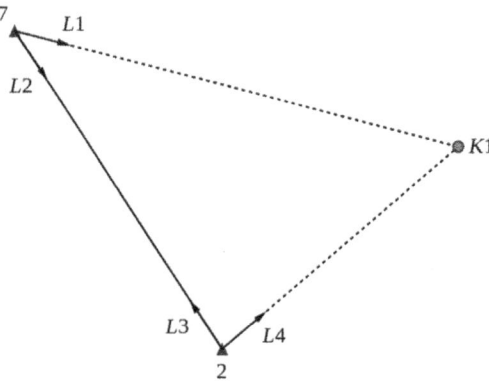

Table 11.42 Field measurements

Station	Sighting point	Horizontal direction
7	2	$L2 = 175°\ 15'38''$
	$K1$	$L1 = 132°\ 39'44''$
2	7	$L3 = 49°\ 58'44''$
	$K1$	$L4 = 133°\ 17'49''$

8. Suppose that the coordinates of a pole (Pst) in the vicinity of points ($E1$) and ($E2$) from problem 6, as shown in Fig. 11.40, need to be determined in the site. To do this, the engineer decided to measure the distances from points ($E1$) and ($E2$) to point (Pst) using a tape. measure The measured distances are given in Table 11.43. Calculate the coordinates of point (Pst).

Fig. 11.40 Measurement geometry

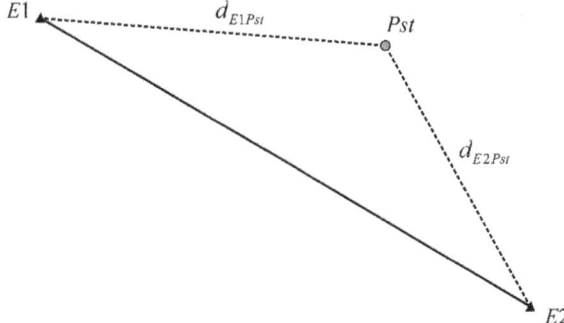

Table 11.43 Field measurements

Station	Sighting point	Horizontal distance [m]
E1	Pst	179.055
E2	Pst	153.829

9. To check the geometry of a retaining wall, a surveying measurement was carried out from station point (S) to points (A), (M) and (B) in the retaining wall, as shown in Fig. 11.41. It is assumed that the geometry of the retaining wall is rectangular at point (M). For the field survey, a total station was set up at point (S) and backsighted to control points (P) and (Q). The coordinates of the control points are given in Table 11.44, and the field measurements are shown in Table 11.45. Calculate the coordinates of points (A), (M) and (B) and check that the lines AM and BM are perpendicular within the allowed tolerance of 1 min of arc.

Fig. 11.41 Measurement geometry

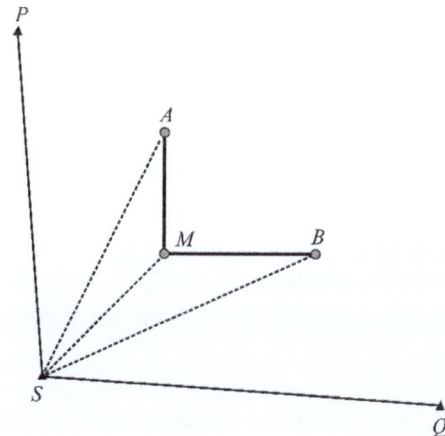

Table 11.44 Control point coordinates

Control point	X [m]	Y [m]
S	10,240.000	10,420.000
P	10,230.000	10,530.000
Q	10,370.000	10,410.000

Table 11.45 Field measurements

Station	Sighting point	Horizontal direction	Horizontal distance [m]
S	P	315° 29′48.0″	–
	Q	55° 05′17.0″	–
	A	347° 15′12″	89.446
	M	5° 41′17″	56.572
	B	26° 43′45″	98.486

10. In the execution of a residential development, a point (A) is to be stakeout in a circular line, as shown in Fig. 11.42. The coordinates of the tangent point (PI) and the centre point (O) of the curve are given in Table 11.46. Given that point (A) is 250 m along the curve, calculate its coordinates.

Fig. 11.42 Circular curve geometry

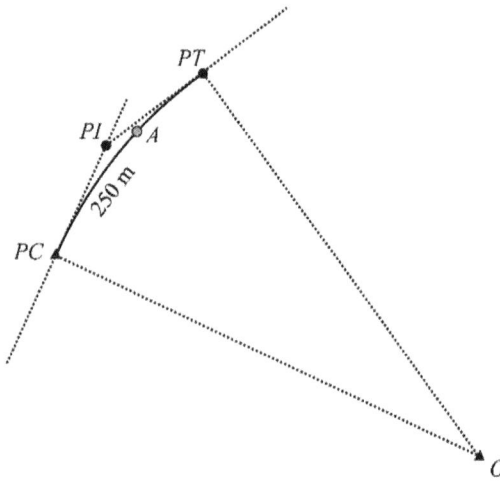

Table 11.46 Coordinates of points (PI) and (O)

Ponto	X [m]	Y [m]
O	10,514.120	10,382.470
PI	9789.074	10,720.565

11. A centre line of a walking track is to be marked every 20 metres, as shown in Fig. 11.43. The four points (1), (2), (3), and (4) of the start and end points of the walking curves are marked on the ground. To carry out the work, the engineer set up a total station at an arbitrary point (STS) and carried out measurements to points (1), (2) and (3). The field measurements are given in Table 11.47. Considering the track geometry and the measurements, calculate the angles and distances to stake out the points (M1) and (M2) with the total station set up at point (STS) and backsighted to point (2).

Fig. 11.43 Track geometry

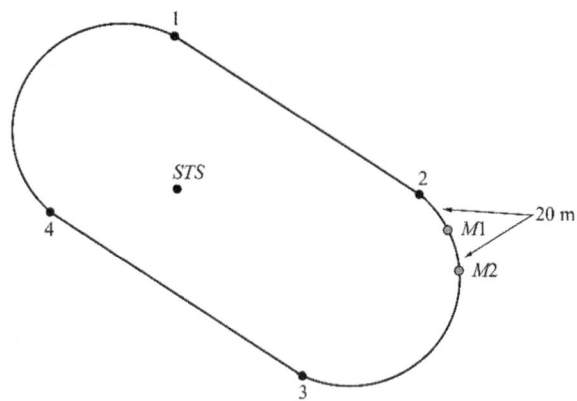

Table 11.47 Field measurements

Station	Sighting point	Horizontal direction	Horizontal distance [m]
Est	1	0° 00′00″	64.839
	2	91° 07′58″	99.701
	3	147° 08′06″	91.353

12. The point (P) on the centreline of a curve, as shown in Fig. 11.44, must be set out using a total station at point (PI) and backsighted to point (P110). Using the coordinates of point (PI) in Table 11.48 and the geometric elements of the curve in Table 11.49, calculate the angle and distance to stakeout point (P).

Fig. 11.44 Geometry of the curve

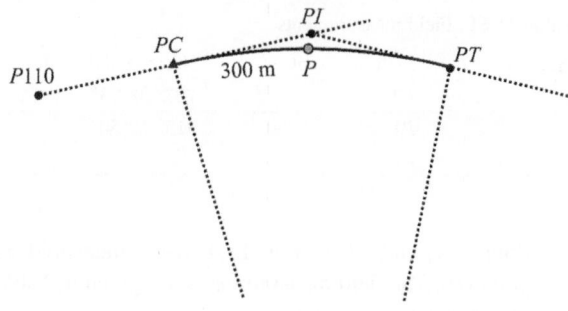

Table 11.48 Coordinates of point (PI)

Point	X [m]	Y [m]
PI	8141.547	13,716.186

Table 11.49 Geometric elements of the curve

Radius	1021.400 m
Arch length	699.170 m
Azimuth of line P110-PI	60° 18′09″

13. Point (P3) of the line P1P3, as shown in Fig. 11.45 must be set out using a total station. The coordinates of points (P1) and (P3) are given in Table 11.50. Considering that points (P1) and (P3) are not visible from each other and that a point (P2) is marked in the ground on the line P1P3, a point (A) was used as an eccentric point to stakeout point (P3). The field measurements are given in Table 11.51. Calculate the angle and distance to stakeout point (P3) with the total station at point (A) and backsighted to point (P1).

Fig. 11.45 Measurement geometry

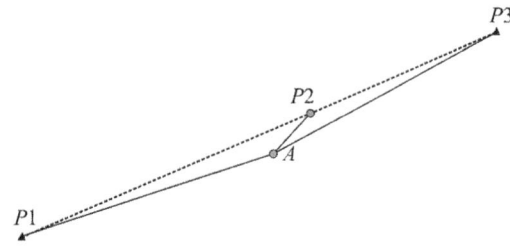

Table 11.50 Coordinates of points (P1) and (P3)

Point	X [m]	Y [m]
P1	6269.292	12,566.067
P3	6686.426	12,740.630

Table 11.51 Field measurements

Station	Sighting point	Horizontal direction	Horizontal distance [m]
A	P2	107° 38′23″	47.468
	P1	315° 12′50″	–
P2	P3	–	177.006

14. Points (A) and (B) in Fig. 11.46 were measured using a total station set up at point (P). The field measurements are given in Table 11.52. Calculate the spatial distance between points (A) and (B).

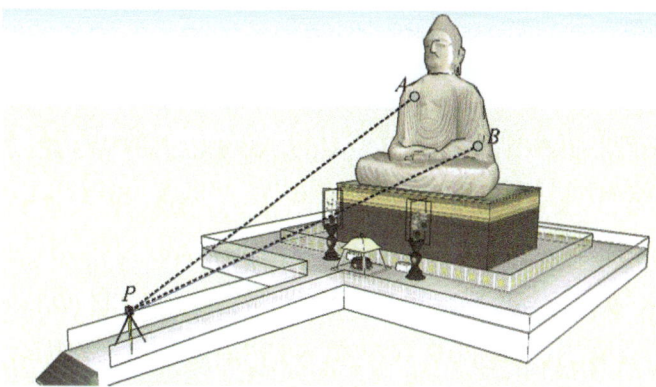

Fig. 11.46 Measurement geometry

Table 11.52 Field measurement

Station	Sighting point	Horizontal direction	Zenith angle	Slope distance [m]
P	A	353° 34′40″	67° 20′08″	36.332
	B	6° 19′14″	73° 38′20″	35.500

15. Using the coordinates of points (P) and (Q) given in Table 11.53, calculate the coordinates of the intersection points of circle 1 with centre at point (P) and radius 416.612 m and circle 2 with centre at point (Q) and radius 481.821 m.

Table 11.53 Field measurement

Point	X [m]	Y [m]
P	11,689.513	12,089.689
Q	12,325.453	11,744.416

References

Anderson, J. M., Mikhail, E. M. (1998). *Surveying, Theory and Practice*. 7th Edition. WCB/ McGraw-Hill, Boston, USA.

Ghilani, C. D., Wolf, P. R. (2007). *Elementary Surveying – An Introduction to Geomatics*. 12 Edition. Pearson Prentice Hall, EUA.

Gonçalves, J. A., Madeira, S., Sousa, J. J. (2012). *Topografia, Conceitos e Aplicações*. 3 Edição. Lidel – Edições Técnicas, Ltda. Lisboa, Portugal.

Mikhail, E. M. (1982). *Observations and Least Squares*. University Press of America. USA. ISBN 9780819123978, 0819123978.

Ogundare, John. (2018). *Understanding Least Squares Estimation and Geomatics Data Analysis*.

Uren, J., Price, B. (2010). *Surveying for Engineers*. 5th Edition. Palgrave Macmillan. Hampshire, England.

Villesuzonne, D. (1988). *Le calcul du géomètre. Eyrolles*, Paris, France.

Chapter 12
Horizontal Control Survey: Traversing

Irineu da Silva and Guilherme Poleszuk dos Santos Rosa

12.1   Introduction

As highlighted several times throughout this book, an engineering project can only be implemented with an accurate framework of control points covering the working area. This framework can be local, covering a small area, such as a construction site, or extending over a large region, encompassing an entire country. Generally, for small areas, the control point framework is based on an engineering datum, as described in *Sect. 4.2.3*, and strategically distributed by the engineer in the working area. It establishes the horizontal reference points and benchmarks for a given construction site. However, for projects in large areas, such as long tunnels, runways, water dams and others, the framework of control points must be based on a *Geodetic Datum*, as described in *Sect. 4.2.4*. In these cases, as the number of geodetic control points is not always densified enough to cover the entire project area, the engineer will need to place new ones according to the regulations and standards specified for the project. The planimetric coordinates of the new control points are then determined based on the horizontal geodetic reference system of the country where the project is being implemented.

Regardless of the type of control point to be established, five control survey methods are currently used to determine the coordinates of new control points.

- GNSS surveys.
- *Triangulation network*, when only horizontal direction observations are performed to form a triangular network of interconnected points.

I. da Silva (✉)
São Carlos School of Engineering, University of São Paulo, São Carlos, SP, Brazil
e-mail: irineu@sc.usp.br

G. Poleszuk dos Santos Rosa
Cartographic Engineer, Researcher at the Faculty of Science and Technology of the São Paulo State University (UNESP), São Paulo, Brazil

© The Author(s), under exclusive license to Springer Nature Switzerland AG 2025
I. da Silva, P. C. L. Segantine, *Geomatics Applied to Civil Engineering*,
https://doi.org/10.1007/978-3-031-75737-2_12

- *Trilateration network*, when only distance observations are performed to form a triangular network of interconnected points.
- *Triangulateration network*, when the horizontal direction and distance observations are performed to form a triangular network of interconnected points.
- *Survey traverse*, when the horizontal direction and distance observations are performed to form a sequence of continuous survey lines, creating a closed or open loop of control points along the route.

Despite their importance in Geomatics, triangulation methods are not presented in this book as they are rarely used in Geomatics applied to Civil Engineering. GNSS survey methods are presented in Chap. 17. The methods for establishing the vertical component of the control points are presented in Chap. 13; therefore, only the two-dimensional horizontal traversing methods for plane surveying will be discussed in this chapter. In the context of Geomatics applied to Civil Engineering, the theories presented in the following sections concern the establishment of a reference frame of control points for:

- Topographic mapping
- Cadastral mapping of small areas
- Subdivision work
- Construction mapping
- Setting out of works
- Earthworks processing
- Geodetic monitoring of structures
- Other small area projects

In addition to the above, a traverse survey has several advantages over other methods that should be highlighted when determining control points in engineering projects, as outlined below:

- Easier field work and office calculations.
- Ease to use, as it is a method that is familiar to almost all professionals involved in Geomatics.
- Less field reconnaissance work, as such a method only requires traversing a route over the terrain.
- Less dependence on terrain conditions, as the route can be easily adapted to the terrain characteristics.
- Independence from predefined geometric figures.
- Easily adapted to project conditions to ensure that control points are close to the geospatial data to be surveyed.

12.2   Traversing

Although the traverse method has been used for a long time in cadastral mapping, the advent of EDM and, more recently, total stations, which have increased the accuracy of observations and greatly simplified fieldwork, has made it the most popular method for establishing control point networks in Civil Engineering. In summary, the fieldwork procedures consist of establishing a series of straight lines connecting successive predefined points along a survey route, defining a traverse. Each point defining the traverse is called a traverse station, which is materialised on the ground according to the technical specifications of the project. As described in the following sections, it becomes a new control point after calculating the traverse. The distances between the traverse stations and the changes in direction along the route are measured with the total station. Depending on the purpose for which the control points are established, the endpoints of the traverse may be tied to an official geodetic control point network, to a predefined GNSS survey point network, or even to a network arbitrarily established by the engineer. The coordinate system of the traverse stations is then set accordingly.

There are two classes of survey traverse: *open* and *closed*. The latter can be categorised as *link traverses* or *polygon (loop) traverses*. Geometric and algebraic details of each of them are presented in the following sections.

12.2.1 Open Traverse

As shown in Fig. 12.1, a survey traverse was run from the control point (A), with an azimuthal orientation at the control point (B), towards the unknown traverse station (P_n). Since it does not close at any control point or allow any computational checking for misclosures, it is called an *open traverse*, which is geometrically and mathematically open. This traverse is composed of two control points, (A) and (B), n free points (P_i) and n traverse lines (courses).

In Fig. 12.1, $L_{i,i+1}$ and $L_{i,i-1}$ are horizontal directions measured in the field, α_i are clockwise horizontal angles calculated considering the forward and backward directions, and $d_{i,i+1}$ are horizontal distances determined from measurements of slope distances $d'_{i,i+1}$ and zenith angles $z_{i,i+1}$ between traverse stations. All angular and slope distance measurements are performed with a total station. The azimuth Az_{BA} is determined from the known coordinates of points (A) and (B). This means the traverse consists of n traverse lines (measured distances) and n horizontal angles.

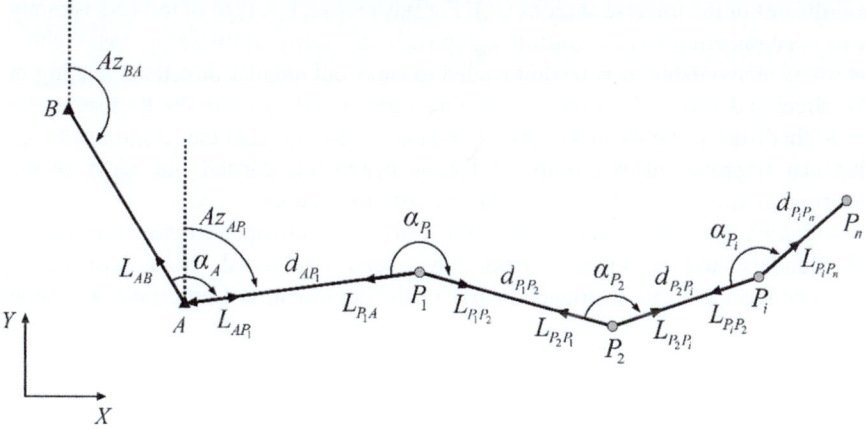

Fig. 12.1 Open traverse

The coordinates of the control points (A) and (B) and the distances $d_{i,i+1}$ can be local or geodetic, depending on the horizontal datum considered for the project, as described in *Sect. 4.2*. In this chapter, all coordinates and distances are referenced to the engineering datum. Details on traversing in other reference systems are given in Chap. 16.

Referring to Fig. 12.1, the coordinates of the traverse stations are calculated from the coordinates of points (A) and (B). Thus, for the first free point (P_1).

$$\alpha_A = L_{AP_1} - L_{AB} \tag{12.1}$$

$$Az_{AP_1} = Az_{BA} + \alpha_A \pm 180° \tag{12.2}$$

$$\begin{aligned} X_{P_1} &= X_A + d_{AP_1} * \sin(Az_{AP_1}) \\ Y_{P_1} &= Y_A + d_{AP_1} * \cos(Az_{AP_1}) \end{aligned} \tag{12.3}$$

For more information on adding or subtracting 180° in Eq. (12.2), refer to *Sect. 11.4*.

Then, following the same reasoning and using the same equations, the coordinates of all the remaining free points (P_i) can be calculated by repeating the process until the last traverse station (P_n) is reached, as shown below:

$$\alpha_i = L_{i,i+1} - L_{i,i-1} \tag{12.4}$$

$$Az_{i,i+1} = Az_{i-1,i} + \alpha_i \pm 180° \tag{12.5}$$

$$\begin{aligned} X_{i+1} &= X_i + d_{i,i+1} * \sin(Az_{i,i+1}) \\ Y_{i+1} &= Y_i + d_{i,i+1} * \cos(Az_{i,i+1}) \end{aligned} \tag{12.6}$$

As this is a sequential process, preparing a calculation table combining all the variables is recommended, as shown in Tables 12.2 and 12.3 of Example 12.1.

When assessing the geometry of the survey traverse shown in Fig. 12.1, it is immediately apparent that its main weakness is the need for more geometric control. Any error in the angular and linear observations will directly affect the quality of the coordinates of the traverse stations (P_i). For this reason, this type of traverse is rarely used in engineering work requiring high positional quality. However, in cases where its use is unavoidable, it is recommended to carry out angular directions reading in the direct and reversed positions of the instrument and measure the backsight and foresight distances between the traverse stations, in addition to other control methods that can improve the reliability of the measurements carried out, such as the determination of angles by closing the horizon, for example.

Although there is no effective error control on an open survey traverse, if all blunders and systematic errors have been eliminated, it is possible to calculate the variance-covariance matrix of the coordinates of any traverse station along the course using the law of propagation of variances. This gives the expected error in the (X) and (Y) directions, which can be used to check the quality of the measurements.

Example 12.1

Using the control point coordinates in Table 12.1 and the field measurements in Table 12.2, calculate the coordinates of each traverse station and their expected error in the (X) and (Y) directions. Assume that the horizontal angles were determined with a standard deviation of $\pm 4''$, that the standard deviation of the azimuth Az_{BA} is equal to $\pm 10''$ and that the traverse lines were measured with a total station with a linear standard deviation of $\pm(2$ mm $+ 2$ ppm). Figure 12.2 shows the geometry of the survey traverse.

Table 12.1 Control point coordinates

Point	X [m]	Y [m]
A	10,794.571 ± 4 mm	12,292.721 ± 4 mm
B	10,609.573	12,839.398

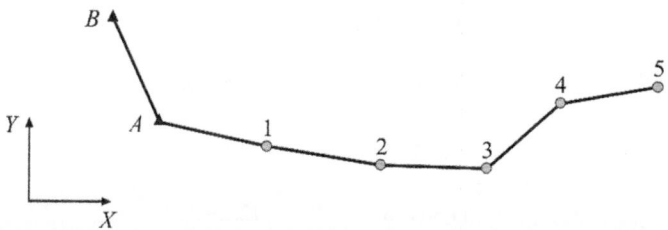

Fig. 12.2 Geometry of the survey traverse

Solution:
The first step in solving this example is calculating the starting azimuth Az_{BA} or Az_{AB} using Eq. (11.4).

$$Az_{BA} = \text{atan}\left(\frac{10,794.571 - 10,609.573}{12,292.721 - 12,839.398}\right) + 180° = 161°18'14.2''$$

$$Az_{AB} = Az_{BA} + 180° = 341°18'14.2''$$

The calculation sequence to obtain the coordinates of the free traverse stations is given in Tables 12.2 and 12.3.

The horizontal distances were calculated using Eq. (12.7).

Table 12.2 Field measurements and computed values

Station	Sighting point	Horizontal direction $(L_{i,j})$	Slope distance [m] $(d'_{i,j})$	Zenith angle $(z_{i,j})$	Horizontal clockwise angle Eq. (12.1)	Azimuth Eq. (11.46)	Horizontal distance [m] Eq. (12.7)
A	B	51° 22′14″	–	–	124° 36′53″	341° 18′14.2″	–
	1	175° 59′07″	245.268	88° 52′47″		105° 55′07.2″	245.222
1	A	333° 15′21″	245.270	91° 07′11″	176° 03′12″		
	2	149° 18′33″	252.783	91° 45′12″		101° 58′19.2″	252.665
2	1	121° 45′34″	252.783	88° 14′50″	171° 33′41″		
	3	293° 19′15″	227.159	90° 57′43″		93° 32′00.2″	227.125
3	2	0° 00′00″	227.156	89° 02′14″	126° 01′20″		
	4	126° 01′28″	256.147	92° 15′48″		39° 33′20.2″	255.945
4	3	0° 00′00″	256.142	87° 44′12″	216° 54′28″		
	5	216° 54′28″	218.600	89° 01′14″		76° 27′48.2″	218.568
						Sum	1,199.525

Table 12.3 Traverse station coordinates

Station	Sighting point	X [m] Eq. (11.21)	Y [m] Eq. (11.21)	Traverse station
A	B 1	10,794.571	12,292.721	A
1	A 2	11,030.389	12,225.463	1
2	1 3	11,277.558	12,173.052	2
3	2 4	11,504.252	12,159.054	3
4	3 5	11,667.245	12,356.389	4
5		11,879.741	12,407.549	5

$$d_{i,i+1} = \frac{d'_{i,i+1} * \sin(z_{i,i+1}) + d'_{i+1,i} * \sin(z_{i+1,i})}{2} \tag{12.7}$$

The standard deviation of the traverse stations can be calculated using Eqs. (11.23) to (11.32). Assuming the standard deviations given below and that the point (A) coordinates are uncorrelated, the matrices for the propagation of variances are as follows:

Given:

$$s_{XA} = s_{YA} = \pm 4 \text{ mm}$$

$$sAz_{AB} = \pm 10''$$

$$s_a = \pm 4''$$

The standard deviation for the horizontal distance is assumed to be equal to ±2.51 mm.

As presented in Sect. 11.3.1, the variance-covariance matrix for the traverse station coordinates can be calculated by applying the law of propagation of variances as follows:

$$\Sigma_{ff} = J^T \Sigma_{ll} J$$

$$J^T = \begin{bmatrix} 1 & 0 & \sin(Az) & d * \cos(Az) \\ 0 & 1 & \cos(Az) & -d * \sin(Az) \end{bmatrix}$$

$$\Sigma_{ll} = \begin{bmatrix} s_X^2 & s_{XY} & 0 & 0 \\ s_{YX} & s_Y^2 & 0 & 0 \\ 0 & 0 & s_d^2 & 0 \\ 0 & 0 & 0 & s_{Az[zad]}^2 \end{bmatrix}$$

$$\Sigma_{ff} = \begin{bmatrix} s_X^2 & s_{XY} \\ s_{YX} & s_Y^2 \end{bmatrix}$$

Then,

For Station 1:

$$J^T = \begin{bmatrix} 1 & 0 & 0.961652 & -6.725770 * 10^4 \\ 0 & 1 & -0.274273 & -2.358184 * 10^5 \end{bmatrix}$$

$$\Sigma_{ll} = \begin{bmatrix} 16.000 & 0 & 0 & 0 \\ 0 & 16.000 & 0 & 0 \\ 0 & 0 & 6.300 & 0 \\ 0 & 0 & 0 & 2.726514 * 10^{-9} \end{bmatrix}$$

$$\Sigma_{ff} = \begin{bmatrix} 34.159827 & 41.582469 \\ 41.582469 & 168.096186 \end{bmatrix} \qquad \begin{bmatrix} s_{XP} \\ s_{YP} \end{bmatrix} = \begin{bmatrix} 5.8 \\ 13.0 \end{bmatrix}_{mm}$$

For Station 2:

$$J^T = \begin{bmatrix} 1 & 0 & 0.978249 & -5.241116 * 10^4 \\ 0 & 1 & -0.207434 & -2.471690 * 10^5 \end{bmatrix}$$

$$\Sigma_{ll} = \begin{bmatrix} 34.159827 & 41.582469 & 0 & 0 \\ 41.582469 & 168.096186 & 0 & 0 \\ 0 & 0 & 6.300 & 0 \\ 0 & 0 & 0 & 3.102585 * 10^{-9} \end{bmatrix}$$

$$\Sigma_{ff} = \begin{bmatrix} 48.711425 & 80.496212 \\ 80.496212 & 357.911980 \end{bmatrix} \qquad \begin{bmatrix} s_{XP} \\ s_{YP} \end{bmatrix} = \begin{bmatrix} 7.0 \\ 18.9 \end{bmatrix}_{mm}$$

Following the same reasoning as for station 2, the variance-covariance and standard deviations for points (3), (4) and (5) are as follows:

For Station 3:

$$\Sigma_{ff} = \begin{bmatrix} 55.669196 & 91.147180 \\ 91.147180 & 536.704151 \end{bmatrix} \qquad \begin{bmatrix} s_{XP} \\ s_{YP} \end{bmatrix} = \begin{bmatrix} 7.5 \\ 23.2 \end{bmatrix}_{mm}$$

For Station 4:

$$\Sigma_{ff} = \begin{bmatrix} 208.331637 & -29.743410 \\ -29.743410 & 642.855994 \end{bmatrix} \qquad \begin{bmatrix} s_{XP} \\ s_{YP} \end{bmatrix} = \begin{bmatrix} 14.4 \\ 25.4 \end{bmatrix}_{mm}$$

For Station 5:

$$\Sigma_{ff} = \begin{bmatrix} 225.359811 & -74.303600 \\ -74.303600 & 834.241567 \end{bmatrix} \qquad \begin{bmatrix} s_{XP} \\ s_{YP} \end{bmatrix} = \begin{bmatrix} 15.0 \\ 28.9 \end{bmatrix}_{mm}$$

The error ellipses for the traverse stations are shown in Fig. 12.3.

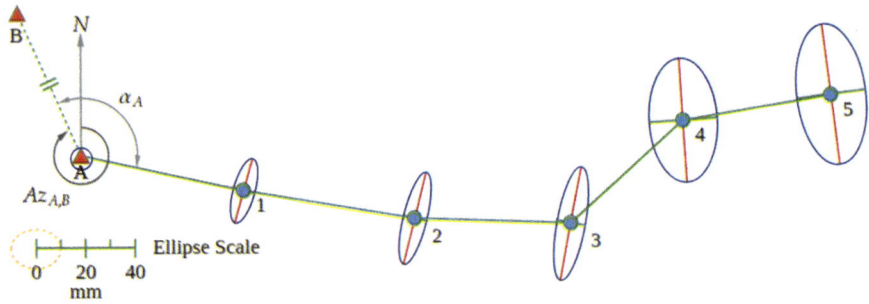

Fig. 12.3 Error ellipses for the calculated open traverse

12.2.2 Link Traverse

To avoid the problem of the lack of geometric control in the open survey traverse, it is recommended to start the route from the control point (A) with an azimuthal orientation at the control point (B) and end it at the control point (M) with an azimuthal orientation at the control point (N), as shown in Fig. 12.4. This traverse is geometrically open but mathematically enclosed in two topographic baselines, as it is restricted to the domain of the four control points already defined in the terrain. For simplicity, this type of traverse is called a *link traverse*. As shown in Fig. 12.4, it consists of four control points, (A), (B), (M) and (N), $n - 1$ free points (P_i), n traverse lines (measured distances) and $n + 1$ calculated angles from the observed horizontal directions. This type of traverse is recommended for linear engineering works such as roads, pipelines, tunnels, power lines and others, depending on the availability of control points. Due to the ease of use of GNSS point determination, it is the most widely used traversing method, and as will be discussed later, it is also the most reliable.

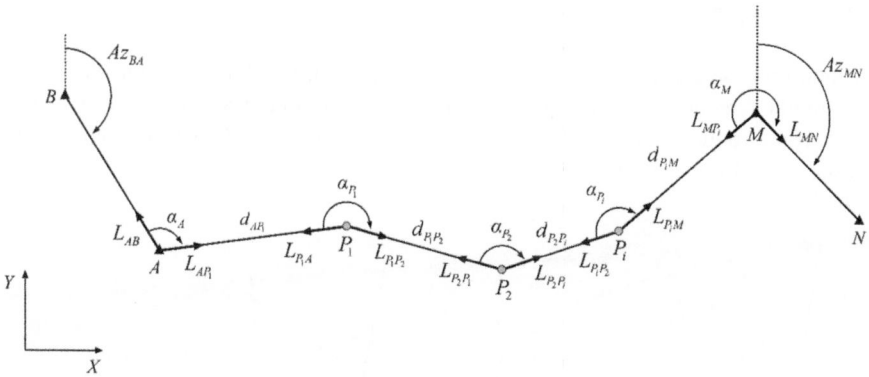

Fig. 12.4 Link traverse

12.2.3 Loop Traverse

A *loop traverse* is a link traverse where the start and end points are the same, as shown in Fig. 12.5. In this case, the control points (A) and (B) are the start and end points for closing the traverse loop. The number of free points, traverse lines and angles are the same as for the link traverse.

As shown in Fig. 12.5, due to the geometry of the traverse, there are two clockwise horizontal angles α_A around point (A). In this book, they are referred to as a α_{start} for the start angle and α_{end} for the closing angle. This type of traverse is often used to establish peripheral reference points on large sites, depending on the availability of control points.

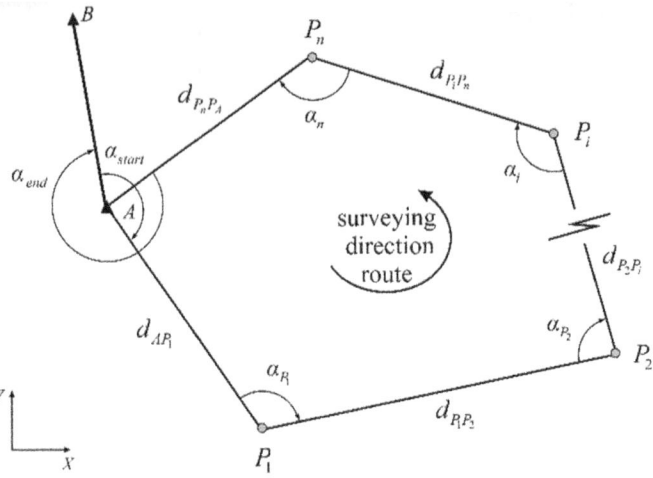

Fig. 12.5 Loop traverse

12.2.3.1   Types of Loop Traverse

A loop traverse can also be described by the direction in which it is run. Since the horizontal direction is clockwise, the traverse is considered an *interior-angle traverse* if the route is counter-clockwise and an *angle to the right traverse* otherwise, as shown in Fig. 12.6a, b.

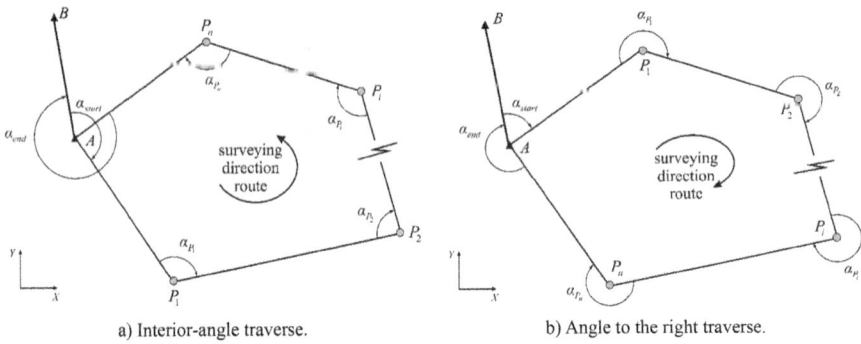

a) Interior-angle traverse.   b) Angle to the right traverse.

Fig. 12.6 Angle to the right and internal-angle traverses.

A loop traverse can also be classified as georeferenced or independent. A *georeferenced loop traverse* is tied to a control point baseline, as shown in Fig. 12.5. Conversely, an *independent loop traverse* does not use control points along the route. As shown in Fig. 12.7, point (P_1) is the start and the endpoint with assumed coordinates provided by the engineer. The azimuth reference point is usually the last point (P_n) on the route to which the first horizontal angle of the traverse is referenced. It can be materialised on the ground before or at the end of the run.

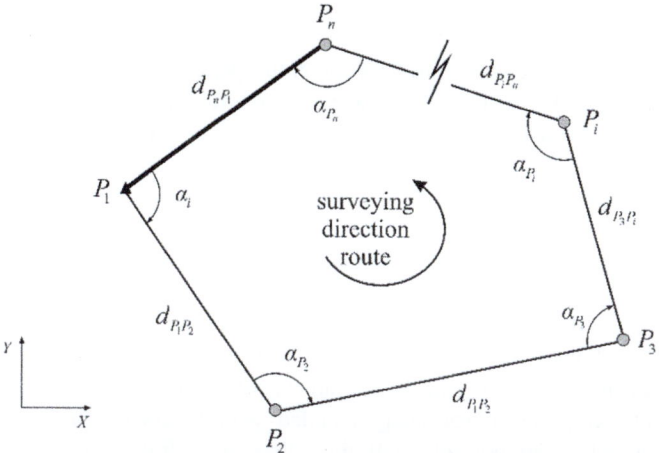

Fig. 12.7 Independent loop traverse

12.2.4 Traverse Field Procedure

The fieldwork procedures for establishing a framework of control points by the traverse method are pretty simple, regardless of the type of traverse used. In Geomatics applied to Civil Engineering, the most commonly used instrument is the total station and its accessories, combined with a set of survey pegs (station marks) to define the traverse stations. Choosing the class of total station and the type of survey marks depends on the accuracy required for the project and how long the station mark will be maintained as a control point. There are no standard international guidelines specifying which tools and accessories should be used. They are, therefore, usually specified by project requirements or based on classical surveying standards adopted by some countries, such as the National Geodetic Survey (NGS) in the USA. Unless otherwise specified, it is advisable to use metal plates and concrete blocks set in a solid, flat and vibration-free ground to facilitate and secure the installation of the instrument. Wooden stakes and stainless-steel nails can also be used in temporary survey traverses set up for cadastral surveys of small areas. A discussion of these guidelines is beyond the scope of this book, and the reader is encouraged to consult specialised references for further information.

The recommended steps for establishing a framework of control points using the traverse method, regardless of the type of project development, are presented below.

12.2.4.1 Field Reconnaissance

Field reconnaissance work is an essential part of any control point survey project. It determines the overall feasibility of the fieldwork and how it should be carried out. The main objective of this stage is to locate suitable positions for the traverse stations, considering their use over the project's lifetime and the ease of executing the route. This means that all adjacent traverse stations must be mutually visible and that the traverse must be maintained for the entire period of use of the materialised points. In addition, the traverse must have as few traverse stations as possible, and short lines must be avoided to minimise the effect of the centring error on the instrument and the prism pole. The lengths of the lines, including the distance between the orientation points, should be approximately equal so that the distribution of linear errors is uniform.

The engineer should also remember that a network of control points will be used to collect the terrain's feature details. These points must, therefore, be conveniently located to ensure the best possible view of the project area for the geospatial data collection. In addition, if they are also to be used for setting out, they must be positioned so that the setting-out points are visible as the work progresses. New control points must be appropriately labelled with letters or numbers for easy identification and, where possible, with their class or quality order.

To start the field reconnaissance, it is advisable to gather all the relevant information about the survey area, which can be done quickly using various web platforms such as Google Earth®. At this stage, it is also important to assess the existing control points, their coordinate values, their physical condition and accessibility. If GNSS surveying is to be used over the traverse points, it is also essential that the sky visibility allows for GNSS data collection. The next step is to visit the survey site to determine the final position of each traverse station. Finally, upon agreement of all supervisors, the traverse marks need to be monumented prior to fieldwork in accordance with the recommendations and regulations specified for the project.

12.2.4.2   Fieldwork Procedures

Considering that a total station will be used for traversing and referring to Fig. 12.4, the following steps must be carried out in the field:

1. Store all control point coordinate values in the instrument's memory.
2. Configure the instrument parameters according to the type of work to be carried out, such as atmospheric corrections, type of reflector prism, type of point numbering, type of code list, and display modes.
3. Install the instrument over the control point (A). If elevation values are also to be determined, the instrument and reflector prism heights must be set accordingly.
4. Take a backsight on the azimuthal control point (B). If the mark in the ground is visible, aim at the mark and take a horizontal direction observation; otherwise, place the prism pole over the ground mark to define the line of sight. Although it is not necessary to measure distances, it is recommended to do so in order to check the coordinates or measurement parameters if necessary.
5. Place the prism pole over the traverse point (P_1), take a foresight horizontal direction observation and measure the distance.

6. Install the instrument over the traverse station point (P_1), place the prism pole over the control point (A) and repeat steps 4 and 5 from the traverse point (P_1) to the control point (M). However, it is now recommended that distance measurements are also taken on the backsight. It is not necessary to measure the distance between control points (M) and (N). However, it is advisable to do so if the algorithm used for the calculation has the option to check the distance to ensure that the correct points have been used in the measurement.
7. Wherever possible, angular observations should be made in double-face readings. Next, slope distances should be measured and stored in the instrument's memory.
8. Some instruments have built-in traverse routines to guide the user during fieldwork. In this case, the traverse computation is performed directly on the instrument at the end of the measurement. It suggests that the user consult the application documentation to define the parameters and format files for the export of the personalised report.

12.2.5   Traverse Computation

Mathematically, a traverse is a geometrically well-defined figure constrained by mathematical rules. However, when measured in the field, due to observational errors, it must be mathematically evaluated to determine the consistency between the measured and geometric values. In other words, all the directions (converted into angles) and distances observed in the field must be checked for any *errors of closure*, which, if smaller than the allowable values, can be distributed among the observations to make the survey traverse geometrically consistent.

Currently, two methods of traverse computation are used in Geomatics applied to Civil Engineering: (1) the conventional method of angle and distance balancing and (2) the adjustment computation by Least Squares. Since angles and distances are involved in traverse measurement, there are two errors of closure to consider: *angular misclosure* and *linear misclosure*. Each is treated mathematically differently, depending on the type of traverse and the adjustment method to be used. They are detected during the traverse calculation and must be adjusted to make the surveyed polygon mathematically consistent so that each traverse station can be used as a control point.

12.2.5.1   Conventional Traverse Adjustment by Angle and Distance Balancing

The first step in the traverse computation by angle and distance balancing is determining the angular error of closure. After that, the linear error of closure is checked and adjusted, as described in the following sections.

12.2.5.1.1 Link Traverse Angle Adjustment

Based on the geometric elements shown in Fig. 12.4, the link traverse angle adjustment is made considering the azimuthal orientations of the initial and final control points, *AB* and *MN*. Under this condition, the following sequence of equations can be written as

$$Az_{BA} = \text{atan}\left(\frac{X_A - X_B}{Y_A - Y_B}\right) + \alpha_{\text{quad}} \tag{12.9}$$

As presented in *Sect. 11.2*, the value of α_{quad} can be $(0\,°, 180° \text{ or } 360°)$ depending on which quadrant the *BA* line lies.

$$\alpha'_i = L_{i,i+1} - L_{i,i-1} \tag{12.10}$$

$$
\begin{aligned}
Az'_{AP_1} &= Az_{BA} + \alpha'_A - 180\,° \\
&\ \vdots \qquad\qquad\quad \vdots
\end{aligned} \tag{12.11}
$$

$$Az'_{P_iM} = Az'_{P_1P_i} + \alpha'_{P_i} - 180\,°$$

$$Az'_{MN} = Az'_{P_iM} + \alpha'_M - 180\,° \tag{12.12}$$

which, generalising, leads to the conclusion that

$$Az'_{MN} = Az_{BA} + \sum \alpha'_i - (n+1) * 180\,° \tag{12.13}$$

where

Az'_{MN} = preliminary azimuth of direction M
Az_{BA} = computed azimuth of direction BA
$Az'_{P_i,P_{i+1}}$ = preliminary azimuth of direction $(i,i+1)$
α'_i = preliminary clockwise horizontal angle around the traverse station (i)
n = number of traverse lines

Eventually, the value of Az'_{MN} must be corrected by $\pm 360°$ to keep it between $0°$ and $360°$.

The angular misclosure e_α is obtained by comparing the preliminary azimuth Az'_{MN} with the computed azimuth Az_{MN}, as given by Eq. (12.14).

$$e_\alpha = Az'_{MN} - Az_{MN} \tag{12.14}$$

The value of e_α must then be compared with the allowable value described in *Sect. 12.4*. If it is acceptable (less than the allowable value), it must be distributed among the preliminary values using one of the alternatives below. Otherwise, the observations should be checked or repeated.

1. Apply an average correction c_α to each angle, as given by Eq. (12.15), assuming that the lengths of the traverse lines are approximately equal.

$$c_\alpha = -\frac{e_\alpha}{n+1} \qquad (12.15)$$

where

n = number of traverse lines

The adjusted angle value is then calculated by adding the angle correction to the preliminary angle using Eq. (12.16).

$$\alpha_i = \alpha_i' + c_\alpha \qquad (12.16)$$

where

α_i = adjusted angle around the traverse station (i)

2. Apply a correction value weighted by the inverse of the traverse line length. Thus, given that there are two observed distances at traverse station (i), the weight to be assigned can be calculated using Eq. (12.17).

$$p_i = \frac{1}{d_{i,i-1} + d_{i,i+1}} \qquad (12.17)$$

where

p_i = weight value at traverse station (i)
d_{i-1} = backsight distance
d_{i+1} = foresight distance

The value of the angular correction c_{α_i} to be applied to the traverse station (i) is given by Eq. (12.18).

$$c_{\alpha_i} = -\frac{e_\alpha}{\sum p_i} * p_i \qquad (12.18)$$

The adjusted angle value is then calculated by adding the angle correction to the preliminary angle using Eq. (12.19).

$$\alpha_i = \alpha_i' + c_{\alpha_i} \qquad (12.19)$$

After the adjustment, the reader should note that considering Eqs. (12.13) and (12.16) or (12.19) gives

$$Az_{MN} = Az_{BA} + \sum \alpha_i - (n+1) * 180° \qquad (12.20)$$

In other words, the difference between the final and initial azimuths must equal the sum of the adjusted angles minus $(n + 1) * 180°$. Finally, the adjusted azimuths can be calculated using Eq. (12.21).

$$
\begin{aligned}
Az_{AP_1} &= Az_{BA} + \alpha_A \pm 180° \\
&\vdots \qquad\qquad \vdots \\
Az_{P_iM} &= Az_{P_iP_{i-1}} + \alpha_{P_i} \pm 180°
\end{aligned}
\qquad (12.21)
$$

12.2.5.1.2 Loop Traverse Angle Adjustment

The loop traverse angle adjustment is similar to the link traverse case. See Fig. 12.8. However, two differences should be highlighted: the first is that the sum of the preliminary angles must consider the two angles observed around the control point (A), denoted as (α_{start}) and (α_{end}); the second is that the azimuthal orientations at the start and end control points (A) and (B) differ by 180°, as given by Eq. (12.22).

$$
Az'_{AB} = Az_{BA} + \sum \alpha'_i - (n + 1) * 180° \qquad (12.22)
$$

where

Az'_{AB} = preliminary azimuth of direction AB
Az_{BA} = computed azimuth of direction BA
$\sum \alpha'_i = \alpha'_{P_i} + \alpha'_{start} + \alpha'_{end}$ = sum of preliminary angles
n = number of traverse lines

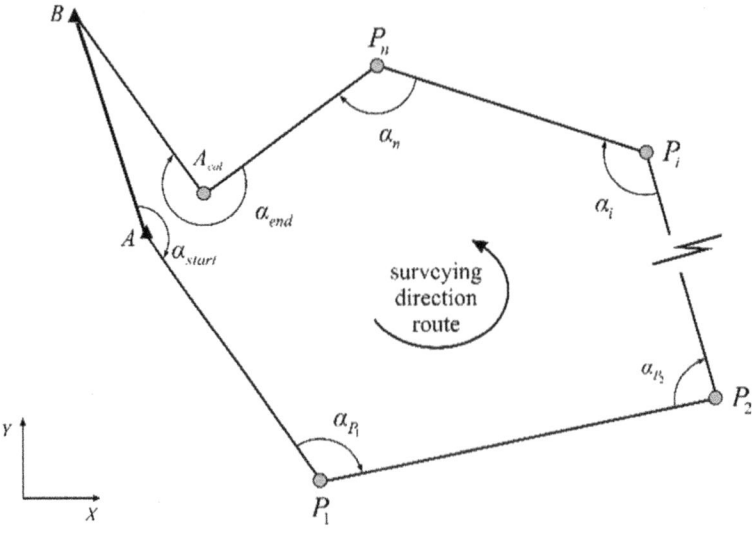

Fig. 12.8 Loop angular misclosure

The angular misclosure e_α is obtained by comparing the preliminary azimuth Az'_{AB} with the computed azimuth Az_{AB}, as given by Eq. (12.23).

$$e_\alpha = Az'_{AB} - Az_{AB} \tag{12.23}$$

The remaining calculations are the same as for the link traverse angle adjustment.

In the case of an independent loop traverse, as shown in Fig. 12.7, the angular error of closure can be calculated considering the geometric condition of the sum of a polygon's internal or external angles, as given by Eqs. (12.24) and (12.25).

For internal angles:

$$\sum \alpha_i = (n-2) * 180° \tag{12.24}$$

For external angles:

$$\sum \alpha_i = (n+2) * 180° \tag{12.25}$$

The angular misclosure e_α is obtained as follows:
For internal angles:

$$e_\alpha = \sum \alpha_i - (n-2) * 180° \tag{12.26}$$

For external angles:

$$e_\alpha = \sum \alpha_i - (n+2) * 180° \tag{12.27}$$

where

$\alpha_i =$ internal or external angle, depending on the type of loop traverse
$n =$ number of traverse lines (in this case, equal to the number of traverse stations)

The remaining calculations are the same as for the loop traverse angle adjustment.

12.2.5.1.3   Link Traverse Linear Adjustment

In traditional adjustment methods, two terms are considered when performing the traverse linear adjustment: *departure* and *latitude*. As shown in Fig. 12.9, the term departure of a traverse line is its orthogonal projection on the X-axis of the coordinate system, which can be expressed as

$$\Delta X = d_{PQ} * \sin(Az_{PQ}) \tag{12.28}$$

As also shown in Fig. 12.9, the term latitude of a traverse line is its orthogonal projection on the Y-axis of the coordinate system, which can be expressed as

$$\Delta Y = d_{PQ} * \cos (Az_{PQ}) \tag{12.29}$$

where

d_{PQ} = length of the traverse line PQ
Az_{PQ} = adjusted azimuth of the traverse line PQ

Fig. 12.9 Departure and latitude of a line

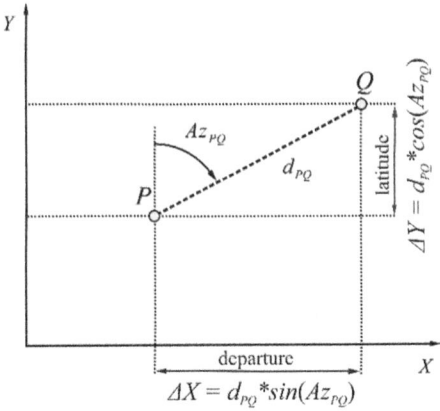

As shown in Fig. 12.10, linear adjustment of the link traverse can be calculated considering the sum of the preliminary latitudes and departures from point (A) to point (M), as given by Eq. (12.30).

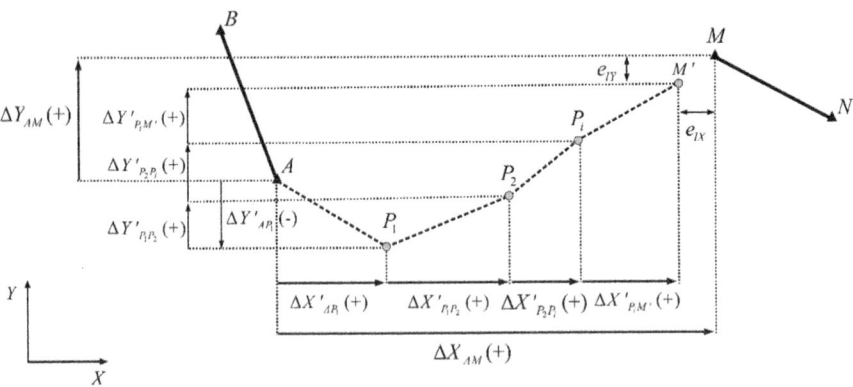

Fig. 12.10 Linear closure of a link traverse

$$X'_M = X_A + \sum \Delta X'_{i,i+1} = X_A + \sum [d_{i,i+1} * \sin(Az_{i,i+1})]$$
$$Y'_M = Y_A + \sum \Delta Y'_{i,i+1} = Y_A + \sum [d_{i,i+1} * \cos(Az_{i,i+1})]$$

(12.30)

The linear misclosures concerning the (X, Y) axes e_{pX} and e_{pY} are given by Eqs. (12.31) or (12.32).

$$e_{Px} = X'_M - X_M$$
$$e_{Py} = Y'_M - Y_M$$

(12.31)

Or,

$$e_{Px} = \sum \Delta X'_{i,i+1} - (X_M - X_A) = \sum \Delta X'_{i,i+1} - \Delta X_{AM}$$
$$e_{Py} = \sum \Delta Y'_{i,i+1} - (Y_M - Y_A) = \sum \Delta Y'_{i,i+1} - \Delta Y_{AM}$$

(12.32)

Finally, the linear misclosure e_p of the traverse and its orientation θ can be calculated using Eqs. (12.33) and (12.34).

$$e_p = \sqrt{e_{Px}^2 + e_{Py}^2}$$

(12.33)

$$\theta = \text{atan}\left(\frac{e_{Px}}{e_{Py}}\right) + \alpha_{\text{quad}}$$

(12.34)

The *relative precision* of the traverse is expressed as a fraction comparing the linear misclosure and the total traverse length, as given by Eq. (12.35).

$$\text{Relative precision} = \frac{e_p}{\sum d_{i,i+1}}$$

(12.35)

where

$\sum d_{i,i+1}$ = total traverse length

Usually, for standardisation, Eq. (12.35) is expressed in ratio form as 1:M. Then, similarly to the angular error of closure, it should be compared with the allowable value, as described in *Sect. 12.4*. It is distributed among the preliminary values if it is acceptable (less than the allowable). Three traditional linear adjustment methods are currently used: the *transit rule*, the *Bowditch rule* and *Crandall's rule*, of which the Bowditch rule is the most used.

The Bowditch rule adjusts the departure and the latitude of the traverse lines in the same proportion as each traverse line length is to the total traverse length, as given by Eq. (12.36).

$$
\begin{aligned}
cP_{X_{i,i+1}} &= -\frac{e_{px}}{\sum d_{i,i+1}} * d_{i,i+1} = -k_X * d_{i,i+1} \\
cP_{Y_{i,i+1}} &= -\frac{e_{py}}{\sum d_{i,i+1}} * d_{i,i+1} = -k_Y * d_{i,i+1}
\end{aligned}
\tag{12.36}
$$

where

$cP_{X_{i,i+1}}, cP_{Y_{i,i+1}}$ = correction in departure $\Delta X'_{i,j}$ and latitude $\Delta Y'_{i,j}$
e_{px}, e_{py} = linear misclosure in (X, Y) axes
$d_{i,i+1}$ = horizontal distance of traverse line $(i,i+1)$
k_X, k_Y = multiplication constant in (X, Y) axes

The adjusted traverse line length is then calculated using Eq. (12.37), that is by adding the linear correction in departure and latitude to the corresponding prelimi-nary traverse line departure and latitude.

$$
\begin{aligned}
\Delta X_{i,i+1} &= \Delta X'_{i,i+1} + cP_{X_{i,i+1}} \\
\Delta Y_{i,i+1} &= \Delta Y'_{i,i+1} + cP_{Y_{i,i+1}}
\end{aligned}
\tag{12.37}
$$

Finally, the adjusted coordinates of the traverse station $(i + 1)$ can be calculated using Eq. (12.38).

$$
\begin{aligned}
X_{i+1} &= X_i + \Delta X_{i,i+1} \\
Y_{i+1} &= Y_i + \Delta Y_{i,i+1}
\end{aligned}
\tag{12.38}
$$

12.2.5.1.4   Loop Traverse Linear Adjustment

The linear adjustment calculation is carried out for a loop traverse, considering Fig. 12.11. In this case, as the start and end points are the same, the linear misclosure is calculated by assuming that the sum of all traverse line departures and all traverse line latitudes must be equal to zero, as expressed by Eq. (12.39).

$$
\begin{aligned}
\sum \Delta X'_{i,i+1} &= 0 \\
\sum \Delta Y'_{i,i+1} &= 0
\end{aligned}
\tag{12.39}
$$

Due to observational errors, Eq. (12.39) are hardly satisfied, and the linear misclosures e_{pX} and e_{pY} can be calculated by taking the algebraic difference of the sum of traverse line departures and latitudes, as expressed by Eq. (12.40).

$$
\begin{aligned}
e_{px} &= \sum \Delta X'_{i,j} \\
e_{py} &= \sum \Delta Y'_{i,j}
\end{aligned}
\tag{12.40}
$$

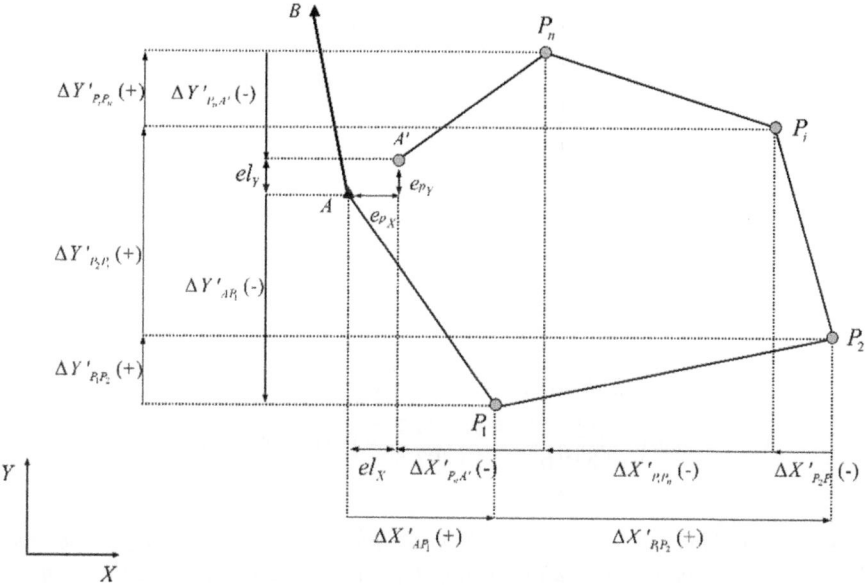

Fig. 12.11 Geometric relations of the linear closure of a loop traverse

As this is a repetitive process, it is recommended to use a spreadsheet, as shown in the following examples of traverse computations.

Example 12.2

Considering that the open traverse from Example 12.1 closes at control points (*M*) and (*N*), as shown in Fig. 12.12, calculate the errors of closure and the adjusted coordinates of the traverse stations. The field measurements are given in Table 12.5, and the control points coordinates are shown in Table 12.4.

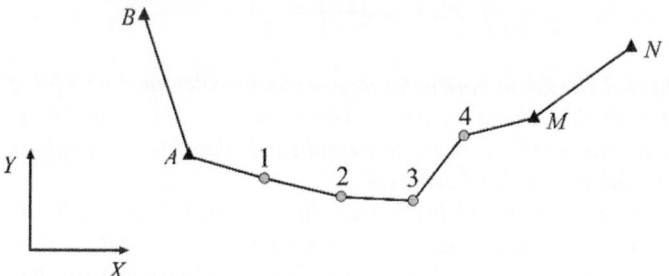

Fig. 12.12 Geometry of the link survey traverse measured in the field

Table 12.4 Control point coordinates

Point	X [m]	Y [m]
A	10,794.571	12,292.721
B	10,609.573	12,839.398
M	11,879.715	12,407.500
N	12,112.239	12,535.602

Solution:
Using Eq. (11.4),

$$Az_{BA} = \text{atan}\left(\frac{10,794.571 - 10,609.573}{12,292.721 - 12,839.398}\right) + 180° = 161° \, 18' 14.2''$$

$$Az_{MN} = \text{atan}\left(\frac{12,112.239 - 11,879.715}{12,535.602 - 12,407.500}\right) = 61° \, 08' 55.6''$$

Next, the preliminary clockwise horizontal angles can be calculated from the horizontal direction observations using Eq. (12.10). The results are given in Table 12.5. Given the azimuth values calculated above and the preliminary angle values determined from the horizontal direction observations, it is possible to calculate the angular error of closure using Eqs. (12.13) and (12.14). The sum of the preliminary angles is also given in Table 12.5.

$$Az'_{MN} = 161° \, 18'14.2'' + 979° \, 50'51.0'' - 1,080° \, 00'00'' = 61° \, 09'05.2''$$

$$e_\alpha = 61° \, 09'05.2'' - 61° \, 08'55.6'' = 00° \, 00'09.6''$$

Assuming that the angular misclosure value is less than the allowable value, the angular correction can be calculated using Eq. (12.15).

$$C_\alpha = -\frac{9.6''}{5 + 1} = -1.6''$$

The adjusted clockwise horizontal angles can be calculated by adding the angle correction to each preliminary angle. The results are shown in Table 12.5. The adjusted azimuths of each traverse line can be calculated from the adjusted angles, as shown in Table 12.5. See Sect. 11.4.

Once the traverse is angularly adjusted, the next step is to adjust the distances. To do this, it is necessary to calculate the horizontal distances and the departure and latitude of each traverse line. The average horizontal distance can be calculated using Eq. (12.7). The results are shown in Table 12.6. The linear error of closure and the relative precision are calculated using Eqs. (12.31), (12.33) and (12.35).

$$X'_M = 10,794.571 + 1,085.161 = 11,879.732\text{m}$$
$$Y'_M = 12,292.721 + 114.852 = 12,407.573\text{m}$$

$$e_{p_X} = 11,879.732 - 11,879.715 = 0.017\text{m}$$
$$e_{p_Y} = 12,407.573 - 12,407.500 = 0.073\text{m}$$

$$e_p = \sqrt{0.017^2 + 0.073^2} = 0.075\text{m}$$

$$\sum d_{ij} = 1199.523\,\text{m}$$

$$\text{Relative precision} = \frac{0.075}{1199.523} \approx \frac{1}{16,035}$$

Assuming that the linear misclosure value is less than the allowable value, the linear corrections can be calculated using Eq. (12.36) and the multiplication constants given below.

$$k_X = -\frac{0.017}{1199.523} = -1.45755 * 10^{-5} \quad k_Y = -\frac{0.073}{1199.523} = -6.06360 * 10^{-5}$$

The results are given in Table 12.6.

The adjusted values can then be calculated using Eqs. (12.37) and (12.38). The values of the final results are shown in Table 12.7.

Table 12.5 Field measurements and calculated partial values

Station	Sighting point	Horizontal direction $(L_{i,j})$	Slope distance [m] $(d'_{i,j})$	Zenith angle $(z_{i,j})$	Preliminary clockwise horizontal angle Eq. (12.1)	Adjusted clockwise horizontal angle Eq. (12.19)	Adjusted azimuth Eq. (12.21)
A	B	51° 22′14″	–		124° 36′53″	124° 36′51.4″	341° 18′14.2″
	1	175° 59′07″	245.268	88° 52′47″			105° 55′05.6″
1	A	333° 15′21″	245.270	91° 07′11″	176° 03′12″	176° 03′10.4″	
	2	149° 18′33″	252.783	91° 45′12″			101° 58′16.0″
2	1	121° 45′34″	252.783	88° 14′50″	171° 33′41″	171° 33′39.4″	
	3	293° 19′15″	227.159	90° 57′43″			93° 31′55.4″
3	2	0° 00′00″	227.156	89° 02′14″	126° 01′20″	126° 01′18.4″	
	4	126° 01′20″	256.147	92° 15′48″			39° 33′13.8″
4	3	0° 00′00″	256.142	87° 44′12″	216° 54′28″	216° 54′26.4″	
	M	216° 54′28″	218.596	89° 01′14″			76° 27′40.2″
M	4	17° 56′58″	218.600	90° 58′45″	164° 41′17″	164° 41′15.4″	
	N	182° 38′15″					61° 08′55.6″
				Sum	979° 50′51″	979° 50′41.4″	

Table 12.6 Distances, preliminary departures and latitudes, and linear corrections

Station	Sighting point	Horizontal distance [m] Eq. (12.7)	$\Delta X'_{i,j}$ [m] Eq. (12.28)	$\Delta Y'_{i,j}$ [m] Eq. (12.29)	$c_{P_{X_{i,j}}}$ [m] Eq. (12.36)	$C_{P_{Y_{i,j}}}$ [m] Eq. (12.36)
A	B					
	1	245.222	235.819	−67.256	−0.004	−0.015
1	A					
	2	252.665	247.170	−52.407	−0.004	−0.015
2	1					
	3	227.125	226.694	−13.993	−0.003	−0.014
3	2					
	4	255.945	162.986	197.340	−0.004	−0.016
4	3					
	M	218.566	212.492	51.167	−0.003	−0.013
M	4					
	N					
Sum		1199.523	1085.161	114.852		

Table 12.7 Adjusted departures and latitudes, and final coordinates

Station	Sighting point	Adjusted ΔX_{ij} [m] Eq. (12.37)	Adjusted ΔY_{ij} [m] Eq. (12.37)	X [m] Eq. (12.38)	Y [m] Eq. (12.38)	Traverse station
A	B			10,794.571	12,292.721	A
	1	235.815	−67.271			
1	A			11,030.386	12,225.450	1
	2	247.166	−52.423			
2	1			11,277.552	12,173.028	2
	3	226.691	−14.006			
3	2			11,504.243	12,159.021	3
	4	162.983	197.325			
4	3			11,667.226	12,356.346	4
	M	212.489	51.154			
M	4			11,879.715	12,407.500	M
	N					
Sum		1085.144	114.779			

Example 12.3

Using the coordinates of control points (A) and (B) in Table 12.8 and the field survey data in Table 12.9, calculate the errors of closure, the relative precision of the survey traverse, and the adjusted coordinates of the traverse stations. Figure 12.13 shows the geometry of the georeferenced loop traverse. Note that, for simplicity, the distances given in Table 12.8 are the horizontal distances calculated from the corresponding slope distances and zenith angles measured in the field.

Table 12.8 Control point coordinates

Point	X [m]	Y [m]
A	9701.147	12,199.247
B	9601.587	12,302.258

Fig. 12.13 Geometry of the georeferenced loop traverse measured in the field

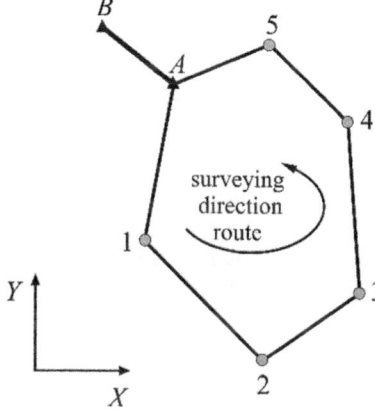

Solution:
Start and end azimuth calculation using Eq. (11.4).

$$Az_{BA} = \text{atan}\left(\frac{9701.147 - 9601.587}{12,199.247 - 12,302.258}\right) + 180° = 135°58'33.6''$$

$$Az_{AB} = 135°58'33.6'' + 180° = 315°58'33.6''$$

Next, the preliminary clockwise horizontal angles can be calculated from the horizontal directions measured in the field. The results are given in Table 12.9. Given the azimuth values calculated above and the preliminary angle values, it is possible to calculate the angular error of closure using Eqs. (12.22) and (12.23). The sum of the preliminary angle values is also given in Table 12.9.

$$Az'_{AB} = 135°58'33.6'' + 1,080°00'06.0'' - 1,260°00'00'' = 315°58'39.6''$$

$$e_\alpha = 315°58'39.6'' - 315°58'33.6'' = 00°00'06''$$

Assuming that the angular misclosure value is less than the allowable value and given that the lengths of the traverse lines are not equivalent, the angular adjustment must be carried out using Eqs. (12.17) and (12.18), i.e., using the correction values weighted by the inverse of each traverse line length. The calculation example for traverse station (A) is given below. The remaining values are shown in Table 12.9.

$$p_A = \frac{1}{340.371+143.260} = 0.00207 \text{ (Note that the distance of 143.260 is calculated}$$
from the coordinates of points (A) and (B).

With $\sum p_i = 0.01400$ given in Table 12.9.

$$c_{\alpha_A} = -\frac{6''}{0.01400} * 0.00207 = -0.88613''$$

The next step is to calculate the adjusted azimuths of the traverse lines, considering the start azimuth and adding the corresponding adjusted horizontal angles. The results are shown in Table 12.10. The adjusted distances are calculated using the departure and latitude values. The results are given in Table 12.11.

The linear error of closure is calculated using Eq. (12.40), and the relative precision is calculated using Eqs. (12.33) and (12.35).

$$e_{P_X} = \sum \Delta X'_{i,+1} = 0.063 \text{ m}$$

$$e_{P_Y} = \sum \Delta Y'_{i,j+1} = 0.037 \text{ m}$$

$$e_p = \sqrt{0.063^2 + 0.037^2} = 0.073 \text{ m}$$

$$\sum d_{ij} = 1699,277 \text{ m}$$

$$\text{Relative precision} = \frac{0.073}{1699.227} \approx \frac{1}{23,409}$$

Assuming that the linear misclosure value is less than the allowable value, the linear corrections are calculated using Eq. (12.36) and the multiplication constants given below.

$$k_X = -\frac{0.063}{1699.227} = -3.69105 * 10^{-5} \qquad k_Y = -\frac{0.037}{1699.227} = -2.15067 * 10^{-5}$$

The results are shown in Table 12.10.

The adjusted values are then calculated using Eqs. (12.37) and (12.38). The results are shown in Table 12.11.

The reader should note that Tables 12.9 to 12.11 are linked together but are presented here individually due to space limitations on a book page. When using an electronic spreadsheet, they must be linked.

Table 12.9 Field measurements and calculated partial values

Station	Sighting point	Horizontal direction ($L_{i,j}$)	Average horizontal distance [m]	Preliminary horizontal angle Eq. (12.1)	Weight Eq. (12.17)	Angular correction Eq. (12.18)	Adjusted angle Eq. (12.19)
A	B	0° 00′00″		232° 27′12″	0.00207	−0.88613″	232° 27′11.1″
	1	232° 27′12″	340.371				
1	A	0° 00′00″		132° 53′44″	0.00148	−0.63347″	132° 53′43.4″
	2	132° 53′44″	336.146				
2	1	0° 00′00″		89° 03′12″	0.00177	−0.75741″	89° 03′11.2″
	3	89° 03′12″	229.679				
3	2	0° 00′00″		126° 07′55″	0.00165	−0.70837″	126° 07′54.3″
	4	126° 07′55″	375.315				
4	3	0° 00′00″		143° 13′30″	0.00167	−0.71427″	143° 13′29.3″
	5	143° 13′30″	224.682				
5	4	0° 00′00″		101° 20′13″	0.00239	−1.02597″	101° 20′12.0″
	A	101° 20′13″	193.034				
A	5	0° 00′00″		254° 54′20″	0.00297	−1.27437″	254° 54′18.7″
	B	254° 54′20″	(143.260)				
		Sum	1699.227	1080° 00′06″	0.01400	−6.0″	1080° 00′00.0″

Table 12.10 Azimuths, preliminary departures and latitudes, and linear corrections

Station	Sighting point	Adjusted azimuth Eq. (11.46)	$\Delta X'_{i,j}$ [m] Eq. (12.28)	$\Delta Y'_{i,j}$ [m] Eq. (12.29)	$c_{P_{X_{i,j}}}$ [m] Eq. (12.36)	$c_{P_{Y_{i,j}}}$ [m] Eq. (12.36)
A	B	315° 58′33. 6″				
	1	188° 25′44. 7″	−49.893	−336.694	−0.013	−0.007
1	A					
	2	141° 19′28. 1″	210.061	−262.428	−0.012	−0.007
2	1					
	3	50° 22′39. 3″	176.913	146.472	−0.008	−0.005
3	2					
	4	356° 30′33. 6″	−22.851	374.619	−0.014	−0.008
4	3					
	5	319° 44′029″	−145.220	171.444	−0.008	−0.005
5	4					
	A	241° 04′14. 9″	−168.947	−93.376	−0.007	−0.004
A	5					
	B	315° 58′33. 6″				
		Sum	0.063	0.037	−0.063	−0.037

Table 12.11 Adjusted departures and latitudes, and final coordinates

Station	Sighting point	Adjusted ΔX_{ij} [m] Eq. (12.37)	Adjusted ΔY_{ij} [m] Eq. (12.37)	X [m] Eq. (12.38)	Y [m] Eq. (12.38)	Traverse station
A	B			9701.147	12,199.247	A
	1	−49.906	−336.702			
1	A			9651.241	11,862.545	1
	2	210.048	−262.436			
2	1			9861.289	11,600.110	2
	3	176.905	146.467			
3	2			10,038.194	11,746.577	3
	4	−22.865	374.611			
4	3			10,015.329	12,121.188	4
	5	−145.228	171.440			
5	4			9870.101	12,292.627	5
	A	−168.954	−93.380			
Sum		0.000	0.000	9701.147	12,199.247	A

Example 12.4

Given the field data in Table 12.12, calculate the errors of closure, the relative precision and the final coordinates of the traverse stations of the independent loop traverse shown in Fig. 12.14. Take the traverse line A-5 as the reference line. Similar to the previous example, the distances given in Table 12.12 are the horizontal distances previously calculated from the corresponding slope distances and zenith angles measured in the field. Assume that point (A) coordinates are (5000.000 m; 10,000.000 m).

Fig. 12.14 Geometry of the independent loop traverse measured in the field

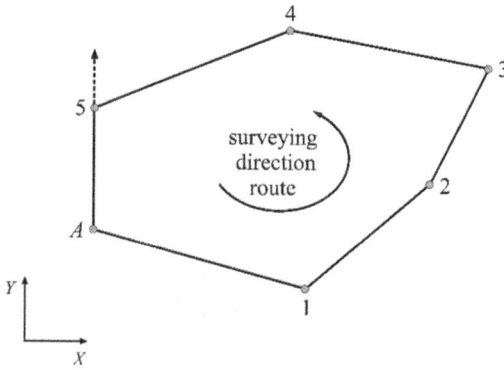

Solution:
The angular error of closure for an interior-angle traverse is calculated using Eq. (12.26) and the values in Table 12.12.

$$e_\alpha = \sum \alpha_i - (n-2) * 180° = 719°59'54'' - (6-2)*180° = -0°00'6''$$

Assuming that the angular error of closure is less than the allowable value and disregarding the difference between the traverse lines, the angular correction can be calculated using Eq. (12.18).

$$C_\alpha = -\frac{(-6'')}{6} = 1.0''$$

The adjusted horizontal angles are calculated by adding the angular correction to each preliminary angle. The adjusted azimuth of each traverse line can then be calculated considering the starting azimuth equal to $0°00'00''$. The results are shown in Table 12.12.

Once the traverse is angularly adjusted, the linear error of closure can be calculated using Eq. (12.40) and the relative precision using Eqs. (12.33) to (12.35).

$$e_{P_X} = \sum \Delta X' = -0.011\,\text{m} \quad e_{P_Y} = \sum \Delta Y' = 0.063\,\text{m}$$

$$e_P = \sqrt{(-0.011)^2 + (0.063)^2} = 0.064\,\text{m}$$

$$\text{Relative precision} = \frac{0.064}{2092.157} \approx \frac{1}{32,748}$$

See page 470 for the values of k_x and k_y.

Table 12.12 Field measurements and calculated partial values

Station	Sighting point	Horizontal direction ($L_{i,j}$)	Average Hz. distance [m]	Preliminary Hz. angle Eq. (12.1)	Adjusted Hz. angle Eq. (12.16)	Adjusted azimuth Eq. (11.46)	$\Delta X'_{i,i+1}$ [m] Eq. (12.28)	$\Delta Y'_{i,i+1}$ [m] Eq. (12.29)
A	5	0° 00′00″				0° 00′00″		
	1	142° 47′33″	273.104	142° 47′33″	142° 47′34″	142° 47′34″	165.146	−217.515
1	A	0° 00′00″						
	2	50° 54′16″	402.612	50° 54′16″	50° 54′17″	13° 41′51″	95.337	391.161
2	1	0° 00′00″						
	3	172° 26′30″	434.452	172° 26′30″	172° 26′31″	6° 08′22″	46.464	431.960
3	2	0° 00′00″						
	4	124° 42′13″	205.325	124° 42′13″	124° 42′14″	310° 50′36″	−155.328	134.281
4	3	0° 00′00″						
	5	76° 28′03″	330.476	76° 28′03″	76° 28′04″	207° 18′40″	−151.630	−293.637
5	4	0° 00′00″						
	A	152° 41′19″	446.188	152° 41′19″	152° 41′20″	180° 00′00″	0.000	−446.188
		Sum	2,092.157	719° 59′54″	720° 00′00″		−0.011	0.063

Table 12.13 Linear corrections, adjusted departures and latitudes, and final coordinates

Station	Sighting point	$c_{P_{X_{ij}}}$ [m] Eq. (12.36)	$c_{P_{Y_{ij}}}$ [m] Eq. (12.36)	Adjusted $\Delta X_{i,j}$ [m] Eq. (12.37)	Adjusted $\Delta Y_{i,j}$ [m] Eq. (12.37)	X [m] Eq. (12.38)	Y [m] Eq. (12.38)	Traverse station
A	5					5000.000	10,000.000	A
1	1	0.0015	−0.0082	165.147	−217.523	5165.147	9782.477	1
	A							
2	2	0.0022	−0.0121	95.339	391.149	5260.486	10,173.627	2
	1							
3	3	0.0024	−0.0131	46.466	431.947	5306.953	10,605.574	3
	2							
4	4	0.0011	−0.0062	−155.327	134.275	5151.625	10,739.849	4
	3							
5	5	0.0018	−0.0099	−151.628	−293.647	4999.998	10,446.201	5
	4							
	A	0.0024	−0.0134	0.002	−446.201	5000.000	10,000.000	A
	Sum			0.000	0.000			

Assuming that the linear misclosure is less than the allowable value, the linear corrections can be calculated using Eq. (12.36). The adjusted values can then be calculated using Eqs. (12.37) and (12.38). All the results of the linear adjustment, the adjusted orthogonal projections and the final adjusted coordinates are shown in Table 12.13.

The values of the multiplication constants are given below:

$$k_X = 5.45789 * 10^{-6} \qquad k_Y = -3.00443 * 10^{-5}$$

12.2.5.2 The Effect of the Traverse Adjustment Procedure by Angle and Distance Balancing

As mentioned earlier, the reader should note that the traverse is not improved by adjusting for angular and linear errors of closure. It only makes the traverse figure geometrically consistent. For example, in the case of a link traverse, it is worth noting that the linear error of closure distribution shifts all the traverse stations cumulatively along the direction of the linear error of closure e_p, as shown in Fig. 12.15. Traverse lines such as $P3P4$, parallel to the direction of the linear error of closure e_p, change linearly but not angularly. In turn, traverse lines such as $P1P2$, perpendicular to the e_p direction, are the opposite. The other traverse lines in other conditions undergo angular and linear changes.

The adjustment procedure, therefore, affects the traverse lines differently depending on their orientation with respect to the orientation of the traverse error of closure. As all measurements are, in principle, of the same accuracy, this type of adjustment is illogical.

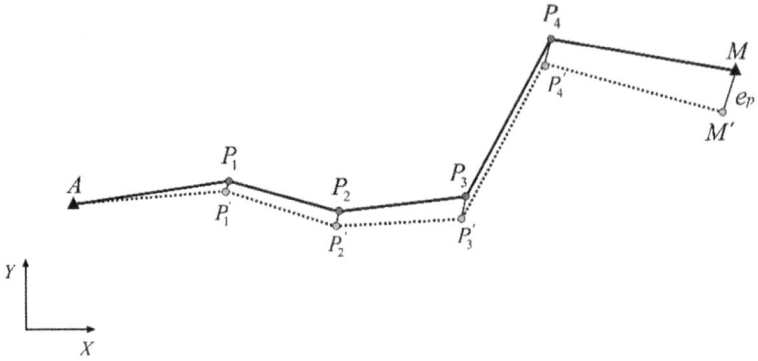

Fig. 12.15 Displacements of the traverse stations due to linear balancing

One of the solutions for this case is to run the traverse so that it is as straight as possible. Another solution is to adjust the traverse using a different adjustment procedure, such as the Similitude method (not presented in this book) or the Parametric adjustment computation model, as described later in this chapter.

In the case of a loop traverse, although it is geometrically well-defined, it is only referenced to an azimuthal orientation. Therefore, if there is an observation error in this orientation, all traverse stations will be rotated accordingly. Furthermore, if there is a systematic scaling error in the distances measured during the field survey, it will not be revealed by the linear error adjustment. In both cases, it is the same geometric figure but with the wrong orientation and scale.

To reduce the effect of orientation error, a common solution is to perform the referencing using two or more orientation azimuths, as shown in Fig. 12.16. Another helpful solution is to include a baseline of reference points in the traverse route, as shown in Fig. 12.17. In the latter case, the orientation of the traverse is defined by the traverse lines AB and BA.

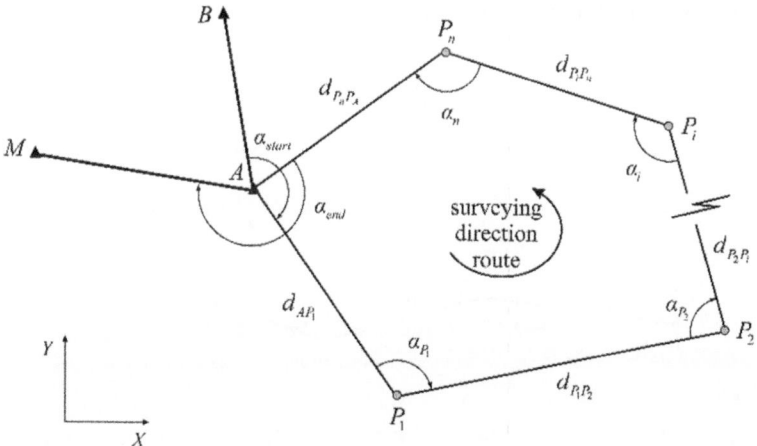

Fig. 12.16 Loop traverse referenced in two orientation azimuths

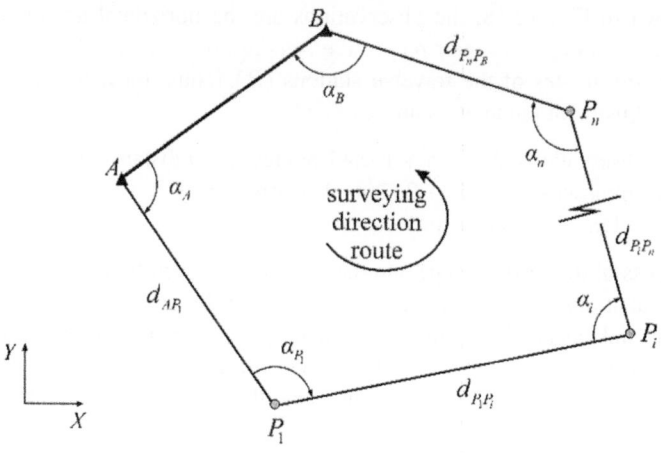

Fig. 12.17 Loop traverse in which the baseline of control points is a traverse line

12.2.5.3   Traverse Computation by Least Squares Adjustment

Although widely used, the traditional adjustment methods presented in the previous sections are considered non-rigorous, as they do not consider the quality of the observations in the adjustment procedure. Least-squares traverse adjustment is a better solution as angles and distances are adjusted simultaneously. In addition, observations can be weighted according to instrument accuracies and adjusted to comply with the geometrical conditions of the traverse route. As a result, the coordinates and the variance-covariance matrix of the traverse stations are computed together with the variance-covariance matrix of the quantities involved in the adjustment computation, allowing the complete data set to be verified.

A traverse can be adjusted using any of the three adjustment computation models presented in Chap. 3. However, the Parametric adjustment model is currently the most widely used and is, therefore, presented in this chapter. In these terms, consider the geometry of the link traverse shown in Fig. 12.18.

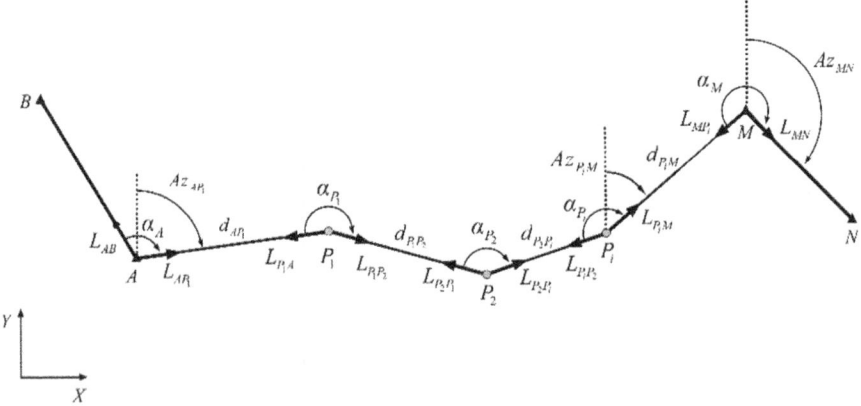

Fig. 12.18 Geometry of the link traverse

As shown in Fig. 12.18, the observations are the horizontal angles α_i and the distances $d_{i,i+1}$, along with their respective standard deviations. The unknowns are the (X, Y) coordinates of the traverse stations (P_i). Thus, for n traverse lines, the following adjustment parameters are revealed.

Number of observations: $N = 2n + 1$ ($n+1$ angles and n distances)
Number of unknowns: $u = 2n - 2$ ($2(n-1)$ coordinates)
Number of redundant observations: $r = N-u = 3$

Regardless of the traverse type and the total traverse length, there will always be three redundant observations. It is also worth noting that, for the sake of simplicity, the angular and linear observations shown in Fig. 12.18 are clockwise horizontal angles and horizontal distances calculated from the corresponding horizontal direction and slope distance observations. The azimuths Az_{AP1} and Az_{P3M} are calculated from the known coordinates of control points A, B, M, and N and the angles α_A and α_{P3}. The weighting strategy must then consider the appropriate error propagation.

An error equation must be written for each observable azimuth, angle and distance when adjusting a traverse using the parametric adjustment model. To do this, it is first necessary to determine the approximate coordinates of each free point (P_i), which can be done using the free traverse calculation strategy. The linearised equations are obtained from those developed earlier in Chap. 11.

For the azimuths, considering Fig. 12.18, the linearised forms of the error equations are developed as follows:

$$Az_{AP_1} = \text{atan}\left(\frac{X_{P_1} - X_A}{Y_{P_1} - Y_A}\right) + \alpha_{\text{quad1}} \tag{12.41}$$

$$Az_{P_3M} = \text{atan}\left(\frac{X_M - X_{P_3}}{Y_M - Y_{P_3}}\right) + \alpha_{\text{quad2}} \tag{12.42}$$

And then,

$$vAz_{AP_1} = \left(\frac{\partial Az_{AP_1}}{\partial X_{P_1}}\right)_0 \delta X_{P_1} + \left(\frac{\partial Az_{AP_1}}{\partial Y_{P_1}}\right)_0 \delta Y_{P_1} + 0\delta X_{P2} + 0\delta Y_{P2} + 0\delta X_{P3}$$
$$+ 0\delta Y_{P3} - \left(Az_{\text{Obs}_{AP1}} - Az_{AP_{1_0}}\right)$$

$$\tag{12.43}$$

$$vAz_{P3M} = 0\delta X_{P1} + 0\delta Y_{P1} + 0\delta X_{P2} + 0\delta Y_{P2} + \left(\frac{\partial Az_{P_3M}}{\partial X_{P_3}}\right)_0 \delta X_{P_3} + \left(\frac{\partial Az_{P_3M}}{\partial Y_{P_3}}\right)_0 \delta Y_{P_3}$$
$$- \left(Az_{\text{Obs}_{P3M}} - Az_{P_{3_0}M}\right)$$

$$\tag{12.44}$$

where

$$\left(\frac{\partial Az_{AP1}}{\partial X_{P1}}\right)_0 = \frac{Y_{P1_0} - Y_A}{d^2_{AP1_0}}$$
$$\left(\frac{\partial Az_{AP1}}{\partial Y_{P1}}\right)_0 = \frac{X_A - X_{P1_0}}{d^2_{AP1_0}} \tag{12.45}$$

$$\left(\frac{\partial Az_{P3M}}{\partial X_{P3}}\right)_0 = \frac{Y_{P3_0} - Y_M}{d^2_{P3_0M}}$$
$$\left(\frac{\partial Az_{P3M}}{\partial Y_{P3}}\right)_0 = \frac{XM - X_{P3_0}}{d^2_{P3_0M}} \tag{12.46}$$

For the distances, the linearised forms of the error equations are developed as follows:

$$d_{i,i+1} = \sqrt{(X_{i+1} - X_i)^2 + (Y_{i+1} - Y_i)^2} \qquad (12.47)$$

$$vd_{AP1} = \left(\frac{\partial d_{P1A}}{\partial X_{P1}}\right)_0 \delta X_{P1} + \left(\frac{\partial d_{P1A}}{\partial Y_{P1}}\right)_0 \delta Y_{P1} + 0\delta X_{P2} + 0\delta Y_{P2} + 0\delta X_{P3}$$

$$+ 0\delta Y_{P3} - \left(d_{\mathrm{Obs}_{AP_1}} - d_{AP1_0}\right) \qquad (12.48)$$

$$vd_{P1P2} = \left(\frac{\partial d_{P1P2}}{\partial X_{P1}}\right)_0 \delta X_{P1} + \left(\frac{\partial d_{P1P2}}{\partial Y_{P1}}\right)_0 \delta Y_{P1} + \left(\frac{\partial d_{P1P2}}{\partial X_{P2}}\right)_0 \delta X_{P2} + \left(\frac{\partial d_{P1P2}}{\partial Y_{P2}}\right)_0 \delta Y_{P} +$$

$$0\delta X_{P3} + 0\delta Y_{P3} - (d_{\mathrm{Obs}_{P1P2}} - d_{P1_0 P2_0}) \qquad (12.49)$$

$$vd_{P2P3} = 0\delta X_{P1} + 0\delta Y_{P1} + \left(\frac{\partial d_{P2P3}}{\partial X_{P2}}\right)_0 \delta X_{P2} + \left(\frac{\partial d_{P2P3}}{\partial Y_{P2}}\right)_0 \delta Y_{P2} + \left(\frac{\partial d_{P2P3}}{\partial X_{P3}}\right)_0 \delta X_{P3} +$$

$$\left(\frac{\partial d_{P2P3}}{\partial Y_{P3}}\right)_0 \delta Y_{P3} - \left(d_{\mathrm{Obs}_{P2P3}} - d_{P2_0 P3_0}\right) \qquad (12.50)$$

$$vd_{P3M} = 0\delta X_{P1} + 0\delta Y_{P1} + 0\delta X_{P2} + 0\delta Y_{P2} + \left(\frac{\partial d_{P3M}}{\partial X_{P3}}\right)_0 \delta X_{P3} + \left(\frac{\partial d_{P3M}}{\partial Y_{P3}}\right)_0 \delta Y_{P3} -$$

$$(d_{\mathrm{Obs}_{P3M}} - d_{P3_0 M}) \qquad (12.51)$$

where

$$\left(\frac{\partial d_{i,i+1}}{\partial X_{i+1}}\right)_0 = \frac{X_{i+1} - X_i}{d_{i,i+1}}$$

$$\left(\frac{\partial d_{i,i+1}}{\partial Y_{i+1}}\right)_0 = \frac{Y_{i+1} - Y_i}{d_{i,i+1}}$$

$$\left(\frac{\partial d_{i,i+1}}{\partial X_i}\right)_0 = \frac{X_i - X_{i+1}}{d_{i,i+1}} \qquad (12.52)$$

$$\left(\frac{\partial d_{i,i+1}}{\partial Y_i}\right)_0 = \frac{Y_i - Y_{i+1}}{d_{i,i+1}}$$

For the angles, the linearised forms of the error equations are developed as follows:

$$\alpha_i = \mathrm{atan}\left(\frac{X_{i+1} - X_i}{Y_{i+1} - Y_i}\right) + \alpha_{\mathrm{quad1}} - \mathrm{atan}\left(\frac{X_{i-1} - X_i}{Y_{i-1} - Y_i}\right) + \alpha_{\mathrm{quad2}} \qquad (12.53)$$

$$v\alpha_{P1} = \left(\frac{\partial \alpha_{P1}}{\partial X_{P1}}\right)_0 \delta X_{P1} + \left(\frac{\partial \alpha_{P1}}{\partial Y_{P1}}\right)_0 \delta Y_{P1} + \left(\frac{\partial \alpha_{P1}}{\partial X_{P2}}\right)_0 \delta X_{P2} + \left(\frac{\partial \alpha_{P1}}{\partial Y_{P2}}\right)_0 \delta Y_{P2} +$$
$$0\delta X_{P3} + 0\delta Y_{P3} - (\alpha_{\mathrm{Obs}_{P1}} - \alpha_{P1_0})$$

$$(12.54)$$

$$v\alpha_{P2} = \left(\frac{\partial \alpha_{P2}}{\partial X_{P1}}\right)_0 \delta X_{P1} + \left(\frac{\partial \alpha_{P2}}{\partial Y_{P1}}\right)_0 \delta Y_{P1} + \left(\frac{\partial \alpha_{P2}}{\partial X_{P2}}\right)_0 \delta X_{P2} + \left(\frac{\partial \alpha_{P2}}{\partial Y_{P2}}\right)_0 \delta Y_{P2} +$$
$$\left(\frac{\partial \alpha_{P2}}{\partial X_{P3}}\right)_0 \delta X_{P3} + \left(\frac{\partial \alpha_{P2}}{\partial Y_{P3}}\right)_0 \delta Y_{P3} - (\alpha_{\mathrm{Obs}_{P2}} - \alpha_{P2_0})$$

$$(12.55)$$

$$v\alpha_{P3} = 0\delta X_{P1} + 0\delta Y_{P1} + \left(\frac{\partial \alpha_{P3}}{\partial X_{P2}}\right)_0 \delta X_{P2} + \left(\frac{\partial \alpha_{P3}}{\partial Y_{P2}}\right)_0 \delta Y_{P2} + \left(\frac{\partial \alpha_{P3}}{\partial X_{P3}}\right)_0 \delta X_{P3} +$$
$$\left(\frac{\partial \alpha_{P3}}{\partial Y_{P3}}\right)_0 \delta Y_{P3} - (\alpha_{\mathrm{Obs}_{P3}} - \alpha_{P3_0})$$

$$(12.56)$$

where

$$\left(\frac{\partial \alpha_i}{\partial X_{i-1}}\right)_0 = \frac{Y_i - Y_{i-1}}{d_{i,i-1}^2}$$

$$\left(\frac{\partial \alpha_i}{\partial Y_{i-1}}\right)_0 = \frac{X_{i-1} - X_i}{d_{i,i-1}^2}$$

$$\left(\frac{\partial \alpha_i}{\partial X_i}\right)_0 = \frac{Y_i - Y_{i+1}}{d_{i,i-1}^2} - \frac{Y_i - Y_{i-1}}{d_{i,i+1}^2}$$

$$\left(\frac{\partial \alpha_i}{\partial Y_i}\right)_0 = \frac{X_i - X_{i-1}}{d_{i,i-1}^2} - \frac{X_i - X_{i+1}}{d_{i,+1}^2}$$

$$\left(\frac{\partial \alpha_i}{\partial X_{i+1}}\right)_0 = \frac{Y_{i+1} - Y_i}{d_{i,i+1}^2}$$

$$\left(\frac{\partial \alpha_i}{\partial Y_{i+1}}\right)_0 = \frac{X_i - X_{i+1}}{d_{i,i+1}^2}$$

$$(12.57)$$

Each observation produces an error equation whose unknown parameters are the coordinate variations δX_{Pi} and δY_{Pi}. For n observations, the error equations can be written in matrix form as $v = Ax - l_0$ (Eq. 3.79), where

$$A = \begin{bmatrix} \left(\dfrac{\partial Az_{AP_1}}{\partial X_{P_1}}\right)_0 & \left(\dfrac{\partial Az_{AP_1}}{\partial Y_{P_1}}\right)_0 & 0 & 0 & 0 & 0 \\[2mm] 0 & 0 & 0 & 0 & \left(\dfrac{\partial Az_{P3M}}{\partial X_{P_3}}\right)_0 & \left(\dfrac{\partial Az_{P3M}}{\partial Y_{P_3}}\right)_0 \\[2mm] \left(\dfrac{\partial d_{AP1}}{\partial X_{P1}}\right)_0 & \left(\dfrac{\partial d_{AP1}}{\partial Y_{P1}}\right)_0 & 0 & 0 & 0 & 0 \\[2mm] \left(\dfrac{\partial d_{P1P2}}{\partial X_{P1}}\right)_0 & \left(\dfrac{\partial d_{P1P2}}{\partial Y_{P1}}\right)_0 & \left(\dfrac{\partial d_{P1P2}}{\partial X_{P2}}\right)_0 & \left(\dfrac{\partial d_{P1P2}}{\partial Y_{P2}}\right)_0 & 0 & 0 \\[2mm] 0 & 0 & \left(\dfrac{\partial d_{P2P3}}{\partial X_{P2}}\right)_0 & \left(\dfrac{\partial d_{P2P3}}{\partial Y_{P2}}\right)_0 & \left(\dfrac{\partial d_{P2P3}}{\partial X_{P3}}\right)_0 & \left(\dfrac{\partial d_{P2P3}}{\partial Y_{P3}}\right)_0 \\[2mm] 0 & 0 & 0 & 0 & \left(\dfrac{\partial d_{P3M}}{\partial X_{P3}}\right)_0 & \left(\dfrac{\partial d_{P3M}}{\partial Y_{P3}}\right)_0 \\[2mm] \left(\dfrac{\partial \alpha_{P1}}{\partial X_{P1}}\right)_0 & \left(\dfrac{\partial \alpha_{P1}}{\partial Y_{P1}}\right)_0 & \left(\dfrac{\partial \alpha_{P1}}{\partial X_{P2}}\right)_0 & \left(\dfrac{\partial \alpha_{P1}}{\partial Y_{P2}}\right)_0 & 0 & 0 \\[2mm] \left(\dfrac{\partial \alpha_{P2}}{\partial X_{P1}}\right)_0 & \left(\dfrac{\partial \alpha_{P2}}{\partial Y_{P1}}\right)_0 & \left(\dfrac{\partial \alpha_{P2}}{\partial X_{P2}}\right)_0 & \left(\dfrac{\partial \alpha_{P2}}{\partial Y_{P2}}\right)_0 & \left(\dfrac{\partial \alpha_{P2}}{\partial X_{P3}}\right)_0 & \left(\dfrac{\partial \alpha_{P2}}{\partial Y_{P3}}\right)_0 \\[2mm] 0 & 0 & \left(\dfrac{\partial \alpha_{P3}}{\partial X_{P2}}\right) & \left(\dfrac{\partial \alpha_{P3}}{\partial Y_{P2}}\right)_0 & \left(\dfrac{\partial \alpha_{P3}}{\partial X_{P3}}\right)_0 & \left(\dfrac{\partial \alpha_{P3}}{\partial Y_{P3}}\right)_0 \end{bmatrix}$$

$$v = \begin{bmatrix} vAz_{AP1} \\ vAz_{P3M} \\ vd_{AP1} \\ vd_{P1P2} \\ vd_{P2P3} \\ vd_{P3M} \\ v\alpha_{P1} \\ v\alpha_{P2} \\ v\alpha_{P3} \end{bmatrix} \quad l_0 = \begin{bmatrix} Az_{Obs_{AP1}} - Az_{AP1_0} \\ Az_{Obs_{P3M}} - Az_{P3_0 M} \\ d_{Obs_{AP1}} - d_{AP1_0} \\ d_{Obs_{P1P2}} - d_{P1_0 P2_0} \\ d_{Obs_{P2P3}} - d_{P2_0 P3_0} \\ d_{Obs_{P3M}} - d_{P3_0 M} \\ \alpha_{Obs_{P1}} - \alpha_{P1_0} \\ \alpha_{Obs_{P2}} - \alpha_{P2_0} \\ \alpha_{Obs_{P3}} - \alpha_{P3_0} \end{bmatrix} \quad x = \begin{bmatrix} \delta X_{P1} \\ \delta Y_{P1} \\ \delta X_{P2} \\ \delta Y_{P2} \\ \delta X_{P3} \\ \delta Y_{P3} \end{bmatrix}$$

Regarding the weights, they are calculated inversely proportional to the standard deviation of the distances, azimuths, and angles (converted into radians), as given by Eq. (12.58). Generally, they are uncorrelated, and the matrix P is diagonal.

$$p_i = \frac{1}{s_i^2} \qquad (12.58)$$

where

s_i = standard deviation of the observation i

Adjustment computation can then be performed using Eqs. (3.84) to (3.87), from which the residual errors and the unknown parameters dX_{Pi} and dY_{Pi} are obtained, followed by the adjusted coordinates of the traverse stations (P_i). As there are three redundant observations, the standard deviations of the unit weight a posteriori can be calculated using Eq. (12.59).

$$s_0^2 = \frac{v^T P v}{3} \qquad (12.59)$$

Finally, the remaining variance-covariance matrices can be explored by considering the hypermatrix (3.91).

Example 12.5

Using the field measurements in Example 12.2, calculate the adjusted coordinates of the traverse stations using the Parametric adjustment model, assuming standard deviations of azimuths $Az_{1A} = Az_{4M} = \pm 10''$, standard deviation of the horizontal distances $d_{i,i+1} = \pm (2 \text{ mm} + 2 \text{ ppm})$ and standard deviations of the angles $s\alpha_i = \pm 4''$.

Solution:
For the solution of the problem, the following equations are used:

Azimuths:

$$Az_{A,1} = \text{atan}\left(\frac{X_1 - X_A}{Y_1 - Y_A}\right) + \alpha_{\text{quad1}} \qquad Az_{4,1} = \text{atan}\left(\frac{X_M - X_4}{Y_M - Y_4}\right) + \alpha_{\text{quad2}} \quad (12.60)$$

Horizontal clockwise angles:

For simplicity, the quadrant compatibility angles are not included in the equations, but they must of course be considered in the calculation.

$$\alpha_1 = Az_{1,2} - Az_{1,A} = \text{atan}\left(\frac{X_2 - X_1}{Y_2 - Y_1}\right) - \text{atan}\left(\frac{X_A - X_1}{Y_A - Y_1}\right)$$

$$\alpha_2 = Az_{2,3} - Az_{2,1} = \text{atan}\left(\frac{X_3 - X_2}{Y_3 - Y_2}\right) - \text{atan}\left(\frac{X_1 - X_2}{Y_1 - Y_2}\right)$$

$$\alpha_3 = Az_{3,4} - Az_{3,2} = \text{atan}\left(\frac{X_4 - X_3}{Y_4 - Y_3}\right) - \text{atan}\left(\frac{X_2 - X_3}{Y_2 - Y_3}\right) \qquad (12.61)$$

$$\alpha_4 = Az_{4,M} - Az_{4,3} = \text{atan}\left(\frac{X_M - X_4}{Y_M - Y_4}\right) - \text{atan}\left(\frac{X_3 - X_4}{Y_3 - Y_4}\right)$$

Distances:

$$d_{A,1} = \sqrt{(X_1 - X_A)^2 + (Y_1 - Y_A)^2}$$

$$d_{1,2} = \sqrt{(X_2 - X_1)^2 + (Y_2 - Y_1)^2}$$

$$d_{2,3} = \sqrt{(X_3 - X_2)^2 + (Y_3 - Y_2)^2} \qquad (12.62)$$

$$d_{3,4} = \sqrt{(X_4 - X_3)^2 + (Y_4 - Y_3)^2}$$

$$d_{4,M} = \sqrt{(X_M - X_4)^2 + (Y_M - Y_4)^2}$$

The coordinates of the control points (A) and (M) and the approximate coordinates of points (1 to 4) taken from Examples 12.2 are given in Table 12.14.

From Tables 12.5 and 12.14, and using Eq. 12.53, the observation and preliminary angles are given in Table 12.15

From the coordinates of control points (A) and (M) and observation angles in Table 12.5, the preliminary azimuths are given in Table 12.16.

From Tables 12.6 and 12.14, the observation and preliminary distances are given in Table 12.17.

Table 12.14 Control points and traverse free points preliminary coordinates

Station	X [m]	Y [m]
A	10,794.571	12,292.721
1	11,030.386	12,225.450
2	11,277.552	12,173.028
3	11,504.243	12,159.021
4	11,667.226	12,356.346
M	11,879.715	12,407.500

Table 12.15 Observation and preliminary angles

Station	Angle	
	Observation	Preliminary
1	176° 03′12″	176° 03′9.6″
2	171° 33′41″	171° 33′40.3″
3	126° 01′20″	126° 01′10.8″
4	216° 54′28″	216° 54′32.1″

Table 12.16 Observation and preliminary azimuths

Line	Azimuth	
	Observation	Preliminary
A-1	105° 55′7.2″	105° 55′18.7″
4-M	76° 27′38.6″	76° 27′51.6″

Table 12.17 Observation and preliminary distances

Line	Distance [m]	
	Observation	Preliminary
A-1	245.222	245.223
1-2	252.665	252.664
2-3	227.125	227.123
3-4	255.945	255.931
4-M	218.566	218.560

Using the given standard deviation of observations and using Eq. (12.58), the diagonal values of the weight matrix (with angles in radians) are given as follows:

$$P = \begin{bmatrix} 4.254517 * 10^8 & 4.254517 * 10^8 & 1.612302 * 10^5 & 1.593199 * 10^5 & 1.660208 * 10^5 & \ldots \end{bmatrix}$$
$$\begin{bmatrix} 1.584889 * 10^5 & 1.683612 * 10^5 & 2.659073 * 10^9 & 2.659073 * 10^9 & 2.659073 * 10^9 & 2.659073 * 10^9 \end{bmatrix}$$

For the application of the Parametric adjustment model, using Eq. (3.79), the above equations can be written in matrix form as $v = Ax - l_0$ (Eq. 3.79), where

$$A = \begin{bmatrix}
\frac{\partial Az_{A1}}{\partial X_1} & \frac{\partial Az_{A1}}{\partial Y_1} & 0 & 0 & 0 & 0 & 0 & 0 \\[2mm]
0 & 0 & 0 & 0 & 0 & 0 & \frac{\partial Az_{4M}}{\partial X_4} & \frac{\partial Az_{4M}}{\partial Y_4} \\[2mm]
\frac{\partial d_{A1}}{\partial X_1} & \frac{\partial d_{A1}}{\partial Y_1} & 0 & 0 & 0 & 0 & 0 & 0 \\[2mm]
\frac{\partial d_{12}}{\partial X_1} & \frac{\partial d_{12}}{\partial Y_1} & \frac{\partial d_{12}}{\partial X_2} & \frac{\partial d_{12}}{\partial Y_2} & 0 & 0 & 0 & 0 \\[2mm]
0 & 0 & \frac{\partial d_{23}}{\partial X_2} & \frac{\partial d_{23}}{\partial Y_2} & \frac{\partial d_{23}}{\partial X_3} & \frac{\partial d_{23}}{\partial Y_3} & 0 & 0 \\[2mm]
0 & 0 & 0 & 0 & \frac{\partial d_{34}}{\partial X_3} & \frac{\partial d_{34}}{\partial Y_3} & \frac{\partial d_{34}}{\partial X_4} & \frac{\partial d_{34}}{\partial Y_4} \\[2mm]
0 & 0 & 0 & 0 & 0 & 0 & \frac{\partial d_{4M}}{\partial X_4} & \frac{\partial d_{4M}}{\partial Y_4} \\[2mm]
\frac{\partial \alpha_1}{\partial X_1} & \frac{\partial \alpha_1}{\partial Y_1} & \frac{\partial \alpha_1}{\partial X_2} & \frac{\partial \alpha_1}{\partial Y_2} & 0 & 0 & 0 & 0 \\[2mm]
\frac{\partial \alpha_2}{\partial X_1} & \frac{\partial \alpha_2}{\partial Y_1} & \frac{\partial \alpha_2}{\partial X_2} & \frac{\partial \alpha_2}{\partial Y_2} & \frac{\partial \alpha_2}{\partial X_3} & \frac{\partial \alpha_2}{\partial Y_3} & 0 & 0 \\[2mm]
0 & 0 & \frac{\partial \alpha_3}{\partial X_2} & \frac{\partial \alpha_3}{\partial Y_2} & \frac{\partial \alpha_3}{\partial X_3} & \frac{\partial \alpha_3}{\partial Y_3} & \frac{\partial \alpha_3}{\partial X_4} & \frac{\partial \alpha_3}{\partial Y_4} \\[2mm]
0 & 0 & 0 & 0 & \frac{\partial \alpha_4}{\partial X_3} & \frac{\partial \alpha_4}{\partial Y_3} & \frac{\partial \alpha_4}{\partial X_4} & \frac{\partial \alpha_4}{\partial Y_4}
\end{bmatrix}$$

Numerically,

$$
A =
\begin{bmatrix}
-1.118683*10^{-3} & -3.921485*10^{-3} & 0 & 0 \\
0 & 0 & 0 & 0 \\
9.616367*10^{-1} & 2.743263*10^{-1} & 0 & 0 \\
-9.782399*10^{-1} & 2.074771*10^{-1} & 9.782399*10^{-1} & -2.074771*10^{-1} \\
0 & 0 & -9.980965*10^{-1} & 6.167134*10^{-2} \\
0 & 0 & 0 & 0 \\
0 & 0 & 0 & 0 \\
1.939841^{*}*10^{-3} & 7.793188*10^{-3} & -8.211582*10^{-4} & -3.871703*10^{-3} \\
-8.211582*10^{-3} & -3.871703*10^{-3} & 1.092691*10^{-3} & 8.266216*10^{-3} \\
0 & 0 & -2.715324*10^{-4} & -4.394513*10^{-3}
\end{bmatrix}
$$

$$
\times
\begin{bmatrix}
0 & 0 & 0 & 0 \\
0 & 0 & 1.070877*10^{-3} & 4.448326*10^{-3} \\
0 & 0 & 0 & 0 \\
0 & 0 & 0 & 0 \\
9.980965*10^{-1} & -6.167134*10^{-2} & 0 & 0 \\
-6.368243*10^{-1} & -7.710090*10^{-1} & 6.368243*10^{-1} & 7.710090*10^{-1} \\
0 & 0 & -9.722244*10^{-1} & -2.340506*10^{-1} \\
0 & 0 & 0 & 0 \\
-2.715324*10^{-4} & -4.394513*10^{-3} & 0 & 0 \\
-2.741035*10^{-3} & 6.882780*10^{-3} & 3.012567*10^{-3} & -2.488267*10^{-3} \\
3.012567*10^{-3} & -2.488267*10^{-3} & -4.083445*10^{-3} & 6.936593*10^{-3}
\end{bmatrix}
$$

$$l_0 = \begin{bmatrix} -5.59257249 * 10^{-5} \\ -6.30922758 * 10^{-5} \\ -5.55377763 * 10^{-4} \\ 9.98527690 * 10^{-4} \\ 1.67330281 * 10^{-3} \\ 1.41227812 * 10^{-2} \\ 6.3815042 * 10^{-3} \\ 1.14893445 * 10^{-5} \\ 3.20576418 * 10^{-6} \\ 4.45558984 * 10^{-5} \\ -1.98754447 * 10^{-5} \end{bmatrix} \qquad x = \begin{bmatrix} \delta X_1 \\ \delta Y_1 \\ \delta X_2 \\ \delta Y_2 \\ \delta X_3 \\ \delta Y_3 \\ \delta X_4 \\ \delta Y_4 \end{bmatrix}$$

Considering the weight matrix given above and using Eqs. (3.84) to (3.87).

$$N = \begin{bmatrix} 3.138901 * 10^5 & -2.434983 * 10^4 & -1.590833 * 10^5 & -5.684429 * 10^3 \\ -2.434983 * 10^4 & 2.268895 * 10^5 & 4,069972 * 10^3 & -1.721920 * 10^5 \\ -1.590833 * 10^5 & 4.069972 * 10^3 & 3.230150 * 10^5 & -6.910447 * 10^3 \\ -5.684429 * 10^3 & -1.721920 * 10^5 & -6.910447 * 10^3 & 2.803961 * 10^5 \\ 5.928964 * 10^2 & 2.795464 * 10^3 & -1.641992 * 10^5 & 3.628073 * 10^4 \\ 9.595507 * 10^3 & 4.524213 * 10^4 & -7.518756 * 10^3 & -1.776525 * 10^5 \\ 0 & 0 & -2.175147 * 10^3 & -3.520285 * 10^4 \\ 0 & 0 & 1.796590 * 10^3 & 2.907622 * 10^4 \end{bmatrix}$$

$$\times \begin{bmatrix} 5.928964 * 10^2 & 9.595507 * 10^3 & 0 & 0 \\ 2.795464 * 10^3 & 4.524213 * 10^4 & 0 & 0 \\ -1.641992 * 10^5 & -7.518756 * 10^3 & -2.175147 * 10^3 & 1.796590 * 10^3 \\ 3.628073 * 10^4 & -1.776525 * 10^5 & -3.520285 * 10^4 & 2.907622 * 10^4 \\ 2.739707 * 10^5 & 6.727948 * 10^2 & -1.189428 * 10^5 & -4.115063 * 10^3 \\ 6.727948 * 10^2 & 2.886282 * 10^5 & 4.335896 * 10^3 & -1.856501 * 10^5 \\ -1.189428 * 10^5 & 4.335896 * 10^3 & 2.923720 * 10^5 & 1.885008 * 10^4 \\ -4.115063 * 10^3 & -1.856501 * 10^5 & 1.885008 * 10^4 & 2.562643 * 10^5 \end{bmatrix}$$

$$
n = \begin{bmatrix} -92.258880 \\ 339.359466 \\ -294.189375 \\ -569.329213 \\ -1,525.672307 \\ -770.624938 \\ 938.040708 \\ 650.884726 \end{bmatrix} \quad x = \begin{bmatrix} -0.00537 \\ -0.0056 \\ -0.0090 \\ -0.0125 \\ -0.0105 \\ -0.0132 \\ -0.0019 \\ -0.0054 \end{bmatrix}_m
$$

The adjusted coordinates of points (1), (2), (3) and (4) are calculated using Eq. (3.80). The results are given in Table 12.18.

Table 12.18   Adjusted coordinates

Points	Preliminary coordinates [m]	Correction [m]	Adjusted coordinates [m]
X_1	11,030.386	−0.0053	11,030.381
Y_1	12,225.450	−0.0056	12,225.444
X_2	11,277.552	−0.0090	11,277.543
Y_2	12,173.028	−0.0125	12,173.015
X_3	11,504.243	−0.0105	11,504.233
Y_3	12,159.021	−0.0132	12,159.008
X_4	11,667.226	−0.0019	11,667.224
Y_4	12,356.346	−0.0054	12,356.341

The residual error vector can be calculated using Eq. (3.79).

$$
v^T = \begin{bmatrix} 17.3'' & 8.5'' & -3.0\,mm & -3.1\,mm & -3.2\,mm & -2.7\,mm & -3.2\,mm & -1.9'' & -6.1'' & -8.6'' & -1.7'' \end{bmatrix}
$$

The standard deviations of the unit weight a posteriori can be calculated using Eq. (12.59).

$$
s_0^2 = 66.21479329
$$

Finally, using Eq. (11.74) and the value of s_0^2 calculated above, the variance-covariance matrix for the parameters is given as follows:

$$\Sigma_{xx} = \begin{bmatrix}
3.33734 * 10^{-5} & 5.71887 * 10^{-6} & 2.58709 * 10^{-5} & 3.86785 * 10^{-6} \\
5.71887 * 10^{-6} & 8.82009 * 10^{-5} & -1.37280 * 10^{-6} & 9.82521 * 10^{-5} \\
2.58709 * 10^{-5} & -1.37280 * 10^{-6} & 5.12442 * 10^{-5} & -5.10368 * 10^{-6} \\
3.86785 * 10^{-6} & 9.82521 * 10^{-5} & -5.10368 * 10^{-6} & 1.64420 * 10^{-4} \\
1.84001 * 10^{-5} & -1.31168 * 10^{-5} & 3.80759 * 10^{-5} & -2.37101 * 10^{-5} \\
5.38570 * 10^{-7} & 7.30000 * 10^{-5} & -5.57868 * 10^{-6} & 1.37056 * 10^{-4} \\
8.17027 * 10^{-6} & 2.73604 * 10^{-6} & 1.56206 * 10^{-5} & 2.91794 * 10^{-6} \\
-5.35574 * 10^{-7} & 4.13346 * 10^{-5} & -4.35923 * 10^{-6} & 8.00750 * 10^{-5} \\
1.84001 * 10^{-5} & 5.38570 * 10^{-7} & 8.17027 * 10^{-6} & -5.35575 * 10^{-7} \\
-1.31168 * 10^{-5} & 7.30000 * 10^{-5} & 2.73604 * 10^{-6} & 4.13346 * 10^{-5} \\
3.80759 * 10^{-5} & -5.57868 * 10^{-6} & 1.56206 * 10^{-5} & -4.35923 * 10^{-6} \\
-2.37101 * 10^{-5} & 1.37056 * 10^{-4} & 2.91790 * 10^{-6} & 8.00750 * 10^{-5} \\
5.82892 * 10^{-5} & -2.15842 * 10^{-5} & 2.23594 * 10^{-5} & -1.39220 * 10^{-5} \\
-2.15842 * 10^{-5} & 1.57667 * 10^{-4} & -1.00499 * 10^{-6} & 9.84373 * 10^{-5} \\
2.23594 * 10^{-5} & -1.00450 * 10^{-6} & 3.10345 * 10^{-5} & -3.09242 * 10^{-6} \\
-1.39220 * 10^{-5} & 9.84373 * 10^{-5} & -3.09242 * 10^{-6} & 8.65132 * 10^{-5}
\end{bmatrix}$$

Using the diagonal values in the variance-covariance matrix for the parameters, the standard deviations of the adjusted coordinates are given as follows:

$$X_1 = 11{,}030.381 \text{ m} \pm 5.8 \text{ mm} \quad Y_1 = 12{,}225.444 \text{ m} \pm 9.4 \text{ mm}$$

$$X_2 = 11{,}277.543 \text{ m} \pm 7.2 \text{ mm} \quad Y_2 = 12{,}173.015 \text{ m} \pm 12.8 \text{ mm}$$

$$X_3 = 11{,}504.233 \text{ m} \pm 7.6 \text{ mm} \quad Y_3 = 12{,}159.008 \text{ m} \pm 12.6 \text{ mm}$$

$$X_4 = 11{,}667.224 \text{ m} \pm 5.6 \text{ mm} \quad Y_4 = 12{,}356.341 \text{ m} \pm 9.3 \text{ mm}$$

The absolute and relative error ellipses for the traverse stations are shown in Fig. 12.19.

Using Eq. (3.73) and the matrix of variance-covariance for the adjusted observations calculated using Eq. (3.94) and the value of s_0^2 calculated above, the following results are obtained for the adjusted observations.

$Az_{A1} = 105°55'24.48'' \pm 7.8''$ $Az_{4M} = 76°27'47.09'' \pm 8.7''$ $d_{A,1} = 245.219\text{m} \pm 5.9\text{mm}$

$d_{1,2} = 252.662\text{m} \pm 5.9\text{mm}$ $d_{2,3} = 227.122\text{m} \pm 5.8\text{mm}$ $d_{3,4} = 255.942\text{m} \pm 6.1\text{mm}$

$d_{4,M} = 218.563\text{m} \pm 5.7\text{mm}$ $\alpha_1 = 176°03'10.05'' \pm 9.4''$ $\alpha_2 = 171°33'34.86'' \pm 8.9''$

$\alpha_3 = 126°01'11.43'' \pm 8.3''$ $\alpha_4 = 216°54'26.26'' \pm 9.5''$

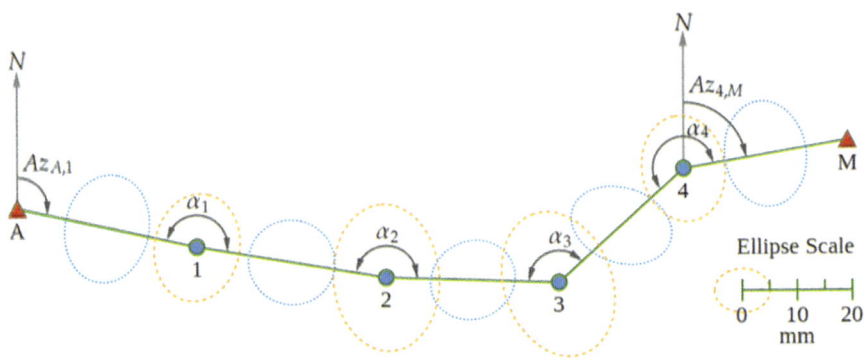

Fig. 12.19 Absolute and relative error ellipses

12.3 Sources of Error in Traversing

The most common sources of error in traversing are related to the use of the surveying instrument, as described in Chap. 10. Among them, the centring error is the most relevant. It is, therefore, advisable to avoid short traverse lines whenever possible. However, if short traverse lines cannot be avoided or the project requires high-accuracy traversing measurements, it is recommended to use the forced centring technique, as described in *Sect. 9.2.5.*

The forced centring technique used for traversing measurements uses a total station, three tripods, three tribraches and two reflector prisms with a prism support assembly, as shown in Fig. 12.20.

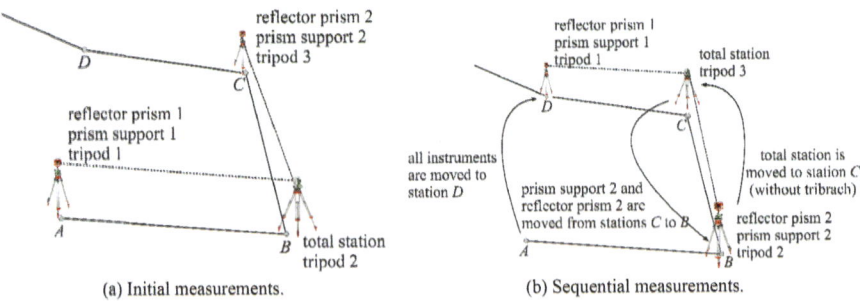

(a) Initial measurements.　　　　　　　(b) Sequential measurements.

Fig. 12.20 Sequence of field procedures for traversing with forced centring

As shown in Fig. 12.20, the measurement procedure starts by setting up the tripods assembled with tribraches at traverse stations *A*, *B* and *C*, with the tribrach levelled and centred over each station. The reflector prism mounted on the prism support is then fixed at stations A and C, while the total station is installed over station B. Once the horizontal directions and slope distances have been measured to points A and C, the tripod, tribrach and prism assembly are moved from station A to station D, the total station without tribrach is moved to station C, and the prism support and reflector prism are moved from station C to station B. The procedure is repeated until reaching the last traverse station. The same process can be performed using four tripods, which can be faster than three.

12.4   Survey Standards Used for Traversing Specification

In Geomatics, survey standards are defined as the minimum accuracy required to meet specific requirements for indicating the positional accuracy of spatial data. These are the operational and accuracy standards for data collection, defined by international criteria, such as the *ISO 19157:2013—Geographic Information—Data quality* standard, which establishes the principles for describing geospatial data quality.

Two survey accuracies are defined by the ISO 19157:2013 standard, commonly used by individual standardisation agencies: *absolute* and *relative* accuracies, defined by some authors as *positional* and *relative closure* ratios (*proportional accuracy*), respectively. Regarding the traverse accuracy, positional accuracy is recommended when the traverse computation is performed by Least Squares adjustment and proportional accuracy is recommended for angular and linear misclosure balancing.

The standard for reporting position accuracy is defined in terms of horizontal and vertical components. As this section deals with 2D traversing, only the horizontal component is of interest.

12.4.1   Horizontal Positional Standard

The standard for the horizontal positional accuracy value represents the uncertainty due to random errors in positioning a traverse station at the 95% confidence level. The statistic to represent it depends on whether it is considered an absolute or relative positioning. Absolute positioning is based on the radius of the 95% confidence standard error ellipse[1] resulting from the least-squares adjustment. Relative positioning is based on the relative error ellipse and indicates the precision of a traverse

---

[1] To convert the standard error ellipse into a 95% confidence error ellipse, the computed values of the ellipse semiaxes (a) and (b) must be multiplied by the appropriate expansion factor, as shown in Table 3.5.

station relative to the adjacent traverse point at the 95% confidence interval. Based on these assumptions, allowable linear unit values are defined by regulatory authorities or specified for individual projects.

The accuracy standard for horizontal control points is sometimes expressed as a distance accuracy ratio of 1:r, as given by Eq. (12.63).

$$r = \frac{d}{s_d} \tag{12.63}$$

where s_d is the propagated standard deviation of the distance between adjacent traverse stations obtained from the least squares adjustment; and d is the distance between the traverse stations considered. In this context, Table 12.19 shows an example of the accuracy standards of horizontal control stations adopted by the Geodetic Survey Section of the Survey and Mapping Office Land Department of Hong Kong.

Table 12.19 Accuracy standards of horizontal control

		Assessment criteria for Least Square Adjustment	
Class	Class description	Allowable residual of distance measurement	Allowable residual of angular measurement
H4.1	Minor control transverse (class 4.1)	1:15,000 or 5 mm (minimum)	$10''$
H4.2	Minor control transverse (class 4.2)	1:15,000 or 5 mm (minimum)	$10''$
	Note:		
	The origin of Class 4.2 is Class 4.1 station		
H5	Traverse (Class 5)	1:10,000 or 10 mm (minimum)	$20''$
H6	Traverse (Class 6)	1:7500 or 10 mm (minimum)	$30''$

12.4.2 Relative Closure Ratio Accuracy

In Geomatics applied to Civil Engineering, the proportional accuracy for traverse surveys is generally specified and classified considering angular and linear misclosures, as explained earlier in the chapter. To be accepted and distributed, they should be smaller than the allowable values specified in technical standards or the specifications for contracting services. These are usually established according to the type of fieldwork and instrument used in the project, as shown in the following sections.

12.4.2.1   Allowable Angular Misclosure

According to the general law of propagation of variances-covariances, if all the horizontal directions are observed with the same instrument and method of measurement, the angle accuracy is considered to have the same standard deviation s_α. In this case, the sum w of the n angles α_i, as given by Eq. (12.64), will have the standard deviation s_w, as given by Eq. (12.65).

$$w = \sum \alpha_i = \alpha_1 + \alpha_2 + \ldots.. + \alpha_n \tag{12.64}$$

Assuming $s_{\alpha 1} = s_{\alpha 2} = \ldots\ldots = s_{\alpha_n} = s_\alpha$:

$$s_w^2 = n * s_\alpha^2 \quad \rightarrow \quad s_w = s_\alpha * \sqrt{n} \tag{12.65}$$

Therefore, s_w represents the maximum angular misclosure expected for the traverse. However, due to the uncertainties of the discrepancies between the measured values, it is not an absolute value but a value comprised in a confidence interval statistically determined by a confidence level. For example, assuming a 95% confidence level, there is a 95% probability that the maximum angular misclosure is within the confidence interval given by Eq. (12.66).

$$P = \left[(-1.96 * s_\alpha * \sqrt{n}) < s_w < (+1.96 * s_\alpha * \sqrt{n}) \right] = 0.95 \tag{12.66}$$

For example, considering a ten-angle link traverse surveyed with a total station with an angular precision of $\pm 5''$, the maximum angular misclosure confidence interval is calculated as shown below.

$$S_w = 5'' * \sqrt{10} = 15.8''$$
$$P = (-31'' < s_w < +31'') = 0.95.$$

Thus, the allowable value T_α for the angular misclosure, at a 95% confidence level, is given by Eq. (12.67):

$$T_\alpha = \pm 1.96 s_\alpha * \sqrt{n} \tag{12.67}$$

However, technical standards (agency, state and local surveying standards) classify the allowable angular misclosure according to the type of surveying work, usually specified by orders of horizontal accuracy standards. In this case, the allowable value T_α is based on a multiplication factor k, as given by Eq. (12.68).

$$T_\alpha = k * s_\alpha * \sqrt{n} \tag{12.68}$$

For example, Table 12.20 shows the angular closure standards for Engineering and Construction Control Surveys proposed by the *Federal Geographic Data Committee* (FGDC), USA.

Table 12.20 Minimum angular closure standards for Engineering and Construction Control Surveys—FGDC

Classification order Engr and Const. control	Closure standard Angle (secs)
Second-order, Class I	$3\sqrt{n}$
Second-order, Class II	$5\sqrt{n}$
Third-order, Class I	$10\sqrt{n}$
Third-order, Class II	$20\sqrt{n}$
Construction (fourth-order)	$60\sqrt{n}$

12.4.2.2   Allowable Linear Misclosure

The most common way of expressing the linear precision of a survey traverse is by relative precision, as given by Eq. (12.35). In this case, the allowable value (the minimum closure accuracy) is given as the ratio 1:M, which is also specified by orders of horizontal accuracy standards. For example, Table 12.21 shows the linear closure standards for Engineering and Construction Control Surveys proposed by the *Federal Geographic Data Committee* (FGDC), USA. Readers should consult Appendix A.1 of the FGDC Geospatial Positioning Accuracy Standards, Part 4, for a more detailed approach.

Table 12.21 Minimum angular closure standards for Engineering and Construction Control Surveys—FGDC

Classification order Engr and Const. control	Closure standard Distance (ratio)
Second-order, Class I	1:50,000
Second-order, Class II	1:20,000
Third-order, Class I	1:10,000
Third-order, Class II	1:5000
Construction (fourth-order)	1:2500

12.5   Detecting Blunders in a Survey Traverse

Suppose that the angular or linear errors of closure exceed the tolerances (allowable values). In this case, the engineer can use geometric analysis to find out which traverse station or line could possibly contain a blunder, provided that only one (angular or linear) has occurred and, in this case, that it has occurred on only one traverse station or line.

This section presents one method for detecting blunders in angular closure and another for detecting blunders in linear closure.

12.5.1 Angular Closure Blunder Detection

For the case of a link traverse, the geometric procedure for detecting the angular closure blunder is to draw the traverse starting from opposite sides, i.e., from (A) to (M) and then from (M) to (A), as shown in Fig. 12.21. When carrying out this procedure, there must be a traverse station at the intersection of the traverses drawn in opposite directions. This must be the traverse station affected by the blunder.

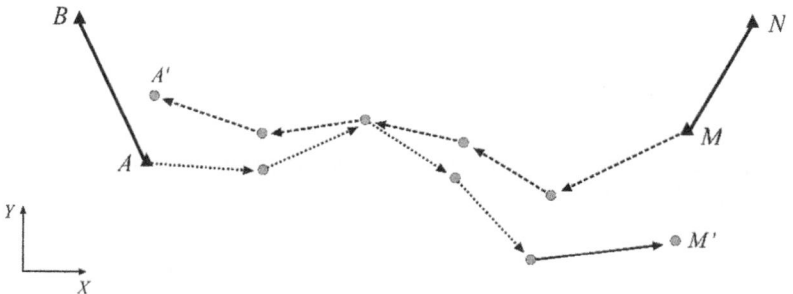

Fig. 12.21 Survey traverses drawn in opposite directions

There is also a mathematical model known as the *Broennimann equation*, which makes it possible to calculate the coordinates of the traverse station affected by the angular blunder, as given by Eqs. (12.69) and (12.70).

$$X = \frac{X_C + X'_C}{2} - \frac{Y_C - Y'_C}{2} * \cot\left(\frac{e_a}{2}\right) \qquad (12.69)$$

$$Y = \frac{Y_C + Y'_C}{2} + \frac{X_C - X'_C}{2} * \cot\left(\frac{e_a}{2}\right) \qquad (12.70)$$

where

$X, Y =$ coordinates of the traverse station affected by the angular measurement blunder

$X_C, Y_C =$ coordinates of the final traverse station

$X'_C, Y'_C =$ preliminary coordinates calculated for the final traverse station

$e_a =$ angular misclosure

Example 12.6
Referring to the data from Table 12.2, consider that there was an error in writing the horizontal direction observation around traverse station 2 in the field book. Instead of the horizontal direction $293° 19' 15''$ for the traverse line 2–3, the value $295° 19' 15''$ was written. Based on the new value of the observed horizontal direction, apply the angular closure blunder detection concept presented in the previous section.

Solution:
When calculating the angular closure of the traverse with the new value of the direction of line 2-3, it is verified that there was an angular error of closure equal 2° 00′ 09.6″. As this error exceeds the allowable angular error for this type of survey traverse, the angular closure blunder detection method can be applied. To do this, the coordinates of the final traverse station must be calculated without using angular or linear corrections, thus obtaining the values shown in Table 12.22.

Table 12.22 Calculated and original coordinates of the traverse station (M)

Traverse station	X [m]	Y [m]
M′	11,887.556	12,386.390
M	11,879.715	12,407.500

Using Eqs. (12.69) and (12.70), the coordinates of the traverse station where the blunder probably occurred are obtained.

$$X = 11,883.635 - 10.555 * \cot g \left(\frac{2°\,00′09.6″}{2} \right) = 11,279.741\,m$$

$$Y = 12,396.945 - 3.920 * \cot g \left(\frac{2°\,00′09.6″}{2} \right) = 12,172.642\,m$$

Based on the results, it is concluded that traverse station 2 is the probable station where the blunder occurred.

12.5.2   Linear Closure Blunder Detection

In the case of detecting linear closure blunders, the procedure is based on calculating the azimuth of the vector resulting from the linear misclosure e_p. It is considered that the traverse line with the length measurement affected by the blunder is the one whose direction is parallel to the direction of the linear misclosure vector.

$$Az_{ep} = \operatorname{atan}\left(\frac{e_{px}}{e_{py}} \right) + \alpha_{quad} \tag{12.71}$$

Once the direction of the linear misclosure vector is known, it is possible to search for the traverse line with a direction parallel to it.

Example 12.7
Still referring to the data from Table 12.2, now consider that no blunder was made in the angular measurements, but in measuring the distance between the traverse stations 2 and 3. Instead of 227.159 m and 227.156 m, the values 237.159 m and 237.156 m were recorded in the field book, respectively. Based on the new value of the measured slope distance, apply the linear closure blunder detection concept presented in the previous section.

Solution: As there was no blunder in the angular measurements, the angular error of closure is the same as that obtained in the solution of Example 12.2. However, since there was an error in reading or writing the distance between the traverse stations 2 and 3 when calculating the linear error of closure, the values $\Delta X = 10.004\,m$ *and* $\Delta Y = -0.568\,m$ *are obtained, i.e., a relative precision of 1:121, which is much lower than the allowable precision for this type of traverse. Therefore, using Eq. (12.71).*

$$Az_{ep} = \operatorname{atan}\left(\frac{9.997}{-0.543}\right) + 180° = 93°06'37.6''$$

The value obtained for the azimuth of the linear misclosure vector is very close to the azimuth of the 2-3 traverse line, which shows that the blunder in the distance measurement must have occurred when measuring or writing down the length of this line.

12.6   Problems

1. Using the values in Fig. 12.22, complete the missing values in Table 12.23.

Fig. 12.22 Geometry of the traverse

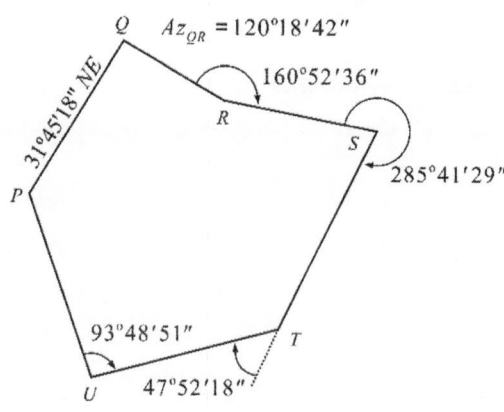

Table 12.23 Azimuth and horizontal angles

Line	Azimuth	Horizontal internal angle	Horizontal external angle	Deflection angle
P-Q				
Q-R	120° 18′42″			
R-S			160° 52′36″	
S-T			285° 41′29″	
T-U				47° 52′188″
U-P		93° 48′51″		

2. Using the values calculated in the previous problem, check the results by calculating the angular error of closure of the traverse as a function of the interior and exterior angles.
3. Using the values for the interior angles of the loop traverse given in Table 12.24, calculate the azimuth of each traverse line and check that the start and end azimuths are the same. Assume that the azimuth of line 29–23 is 41° 10′35″.

Table 12.24 Interior angles of the loop traverse

Traverse station	Internal angle (α_i)
23	261° 31′26″
24	98° 03′59″
25	93° 02′06″
26	121° 44′12″
27	97° 10′07″
28	196° 04′10″
29	32° 24′00″

4. A loop traverse was run, as shown in the traverse sketch of Fig. 12.23. The coordinates of the control points are given in Table 12.25, and the field measurements are given in Table 12.26. Calculate the angular error of closure of the traverse and classify the error of closure according to the closure standard given in Table 12.21.

Fig. 12.23 Traverse sketch

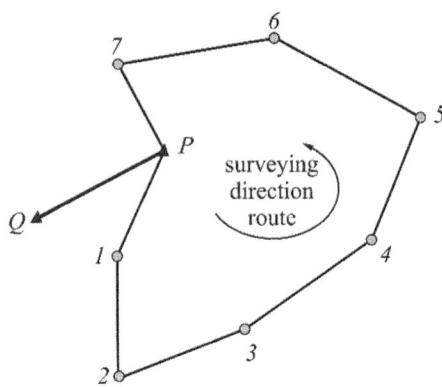

Table 12.25 Coordinates of the control points

Point	X [m]	Y [m]
P	14,550.798	11,159.290
Q	13,937.796	10,844.072

Table 12.26 Field measurements

Station	Sighting point	Horizontal direction
P	Q	12° 44′56″
	1	335° 16′01″
1	P	322° 34′56″
	2	117° 16′40″
2	1	33° 51′02″
	3	103° 07′47″
3	2	192° 11′00″
	4	358° 14′04″
4	3	44° 54′32″
	5	192° 13′05″
5	4	65° 22′43″
	6	160° 25′02″
4	5	21° 54′32″
	7	165° 13′10″
7	6	0° 00′54″
	P	68° 54′53″
P	7	261° 00′10″
	Q	173° 54′09″

5. Using the results of problem 4, calculate the adjusted angles in the traverse using one of the angle adjustment methods presented in *Sect. 12.2.5.1.1*.
6. Using the adjusted angles from problem 5 and the horizontal distances given in Table 12.27, calculate the linear error of closure and the relative precision of the traverse. Classify the error of closure according to the closure standard given in Table 12.20.

Table 12.27 Horizontal distances

Station	Sighting point	Horizontal distance [m]
P	1	538.271
1	2	551.699
2	3	633.559
3	4	727.497
4	5	597.854
5	6	779.570
6	7	740.085

7. Using the results of problem 6, perform the linear adjustment of the traverse and calculate the adjusted coordinate of each traverse station.
8. A link traverse was run between points P1 and P3, as shown in the traverse sketch in Fig. 12.24. The coordinates of the control points are given in Table 12.28, and the field measurements are given in Table 12.29. Calculate the adjusted coordinates of each traverse station using the angle and distance balancing method.

Fig. 12.24 Traverse sketch

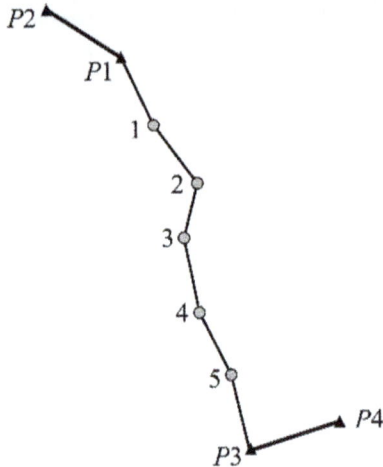

Table 12.28 Control point coordinates

Ponto	X [m]	Y [m]
P1	18,225.268	12,651.965
P2	17,496,.76	13,107.972
P3	19,568.152	8661.903
P4	20,486.246	8943.941

Table 12.29 Field measurements

Station	Sight point	Horizontal clockwise angle	Horizontal distance [m]
P1	1	211 ° 58′40″	741.402
1	2	169 ° 05′07″	789.695
2	3	230 ° 15′12″	531.784
3	4	155 ° 11′23″	751.587
4	5	165 ° 47′22″	729.665
5	P3	191 ° 09′34″	806.148
P3	P4	87 ° 5′33″	

9. An independent loop traverse was run, as shown in the traverse sketch in Fig. 12.25. The field measurements are given in Table 12.30. Assuming that the coordinates of point (1) are $X = 5000.000$ m and $Y = 10,000.000$ m, calculate the adjusted coordinates of each traverse station using the angle and distance balancing method.

Fig. 12.25 Traverse sketch

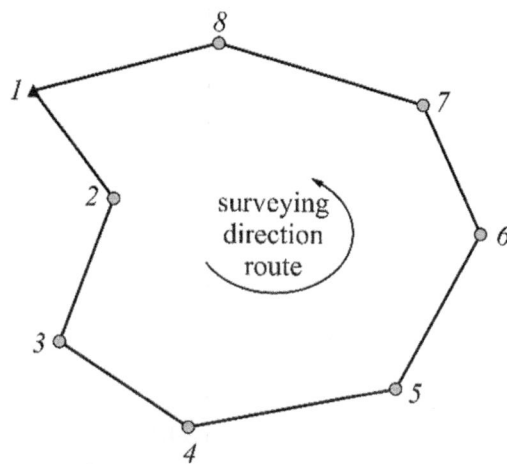

Table 12.30 Field measurements

Station	Sighting point	Horizontal clockwise angle	Horizontal distance [m]
1	2	67 ° 17′38″	565.381
2	3	237 ° 59′45″	650.352
3	4	101 ° 13′23″	661.205
4	5	136 ° 40′14″	905.311
5	6	130 ° 02′04″	729.936
6	7	126 ° 38′14″	597.415
7	8	131 ° 01′04″	901.158
8	1	149 ° 07′22″	822.826

10. A link traverse was run between points A and B, as shown in the traverse sketch in Fig. 12.26. The coordinates of the control points are given in Table 12.31, and the field measurements are given in Table 12.32. Calculate the adjusted coordinates of the traverse stations using the Parametric adjustment model, assuming standard deviations of azimuths $Az_{1-A} = Az_{3-C} = \pm 10''$, standard deviation of the horizontal distances $d_{i,i+1} = \pm (2 \text{ mm} + 2 \text{ ppm})$ and standard deviations of the angles $s\alpha_i = \pm 4''$.

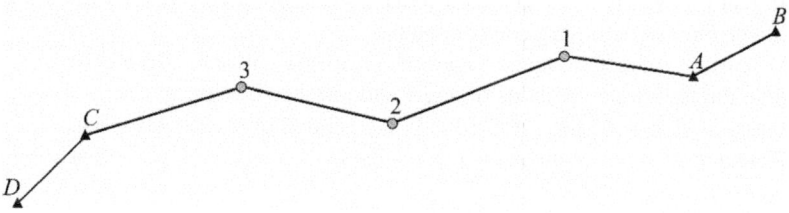

Fig. 12.26 Traverse sketch

Table 12.31 Control point coordinates

Ponto	X [m]	Y [m]
A	11,178.977	12,504.330
B	11,656.792	12,775.934
C	8,600.622	12,453.243
D	8,261.894	12,154.893

Table 12.32 Field measurements

Station	Sighting point	Horizontal clockwise angle	Horizontal distance [m]
A	1	220 ° 35'07"	562.214
1	2	159 ° 59'53"	734.041
2	3	197 ° 52'50"	641.964
3	C	159 ° 09'55"	681.930
C	D	150 ° 36'35"	

12.7   Review Questions

1. Discuss the different surveying methods that can be used to determine control points in Geomatics and which are commonly used in current Civil Engineering works.
2. Discuss in which situations the survey traverse method may be more appropriate than GNSS surveying for the determination of control points for Civil Engineering works.
3. Explain how to carry out link traverse surveying in the field using a total station.
4. Discuss the situations in which using an open traverse is acceptable in civil engineering works.
5. Discuss how an open traverse can be checked for acceptability in Civil Engineering work.
6. Discuss the types of Civil Engineering projects for which you would recommend the use of a link, a loop or an independent survey traverse.
7. Discuss what are the errors of closure for a survey traverse and how they are determined and adjusted.
8. Discuss when you would recommend using forced centring for traverse surveying.
9. Explain why a link traverse is a more reliable surveying method than a loop traverse for determining control points.
10. Discuss why the least-squares method of traverse calculation is a better solution than the traditional methods of angle and distance balancing.
11. Discuss what a closure standard is in the context of a survey traverse.
12. Research the positioning standards available in your region for engineering and construction control surveys.
13. Discuss how control points should be monumented for use in a Civil Engineering project to ensure their availability throughout the project's lifetime.

References

ASPRS (2015). *Positional Accuracy Standards for Digital Geospatial Data. Edition* 1, Version 1.0. Photogrammetric Engineering & Remote Sensing. Vol 18, N.3, March, pages A1-A26.

Bird, R. G. (1989). *EDM Traverse – Measurement, computation and Adjustment.* Longman Scientific & Technical. ISBN 0-470-21318-3. England.

Brabant, M. (2012). *Topographie opérationnelle. Groupe Eyrolles.* ISBN 978-2-212-12847-5. France.

Federal Geographic Data committee (FGDC). FGDC-STD-007.4-2002. *Geospatial Positioning Accuracy Standards PART 4: Standards for Architecture, Engineering, Construction (A/E/C) and Facility Management.* Reston, Virginia, USA.

Geodetic Survey Section Survey and Mapping Office Lands Department (2010). *Accuracy Standards of Control Survey (Version 2.0).* Hong Kong.

Ghilani, C. D., Wolf, P. R. (2007). *Elementary Surveying – An Introduction to Geomatics.* 12 Edition. Pearson Prentice Hall, EUA.

Guyer, J. P, Fellow, R. A. (2017). *An Introduction to Accuracy Standards for Land Surveys.* The Clubhouse Press, El Macero, California, USA.

ISO 19157:2013. *Geographic information — Data quality.*

NBR 13.133 (1994). *Execução de levantamento topográfico.* Associação Brasileira de Normas Técnicas (ABNT), Rio de Janeiro, 35 pag.

NBR 13.133 (2021). *Execução de levantamento topográfico.* Associação Brasileira de Normas Técnicas (ABNT), Rio de Janeiro, 45 pag.

Ogundare, J. O. (2015). *Precision Surveying: The Principles and Geomatics Practice.* John Wiley & Sons Inc. USA. ISBN 978-1-119-10251-9.

Ogundare, J. O. (2018). *Understanding Least Squares Estimation and Geomatics Data Analysis.* John Wiley & Sons Inc. USA. ISBN 978-1119501398.

Paar, Rinaldo, Novakovic, Gorana, Zulijani, Emili. (2009). *Positioning Accuracy Standards for Geodetic Control.* Allgemeine Vermessungs-Nachrichten. D 1103. 280–287.

Chapter 13
Vertical Control Survey: Levelling

Irineu da Silva and Marcelo Monari

13.1 Introduction

Vertical control refers to surveying techniques that provide measurements to determine the elevation (or height) of geospatial data. In Civil Engineering, the relief of the terrain is represented by differences in elevation, making this measurement an essential process in the production of vertical data for mapping, engineering design and construction. The process of measuring vertical distances to determine the elevations of given points relative to a given or assumed datum, or to establish points at a given elevation, is called levelling. For Geomatics applied to Civil Engineering, levelling is a typical operation used in the design of roads, railways, tunnels and canals, grade setting works, volume calculations, surveying and implementation of sewerage and drainage systems, and many other types of Civil Engineering works where elevation values need to be determined. In this sense, this chapter aims to examine the methods and techniques of levelling used in Civil engineering.

It is also important to note that in the case of Geomatics applied to Civil Engineering, some projects require the elevations of points to be referenced to an official reference datum. In most cases, however, the Civil Engineer is concerned with the relative height from one point to another rather than with an official datum. It is, therefore, common practice to adopt an arbitrary vertical reference datum defined by a stable mark within the boundaries of the survey area.

I. da Silva (✉)
São Carlos School of Engineering, University of São Paulo, São Carlos, SP, Brazil
e-mail: irineu@sc.usp.br

M. Monari
Civil Engineer, Assistant Professor in the Department of Civil Engineering at the Federal University of São Carlos (UFScar), São Paulo, Brazil
e-mail: marcelo.monari@ufscar.br

© The Author(s), under exclusive license to Springer Nature Switzerland AG 2025
I. da Silva, P. C. L. Segantine, *Geomatics Applied to Civil Engineering*,
https://doi.org/10.1007/978-3-031-75737-2_13

For readers less familiar with levelling concepts, the meaning of some key terms used throughout this book is explained below.

Height: The vertical distance between the top and bottom of an object. Also, the vertical distance of an object or a point above an arbitrary reference level.

Elevation: The vertical distance from a datum to a point or object on the Earth's surface.

Gradient (i%): The rate at which the vertical distance ΔH changes for every 100 m change in the horizontal distance d.

$$i\% = \frac{\Delta H}{d} * 100 \tag{13.1}$$

Level surface: A curved surface where each point is normal to the vertical line (direction of the plumb line) at that point. Every point on it is at the same elevation relative to a specific reference level.

Level line: A line on a level surface which has a constant elevation across its length. It is, therefore, normal to the plumb line at all points.

Reference plane: The horizontal plane adopted as the horizontal reference for determining elevations. The reference plane at a point is a plane tangent to the level surface at that point.

Horizontal line: The straight-line tangent to the level line at a point.

Vertical datum: The assumed or fixed elevation of a given point or level surface, which is the reference for vertical point positioning. See Chap. 4 for more details.

Relief: The variation in the elevation of the ground surface.

Benchmark (BM): A permanent reference point (materialised in the ground) for the vertical point positioning in relation to an official reference datum. In Civil Engineering applications, various objects are used to represent a benchmark, including metal rods, bolts, chiselled marks, and stone monuments. In all cases, they must be robust enough to be repeatedly and consistently used without changing their elevation.

Temporary benchmark (TBM): A temporary point used as a reference for level control during construction and surveying works.

Reduced level: Height of a point referenced to a surface of assumed level.

13.2   Vertical Distance Measurement

As mentioned above, in Geomatics, performing a vertical distance measurement involves determining the difference in elevation between two level surfaces. However, it is essential to remember that level surfaces are surfaces of constant gravitational potential energy,[1] which brings them closer to the shape of the geoid. Therefore, due to the difficulty of rigorously modelling each level surface, for practical applications, the difference in elevation is determined by establishing a horizontal reference plane tangent to the level surface[2] at the measurement location, as shown in Fig. 13.1.

---

[1] Equipotential surface.

[2] Perpendicular to the plumb line.

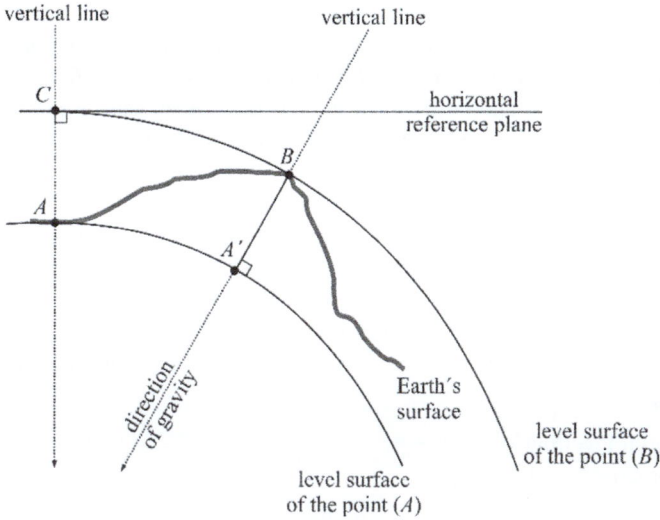

Fig. 13.1 Horizontal reference plane and level surfaces

Since level surfaces are gravity fields, the only possible and affordable way to establish a horizontal reference plane anywhere on the Earth's surface for Geomatics applied to Civil Engineering is by referencing them to the vertical line determined by the level bubble of a surveying instrument or the plumb line, as presented in Chap. 9.

The use of a horizontal reference plane rather than a level surface to determine point elevation obviously limits the range of sighting distances and distorts the calculated values due to the gradual separation of the horizontal plane from the level surface. Add to this the effect of vertical atmospheric refraction in defining the horizontal plane and the non-parallelism of level surfaces, as shown in Fig. 13.2, and the result is a complex physical model for determining the elevation of points on the Earth's surface. However, for Civil Engineering applications, the small size of the working areas allows some simplifications to be made without causing major changes in the values obtained, making the mathematical models easier to apply.

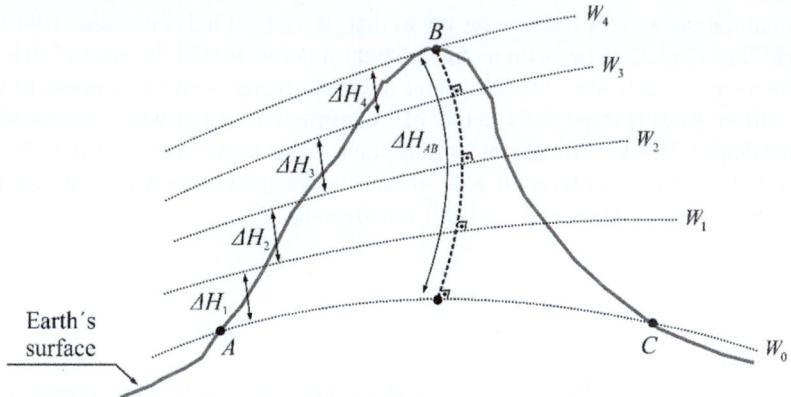

Fig. 13.2 Non-parallelism of level surfaces

The main simplification adopted in Civil Engineering is to ignore the effect of the non-parallelism of the level surfaces. As shown in Fig. 13.2, this causes the sum of the partial differences in elevations ΔH_i between points (A) and (B) to differ from ΔH_{AB}, making the calculated difference in elevation dependent on the path. However, this effect is minimised for short distances such as those used in Civil Engineering projects.

The effects of Earth's curvature and vertical refraction on levelling results are discussed throughout this book. However, the effects of the non-parallelism of equipotential surfaces are beyond the scope of this book. Therefore, all levelling methods presented in this chapter, as they relate to current Civil Engineering applications, ignore the non-parallelism of the level surfaces.

Four levelling methods are currently used in Geomatics applied to Civil Engineering.

- Differential levelling[3]
- Trigonometric levelling
- GNSS levelling
- Laser levelling

13.3   Differential Levelling

Differential levelling is the method of determining the difference in elevation between points on the Earth's surface by measuring their vertical distance from the ground to a horizontal reference plane. It is, therefore, necessary to use an instrument capable of generating such a plane and a measuring rod for vertical measurements. The instrument for this purpose is the *surveyor's level*, as described in Sect. 9.5 and the measuring rod is the *levelling staff*, as described in Sect. 9.6. As shown in Fig. 13.3a, the relative elevations of the ground points can be determined by the intersection of the horizontal line of sight defined by the surveyor's level telescope and the levelling staff (see Fig. 13.3b). Differential levelling is the most accurate method of determining levels and is most commonly used by Civil Engineers.

When the surveyor's level is set up so that its vertical axis coincides with the vertical line, the telescope, with its line of sight rotating around the vertical axis of the instrument, establishes the horizontal reference plane, which is tangent to the level surface passing through the centre of the instrument. In this way, together with the levelling staff held vertically at various points in the survey area, it is possible to measure differences in elevation with millimetre accuracy, which is sufficient for most applications of Geomatics in Civil Engineering.

---

[3] Also referred as direct levelling, spirit levelling and geometric levelling by some authors.

a)

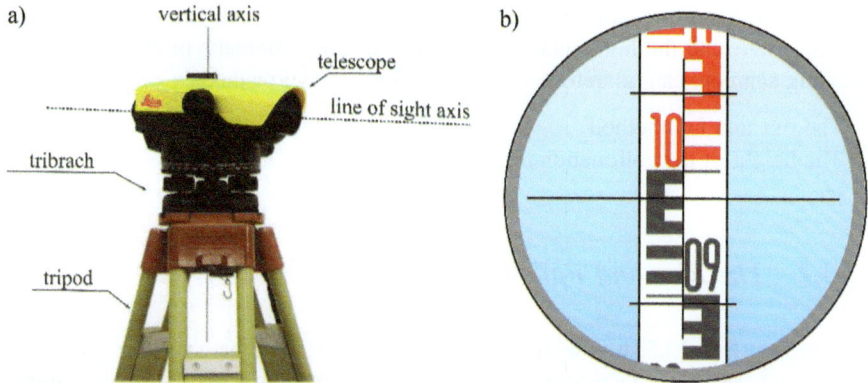

b)

Fig. 13.3 Surveyor's level and levelling staff for differential levelling. (**a**) Surveyor's level for differential levelling. (Courtesy: Leica Geosystems. (**b**) Example of readings taken on the levelling staff)

As shown in Fig. 13.4, the difference in elevation between points (A) and (B) is calculated as the difference between the vertical distances determined by the values of the readings at the intersections of the horizontal plane and the vertically held levelling staff installed at these points.

The field procedure for differential levelling measurements, as shown in Fig. 13.4, is to set up the surveyor's level correctly in a position approximately halfway between points (A) and (B) on the ground (not necessarily in AB line), aim the levelling staff at point (A) and take a sight reading. The levelling staff must then be placed over point (B) for a new reading. If the first reading is taken at a benchmark or at a point whose elevation has already been determined, it is called a Backsight Reading (BS). If it is taken at a point that ends a levelling segment, it is called a Foresight Reading (FS).

Fig. 13.4 Principle of differential levelling

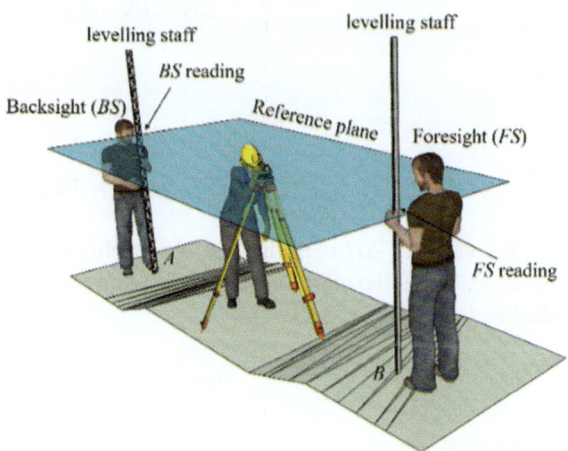

The readings correspond to the vertical distances between the bottom of the levelling staff and the horizontal reference plane. The difference in elevation for a levelling segment can, therefore, be calculated using two methods.

- The rise and fall method
- The height of the collimation method

13.3.1 The Rise and Fall Method

Considering Fig. 13.5, it is worth noting that the difference between the backsight (*BS*) and foresight (*FS*) readings, or the previous reading minus the subsequent reading, gives the difference in elevation (ΔH) *between the foresight point and the backsight point*, as given by Eq. (13.2).

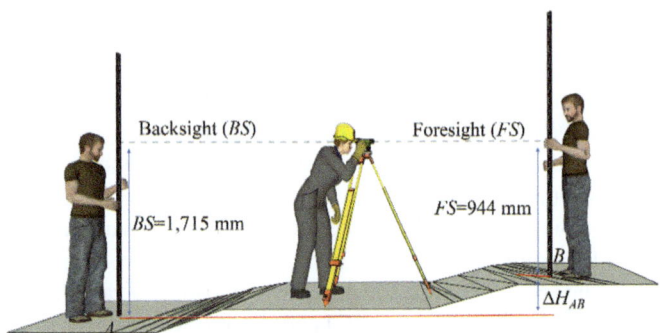

Fig. 13.5 Backsight and foresight readings

$$\Delta H = BS - FS \tag{13.2}$$

Or,

$$\Delta H_{AB} = BS_A - FS_B \tag{13.3}$$

$$\Delta H_{BA} = BS_B - FS_A \tag{13.4}$$

If the value of ΔH is positive, a *rise* is obtained; if negative, a *fall* is obtained. Therefore, if the elevation of point (*A*) is known, the elevation of point (*B*) can be calculated using Eq. (13.5).

$$H_B = H_A + \Delta H_{AB} \tag{13.5}$$

Likewise,

$$\Delta H_{AB} = H_B - H_A = BS_A - FS_B \tag{13.6}$$

where

H_A = elevation of point A

H_B = elevation of point B

13.3.2 The Height of Collimation Method

The *height of collimation method* is based on determining the elevation of the plane of collimation[4] (*HPC*) by adding the staff reading to the elevation of the backsight point, as given by Eq. (13.7). Similarly, the elevation of any subsequent point whose reading is taken from the same instrument setup is calculated by subtracting that reading from the *HPC*, as given by Eq. (13.8). The reader should note that the elevation of the plane of collimation corresponds to the elevation of the line of sight of the instrument, which some authors refer to as the *height of the instrument*[5] (*H.I.*). Hence, the name *height of the instrument method*.

$$HPC = H_A + BS_A \tag{13.7}$$

$$H_B = HPC - FS_B \tag{13.8}$$

In practice, in Geomatics applied to Civil Engineering, the levelling staff readings are taken to the nearest millimetre, requiring that the distance between the instrument and the levelling staff does not exceed 40–50 m. See Sect. 9.6 for more information.

Example 13.1 Using the readings shown in Fig. 13.5 and assuming that the reduced level (*RL*) of point (*A*) is equal to 785.147 m, calculate the reduced level (*RL*) of point (*B*) using the *rise and fall* and *height of collimation* methods.

Solution:

Using the rise and fall method, the reduced level of point (B) can be calculated using Eqs. (13.3) and (13.5).

$$\Delta H_{AB} = BS_A - FS_B = 1,715 - 944 = 771 \text{ mm} = 0.771 \text{ m}$$
$$H_B = 785.147 + 0.771 = 785.918 \text{ m}$$

Using the height of the collimation method, the reduced level of point (B) can be calculated using Eqs. (13.7) and (13.8).

---

[4]Horizontal reference plane at the elevation of the line of sight.

[5]Note that the concept of instrument height is different for levelling than for total stations.

$$HPC = 785.147 + 1.715 = 786.862 \text{ m}$$
$$H_B = 786.862 - 0.944 = 785.918 \text{ m}$$

In this book, differential levelling is classified as follows to facilitate its practical application in Civil Engineering projects:

- Simple levelling
- Line levelling
- Precision levelling
- Network levelling

13.3.3 Simple Levelling

Differential levelling is called *simple levelling* when the difference in elevation between ground points is determined from a single instrument setup. Reduced levels are calculated by taking a backsight reading to a point of assumed reduced level and subsequent readings at each point where the reduced level is to be found. A typical application of this levelling method is to determine the difference in elevation between a series of points distributed over a small construction site or to determine the cross-section of an alignment, as shown in the following examples.

Example 13.2 The readings given in Table 13.1 were taken to determine the difference in elevation between points (A), (B), (C) and (D), as shown in Fig. 13.6. Calculate the corresponding values using the height of collimation method.

Fig. 13.6 Geometry of the levelling measurement

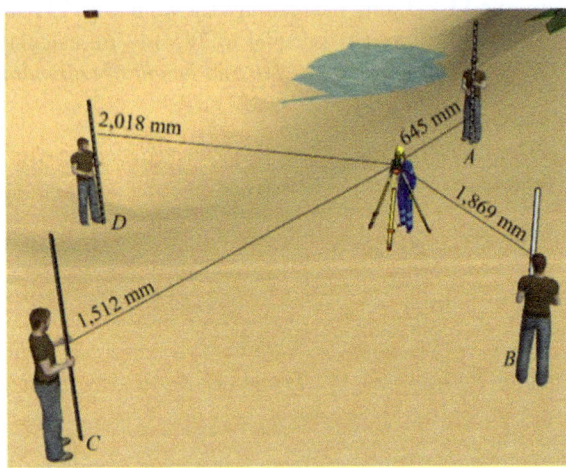

Solution:

   As the four points can be sighted from a simple instrument setup, the surveyor's level was placed within the survey area, and successive readings were taken at points (A), (B), (C) and (D). The reduced level of point (A) was assumed to be 100.000 m. The HPC value was then calculated using Eq. (13.7), and the reduced levels of points (B), (C) and (D) using Eq. (13.8). The results are shown in the three columns on the right side of Table 13.1.

Table 13.1 Reading values and calculation results

Sighting point	Reading values		Calculation results		
	Backsight (BS) [mm]	Foresight (FS) [mm]	HPC [m] (Eq. 13.7)	Reduced level [m] (Eq. 13.8)	Difference in elevation [m]
A	645		100.645	TBM = 100.000	
B		1869		98.776	−1.224
C		1512		99.133	−0.867
D		2018		98.627	−1.373

Example 13.3 A simple levelling has been carried out to determine the cross-section of a road, as shown in Fig. 13.7. Calculate the reduced level (*RL*) of each point along the cross-section using the rise and fall method. Consider the reduced level of the TBM to be 815.325 m. The readings are given in Table 13.2.

Fig. 13.7 Geometry of the levelling measurement for the cross-section determination

Table 13.2 Reading values and calculation results

Sighting point	Reading values		Calculation results	
	Backsight (BS) [mm]	Foresight (FS) [mm]	Rise and fall [m] (Eq. 13.6)	RL [m] (Eq. 13.5)
TBM	337			TBM = 815.325
A		1051	−0.714	814.611
B		1613	−1.276	814.049
C		1642	−1.305	814.020
D		1640	−1.303	814.022
E		1101	−0.764	814.561
F		259	0.078	815.403

Solution:

As all the points belonging to the cross-section can be sighted from a simple instrument setup, the readings were obtained by placing the levelling staff first on the TBM and then on each point from (A) to (F). The calculation results are shown in the two columns on the right side of Table 13.2.

13.3.4 Line Levelling

Differential levelling is classified as line levelling when the levelling operation requires a series of instrument setups along a route consisting of several levelling spans.[6] Each span refers to a segment of the total levelling line between two successive points where readings are taken. It is generally used when the terrain has steep gradients, or the endpoints of the levelling course are far apart. Typical Civil Engineering applications of this type of levelling include roads, bridges, railways, tunnels, airports, pipelines, water supply, sewerage, etc.

Figure 13.8 shows a typical example of line levelling. The field procedure, in this case, is to place a levelling staff over point (A) and set up the surveyor's level in a suitable position to carry out the levelling in the direction from (A) to (N). A backsight reading (BS) is taken at point (A), followed by a foresight reading (FS) at the next point (B) along the route. The instrument is then moved forward so that point (B) can be sighted for a new backsight reading, and the measurement process is repeated until the course is completed at point (N). As shown in Fig. 13.8, there are several intermediate points between the endpoints points (A) and (N), where backsight and foresight readings are taken from different instrument setups (points (B), (C) and (N-1) in Fig. 13.8). Such a point is called *a turning point (TP)* or a *change point (CP)*. They are the connecting points between successive levelling sections, so they must be stable and have a well-defined high point or be marked at the exact point of contact of the levelling staff. They should not be removed after use for precise levelling works but left in place to provide a check for blunders or excessive misclosures.

As shown in Fig. 13.8, the reader should note that for this type of levelling, it is not necessary to align the intermediate levelling staff with the instrument setup. The positions should only ensure that successive levelling staffs are visible and that the distances between the instrument and the backsight and foresight readings are approximately the same.

[6]Each span consists of a backsight (BS) reading and a foresight (FS) reading, taken from a single instrument setup.

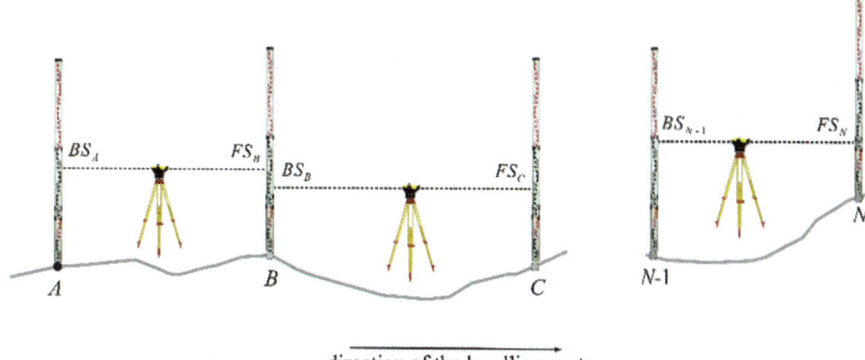

direction of the levelling route

Fig. 13.8 Principle of line levelling

Considering Fig. 13.8, the line levelling calculation can be performed as follows:

$$\Delta H_{AB} = BS_A - FS_B$$
$$\Delta H_{BC} = BS_B - FS_C$$
$$\Delta H_{CD} = BS_C - FS_D$$
$$\cdots\cdots\cdots\cdots\cdots\cdots\cdots \qquad (13.9)$$
$$\Delta H_{N-1,N} = BS_{N-1} - FS_N$$
$$\Delta H_{AN} = \sum BS - \sum FS$$

Finally, if the elevation of point (A) is known, the elevation of point (N) can be calculated using Eq. (13.10).

$$H_N = H_A + \Delta H_{AN} \qquad (13.10)$$

The calculation table for the rise and fall method can be organised as shown in Table 13.3.

Table 13.3 Example of calculation table for the rise and fall method

Sighting point	Reading values		Calculation results	
	BS [mm]	FS [mm]	Rise and fall [mm] (Eq. 13.2)	Elevation [m] (Eq. 13.5)
A				500.000
B	1947	913	1947–913 = 1034	500.000 + 1.034 = 501.034
C	1987	845	1987–845 = 1142	501.034 + 1.142 = 502.176
D	539	1807	539–1807 = −1268	502.176–1.268 = 500.908
	$\sum BS = 4,473$	$\sum FS = 3,565$		

An arithmetic check of the final calculated values is possible and advisable at the end of the calculation, as given by Eq. (13.11).

$$\sum BS - \sum FS = \sum \text{rises} - \sum \text{falls} = \text{Last elevation} - \text{First elevation} \quad (13.11)$$

Example 13.4 Given the line levelling geometry shown in Fig. 13.9 and the field notes in Table 13.4, calculate the difference in elevation between point (*C*) and the *TBM*. Assuming that the reduced level of the *TBM* is 843.871 m, calculate the reduced level of point (*C*).

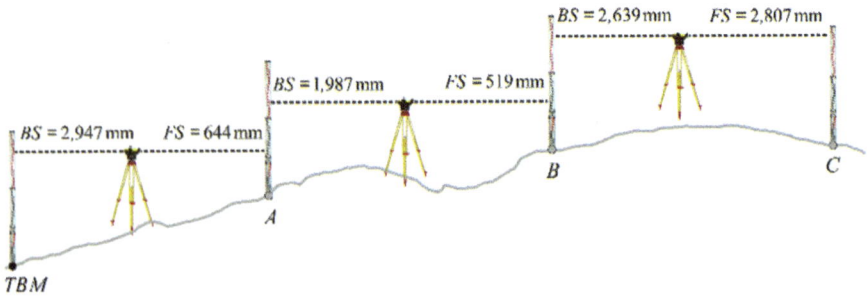

Fig. 13.9 Line levelling geometry

Table 13.4 Reading values and calculation results

	Reading values		Calculation results	
Sighting point	BS [mm]	FS [mm]	Rise and fall [m] (Eq. 13.2)	Elevation [m] (Eq. 13.5)
TBM	2947			843.871
A	1987	644	2.303	846.174
B	2639	519	1.468	847.642
C		2807	−0.168	847.474
	$\sum BS = 7,573$	$\sum FS = 3,970$		

Solution:
The results of the calculations are shown in the two columns on the right side of Table 13.4.

The difference in elevation between point (C) and the TBM can be calculated by summing the rise and fall values as follows:

$$\Delta H_{TRM-C} = 2.303 + 1.468 - 0.168 = 3.603 \text{ m}$$

The arithmetic check can be applied using Eq. (13.11).

$$7.573 - 3.970 = 2.303 + 1.468 - 0.168 = 3.603 \text{ m}$$

The reduced level of point (C) is 843.871 + 3.603 = 847.474 m.

In some cases, line levelling may include *intermediate sights (IS),*[7] as shown in Fig. 13.10, which are readings taken at ground points between the *BS* and the *FS* i.e, withing a levelling span) but not part of the levelling route. In the example in Fig. 13.10, BM_1 and BM_2 are the initial and final benchmarks; points (A), (B), (C), (D) and (E) are the turning points and points (1) to (7) are the intermediate sights. In this case, it is important to note that only the initial, final and turning points are considered for calculating the line levelling.

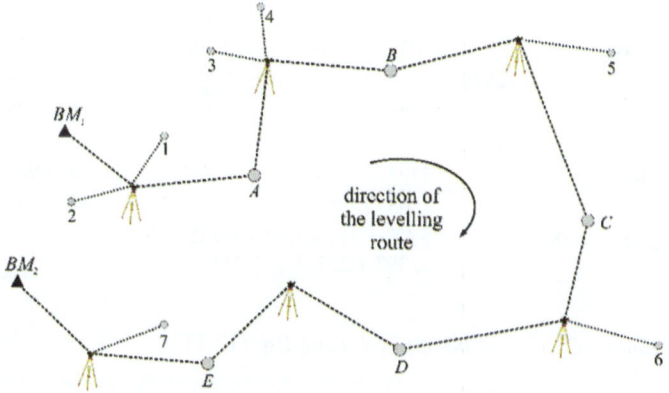

Fig. 13.10 Levelling route including intermediate sights

Example 13.5 Using the line levelling geometry shown in Fig. 13.10 and the readings given in Table 13.5, calculate the elevation of each observed point. The elevation of BM_1 is 815.439 m.

Solution:
    The calculation results are presented in the two columns on the right side of Table 13.5.
    The observed difference in elevation between BM_1 and BM_2 can be calculated using Eq. (13.6).

$$\Delta H_{BM1-BM2} = 827.549 - 815.439 = 12.110 \text{ m}$$

---

[7] Sideways or sideshots.

Table 13.5 Reading values and calculation results

Sighting point	Reading values				Calculation results	
	BS [mm]	IS [mm]	FS [mm]	Span length [m]	Rise and fall [m] (Eq. 13.2)	Elevation [m] (Eq. 13.5)
BM_1	3817					815.439
1		2412			1.405	816.844
2		1509			2.308	817.747
A	3751		344	91.40	3.473	818.912
3		2914			0.837	819.749
4		2564			1.187	820.099
B	3804		1095	69.90	2.656	821.568
5		1932			1.872	823.440
C	3566		902	100.40	2.902	824.470
6		2928			0.638	825.108
D	3727		438	90.20	3.128	827.598
7		1912			1.815	829.413
E	1186		2397	70.00	1.330	828.928
BM_2			2565	90.00	−1.379	827.549
	$\sum BS = 19,851$		$\sum FS = 344 + 1,095 + 902 + 438 +$ $2,397 + 2,565 = 7,741$			

The arithmetic check can be applied using Eq. (13.11).

$$(3,817 + 3,571 + 3,804 + 3,566 + 3,727 + 1,186)$$
$$- (344 + 1,095 + 902 + 438 + 2,397 + 2,565) = 19,851 - 7,741$$
$$= 12,110 \; mm = 12.110 \; m$$

As with a survey traverse, a line levelling course must be checked by running closed circuits. This can be done by returning to the starting benchmark (*loop levelling* or *closed-circuit levelling*) or by ending the course at another benchmark[8] (*link levelling*). In closed-circuit levelling, an *error of closure* or *levelling misclosure* is detected each time the known elevation of the benchmark is compared with the observed (calculated) elevation. If it is lower than the tolerance, it must be distributed over the observations to make the levelling geometrically coherent.

13.3.4.1 Loop Levelling (Circuit Levelling)

As the name suggests, the *loop levelling* (or *circuit levelling*) technique consists of carrying out a closed circuit, that is starting the measurements at a point of known or assumed elevation (*BM*), reaching the endpoint (*N*) of the levelling work and returning to the same starting point, as shown in Fig. 13.11. The difference in

---

[8] Also known as a levelling section (a series of setups between two permanent benchmarks).

Fig. 13.11 Loop levelling (circuit levelling)

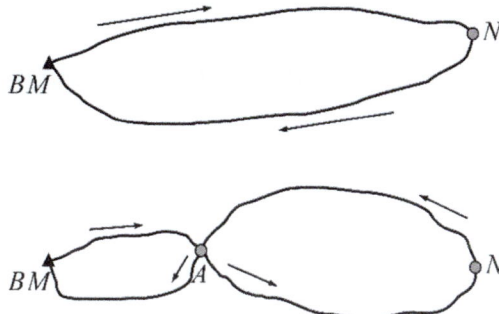

elevation between the calculated (BM_{calc}) and the known (BM_{known}) benchmarks gives the error of closure e_n as follows:

$$e_n = BM_{calc} - BM_{known} \qquad (13.12)$$

As with the survey traverse, the error of closure en must be less than a predefined allowable value for the levelling result to be accepted. Otherwise, field operations must be repeated.

The reader should note that, as shown in Fig. 13.11, the routes taken from (BM) to (N) and from (N) to (BM) do not have to be the same, that is they do not necessarily have to pass through the same points. However, the lengths of the two routes must be as close as possible.

Example 13.6 Given the readings shown in Table 13.6, calculate the elevation of each observed point and the levelling misclosure. The elevation of BM_1 is 785.547 m.

Table 13.6 Reading values and calculation results

| Sighting point | Reading values | | | Calculation results | |
	BS [mm]	IS [mm]	FS [mm]	Rise and fall [m] (Eq. 12.2)	Elevation [m] (Eq. 12.5)
BM_1	1820				**785.547**
1		3725		−1.905	783.642
A	833		3749	−1.929	783.618
2		2501		−1.668	781.950
3		2034		−1.201	782.417
4		3686		−2.853	780.765
B	3460		3990	−3.157	780.461
C	2869		305	3.155	783.616
BM_1			934	1.935	785.551
	$\sum BS = 8,982$		$\sum FS = 8,978$		

Solution:

The results of the calculations are shown in the two columns on the right side of Table 13.6.

The levelling misclosure can be calculated using Eq. (13.12).

$$e_n = BM_{calc} - BM_{known} = 785.551 - 785.547 = 4 \text{ mm}$$

13.3.4.2   Link Levelling

The levelling misclosure can also be calculated if the route starts at a point of known or assumed elevation (BM_i) and ends at another point of known elevation (BM_f), passing through the points whose elevations need to be determined, such as point (A) in Fig. 13.12. This type of levelling technique is called *link levelling*. The difference between the calculated and known elevations of the final benchmark gives the error of closure e_n as follows:

$$en = BM_{f(calc)} - BM_{f(known)} \qquad\qquad (13.13)$$

Fig. 13.12 Link levelling

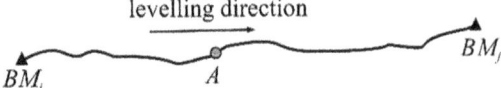

Example 13.7 Given the calculation results from Example 13.5 and knowing that the elevation of BM_2 is 827.544 m, calculate the levelling misclosure.

Solution:
   Using Eq. (13.13).

$$e_n = 827.549 - 827.544 = 5 \text{ mm}$$

13.3.4.3   Field Procedure for Line Levelling

Fieldwork procedures for line levelling are quite simple, regardless of the type of measurement method used. It is usually applied to establish a new benchmark or a preferred ground point in addition to an existing one or to determine elevations of a series of ground points required for a Civil Engineering project. In the first case, the levelling course starts from a benchmark and runs towards the benchmark (or the ground point) to be established. When the new benchmark is reached, a new course must be levelled back to the first benchmark, thus establishing a loop levelling. Because it consists only of taking backsights and foresights, without intermediate sights, it is called *fly levelling* by some authors. In the second case, the main objective is to determine the elevations of essential features for the development of the project. Line levelling is then used to establish temporary vertical control points for feature point sideshots and must, therefore, be closed by a loop or link levelling.

The field levelling procedure for fly levelling is as follows:

(a) First, study the relief of the area to be levelled, check the stability of the starting benchmark, and define approximatively the locations of the turning points.
(b) Set the surveyor's level on stable ground approximately halfway between the benchmark and the first turning point. Take a backsight reading on the levelling staff held vertically at the benchmark.
(c) Move the levelling staff to the turning point and take a foresight reading.
(d) Move the instrument to the next levelling section and repeat steps (b) and (c) until the second benchmark or the desired ground point is reached.
(e) Repeat the process until the fly levelling is completed at the starting benchmark.

The reader must take all levelling precautions described in Chap. 9 and record the readings on an appropriate worksheet or in the instrument memory.

In the case of levelling feature points, the field levelling procedure is as follows:

(a) First, study the relief of the area to be levelled, check the stability of the benchmarks to be used and approximatively define the levelling course considering the feature points to be levelled.
(b) Set the surveyor's level on stable ground approximately halfway between the benchmark and the turning point. Take a backsight reading on the levelling staff held vertically at the benchmark.
(c) Move the reading staff to the turning point and take a foresight reading.
(d) Move the levelling staff to each desired feature point and take readings at each.
(e) Move the instrument to the next levelling section and repeat Steps (b) to (d) until reaching the second benchmark or returning to the first one.

The planimetric position of the feature points to be levelled may already be defined in the project database or may need to be located during the levelling process. In the first case, the operator must identify the feature point before taking the reading on the levelling staff. In the second case, the planimetric position of each feature point must be determined using an appropriate surveying technique, as described in Chap. 11. Levelling procedures for some specific type of feature point elevation determination used in Civil Engineering projects are presented in Chap. 14.

13.3.4.4 Double-Run Levelling

Closed-link levelling can also be carried out by double-run levelling, which consists of simultaneously levelling two parallel and independent paths starting from point (A) and ending at point (B), as shown in Fig. 13.13. To ensure accuracy, two different teams of foremen should be used with their equipment. Steel turning plates for precise levelling should also be used, as shown in Sect. 10.4.1.4, which must be painted in different colours for each team.

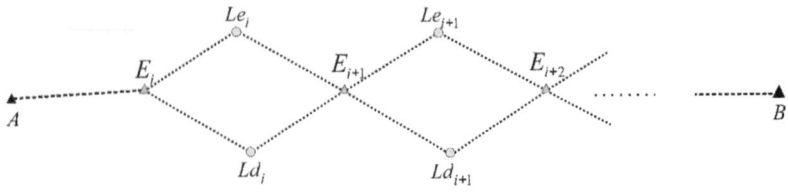

Fig. 13.13 Principle of double-run levelling using two levelling staffs

This levelling method is also called the Cholesky method, in honour of its creator. The field procedure for its application is as follows:

Points (Le_i) on the left and the points (Ld_i) on the right, according to the direction of the course, are positioned approximately 1.0 m apart. Points $(Lei-1)$, $(Ldi-1)$ and (Lei), (Ldi) are then read simultaneously at each position of the surveyor's level (Ei). The endpoints points (A) and (B) are read separately with the two levelling staffs to ensure independent readings. The absolute deviation between the two independent level differences between points (A) and (B), calculated for the left and right paths, gives the levelling error of closure. If this error is less than the preset tolerance, the arithmetic mean of the two differences in level is taken as the levelling result. Otherwise, the field procedure must be repeated.

This method should be used cautiously, as there is no control over the readings taken along the way. The equidistance between the sights from the same instrument setup must also be strictly observed. It is therefore advisable to use digital levels with distance metre to check the length of the sights.

A variation of the method described is to use two surveyor's levels, (Ei) and (Di), and a single levelling staff (L), as shown in Fig. 13.14. The field procedure and the final levelling calculation are self-explanatory, as shown in Fig. 13.14.

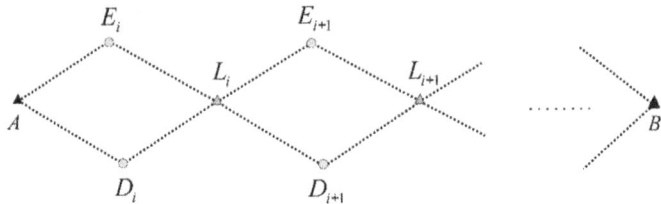

Fig. 13.14 Principle of double-run levelling using two surveyor's levels

13.3.4.5   Inverted Levelling Staff

When levelling underground, the levelling staff is often positioned on the roof of a particular structure rather than on the ground, as shown in Fig. 13.15. In these cases, the staff is placed upside down above the reference plane. Therefore, the readings must be stored as negative, and the calculations performed considering such algebraic signs. In Fig. 13.15, the levelling result will be negative because $BS < FS < 0$.

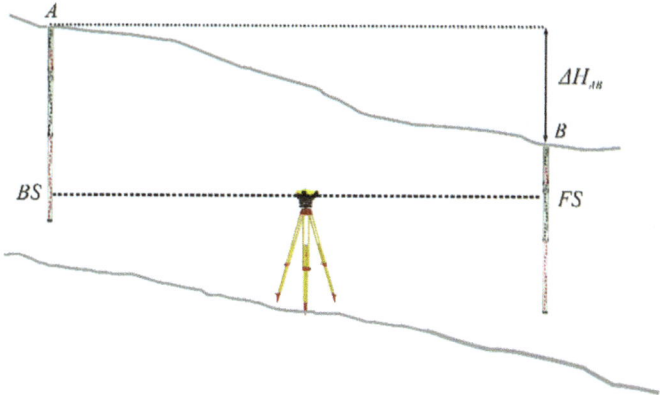

Fig. 13.15 Principle of inverted levelling staff measurement

Example 13.8 Considering the measurement principle shown in Fig. 13.15 and the readings given in Table 13.7, calculate the difference in elevation between points (A) and (B).

Table 13.7 Reading values

Sighting point	BS [mm]	FS [mm]
A	−2915	
B		−912

Solution:
   Using Eq. (13.2).

$$\Delta H_{AB} = BS_A - FS_B = -2,915 - (-912) = -2,003 \text{ mm} = -2.003 \text{ m}$$

13.3.4.6   Reciprocal Levelling

As already explained, ensuring short and approximately equal sight lengths is essential to ensure accurate results in differential levelling. Occasionally, however, it may not be possible to carry out levelling by keeping the back and foresight distances equal, for example, when crossing a valley or river. In these cases, the solution is to carry out simultaneous reciprocal differential levelling, as shown in Fig. 13.16. This figure shows two points (A) and (B) to be levelled on opposite sides of a valley.

   The levelling procedure, in this case, is to set up a surveyor's level at (P_1), close to point (A) and place a levelling staff at point (A) on the left side of the valley. At the same time, a level is set up at (P_2), close to point (B), and a levelling staff is placed at point (B) on the right side of the valley. Simultaneously, both instruments take

backsight and foresight readings at points (A) and (B). The elevation difference can
then be calculated using Eq. (13.14).

$$\Delta H_{AB} = \frac{(FS_B - BS_A)_{\text{From } P1} + (BS_B - FS_A)_{\text{From } P2}}{2} \qquad (13.14)$$

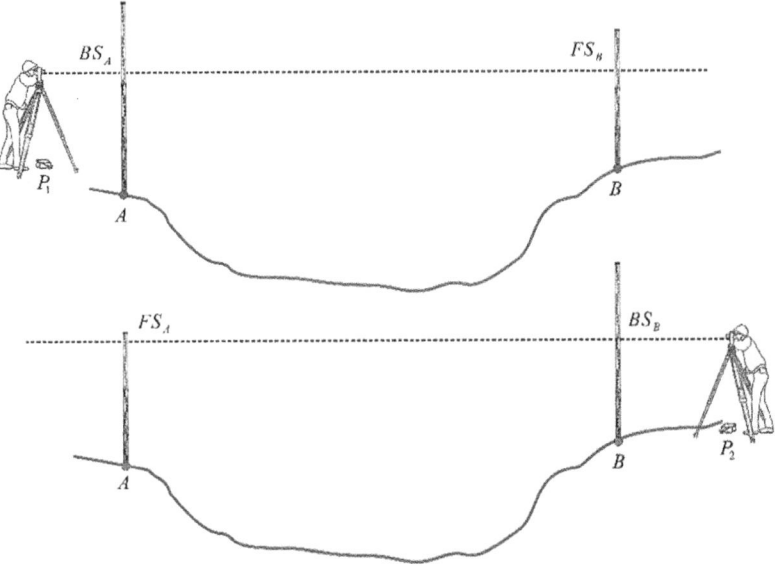

Fig. 13.16 Principle of reciprocal levelling

As this procedure does not allow the results to be checked, it is advisable to repeat
it several times, changing the position of the instruments. The arithmetic mean of the
partial results gives the final result. In cases where two instruments cannot be used
simultaneously, readings should be taken as soon as possible to minimise the effect
of refraction variation. Even in cases where two surveyor's levels are used, it is
recommended that the readings be taken with the staff in alternate positions to
minimise the effects of their different collimation errors.

Example 13.9 Considering the measurement principle shown in Fig. 13.16 and the
readings given in Table 13.8, calculate the difference in elevation between points (A)
and (B). Assuming that the elevation of point (B) is 811.060 m, calculate the
elevation of point (A).

Table 13.8 Reading values

	BS [mm]		FS [mm]	
Instrument setup	A	B	A	B
P_1	1361			2117
P_2		1537	783	

Solution:

Using Eq. (13.14),

$$\Delta H_{AB} = \frac{(2.117 - 1.361) + (1.537 - 0.783)}{2} = 0.755 \text{ m}$$

The elevation of point (A) can be calculated using Eq. (13.5).

$$H_A = H_B - \Delta H_{AB} = 811.060 - 0.755 = 810.305 \text{ m}$$

13.3.5 Precision Levelling

For some specific Civil Engineering applications, such as structural geodetic monitoring, water dam construction, tunnelling, long-span bridges and the like, the engineer needs to carry out levelling operations that are considered to be of *high precision*. Measurements in these cases must be carried out using a specific levelling instrument, called a *precision level*, and a special levelling staff, such as the invar levelling staff, described in Sect. 9.6. Fieldwork must also be performed with care to ensure the required accuracy. The first precaution is to ensure the equidistance between the back and foresights, which should be within a few centimetres. Long lines of sight should also be avoided, keeping them between 25 and 50 m long and never performing observations close to the ground level (less than 50 cm). It is also advisable to avoid measurements on steep slopes to avoid the effect of different vertical atmospheric refraction gradients. Furthermore, during the field measurement process, certain events must be avoided, such as:

- Vibrations in the vicinity of the installation site.
- For instruments with pendulum compensator with magnetic damping, strong magnetic fields in the immediate vicinity (e.g. transformers, melting furnaces...) can affect the compensator and lead to measurement errors.
- Instability of turning points.
- Network levelling instead of line levelling, whenever possible.
- Excessive exposure of the instrument and tripod to sunlight. Heat can cause differential deformation in optical and metallic components of the levelling instrument. An umbrella should be used in this case, as shown in Fig. 13.17.

It is also recommended to carry out a series of observations of the type *BSFSBSFS, BSFSFSBS, BSFSFSBSBSFSFSBS* and *BSFSBSFSBSFS* to control the quality of the readings. If the differences are within the tolerance specified for the project, the mean is calculated and adopted as the final reading. In practice, the *BSFSFSBS* series is the most commonly used.

Detailed information on high-precision levelling instruments is beyond the scope of this book. Readers are advised to refer to specialised references for further details.

Fig. 13.17 Surveying
instrument protected from
sunlight

13.3.6   Levelling Network

A levelling network is configured when several levelling loops are interconnected to
form a 1-*D* network, as shown in Fig. 13.18. This levelling procedure allows each
observed point to benefit from multiple observations with consequent redundancy.
Instead of calculating the individual line levelling circuit, all observations within the
levelling network can be adjusted simultaneously by a least squares adjustment
method, as presented in Sect. 13.4.2. This type of levelling procedure is useful for
benchmark determination on construction sites, water supply, sewerage and drainage
networks and many other Civil Engineering projects where a series of benchmarks
must be established over a construction area.

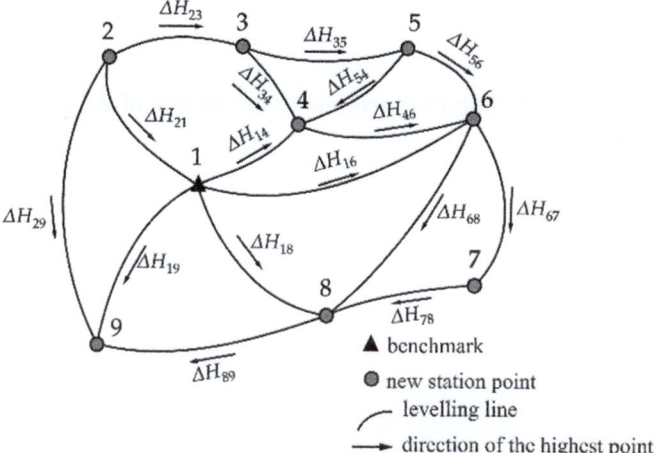

Fig. 13.18 Levelling network

13.4   Levelling Adjustment

Two levelling adjustment methods are currently used in Geomatics applied to Civil Engineering, which are both described in the following subsections: the conventional error distribution method and the least squares adjustment method. The conventional error distribution method is the most widely used for closed-circuit line levelling, while the least squares adjustment method is preferred for levelling networks.

13.4.1   Line Levelling Adjustment by Error Distribution

In the case of line levelling, the adjustment is performed by applying corrections to the observed rises and falls or instrument heights, considering the magnitude of the error of closure for the points of the levelling line, i.e., $H_i = H_{i-1} + \Delta H_{i-1,i} + C_{Hi}$. There is no correction for the intermediate sights as there is no control over their measurements. Their elevation is calculated based on the corrected value of the elevation of the backsight point, i.e., $H_{IS} = H_{BS} + \Delta H_{BS,\,IS}$. In this respect, three error distribution methods are currently used in Civil Engineering.

13.4.1.1   Error Distribution Proportional to the Number of Levelling Sections

The correction value C_{H_i} for each levelling section i is calculated by dividing the error of closure e_n by the number of levelling sections n, according to Eq. (13.15).

$$CH_i = \frac{-e_n}{n} \qquad (13.15)$$

13.4.1.2   Error Distribution Proportional to the Length of the Levelling Section

In this case, it is assumed that the greater the length d_i of the levelling section i, the greater the correction value C_{H_i}, as given by Eq. (13.16). Note that to apply this adjustment method, it is necessary to know the length of all the levelling sections.

$$CH_i = -e_n \frac{d_i}{\sum d_i} \qquad (13.16)$$

13.4.1.3   Error Distribution Proportional to the Absolute Value of the Difference in Elevation of the Levelling Section

In this case, the correction value C_{H_i} to be applied for a given levelling section i depends on its difference in elevation ΔH_i, according to Eq. (13.17).

$$C_{H_i} = -e_n \frac{|\Delta H_i|}{\sum |\Delta H_i|} \qquad (13.17)$$

13.4.1.4   Elevation Correction

The elevation correction for each levelled point is carried out by adding the correction value C_H to the reduced levels calculated from each instrument position, as given by Eq. (13.18) in the case of the rise and fall calculation method and Eq. (13.19) in the case of the height of collimation method. This means that all reduced levels, including the intermediate sights, are adjusted. See Example 13.10.

$$H_i = H_{i-1} + \Delta H_{i-1,i} + CH_i \qquad (13.18)$$
$$H_i = HPC_i + CH_i\text{-}FS_i, \qquad (13.19)$$

Example 13.10 Considering the data from Example 13.5, calculate the adjusted elevation of each observed point using Eqs. (13.15), (13.16) and (13.17).

Solution:

The calculation results in **Example 13.5** *shows that a closing error of $+5$ mm has occurred. Therefore, a correction of $CH = -5$ mm should be applied.*

(a) *Error distribution proportional to the number of levelling sections n.*

$$C_{H_i} = \frac{C_H}{n} = \frac{-5}{6} = -0.83 \text{ mm}$$

(b) *Error distribution proportional to the length of the levelling section d_i.*

$$C_{H_{BM1-A}} = -5 * \frac{91.4}{511.9} = -0.89 \text{ mm} \quad C_{H_{AB}} = -5 * \frac{69.9}{511.9} = -0.68 \text{ mm}$$

$$C_{H_{BC}} = -5 * \frac{100.4}{511.9} = -0.98 \text{ mm} \quad C_{H_{CD}} = -5 * \frac{90,2}{511.9} = -0.88 \text{ mm}$$

$$C_{H_{DE}} = -5 * \frac{70}{511.9} = -0.68 \text{ mm} \quad C_{H_{E-BM2}} = -5 * \frac{90}{511.9} = -0.88 \text{ mm}$$

(c) *Error distribution proportional to the absolute value of the difference in elevation of the levelling section $|\Delta H_i|$.*

From **Table 13.5.**

$$\sum |\Delta H_i| = 3,473 + 2,656 + 2,902 + 33,128 + 1,330 + 1,379 = 14.868 \text{ mm}$$

$$CH_{BM1-A} = -5 * \frac{3.473}{14.868} = -1.17 \text{ mm} \quad CH_{AB} = -5 * \frac{2.656}{14.868} = -0.89 \text{ mm}$$

$$CH_{BC} = -5 * \frac{2.902}{14.868} = -0.98 \text{ mm} \quad CH_{CD} = -5 * \frac{3.128}{14.868} = -1.05 \text{ mm}$$

$$CH_{DE} = -5 * \frac{1.330}{14.868} = -0.45 \text{ mm} \quad CH_{E-BM2} = -5 * \frac{1.379}{14.868} = -0.46 \text{ mm}$$

The adjusted elevations using the above correction values are given in **Table 13.9.** Note that no correction has been applied to the rise and fall values of the intermediate points.

As can be seen from the results in **Table 13.9,** there are small differences between the methods used.

Table 13.9 Calculation results

Point	Calculated elevation [m]	Adjusted elevation [m]		
		$C_{Hi} = \frac{C_H}{n}$	$C_{Hi} = C_H \frac{d_i}{\sum d_i}$	$C_{Hi} = C_H \frac{\|\Delta H_i\|}{\sum \|\Delta H_i\|}$
BM_1	815.439	**815,439**	815,439	815,439
1	816.844	816,844	816,844	816,844
2	817.747	817,747	817,747	817,747
A	818.912	**818,911**	**818,911**	**818,911**
3	819.749	819,748	819,748	819,748
4	820.099	820,098	820,098	820,098
B	821.568	**821,566**	**821,566**	**821,566**
5	823.440	823,438	823,438	823,438
C	824.470	**824,468**	**824,467**	**824,467**
6	825.108	825,106	825,105	825,105
D	827.598	**827,595**	**827,595**	**827,594**
7	829.413	829,410	829,410	829,409
E	828.928	**828,924**	**828,924**	**828,923**
BM_2	827.549	827,544	827,544	827,544

13.4.2   Levelling Computation by Least Squares Adjustment

Least squares levelling computation provides the most rigorous solution for network levelling. Therefore, as described in the following subsections, this is the levelling calculation method used for network levelling and sometimes even for line levelling.

A levelling network can be adjusted using either the Conditional or Parametric adjustment models presented in Chap. 3, although the latter is currently the most used and is, therefore, presented in this chapter.

13.4.2.1   Line Levelling Adjustment by Least Squares Adjustment

To apply the Parametric adjustment model to a line levelling, consider the closed-link levelling geometry shown in Fig. 13.19.

In Fig. 13.19, the start and end points have elevations BM_1 and BM_2, respectively; ΔH_i is the observed difference in elevation for the levelling section i; and (H_i) is the unknown elevation of point (i). The arrow indicates the direction of the highest point of the course.

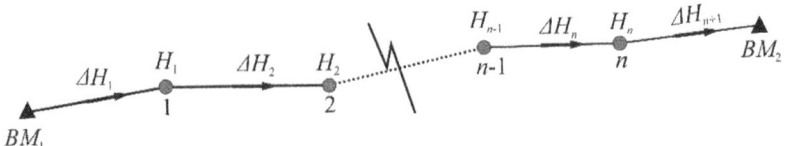

Fig. 13.19 Line levelling geometry

In this case, the observables are the differences in elevation ΔH_i given by Eq. (13.20).

$$\Delta H_i = H_{i+1} - H_i \tag{13.20}$$

Since Eq. (13.20) is linear, the error equation considering the Parametric adjustment model can be written as follows:

$$v\Delta H_i = H_{i+1} - H_i - \Delta H_i \tag{13.21}$$

where

$v\Delta H_i$ = residual error of the difference in elevation for levelling section (i)
ΔH_i = observed difference in elevation for levelling section (i)

Each observation produces an error Eq. (13.21), whose unknown parameters are the elevations H_i. For $n + 1$ observations, Eq. (13.21) can be written as follows:

$$\begin{aligned}
v\Delta H_1 &= H_1 \quad \ldots \quad -(\Delta H_1 + BM_1) \\
v\Delta H_2 &= -H_1 + H_2 \quad \ldots \quad -\Delta H_2 \\
v\Delta H_n &= \ldots -H_{n-1} + H_n - \Delta H_n \\
v\Delta H_{n+1} &= \quad \ldots \quad -H_n - (\Delta H_n BM_2)
\end{aligned} \tag{13.22}$$

Equation (13.22) can be written in matrix form as $v = Ax - l_0$ (Eq. 3.79), where

$$A = \begin{bmatrix} 1 & 0 & \ldots & 0 & 0 \\ -1 & 1 & & 0 & 0 \\ \vdots & \vdots & \ddots & \vdots & \vdots \\ 0 & 0 & & -1 & 1 \\ 0 & 0 & & 0 & -1 \end{bmatrix} \quad v = \begin{bmatrix} v\Delta H_1 \\ v\Delta H_2 \\ \vdots \\ v\Delta H_{n+1} \end{bmatrix}$$

$$l_0 = \begin{bmatrix} \Delta H_1 + BM_1 \\ \Delta H_2 \\ \vdots \\ \Delta H_n \\ \Delta H_n - BM_2 \end{bmatrix} \quad x = \begin{bmatrix} H_1 \\ H_2 \\ \vdots \\ H_{n-1} \\ H_n \end{bmatrix}$$

Note that in this case the approximate value δH_i is assumed equal to zero.

In terms of weighting, instead of using variances, it is generally assumed that each observation has a weight proportional to the length of the levelling section, in km, as given by the weight matrix (13.23).

$$P = \begin{bmatrix} 1/d_1 & 0 & \cdots & 0 & 0 \\ 0 & 1/d_2 & \cdots & 0 & 0 \\ \vdots & \vdots & \ddots & \vdots & \vdots \\ 0 & 0 & \cdots & 1/d_n & 0 \\ 0 & 0 & \cdots & 0 & 1/d_{n+1} \end{bmatrix} \qquad (13.23)$$

The adjustment computation can then be performed using Eqs. (3.84), (3.85), (3.86) and (3.87), from which H_i is obtained, followed by the observation residuals using Eq. (3.79), the adjusted observations $\Delta \overline{H}$ using Eq. (13.24) and, finally, the adjusted elevations using Eq. (13.20).

$$\Delta \overline{H}_i = \Delta H_i + v\Delta H_i \qquad (13.24)$$

As a total of $n + 1$ observations and n unknowns are considered, the standard deviations of the unit weight can be calculated using Eq. (13.25) and the variance-covariance matrix of the parameters (elevations) using Eq. (11.74). The remaining statistical parameters can be determined as indicated in Sect. 3.9.1.1.

$$s_0^2 = v^T P v \qquad (13.25)$$

13.4.2.2   Network Levelling by Least Squares Adjustment

To apply the Parametric adjustment model to a network levelling, consider the levelling geometry shown in Fig. 13.20.

Fig. 13.20 Example of a levelling network

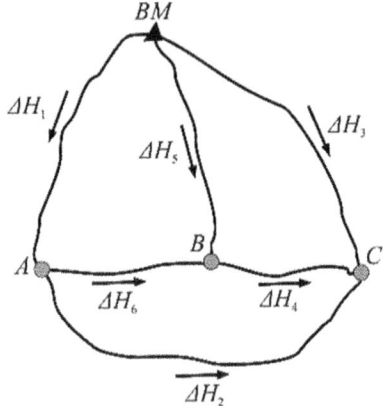

In Fig. 13.20, BM is a benchmark, ΔH_i is the observed difference in elevation for the levelling section i, and (A), (B) and (C) are ground points to be levelled. The arrow indicates the direction of the highest point of the course. The unknowns are the elevations of the ground points (A), (B) and (C).

The reasoning for adjusting the network levelling is similar to the computation of the line levelling adjustment presented in the previous section. The difference is that instead of interconnected line runs, there are interconnected closed circuits that form a levelling network.

In this example, six levelling runs and three circuits have been levelled, and the observations are the differences in elevation of each. The elevations (H_A), (H_B) and (H_C) are the unknowns. There are, therefore, six error equations, which can be written as follows:

$$
\begin{aligned}
v\Delta H_1 &= H_A + 0H_B + 0H_C - (\Delta H_1 + BM) \\
v\Delta H_2 &= -H_A + 0H_B + H_C - \Delta H_2 \\
v\Delta H_3 &= 0H_A + 0H_B + H_C - (\Delta H_3 + BM) \\
v\Delta H_4 &= 0H_A - H_B + H_C - \Delta H_4 \\
v\Delta H_5 &= 0H_A + H_B + 0H_C - (\Delta H_5 + BM) \\
v\Delta H_6 &= -H_A + H_B + 0H_C - \Delta H_6
\end{aligned}
\tag{13.26}
$$

Equations (13.26) can be written in matrix form as $v = Ax - l_0$ (Eq. 3.79), where

$$
A = \begin{bmatrix} 1 & 0 & 0 \\ -1 & 1 & 0 \\ 0 & 0 & 1 \\ 0 & -1 & 1 \\ 0 & 1 & 0 \\ -1 & 1 & 0 \end{bmatrix} \quad
v = \begin{bmatrix} v\Delta H_1 \\ v\Delta H_2 \\ v\Delta H_3 \\ v\Delta H_4 \\ v\Delta H_5 \\ v\Delta H_6 \end{bmatrix} \quad
l_0 = \begin{bmatrix} \Delta H_1 + BM \\ \Delta H_2 \\ \Delta H_3 + BM \\ \Delta H_4 \\ \Delta H_5 + BM \\ \Delta H_6 \end{bmatrix} \quad
x = \begin{bmatrix} H_A \\ H_B \\ H_C \end{bmatrix}
$$

Assuming that weights are proportional to the length of the levelling line, the matrix of weights is given as follows:

$$
P = \begin{bmatrix}
1/d_1 & 0 & \cdots & 0 & 0 \\
0 & 1/d_2 & \cdots & 0 & 0 \\
\vdots & \vdots & \ddots & \vdots & \vdots \\
0 & 0 & \cdots & 1/d_n & 0 \\
0 & 0 & \cdots & 0 & 1/d_{n+1}
\end{bmatrix}
\tag{13.27}
$$

As for the level line adjustment, the adjustment computation can then be performed using Eqs. (3.84), (3.85), (3.86) and (3.87), from which H_i is obtained, followed by the observation residuals using Eq. (3.79), the adjusted observations $\Delta \overline{H}$ using Eq. (13.24) and, finally, the adjusted elevations using Eq. (13.20).

As a total of six observations and three unknowns are considered, the standard deviations of the unit weight can be calculated using Eq. (13.28) and the variance-covariance matrix of the parameters (elevations) using Eq. (11.74). The remaining statistical parameters can be determined as indicated in Sect. 3.9.1.1.

$$s_0^2 = \frac{v^T P v}{3} \qquad (13.28)$$

Example 13.11 Using the data from Example 13.5, summarised in Table 13.10 and Fig. 13.21, calculate the adjusted elevation of each levelling point using the Parametric adjustment model. For the calculation, assume $BM_1 = 815.439$ m and $BM_2 = 827.544$ m.

Table 13.10 Difference in elevation between levelled points

Levelling section	Difference in elevation (ΔH_i) [m]	Section length [m]
(1) BM_1 - A	3.473	91.4
(2) A-B	2.656	69.9
(3) B-C	2.902	100.4
(4) C-D	3.128	90.2
(5) D-E	1.330	70.0
(6) E - BM_2	−1.379	90.0

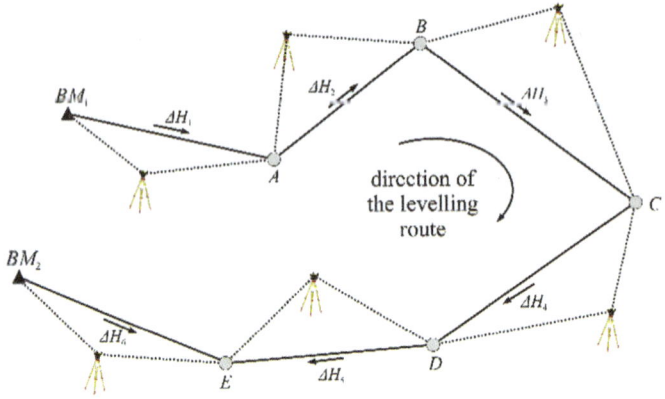

Fig. 13.21 Closed-link levelling

Solution:

Using Fig. 13.21, *the following observation equations can be written as*

$$\Delta H_1 + v_{\Delta H_1} = H_A - BM_1$$
$$\Delta H_2 + v_{\Delta H_2} = H_B - H_A$$
$$\Delta H_3 + v_{\Delta H_3} = H_C - H_B$$
$$\Delta H_4 + v_{\Delta H_4} = H_D - H_C$$
$$\Delta H_5 + v_{\Delta H_5} = H_E - H_D$$
$$\Delta H_6 + v_{\Delta H_6} = BM_2 - H_E$$

There are six equations, five unknowns and one redundancy. To apply the Parametric adjustment model, the following error equations can be written as

$$v\Delta H_1 = 1H_A + 0H_B + 0H_C + 0H_D + 0H_E - (\Delta H_1 + BM_1)$$
$$v\Delta H_2 = -1H_A + 1H_B + 0H_C + 0H_D + 0H_E - \Delta H_2$$
$$v\Delta H_3 = 0H_A - 1H_B + 1H_C + 0H_D + 0H_E - \Delta H_3$$
$$v\Delta H_4 = 0H_A + 0H_B - 1H_C + 1H_D + 0H_E - \Delta H_4$$
$$v\Delta H_5 = 0H_A + 0H_B + 0H_C - 1H_D + 1H_E - \Delta H_5$$
$$v\Delta H_6 = 0H_A + 0H_B + 0H_C + 0H_D - 1H_E - (\Delta H_6 - BM_2)$$

From the error equations above and using Eq. (3.79).

$$A = \begin{bmatrix} 1 & 0 & 0 & 0 & 0 \\ -1 & 1 & 0 & 0 & 0 \\ 0 & -1 & 1 & 0 & 0 \\ 0 & 0 & -1 & 1 & 0 \\ 0 & 0 & 0 & -1 & 1 \\ 0 & 0 & 0 & 0 & -1 \end{bmatrix} \qquad x = \begin{bmatrix} H_A \\ H_B \\ H_C \\ H_D \\ H_E \end{bmatrix}$$

$$l_0 = \begin{bmatrix} \Delta H_1 + BM_1 \\ \Delta H_2 \\ \Delta H_3 \\ \Delta H_4 \\ \Delta H_5 \\ \Delta H_6 - BM_2 \end{bmatrix} = \begin{bmatrix} 818.912 \\ 2.656 \\ 2.902 \\ 3.128 \\ 1.330 \\ -828.923 \end{bmatrix}_m$$

Assuming the weighting model based on the inverse of the levelling section's length, the weight matrix P is given as follows:

$$
P = \begin{bmatrix}
10.9409 & 0 & 0 & 0 & 0 & 0 \\
0 & 14.3062 & 0 & 0 & 0 & 0 \\
0 & 0 & 9.9602 & 0 & 0 & 0 \\
0 & 0 & 0 & 11.0865 & 0 & 0 \\
0 & 0 & 0 & 0 & 14.2857 & 0 \\
0 & 0 & 0 & 0 & 0 & 11.1111
\end{bmatrix}
$$

Considering the weight matrix given above and using Eqs. (3.84), (3.85), (3.86) and (3.87).

$$
N = \begin{bmatrix}
25.24707 & -14.30615 & 0 & 0 & 0 \\
-14.30615 & 24.26631 & -9.96016 & 0 & 0 \\
0 & -9.96016 & 21.04663 & -11.08647 & 0 \\
0 & 0 & -11.08647 & 25.37219 & -14.28571 \\
0 & 0 & 0 & -14.28571 & 25.39683
\end{bmatrix}
$$

$$
n = \begin{bmatrix}
8,921.65275 \\
9.09276 \\
-5.77411 \\
15.67849 \\
9,229.25556
\end{bmatrix}
\qquad
x = \begin{bmatrix}
818.9111 \\
821.5664 \\
824.4674 \\
827.5946 \\
828.9239
\end{bmatrix}_m
$$

Using Eq. (3.79),

$$
v^{T} = [-0.89 \;\; -0.68 \;\; -0.98 \;\; -0.88 \;\; -0.68 \;\; -0.88]_{mm}
$$

In this example, since the approximate values are equal to zero, the adjusted values of the unknowns are equal to the correction values given in the x vector. Using Eq. (13.28).

$$
s_0^2 = 4.884 * 10^{-5}\,m^2 = 48.84\,mm^2
$$
$$
s_0 = \pm 6.99\,mm
$$

Using Eq. (11.74) and the value of s_0^2 calculated above, the variance-covariance matrix for the parameters is given as follows:

$$\Sigma_{xx} = s_0^2 N^{-1} = \begin{bmatrix} 3.6668 & 3.0572 & 2.1817 & 1.3952 & 0.7848 \\ 3.0572 & 5.3953 & 3.8503 & 2.4622 & 1.3850 \\ 2.1817 & 3.8503 & 6.2468 & 3.9948 & 2.2471 \\ 1.3952 & 2.4622 & -11.0865 & 5.3717 & 3.0216 \\ 0.7848 & 1.3850 & 2.2471 & 3.0216 & 3.6226 \end{bmatrix}$$

Using the diagonal values in the variance-covariance matrix for the parameters, the standard deviations of the adjusted coordinates are given as follows.

$H_A = 818.9111$ m $\pm$ 1.9 mm $H_B = 821.5664$ m $\pm$ 2.3 mm $H_C = 824.4674$ m $\pm$ 2.5 mm
$H_D = 827.5946$ m $\pm$ 2.3 mm $H_E = 828.9239$ m $\pm$ 1.9 mm

Using Eqs. (13.20) and (3.94), the following results are obtained for the adjusted observations.

$\Delta H_1 = 3.4721$ m $\pm$ 1.91 mm $\Delta H_2 = 2.6553$ m $\pm$ 1.72 mm $\Delta H_3 = 2.9010$ m $\pm$ 1.99 mm
$\Delta H_4 = 3.1271$ m $\pm$ 1.90 mm $\Delta H_5 = 1.3293$ m $\pm$ 1.72 mm $\Delta H_6 = -1.3799$ m $\pm$ 1.90 mm

Example 13.12 Using the levelling network shown in Fig. 13.20 and the readings given in Table 13.11, calculate the adjusted elevations of points (A), (B), and (C). Assume that the elevation of the BM is equal to 815.042 m.

Table 13.11 Field observations

Levelling run	Difference in elevation (ΔH_i) [m]	Levelling run length [km]
(1) BM-A	1.0154	1.21
(2) B-BM	11.5610	1.11
(3) B-A	12.5750	0.65
(4) B-C	6.4145	0.75
(5) C-A	6.1610	1.32
(6) C-BM	5.1410	1.23

Solution:

Using Fig. 13.20, the following observation equations can be written.

$$\Delta H_1 + v\Delta H_1 = H_A - BM$$
$$\Delta H_2 + v\Delta H_2 = BM - H_B$$
$$\Delta H_3 + v\Delta H_3 = H_A - H_B$$
$$\Delta H_4 + v\Delta H_4 = H_C - H_B$$
$$\Delta H_5 + v\Delta H_5 = H_A - H_C$$
$$\Delta H_6 + v\Delta H_6 = BM - H_C$$

So, there are six equations, three unknowns and three redundancies. To apply the Parametric adjustment model, the following error equations can be written as

$$v\Delta H_1 = 1H_A + 0H_B + 0H_C - (\Delta H_1 + BM)$$
$$v\Delta H_2 = 0H_A - 1H_B + 0H_C - (\Delta H_2 - BM)$$
$$v\Delta H_3 = 1H_A - 1H_B + 0H_C - \Delta H_3$$
$$v\Delta H_4 = 0H_A - 1H_B + 1H_C - \Delta H_4$$
$$v\Delta H_5 = 1H_A + 0H_B - 1H_C - \Delta H_5$$
$$v\Delta H_6 = 0H_A + 0H_B - 1H_C - (\Delta H_6 - BM)$$

From the error equations above and using Eq. (3.79).

$$
A = \begin{bmatrix} 1 & 0 & 0 \\ 0 & -1 & 0 \\ 1 & -1 & 0 \\ 0 & -1 & 1 \\ 1 & 0 & -1 \\ 0 & 0 & -1 \end{bmatrix}
\quad x = \begin{bmatrix} H_A \\ H_B \\ H_C \end{bmatrix}
\quad l_0 = \begin{bmatrix} \Delta H_1 + BM_1 \\ \Delta H_2 - BM \\ \Delta H_3 \\ \Delta H_4 \\ \Delta H_5 \\ \Delta H_6 - BM \end{bmatrix}
= \begin{bmatrix} 816.0574 \\ -803.4810 \\ 12.5750 \\ 6.4145 \\ 6.1610 \\ -809.8980 \end{bmatrix}_m
$$

Assuming the weighting model based on the inverse of the levelling run length, the weight matrix **P** *is given as follows:*

$$
P = \begin{bmatrix}
0.826446 & 0 & 0 & 0 & 0 & 0 \\
0 & 0.900901 & 0 & 0 & 0 & 0 \\
0 & 0 & 1.538462 & 0 & 0 & 0 \\
0 & 0 & 0 & 1.33333 & 0 & 0 \\
0 & 0 & 0 & 0 & 0.757576 & 0 \\
0 & 0 & 0 & 0 & 0 & 0.813008
\end{bmatrix}
$$

Considering the weight matrix given above and using Eqs. (3.84), (3.85), (3.86) *and* (3.87).

$$N = \begin{bmatrix} 3.122484 & -1.538462 & -0.757576 \\ -1.538462 & 3.772696 & -1.333333 \\ -0.757576 & -1.333333 & 2.903917 \end{bmatrix} \quad n = \begin{bmatrix} 698.4412 \\ 695.9579 \\ 662.3389 \end{bmatrix}$$

$$x = \begin{bmatrix} 816.0573 \\ 803.4820 \\ 809.8969 \end{bmatrix}_m$$

Using Eq. (3.79),

$$v^T = \begin{bmatrix} -0.05 & -1.04 & 0.30 & 0.36 & -0.55 & 1.10 \end{bmatrix}_{mm}$$

Using Eq. (13.28),

$$s_0^2 = 8.366267 * 10^{-7} \text{ m}^2 = 0.8366267 \text{ mm}^2$$
$$s_0 = \pm 0.91 \text{ mm}$$

The variance-covariance matrix of the parameters is given as follows:

$$\Sigma_{xx} = s_0^2 N^{-1} = \begin{bmatrix} 0.465043 & 0.277555 & 0.248760 \\ 0.277555 & 0.430369 & 0.270013 \\ 0.248760 & 0.270013 & 0.476976 \end{bmatrix}_{mm^2}$$

Using the diagonal values in the variance-covariance matrix of the parameters, the standard deviations of the adjusted coordinates are given as follows:

$$H_A = 816.0573 \text{ m} \pm 0.68 \text{ mm}$$
$$H_B = 803.4820 \text{ m} \pm 0.66 \text{ mm}$$
$$H_C = 809.8969 \text{ m} \pm 0.69 \text{ mm}$$

Using Eqs. (13.20) and (3.94), the following results are obtained for the adjusted observations.

$$\Delta H_1 = 1.01535 \text{ m} \pm 0.68 \text{ mm}$$
$$\Delta H_2 = 11.55996 \text{ m} \pm 0.66 \text{ mm}$$
$$\Delta H_3 = 12.5730 \text{ m} \pm 0.58 \text{ mm}$$
$$\Delta H_4 = 6.41486 \text{ m} \pm 0.61 \text{ mm}$$
$$\Delta H_5 = 6.16045 \text{ m} \pm 0.67 \text{ mm}$$
$$\Delta H_6 = 5.14510 \text{ m} \pm 0.69 \text{ mm}$$

13.5   Sources of Error in Differential Levelling

Errors that occur in differential levelling come from a variety of sources and, like all measurement errors, can be classified as blunders, systematic or random errors. The reliability of the results obtained will be greater if there are fewer blunders and systematic errors. They must therefore be avoided and, if they persist, corrected.

It is, therefore, necessary to establish measurement control routines and have considerable knowledge of the relationships between the physical and mathematical models used for differential levelling to avoid blunders and systematic errors. In addition, there are some measurement procedure recommendations that can increase confidence in the quality of the results, as shown below.

Procedures for checking systematic and random errors associated with surveying instruments are presented in Chap. 10. Details on the blunders that can occur during differential levelling and the systematic errors associated to environmental conditions affecting this type of surveying work are given below.

13.5.1   Blunders

The most common blunders in differential levelling are related to instrument setup and field readings, as shown below:

- Instrument incorrectly positioned, that is instrument not levelled or in an unstable position.
- Incorrect reading on the levelling staff.
- Incorrect reading annotation.
- Errors due to the incorrect assembly of the moving parts (sliding joints) of the levelling staff.

13.5.2   Systematic Errors Related to Environmental Conditions

13.5.2.1   Error Due to the Curvature of the Earth

Assuming for simplicity that the level surfaces are concentric spheres due to the curvature of the Earth, there will be an error δR_0 in the sight readings, as shown in Fig. 13.22.

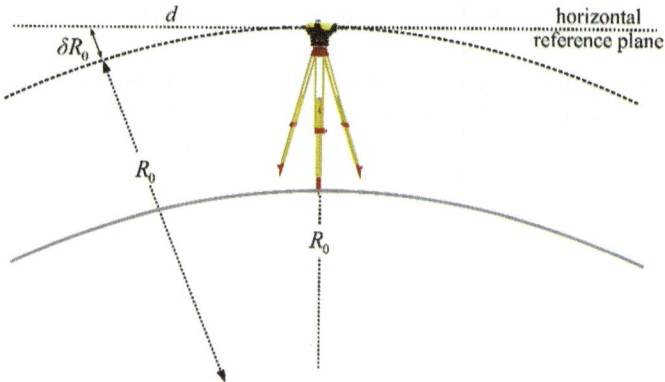

Fig. 13.22 Influence of Earth's curvature on differential levelling

R_0 = local mean radius of the Earth
d = sighting distance
δR_0 = correction due to the Earth's curvature

From Fig. 13.22.

$$R_0^2 + d^2 = (R_0 + \delta R_0)^2 = R_0^2 + (2R_0 * \delta R_0) + \delta R_0^2 \qquad (13.29)$$

Resulting in

$$\delta R_0 = \frac{d^2}{2R_0} - \frac{\delta R_0^2}{2R_0} \qquad (13.30)$$

Because the second term of Eq. (13.30) is very small compared to the first, the value of δR_0 can be calculated using Eq. (13.31).

$$\delta R_0 = \frac{d^2}{2R_0} \qquad (13.31)$$

For example, considering a region where the local mean radius of the Earth is of the order of 6,362,000 m, the expected error due to the curvature of the Earth for differential levelling at a sighting distance of 50 m is approximately 0.2 mm.

It is important to point out that the error δR_0 is the same for the backsight and foresight readings. So, they cancel each other out if the readings are taken at the same sighting distance from the instrument.

13.5.2.2   Error Due to the Vertical Atmospheric Refraction

As explained in Chap. 7, due to vertical atmospheric refraction, the light beam propagates according to a curve in the atmosphere, and this causes the readings on the staff to be lower than the expected readings, as shown in Fig. 13.23.

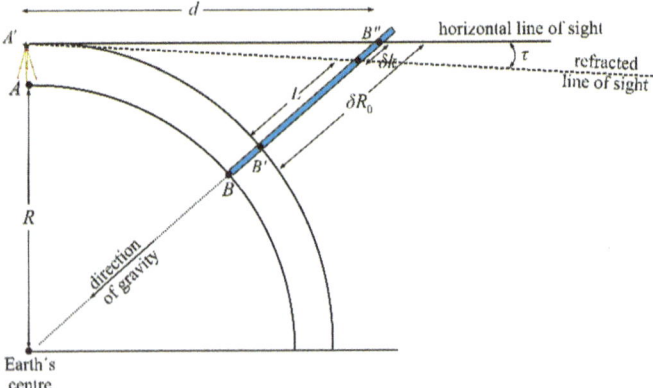

Fig. 13.23 Effects of vertical atmospheric refraction and Earth's curvature on differential levelling

d = distance between the instrument and the levelling staff
τ = angle of refraction
δk = correction due to vertical atmospheric refraction
δR_0 = correction due to the Earth's curvature

Due to the Earth's curvature, point (B'), which is at the same elevation as point (A'), would appear on the line of sight at height (B''). In practice, however, the line of sight is refracted due to atmospheric refraction, and point (B'') is read on the levelling staff at height (L). Therefore, a deviation equal to δk is observed, which, being a small value, allows consideration of the mathematical relationship indicated in Eq. (13.32).

$$\tau = \frac{\delta k}{d} \tag{13.32}$$

Replacing Eq. (7.28) in (13.32).

$$\frac{\delta k}{d} = k * \frac{d}{2R_0} \quad \rightarrow \quad \delta k = k * \frac{d^2}{2R_0} = k * \delta R_0 \tag{13.33}$$

Resulting in

$$(\delta R_0 - \delta k) = \delta R_0 * (1 - k) = \frac{d^2}{2R_0} * (1 - k) \tag{13.34}$$

This correction is very small, especially for short sights. It is also eliminated if the readings are taken at the same sighting distance from the instrument. In any case, combining the error δR_0 due to the Earth's curvature with the error δk due to the vertical atmospheric refraction, the reading taken in the field on the levelling staff must be corrected by a value equal to $(\delta R_0 - \delta k)$. This value, for $k = 0.13$ and $R_0 = 6,$ $371,000$ m, equals $0.00006828d^2$ [mm] for (d) in metres, i.e., an error of 1.0 mm for a sighting distance of the order of 122 m. This also means that for current engineering levelling, at distances of 30–40 m, it is of the order of a tenth of a millimetre. However, to minimise the influence of these errors, it is recommended that the sum of the distance differences between the forward and backward readings should not exceed 5 m for high-precision levelling, although this sum can reach 20 m, for example, for current Civil Engineering levelling. In the latter case, this means a variation between 30 cm and 1.25 m in each levelling section in a line levelling length of 1 km (considering 16 sections/km); therefore, distance estimation by optical measurement method or pacing is sufficient for most levelling works.

In addition to eliminating the aforementioned blunders and systematic errors, it is also recommended to follow some field procedures that may help to improve the levelling results.

- Whenever possible, carry out levelling with a series of measurements.
- For high-precision levelling, it is recommended to check the closure of each levelling section individually and, if necessary, perform more than one measurement in each of them, changing the position of the surveyor's level.
- Perform maximum sightings between 40 and 50 m, depending on the class of the instrument used.
- If a digital level is used, as shown in Sect. 9.5.2, the sighting distances may be greater (approximately 100 m). In this case, it is recommended to take at least three readings with a standard deviation less than or equal to 1.0 mm.
- When using a surveyor's level whose telescope has three reticules (upper, middle, lower), it is recommended to take readings on all three and average the readings on the upper and lower to verify the reading on the middle crosshair. Reading the three crosshairs allows the approximate calculation of the distance between the surveyor's level and the levelling staff, which is sufficient to maintain the recommended distance between sights.
- Carefully check the stability of the deployed benchmarks to ensure that they can be used reliably.
- Use levelling staffs made of invar whenever possible.
- Use steel turning plates for staff support in high-precision levelling.
- Secure the level bubble to the levelling staff.
- Use an umbrella to protect the surveying instrument and its tripod whenever possible.
- Avoid working during periods of high line-of-sight luminous fluctuation to avoid errors in reading the levelling staff.
- Calibrate the instrument and levelling staff regularly.

- Avoid using different units of measurement. In general, millimetres are used for levelling staff readings and metres for distances. Some surveyors recommend adding a zero to the left of values less than a thousand millimetres when reading values on the levelling staff.

13.6   Survey Standards Used for Levelling Specification

In the case of the line levelling in Example 13.4, the levelling work started at *BM* and ended at point (*C*) without any known control points other than the *BM* included in the run. However, due to various factors inherent to the measurement process, as indicated in Sect. 13.5, there are several errors that can affect the result of the survey. For this reason, as already mentioned, it is recommended for most levelling works to carry out a loop or link levelling procedure so that the closing error can be evaluated and, thus, a parameter indicative of the levelling quality can be obtained. Levelling without closing control should be avoided and should only be carried out in cases of extreme necessity. The use of network levelling should also be prioritised whenever the configuration of the work allows it.

The accuracy of levelling lines in engineering works is usually specified, classified and reported on the basis of height differences at the 95% confidence level, that is the standard for the vertical component of spatial data is a linear value specified so that the correct or theoretical location of the point lies within +/− this value 95% of the time. Allowable values for errors of closure are then specified based on this assumption, as shown in the following subsection.

13.6.1   Allowable Error of Closure

The allowable (or standard) error of closure for conventional line levelling is based on the error propagation theory considering the random errors that occur in reading a levelling staff, including compensator and pointing errors, and the reading error resulting from the levelling staff not being held vertically, as described in Sect. 10. 4.2. As discussed in the previous sections, it is assumed that blunders and systematic errors have been eliminated. In this situation, since a line levelling consists of a series of simple, consecutive levelling operations, as given by Eq. (13.9), for a sequence of *n* instrument setups, the standard deviation of the total level difference is given by Eq. (13.35).

$$s_{\Delta H(\text{total})} = \pm s_{sl} * \sqrt{2n} \qquad (13.35)$$

where

s_{sl} = standard deviation of a single levelling.

Assuming a 95% confidence level, the maximum elevation misclosure is given by Eq. (13.36).

$$s_{\Delta H(\text{total})} = \pm 1.96 s_{sl} * \sqrt{2n} \tag{13.36}$$

Considering that the standard deviation of a single levelling observation is composed of several random errors, as shown below, applying the general law of propagation of variances-covariances gives the standard deviation for a sighting length of 30 m, as shown in the following mathematical development.

1. Compensator error e_n for a compensator accuracy of 0.5″.

$$e_n = \pm \frac{0.5''}{206,264.8062''} * 30,000 = \pm 0.073 \text{ mm}$$

2. Levelling staff verticality error e_m considering a circular bubble level accuracy of 10′ and 2 m levelling staff reading.

$$e_m = \pm \frac{2,000}{2} * \left(\frac{10'}{3,437.7467708'}\right)^2 = \pm 0.008 \text{ mm}$$

3. Pointing error e_p considering a telescope magnification power of 32x.

$$e_p = \pm \frac{45''*30,000}{32 * 206,264.8062''} = \pm 0.204 \text{ mm}$$

The standard deviation of a single levelling observation, for a sighting length of 30 m is then given as follows:

$$s_{sl} = \sqrt{0.073^2 + 0.008^2 + 0.204^2} = \pm 0.217 \text{ mm}$$

Likewise, for the measurement conditions described above, considering Eq. (13.36), the estimated standard deviation for a 1 *km* line levelling is given as follows:

$$s_{\Delta H(1km)} = \pm 1.96 * 0.217 * \sqrt{\frac{2 * 1,000}{30}} = \pm 3.5 \text{ mm}$$

However, most line levelling standards are based on the standard error of differential levelling over a distance (d [Km]), where d is the one-way length of a link levelling, or the total loop length of a double run. In this case, the allowable error of closure Te for the levelling of the link or loop is specified according to the class of survey work, given by a multiplication factor k and the length d of the line levelling, as follows:

$$Te = \pm k * \sqrt{d} \tag{13.37}$$

For example, Table 13.12 shows the accuracy standards for vertical control of line levelling proposed by the Federal Geographic Data Committee (FGDC), USA.

Table 13.12 Minimum elevation closure standards for vertical control surveys (FDGC)

Order of accuracy	Elevation closure standard (mm) where d [km]
First order, class I	$3\sqrt{d}$
First order, class II	$4\sqrt{d}$
Second order, class I	$6\sqrt{d}$
Second order, class II	$8\sqrt{d}$
Third order	$12\sqrt{d}$
Construction layout	$24\sqrt{d}$

Typical tolerance limits range from $\pm 3\sqrt{d}$ to $\pm 4\sqrt{d}$ mm for precision levelling, and from $\pm 6\sqrt{d}$ mm to $\pm 8\sqrt{d}$ mm for construction sites, where d is given in km.

In the case of line levelling, the quality of the levelling is assessed by directly comparing the closing error e_n with an allowable value Te. If the error of closure is within the tolerance, the work is accepted, and the error must be distributed over the levelling sections.

In the case of the calculation presented above, assuming that the calculated result represents a 1 km one-way link levelling, the calculated standard deviation must be multiplied by $\sqrt{2}$ and the survey work would be classified as second order, class I, as given below.

$$s_{\Delta H(1km)} = \pm \sqrt{2} * 3.5 = \pm 4.9 \text{ mm} < \pm 6 \text{ mm}$$

In the case of vertical control standards for network levelling, the allowable value is generally based on the propagated standard deviation of the elevation difference in millimetres for a sample of adjusted control points (pairs) obtained from least squares adjustment, divided by the approximate distance in kilometres between the positions of the control point pairs, as given by Eq. (13.38).

$$Te = \pm \frac{s_{\Delta H}}{\sqrt{d}} \tag{13.38}$$

where

Te = allowable elevation difference
$s_{\Delta H}$= propagated standard deviation between a pair of control points
d = distance between control points, in kilometres

Table 13.13 shows an example of accuracy standards for vertical control of network levelling proposed by the Federal Geographic Data Committee (FGDC), USA.

Table 13.13 Accuracy standards for vertical control of directly connected points of levelling networks (FGDC)

Order of accuracy	Relative accuracy between directly connected points of benchmarks (standard deviation of elevation difference in mm)
First order, class I	$0.5\sqrt{d[\,km]}$
First order, class II	$0.7\sqrt{d[\,km]}$
Second order, class I	$1.0\sqrt{d[\,km]}$
Second order, class II	$1.3\sqrt{d[\,km]}$
Third order	$2.0\sqrt{d[\,km]}$

Example 13.13 Using the results of Example 13.12, check the order of accuracy of the levelling network using the accuracy standards proposed by the FGDC in Table 13.13.

Solution:

As stated by the FGDC standards for vertical control of directly connected points of the levelling network, the allowable value is based on the propagated standard deviation between a pair of control points, as given by Eq. (13.38). Thus, using the results from Example 13.12, the following standard deviations were obtained for control points (A), (B) and (C):

$$SH_A = \pm 0.68 \text{ mm} \quad SH_B = \pm 0.66 \text{ mm} \quad SH_C = \pm 0.69 \text{ mm}$$

Using the law of propagation of variances.

$$s_{\Delta H_{AB}} = \pm \sqrt{0.68^2 + 0.66^2} = \pm 0.95 \text{ mm} \quad \rightarrow \quad T_e = \pm \frac{0.95}{\sqrt{0.65}} = \pm 1.17 \text{ mm}$$

$$s_{\Delta H_{AC}} = \pm \sqrt{0.68^2 + 0.69^2} = \pm 0.97 \text{ mm} \quad \rightarrow \quad T_e = \pm \frac{0.97}{\sqrt{1.32}} = \pm 0.84 \text{ mm}$$

$$s_{\Delta H_{AB}} = \pm \sqrt{0.66^2 + 0.69^2} = \pm 0.95 \text{ mm} \quad \rightarrow \quad T_e = \pm \frac{0.95}{\sqrt{0.75}} = \pm 1.10 \text{ mm}$$

Compared to the values in Table 13.13, *the network adjustment is considered as a second order class II.*

13.7   Determining the Precision of Levelling Instruments

As with other measuring instruments, levelling instruments should be checked periodically or before carrying out major levelling work. It is, therefore, recommended to adopt the specifications indicated in the ISO 17.123-2, which specifies field procedures for determining and evaluating the precision of surveyor's levels used in Civil Engineering construction and surveying measurements. As indicated in the text of the ISO standard, *the proposed tests are intended to be field verifications of the suitability of a particular instrument for the immediate task at hand and to satisfy the requirements of other standards.*

The ISO standard text presents two field procedures: the simplified and the full test. The former is intended to verify the precision of a surveyor's level to be used for area levelling applications where measurements with unequal lengths are common, such as in Civil Engineering construction sites. The full test procedure is used to determine the best possible precision for precise levelling in linear applications, as in the case of big Civil Engineering projects. The measurement precision achieved in the full test procedure is expressed in terms of the experimental standard deviation of a 1 km double-run levelling in accordance with the conditions specified by the equipment manufacturers. The field procedures for both tests are fully described in the text of the ISO standard. Readers are encouraged to consult it for further details.

13.8   Trigonometric Levelling

In Geomatics applied to Civil Engineering, there are some levelling works where the terrain conditions, such as highly variable terrain topography and the need for cost-effective and rapid data acquisition, make conventional differential levelling methods inefficient, empathising the use of a levelling approach called *trigonometric levelling*.

The determination of the difference in elevation between two points, in this case, is carried out by measuring the slope distance and the vertical angle between them using a total station, as shown in Fig. 13.24. Thus, as targets can be placed at different heights above the ground, it is possible to extend the range to several hundred metres.

Three types of trigonometric levelling techniques are currently used in Geomatics applied to Civil Engineering: *unidirectional levelling, leap-frog levelling*, and *reciprocal levelling*, as described in the following subsections.

13.8.1   Unidirectional Trigonometric Levelling for Short Lines

The simplest levelling technique for short lines (300 m or less) is the unidirectional trigonometric levelling, which requires a tripod-mounted total station at one of the ground points and a reflector prism installed at the other. The total station is set up at only one point, and observations are performed in only one direction. From this setup, as shown in Fig. 13.24, the operator needs to measure the height of the instrument h_i, the height of the reflector prism h_r, the slope distance d'_{PQ} and the vertical angle β_{PQ} or z_{PQ} from (P) to (Q). The calculations, in this case, do not consider the curvature of the Earth and vertical atmospheric refraction.

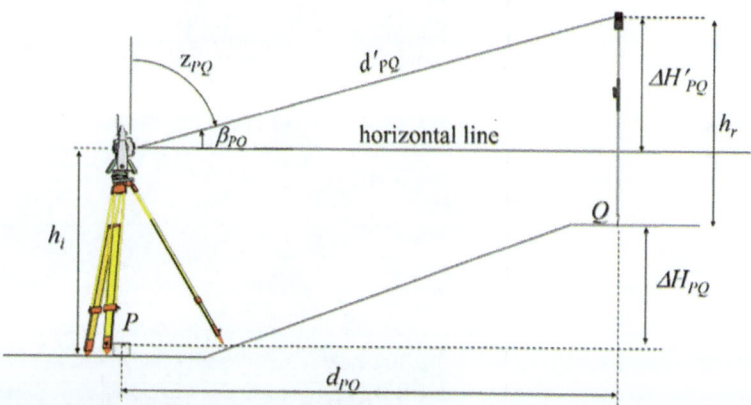

Fig. 13.24 Principle of trigonometric levelling

In Fig. 13.24:

$\Delta H'_{PQ}$ = vertical distance from the centre of the total station and the centre of the reflector prism

ΔH_{PQ} = difference in elevation between points (Q) and (P)

d'_{PQ} = slope distance for sight PQ

d_{PQ} = horizontal distance between points (P) and (Q)

β_{PQ} = vertical altitude angle

z_{PQ} = vertical zenith angle

h_i = height of the instrument

h_r = height of the reflector prism

The height of the instrument is usually measured with a steel measuring tape, considering the distance between the top of the ground mark and the mark on the side of the instrument, as shown in Fig. 13.25. Some instrument manufacturers provide special accessories to assist in measuring this distance (see Fig. 13.26), and others allow it to be measured directly using a laser distance metre inserted into the instrument's alidade.

The height of the reflector prism is the distance from the bottom of the prism pole to the centre of the reflector prism. Numerical values are printed on the body of the extendable pole, indicating the height of the reflector prism as it is extended. See Fig. 13.27.

Fig. 13.25 Measurement of the height of the instrument using a measuring tape

Fig. 13.26 Instrument height metre. (Courtesy Leica Geosystems AG)

Fig. 13.27 Pole height metre. (Courtesy Seco Manufacturing Co)

As presented in Chap. 7, the following equations can be used to calculate the elevation difference by unidirectional trigonometric levelling for short lines:

$$\Delta H'_{PQ} = d'_{PQ} * \sin(\beta_{PQ}) \text{ or } \Delta H'_{PQ} = d'_{PQ} * \cos(z_{PQ}) \text{ as given by Eq. (7.3)}$$
$$d_{PQ} = d'_{PQ} * \cos(\beta_{PQ}) \text{ or } d_{PQ} = d'_{PQ} * \sin(z_{PQ}) \text{ as given by Eq. (7.1)}$$

The difference in elevation ΔH_{PQ} between points (Q) and (P) is given by Eq. (13.39).

$$\Delta H_{PQ} = h_i + \Delta H'_{PQ} - h_r = \Delta H'_{PQ} + (h_i - h_r) = d'_{PQ} * \cos(z_{PQ}) + (h_i - h_r)$$
$$(13.39)$$

Thus, if the elevation of the points (P) or (Q) is known, the elevation of the other can be calculated using Eqs. (13.40) and (13.41).

$$H_Q = H_P + \Delta H_{PQ} \qquad (13.40)$$

$$H_P = H_Q - \Delta H_{QP} \qquad (13.41)$$

The above equations are general and valid for any values of β or z. However, the reader should note that if the angle β is used, it is necessary to consider its algebraic sign $(+)$ or $(-)$ to calculate the vertical distance $\Delta H'$. On the other hand, if the angle

z is used, the value of the vertical distance will already have the algebraic sign corresponding to the value of the z angle.

Trigonometric levelling is simpler and faster to use than differential levelling. However, it is difficult to achieve a comparable level of accuracy because the number of variables that can affect the quality of the measurements is much greater than for differential levelling, as described in the following subsections.

Example 13.14 Two sets of measurements were taken from point (P) to determine the elevations of points (Q) and (S) by trigonometric levelling. The results are given in Table 13.14. Assuming that the elevation of point (P) is 821.175 m, the height of the instrument is 1.561 m, and the height of the reflector prism is 1.900 m, calculate the elevations of points (Q) and (S).

Table 13.14 Field measurement

Station	Sighting point	Slope distance [m]	Zenith angle
P	Q	245.732	87° 34′29″
	S	292.146	92° 25′31″

Solution:

Using Eq. (13.39),

$$\Delta H_{PQ} = 245.732 * \cos(87° 34' 29'') + (1.561 - 1.900) = 10.060 \text{ m}$$
$$\Delta H_{PS} = 292.146 * \cos(92° 25' 31'') + (1.561 - 1.900) = -12.702\text{m}$$

Using Eqs. (13.40) and (13.41),

$$H_Q = 821.175 + 10.060 = 831.235 \text{ m}$$
$$H_S = 821.175 - 12.702 = 808.473 \text{ m}$$

13.8.2   Unidirectional Trigonometric Levelling for Long Lines

If the levelling distance is long enough to require corrections for Earth's curvature and vertical atmospheric refraction, the geometry shown in Fig. 13.28 and the corresponding equations must be considered.

Note from Fig. 13.28 that, due to vertical atmospheric refraction, although the instrument's telescope is pointing at point (B'), the observer sees point (B). It is, therefore, necessary to apply corrections due to the curvature of the Earth and vertical atmospheric refraction, as described below.

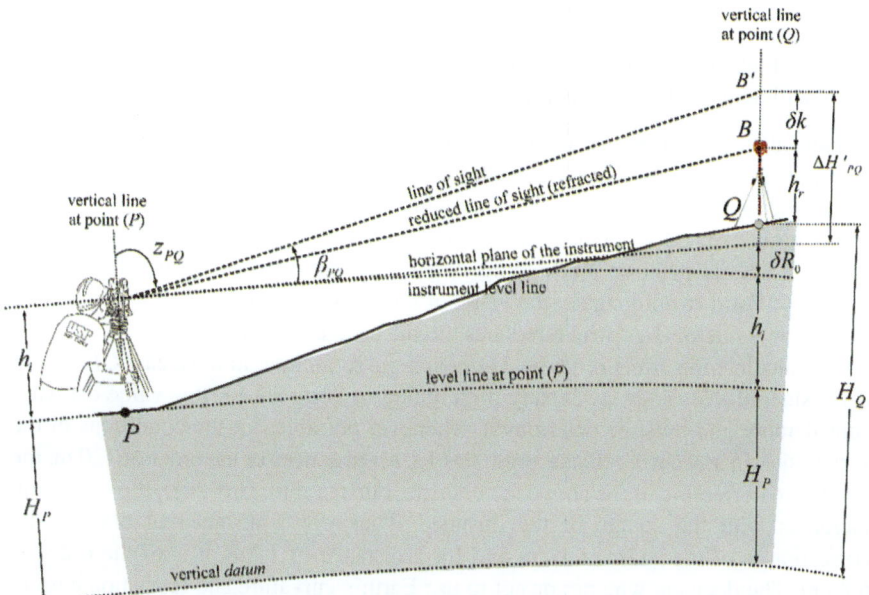

Fig. 13.28 Effects of vertical atmospheric refraction and Earth's curvature on trigonometric levelling. (Adapted from Uren and Price 2010)

$$H_Q = H_P + h_i + \delta R_0 + \Delta H'_{PQ} - \delta k - h_r = H_P + \Delta H'_{PQ} + (h_i - h_r) + (\delta R_0 - \delta k) \tag{13.42}$$

Thus,

$$\Delta H_{PQ} = \Delta H'_{PQ} + (h_i - h_r) + (\delta R_0 - \delta k) \tag{13.43}$$

Then, considering Eqs. (13.31) and (13.34),

$$(\delta R_0 - \delta k) = (1 - k) * \frac{\left[d'_{PQ} * \sin(z_{PQ}) \right]^2}{2R_0} \tag{13.44}$$

Finally,

$$HQ = HP + d'_{PQ} * \cos(z_{PQ}) + (hi - hr) + (1 - k) * \frac{\left[d'_{PQ} * \sin(z_{PQ}) \right]^2}{2R0} \tag{13.45}$$

where

R_0= local mean radius of the Earth
k = coefficient of atmospheric refraction

Note that the difference between considering the curvature of the Earth and the vertical atmospheric refraction into account or not is given by the correction $\delta R_0 - \delta k$, also called $(c\text{-}r)$ in many surveying books.

Equation (13.45) is valid for uphill and downhill sights. For uphill sights, the value of $\Delta H'$ is positive and adding $\delta R_0 - \delta k$ increases the total difference in elevation. For downhill sights, the value of $\delta R_0 - \delta k$ is also added, but this time to a negative value of $\Delta H'$, which reduces the difference in elevation.

To calculate the effect of vertical atmospheric refraction, in most cases, $k = 0.13$ is considered. However, as already highlighted in Sect. 7.2.11, this value can vary considerably and must be determined, whenever possible, for the conditions of the survey site. In any case, considering $k = 0.13$, for distances of the order of 120 m, the combined correction of the above factors is less than 1 mm. However, the correction increases with the square of the distance. Thus, for a distance of 400 m, the correction is of the order of 1 cm and for a distance of 1 km, it is of the order of 6.8 cm. The decision whether or not to use Earth's curvature and refraction corrections depends on the accuracy expected for the levelling work.

13.8.2.1 Error Propagation in Unidirectional Trigonometric Levelling for Long Lines

Applying the general law of propagation of variances to Eq. (13.43), the standard deviation of the difference in elevation $s_{\Delta H}$ is given as follows:

$$s_{\Delta H}^2 = J^T \Sigma_{ll} J \tag{13.46}$$

where Σ_{ll} is given by Eq. (13.47).

$$\Sigma_{ll} = \begin{bmatrix} s_{d'}^2 & 0 & 0 & 0 & 0 \\ 0 & s_z^2 & 0 & 0 & 0 \\ 0 & 0 & s_k^2 & 0 & 0 \\ 0 & 0 & 0 & s_{hi}^2 & 0 \\ 0 & 0 & 0 & 0 & s_{hr}^2 \end{bmatrix} \tag{13.47}$$

where

s_d= standard deviation of the slope distance
s_z=standard deviation of the vertical zenith angle
s_k= standard deviation of the coefficient of atmospheric refraction

s_{hi}= standard deviation of the height of the instrument
s_{hr}= standard deviation of the height of the reflector prism.

Likewise,

$$J^T = \left[\left(\frac{\partial \Delta H}{\partial d'} \right)_0 \quad \left(\frac{\partial \Delta H}{\partial z} \right)_0 \quad \left(\frac{\partial \Delta H}{\partial k} \right)_0 \quad \left(\frac{\partial \Delta H}{\partial h_i} \right)_0 \left(\frac{\partial \Delta H}{\partial h_r} \right)_0 \right] \qquad (13.48)$$

$$\left(\frac{\partial \Delta H}{\partial d'} \right)_0 = \cos(z_{PQ}) + \frac{1-k}{R_0} * d'_{PQ} * \sin^2(z_{PQ}) \qquad (13.49)$$

$$\left(\frac{\partial \Delta H}{\partial z} \right)_0 = -d'_{PQ} * \sin(z_{PQ}) + \frac{1-k}{R_0} * \left(d'_{PQ} \right)^2 * \sin(z_{PQ}) * \cos(z_{PQ}) \qquad (13.50)$$

$$\left(\frac{\partial \Delta H}{\partial k} \right)_0 = \frac{- \left[d'_{PQ} * \sin(z_{PQ}) \right]^2}{2R_0} \qquad (13.51)$$

$$\left(\frac{\partial \Delta H}{\partial h_i} \right)_0 = 1 \qquad (13.52)$$

$$\left(\frac{\partial \Delta H}{\partial h_r} \right)_0 = -1 \qquad (13.53)$$

For example, in the case of a trigonometric levelling of a 100 m span carried out with a total station with linear precision of ±(2 mm + 2 ppm) and an angular precision of ±5″, if the precision of the coefficient of refraction is equal to ±0.1, the instrument and reflector prism height precisions are ±3 mm and ±1 mm, respectively, and the measured vertical zenith angle is 89°00′00″, a levelling precision of ±4.0 mm is expected, as long as the effects of Earth's curvature and vertical atmospheric refraction are taken into consideration.

Note that the precision of the distance has little influence on the levelling precision. If the heights of the instrument and the reflector prism are well-measured, the precision of the angle measurement is decisive, so it is recommended to use calibrated instruments and carry out a series of measurements on both sides of the telescope. Therefore, for engineering work where a precision of ±1 cm is acceptable, unidirectional trigonometric levelling can be used for distances less than 220 m, ignoring the effects of Earth's curvature and atmospheric refraction.

The reader should remember the difficulty of correctly determining the value of the atmospheric refraction coefficient, which may affect the precision of the trigonometric levelling by ±1 to ±2 mm. To avoid the effects of the Earth's curvature and vertical atmospheric refraction, it is advisable to carry out trigonometric levelling, wherever possible, using the leap-flog technique, as described in Sect. 13.8.3, or using reciprocal and simultaneous observations, as described in Sect. 13.8.4.

Example 13.15 Using the data from Example 7.5, calculate the elevation of the point (Q) considering the curvature of the Earth and the vertical atmospheric refraction. Assuming that the instrument used has a linear precision of ±(1 mm + 1 ppm), an angular precision of ±1″ (specified by the manufacturer), a precision of the vertical atmospheric refraction coefficient of ±0.04 and that the heights of the instrument and the reflector prism were measured with a precision of the order of ±1 mm, calculate the standard deviation of the levelling work.

Solution:
   From the data in Example 7.5.

$d'_{PO} = 1,045.022\,\text{m}$ $z_{PO} = 84°32'59''$ $H_P = 700.456\,\text{m}$
$h_i = 1.269\,\text{m}$ $h_r = 1.273\,\text{m}$ $k = 0.380$ $R_0 = 6,362,745\,\text{m}$

Using Eq. (13.45),

$$H_Q = 700.456 + 1,045.022 * \cos(84°32'59'') + (1.269 - 1.273)$$
$$+ (1 - 0.380) * \frac{[1,045.022 * \sin(84°32'59'')]^2}{2 * 6,362,745}$$
$$H_Q = 799.768\,\text{m}$$

Using the additive calculation for the standard deviation of the slope distance.

$$s_{d'_{PQ}} = \pm(1 + 1 * 1.045022) = \pm 2.045\,\text{mm}$$

Considering that the zenith angle was measured in a single face reading.

$$sz = 2'' = 9.696274 * 10^{-6}\,\text{rad}$$

Considering the given standard deviations, the matrix of variance-covariance of observations is given as follows:

$$\boldsymbol{\Sigma}_{ll}=\begin{bmatrix} 4.182115*10^{-6} & 0 & 0 & 0 & 0 \\ 0 & 9.401772*10^{-11} & 0 & 0 & 0 \\ 0 & 0 & 1.60*10^{-3} & 0 & 0 \\ 0 & 0 & 0 & 1.00*10^{-6} & 0 \\ 0 & 0 & 0 & 0 & 1.00*10^{-6} \end{bmatrix}$$

Using Eq. (13.48),

$$\boldsymbol{J}^T = [0.09508281 \quad -1,040.2287494 \quad -0.085043 \quad 1 \quad -1]$$

Using Eq. (13.46),

$$s_{\Delta H}^2 = 1.153554*10^{-4}\,\mathrm{m}^2$$

$$s_{\Delta H} = \pm 0.01074\,\mathrm{m} = \pm 10.7\,\mathrm{mm}$$

Finally,

$$H_Q = 799.768\,\mathrm{m} \pm 10.7\,\mathrm{mm}$$

13.8.3   Leap-Frog Trigonometric Levelling

Trigonometric levelling can also be used for line levelling using a field measurement method called the *leap-frog* technique. In this case, the observations are performed in a similar way to differential levelling, i.e., taking backsight and foresight readings at each total station setup between turning points, as shown in Fig. 13.29.

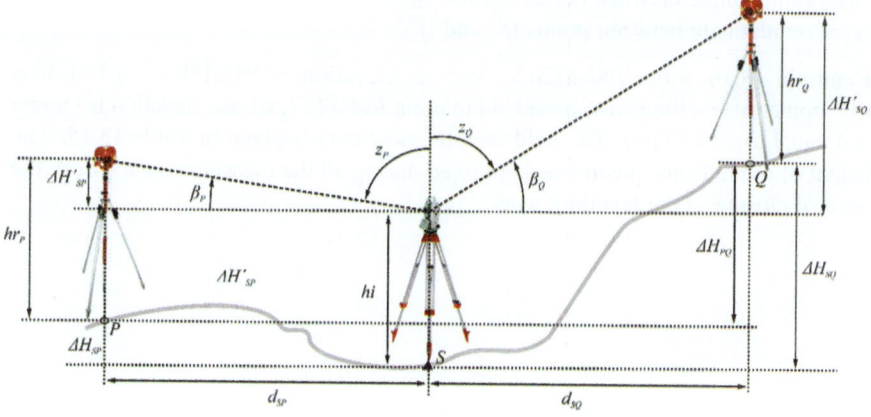

Fig. 13.29 Illustration of leap-frog trigonometric levelling

As shown in Fig. 13.29, a total station is set up over point (S) approximately halfway between ground points (P) and (Q) (not necessarily on the PQ line), sighting the reflector prism placed at point (P) and carrying out a vertical angle and slope distance observations. The reflector prism must then be placed at point (Q) for a new observation.

According to Fig. 13.29 and Eq. (13.39), the following equations can be written for short line observations:

From point (S) to point (P):

$$\Delta H_{SP} = \Delta H'_{SP} + (h_i - h_{rp}) \tag{13.54}$$

From point (S) to point (Q):

$$\Delta H_{SQ} = \Delta H'_{SQ} + (h_i - h_{rQ}) \tag{13.55}$$

Then,

$$\Delta H_{PQ} = \Delta H_{SQ} - \Delta H_{SP} = \left(\Delta H'_{SQ} - \Delta H'_{SP}\right) - \left(h_{rQ} - h_{rp}\right) \tag{13.56}$$

If the target heights are the same in the foresight and backsight observations, Eq. (13.56) can be written as follows:

$$\Delta H_{PQ} = d'_{SQ} * \cos(z_{SQ}) - d'_{SP} * \cos(z_{SP}) \tag{13.57}$$

where

ΔH_{PQ} = difference in elevation between points (P) and (Q)
d'_{SQ} = slope distance for sight SP
d'_{SP} = slope distance for sight SQ
z_{SQ} = zenith angle between points (S) and (Q)
z_{SP} = zenith angle between points (S) and (P)

Example 13.16 From *BM-EESC1*, with an elevation of 845.150 m, a leap-frog trigonometric levelling was carried out to point *BM-STT1*, whose elevation is known and equal to 815.592 m. The field measurement data is given in Table 13.15. The height of the reflector prism was kept fixed during all the measurements. Check the error of closure of the levelling work.

Table 13.15 Field measurement and calculation results

Sighting point	BS Zenith angle	BS Slope distance [m]	FS Zenith angle	FS Slope distance [m]	Difference in elevation Eq. 13.55 [m]	Elevation [m]
BM-EESC1	84° 51′52″	23.5355	–	–	–	**845.150**
1	86° 44′31″	25.9653	91° 49′08″	24.0742	−2.8708	842.279
2	87° 28′52″	53.0878	90° 38′55″	25.5520	−1.7649	840.514
3	87° 04′52″	55.8535	91° 39′58″	53.1692	−3.8790	836.635
4	87° 04′57″	65.6203	91° 34′51″	54.0200	−4.3344	832.301
5	85° 42′03″	51.5500	92° 30′35″	65.4459	−6.2057	826.095
6	87° 46′34″	87.1178	91° 35′43″	51.4898	−5.2978	820.797
7	89° 05′42″	19.7056	91° 36′11″	72.2160	−5.4008	815.396
BM-STT1			88° 45′51″	22.5358	0.1748	**815.571**

Solution:

The solution to this problem is given by Eq. (13.57). *The calculation results are shown in the highlighted columns of* Table 13.15. *Note that the calculation ignores the Earth's curvature and atmospheric refraction due to the small and approximately equal levelling distances.*

Comparing the levelling result with the known elevation of the BM-STT1, it is verified that there was an error of closure of − 21 mm.

13.8.3.1 Error Propagation in Leap-Frog Trigonometric Levelling

Applying the general law of propagation of variances to Eq. (13.57), the standard deviation of the difference in elevation $s_{\Delta H}$ can be calculated using Eqs. (13.46) and (13.58), (13.59), (13.60), (13.61), (13.62) and (13.63).

$$\Sigma_{ll} = \begin{bmatrix} Sd'^2_{SQ} & 0 & 0 & 0 \\ 0 & Sz^2_{SQ} & 0 & 0 \\ 0 & 0 & Sd'^2_{SP} & 0 \\ 0 & 0 & 0 & Sz^2_{SP} \end{bmatrix} \tag{13.58}$$

with

$S_{d'}$ = standard deviation of the slope distance
S_z = standard deviation of the zenith angle

Likewise,

$$J^T = \left[\left(\frac{\partial \Delta H}{\partial d'_{SQ}} \right)_0 \quad \left(\frac{\partial \Delta H}{\partial z_{SQ}} \right)_0 \quad \left(\frac{\partial \Delta H}{\partial d'_{SP}} \right)_0 \quad \left(\frac{\partial \Delta H}{\partial z_{SP}} \right)_0 \right] \qquad (13.59)$$

$$\left(\frac{\partial \Delta H}{\partial d'_{SQ}} \right)_0 = \cos(z_{SQ}) \qquad (13.60)$$

$$\left(\frac{\partial \Delta H}{\partial z_{SQ}} \right)_0 = -d'_{SQ} * \sin(z_{SQ}) \qquad (13.61)$$

$$\left(\frac{\partial \Delta H}{\partial d'_{SP}} \right)_0 = -\cos(z_{SP}) \qquad (13.62)$$

$$\left(\frac{\partial \Delta H}{\partial z_{SP}} \right)_0 = d'_{SP} * \sin(z_{SP}) \qquad (13.63)$$

The standard deviation of a 1 km line levelling is calculated as:

$$s^2_{\Delta H(1km)} = \frac{1,000 s^2_{\Delta H}}{2d'} \qquad (13.64)$$

Example 13.17 Using the data from the first levelling section in Example 13.16 and assuming that the observations were carried out with single face reading using a total station with a linear precision of $\pm(1 \text{ mm} + 1 \text{ ppm})$ and angular precision of $\pm 1''$ (specified by the manufacturer), calculate the standard deviation of this levelling section.

Solution:
From the data in Example 13.16.

BS zenith angle $= 1.481163973$ rad *FS zenith angle* $= 1.602541927$ rad
BS slope distance $= 23.5355$ m   FS slope distance $= 24.0742$ m
Using the additive calculation for the standard deviation of the slope distance.

$$sd'_{SP} = \pm 1.0235 \text{ mm}$$
$$sd'_{SQ} = \pm 1.0241 \text{ mm}$$

Considering that the zenith angle was measured in a single face reading.

$$sz = 2'' = 9.696274 * 10^{-6} \text{rad}$$

Using Eq. (13.58),

$$\Sigma_{ll} = \begin{bmatrix} 1.048728 * 10^{-6} & 0 & 0 & 0 \\ 0 & 9.401772 * 10^{-11} & 0 & 0 \\ 0 & 0 & 1.047625 * 10^{-6} & 0 \\ 0 & 0 & 0 & 9.696274 * 10^{-6} \end{bmatrix}$$

Using Eq. (13.59),

$$J^T = [\,-0.031740 \quad -24.062070 \quad -0.089512 \quad 23.441022\,]$$

Finally, using Eq. (13.46),

$$s^2_{\Delta H} = 5.327987 * 10^{-3}\ \text{m}^2$$
$$s_{\Delta H} = \pm 0.072993\ \text{m} = \pm 7.29\ \text{cm}$$

And

$$\Delta H_{PQ} = -2.8708\ \text{m} = \pm 7.29\ \text{cm}$$

13.8.4   Simultaneous-Reciprocal Trigonometric Levelling

Like differential levelling, trigonometric levelling can be performed using reciprocal and simultaneous observations to eliminate the effects of Earth's curvature and vertical refraction. The measurement principle of this technique is shown in Fig. 13.30.

In this case, the field procedure requires the installation of two total stations, one at each end of the levelling section, allowing simultaneous observations in opposite directions.

In Fig. 13.30, h_{i_A} is the height of the instrument at point (A), and h_{r_A} is the height of the target at the left side. Such a configuration is repeated on the reciprocal side.

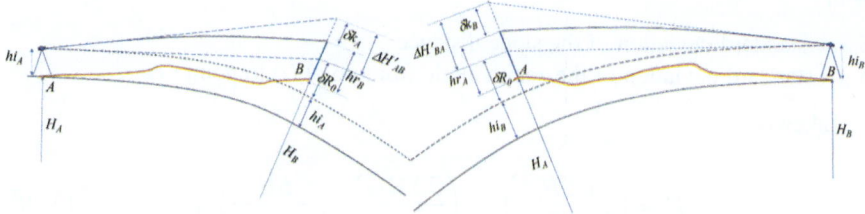

Fig. 13.30 Principle of simultaneous-reciprocal trigonometric levelling

The difference in elevation ΔH_{AB} is then formulated as follows:

$$H_B = H_A + h_{i_A} + \delta R_0 + \Delta H'_{AB} - \delta k_A - h_{r_B} \qquad (13.65)$$

$$H_A = H_B + h_{i_B} + \delta R_0 + \Delta H'_{BA} - \delta k_B - h_{r_A} \qquad (13.66)$$

The elevation differences from (A) to (B) and from (B) to (A) are formulated separately, so the arithmetic mean is given by Eq. (13.67).

$$\Delta H_{AB} = \frac{1}{2}\left[\left(h_{i_A} - h_{i_B}\right) + \left(\Delta H'_{AB} - \Delta H'_{BA}\right) + \left(h_{r_A} - h_{r_B}\right) + \left(\delta R_0 - \delta k_A\right) - \left(\delta R_0 - \delta k_B\right)\right]$$
$$(13.67)$$

On the reasonable assumption that $(\delta R_0 - \delta k_A) = (\delta R_0 - \delta k_B)$, the difference in elevation between station points (A) and (B) for simultaneous-reciprocal trigonometric levelling is given as follows:

$$\Delta H_{AB} = \frac{d'_{AB} * [\cos(z_A) - \cos(z_B)]}{2} \qquad (13.68)$$

13.8.4.1   Error Propagation in Simultaneous-Reciprocal Trigonometric Levelling

Applying the general law of propagation of variances to Eq. (13.68), the standard deviation of the difference in elevation $s_{\Delta H}$ can be calculated using Eqs. (13.46), (13.69), (13.70), (13.71), (13.72) and (13.73).

$$\Sigma_{ll} = \begin{bmatrix} Sd'^2 & 0 & 0 \\ 0 & Sz^2_{AB} & 0 \\ 0 & 0 & Sz^2_{BA} \end{bmatrix} \qquad (13.69)$$

with

$S_{d'}$ = standard deviation of the slope distance
$S_{z_{ij}}$ = standard deviation of the zenith angle from point (i) to point (j)

Likewise,

$$J^T = \left[\left(\frac{\partial \Delta H_{AB}}{\partial d'_{AB}}\right)_0 \quad \left(\frac{\partial \Delta H_{AB}}{\partial z_{AB}}\right)_0 \quad \left(\frac{\partial \Delta H_{AB}}{\partial z_{BA}}\right)_0 \right] \qquad (13.70)$$

where

$$\left(\frac{\partial \Delta H_{AB}}{\partial d'_{AB}}\right)_0 = \frac{\cos(z_{AB}) - \cos(z_{BA})}{2} \tag{13.71}$$

$$\left(\frac{\partial \Delta H}{\partial z_A}\right)_0 = -\frac{d'_{AB} * \sin(z_A)}{2} \tag{13.72}$$

$$\left(\frac{\partial \Delta H}{\partial z_B}\right)_0 = \frac{d'_{AB} * \sin(z_B)}{2} \tag{13.73}$$

With regard to accuracy standards for trigonometric levelling, the reader will find standard specifications ranging from $12\sqrt{d}$ mm to $100\sqrt{d}$ mm, for d in km, depending on the class or order of work.

Example 13.18 Using the data from Example 7.5, consider the simultaneous and reciprocal observations between points (P) and (Q) and calculate the difference in elevation between the two points and its standard deviation. Assume that the observations were carried out with readings on a single face reading using a total station with linear and angular standard deviations of $\pm(1 \text{ mm} + 1 \text{ ppm})$ and $\pm1''$, as specified by the manufacturer.

Solution:

From the data in Example 7.5:

$d'_{PQ} = 1,045.022$ m $z_{PQ} = 84°32'59.0''$ $z_{QP} = 95°27'22.0''$

From the instrument specifications:

$$sd'_{PQ} = \pm 2.045 \text{ mm}$$

$$sz = 2'' = 9.696274 * 10^{-6} \text{rad}$$

Using Eq. (13.68),

$$\Delta H_{PQ} = \frac{1,045.022 * [\cos(84°32'50.0'') - \cos(95°27'20.0'')]}{2} = 99.311 \text{ m}$$

Using Eq. (13.69),

$$\Sigma_{ll} = \begin{bmatrix} 4.182115 * 10^{-6} & 0 & 0 \\ 0 & 9.401772 * 10^{-11} & 0 \\ 0 & 0 & 9.401772 * 10^{-11} \end{bmatrix}$$

Using Eq. (13.70),

$$J^T = [0.095033 \quad -520.148728 \quad 520.143672]$$

Finally, using Eq. (13.46),

$$s^2_{\Delta H} = 5.091115 * 10^{-5} \text{ m}^2$$
$$s_{\Delta H} = \pm 0.007135 \text{ m} = \pm 7.14 \text{ mm}$$

And

$$\Delta H_{PQ} = 99.311 \text{ m} = \pm 7.14 \text{ mm}$$

13.9   Levelling with GNSS Technology

Levelling with GNSS technology is the name given to the levelling work carried out using GNSS surveys in relative mode (post-processed or RTK). In this case, the field procedure consists of installing the reference receiver antenna on a point with known coordinates (X, Y, h) and moving the remote receiver antenna over the points whose coordinates are to be determined. In this case, the heights obtained for the levelled points are the ellipsoidal heights, whose values are determined from the ellipsoidal height of the reference point. Although the accuracy of GNSS levelling is still well below that of differential levelling methods, especially for short and medium distances, it is sufficient for lower-order levelling applications in Civil Engineering projects. Under favourable conditions, centimetre accuracy can be achieved with this technology. This type of levelling is extremely fast compared to others, mainly when carried out in differential RTK mode. For more information on this subject, the reader is suggested to consult Chap. 17.

However, as shown in Fig. 13.31, it is important to note that GNSS observations determine the difference between ellipsoidal heights $\Delta h_{PQ} - h_Q - h_P$, while differential levelling results in the determination of the difference between orthometric altitudes $\Delta H_{PQ} = H_Q - H_P$. Therefore, for using GNSS technology for levelling applications, it is required to know the geoid undulation for points (P) and (Q). In this case, three solutions can be used, as described below:

1. *Calculate the value of the geoid undulation for each levelled point and correct the ellipsoidal heights to the orthometric altitudes.* The details of this type of correction are described in Sect. 4.4.1.
2. *Determine the geoid undulation of the region or the geoid height for each levelling point.* This type of action requires that gravimetric measurements be taken in the region of interest.
3. *If the orthometric altitude of the point where the base GNSS antenna will be installed is known, this value can be specified as the ellipsoidal height (h). The other points measured with the remote receiver will then have their heights related to the height value given for the base antenna.* This method is suitable for small area projects where the variation of the geoid undulation is minimal. However, care should be taken not to apply it indiscriminately. In this case, as shown in Fig. 13.31, the following mathematical relationships can be used.

$$H_Q - H_P \cong h_Q - h_P \tag{13.74}$$

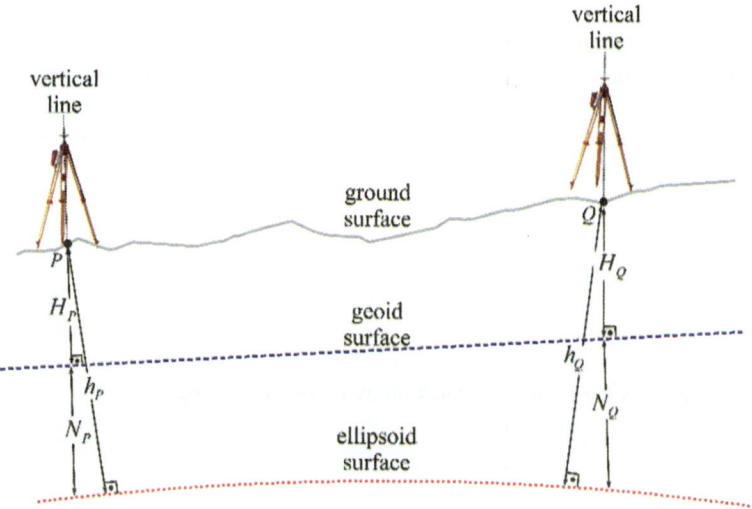

Fig. 13.31 Principle of GNSS levelling

GNSS processing software usually has a built-in geoid model, which allows to obtain the values of the geoid undulations and, from them, the orthometric altitudes of the points to be levelled. The problem, however, is that the embedded geoid models are often global and may generate inconsistent values for specific regions. However, it is important to note that, some processing software allows the user to import a personalised local geoid model.

Another procedure that can help to improve the quality of GNSS levelling is to measure the orthometric altitudes (H) of some points in the levelling region using a conventional differential levelling, determine the variation of the geoid undulation in that region and consider it constant (linear) for the other points. The geoid undulations of other points can then be calculated through linear interpolation based on the points of known altitudes.

Regarding accuracy standards, not so many standards are available. The reader is advised to consult regional and international standardisation agencies for further information.

13.10   Levelling with a Laser Level

Laser levelling can be considered a particular case of differential levelling. In this case, the difference is that the optical level is replaced by a laser level, which generates a levelling plane using a visible laser beam that rotates continuously around the vertical axis of the instrument, as shown in Fig. 13.32. Using a laser beam sensor attached to a levelling staff, the operator determines, for example, the height differences between any points on a construction site.

A laser level is generally used for vertical control of Civil Engineering works. The instrument is installed in a safe and suitable place on the site and continuously produces a horizontal plane throughout the working period, as shown in Fig. 13.32a. Any operator with access to a laser beam sensor can use it to level elements on site. The laser level can also be positioned to create a vertical or tilted plane, making it suitable for many other applications. The interested reader can find guidelines for use in various fields of engineering and architecture on specialised websites or from the manufacturers of the instruments.

Laser levels can also be used to control Earth-moving machinery. In this case, the sensors are correctly installed to control the vertical movements of the blade or arm of the machines, as shown in Fig. 13.32b. Applications in this area of Civil engineering have increased in recent years. Interested readers are advised to consult specialist websites for regular updates on available technologies.

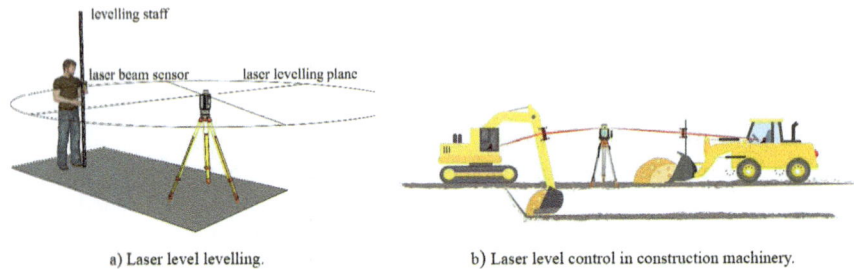

a) Laser level levelling.   b) Laser level control in construction machinery.

Fig. 13.32 Laser levelling in Civil Engineering works. (a) Laser level levelling. (b) Laser level control in construction machinery

13.11 Review Questions

1. Explain which are the levelling methods that can be used in Geomatics applied to Civil engineering and the level of accuracy expected from each when used on a construction site.
2. Briefly explain how differential levelling is carried out to determine the difference in elevation for a levelling segment between two points.
3. Briefly explain the difference between the rise and fall and height of collimation differential levelling calculation procedures.
4. Explain the sources of errors that can occur in differential levelling and describe how they can be minimised.
5. Explain why it is important to have an equidistant sight during a differential levelling measurement.
6. Briefly explain the difference between line levelling and network levelling techniques.
7. Briefly explain the difference between loop levelling and link levelling.

8. Briefly explain the difference between differential, trigonometric and leap-frog levelling.
9. List the equipment required for carrying out a trigonometric levelling.
10. Explain when you would suggest using a trigonometric levelling method instead of a differential levelling method on a construction site.
11. Briefly explain the advantages of simultaneous-reciprocal trigonometric levelling compared to unidirectional trigonometric levelling.
12. Under what conditions would you accept GNSS levelling on a construction site?
13. What type of construction site levelling work would you recommend using a laser level for?

13.12 Problems

1. Using the differential levelling observations given in Table 13.16 and assuming that the elevation of point (P) is 845.149 m, calculate the elevation of point (Q) using the rise and fall and the heigh of collimation methods.

Table 13.16 Field observations

Sighting point	BS [mm]	FS [mm]
P	2574	
Q		857

2. Several levels have been designed for a Civil Engineering project, as shown in Fig. 13.33. Using the field measurements given in Table 13.17, calculate the heights of points (2), (4) and (5), the value the operator needs to read from the levelling staff to set out point (3), and the gradients of ramps 1, 2 and 3. The heights of the access steps to ramp 2 are all equal to 17.0 cm.

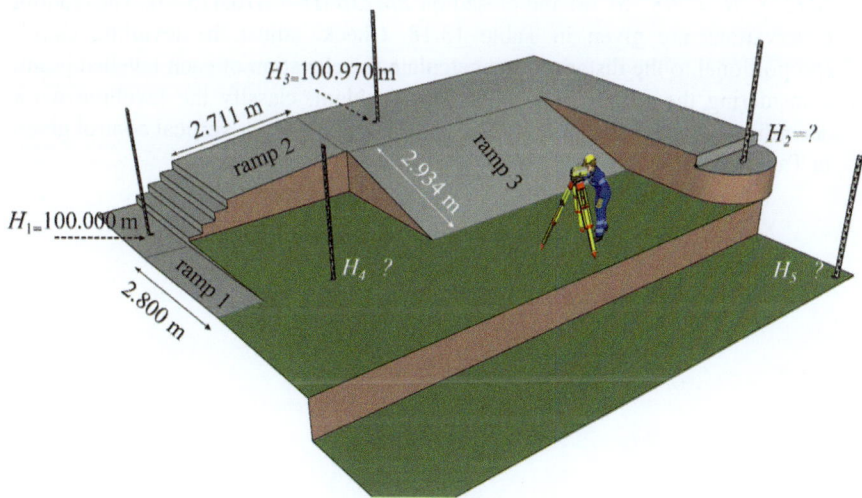

Fig. 13.33 Platform levels

Table 13.17 Field observations

Sighting point	BS [mm]	FS [mm]
P1	1681	
P2		747
P4		1580
P5		2332

3. Prepare the levelling field notes for the field measurement sketch shown in Fig. 13.34 and calculate the elevation of point (C).

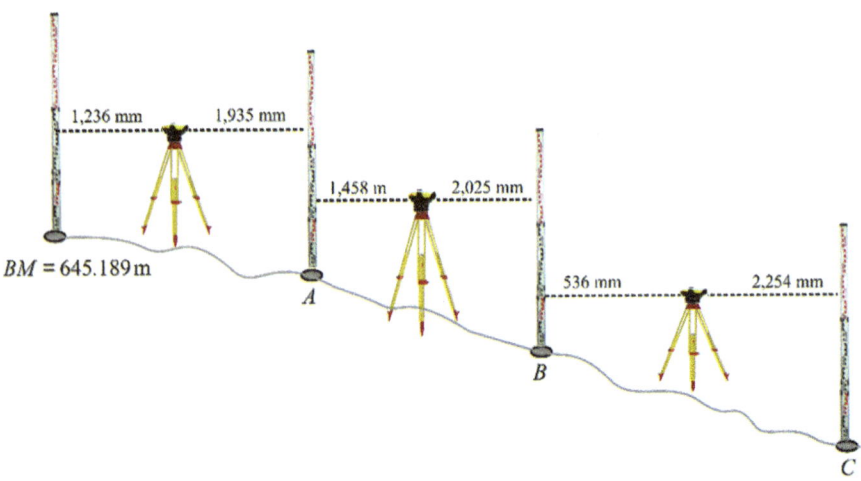

Fig. 13.34 Field measurement sketch

4. A differential levelling circuit with intermediate sights (*1*) to (*7*) started on BM235 (H = 798.561 m) and closed on BM236 (H = 810.415 m). The reading observations are given in Table 13.18. Check, adjust the levelling circuit (proportional to the distance) and calculate the elevation of each levelled point.
5. Considering the results of the previous problem, classify the levelling work according to the minimum elevation closure standards for vertical control given in Table 13.12.

Table 13.18 Field observations

Instrument position	Sighting point	Levelling staff reading BS [mm]	FS [mm]	Back + Fore sighting distance [m]
I1	BM235	3813		
	1		2398	
	2		1789	
	A	3755	358	64.50
I2	3		2876	
	4		2357	
	B	3821	1145	60.35
I3	5		1859	
	C	3258	932	50.75
I4	6		2892	
	D	3598	452	80.60
I5	7		1947	
	E	1327	2351	74.00
I6	BM236		2473	36.90

6. To determine the elevation of a water tank, a differential loop levelling was carried out beginning and closing on *BM100* ($H = 857.145$ m). Using the readings given in Table 13.19, check, adjust the levelling loop and calculate the elevation of the water tank.

Table 13.19 Field observations

Sighting point	Levelling staff reading BS [mm]	FS [mm]	Section length [m]
BM100	**1167**		
1	437	1215	67.50
2	822	1032	62.20
3	657	1388	74.60
4	1105	1721	80.10
5	2002	1814	77.80
6	1745	815	59.90
7	992	1075	65.55
8	601	418	45.00
9	2113	1234	59.80
10	1510	525	60.10
Water tank	**1611**	**515**	64.50
11	1513	2348	67.30
12	1312	2803	63.65
13	1615	634	63.95
14	418	2865	70.60
15	2880	2004	58.00
16	1327	2638	83.40
17	2890	1990	87.90
18	1184	543	86.10
19	965	574	80.80
BM100		**703**	59.90

7. Given the levelling network shown in Fig. 13.35, write the error and matrix equations for calculating H_P and its standard deviation. The arrow indicates the direction of the highest point of the course.

Fig. 13.35 Levelling network sketch

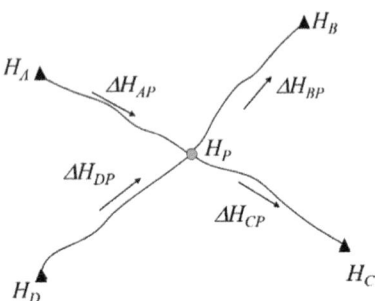

8. To determine the elevations of points (P), (Q) and (R) from benchmarks $H_A = 421.362$ m, $H_B = 406.713$ m and $H_C = 430.844$ m, a levelling network with seven differential levelling sections was measured as shown in Fig. 13.36. Using the field observations given in Table 13.20, calculate the unknown elevations, their standard deviations and the adjusted values and standard deviation of the observations. The arrow indicates the direction of the highest point of the course. Assume that the observation weight P is the inverse of the distance and $p = 1$ for 1 km levelling.

Fig. 13.36 Levelling network sketch

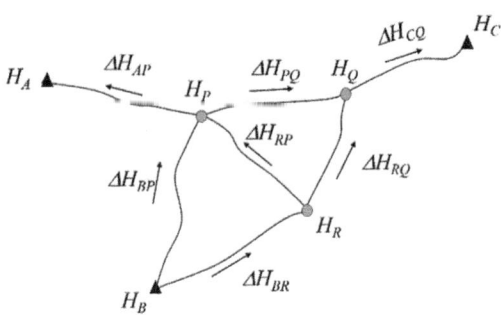

Table 13.20 Filed observations

Levelling section	Difference in elevation [m]	Section length [km]
AP	6.773	1.4
PQ	6.752	0.6
CQ	11.515	0.8
BP	5.865	1.2
PR	2.944	0.9
QR	9.651	0.8
BR	2.948	0.9

9. A double-run differential levelling consisting of five levelling runs was carried out, as shown in Table 13.21. Calculate:

   (a) The standard deviation of 1 km single run levelling.
   (b) The standard deviation of the 1 km double-run levelling.
   (c) The residual error of each forward levelling.
   (d) The residual error of each forward/backward levelling.
   (e) The residual error of the total forward/backward levelling.

Table 13.21 Filed observations

Sighting point	Levelling run length [km]	Forward run [m]	Backward run [m]
1	3.1	6.471	6.468
2	4.6	6.035	6.039
3	4.3	2.498	2.500
4	6.6	5.604	5.608
5	6.5	0.490	0.485

10. A trigonometric levelling was carried out to determine the difference in elevation between points ($Q1$) and ($Q2$), as shown in Fig. 13.37. The field observations were carried out using a total station with a linear standard deviation of $\pm(2 \text{ mm} + 2 \text{ ppm})$ and an angular standard deviation of $\pm 3''$. Assuming that the instrument height at point (P) is 1.61 m and the height of the reflector prism at the target points is 1.90 m, calculate the difference in elevation and its standard deviation. The field observations are given in Table 13.22.

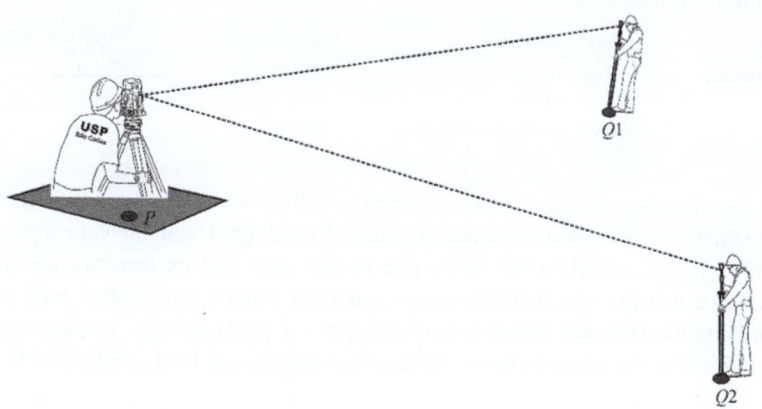

Fig. 13.37 Measurement geometry

Table 13.22 Field
observations

Station	Sighting point	Zenith angle	Slope distance [m]
P	Q1	88° 14′ 29″	451.369
	Q2	91° 55′ 31″	397.128

11. A total station was used to determine the clearance of an overpass, as shown in Fig. 13.38. Using the readings given in Table 13.23, calculate the clearance value.

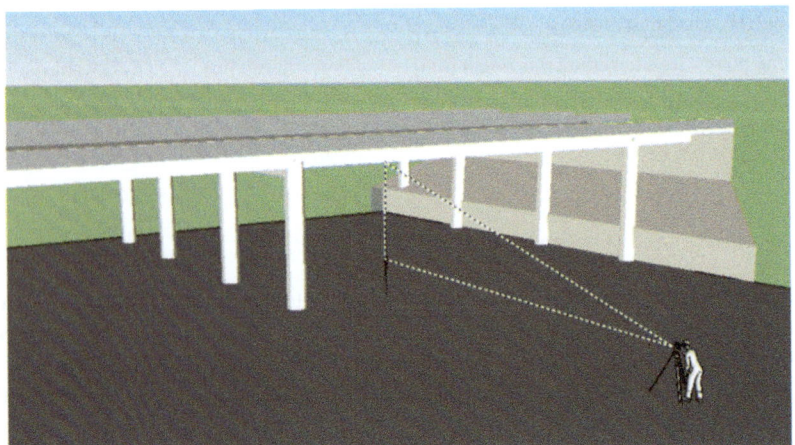

Fig. 13.38 Measurement geometry

Table 13.23 Field observations

Station	Sighting point	Zenith angle	Slope distance [m]
Total station	Reflector prism $h_r = 1.10$ m	90° 25′ 00″	57.430
	Overpass undersurface	87° 18′ 10″	–

12. A simultaneous-reciprocal trigonometric levelling was carried out to determine the difference in elevation between points (P) and (Q). Using the values given in Table 13.24, calculate the difference in elevation and its standard deviation. Assume that the observations were carried out using a total station with linear and angular standard deviations of ±(1 mm + 1 ppm) and ±1″ respectively, as specified by the manufacturer. All readings were taken in single-face mode.

Table 13.24 Field
observations

Station	Sighting point	Zenith angle	Slope distance [m]
P	Q	85° 32′ 21″	1827.430
Q	P	94° 26′ 33″	1827.425

13. A student exercise has been proposed to compare the results of differential, trigonometric and leap-frog levelling methods. Using the readings given in Tables 13.25, 13.26 and 13.27, check, adjust the levelling loop and calculate the difference in elevation between points (A) and (B) for each levelling method and the precision of these differences. Compare the results. The total station used for the trigonometric and leap-frog levelling has a linear standard deviation of ±(1mm + 1 ppm) and an angular standard deviation of 1″. All readings were taken in single-face mode.

Table 13.25 Differential levelling observations

Sighting point	Levelling staff reading	
	BS [mm]	FS [mm]
A	1516.2	
1	2257.2	523.6
2	2410.1	864.3
B	342.2	363.4
3	972.0	1777.1
4	819.3	2564.3
A		2227.8

Table 13.26 Trigonometric levelling observations

Station	Sighting point	Slope distance [m]	Zenith angle
A	B	137.4674	88° 09′ 19.6″

Table 13.27 Leap-frog levelling observations

Station	Sighting point	Slope distance [m]	Zenith angle
M	B	68.0887	88° 13′ 29.6″
	A	69.4475	91° 55′ 02.3″

References

Bomford, G. (1980). *Geodey*. 4 Ed. Oxford University Press. England.

Brabant, M. (2012). *Topographie opérationnelle*. Eyrolles Paris, France.

Ceylan, A. (2008). *Precise height determination using simultaneous-reciprocal trigonometric levelling*. Survey Review. https://doi.org/10.1179/003962608X290997.

Ceylan, A., Baykal, O. (2006). *Precise height determination using leap-flog trigonometric levelling*. Journal of Surveying Engineering © ASCE. https://doi.org/10.1061/(ASCE)0733-9453 (2006)132:3(118).

Federal Geographic Data Committee's *Geospatial Positioning Accuracy Standards*, FGDC-STD-007.1-1998 (Part 1: Reporting Methodology), FGDC-STD-007.2-1998 (Part 2: Standards for Geodetic Networks), and FGDC-STD-007.4-2002 (*Part 4: Architecture, Engineering, Construction, and Facilities Management*).

ISO 17.123-2:2001. *Optics and optical instruments — Field procedures for testing geodetic and surveying instruments — Part 2: Levels.*

Milles, S., Lagofun, J. (2011). *Topographie et topométrie modernes.* Eyrolles Paris, France.

Rüeger J. M. (1997). *Staff Errors in and the Adjustment of Ordinary Levelling Runs,* Australian Surveyor, 42:1, 16–24, https://doi.org/10.1080/00050342.1997.10558662.

USGS (2020). *Procedures and Best Practices for Trigonometric Levelling in the U.S. Geological Survey.* Virginia, USA.

Uren, J., Price, B. (2010). *Surveying for Engineers.* 5th Edition. Palgrave Macmillan. Hampshire, England.

Chapter 14
Levelling Applications for Relief Representations

Irineu da Silva and Paulo C. L. Segantine

14.1 Introduction

In order to develop or design an engineering project, it is often necessary to know the relief of the construction site. The design of a highway, a water dam, an energy transmission line, a water distribution network, a sewage network or a subdivision, to name only the most expressive ones, can only be developed if the engineer has at his disposal, in addition to the location of the feature elements of the terrain, a representation of the relief of the area in which it is to be designed, showing the elevations, depressions and other relief features. In this way, over the years, different forms of relief representation have been developed, allowing professionals in the field of geosciences and engineering to use them according to the needs of their projects. In this sense, to help Civil Engineers develop their projects, this chapter aims to study different types of relief representations for use in Civil Engineering projects.

14.2 Profile Levelling

Profile levelling is a measurement technique used in Civil Engineering to produce ground elevation values along a predefined alignment. The resulting graph of these measured values is called a *longitudinal profile*, in which the elevations of the points along the alignment and the horizontal distances between them are plotted on the vertical and horizontal axes of a vertical Cartesian plane view, as shown in Fig. 14.1. In other words, it is the graphical representation of the vertical section of the terrain along the levelling path.

I. da Silva (✉) · P. C. L. Segantine
São Carlos School of Engineering, University of São Paulo, São Carlos, SP, Brazil
e-mail: irineu@sc.usp.br; pclsegantine@usp.br

© The Author(s), under exclusive license to Springer Nature Switzerland AG 2025
I. da Silva, P. C. L. Segantine, *Geomatics Applied to Civil Engineering*,
https://doi.org/10.1007/978-3-031-75737-2_14

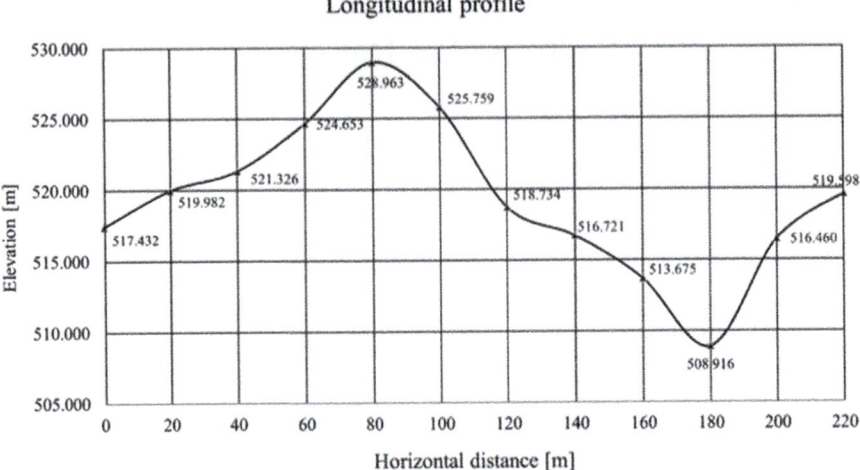

Fig. 14.1 Longitudinal profile of the terrain

In many Civil Engineering projects, the longitudinal profile may not be sufficient to represent the relief of the terrain, requiring, in this case, a cross-sectional levelling at 90° to the longitudinal line, as shown in Fig. 14.2. The length and spacing of the cross-sections will depend on the project. They are usually placed every 20 m and extend 20-30 m on either side of the alignment.

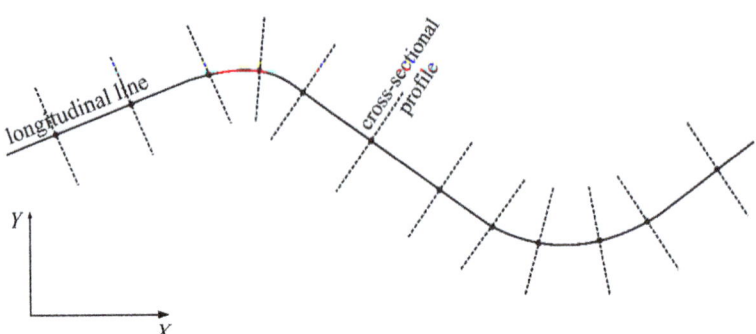

Fig. 14.2 Plan illustration of an alignment with cross-sections

Like the longitudinal profile, the *cross-sectional profile* is also graphically represented in a vertical Cartesian plane view, as shown in Fig. 14.2.

Several engineering projects define their geometric guidelines using this type of relief representation. A classic example is its use in the design of land transportation routes, gas pipelines, water mains, and road networks in subdivision projects.

The determination of longitudinal and cross-sectional profiles can be performed by three profiling techniques, namely:

- Field levelling
- Point interpolation over a cartographic representation with contour lines
- Digital terrain modelling

14.2.1   Profile Determination by Field Levelling

Profile levelling consists of determining the elevations of specific ground points defined along the centre line of a track of land on which a linear structure is to be constructed, using a levelling method, as described in Chap. 13. The field operation is relatively simple, consisting only of establishing an alignment in the terrain and measuring the difference in elevation of specific relief points along it. The choice of levelling method depends on the accuracy required for the project. It is usually carried out using differential line levelling with intermediate sights (IS). However, there are cases where trigonometric levelling or levelling using GNSS technology can also meet the requirements of specific projects.

The choice of points to be levelled depends on the purpose for which the profile (or cross-section) is being surveyed. If the project requires a full description of the relief variations along the alignment, the engineer should choose to measure the elevations of all points that indicate such variations. In other cases, it is sufficient to measure the elevations of points at regular intervals along the alignment, usually every 20 m, to represent the profile.

The field procedure for levelling the longitudinal profile or cross-sections is practically the same, regardless of the method used. First, the engineer must define the alignment in the field, using a surveying method that best suits the shape of the terrain and the surveying instrument available. Next, the field assistant must go through the alignment, positioning the levelling staff, in the case of differential levelling, or the prism pole, in the case of trigonometric levelling, on the relevant points of the relief or the equidistant points of the profile.

The instrument operator then performs the levelling of the points occupied by the assistant, always considering the prerogatives regarding the levelling methods presented in Chap. 13. If the project uses the levelling method with GNSS technology, all the instrument operator has to do is walk along the alignment or cross-section and position the GNSS receiving antenna over the points of interest. In post-processing mode, the operator needs to be guided along the alignment. In RTK mode, on the other hand, the graphical interface of the GNSS receiver (data collector) shows the path to be followed and, in the case of equidistant point measurements, the points to be levelled. For more information on GNSS measurements, the reader is referred to Chap. 17.

A typical example of field-measured values for profiling a terrain cross-section for a highway design is shown in Table 14.1. In this case, the cross-section extends 50 m on each side of the alignment axis, and specific ground points are levelled. Figure 14.3a and b show cross-section drawings in plan and profile views, respectively.

Table 14.1 Field measurements along the cross-section

Left side [m]			Stake	Right side [m]				
$-\frac{0.80}{19.00}$,	$+\frac{1.20}{19.00}$,	$+\frac{0.60}{12.00}$,	**[112]** **Elevation = 845.236 m**	$+\frac{0.80}{10.00}$,	$+\frac{0.65}{9.00}$,	$+\frac{1.40}{13.00}$,	$+\frac{0.35}{9.00}$,	$+\frac{1.20}{9.00}$

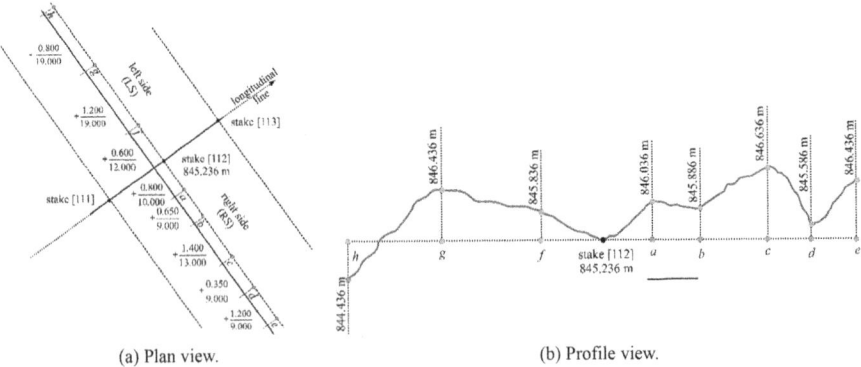

(a) Plan view.   (b) Profile view.

Fig. 14.3   Cross-section drawing in plan and profile views

14.2.2   Profile Determination by Point Interpolation Over a Paper Map with Contour Lines

In this case, the procedure for determining the profile (or cross-section) is to draw the alignment over a cartographic representation with contour lines and manually interpolate the elevation values of equidistant points along the alignment or extract the integer elevation values of contour lines that intersect the alignment. The profile is then drawn using the integer or interpolated elevations with the corresponding horizontal distances. Figure 14.4 shows an example of a longitudinal line with a cross-section drawn over the contour lines and the profile of the corresponding cross-section.

At present, this type of terrain profile representation is rarely used. Its use is recommended only for special cases of preliminary projects where only paper maps are available. Even then, it is preferable to scan the map and convert its contours into a set of points of known elevation to be used in a digital terrain model that can automatically determine the profile.

The accuracy of interpolated elevations on a contour map depends on the quality of the graphical representation. Typically, a lower-quality profile representation is obtained compared to other profile determination methods.

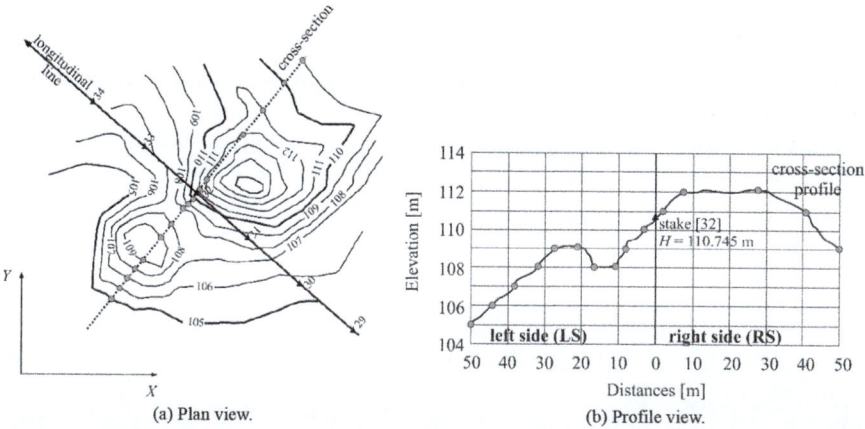

Fig. 14.4 Profile determination by point interpolation over a cartographic representation with contour lines

14.2.3 Profile Determination by Digital Terrain Modelling

Determining profiles and cross-sections has become a relatively simple process with digital terrain modelling techniques. The engineer simply selects or (draws) the profile alignment in the plan view of the graphical representation of the modelled surface for the profile to be generated. Details of this type of profile determination are given in Sect. 15.7.2.

14.2.4 Profile Plotting

Another critical element to consider when plotting profiles is the choice of the drawing scale. Commonly, to highlight the relief variations, different scales are used for the distances on the horizontal axis and the elevations on the vertical axis. The most common solution is to represent the vertical with a ten-fold exaggeration of the horizontal, for example, a horizontal scale of 1:10,000 and a vertical scale of 1: 1000.

14.3 Representation of Relief Features by Spot Elevations

A *Spot Elevation* is a ground point that identifies the position and elevation (*X, Y, H*) of a given point on a map, as shown in Fig. 14.6. In this case, points are positioned and levelled using surveying techniques described in the previous chapters.

Differential, trigonometric, and GNSS levelling can determine elevations for small areas. However, depending on the accuracy required, laser scanning or photogrammetric modelling is recommended for a high density of points, as described in the chapters on these two technologies.

Typically, spot elevations indicate relevant points in a project, such as the tops of hills and slopes, the bottoms of valleys, river banks and so on. However, as discussed in Chap. 15, it can also be a point cloud for surface modelling. They are usually represented by a dot with a height value, as shown in Fig. 14.5.

Fig. 14.5 Spot points over an orthophoto. (Source: Digital map of the city of São Paulo, Brazil. *Location*: https://download.geosampa.prefeitura.sp.gov.br/PaginasPublicas/_SBC.aspx)

14.4 Contouring

A contour line (or contour) is an imaginary *isoline of height*, i.e., a line on a map representing a series of points at the same height. It can be considered the trace of the extremities of several horizontal surfaces crossing the terrain at different elevations, as shown in Fig. 14.6.

The vertical distance between two successive horizontal planes must be constant. It determines the equidistance between contour lines on the map, which depends on the design guidelines for which the contours are being drawn and the scale of the map. Contour lines are generally drawn at intervals of 1, 5, 10, 25 or 50 m, with elevation values (index) marked on every fifth contour line (major contour line) and highlighted on the map with a coloured or thicker line.

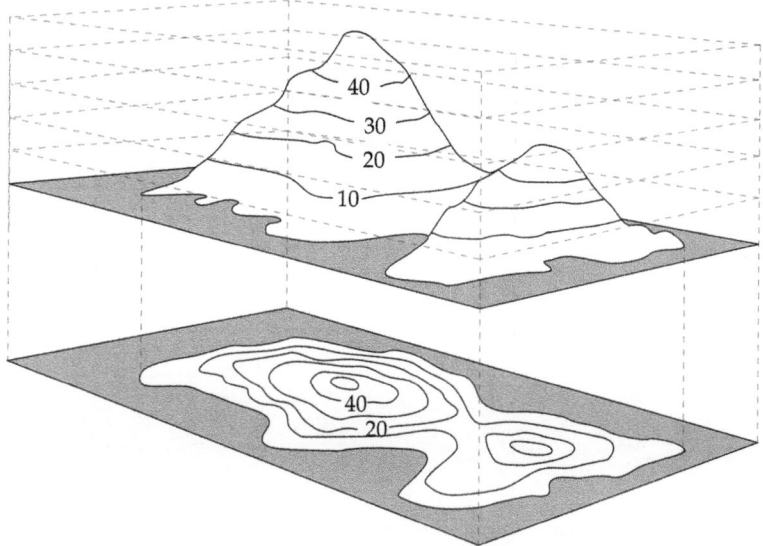

Fig. 14.6   Illustration of the graphic generation of terrain relief contour lines

In some mapping applications, in addition to the contour lines, it is also necessary to show some spot elevations, such as peaks, valleys and other points of interest for the project. Figure 14.7 shows a typical example of a contour map with spot elevations.

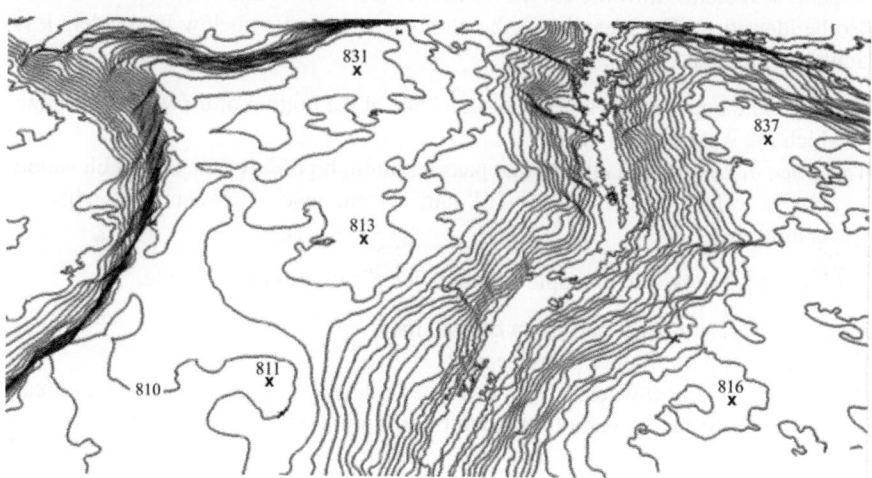

Fig. 14.7   Contour map

The sequence of elevation values labelled on the contour lines indicates whether they represent an elevation or a depression. If the smaller values surround the larger ones, there is an elevation; otherwise, there is a depression. See Fig. 14.8.

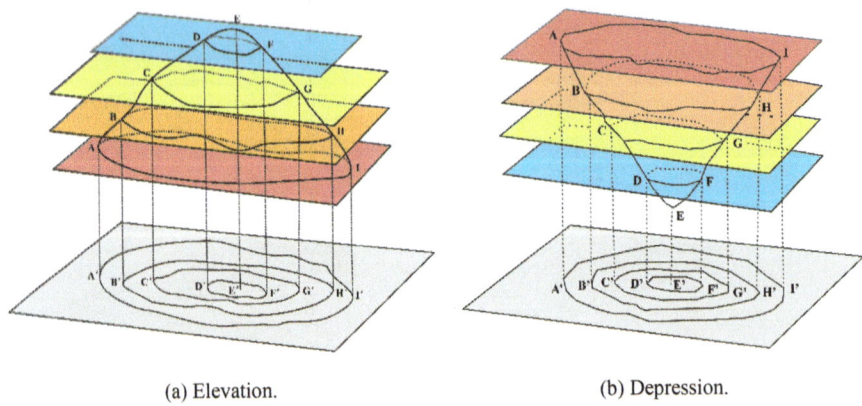

(a) Elevation.   (b) Depression.

Fig. 14.8 Interpretation of contour lines

In order for the graphical representation using contour lines to reflect the altimetric conditions of the terrain as closely as possible, the engineer must identify some elementary features of the relief during the field survey, such as slopes, ridges, watersheds, foothills and embankment crests, among others. Identifying these relief elements will dictate how the curves are drawn. Dead zones and retaining walls must also be identified. The meaning of the above terms is given below for readers less familiar with them.

Thalweg: The opposite of ridge. A sinuous line at the bottom of the valleys through which the watercourses flow.

Watershed divide: An imaginary line passing through points of maximum elevation along a ridge, separating rainfall and watercourses between two adjacent catchments.

Figure 14.9 shows an example of the thalweg and watershed divide.

Toe of slope: The lowest part of an embankment.

Crest: Highest part of an embankment.

Dead zone: Land area where it is impossible to display contour lines, such as areas covered by dense vegetation, flooded or built-up areas.

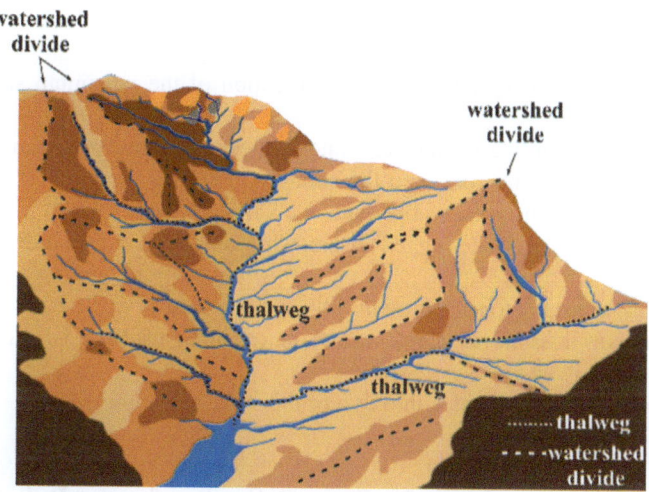

Fig. 14.9 3D view of relief with the indication of landforms

Figure 14.10 shows an example of a toe of slope and a crest of a road-cut embankment.

Fig. 14.10 Toe of slope and crest of a road-cut-slope

14.4.1 Properties of Contour Lines

For contour lines to be correctly displayed on a map, they must have the following characteristics:

1. Each contour line must be self-contained, although this is often outside the extent of the map on which it is drawn.
2. Contour lines are perpendicular to the direction of the maximum slope of the terrain.
3. The slope between two successive contour lines is assumed to be uniform.
4. The horizontal distance between the contours indicates the slope of the terrain. Greater distances mean less slope and vice versa.
5. Irregular contours indicate rugged terrain, while smooth contours indicate smooth terrain.
6. Concentric close contours represent hills or depressions.
7. Two contour lines never meet, except in retaining walls or in rare cases of cavities.
8. A contour line never divides into two of the same elevation.

14.4.2   Field Procedures for Data Collection for the Relief Representation Using Contour Lines

There are two field procedures for collecting data to represent the relief using contour lines:

- *Regular grid* data collection
- *Irregular grid* data collection

A regular grid survey consists of establishing contiguous regular squares on the ground, referenced to a particular element of the survey or to linear features of the survey area for grid-to-map referencing. Once the reference line has been established, the grid intersections (vertices) are established on the ground, generally with equidistant points every 20 m. The field procedure for determining these intersections depends on the type of surveying instrument available. Typically, a theodolite or total station will allow the work to be carried out quickly and accurately. Field surveying using GNSS technology in RTK mode is another viable option, depending on the suitability of the site for GNSS measurements and the accuracy required for the project.

After steaking out the vertices of the grid on the terrain, each point must be levelled using one of the levelling methods indicated in Chap. 13. In some cases, for a better representation of the relief, it may also be necessary to level feature points of the terrain other than the grid vertices, as in the case of the existence of any of the geographical features mentioned in the previous section. Figure 14.11 shows the field procedures for levelling the vertices of a regular grid using differential levelling.

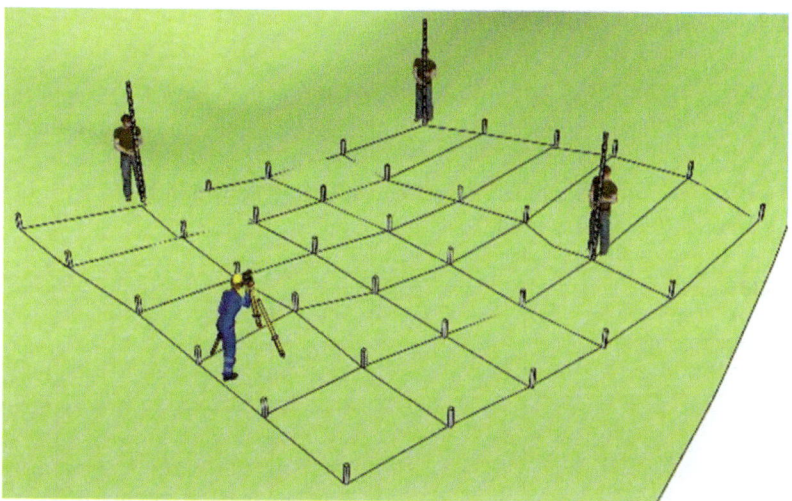

Fig. 14.11 Differential levelling over a regular grid for contour determination

The levelling procedure for determining contours over an irregular grid is to determine the difference in elevation between a series of ground points irregularly distributed over the terrain, provided that all relevant feature points are surveyed, as shown in Fig. 14.12.

In this case, the field procedure involves levelling the feature points of the terrain using one of the levelling methods indicated in Chap. 13. Usually, trigonometric or GNSS-RTK levelling is used. In the first case, the field procedure involves materialising a network of traverse stations on the ground and levelling the terrain feature points, taking the traverse stations as a reference, as shown in Fig. 14.12a. When using a GNSS-RTK levelling, the field procedure is to move the GNSS antenna around the terrain in a planned manner and conveniently measure the feature points, as shown in Fig. 14.12b.

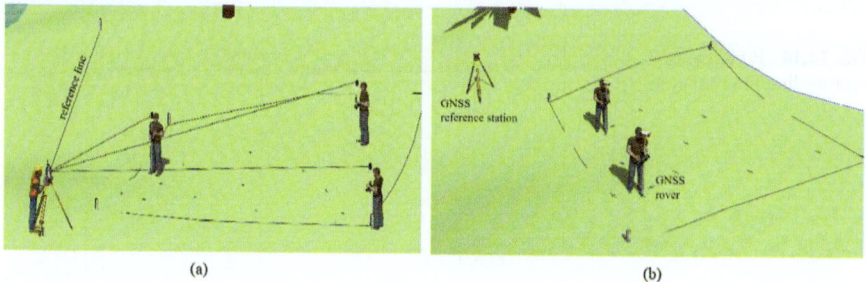

Fig. 14.12 Procedure for levelling an irregular grid of elevation points using total station and GNSS-RTK receivers. (**a**) Trigonometric levelling for determining an irregular grid of elevation points. (**b**) GNSS-RTK levelling for determining an irregular grid of elevation points

14.4.3 Contour Line Plotting

Contour line plotting can be performed manually or through a DTM application program, as described in Sect. 15.7.1. In this section, only manual plotting procedures are discussed.

The first step is to determine the grid elevations, as shown in Fig. 14.13. In this case, a regular 20 m grid is used.

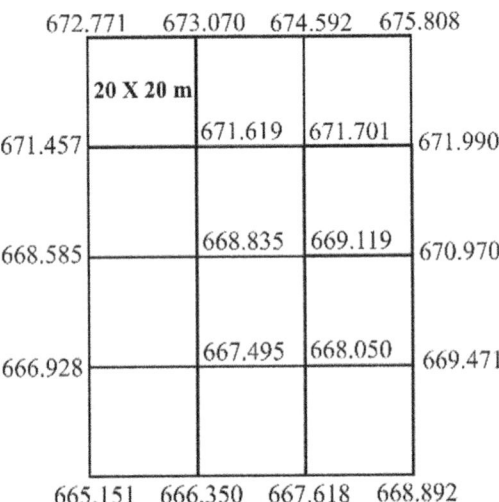

Fig. 14.13 Regular grid of points to be interpolated for contour lines determination

The drawing process consists of determining the location of points with integer elevation values by linear interpolation between the elevation values of their neighbours. Each interpolated point should then be connected by a sinuous line, representing the contour line, as shown in Fig. 14.15.

To illustrate the process of locating points with integer elevation values, consider the grid shown in Fig. 14.14.

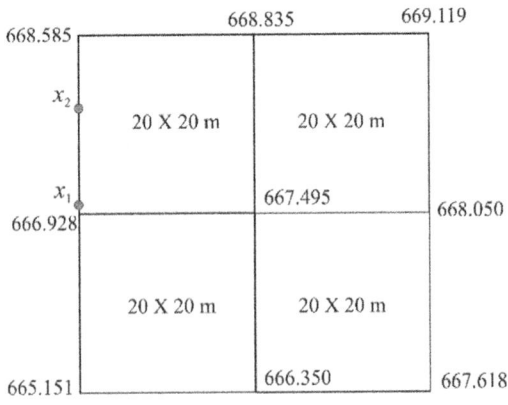

Fig. 14.14 Principle of contour line interpolation

Typically, values are interpolated concerning the grid lines, for example, between the vertices with elevations of 668.585 m and 666.928 m. In this case, the integer elevation values to be interpolated are 667.000 m and 668.000 m, located on the corresponding grid line. Thus, for elevation 667.000 m, the following calculation can be performed:

$$\frac{668.585 - 666.928}{20.000} = \frac{667.000 - 666.928}{x_1} \rightarrow x_1 = 0.87m$$

The following calculation must be applied for a point at an elevation of 668.000 m.

$$\frac{668.585 - 666.928}{20.000} = \frac{668.000 - 666.928}{x_2} \rightarrow x_2 = 12.94m$$

In other words, the contour lines at 667.000 m and 668.000 m are 0.87 m and 12.94 m, respectively, from the grid point with an elevation of 666.928 m. Points with elevation values calculated in this way must be plotted on the appropriate edge of the grid. The calculation process must then be repeated for all grid lines.

Contour lines are then drawn manually, connecting all points of the same elevation, as shown in Fig. 14.15.

Fig. 14.15 Manually interpolated contour lines

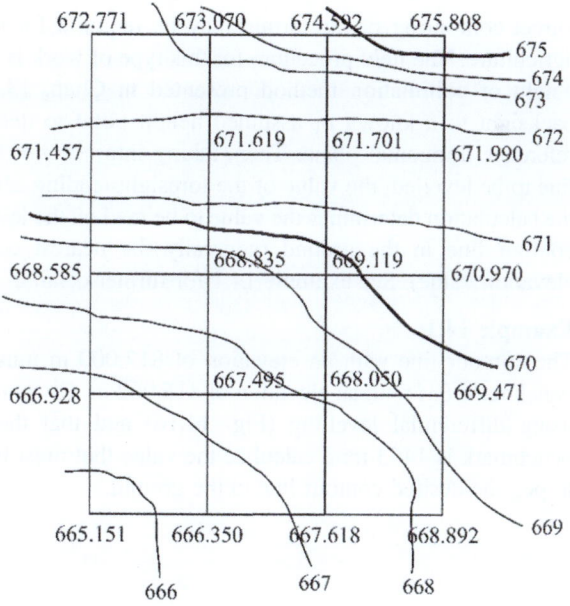

For a long time, contour lines were drawn manually and required skill from the designer. As mentioned, with the advent of application programs for DTM systems in CAD or GIS environments, contour lines are generated automatically. Therefore, it no longer makes sense to worry about the details of manual contouring. Nevertheless, if necessary, the principle must be known even for analysing the results of automatic plotting.

The vertical equidistance between contour lines must be selected according to the project for which they are being generated and the scale of the plot. Table 14.2 shows suggested equidistance values for different plot scales.

Table 14.2 Relationship between drawing scale and vertical equidistance between contour lines

Scale	Equidistance
1:500	0.25 m a 0.50 m
1:1000	1.00 m
1:2000	2.0 m
1:5000	5.0 m
1:10,000	10.0 m
1:50,000	20.0 m
1:100,000	50.0 m

14.4.4 Direct Contouring on the Terrain

Direct contouring of the terrain may be required for special applications such as agriculture. The field procedure for this type of work is based on levelling using the height of collimation method presented in Chap. 13. Firstly, the operator must backsight to a known or assumed height point to determine the elevation of the reference horizontal plane. Then, taking into account the elevation of the contour line to be levelled, the value of the foresight reading can be calculated. In this case, the calculation determines the value to be read on the levelling staff to set the desired contour line in the ground (generally the nearest contour line with an integer elevation value). See example 14.1 for further details.

Example 14.1
The contour line with an elevation of 817.000 m must be set in the field from a benchmark *BM* with an elevation of 815.962 m. Given that the process is carried out using differential levelling (Fig. 14.16) and that the backsight reading on the benchmark is 1463 mm, calculate the value that must be read on the foresight staff to peg the desired contour line in the ground.

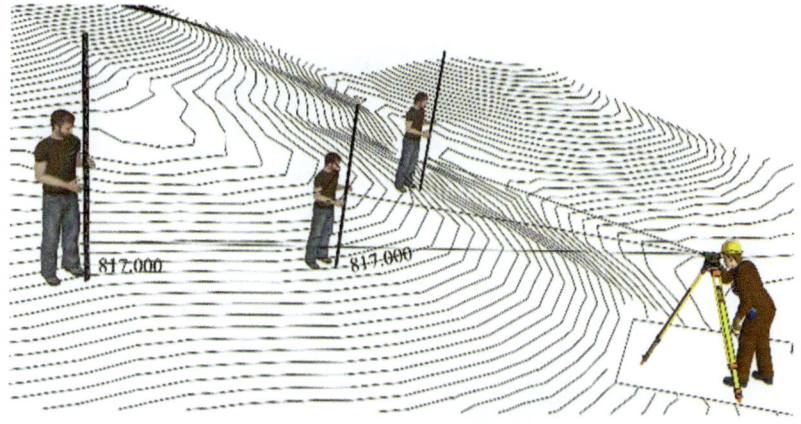

Fig. 14.16 Direct contouring on the terrain

Solution:
  Considering

$$RL = H_P + BS_P = 815.962 + 1.463 = 817.425 \text{ m}$$

$$H_Q = RL - FS_Q$$
$$817.000 = 817.425 - FS_Q$$
$$\therefore \ FS_Q = 425 \text{ mm}$$

  Thus, whenever the value read on the levelling staff equals 425 mm, *the point's elevation on the ground is* 817.000 m.

14.5 Perspective View

Relief representation for Geomatics applications is slowly moving towards a 3D view, as it can better represent the real world. It is intuitive and informative. However, its use in Civil Engineering projects is still limited to a few applications, such as urban planning projects and transportation routes.

The perspective view of the relief uses 3D computer graphics (perspective techniques) based on the digital terrain model of the area to be modelled, as shown in Fig. 14.17. Several 3D visualisation models are currently available. They range from simple wireframe 3D views to high-quality textured 3D digital surface models. Virtually all DTM or DSM programs can generate a 3D visualisation of the digital model. The reader must consult specialised references for more information on this subject, mainly in the computer graphics sciences.

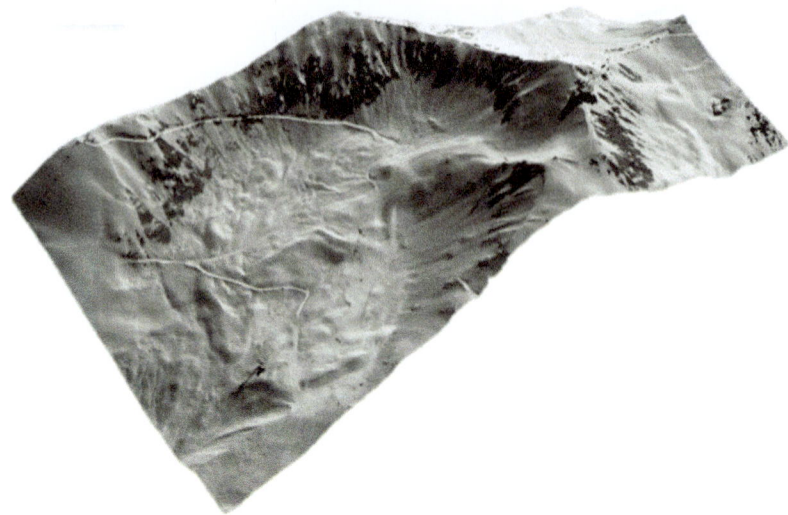

Fig. 14.17 Relief representation by the perspective view

14.6 Watershed

Watershed is the name given to the area of land that drains all watercourses (streams, rivers, etc.) and rainfall to a common outlet, as shown in Fig. 14.18. The dotted line indicates the watershed boundary, that is the drainage basin boundary.

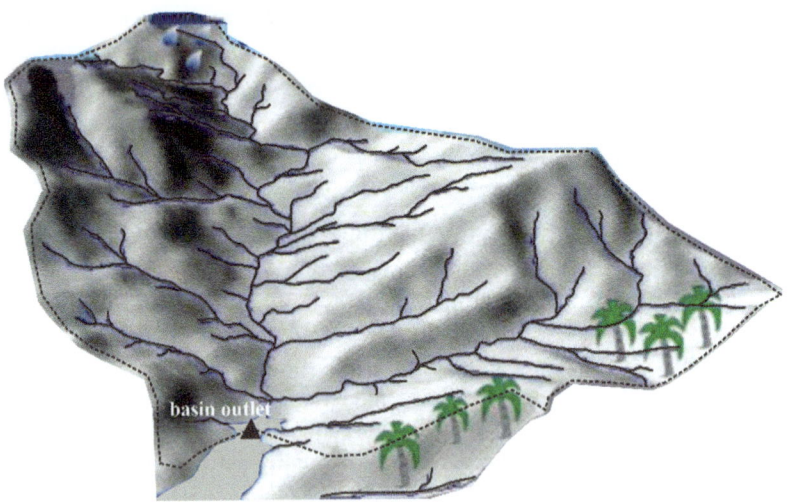

Fig. 14.18 Illustration of a watershed

Watersheds are used in engineering as a starting point for the development of drainage projects or when it is necessary to determine the flow of rainwater at specific points within a project area (outfall points), such as reservoirs, river crossings and culverts. The correct delineation of watersheds is therefore essential for the preservation, conservation, development and management of natural resources, as well as for disaster management and the dimensioning of transport drainage projects.

The use of a watershed in engineering projects is a subject covered in disciplines related to hydrology. This book discusses only the essential elements of a watershed and the recommendations for drawing its boundaries, that is the geometric determination of the watershed.

A watershed can be delineated manually on a topographic map using contour lines or automatically using a DTM. In the first case, the procedure is based on the following steps:

1. Have a topographical map with contour lines of the region where the watershed is to be delineated.
2. Study the map carefully to get all the essential information about the relief of the terrain, the contour lines and the direction of the water flow.
3. Define the basin outlet, that is the endpoint of the stream from which the watershed is to be delineated. This point is always over a watercourse in the lowest part of the catchment.
4. Highlight the corresponding watercourse on the topographic map up to its source.
5. Highlight the tributaries of the main watercourse.
6. Identify and highlight the highest points around the watercourse and its tributaries identified in Steps 4 and 5. These points are usually indicated on topographic maps by spot heights.
7. Draw the watershed boundaries starting and ending at the outfall identified in Step 3. To do this, look for paths to the highest points highlighted in the previous step and consider the highest points of the terrain between adjacent catchments. The catchment boundaries surround the watercourse and the sources of its tributaries, as shown in Fig. 14.19. The dividing line is drawn perpendicular to the contour lines.
8. On completion of the above steps, carry out a thorough review of the work.
   According to the above, a watershed has the following characteristics:

   - The dividing line between adjacent basins never intersects the watercourse or its tributaries, except at the outlet.
   - The watershed passes through the highest regions of the basin, and there may be elevated points within the basin that are not part of its boundaries.
   - Water flow is always perpendicular to the contours of the watercourse.
   - Terrain slopes opposite to the identified drainage lines indicate regions belonging to other basins.
   - The catchment boundary line must pass through the outlet.

Figure 14.19 shows an example of watershed delineation according to the steps indicated.

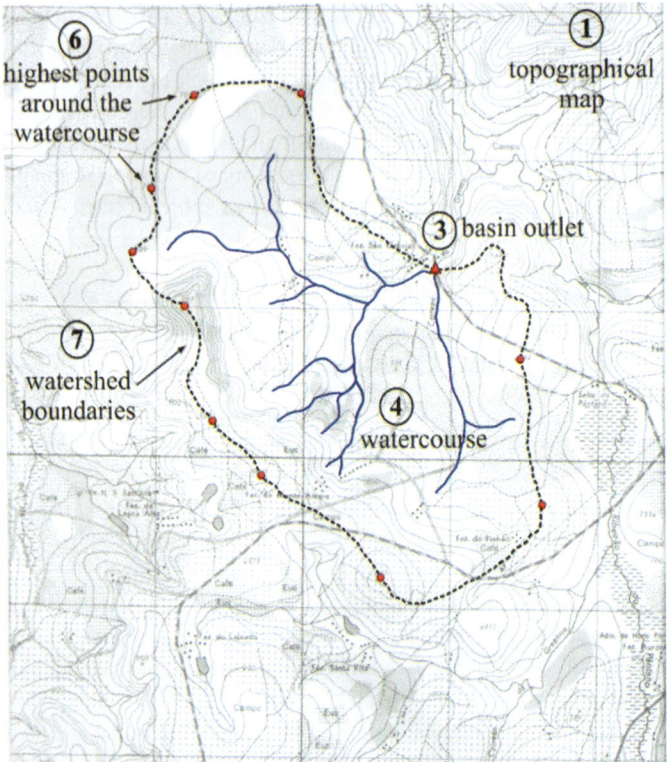

Fig. 14.19 Steps for the delimitation of a watershed

The reader must look for specific application programs when using a DTM to delineate a watershed. In this case, the drawing is carried out automatically, considering the slopes of the terrain generated by the interpolation method used in the numerical modelling of the terrain. However, the user must carefully check that the layout of the basin is consistent with the relief and, if necessary, edit the final product generated by the software.

14.7   Review Questions

1. Explain the difference between longitudinal and cross-sectional profiles.
2. Discuss which levelling methods can be used to determine profiles and in what circumstances would you recommend their use in Civil Engineering works.
3. Explain under what conditions the determination of profiles by point interpolation over a cartographic map with contour lines can be used in a Civil Engineering project.
4. Explain why profile plotting is usually carried out with vertical exaggeration.

5. Explain how contouring is currently carried out for a Civil Engineering project. Discuss the field and office procedures used for the different data collection methods.
6. Discuss for which type of Civil Engineering project you would suggest using differential levelling, trigonometric levelling, GNSS, airborne laser scanning and airborne photogrammetry to collect data for relief determination.
7. Briefly explain how a relief perspective view can be generated from surveying engineering data collection.
8. Discuss the importance of watershed determination for Civil Engineering and how it can be determined.

Reference

Barras, V (2015). *Les relevés de terrain en 3D, Les profils, les plans topographiques, les MNTs.* heig-vd - Haute Ecole d'Ingénierie et de Gestion du Canton de Vaud. Yverdon-les-Bains. Switzerland.

Chapter 15
Digital Terrain Modelling

Irineu da Silva

15.1 Introduction

The ability to describe and represent terrain relief through mathematical modelling has been an invaluable advance in geosciences and has changed how Civil Engineering projects are designed. Since its creation in the 1950s in the USA to support the automation of road projects, it has been used in a wide range of engineering applications requiring elevation information on the relief of the terrain. The list of products derived from such models is extensive, including contours, terrain profiles, volume calculations, earthworks, geometric design of transport routes, pipeline design, electrical transmission lines, slope maps, drainage networks, landscape perspective views, support for the production of orthophotos in photogrammetry, and many others.

Mathematical modelling of a terrain surface means approximating part or all of the continuous surface of the terrain by mathematical functions capable of describing the geometric shape of the topographic relief based on a set of discrete 3D points with unique elevation values. In this case, the mathematical functions are interpolation functions from which a mathematically modelled surface is generated. As a result, the elevation value of any point in the model can be extracted from the DEM through this surface.

The set formed by the variables to be modelled, plus the interpolation functions and the resources for practical use of the modelled surface, is given the generic name of the *Digital Elevation Model* (DEM). Under the umbrella of DEM, two specific terms are used to describe the concept of terrain surface modelling: the *Digital Terrain Model* (DTM) and the *Digital Surface Model* (DSM). A DTM represents the topographic surface of the Earth's bare ground, excluding trees, buildings, and any other surface objects. Conversely, a DSM includes vegetation heights and artificial features, as shown in Fig. 15.1. In any case, the reader needs to understand that a

I. da Silva (✉)
São Carlos School of Engineering, University of São Paulo, São Carlos, SP, Brazil
e-mail: irineu@sc.usp.br

DEM is a digital spatial representation of terrain elevation consisting of 2D positioning plus a mathematical set of equations that describe the elevation of any point in a 3D geographic space.

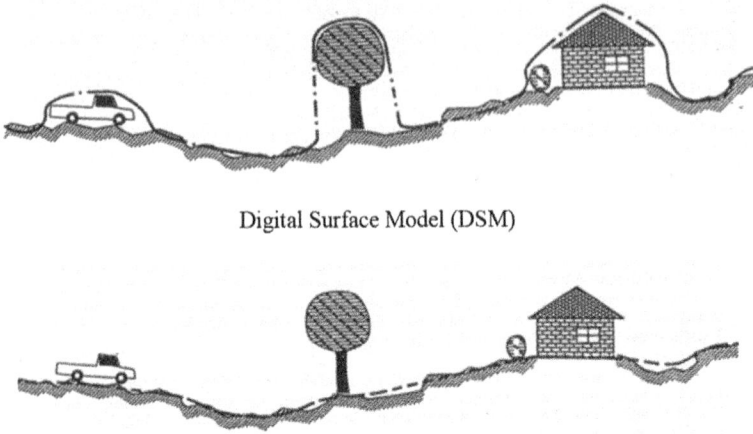

Digital Surface Model (DSM)

Digital Terrain Model (DTM)

Fig. 15.1 DSM and DTM concepts

Generally, a DSM is obtained when the surface is modelled by point clouds generated by airborne laser scanners or photogrammetric image correlation techniques. Therefore, to obtain a DTM, filtering techniques are used to filter out all non-relief elements, as shown in the perspective views in Fig. 15.2.

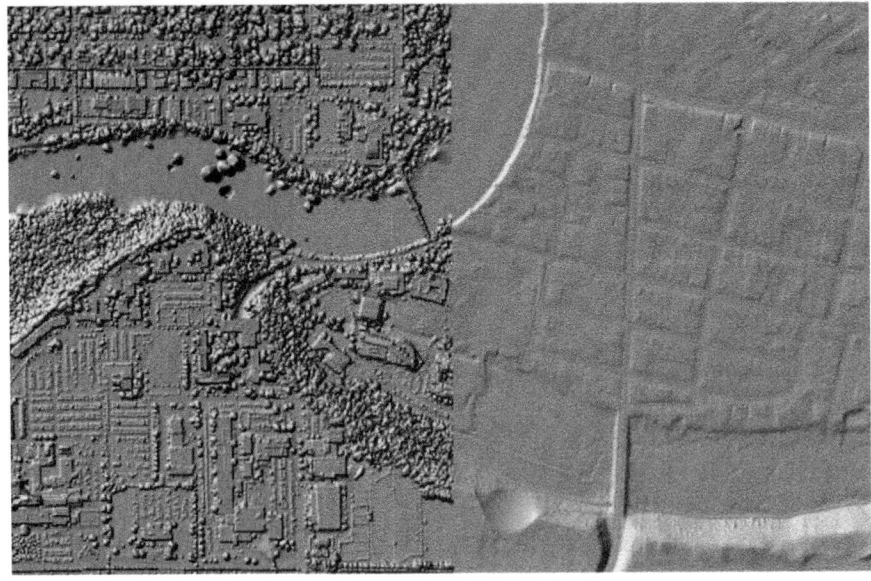

Fig. 15.2 Perspective views of the DSM (on the left) and the filtered DTM (on the right)

Another aspect to consider in digital elevation modelling is the data structure, which can be raster or vector. The raster data structure is a grid of uniform, continuous, digital square cells covering an area of the Earth's surface, with each cell assigned an elevation value. The vector data structure consists of discrete 3D points (nodes) representing the terrain surface and lines or polygons representing terrain features. The advantage of vector data is that it is more accurate than raster data. However, a raster model is easier and faster to handle. However, since in most current Civil Engineering applications, the elevation data to be modelled correspond to discrete 3D points on a terrain surface, only the concepts relating to the structure of vector data are covered in this chapter. Figure 15.3 shows an example of both DEM data structures.

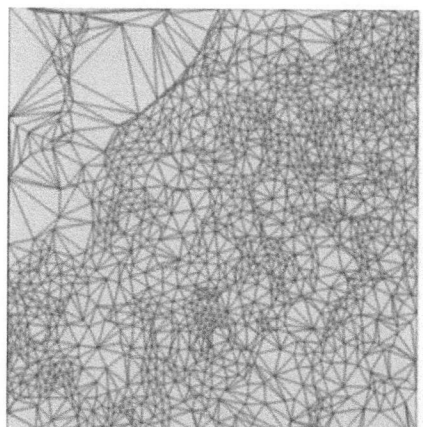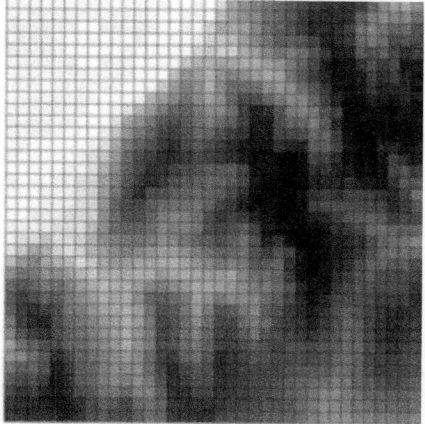

Fig. 15.3 Vector and raster DEM data structures (Hengl and Evans 2009)

In most cases, Civil Engineering DEMs are limited to a specific area related to a project and are of high spatial resolution and accuracy. However, Civil Engineers can also use global DEM data sources from international institutes for preliminary design studies. They can be published under a free or restricted licence and cover the whole world or specific regions of a country.

The Shuttle Radar Topography Mission (SRTM), carried out in February 2000 by the US National Aeronautics and Space Administration (NASA), the German Aerospace Center (DLR) and the Italian Space Agency (Agenzia Spaziale Italiana, ASI), was a pioneering effort to produce a DEM on a near-global scale. It was based on a technique known as Synthetic Aperture Radar (SAR) interferometry, i.e., a specially modified radar system based on two SAR antennas (flown aboard the Space Shuttle Endeavor) separated by a 60-m extender, covering about 80% of the Earth's surface (from 56° S to 60° N). Today, the 1-arc second global DEM (30 m along the equator) is available from the United States Geological Survey website.

Although the SRTM DEM is one of the most widely used and perhaps the most popular among Geomatics professionals, it is not the only high-resolution digital topographic database of the Earth available to the community. Others, such as the 2020 NASADEM, TanDEM-X, *Copernicus*-DEM and *Forest and Buildings removed Copernicus*-DEM (FABDEM), are examples of DEMs produced from radar interferometry. Other examples are the *Advanced Spaceborne Thermal Emission and Reflection Radiometer* (ASTER) and the ALOS World 3D-30 m (AW3D30), produced using photogrammetry and available in the public domain.

Table 15.1 lists some currently available global DEM data sources for Geomatics applications.

Table 15.1 Global DEM data sources

Name	Region	License	Resolution
SRTM 3-Arc seconds	World from 56° S to 60° N	Public domain	3 Arc seconds
SRTM 1-Arc second	World from 56° S to 60° N, except for the Middle East (map)	Public domain	1 Arc second
GTOPO30	World	Public domain	30 Arc seconds
GLOBE	World	Public domain	30 Arc seconds
CGIAR	World from 56° S to 60° N	Non-commercial	Diverse
MERIT DEM	World from 60° S to 90° N	Non-commercial	3 Arc seconds
Viewfinder Panoramas enhanced SRTM data	World	Research and private use	Diverse
Allows world 3D—30 m	World	Terms of use	30 m
govdata Germany	Germany	Variable	Diverse
gis.arso.gov.si	Slovenian	Public domain-like	1 m
Radarsat Antarctic Mapping Project	Antarctica	Citation	Diverse
CDED	Canada	Public domain	Diverse
EU-DEM	Europe	Terms of use	1 Arc second
Opendata Portal Europe	Europe	Legal notice	Diverse
NED Alaska	Alaska	Public domain	Diverse

A DEM is always used through a computer application program developed for this purpose. Such an application must include modules for data acquisition, surface modelling and the generation of derived products. In addition, to be considered a DEM, it must be able to consider the geographical features of the relief, such as slopes, watersheds, ridges, valleys, river banks, retaining walls, built-up areas and others.

Another aspect of DEM that should be highlighted is the frequent misuse of the term. It is common to find references to the term when describing cloud points, especially when they are generated by laser scanners. In these cases, the point cloud is incorrectly described as a DEM when it is only a sample of heights at discrete locations and not a land surface model. Another common mistake is to call a perspective view of terrain a DEM when it is not. A perspective view may have been generated from the modelled surface but is not the model itself. It is also common to say that a DEM is a contouring application. This is true, but contouring is only one of the products generated by a DEM and not even the most important.

As shown in Fig. 15.4, the production of a DEM involves several interrelated functional tasks, from field data collection to practical application. Therefore, a DEM must be interpreted as much more than a simple application program, as the quality of the final product depends on the quality of each step.

Fig. 15.4 Functional tasks to generate a DEM. (Adapted from Hutchinson and Gallant 2000, p. 30)

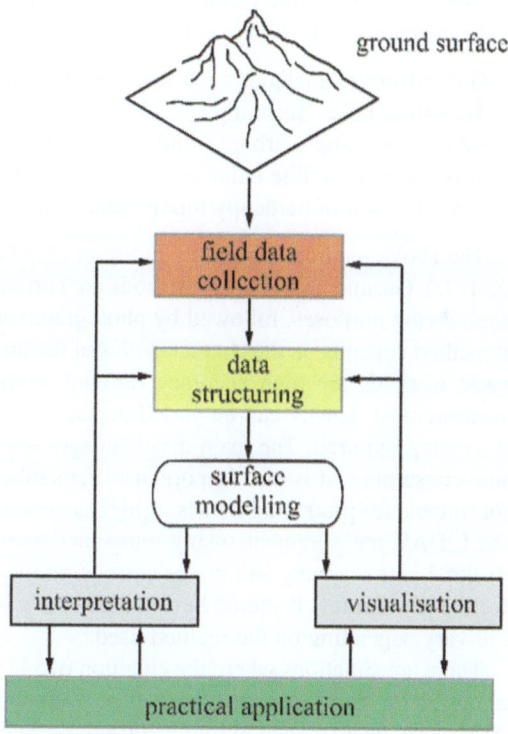

Thus, according to Fig. 15.4, generating a DEM comprises the following functional tasks:

- Field data collection
- Data structuring
- Surface modelling
- Model visualisation and interpretation
- Practical application

15.2   Field Data Collection

Data acquisition is the main task in digital elevation modelling. It consists of two steps: sampling and measurement. Sampling refers to the selection of the feature location, while measurement determines the 3D coordinates of the sampling points. It can be considered the fundamental step in the process of generating a DEM, which is based on the use of the various spatial coordinate determination methods discussed in the chapters of this book, as follows:

- Ground survey methods such as differential, trigonometric and GNSS levelling
- Terrestrial Laser Scanning
- Remote Sensing (airborne and satellite imagery, airborne laser systems, and airborne and satellite radar using interferometry)
- Digitalization of hardcopy topographic maps

The above methods produce a group or cloud of points with known coordinates (*X, Y, H*). Ground and LiDAR methods are currently the most widely used for Civil Engineering purposes, followed by photogrammetric image correlation. The choice of method depends on the characteristics of the project. Some advantages of ground-based methods are high accuracy (around or even better than 1 cm), flexibility (measurement density can be varied depending on the terrain) and very little post-processing required. The main disadvantages are that they are labour-intensive and time-consuming. It is also important to remember that ground survey methods are not suitable for producing DSMs. If high accuracy is not a priority, photogrammetry and LiDAR measurements using unmanned aerial vehicles can also provide good results. Laser scanning and image correlation may be the choice for large areas and decimetre accuracy. It should be noted that the amount of data and spatial resolution will vary depending on the method used.

There are situations where the creatiion of a DTM from surveying methods is not an option. In such cases, if topographic maps with contours are available, suitable DTMs can be extracted through an appropriate digitisation process. The reader should note, however, that these methods produce very poor results and are only suitable for preliminary studies in Civil Engineering projects.

For DTM generation, ground points can be classified into individual and feature-specific points. Individual points are primarily used as spot elevations for mathematical interpolation, while feature-specific points, when connected by lines, are used to describe topographic elements of the terrain. Lines connecting feature-specific points are called feature-specific lines and are used to define or constrain interpolation regions in the surface modelling process. They are mainly borderlines, which indicate thresholds for surface modelling; dead areas, which are surface regions without elevation information i.e., water surfaces); and break lines, which are sudden changes in slope in the terrain (river banks, ridge lines, slope picks, retaining walls, etc.). Typically, a feature-specific line is manually labelled during data acquisition.

DTM surface modelling is then performed by selecting individual points and feature-specific lines separately. Figure 15.5 shows feature-specific lines represented in a 2D terrain view. The vertices of the triangle are individual points of known elevation.

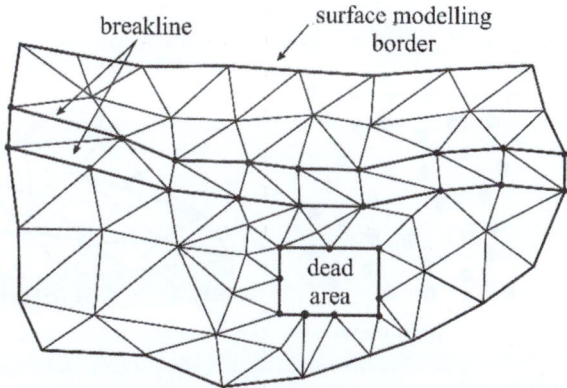

Fig. 15.5 Feature-specific lines for DEM surface modelling

From a geometric point of view, a terrain surface can be represented by two geometric patterns: *regular* or *irregular*. The regular pattern can be a regular grid or a series of contiguous equilateral triangles, hexagons or other regularly shaped geometric figures. For Geomatics applied to Civil Engineering, however, the regular grid is the most commonly used. This pattern can be achieved by defining the point intervals in both directions (X, Y) to form the plane grid. The elevations of all points at the grid nodes are then determined, as shown in Fig. 15.6. This pattern requires special care to define the terrain points coherently so that there is no excessive loss of feature-specific points.

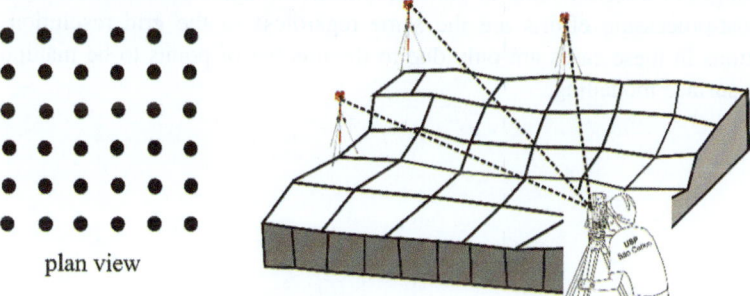

Fig. 15.6   Field data collection by regular grid points

In the case of field data collection for Civil Engineering purposes, the elevations of regular grid points are mainly determined by differential levelling. In this case, the field procedure is the same as described in Sect. 14.4.2.

Due to levelling limitations, this type of data sampling is recommended for small areas and smooth terrain.

In cases where the terrain surface has too many feature-specific lines, one solution is to combine regular grid sampling with selective feature-specific points. In practice, this means that more than one ground levelling method or angle/distance measurement method must be used for 2D point positioning of the feature-specific points.

The irregular pattern often used in Geomatics applied to Civil Engineering is the random data collection method, which means that the measured points are randomly distributed over the terrain, as shown in Fig. 15.7. Any of the 3D positioning methods mentioned above can generate this geometric pattern.

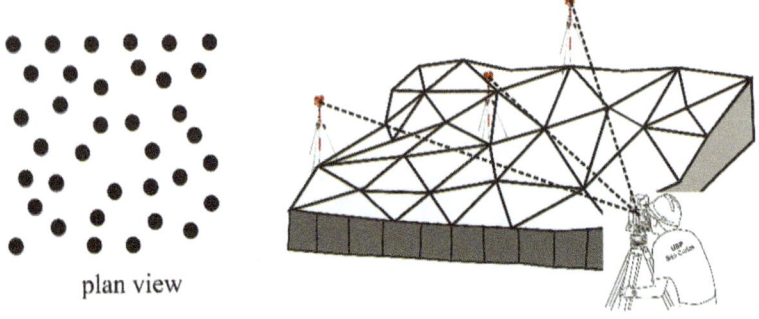

plan view

Fig. 15.7 Field data collection by irregular grid points

Using a ground survey method, the operator can choose which random, feature-specific points to measure best represent the terrain surface.

Whether laser scanning or photogrammetric correlation techniques are applied, random data is collected automatically, and feature-specific points are detected during the data post-processing phase.

Another important issue to consider when generating the DEM is the resolution of the sampling grid. The values vary depending on the intended use of the DEM. In any case, it is only of great interest if the fieldwork constraints require land surveying. For laser scanning or photogrammetric image correlation, the fieldwork and post-processing efforts are the same regardless of the grid resolution. The limitations in these cases are only due to the number of points to be manipulated during surface modelling.

15.3   Data Structuring

The dataset of measured points must be appropriately structured before subsequent digital elevation modelling operations are applied. Data structuring, in the case of vector configuration, consists of generating a grid (regular or irregular) of discrete 3D points suitable for applying surface modelling interpolation functions. It is at this stage that it is necessary to evaluate: (1) the accuracy with which the data represent surface roughness, (2) whether all geographic elements are represented consistently, and (3) whether there is geometric consistency in the generated grid. In short, it is at this stage that the positional and distributional quality of the data collected in the field must be evaluated.

15.3.1   Data Structuring by Regular Grid Points

The data structuring by regular grid points can result directly from the geometric distribution of the data collected in the field (points already distributed in a regular pattern, as indicated in the previous section) or, in the case of a random data collection method (irregular pattern), from an interpolation method based on points neighbouring the vertices of the grid to be generated, as shown in Fig. 15.8. In the latter case, the height values (HP) of the grid vertices are interpolated as a function of the position (Xi, Yi) and height (Hi) of the neighbouring points. This type of data structuring results in a regular grid with vertices of known coordinates (X, Y, H).

As mentioned above, structuring the data into a regular grid has the drawback that it does not always match the complexity of the terrain relief, requiring refinements to the grid to improve the accuracy of the final model.

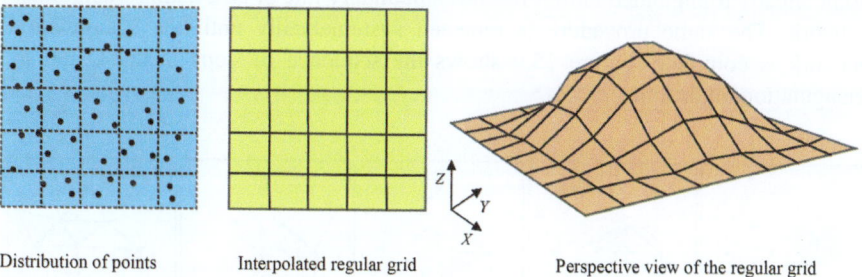

Distribution of points   Interpolated regular grid   Perspective view of the regular grid

Fig. 15.8 Data structuring by regular grid points

15.3.2   Data Structuring by Irregular Grid Points

Another alternative to data structuring is to use an irregular triangulation method. In this case, the data collected in the field is joined three to three to form a 2D network of triangles connecting all the points available for the model, as shown in Fig. 15.9b.

The resulting triangle-based approach is called a Triangular Irregular Network (TIN).

For the network of triangles to be able to generate a consistent DEM, three conditions need to be met:

- For each set of points, the generated irregular network (TIN) must be unique,
- The geometric shape of the triangles must be optimised so that each triangle is as equilateral as possible, and.
- Each triangle must be formed by the nearest neighbours, i.e. the sum of the three sides must be as small as possible.

There are several algorithms available for generating the TIN. Of these, the only one that satisfies the three conditions above is the Delaunay triangulation algorithm, briefly described below. Delaunay triangles are inter-connected figures where the vertices are the sample points. The triangles are created to be as equilateral as possible, and there is only one possible optimal triangulation for a data set. For more details on this triangulation algorithm, the reader is advised to consult specialised literature.

The first step in the Delaunay triangulation algorithm is the selection of the start point (also known as the rotation point). Although any point can be chosen, it is recommended to start from a point in the geometric centre of the dataset. Once the start point has been selected, the next step is to find the closest point to form the base of the first triangle. A circle is then drawn with a diameter equal to the distance between the two points, and the number of points inside the circle is checked. If there are no data points within the circle, it is progressively enlarged until one or more points are found. If only one point is found, it is selected as the third vertex of the triangle. If there is more than one point inside the circle, they are tested to see which meets Delaunay's criteria. Once this is achieved, the search for the next neighbour continues until a board line is reached. The triangulation procedure continues from a point already triangulated until it reaches a boundary line or an existing triangulated network. The same procedure is repeated systematically until the triangulation network is complete. Figure 15.9 shows the sequence of steps in the Delaunay triangulation algorithm.

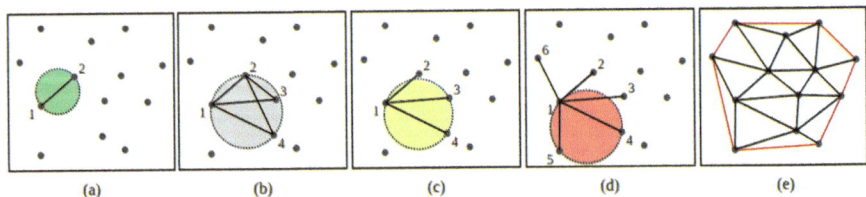

Fig. 15.9 Sequence of steps of the Delaunay Triangulation algorithm

Steps:

1. Find the nearest neighbour to point 1.
2. Create a circle with a diameter equal to the distance between points 1 and 2.
3. Extend the circle.
4. Identify the points inside the circle (3 and 4) and choose the one that forms the largest angle with the base (3).
5. Create a circle through points 1 and 3.
6. Repeat the process for point 2 and so on.

Figure 15.10 shows a top view and a full perspective view of the irregular network (TIN) generated from the set of points with known coordinates (X, Y, H). Each vertex of the grid corresponds to a point measured on the terrain.

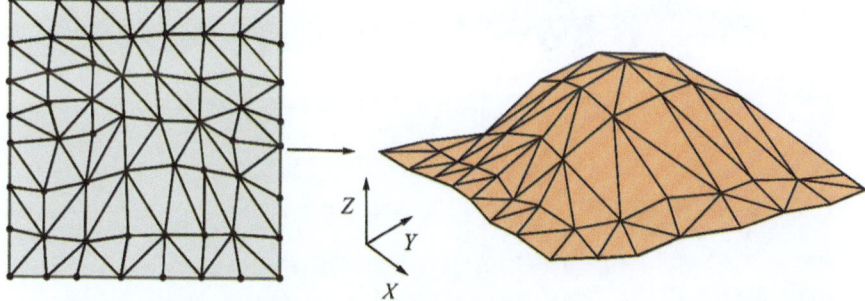

Fig. 15.10 Data structured according to an irregular grid - TIN

Once the triangle mesh (TIN) has been created, it must be checked for possible edits and resampling. This step involves inserting and/or editing the feature-specific lines of the model and filtering out values considered 'suspect' for representing the surface. As mentioned earlier, feature-specific lines are generated from feature-specific points collected in the field. In this context, some DTM generation applications allow the feature-specific points collected in the field to be automatically inserted into the grid pattern during the data download. Others require this to be done manually. Whichever method is used, the result is the generation of an irregular grid whose feature-specific lines are included as sides of the triangles, i.e. no triangle side should cross it.

Figure 15.11 shows an example of editing a *break line* to be inserted into the grid, and Fig. 15.12 shows a 3D view of a DTM with its *break lines*.

The TIN structure becomes more complex if feature-specific lines need to be included. However, the greater the number of these lines, the more accurate the model will be as it will better represent the irregularities of the relief.

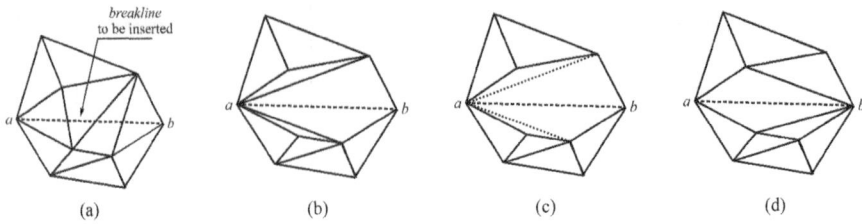

Fig. 15.11 Example of including a breakline in the Delaunay Triangulation

Fig. 15.12 Perspective view of a DTM with its break lines

15.4   Surface Modelling: Interpolation Function

An interpolation function is, by definition, a continuous function that approximately describes the mathematical behaviour of a phenomenon for which only discrete values are known. In the case of a three-dimensional surface, it consists of determining the surface that approximately describes the behaviour of a set of discrete points defined by their three coordinates (X, Y, H). It is generally assumed that the coordinates (X, Y) are the independent variables, and the elevation (H) is the dependent variable, so that

$$H = f(X, Y) \tag{15.1}$$

In other words, the value of the variable (H) can be calculated for all values of (X, Y) that fall within the domain of the function.

In the case of a DEM, the interpolation functions are used to determine the height (H_P) of any point (P) from the known values of the heights (H_i) of its neighbours. In addition, it must consider that the surface of a terrain has some peculiar characteristics that impose some constraints on the mathematical model of the interpolation, which are:

- The terrain surface is continuous and smooth
- There is always a high correlation between adjacent points

In these terms, the interpolation methods used to generate a DEM can be classified into

- Point-based methods
- Grid-based (or area-based) methods

15.4.1 Point-Based 3D Interpolation Methods

A point-based interpolation method determines the height (H) of an output point in the model from its nearest neighbour dataset of measured points with known (X, Y, H) coordinates.

There are several point-based 3D interpolation methods available in the literature. Typically, they are all very simple to use and can produce good results for dense and well distributed point sets. Most are based on averaging techniques, as presented in the following subsections.

15.4.1.1 Moving Average Interpolation Method

The Moving Average interpolation method assigns an elevation value (H_P) to an output point (P) by averaging the (H_i) known elevation values of n points in a surrounding search area, as given by Eq. (15.2).

$$H_P = \frac{1}{n}\sum_{i=1}^{n} H_i \qquad (15.2)$$

It is a simple method to apply but has the disadvantage of being empirical and requiring the definition of a neighbourhood for interpolation.

15.4.1.2 Weighted Moving Average Interpolation Method

The Weighted Moving Average interpolation method assigns a weighted average of the surrounding points (i) to the output point (P), as given by Eq. (15.3).

$$H_P = \sum_{i=1}^{n} w_i * H_i \tag{15.3}$$

where w_i is the weight assigned to point (i).

The weight calculation is usually based on the inverse of the distance between the surrounding points (i) and the output point (P), i.e., the shorter the distance, the greater the weight, as given by Eq. (15.4).

$$w_i = \frac{\frac{1}{d_i^x}}{\sum_{i=1}^{n} \frac{1}{d_i^x}} \tag{15.4}$$

where

d_i = relative distance between the point (i) and the output point (P)
x = weight adjustment coefficient

Although simple, these interpolation methods are not currently used for DTM generation, mainly due to the resulting discontinuities on its surface and the need to define a search area. However, they can be used for specific applications such as volume calculation.

Example 15.1
Given the Cartesian spatial coordinates of the points indicated in Table 15.2, calculate the elevation of the point (P) using the Weighted Moving Average interpolation method with a weight adjustment coefficient equal to 2.

Table 15.2 Known coordinates

Point	X [m]	Y [m]	H [m]
P	11,556.118	15,468.963	–
A	11,588.256	15,511.332	815.325
B	11,599.247	15,450.458	817.269
C	11,520.053	15,436.214	816.567
D	11,527.000	15,510.000	817.864
E	11,560.000	15,410.000	815.943

Solution:
To solve this example, it is first necessary to assign weights to the elevations of the neighbours of the point (P), which are determined as a function of the distances

between the point (P) and each one of them. The results of the calculations are given in Table 15.3.

Table 15.3 Partial values for determining the weights

Line	Distance [m]	H [m]	$\frac{1}{d^2}$	w_i	$w_i * H_i$
PA	53.179	815.325	0.0004	0.185	150.916
PB	46.931	817.269	0.0005	0.238	194.236
PC	48.715	816.567	0.0004	0.221	180.115
PD	50.318	817.864	0.0004	0.207	169.090
PE	59.091	815.943	0.0003	0.150	122.321
			$\sum \frac{1}{d^2} = 0.0019$		

From the partial results presented in Table 15.3 *and using* Eq. (15.3), *the elevation of the point (P) is obtained as follows:*

$$H_P = 150.916 + 194.236 + 180.115 + 169.090 + 122.321 = 816.679\,\text{m}$$

15.4.2 Grid-Based 3D Interpolation Methods

A grid-based interpolation method determines the elevation (H) of an output point in the model using a surface interpolation math function from a grid of datasets of measured points with (X, Y, H) coordinates. If the interpolation surface is a single complex 3D surface from the entire dataset of measured points, it is called a *Global Interpolation Method*. Suppose the domain of the interpolation surface is divided into several small patches so that an interpolation surface can be generated for each patch. In that case, it is called a *Local, Griding or Patchwise* interpolation method.

The Global interpolation method consists of fitting a single 3D surface defined by a high-degree polynomial function to describe the entire behaviour of the interpolation surface, as given by Eq. (15.5).

$$H(x, y) = \sum_{i,j=0}^{n} a_{i,j} x^i y^j \tag{15.5}$$

where

$a_{i, j}$= polynomial coefficients
x, y= polynomial variables
n= degree of the polynomial

The problem with using this interpolation method is that, as the degree of the polynomial increases, the interpolated surface quickly becomes unrealistic, making the use of the method unfeasible for digital terrain modelling. To avoid such an occurrence, it is suggested to use Local interpolation methods, as presented in the following subsections.

15.4.2.1   Modelling of Irregular Surfaces by a Linear Function

Triangle-based surface modelling is the simplest way to represent a DEM in an irregular geometric pattern. Each triangle represents a patch in a pattern that can be described by a single interpolation surface. In this case, the most common interpolation method is linear interpolation, from which a plane containing the three vertices of the triangle, as shown in Fig. 15.13, is defined in a rectangular coordinate system as given by Eq. (15.6).

Fig. 15.13 Linear interpolation surface

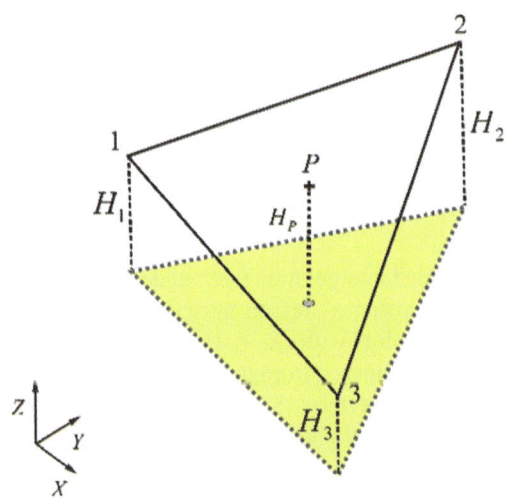

$$H = a_0 + a_1 x + a_2 y \qquad\qquad (15.6)$$

where

$a_i =$ polynomial coefficients
$x, y, H =$ coordinates of a surface point

The values of the three coefficients a_i of the polynomial are calculated as a function of the known coordinates (x, y, H) for each vertex of the triangle, that is, $P_1(x_1, y_1, H_1)$, $P2\ (x_2, y_2, H_2)$ and $P_3\ (x_3, y_3, H_3)$. Thus, the following set of equations can be written for each triangle:

$$H_1 = a_0 + a_1 x_1 + a_2 y_1$$
$$H_2 = a_0 + a_1 x_2 + a_2 y_2 \qquad (15.7)$$
$$H_3 = a_0 + a_1 x_3 + a_2 y_3$$

Or, in matrix form:

$$\begin{bmatrix} H_1 \\ H_2 \\ H_3 \end{bmatrix} = \begin{bmatrix} 1 & x_1 & y_1 \\ 1 & x_2 & y_2 \\ 1 & x_3 & y_3 \end{bmatrix} \begin{bmatrix} a_0 \\ a_1 \\ a_2 \end{bmatrix} \qquad (15.8)$$

$$H = XA \qquad (15.9)$$

From this, the values of the polynomial coefficients can be calculated using Eq. (15.10).

$$A = X^{-1}H \qquad (15.10)$$

In this way, the values of the coefficients a_0, a_1 and a_2 are obtained and, consequently, the value (H_P) of any point (P) with known coordinates (X_P, Y_P) inside the triangle, simply by substituting its coordinates (X_P, Y_P) into Eq. (15.6). It is worth noting that the polynomial coefficients a_i must be solved for each triangle, which makes this interpolation method completely local.

15.4.2.2   Regular Grid Surface Modelling by Bilinear Function

In the case of a regular grid of measured points, the interpolation can be performed using a bilinear polynomial function, as given by Eq. (15.11).

$$H = a_0 + a_1 x + a_2 y + a_3 xy \qquad (15.11)$$

where

a_i = polynomial coefficients
x, y, H = coordinates of a surface point

As with triangle-based linear interpolation, the values of the four coefficients a_i of the polynomial are calculated as a function of the known coordinates (X, Y, H) for each vertex of the grid cell, that is, P_1 (x_1, y_1, H_1), P_2 (x_2, y_2, H_2), P_3 (x_3, y_3, H_3) and P_4 (x_4, y_4, H_4). Thus,

$$\begin{bmatrix} a_0 \\ a_1 \\ a_2 \\ a_3 \end{bmatrix} = \begin{bmatrix} 1 & x_1 & y_1 & x_1y_1 \\ 1 & x_2 & y_2 & x_2y_2 \\ 1 & x_3 & y_3 & x_3y_3 \\ 1 & x_4 & y_4 & x_4y_4 \end{bmatrix}^{-1} \begin{bmatrix} H_1 \\ H_2 \\ H_3 \\ H_4 \end{bmatrix} \tag{15.12}$$

In this way, the values of the coefficients a_0, a_1, a_2 and a_3 are obtained and, consequently, the value (H_P) of any point (P) with known coordinates (X_P, Y_P) inside the grid cell simply by substituting its coordinates (X_P, Y_P) in Eq. (15.11).

However, as described above, using linear or bilinear interpolation functions presents a problem of smoothness between patches. Because they are flat facets, they are not smooth, as shown in Fig. 15.14. To solve this problem, it is necessary to use higher-degree polynomials, such as polynomials with 8, 12 and even 16 terms, as shown in the sequence.

Fig. 15.14 Discontinuity of flat facets of a triangular network (Zhou 2017)

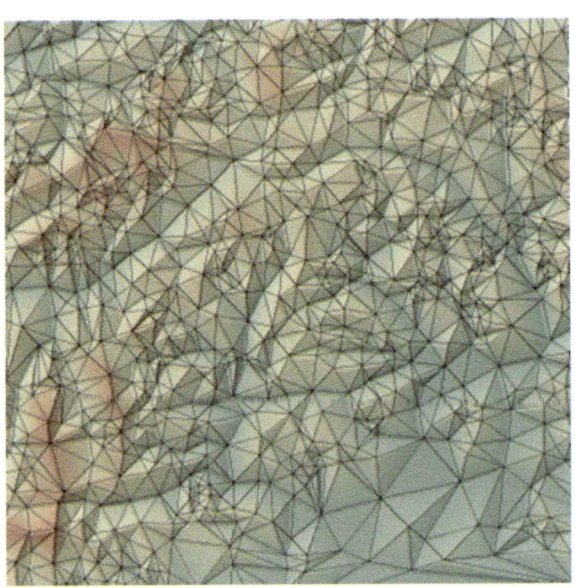

In the case of using an 8-term polynomial, the bi-quadratic interpolation polynomial is used as described in Eq. (15.13).

$$H = a_0 + a_1x + a_2y + a_3xy + a_4x^2 + a_5y^2 + a_6x^2y + a_7xy^2 \tag{15.13}$$

The terms x^3 and y^3 have been neglected for ease of calculation. There are, therefore, eight coefficients to calculate. The solution is to use a regular grid with eight reference points, i.e., to add four new points on the sides of each grid cell, e.g., a two-dimensional cubic interpolation.

For the regular grid, one can also consider using a bi-cubic polynomial with 16 terms, as shown in Eq. (15.14).

$$H = a_0 + a_1 x + a_2 y + a_3 xy + a_4 x^2 + a_5 y^2 + a_6 x^2 y + a_7 xy^2 + a_8 x^2 y^2 +$$
$$a_9 x^3 + a_{10} y^3 + a_{11} x^3 y + a_{12} xy^3 + a_{13} x^2 y^3 + a_{14} x^3 y^2 + a_{15} x^3 y^3 \tag{15.14}$$

There are, therefore, 16 coefficients to calculate. The proposed solution for this case is to use the first and second partial derivatives of the function as continuity conditions for the subdomains. In the case of regular grids, the first derivatives at (X) and at (Y) are considered equal to the tangents of the junctions between two adjacent subdomains in both directions, and the second derivative is equal to the diagonal tangents.

The solution for calculating the 16 coefficients is obtained by considering that four equations are formed by the values of (H_i) of each vertex of the quadrangular element, and the remaining 12 equations are created by the three derivatives in each vertex. Once the values of the coefficients a_i are obtained, the value (H_P) of any point (P) with known coordinates (X_P, Y_P) inside the grid cell can also be determined by simply substituting its coordinates (X_P, Y_P) into Eq. (15.14).

In the case of triangular grids, there are also proposed solutions using bi-cubic polynomials. However, the methodology used requires mathematical calculations that are somewhat more sophisticated than the previous ones. For this reason, they have not been included in this text. The interested reader should consult specialised references.

In addition to the interpolation models mentioned above, the reader will find in the literature several others that can be applied to digital terrain modelling. Among these, the Kriging interpolation method stands out, assuming that the surface generated is based on a spatial correlation structure. This surface has three components: a structural component associated with a constant mean or trend (deterministic function), a stochastic term that varies locally and spatially as a function of the structural component, and an uncorrelated random noise that is normally distributed. It should be noted that this interpolation method can only be applied if the above components are detected. This is done using a variogram graph obtained from the measured points. Although this method is available in many terrain modelling applications, it will not be discussed in this book.

15.5 Surface Model Visualisation

For the DEM to be displayed graphically, it is necessary to use one of the following traditional forms of graphical surface representation:

- Contouring
- Shading
- Three-dimensional perspective view

15.5.1   Surface Model Visualisation by Contouring

As described in Chap. 14, contour line relief representation is the most widely used technique in engineering projects worldwide. Therefore, it is usually the preferred method for displaying DEMs. However, the main disadvantage of contouring, especially for non-specialists, is its inability to provide a visual impression of the terrain relief.

15.5.2   Surface Model Visualization by Shading View

In this case, the representation is flat and carried out using a hypothetical illumination of the terrain surface from a particular azimuth and an altitude of the sun. Figure 15.15 shows a classic example of this type of relief map. This is a convenient form of representation for cartographic inspection of the model, although it can easily occlude small relief areas.

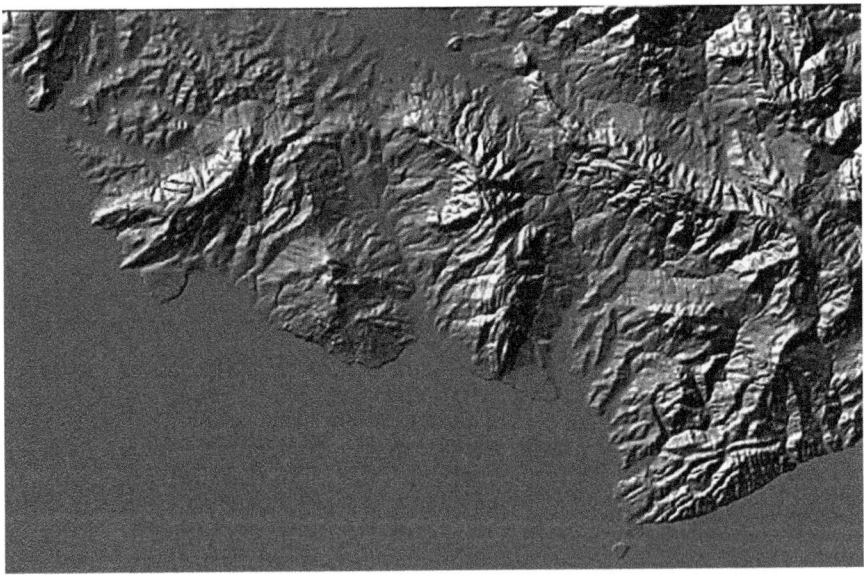

Fig. 15.15 DTM representation by shading view

15.5.3   Surface Model Visualisation by Three-Dimensional Perspective View

The perspective view of a DEM, as shown in Fig. 15.16, is easy for non-professionals to understand. Although it does not provide a complete view of all parts of the terrain, it allows highlighted areas to be viewed as a 3D projection of the surface, thus eliminating any possibility of occluding landscape areas. The ability to rotate the terrain view in any direction in space is its greatest asset.

The 3D perspective view can also be enhanced by rendering techniques that create vivid representations of 3D objects by assigning a grey level or colour to small portions of the surface model based on an illumination model and a user-defined viewpoint.

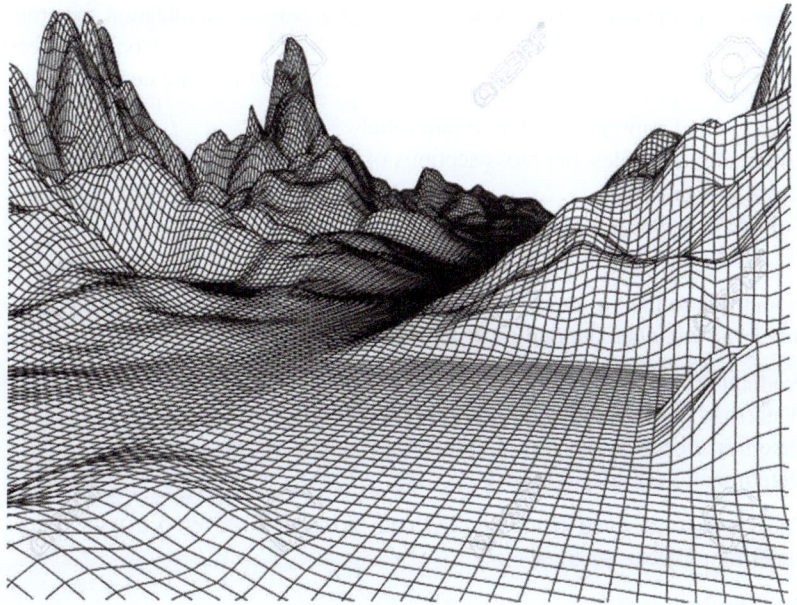

Fig. 15.16 DTM representation by Three-Dimensional wireframe perspective view

15.6   Interpretation of the Results

Interpreting the results of terrain modelling involves examining the model to confirm its correspondence with the real surface modelled and to refine the assumptions made. Therefore, visualisation through perspective views is one way of interpreting the terrain modelling results that must be highlighted. In this case, the interpretation consists of a careful verification of the model's conformity with the relief's roughness through detailed observations of the areas of interest of the model. The interpretation of the conformity of the landforms is the most important.

Another way of interpreting the results is by generating the contours of the model. The contour lines' conformity can indicate the adequacy of the model to the relief represented. In the latter case, peaks or very large flat areas are sought. In both cases, however, the interpretation of the results is carried out visually and manually.

15.7   Applications of a DTM: Derivative Products

The list of products derived from a DTM for engineering purposes is extensive. It is not necessary to be an expert to imagine the benefits of having a model that describes the geometric behaviour of a terrain surface. For this reason, an infinite number of digital terrain modelling applications have been developed for a wide variety of engineering applications. The choice of the most suitable application programme depends on the intended use and the expected results. Usually, for Civil Engineering purposes, a DTM should be able to generate at least the following derived products:

- Contour lines representing the terrain relief
- Longitudinal profiles and cross-sections of alignments drawn on the model
- Elevations of points
- Volume calculation worksheets
- 3D representation of the modelled terrain surface

In addition to the basic products mentioned above, depending on the application required, the engineer can add other applications to the list, such as:

- Production of slope maps
- Watershed delineation
- Determination of water flows, drainage networks, etc.
- Determination of lines of sight

The products derived from a DTM can be used to develop a wide range of engineering projects, such as:

- Road projects
- Pipeline construction projects
- High voltage electrical network projects
- Cell tower installation projects
- Earthworks and dam projects
- Urban projects (subdivisions and others)
- Orthophoto generation
- Hydrological projects
- Storm water drainage network projects
- Drinking water distribution network projects
- Sewage collection network projects
- Automation of earthmoving machines

- Precision agriculture
- Flight simulation
- Others

For each application, there are methods and techniques to extract specific information from the DTM. The reader interested in these methods and techniques should consult other specialised references. This section discusses one of the techniques used to trace contours, the design of terrain profiles using DTMs and the main elements that need to be evaluated when using a DTM in road projects. The technique for calculating volumes using a DTM is described in Sect. 22.7.5.1.

15.7.1   Tracing Contour Lines Through a DTM

The first concept to be considered in this section is the understanding that the contour lines generated using a DTM are just isolines for relief visualisation. The modelled surface is the reference for all data generated from the model. Therefore, there is little point in worrying about the details of displaying contours as if they had been drawn by hand using traditional contour tracing methods.

The technique used for tracing contour lines through a DTM, generated by either a regular grid or a TIN, is practically the same and includes the following:

- Selecting the starting point for tracing the contours
- Interpolating the coordinates of the intersection points between each contour and the edges of the grid or TIN elements
- Tracing the contours
- Smoothing the trace of the contours, if necessary, using interpolation functions

The algebraic details of each of the above steps are presented below.

15.7.1.1   Selecting the Starting Point for Tracing the Contours

The first step in selecting the starting point for tracing the contour lines is to know the values of the maximum and minimum elevations of the terrain to determine which are the contour lines with the highest and the lowest values to draw, as given by Eqs. (15.15) and (15.16).

$$h_{\min} = \left[\text{integer} \left(\frac{H_{\min}}{\Delta h} \right) + 1 \right] * \Delta h \qquad (15.15)$$

$$h_{\max} = \text{integer} \left(\frac{H_{\max}}{\Delta h} \right) * \Delta h \qquad (15.16)$$

where

$h_{\min}$ = lowest value of contour elevation
$h_{\max}$ = highest value of contour elevation

H_{min}= minimum elevation of the terrain
H_{max}= maximum elevation of the terrain
Δh= vertical equidistance between contour lines

The next step is to select the contour line to be drawn and look for all the edges they cross. The following mathematical relationship can be used to know if an edge contains one or more contour lines.

If $H_{P_1} > h > H_{P_2}$ or $H_{P_1} < h < H_{P_2}$ → the contour passes through the edge $\overline{P_1 P_2}$
If $H_{P_1} = h$ or $H_{P_2} = h$ → the contour passes through the vertex (P_1) or (P_2)
Otherwise, the contour will not pass through the border in question.
where

P_1, P_2= vertices of the considered edge
h= elevation of the considered contour
H_{P_i} = elevation of vertex (Pi) of the considered edge

15.7.1.2   Interpolating the Coordinates of the Intersection Points

The intersection of the contour with the edge of the grid is determined by linear interpolation between the values of the vertices of the edge, as given by Eqs. (15.17) and (15.18).

$$X_h = X_A + \left[\left(\frac{h - H_A}{H_B - H_A} \right) * (X_B - X_A) \right] \tag{15.17}$$

$$Y_h = Y_A + \left[\left(\frac{h - H_A}{H_B - H_A} \right) * (Y_B - Y_A) \right] \tag{15.18}$$

Knowing the start point of the contour in the grid edge, the next step is to determine the endpoint in the same grid element and so on, as shown in Fig. 15.17.

Fig. 15.17 Contour determination on a grid of triangular elements

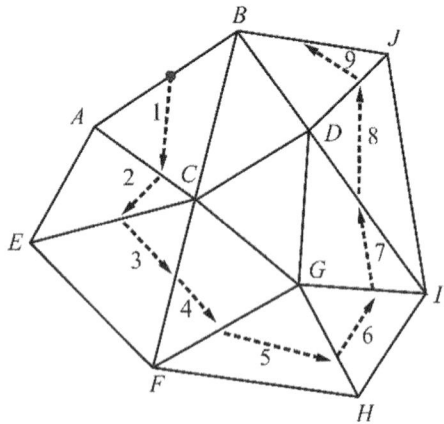

15.7.1.3   Contour Line Drawing

Contour line drawing is based on the start and end points of each contour. In the case of a regular grid, the results may be ambiguous. It is then necessary to check the interpolation on the diagonals of the element. The contour line drawing ends when it reaches the borderline of the model or when it closes to its starting point.

15.7.1.4   Contour Line Smoothing

Contours should be visually smooth and continuous to better represent terrain relief. For this reason, most contours generated from DTM interpolation need to be smoothed. Many different algorithms are currently available for smoothing contours to a high degree of accuracy. The simplest of these are based on dividing the patch element into smaller elements or using mathematical functions or series of functions that describe the geometric nature of isolated lines, such as b-splines, Bézier curves or cubic splines. More rigorous algorithms deal with groups of lines, considering their neighbours. The discussion of such smoothing techniques is beyond the scope of this book, and readers are advised to consult specialist literature for further details.

Figure 15.18 presents an example of a triangular element network (TIN) with the respective superimposed smoothed contours.

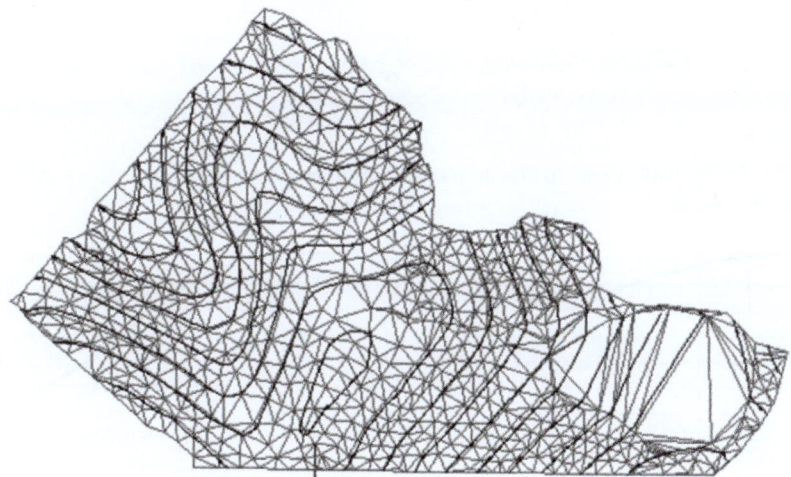

Fig. 15.18 Example of a TIN with overlapping contours

15.7.2   Alignment Profiling from a DTM

Profiling an alignment from a DTM is a simple process of interpolating elevations as a function of known coordinates (X, Y) of predetermined points along that alignment. The steps involved in profiling are as follows:

1. Generate the DTM of the surface of interest.
2. Plot the alignment over the plan view of the modelled surface.
3. Specify profile design settings such as staking, scales and others.
4. Interpolate the elevation values of the feature points over the alignment.
5. Draw the profile view.

Figure 15.19 shows an example of profiling an alignment AB defined over a DTM.

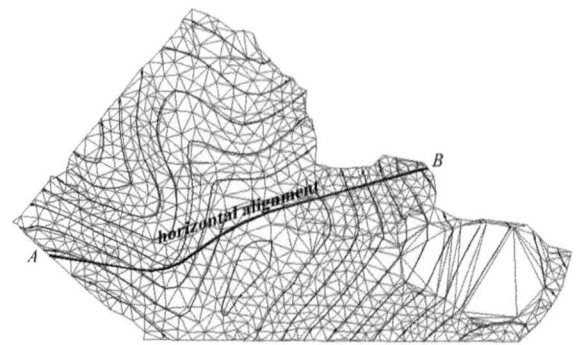

Alignment AB over the plan view of the project area

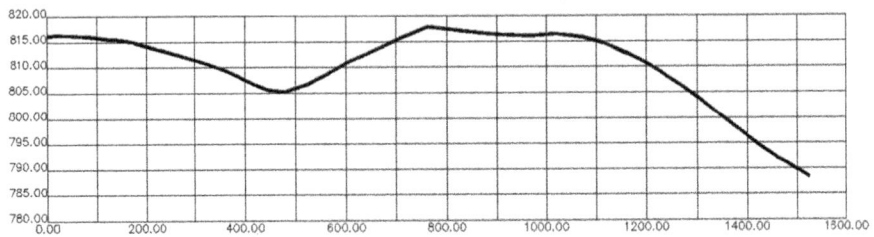

Longitudinal profile of the alignment AB

Fig. 15.19   Example of alignment profiling from a DTM

15.7.3 DTM as an Auxiliary Tool for Road Geometric Design

The use of a DTM as an auxiliary tool for the elaboration of geometric designs of transport routes has been a known subject since the beginning of the application of this technique. Currently, there are practically no land transport routes that are designed without the aid of a DTM. In this context, the main aspects being considered when designing a transport route using a DTM are presented below.

Selecting the layout of a road or a railroad, for example, is usually carried out on a topographic map at an appropriate scale. Next, the relief and details of the terrain surrounding the proposed route are surveyed using a suitable topographic survey and levelling method. From this stage, the application of the DTM begins, the first step being the creation of the mathematical model using the data collected in the field. Then, once the modelling has been completed and verified, the geometric design of the road is developed on the generated model, a stage in which the contour lines representing the terrain surface are usually also drawn to provide visual support for the project guidelines.

Once the geometric guidelines of the project have been defined, the positions of the tangents and horizontal curves of the road are determined in a plan view of the model. The result of this positioning is the longitudinal profile of the terrain through which the road axis will pass. The terrain profile defined in this way is later used in the vertical layout of the road, i.e. the project slopes and vertical curves.

The next stage of the project is to define the stakes for profiling the cross-sections. Once all the transversal elements, such as the roadbed, shoulders, drainage and slopes of the project, have been considered, the project elements are ready to be placed over the DTM.

The project is defined by selecting the best route in the plan and profile, considering the project guidelines, road visibility conditions and the cut and fill volumes generated.

The automation of the project is because the interpolation of the vertical and horizontal axes takes place practically in real-time. Each time the designer changes the road axis, the program interpolates the new horizontal and vertical sections, taking into account all the variants of the new road location, and generates a new numerical model of the road surface, which, when superimposed on the DTM, generates the data of the earthworks, project drawings, 3D views and even a fly-through simulator, depending on the power of the program used. Figure 15.20 shows a perspective view of a motorway designed from a DTM.

Fig. 15.20 Perspective view of a highway designed based on a DTM

15.8   Accuracy of a DTM

Indicating the accuracy of a DTM is still a controversial subject. Although research in this area began in the 1950s, there is still little consensus on the subject. The accuracy of the model depends on the morphology of the terrain, the density of the points, the accuracy of the field surveying and the interpolation methods used. The reader will find several models in the literature that allow the determination of accuracy values, but all require extensive statistical analysis. The simplest method is still the checkpoint method, i.e., a few strategically selected points with known coordinates (X, Y, H) that can be used as reference values. Considering that blunders and systematic errors have been eliminated from the model, the Root Mean Square Error (RMSE) is a good parameter for analysing the DTM accuracy. The decision to accept or reject the generated modelling will depend on the purpose for which the DTM is produced.

15.9   Review Questions

1. Discuss the difference between DTM and DSM and under what circumstances would you recommend using each in a Civil Engineering project.
2. Examine the various global DEMs currently available for mapping and discuss their availability for Civil Engineering projects.
3. Discuss why a DEM must be considered as a system rather than a technique of terrain modelling.
4. Briefly discuss the functional tasks encompassed by a digital terrain modelling determination.
5. Discuss the coordinate determination methods that can be used to collect data for a digital terrain modelling surveying project.

6. Explain the difference between individual and feature-specific points for terrain modelling and how they are used in surface modelling.
7. Explain what break lines are and their importance in digital terrain modelling.
8. Discuss when you would suggest using regular and irregular grid data collection for digital terrain modelling.
9. Discuss how contours are generated from a DTM and explain why the conformation of the contour plot is of little importance for a Civil Engineering project.
10. Explain the importance of TIN for digital terrain modelling.
11. Briefly discuss the interpolation methods currently used for digital terrain modelling and those more commonly used in commercial DTM software.
12. Explain which surface model visualisation can be used for DTM or DSM plotting and which you would suggest for Civil Engineering projects.
13. Discuss how to check the completeness and geometric accuracy of a DEM.
14. List at least ten applications of a DTM for Civil Engineering.
15. Explain how an alignment profile can be generated from a DTM.
16. Briefly discuss how a road is designed using a DTM as a tool for geometric definition and earthwork calculations.

References

Hengl, T., Evans, I. (2009). *Mathematical and Digital Models of the Land Surface.* In: Hengl, T., Reuter, H.I. (Eds.), Geomorphometry –Concepts, Software, Applications. Elsevier, Amsterdam, pp. 31–64. 07 Mar. 2016. (Series Developments in Soil Science, 33).

Hengl, T., Reuter, H.I. (Eds.), (2009). *Geomorphometry: Concepts, Software, Applications.* Elsevier, Amsterdam.

Hirt, C. (2014). *Digital terrain models.* Encyclopedia of Geodesy. https://doi.org/10.1007/978-3-319-02370-0_31-1 # Springer International Publishing, Switzerland.

Hutchinson, M.F. (2008). *Adding the Z-dimension.* In: Wilson, J.P., Fotheringham, A.S. (Eds.), The Handbook of Geographic Information Science. Blackwell Publishers, Oxford, pp. 144–168.

Hutchinson, M.F., Gallant, J.C. (2000). *Digital elevation models and representation of terrain shape.* In: Wilson, J.P., Gallant, J.C. (Eds.), Terrain Analysis: Principles and Applications. John Wiley and Sons, New York, pages 29–50.

Ministério da Defesa. Exército Brasileiro. Departamento de Ciência e Tecnologia. (2016). *Norma para especificação técnica para produtos de conjunto de dados geoespaciais (ET-PCDG).* 2° edição, Brasília.

Shingare, P. P., and Kale, S. S. (2013). *Review on digital elevation model.* International Journal of Modern Engineering Research, 3, 2412–2418.

Silva, I. (1990). *Méthodologie pour le lissage et le filtrage des données altimétriques dérivées de la corrélation d'images.* Doctoral Thesys. EPFL, Switzerland.

Wilson, J. P. (2012). *Digital terrain modelling.* Geomorphology 137, pag 107–121. www. e l s e v i e r. c om / l o c a t e / g e om o r p h. https://doi.org/10.1016/j.geomorph.2011.03.012.

Zhilin Li, Qing Zhu, and Chris Gold (2005). *Digital terrain modelling. Principles and methodology.* CRC Press. USA.

Zhou, Q. (2017). *Digital Elevation Model and Digital Surface Model.* https://doi.org/10.1002/9781118786352.wbieg0768. The International Encyclopaedia of Geography.

Chapter 16
Map Projections

Irineu da Silva and Paulo C. L. Segantine

16.1 Introduction

As previously discussed, geospatial databases are spatially defined on Three-Dimensional representations of the Earth's surface and visualised on Two-Dimensional forms for Civil Engineering projects, such as maps and topographic plans. Furthermore, it is desirable to perform data modelling in a simple Two-Dimensional coordinate system rather than on spherical surfaces whenever possible. For these reasons, whenever *geodetic data* are used for data modelling, specific mathematical functions must be applied to transform the Earth's spherical surface into a flat surface. This is carried out using *Map Projections*, which are defined as a mathematical function (or mapping equations) that transforms geodetic coordinates (ϕ_g, λ_g) on the reference ellipsoid into the Two-Dimensional coordinate system (X, Y) on a plane surface as given by Eq. (16.1). The Two-Dimensional coordinates based on a map projection plane are usually called *grid coordinates*.

$$
\begin{aligned}
X &= f_1\left(\phi_g, \lambda_g\right)\\
Y &= f_2\left(\phi_g, \lambda_g\right)
\end{aligned}
\tag{16.1}
$$

Geometrically, the transformation from geodetic to grid coordinates (a map projection) is carried out by projecting the geodetic coordinates from the reference ellipsoid directly onto a plane surface or a developable surface to obtain a grid coordinate map. Mathematically, this is done by selecting specific functions *f1* and *f2* that minimise the geometric deformations of the geospatial data represented in the plane surface, as described in the following subsections.

I. da Silva (✉) · P. C. L. Segantine
São Carlos School of Engineering, University of São Paulo, São Carlos, SP, Brazil
e-mail: irineu@sc.usp.br; pclsegantine@usp.br

619

Considering the above, map projection modelling is usually carried out in three steps:

1. First, the positions of points (ϕ_g, λ_g, h) on the surface of the Earth (ground surface) are reduced to positions $(\phi_g, \lambda_g, 0)$ on the reference ellipsoid.
2. Second, a map projection is selected to transform the geodetic coordinates (ϕ_g, λ_g) into grid coordinates (X, Y).
3. Third, the grid coordinates of the feature points are calculated and stored for future use.

Several map projection systems have been developed over the years, and many books have been written on the subject. Of these, Transverse Mercator (TM) is the most widely known. For this reason, this chapter deals exclusively with this system, explaining its mathematical basis and how it can be conveniently used in Civil Engineering projects.

16.2 Map Projection Parameters

Map projection parameters are descriptors that define the shape and distortions of the map projection surface and its position and orientation concerning the reference ellipsoid. They also define the location of the Two-Dimensional coordinate origin and its false northing and false easting values. The most important of these for the purposes of this book are presented below.

- *Scale factor*: a positive real number k expressing the ratio of an infinitesimal distance on the map projection surface to the corresponding infinitesimal distance on the reference ellipsoid. If $k < 1$, then the map projection surface (developable surface) is interior to the reference ellipsoid, and feature lengths in the map projection are smaller compared to the reference ellipsoid. If $k > 1$, then the map projection surface is exterior to the reference ellipsoid and feature lengths in the map projection are larger compared to the reference ellipsoid. If $k = 1$, then the map projection and the reference ellipsoid surfaces intersect at a line and feature lengths in the map projection and the reference ellipsoid are equal. See Fig. 16.16 for more details.
- *False easting*: a linear value applied to the origin of the X-grid coordinates.
- *False northing*: a linear value applied to the origin of the Y-grid coordinates.
- *Central meridian*: defines the origin of the X-grid coordinates.
- *Central parallel*: defines the origin of the Y-grid coordinates.

Defining these parameters for specific map projections is described in the following subsections.

16.3 Classification of Map Projections

Map projections are classified according to the geometric and graphical characteristics of their construction. The main classifications for Geomatics applied to Civil Engineering are presented below.

16.3.1 Classification Based on the Preserved Features

The fundamental geometric quantities measured on a topographic representation are distance, area and angle (shape). The ideal map projection would be one that preserves all these quantities simultaneously. However, none of them has this property. The quantities preserved are mutually exclusive. Therefore, map projections can be classified according to the geometric quantities they preserve.

Conformal projection,[1] is those that preserve the angles and, therefore, the shape of small areas represented in the map projection. This is the type of projection used in engineering projects because it allows a direct relationship between the angles measured on the ground (topographic surface) and their values on a flat surface (cartographic projection).

Equal area or equivalent projection, those that preserve areas, i.e., an area determined on the map projection corresponds in scale to its corresponding area on the ground. This is the type of projection used to produce maps at reduced scales, for example, for a geographical atlas.

Equidistant projection preserves distances, i.e., a distance determined on the map projection corresponds in scale to its corresponding distance on the ground.

In addition to the projections mentioned, there are others that are beyond the scope of this book. Readers interested in further information are advised to consult specialised literature.

16.3.2 Classification Based on the Position of the Viewpoint

A map projection can be generated graphically from three viewpoints (or projection points). From the centre of the Earth, from the pole opposite the point of tangency of the projection surface, and from infinity. This results in three basic types of map projection.

---

[1] The conformal projection was developed by the German mathematician and geodesist Carl Friedrich Gauss (1777–1855). It was used in geodesy after its mathematical development by J. H. L. Krüger (1857–1923). In their honour, the Gauss conformal projection and its coordinates were named the Gauss-Krüger projection in Germany in the 1920s and later in other European countries.

Gnomonic map projection: where the source of projection is the centre of the Earth.
Stereographic map projection: where the projection source is the pole opposite the
     point of tangency of the projection surface.
Orthographic map projection: where the projection source is at infinity.

  Figure 16.1 shows the three classes of map projections mentioned.

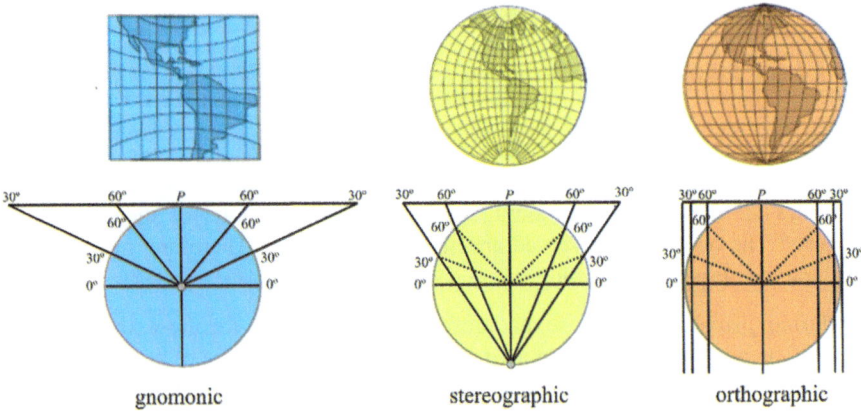

| gnomonic | stereographic | orthographic |

Fig. 16.1   Classes of map projections based on the position of the viewpoint

16.3.3   Classification Based on Developable Surface

In terms of developable surfaces, map projections can be divided into three
categories.

- *Cylindrical map projection:* the one in which the mathematical modelling considers that the surface of the reference ellipsoid is projected onto a cylinder
  tangent or secant to it from the centre of the ellipsoid, which is then developed
  to obtain the flat cartographic representation, as shown in Fig. 16.2.

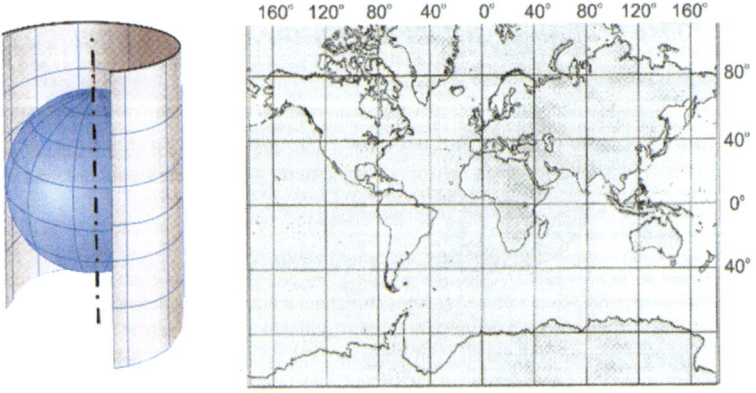

Fig. 16.2   Cylindrical map projection

- *Conical map projection:* the one in which the mathematical modelling considers that the surface of the reference ellipsoid is projected onto a cone tangent or secant to it from the centre of the ellipsoid, which is subsequently developed to obtain the flat cartographic representation, as shown in Fig. 16.3.

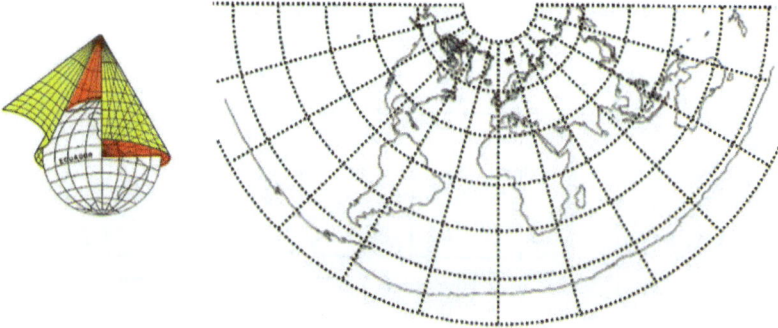

Fig. 16.3 Conical map projection

- *Azimuthal map projection:* the one in which the mathematical modelling considers that the surface of the reference ellipsoid is projected onto a tangent or secant plane to it from a perspective point, as shown in Fig. 16.4.

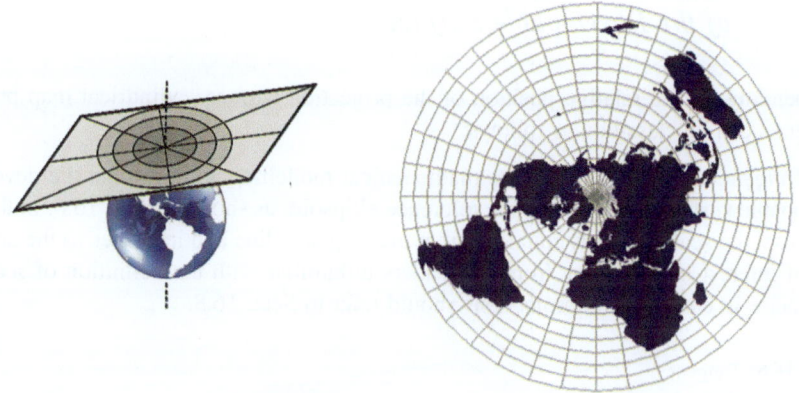

Fig. 16.4 Polar azimuthal equidistant map projection. (M.Galo, UNESP/FCT)

16.3.4 Cylindrical Map Projections: Orientation of the Developable Surface

In terms of the orientation of the developable surface, cylindrical map projections can be divided into three categories.

- *Normal:* when the axis of the cylinder coincides with the axis of rotation of the Earth.

- *Transverse:* when the axis of the cylinder coincides with the plane of the equator.
- *Oblique:* when the axis of the cylinder is inclined to the axis of rotation of the Earth.

Figure 16.5 shows the three classes of map projections mentioned.

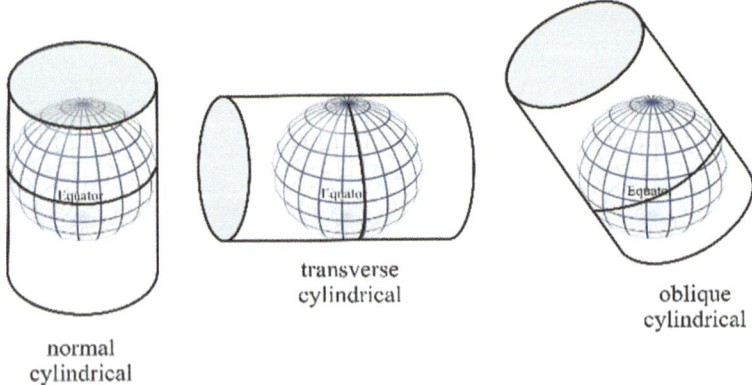

transverse
cylindrical

oblique
cylindrical

normal
cylindrical

Fig. 16.5 Normal, transverse and oblique cylindrical map projections

16.3.5 Cylindrical Map Projections: The Relative Position of the Developable Surface

Depending on the relative position of the projection surface, cylindrical map projections can be classified as follows:

- *Tangent:* the one in which the mathematical modelling assumes that the developable surface is tangent to the reference ellipsoid, as shown in Fig. 16.6. In this case, the scale factor is equal to 1.0 at the tangency line and increases as the area of interest moves away from it. Readers unfamiliar with the definition of scale factor in cartographic projections should refer to Sect. 16.8.

Fig. 16.6 Tangent cylindrical map projection

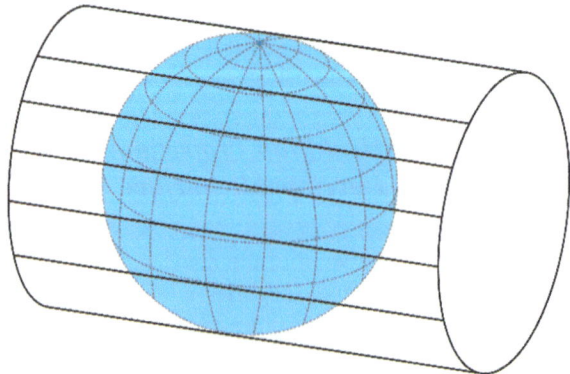

- *Secant:* the one in which the mathematical modelling assumes that the developable surface is secant to the reference ellipsoid, as shown in Fig. 16.7. In this case, the scale factor is less than 1.0 at the centre of the projection, increases to 1.0 near the secant lines and becomes greater than 1.0 as the area of interest moves away from them.

Fig. 16.7 Secant cylindrical map projection

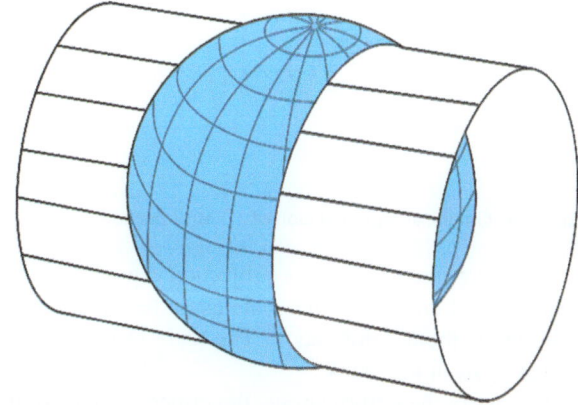

16.4   Universal Transverse Mercator (UTM) Map Projection

The Universal Transverse Mercator (UTM) map projection is the best example of a cylindrical projection. It is based on the modified Gauss-Krueger conformal projection. Its main advantage is that it allows large areas of the Earth's surface to be represented on a plane surface with few deformations and only one set of equations to perform coordinate transformations. It is based on a Two-Dimensional coordinate system, which facilitates its use in civil engineering applications. The International Union of Geological Sciences (IUGS) recommended this projection for topographic mapping in 1951, and it has since become the world's most widely used map projection.

16.4.1   Geometrical Characteristics of UTM Projection

The UTM projection is a cylindrical, conformal, transverse and secant projection defined by a set of equations to transform geodetic coordinates (ϕ_g, λ_g) into grid coordinates (N, E). It can be represented graphically, as illustrated in Fig. 16.8. As this is a secant cylindrical projection, there are two intersection lines between the cylinder and the ellipsoidal surface: AB and CD, in Fig. 16.8.

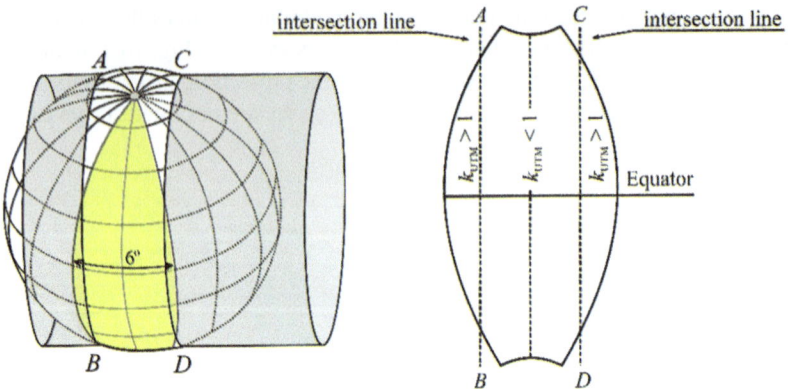

Fig. 16.8 Graphical representation of UTM projection

There are, therefore, three projection regions: two projected with a scale factor greater than 1 and one with a scale factor less than 1. Points on lines AB and CD are projected without deformation. The scale factor of the UTM Projection is abbreviated as k_{UTM} in this book.

To control the deformations, the projection area of the ellipsoidal surface on the cylinder is limited in amplitude to 6°, defining several projection zones, whose position in longitude varies as the cylinder is rotated horizontally. Each zone is represented by a number identifying it or by the longitude of its central meridian, as shown in Fig. 16.9.

Fig. 16.9 UTM zones

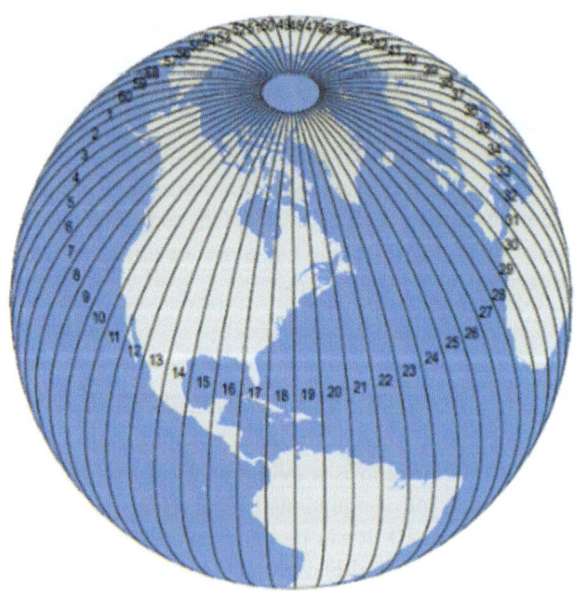

As already mentioned, the graphical representation of the UTM projection system is made on a Two-dimensional coordinate system called the *UTM coordinate system*. To identify it, the abscissa axis is called East, represented by the letter (E), and the ordinate axis is called North and represented by the letter (N). They are commonly referred to as (N, E) UTM coordinates. The rectangular grid of this coordinate system encloses each zone with the ordinate parallel to the central meridian of the zone and the abscissa parallel to the equator, as shown in Fig. 16.10.

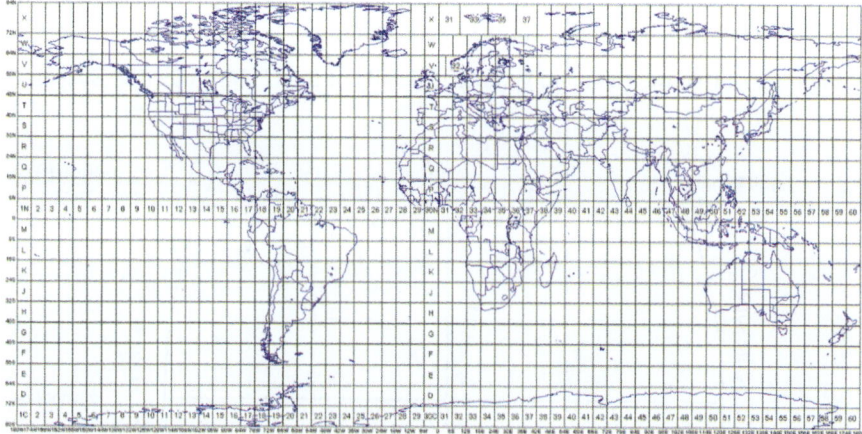

Fig. 16.10 UTM grid zone designations for the world. (Compiled by Alan Mortin (www.dmap. co.uk))

As it is a universal map projection system, the UTM map projection has a series of geometric parameters predefined, as presented below:

- *Zone width*: 6° with 3° on either side of the central meridian.
- *Zone numbering*: 1–60, from the antemeridian (180° longitude) to the East. Thus, zone 1 has a central meridian equal to 177° W, and zone 60 has a central meridian equal to 177° E (see Fig. 16.10).
- The projection zones are limited to 84° N and 80° S to avoid extreme distortions in the polar areas.
- The central parallel (Equator line) and the central meridian are represented as straight lines in the map projection. All the remaining projections of meridians and parallels are curves symmetrical about the central meridian and the nearest pole, respectively.
- Meridians and parallels intersect at right angles in the projection.
- The distance between the meridians in the graphical representation increases as they move away from the central meridian. The north-south scale is also deformed to maintain the proportionality of the conformal projection resulting in a different scale for each point on the same side of the meridian.

- The projection cylinder secant to the reference surface is placed on it so that the secant lines are located 180 km from the central meridian. This gives a scale factor at the central meridian k_0 of 0.9996. This is, therefore, the smallest scale factor of the UTM projection system.
- The maximum scale factor occurs at the edge of the zone and is 1.000981060 for the GRS80 ellipsoid, and
- The rectangular coordinate grid is placed over the zone so that the intersection of the central meridian with the equator has the coordinates ($N = 10,000,000.000$ m, $E = 500,000.000$ m) for the southern hemisphere and ($N = 0,000$ m, $E = 500,000.000$ m) for the northern hemisphere.

Figure 16.11 shows the main geometric details of the UTM Projection System.

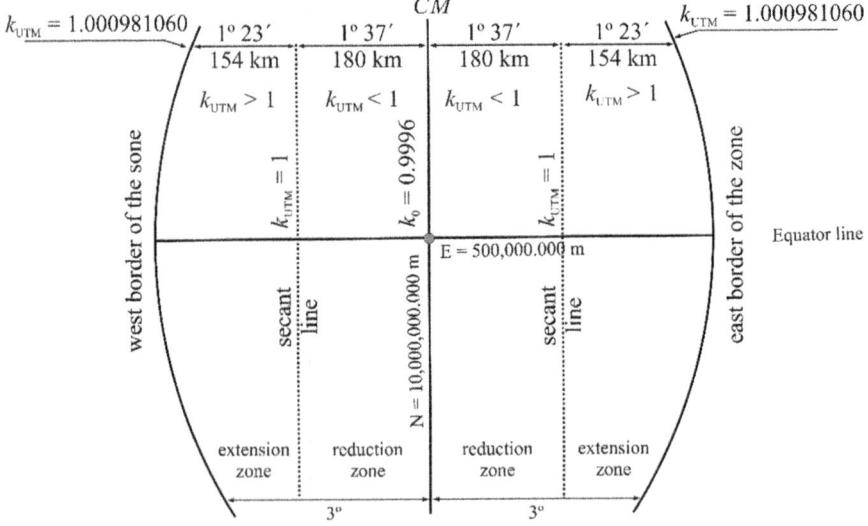

Fig. 16.11 Geometric details of the UTM Projection System. (Adapted from A MIRA (2000))

16.4.2   Determining the UTM Zone and Central Meridian

The value of the central meridian *CM* of the UTM projection is determined by considering that its variation occurs from 6° to 6°. The eastern central meridians have values equal to 3°, 9°, 15°, 21° and up to 177°. To know the value of the longitude of the central meridian of the zone in which a given point of known longitude is located, it is sufficient to determine in which interval of longitudes of 6° the point falls. The relation zone/*CM*, considering the algebraic sign of the value of the central meridian, is given by Eqs. (16.2) and (16.3).

$$CM = 6 * zone - 183° \tag{16.2}$$

$$zone = abs\left(\frac{-CM - 183°}{6}\right) \tag{16.3}$$

where

$zone$ = number of the zone

The value of the central meridian can also be calculated using Eq. (16.4), knowing the longitude λ_g of a point in the project area. In this case, the value of λ_g must include the algebraic sign.

$$CM = 6 * \text{roundup}\left(\frac{\lambda_g}{6}\right) \pm 3° \tag{16.4}$$

Add $3°$ for points west of the Greenwich meridian and subtract for points east of the Greenwich meridian.

Example 16.1

Given that point (E010) has a longitude of $-47°51'37.2336''$, calculate the values of the central meridian and the zone number in which it is located.

Solution:

Using Eq. (16.4),

$$CM = 6 * \text{roundup}\left(\frac{-47°51'37.2336''}{6}\right) \pm 3° = -45°$$

Using Eq. (16.3),

$$\text{Zone} = abs\left[\frac{-(-45°) - 183°}{6}\right] = 23$$

Knowing the value of the zone, it is also possible to calculate the value of the central meridian using Eq. (16.2)

$$CM = 6 * 23 - 183° = -45°$$

16.5 UTM Coordinate Transformation

To use map projection effectively in engineering projects, the engineer must have sufficient knowledge to perform the appropriate coordinate transformations between the reference surfaces. Most major engineering projects are developed based on the coordinate values of the points represented in the map projection. Therefore, the mathematical models used for coordinate transformations between geodetic (ϕ_g, λ_g) and UTM grid coordinates (N, E) are presented below.

16.5.1   Transformation from Geodetic (ϕ_g, λ_g) to UTM Grid Coordinates

Since most engineering projects are based on the two-dimensional coordinate system, the transformation from geodetic to UTM coordinates is the most commonly used in Geomatics applied to Civil Engineering. The rigorous mathematical model used for this transformation is based on the following equations, which must be used with the algebraic signs for latitude and longitude, considering the position of the point on the terrestrial globe.

$$N' = (I) + (II) * p^2 + (III) * p^4 + (A_6) * p^6 \qquad (16.5)$$

$$E' = (IV) * p + (V) * p^3 + (B_5) * p^5 \qquad (16.6)$$

where

Ordinate $N = N'$ (for the northern hemisphere),
Ordinate $N = N'$ (for the northern hemisphere),
Ordinate $N = 10, 000, 000.000 + N'$ (for the south hemisphere), and
Abscissa $E = 500, 000.000 + E'$

$$(I) = k_0 * S \qquad (16.7)$$

$$S = a * (1 - e^2) * \left[A * \phi_g - \frac{B}{2} \sin(2\phi_g) + \frac{C}{4} \sin(4\phi_g) - \frac{D}{6} \sin(6\phi_g) + \frac{E}{8} \sin(8\phi_g) - \frac{F}{10} \sin(10\phi_g) \right]$$
$$(16.8)$$

$$A = 1 + \frac{3}{4} e^2 + \frac{45}{64} e^4 + \frac{175}{256} e^6 + \frac{11,025}{16,384} e^8 + \frac{43,659}{65,536} e^{10} \qquad (16.9)$$

$$B = \frac{3}{4} e^2 + \frac{15}{16} e^4 + \frac{525}{512} e^6 + \frac{2,205}{2,048} e^8 + \frac{72,765}{65,536} e^{10} \qquad (16.10)$$

$$C = \frac{15}{64} e^4 + \frac{105}{256} e^6 + \frac{2,205}{4,096} e^8 + \frac{10,395}{16,384} e^{10} \qquad (16.11)$$

$$D = \frac{35}{512} e^6 + \frac{315}{2,048} e^8 + \frac{31,185}{131,072} e^{10} \qquad (16.12)$$

$$E = \frac{315}{16,384} e^8 + \frac{3,465}{65,536} e^{10} \qquad (16.13)$$

$$F = \frac{693}{131,072} e^{10} \tag{16.14}$$

$$(II) = \frac{N * \sin\left(\phi_g\right) * \cos\left(\phi_g\right) * 10^8 k_0}{2\rho''^2} \tag{16.15}$$

$$(III) = \frac{N * \sin\left(\phi_g\right) * \cos^3\left(\phi_g\right)}{24\rho''^4} *$$
$$\left[5 - \tan^2\left(\phi_g\right) + 9e'^2 * \cos^2\left(\phi_g\right) + 4e'^4 * \cos^4\left(\phi_g\right)\right] * 10^{16} k_0 \tag{16.16}$$

$$(IV) = \frac{N * \cos\left(\phi_g\right) * 10^4 k_0}{\rho''} \tag{16.17}$$

$$(V) = \frac{N * \cos^3\left(\phi_g\right)}{6\rho''^3} * \left[1 - \tan^2\left(\phi_g\right) + e'^2 * \cos^2\left(\phi_g\right)\right] * 10^{12} k_0 \tag{16.18}$$

$$p = 0.0001 * \Delta\lambda'' \tag{16.19}$$

$$\Delta\lambda = \lambda_g - \lambda_{MC} \tag{16.20}$$

$$(A_6) = \frac{N * \sin\left(\phi_g\right) * \cos^5\left(\phi_g\right)}{720\rho''^6} * \begin{bmatrix} 61 - 58\tan^2\left(\phi_g\right) + \tan^4\left(\phi_g\right) + \\ 270e'^2 * \cos^2\left(\phi_g\right) - 330e'^2 * \sin^2\left(\phi_g\right) \end{bmatrix} * 10^{24} k_0 \tag{16.21}$$

$$(B_5) = \frac{N * \cos^5\left(\phi_g\right)}{120\rho''^5} * \begin{bmatrix} 5 - 18\tan^2\left(\phi_g\right) + \tan^4\left(\phi_g\right) + 14e'^2 * \cos^2\left(\phi_g\right) - 58e'^2 * \sin^2\left(\phi_g\right) \end{bmatrix} * 10^{20} k_0 \tag{16.22}$$

where

$\rho'' = 206,264.8062''$ (radian to arc second conversion factor)
ϕ_g = geodetic latitude of the point, with its algebraic sign
λ_g = geodetic longitude of the point, with its algebraic sign
λ_0 = longitude of the central meridian, with its algebraic sign
$\Delta\lambda = \lambda_g - \lambda_0$ = difference between the longitude of the point and the longitude of the central meridian, considering the algebraic signs
$k_0 = 0.9996$
N = radius of curvature of the first vertical calculated for the ϕ_n value, according to Eq. (4.8)
e, e' = first and second eccentricity, according to Eqs. (4.3) and (4.5)

To assist the reader in his calculations, Table 16.1 shows the values of the fixed parameters used to transform geodetic coordinates to UTM coordinates, considering the GRS80 ellipsoid.

Table 16.1 Fixed parameters for transforming geodetic coordinates into UTM coordinates, considering the GRS80 ellipsoid

a	e	e'	A	B
6,378,137.000 m	0.081819191	0.082094438	1.005052502	0.005063109
C	D	E	F	
$1.062759026 * 10^{-5}$	$2.082037857 * 10^{-8}$	$3.932371371 * 10^{-11}$	$7,108453403 * 10^{-14}$	

Example 16.2

Given the geodetic coordinates of the point (E010) in Table 16.2, calculate the corresponding UTM plane coordinates.

Table 16.2 Geodetic coordinates of the point (E010)—SIRGAS2000

Point	Latitude	Longitude	Ellipsoidal height[a] (h) [m]
E010	−22°01′52.5229″	−47°51′37.2336″	866.868

[a]The value of the ellipsoidal height is given in this table because it will be used in later calculations throughout the chapter. It does not affect the calculations of the transformation from geodetic to UTM coordinates

Solution:

Given the parameters in Table 16.1, *and knowing that point (E010) is in the southern hemisphere and west of the Greenwich meridian, the preliminary values given in* Table 16.3 *are obtained.*

The calculated UTM coordinates of the point (E010) shown in Table 16.4 are obtained from the preliminary values in Table 16.3.

Table 16.3 Preliminary values for calculating the coordinates of the point (E010)

N = 6,381,143.1030 m	Δλ = − 10, 297.2336″	p = − 1.029723360″	S = −2,437,261.5097 m
(I) = −2,436,286.6051 m	(II) = −2.606,6090 m	(III) = −21,447 m	(IV) = 286.661,6936 m
(V) = 81,2534 m	(A_6) = − 0, 0016 m	(B_5) = 0.0205 m	
$E^{'}$ = − 295, 270.9823 m			

Table 16.4 UTM coordinates

Point	E [m]	N [m]
E010	204,729.0177	7,560,947.1156

NB: The values calculated in Table 16.4 will be used in the rest of this chapter, rounding to the specified four decimal places

16.5.2   Transformation from UTM (N, E) to Geodetic (ϕ_g, λ_g) Coordinates

The transformation from UTM to geodetic coordinates has little application in Civil Engineering. However, it is important that the engineer has sufficient knowledge to perform it if required.

16.5.2.1   Calculation of the latitude ϕ_g

The calculation of the latitude ϕ_g is an iterative process. The initial latitude ϕ_0 is calculated using Eq. (16.23).

$$\phi_0 = \frac{N' * \rho^{\circ}}{\alpha * k_0} \qquad (16.23)$$

where

$$\alpha = A * a * \left(1 - e^2\right) \qquad (16.24)$$

ϕ_0 = initial latitude, in decimal degrees
a = semi-major axis of the reference ellipsoid
A = parameter calculated using the Eq. (16.9).
N' = parameter calculated using the Eq. (16.5)
k_0 = 0.9996
ρ° = 57.295779513° (radian to decimal degree conversion factor)

With the initial latitude ϕ_0 calculated, the values of the subsequent latitudes are determined using Eq. (16.25).

$$\phi_{i+1} = \frac{\rho^{\circ}}{\alpha}$$
$$* \left[\frac{N'}{k_0} + \beta * \sin(2\phi_i) - \gamma * \sin(4\phi_i) + \delta * \sin(6\phi_i) - \varepsilon * \sin(8\phi_i) + \xi * \sin(10\phi_i)\right]$$
$$(16.25)$$

where

$$\beta = \frac{B * a * \left(1 - e^2\right)}{2} \qquad (16.26)$$

$$\gamma = \frac{C * a * (1 - e^2)}{4} \tag{16.27}$$

$$\delta = \frac{D * a * (1 - e^2)}{6} \tag{16.28}$$

$$\varepsilon = \frac{E * a * (1 - e^2)}{8} \tag{16.29}$$

$$\xi = \frac{F * a * (1 - e^2)}{10} \tag{16.30}$$

Where *B, C, D, E, F* = parameters defined in the previous section.

To help the reader with his calculations, Table 16.5 shows the values of the fixed parameters used to transform UTM coordinates into geodetic coordinates, considering the GRS80 ellipsoid.

Table 16.5 Fixed parameters for transforming UTM coordinates into geodetic coordinates for the GRS80 ellipsoid

α	β	γ	δ	ε	ξ
6,367,449.1458 m	16,038.5087 m	16.8326 m	0.0220 m	$3.1142*10^{-5}$ m	$4.1526*10^{-8}$ m

For Civil Engineering design purposes, iteration of Eq. (16.25) should continue until $\phi_n = \phi_{i+1} - \phi_i \leq 0,0001''$. This condition is usually satisfied after the fifth iteration.

Given the value of ϕ_n, the calculation can be completed as follows:

$$\phi_g = \phi_n - \left[(VII) * q^2 - (VIII) * q^4 + (D_6') * q^6\right]'' \tag{16.31}$$

Note that the second term in Eq. (16.31) is in arc seconds.

$$q = 0.000001E' \tag{16.32}$$

$$(VII) = \frac{\tan\phi_n * \left[1 + e'^{2*}\cos^2(\phi_n)\right] * 10^{12}\rho''}{2N^2 * k_0^2} \tag{16.33}$$

$$(VIII) = \frac{\tan\phi_n}{24N^4} * \begin{bmatrix} 5 + 3\tan^2(\phi_n) + 6e'^2 * \cos^2(\phi_n) - 6e'^2 * \sin^2(\phi_n) - \\ 3e'^4 * \cos^4(\phi_n) - 9e'^4 * \cos^2(\phi_n) * \sin^2(\phi_n) \end{bmatrix} * \frac{10^{24}\rho''}{k_0^4} \tag{16.34}$$

$$(D_6') = \frac{\tan\phi_n}{720N^6} * \left[\begin{array}{l} 61 + 90\tan^2(\phi_n) + 45\tan^4(\phi_n) + 107e'^2 * \cos^2(\phi_n) - \\ 162e'^2 * \sin^2(\phi_n) - 45e'^2 * \tan^2(\phi_n) * \sin^2(\phi_n) \end{array} \right] * \frac{10^{36}\rho''}{k_0^6}$$

$$(16.35)$$

Where N is the radius of curvature of the first vertical calculated for the ϕ_n value, according to Eq. (4.8).

16.5.2.2   Calculation of the longitude λ_g

The following equations can be used to calculate longitude λ_g.

$$\lambda_g = \lambda_0 + \left[(IX) * q - (X) * q^3 + (E_5') * q^5 \right]''$$

$$(16.36)$$

Note that the second term in Eq. (16.36) is in arc seconds.

$$(IX) = \frac{\sec\phi_n * 10^6\rho''}{N * k_0}$$

$$(16.37)$$

$$(X) = \frac{\sec\phi_n}{6N^3} * \left[1 + 2\tan^2(\phi_n) + e'^2 * \cos^2(\phi_n) \right] * \frac{10^{18}\rho''}{k_0^3}$$

$$(16.38)$$

$$(E_5') = \frac{\sec\phi_n}{120N^5} * \left[5 + 28\tan^2(\phi_n) + 24\tan^4(\phi_n) + 6e'^2 * \cos^2(\phi_n) + 8e'^2 * \sin^2(\phi_n) \right] * \frac{10^{30}\rho''}{k_0^5}$$

$$(16.39)$$

where

$N =$ radius of curvature of the first vertical calculated for the ϕ_n value, according to
   Eq. (4.8)
$e'^2 =$ second quadratic eccentricity, according to Eq. (4.6)
$\rho'' = 206.264,8062''$ (radian to arc seconds conversion factor)

Example 16.3
Given the values in Table 16.4, transform the UTM coordinates of the point (*E010*) into geodetic coordinates in the SIRGAS2000 reference system.

Solution:
   The parameters shown in Table 16.6 *are determined using* Eqs. (16.23) to (16.35).

   To calculate the longitude, the sequence of Eqs. (16.36) to (16.39) *must be used to obtain the parameter values shown in* Table 16.7.

Table 16.6 Parameters for calculating the latitude of the point (E010)

$\phi_0 = -21.955944120°$ initial latitude	$\phi_n = -22.056248366°$ 5^a iteration	$q = -0.295270982$ m
$N = 6,381,149.5913$ m	$(VII) * q^2 = -90.059412072''$	$(VIII) * q^4 = -0.088283343''$
$(D_6') * q^6 = -8.845032568 * 10^{-5''}$	$\phi = -22.031256361°$	$\phi = -22°01'52.5229''$

Table 16.7 Parameters used to calculate the longitude of point (E010)

$(IX) * q = -10,302.1382''$	$(X) * q^3 = -4.9086''$	$(E_5') * q^5 = -0.0041''$
$\lambda_g = -45° - \frac{10,297.2336''}{3,600} = -45° - 2.860342668° = -47°51'37.2336''$		

16.6   Convergence of the Meridians (γ)

In the UTM coordinate system, the ordinate axis is the direction taken as the reference for coordinate geometry calculations. To distinguish it from the geodetic reference direction given by the terrestrial meridians, it is called *Grid North (GN)* and is referenced to the grid cell of the UTM coordinate system.

The terrestrial meridians, in turn, represent the true direction of the geodetic north, establishing what is known as the *true north* or *geodetic north (TN)*. As already seen, when projected onto the UTM Projection plane, they are represented by a concave line about the central meridian.

Taking any point in the UTM projection plane, Grid North (GN) is given by the direction of the ordinate axis passing through the point, and True North (TN) is given by the tangent to the meridian passing through the same point. Thus, as shown in Fig. 16.12, there is an angular difference between them. This angle is called the *convergence of the meridians γ,*[2] i.e., the angle at any point in the projection between the north-south grid line and the meridian at that point.

In the southern hemisphere, the value of γ is positive for points west of the meridian and negative for points east of the meridian. As presented below, the mathematical equations used to calculate this value already produce results with their respective signs.

Several equations are available in the literature for calculating the meridian convergence. Two of them, based on the geodetic coordinates, are presented below. There are also other equations based on UTM coordinates. However, they use the latitudes, which must be calculated by a transformation method from UTM coordinates to geodetic coordinates. This adds little advantage over the meridian convergence calculation model based on geodetic coordinates. For this reason, they have not been included in this text.

---

[2]It is also referred to as grid convergence by some authors.

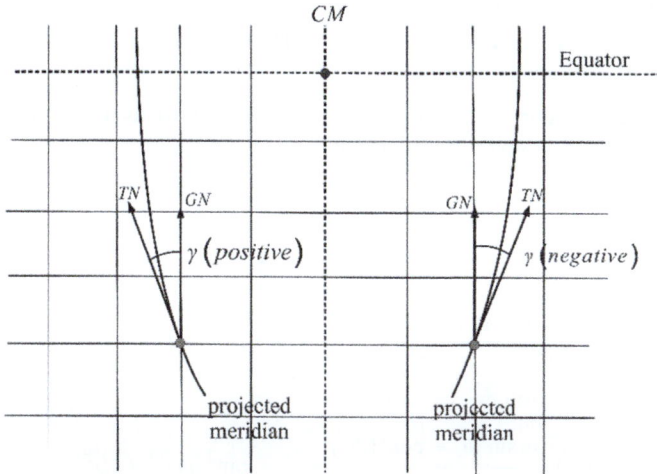

Fig. 16.12 Convergence of the meridians γ

The interest in knowing the value of meridian convergence for engineering applications is because, in some cases, topographic or geodetic surveys have True North (*TN*) as the only reference, such as underground works where gyro-theodolites (theodolites coupled to a gyroscope) are used. Another case is the installation of solar panels, where the direction of the true north is sought to maximise the installation's efficiency.

16.6.1   Approximate Calculation of the Convergence of the Meridians from Geodetic Coordinates

The approximate value for the convergence of the meridians as a function of the geodetic coordinates can be obtained using Eq. (16.40).

$$\gamma = \left(\lambda_g - \lambda_{CM}\right) * \sin\left(\phi_g\right) \tag{16.40}$$

where

γ = convergence of the meridians
λ_g = longitude of the point under consideration
λ_{CM} = longitude of the central meridian
ϕ_g = latitude of the point under consideration

16.6.2 Rigorous Calculation of the Convergence of the Meridians from Geodetic Coordinates

The rigorous calculation of the convergence of the meridians as a function of the geodetic coordinates is given by Eq. (16.41).

$$\gamma = (XII) * p + (XIII) * p^3 + (C_5) * p^5 \tag{16.41}$$

$$(XII) = 10^4 \sin \phi_g \tag{16.42}$$

$$(XIII) = \frac{\sin \phi_g * \cos^2(\phi_g)}{3\rho''^2} * \left[1 + 3e'^2 * \cos^2(\phi_g) + 2e'^4 * \cos^4(\phi_g)\right] * 10^{12} \tag{16.43}$$

$$(C_5) = \frac{\sin \phi_g * \cos^4(\phi_g)}{15\rho''^4} * \left[2 - \tan^2(\phi_g)\right] * 10^{20} \tag{16.44}$$

where

$\gamma =$ convergence of the meridians in arc seconds
$\phi_g =$ geodetic latitude of the point under consideration
$\rho'' = 206, 264.8062''$ (radian to arc second conversion factor)
$p =$ as given by Eq. (16.19)
$e'^2 =$ second quadratic eccentricity, according to Eq. (4.6)

Example 16.4

Given the geodetic coordinates of the point (E010) in Table 16.1, calculate the convergence of the meridians for this location using the approximate equation and the rigorous equation as a function of the geodetic coordinates.

Solution:
   Approximate calculation of the convergence of the meridians from geodetic coordinates:
   Using Eq. (16.40),

$$\gamma = [-47°51'37.2336'' - (-45°)] * \sin(-22°01'52.5229'') = 1°04'22.6194''$$

Rigorous calculation of the convergence of the meridians from geodetic coordinates:
   The values in Table 16.8 *can be calculated using* Eqs. (16.41) to (16.44).

Table 16.8 Parameters for calculating the convergence of the meridians of the point (E010) as a function of the geodetic coordinates

$p = -1.02972336''$	$(XII) * p =$ 3862.6194''	$(XIII) * p^3 = 2.8054''$	$(C_5) * p^5 = 0.0022''$	$\gamma = 3,865.4270''$ $\gamma = 1°04'25.4270''$

16.7 Arc-to-Chord Correction δ

The shortest distance between two points (A) and (B) on the surface of an ellipsoid is given by a curve called a geodesic line or simply *geodesic*. When projected onto the UTM projection plane, this curve is represented by an arc with concavity towards the central meridian, as shown in Fig. 16.13.

Fig. 16.13 Arc-to-chord representation

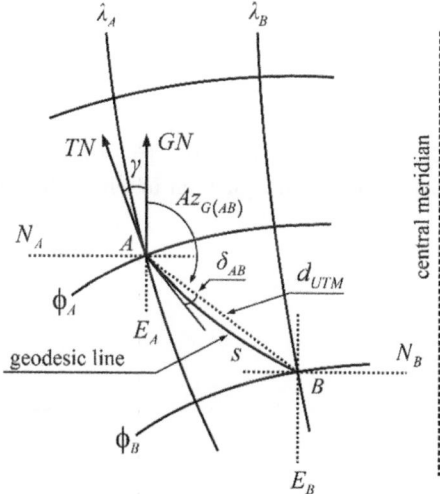

In the UTM Projection, when the coordinates of points (A) and (B) are known, the distance d_{UTM} between them is calculated by Eq. (16.45) and the grid azimuth Az_{GAB} by Eq. (16.46).

$$d_{UTM} = \sqrt{(E_B - E_A)^2 + (N_B - N_A)^2} \qquad (16.45)$$

$$Az_{GAB} = \text{atan}\left(\frac{E_B - E_A}{N_B - N_A}\right) \pm \alpha_{quad} \qquad (16.46)$$

$$\delta_{AB} = \left(\frac{2E'_A + E'_B}{6R_0^2}\right) * (N_B - N_A) * \rho'' \qquad (16.47)$$

$$\delta_{BA} = \left(\frac{E'_A + 2E'_B}{6R_0^2}\right) * (N_A - N_B) * \rho'' \qquad (16.48)$$

where

α_{quad} = quadrant compatibility
δ_{ij} = arc-to-chord, in arc seconds
$E'_i = E - 500, 000.000 \, \text{m}$

N_i, N_j= UTM ordinates of the endpoints of line ij
R_0= local mean radius of the Earth
$\rho'' = 206, 264.8062$ (radian to arc second conversion factor).

Generally, the geodesic curvature is very small for distances measured in engineering surveys and is therefore ignored. The chord is then taken as the angular
reference. For a distance of 2 km, for example, the arc-to-chord correction is one
second, i.e., of the angular accuracy of a high-precision total station. It can, therefore,
be safely discarded for the vast majority of surveying applications.

In cases where the reduction angle cannot be disregarded, it will affect the values
of the directions measured in the field and alter the values of the angles determined
between them, as shown in Fig. 16.14.

The angular conversion between the tangent angle α_t and the chord angle α_c in
these cases is given by Eq. (16.49), where the positive or negative sign must be used
depending on the quadrant in which the line lies.

$$\alpha_t = \alpha_c \pm \delta_{AB} \pm \delta_{AC} \tag{16.49}$$

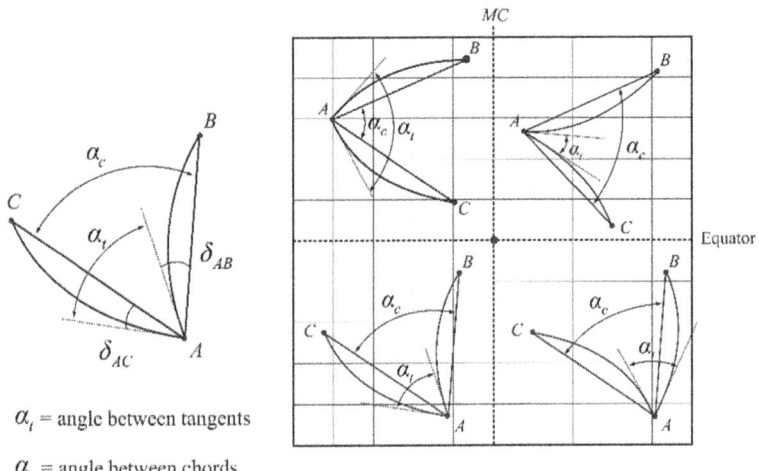

α_t = angle between tangents

α_c = angle between chords

Fig. 16.14 Arc-to-chord corrections

Example 16.5
Given the geodetic coordinates of the point ($E010$) shown in Table 16.1 and the
UTM coordinates of points ($E010$) and ($E014$) in the SIRGAS2000 geodetic system
shown in Table 16.9, calculate the arc-to-chord correction at point ($E010$).

Point	E [m]	N [m]
E010	204,729.0177	7,560,947.1156
E014	203,551.8201	7,560,860.4734

Table 16.9 UTM coordinates

Solution:

To solve this exercise, it is first necessary to calculate the local mean radius of the Earth and the values of $\left(E'_{E010}\right)$ and $\left(E'_{E014}\right)$. Thus, using point (E010) as a reference.

$$M_{E010} = 6,344,401.487\,\text{m}$$

$$N_{E010} = 6,381,143.103\,\text{m}$$

$$R_{0E010} = 6,362,746.000\,\text{m}$$

$$E'_{E010} = 204,729.0177 - 500,000 = -295,270.9823\,\text{m}$$

$$E'_{E014} = 203,551.8201 - 500,000 = -296,448.1799\,\text{m}$$

Then, using Eq. (16.47).

$$\delta_{E010} = \frac{\begin{array}{c}[2*(-295,270.9823) + (-296,448.1799)]\\ *(7,560,860.4734 - 7,560,947.1156)*206,264.8062\end{array}}{6*6,362,746.000^2}$$

$$= 0.0653''$$

16.8   Scale Factor

As explained in previous sections, the UTM distances shown on a map or calculated from UTM coordinates are affected by the scale factor k_{UTM}, which varies according to their location in the projection zone.

The UTM scale factor makes it possible to relate the distances indicated on the UTM projection plane d_{UTM} to their corresponding distances on the ellipsoidal surface d_0, as shown schematically in Fig. 16.15.[3]

Remember that the ellipsoidal distance d_0 is related to the ground distance as a function of the elevation scale factor k_{alt}, as described in Sect. 7.2

---

[3] Note that this is a simplified representation of the plane distance compared to the ellipsoidal and spherical distances. It has no geometrical meaning.

Fig. 16.15 Geometric
reduction for UTM distance
calculation

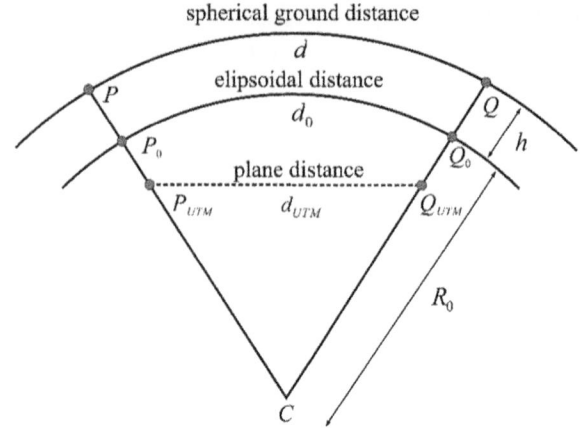

Similar to the calculation of the meridian convergence, there are several equations in the literature for calculating the UTM scale factor k_{UTM}. An approximate calculation based on UTM coordinates and a rigorous calculation based on geodetic coordinates are given below.

16.8.1   Approximate Calculation of the UTM Scale Factor from Grid Coordinates

The approximate value of the scale factor as a function of the UTM coordinates can be obtained using Eq. (16.50).

$$k_{UTM} = k_0 * \left(1 + \frac{E'^2}{2R_0^2} + \frac{E'^4}{24R_0^4} \right) \qquad (16.50)$$

where

$k_0 = 0.9996$
$E' = E{-}500{,}000.000\text{m}$
$R_0 = $ local mean radius of the Earth

16.8.2   Rigorous Calculation of the UTM Scale Factor from Geodetic Coordinates

The rigorous calculation of the scale factor as a function of the geodetic coordinates can be done using Eq. (16.50).

$$k_{UTM} = k_0 * \left[1 + C_2 * L^2 * \left(1 + C_4 * L^2 \right) \right] \tag{16.51}$$

where

$$L = (\lambda g - \lambda_0) * \cos\left(\phi_g\right) \quad \text{(in radians)} \tag{16.52}$$

$$\eta^2 = \frac{e^2 * \cos^2\left(\phi_g\right)}{1 - e^2} \tag{16.53}$$

$$C_2 = \frac{1 + \eta^2}{2} \tag{16.54}$$

$$C_4 = \frac{5 - 4 * \tan^2\left(\phi_g\right) + \eta^2 * \left[9 - 24 * \tan^2\left(\phi_g\right) \right]}{12} \tag{16.55}$$

$e^2 =$ first quadratic eccentricity, according to Eq. (4.4).

As shown in Fig. 16.16, the value of the UTM scale factor at the central meridian is equal to $k_0 = 0.9996$. It then increases both westwards and eastwards until it reaches the value $k_{UTM} = 1.000$ near the values $E = 320,000.000$ m and $E = 680,000.000$ m and continues to increase until it reaches the value $k_{UTM} = 1.000981060$ at the edges of the zone (for the GRS80 ellipsoid).

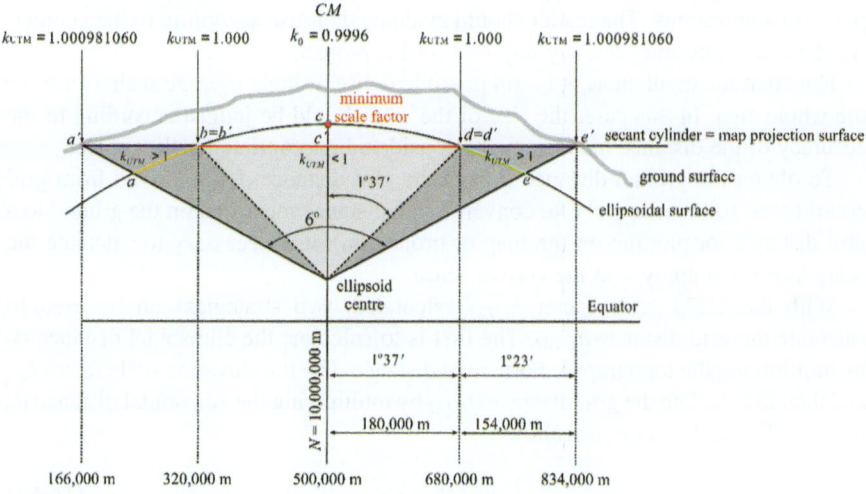

Fig. 16.16 Variation of the UTM scale factor within the UTM zone

As the UTM scale factor is point-based, it varies according to the position of the point on the projection surface. For this reason, to apply it to a distance between two points, some experts consider it appropriate to use the scale factor of any point on the line for distances less than 1 km; to use the average scale factor calculated according to the scale factors of the endpoints for distances between 1 and 4 km; and to use the scale factor calculated by a weighted average using the value of the scale factor of the centre of the line with a weight equal to 4 for distances greater than 4 km.

For distances less than 1 km:

$$k_{UTM} = k_{UTMi} \text{ ou } k_{UTMj} \tag{16.56}$$

For distances between 1 km and 4 km:

$$k_{UTMm} = \frac{k_{UTMi} + k_{UTMj}}{2} \tag{16.57}$$

For distances greater than 4 km:

$$k_{UTMm} = \frac{k_{UTMi} + 4k_{UTMij} + k_{UTMj}}{6} \tag{16.58}$$

where

k_{UTMi}, k_{UTMj} = UTM scale factor of endpoints i, j
k_{UTMm} = mean UTM scale factor
k_{UTMij} = UTM scale factor of the centre point of line ij.

The strategies for weighting the value of the UTM scale factor given are only practical suggestions. The reader should evaluate their use according to the geometric characteristics and accuracy required for his project.

Note that for small areas, it is still possible to use a single average scale factor for the whole area. In this case, the size of the area should be judged according to the accuracy of the distance measurements considered appropriate for the project.

To obtain the ground distance d from the grid distance d_{UTM} derived from grid coordinates; (or alternatively, to convert a true distance measured on the ground to a grid distance for plotting on the map or projection), it is necessary to calculate the scale factor and apply it in the correct sense.

With the UTM scale factor k_{UTM} calculated, two strategies can be used to calculate the grid distance d_{UTM}. The first is to calculate the ellipsoidal distance d_0 by multiplying the topographic horizontal distance d by the elevation scale factor k_{alt} and then to calculate the grid distance d_{UTM} by multiplying the ellipsoidal distance d_0 by the UTM scale factor, as follows:

$$d_0 = k_{alt} * d \tag{16.59}$$

$$d_{UTM} = k_{UTM} * d_0 \tag{16.60}$$

To avoid performing the calculation in two steps, it is recommended to use the Combined Scale Factor (CSF), in this book simplified to k_T, according to Eq. (16.61).

$$CSF = k_T = k_{alt} * k_{UTM} \tag{16.61}$$

In this case, multiplying the combined scale factor k_T by the ground distance d gives the grid distance d_{UTM}. Dividing the grid distance d_{UTM} by the combined scale factor k_T gives the ground distance d, as follows:

$$d_{UTM} = k_T * d \ \text{ or } \ d = \frac{d_{UTM}}{k_T} \tag{16.62}$$

The effect of the elevation scale factor k_{alt} in calculating the grid distance or its inverse is a process that must be carefully observed. Using the same principle of calculating the variation of the ellipsoidal distance as a function of the local mean radius of the Earth and the elevation of the terrain, given in Sect. 7.2.4— shows that a variation of 6 m in elevation causes a variation of 1 ppm in the grid distance (2 ppm for 12 m and 3 ppm for 18 m). These values can indicate when to consider or not the geoid undulation and/or the variation in elevation when calculating the elevation scale factor and, consequently, the combined scale factor. For a project spread over an area of less than 1 km^2, on a terrain with little gradient change, the scale factor can be considered a constant equal to the combined scale factor at the centre of the project as a whole.

Example 16.6
The slope distance between points (E010) and (E014) has been measured in the field, as shown in Table 16.10. Using the field measurements and the known coordinate values from Tables 16.9 and 16.11, calculate the scale factors and UTM grid distances between points (E010) and (E014) for each calculation criteria presented in the previous section.

Table 16.10 Field measurements

Station	Sighting point	Slope distance [m]	Zenith angle
E010	E014	1179.788	90°30′55″

Table 16.11 Known coordinates

Point	ϕ_g	λ_g	H [m]
E010	−22°01′52.5229″	−47°51′37.2336″	866.868
E014	−22°01′54.6192″	−47°52′18.3045″	856.791

Solution:
   To solve this exercise, it is first necessary to calculate the horizontal distance d_{E010} at the level of (E010), using Eq. (7.1).

$$d_{E010} = 1,179.788 * \sin(90°30′55″) = 1,179.740 \text{ m}$$

Note that the effect of the curvature of the Earth's and vertical atmospheric refraction has not been considered.
   As the distance between the points is about 1.2 km, in this exercise, the average elevation between the two points and the local mean radius of the Earth at the latitude of the point (E010) are used to calculate the elevation scale factor.

$$H_{\acute{m}edio} = \frac{866.868 + 856.791}{2} = 861.830 \text{ m}$$

$$R_{0E010} = 6,362,746.000 \text{ m} \quad (as \text{ in example } 16.5)$$

$$kalt_m = \left(\frac{6,362,746.000}{6,362,746.000 + 861.830}\right) = 0.999864569$$

The scale factors and UTM distances are shown in Table 16.12.

Example 16.7
Using the data in Table 16.9, calculate the ground horizontal distance between points (E010) and (E014), considering the k_{UTM} values in Table 16.12.

Table 16.12 Scale factor and UTM distance

Calculation criterion	Point	k_{UTM}	$k_{UTM_{mean}}$	$k_{T_{mean}}$	d_{UTM} [m]
Approximate method using UTM coordinates	E010	1.000676531	1.000680832	1.000545309	1180.384
	E014	1.000685134			
Rigorous method using geodetic coordinates	E010	1.000677393	1.000681697	1.000546174	1180.385
	E014	1.000686002			

Solution:
  Using coordinates given in Table 16.9.

d_{UTM}

$$= \sqrt{(204,729.0177 - 203,551.8201)^2 + (7,560,947.1156 - 7,560,860.4734)^2}$$

$$= 1,180.3817 \text{ m}$$

Using the results of example 16.6, the following ground distances are obtained. Using the approximative method with UTM coordinates:

$$d = \frac{1,180.3817}{1.000545309} = 1,179.7384 \text{ m}$$

Using the Rigorous method with geodetic coordinates:

$$d = \frac{1,180.3817}{1.000546174} = 1,179.7374 \text{ m}$$

16.9   Azimuths to Be Considered in the UTM Projection

As Grid North and True North are involved in the UTM projection, there are two types of azimuths to consider when working with UTM coordinates.

- Plane Azimuth or Grid Azimuth (Az_G), and
- Geodetic Azimuth or True Azimuth (Az_T).

The grid azimuth Az_G is the clockwise angle in the projection between the Grid North and the tangent to the arc of the projection of the line under consideration. It is calculated as a function of the UTM coordinates (N, E) of the endpoints of the line using Eq. (16.63).

$$Az_G = \text{atan}\left(\frac{\Delta E}{\Delta N}\right) \pm \alpha_{quad} \tag{16.63}$$

The geodetic azimuth Az_T is the clockwise angle in projection between the

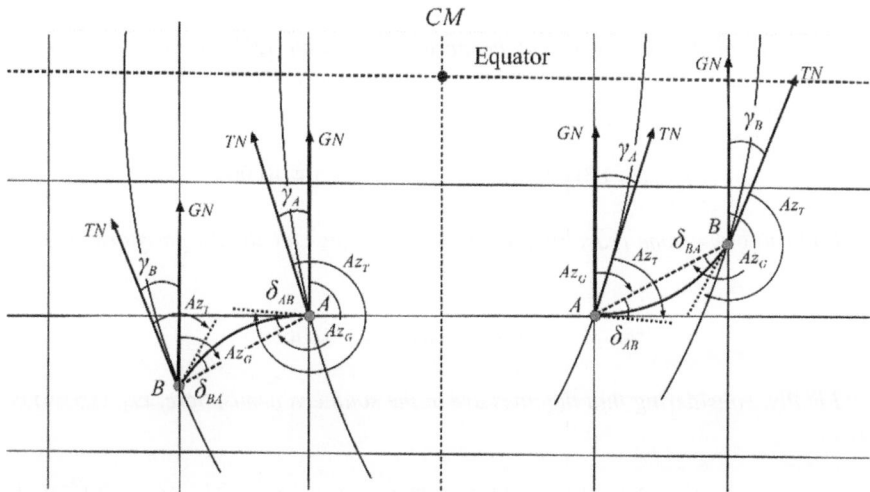

Fig. 16.17 Grid and geodetic azimuths considering the convergence of the meridians and the arc-to-chord correction

tangent to the projected meridian passing through the starting point of the line and the tangent to the arc of the projection of the line under consideration, as shown in Fig. 16.17.

According to the geometry in Fig. 16.17, the geodetic azimuth $Az_{T(AB)}$ and the grid azimuth Az_{GAB} are related as a function of the meridian convergence γ_A and the arc-to-chord correction δ_{AB}.

The equation for relating the two azimuths depends on the quadrant in which the alignment is located. Thus, the following equation can be used for the southern hemisphere and considering the algebraic signs of γ and δ.

$$Az_T = Az_G + \gamma + \delta \tag{16.64}$$

As given by Eq. (16.64), the true azimuth is obtained by calculating the grid azimuth and applying the convergence and arc-to-chord corrections.

Example 16.8

Using points (*E010*) and (*E014*) from Example 16.5, calculate the geodetic azimuth (Az_T) of lines *E010-E014* and *E014-E010*.

Solution:

   To solve this exercise, it is first necessary to calculate the grid azimuths of directions *E010-E014* and *E014-E010*, the convergence of the meridians and the arc-to-chord correction at points (*E010*) and (*E014*).

   Using Eq. (16.63).

$$Az_{GE010-E014} = \text{atan}\left(\frac{203,551.8201 - 204,729.0177}{7,560,860.4734 - 7,560,947.1156}\right) + 180° = 265°47'26.1537''$$

$$Az_{G(E014-E010)} = 265°47'26.1537'' - 180° = 85°47'26.1537''$$

   The value of the convergence of the meridians at point (E010) is given in example 16.4. Using the same calculation procedure, the convergence of the meridians at point (E014) is shown as follows:

$$\gamma_{E010} = 1°04'25.4270'' \qquad \gamma_{E014} = 1°04'40.9644''$$

   Following the same reasoning as above, the values for the arc-to-chord corrections are

$$\delta_{E010} = 0.0653'' \qquad \delta_{E014} = -0.0653''$$

   Finally, considering that the lines are in the southern hemisphere, Eq. (16.64) *is used.*

$$Az_{T(E010-E014)} = 265°47'26.1537'' + 1°04'25.4270'' + 0°00'00.0653'' = 266°51'51.6460''$$

$$Az_{T(E014-E010)} = 85°47'26.1537'' + 1°04'40.9644'' + (-0°00'00.0653'') = 86°52'7.0528''$$

16.10   Transverse Mercator Projection (TM)

Although the UTM projection system is the most widely used, it may be more appropriate to use a Transverse Mercator (TM) projection with local or regional characteristics in specific engineering projects. In this case, the solution is to reduce the size of the zone and increase the value of the scale factor at the central meridian k_0 to reduce the effects of geometric deformations of the objects represented on the projection plane. Examples of TM projections used in some countries include:

Gauss-Krüger projection, which has a 3° zone amplitude, $k_0 = 1$ and central meridian, is defined similarly to the UTM projection system.

Regional TM, which has 2° zone amplitude, $k_0 = 0.999995$ and central meridian
   defined in the odd meridians.
Local TM, as presented in detail below.

- Zone amplitude: 1°
- Central meridians at longitudes of 30′
- Latitude limits: 45 ° N and 45°S
- Scale factor k_0 in the central meridian is equal to 0.999995

Coordinates at the intersection of the central meridian and the Equator equals to
($N = 5,000,000.000$ m; $E = 200,000.000$ m) for the southern hemisphere and
($N = 0.000$ m; $E = 200,000.000$ m) for the northern hemisphere. They are usually
designated as (X, Y) to differentiate them from UTM coordinates.

For the coordinate transformations and the determination of the values of the
geometric elements of the TM projection system, refer to the equations used for the
UTM projection system presented in the previous sections, obviously changing the
value of k_0 (Fig. 16.18).

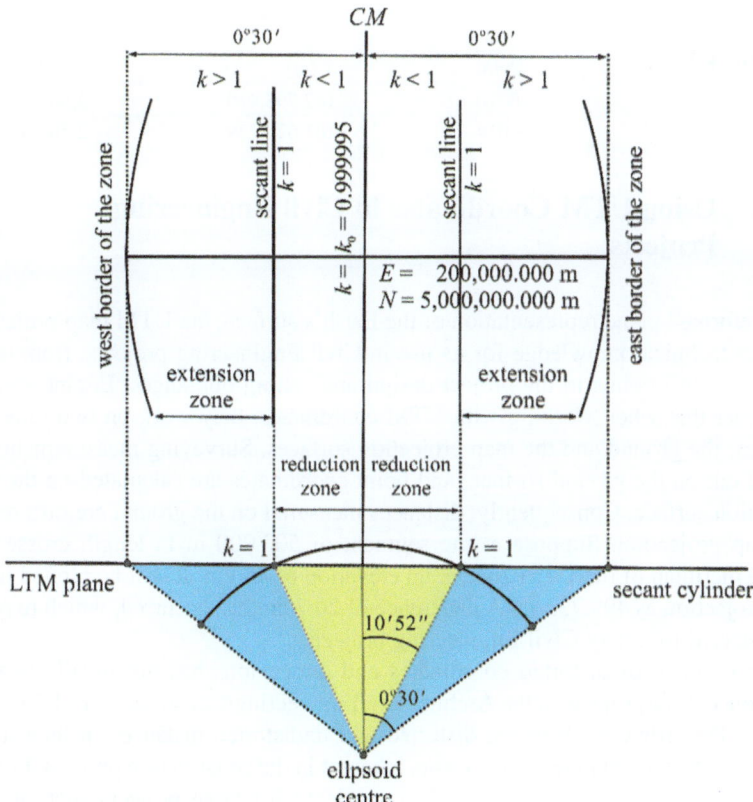

Fig. 16.18 Principle of the LTM projection

Example 16.9

Given the geodetic coordinates of points (*E*010) and (*E*014) in Table 16.11, calculate their LTM plane coordinates in the SIRGAS2000 reference system.

Solution:

Considering that LTM projection is to be centred at the meridian $-47°30'00''$ *and with* $k_0 = 0.999995$, *the LTM plane coordinates can be calculated as given in* Table 16.13 *and using the parameters given in* Table 16.1.

The LTM coordinates calculated for points (E010) and (E014) are given in Table 16.14.

Table 16.13 LTM parameters for calculating the coordinates of points (E010) and (E014)

E010	$S = -2,437,261.5097$ m	$\Delta\lambda = -1,297.2336''$	$p = -0.129723360$	$(I) = -2,437,249.3234$ m
	$(II) = -2607.6390$ m	$(III) = -2.1455$ m	$(IV) = 286,774.9703$ m	$(V) = 81.2855$ m
	$A_6 = -0.0016$ m	$B_5 = 0.0205$ m	$N' = -2,437,293.2057$ m	$E' = -37,201.5902$ m
E014	$S = -2,437,325.9888$ m	$\Delta\lambda = -1,338.304''$	$p = -0.133830450$	$(I) = -2,437,313.8022$ m
	$(II) = -2607.6938$ m	$(III) = -2.1455$ m	$(IV) = 286,773.7976$ m	$(V) = 81.2835$ m
	$A_6 = -0.0016$ m	$B_5 = 0.0205$ m	$N' = -2,437,360.5082$ m	$E' = -38,379.2612$ m

Table 16.14 LTM coordinates

Point	X [m]	Y [m]
E010	162,798.410	2,562,706.794
E014	161,620.739	2,562,639.492

16.11   Using UTM Coordinates in Civil Engineering Projects

As a deformed plane representation of the Earth's surface, the UTM map projection requires technical knowledge for its use in Civil Engineering projects, from determining control points to the project design and setting out stages. Engineers must remember that when working with UTM coordinates, they work on two reference surfaces: the ground and the map projection surfaces. Surveying measurements are carried out on the ground surface, and point coordinates are calculated on the map projection surface. Consequently, distances measured on the ground are different in the map projection. Suppose a line segment of 500.000 m in length crosses the central meridian of the UTM zone at an elevation of 850 m. It will be plotted in the map projection as 499.733 m. A difference of 26.7 cm has occurred, which may not be neglected for many Civil Engineering projects.

The problem of distorted coordinates and dimensions has historically been of little concern for projects in the Architecture/Engineering/Construction (AEC) sector because the difference between distorted and undistorted distances in these cases was insignificant compared to the errors inherent in the construction process. For this reason, distortion concerns were only considered for large projects such as road construction. Today, however, as engineering designs are created in CAD and BIM systems using local Cartesian coordinate system (also called the engineering CS), and topographic modelling and GIS methods are in the geospatial domain (map projection coordinate system), spatial integration between them requires detailed knowledge of coordinate properties when digital topographic and building data need to be integrated during the design and construction phases of the project.

16.11.1 Determining Control Points in UTM Coordinates

For many Civil Engineering projects, the spatial database for project development needs to be established in UTM coordinates linked to a national framework of geodetic reference points. This requires the network of control points to be defined in UTM coordinates. Surveying measurements and calculations can then be carried out using any of the coordinate geometry methods presented in Chap. 11, provided that observation reductions from the ground to the ellipsoid and the projection plane are applied appropriately. This means calculating the elevations and UTM scale factors (average or individual for each measured distance), reducing the ground distances to grid distances, calculating and applying the arc-to-chord correction to each line if necessary, and finally calculating the final UTM coordinates of the control points.

Supposing that the network of control points is to be determined by traverse adjustment, the calculation procedure may include the steps presented below. For this to be done, consider a link traverse composed of four reference points (*P*), (*Q*), (*R*), and (*S*) with known UTM coordinates, as shown in Fig. 16.19. Suppose the traverse lines were measured using a total station, meaning that each line length, after horizontal reduction, is ground distances at the instrument elevation. They need, therefore, to be reduced to the projection plane before traverse computation. The calculation steps are then as follows:

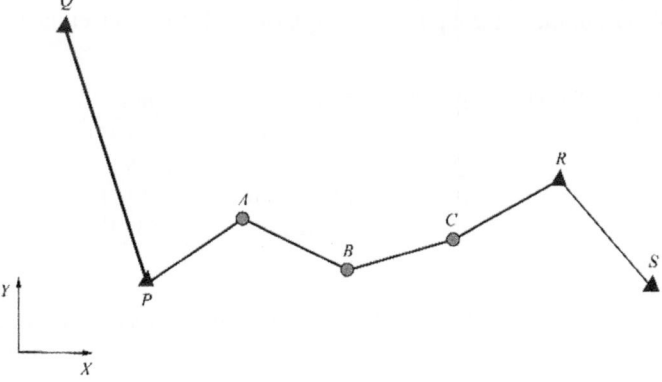

Fig. 16.19 Traverse geometry

1. Calculate starting grid azimuth Az_{GPQ} using the UTM coordinates of points (P) and (Q).
2. Calculate the preliminary azimuth $Az'G_{ij}$ for each traverse line using the field direction observations.
3. Calculate the preliminary grid coordinates (N',E') and preliminary elevations (H') of each traverse station considering the preliminary azimuths and the measured ground distances. Note that these three steps are essentially computing an open traverse without applying traverse adjustment.
4. Once the approximate elevations have been calculated and the elevation of at least one of the traverse stations is known, calculate the approximate elevations (mean or individual for each station) and then the corresponding elevation scale factor. Note that unless high-accuracy results are required or there are significant differences in elevation between the various traverse stations, it is sufficient to use a mean elevation of the area.
5. If necessary, the reduced distances to the ellipsoidal surface should be calculated considering the curvature of the Earth and atmospheric refraction as described in Chap. 7.
6. Based on the preliminary UTM coordinates (N',E'), calculate the UTM scale factor (mean or individual for each line) according to the weighting strategy adopted for the project area.
7. If necessary, calculate and apply the arc-to-chord correction for each traverse line.
8. Reduce ground distances to grid distances based on the calculated scale factors.
9. Considering start and ending UTM azimuths, observational directions and the calculated grid distances, perform traverse adjustment.
10. Finally, calculate UTM coordinates for the control points.

Example 16.10
Using the field measurements in Table 16.17, calculate the UTM coordinates of the linked traverse shown in Fig. 16.19. The latitude of point (P) is $-22°03'53.5660''$. The coordinates of the control points (P), (Q), (R) and (S) are given in Table 16.15.

Table 16.15 Control point UTM coordinates

Point	E [m]	N [m]	(H) [m]
P	217,377.303	7,557,453.190	852.036
Q	216,463.837	7,558,135.327	856.791
R	221,801.561	7,557,717.180	866.842
S	222,968.157	7,557,431.186	869.790

Solution:
   The first step in solving this exercise is to calculate the local mean radius of the Earth at the latitude of (P). The resulting values are given in Table 16.16.

Table 16.16 Mean local radius of the Earth at point (P)

P	$\phi_P = -22°03'53.5660''$	$N = 6{,}381{,}151.833$ m
	$M = 6{,}344{,}427.527$ m	$R_0 \cong 6{,}362{,}763.000$ m

The next step is to calculate the preliminary coordinates of the traverse vertices based on the values measured in the field without applying any corrections. With the preliminary coordinates, the scale factors can be calculated to convert the horizontal ground distances into UTM distances. Therefore, the average elevation of all the traverse points must be considered. All the values calculated so far are shown in Tables 16.17, 16.18 and 16.19.

The traverse computation can be carried out at this stage according to the calculation sequence presented in Sect. 12.2.5. The intermediate values of the calculations are not indicated in the text.

To help the reader check their calculations, Table 16.20 gives the traversing angular and linear errors of closure, the relative linear accuracy and the adjusted UTM coordinates of the traverse stations.

It should be noted that the relative precision of the traverse calculation would be ~1/2, 300 if the conversions from ground to UTM distances were not carried out.

Table 16.17 Field measurements and calculation results

Station	Sighting point	Horizontal direction	Slope distance [m]	Zenith angle	Preliminary horizontal clockwise angle Eq. (12.1)	Preliminary azimuth Eq. (11.46)	Horizontal ground distance [m] Eq. (12.7)
P	Q	66°53'32"	–	–	134°05'26"		
	A	200°58'58"	1.034,623	89°30'00"		80°50' 28.8392"	1034.587
A	P	299°28'17"	1.034,630	90°30'07"	196°12'57"		
	B	135°41'14"	1.165,483	89°52'55"		97°03' 25.8392"	1165.477
B	A	227°34'45"	1.165,475	90°07'10"	168°29'55"		
	C	36°04'40"	1.136,016	89°48'44"		85°33' 20.8392"	1136.004
C	B	148°29'22"	1.136,005	90°11'23"	176°33'18"		
	R	325°02'40"	1.122,301	89°51'46"		82°06' 38.8392"	1122.302
R	C	282°48'23"	1.122,310	90°08'20"	201°40'12"		
	S	124°28'35"				103°46' 50.8392"	

Table 16.18 Preliminary coordinates and scale factor

Station	Sighting point	Preliminary X-coordinate [m] Eq. (11.21)	Preliminary Y-coordinate [m] Eq. (11.21)	E' [m] Eq. (16.6)	k_{UTM} Eq (16.50)
P	$\dfrac{Q}{A}$	217,377.303	7,557,453.190	−282,622.697	1.000586257
A	$\dfrac{P}{B}$	218,398.700	7,557,617.864	−281,601.300	1.000579140 1.000579140
B	$\dfrac{A}{C}$	219,555.347	7,557,474.674	−280,444.653	1.000571112 1.000571112
C	$\dfrac{B}{R}$	220,687.936	7,557,562.701	−279,312.064	1.000563283 1.000563283
R	$\dfrac{C}{S}$	221,799.616	7,557,716.746	−278,200.384	1.000555629

Table 16.19 Scale factors and UTM distances

Station	Sighting point	k_{UTM} Eq. (16.50)	k_{UTM} (mean) Eq. (16.57)	k_{alt} Eq. (7.12)	k_T (mean) Eq. (16.61)	UTM distance [m] Eq. (16.62)
P	$\dfrac{Q}{A}$	1.000586257	1.000582699	0.999864642	1.000447262	1035.050
A	$\dfrac{P}{B}$	1.000579140 1.000579140	1.000575126	0.999864642	1.000439691	1165.989
B	$\dfrac{A}{C}$	1.000571112 1.000571112	1.000567198	0.999864642	1.000431763	1136.495
C	$\dfrac{B}{R}$	1.000563283 1.000563283	1.000559456	0.999864642	1.000424023	1122.778
R	$\dfrac{C}{S}$	1.000555629				

Table 16.20 Errors of closure, and adjusted UTM coordinates

Closing errors		UTM coordinate X [m]	Y [m]	Station
Angular	22.4″	217,377.303	7,557,453.190	P
Linear X [m]	−0.0357	218,399.163	7,557,617.979	1
Linear Y [m]	−0.0764	219,556.333	7,557,474.795	2
Relative precision	~1/53,000	220,689.414	7,557,562.954	3

16.11.2   Zone Transformation for the UTM Projection

Due to the configuration of the UTM projection, each zone always starts at the intersection of the central meridian and the central parallel of the zone. For this reason, the coordinate systems of each projection zone are independent of each other. However, a situation may arise where a surveying measurement extends from one zone to another, as in a road project. Several solutions to this type of problem are given in the literature. This book presents the solution using UTM coordinates of two points on either side of adjacent and overlapping zones, as described below.

Suppose the UTM coordinates of points (P) and (Q) in the western (or eastern) boundary of zone A are known. The UTM coordinate of a new point (R) can be calculated from these two points using coordinate geometry calculations as described in Chap. 11. The coordinates of the point (R) are then referenced to zone A. Suppose point (R) is located on the east boundary (or west boundary) of zone (B). If a new point (S) is calculated from point (R), all four points will have UTM coordinates referenced to zone A. The coordinates of each point can then be transformed into geodetic coordinates in zones A and B by extending the zone widths. The transformed geodetic coordinates can then be used to calculate UTM coordinates in adjacent Zones A and B. This provides four points with known UTM coordinates in both Zones, allowing the traverse measurements to continue.

The reader should note that this zone transformation can also be carried out using GNSS technology, which directly gives the measured geodetic coordinates of each point. The other coordinate transformations remain unchanged.

Example 16.11
Given the UTM coordinates of points $(P005)$ to $(P008)$ in zone 22, given in Table 16.21, determine the coordinates of points $(P007)$ and $(P008)$ in zone 23. See Fig. 16.20.

Table 16.21 UTM coordinates of points $(P005)$ to $(P008)$ in zone 22 (SIRGAS2000)

Point	E [m]	N [m]
P005	809,829.641	7,572,038.287
P006	809,863.375	7,572,152.411
P007	810,166.879	7,571,928.123
P008	810,146.779	7,572,224.971

Fig. 16.20 Point location in zones 22 and 23. (Image from Google Maps)

Solution:

1. *The following geodetic coordinates are obtained Based on the UTM coordinates in* Table 16.21:

$$\phi_{P005} = -21°55'43.2035''$$
$$\lambda_{P005} = -48°00'03.1458''$$

$$\phi_{P006} = -21°55'39.4753''$$
$$\lambda_{P006} = -48°00'02.0492''$$

$$\phi_{P007} = -21°55'46.5672''$$
$$\lambda_{P007} = -47°59'51.3310''$$

$$\phi_{P008} = -21°55'36.9384''$$
$$\lambda_{P008} = -47°59'52.2331''$$

2. *Considering the calculated geodetic coordinates, the UTM coordinates of points (P005) to (P008) in zone 23 are as follows:*

$E_{P005} = 189,989.694\,\text{m}$	$N_{P005} = 7,572,034.751\,\text{m}$	$E_{P006} = 190,018.937\,\text{m}$	$N_{P006} = 7,572,150.108\,\text{m}$
$E_{P007} = 190,330.984\,\text{m}$	$N_{P007} = 7,571,937.867\,\text{m}$	$E_{P008} = 190,299.285\,\text{m}$	$N_{P008} = 7,572,233.700\,\text{m}$

16.11.3   Project Design Based on UTM Control Points

The design of an engineering project based on reference points with UTM coordinates is an operation that requires scientific knowledge and technical care on the part of the designer. He must be aware that the values of the horizontal distances involved in the project are affected by the combined scale factor k_T. Therefore, to carry out his project correctly, he must define a suitable working strategy, agreed upon by all the professionals involved in the project, to determine the scale factors accordingly.

- The first strategy is to develop design elements (roads, buildings, infrastructure) directly in the UTM coordinate system. This means selecting the appropriate UTM zone, collecting and inputting data in UTM coordinates, developing the design in the UTM framework, and ensuring accuracy during field stakeout also in UTM coordinates. Design development in this case is facilitated by the use of CAD software with the ability to set the CAD design project's coordinate system to match the project's UTM zone and datum. The location of key design elements must then be entered into the project using UTM coordinates. The same process must be followed for the stakeout, i.e., the stakeout design point coordinates generated from the UTM grid are loaded in the memory of a GNSS or total station instrument that has been previously set to the correct UTM zone of the project area, and then stake out in the field, as presented in the next subsection.
- A second strategy is to transform the UTM coordinates into ground coordinates (Engineering Datum) and develop the project in this system. In this case, the project is designed without any change in the scale of the values of the horizontal distances of the design elements. The equations for performing these coordinate transformations are presented in Sect. 16.12. Note that this type of coordinate transformation is only suitable for small areas. Deformations can become significant as distances increase. For large areas, the engineer may need to consider the different ground coordinates generated at different altitudes.
- A third solution is to use a single combined scale factor for the whole project area if the project area is small enough or if the required accuracy is not affected by different projection distortions.
- Another solution, used in some countries, is to use the Local Transverse Mercator (LTM) projection. This projection is favored because it produces minimal distortion over the project area, making it highly suitable for engineering tasks in long line projects. By using the LTM projection, both project design and stakeout can be carried out in the same coordinate system, eliminating the need for coordinate transformations. This approach simplifies the workflow, ensures accuracy and reduces the risk of errors during construction.

16.11.4   Point Stakeout from a Project with UTM Coordinates

Stakeout is now a straightforward field operation carried out using total stations and GNSS instruments separately or in combination, depending on the working site environment. The use of these instruments depends on the type of coordinate system used in the project database. If the project has been designed using the UTM map projection, stakeout with GNSS RTK instruments is the best solution. As explained later in Chap. 17, a GNSS RTK instrument will work directly with UTM coordinates if configured. In this case, the operator works with UTM coordinates. However, the GNSS data logger calculates the ground distances, and the points are stakeout directly on the local reference plane in the terrain.

If GNSS-RTK stakeout is not an option, or if GNSS-RTK and total station measurements must be combined in the field, the UTM distances must be corrected by the respective combined scale factor k_T when using a total station. Each distance calculated in the projection plane (grid distance) must then be divided by its combined scale factor to obtain the ground distance to be staked out in the field. It is important to note that some total stations have built-in routines to work directly with UTM coordinates, i.e., the instrument can be configured with the projection's parameters, simplifying field operations. Staking out, in this case, is similar to GNSS RTK instruments, provided the total station has been configured to consider the necessary distance reductions to the local reference plane.

As mentioned above, the combined scale factor is critical for staking out points with UTM coordinates. For this reason, when working with GNSS-RTK or total stations that work directly with UTM coordinates, it is essential to evaluate which calculation method they use to determine the combined scale factor k_T. Using a single combined scale factor for the whole area is always desirable as long as it is acceptable and there are no significant differences in elevation between points in the area where the project is to be set out.

16.12   Coordinate Transformation Between Grid and Ground Coordinates

Although the advantages of working with the UTM coordinate system are obvious, the engineer may prefer to work with local ground-based coordinates (Engineering Datum). As already mentioned, it is easier to design an engineering project in this system, as the designer does not have to worry about the deformation of the geometric figures represented. For these reasons, there are coordinate transformation models that make grid and ground coordinates compatible. The most relevant of these for Geomatics applied to Civil Engineering are presented in the following subsections.

- Transformation between grid and local ground coordinates using the *Two-Dimensional Conformal Coordinate Transformation* presented in Sect. 5.9.1.2.1

- Transformation from Global Geocentric Cartesian to Topocentric Cartesian coordinates, as presented in Sect. 5.9.2.4
- Local ground coordinates from a Tangent Plane
- Local ground coordinates from Low Distortion Projection (LDP)
- Local ground coordinates by geometric reductions

16.12.1   Transformation Between Grid and Ground Coordinates Using the Two-Dimensional Conformal Coordinate Transformation

The equations used in this coordinate transformation are presented in Sect. 5.9.1.2.2.

This type of transformation is often used on construction sites where the project is on the local ground-based plane, and project points are to be stakeout with GNSS RTK instruments. In this case, since GNSS instruments can provide UTM (or TM) coordinates, it is necessary to perform a coordinate transformation. To do this, the engineer must have a network of control points with known coordinates in the local ground-based horizontal plane to be selected for GNSS observations, thus creating a set of homologous points with known coordinates in both systems (grid and local). The set of points thus defined is used to determine the transformation parameters, which, once entered into the data collector of the GNSS instrument, will allow the use of both coordinate systems. It is important to note that the homologous points must be evenly distributed over the project area. The quality of the results will depend on the coordinates of the control points determined in both systems.

As this is a transformation of coordinates between Two-Dimensional Cartesian coordinate systems, the practice has shown that the geometric deformations generated by the model can become unacceptable for areas greater than 10 km × 10 km, depending on whether the project is predominantly north-south or east-west. If it is east-west, the deformations will be more significant. However, the method is highly reliable for small areas and gives better results than other methods for the same area. Experiments carried out by Morais (2019) in an area of 3.7 km × 12 km in the N-S direction gave average results of around ±5 cm at the extremes of the area.

16.12.2   Transformation Between Global Geocentric Cartesian and Topocentric Cartesian Coordinates

The equations used in this coordinate transformation are given in Sect. 5.9.2.4. The dimension of the area for applying this coordinate transformation model is limited to 10 km × 10 km. The experiment above produced average results of about ±9 cm for this transformation model.

16.12.3   Local Ground Coordinate from a Tangent Plane

The calculation of local ground coordinates from a tangent plane is based on a direct transformation from geodetic to ground coordinates (and vice versa), considering a local tangent plane placed at ground level and transferring features from the ellipsoid to the plane. To define the projection, it is first necessary to specify the point of tangency and the orientation of the projection. The tangent point becomes the centre of the projection and is usually chosen near the centre of the project site. The orientation of the tangent plane can be selected to align with the map grid, previous survey plans in the area, or any other convenient meridian. The equations for this coordinate transformation method are presented in the following subsections.

16.12.3.1   Transformation from Geodetic to Ground Coordinates

In this coordinate transformation, the ground points are Two-dimensional coordinates (XL, YL). Thus, given the point (P) with (ϕ_P, λ_P) geodetic coordinates and the tangent reference point (O) with (ϕ_O, λ_O) geodetic and (XL_O, YL_O) ground coordinates (known or assumed), the (XL_P, YL_P) ground coordinates of the point (P) can be calculated as follows:

$$XL_P = XL_O + \Delta x_P \tag{16.65}$$

$$YL_P = YL_O + \Delta y_P \tag{16.66}$$

where

$$\Delta x_P = \frac{\Delta x'_P}{k_{alt}} \tag{16.67}$$

$$\Delta x'_P = \frac{\Delta \lambda_1 * \cos(\phi_P) * N_P}{\rho''} \tag{16.68}$$

$$k_{alt} = \frac{R0}{R0 + H_m} \qquad \text{same as equation (7.12)}$$

$$\Delta \lambda_1 = \Delta \lambda'' * \left[1 - \left(\frac{\Delta \lambda''^2}{6 \rho''^2} \right) \right] \tag{16.69}$$

$$\Delta \lambda'' = \lambda_P - \lambda_0 \tag{16.70}$$

$$\Delta y_P = \frac{\Delta y'_P}{k_{alt}} \tag{16.71}$$

$$\Delta y'_P = \frac{1}{B} * \left[\Delta \phi_1 + C * \Delta x'^2_P + D * (\Delta \phi_1)^2 + E * (\Delta \phi_1) * \Delta x'^2_P + E * C * \Delta x'^4_P \right] \tag{16.72}$$

$$\Delta\phi_1 = \Delta\phi'' * \left[1 - \left(\frac{\Delta\phi''^2}{6\rho''^2}\right)\right] \tag{16.73}$$

$$\Delta\phi'' = \phi_P - \phi_0 \tag{16.74}$$

$$B = \frac{\rho''}{M_0} \tag{16.75}$$

$$C = \frac{\tan(\phi_0) * \rho''}{2M_0 * N_0} \tag{16.76}$$

$$D = \frac{3e^2 * \sin(\phi_0) * \cos(\phi_0)}{\left[1 - e^2 * \sin^2(\phi_0)\right] * 2\rho''} \tag{16.77}$$

$$E = \frac{1 + 3\tan^2(\phi_0)}{6N_0^2} \tag{16.78}$$

where

M_0= radius of curvature of the meridian section of the reference ellipsoid at (O)
   (origin of the system)—Eq. (4.7)
N_0= radius of curvature of the first principal vertical of the reference ellipsoid at
   (O) – (Eq. 4.8)
N_P= radius of curvature of the first principal vertical of the reference ellipsoid at
   (P) – (Eq. 4.8)
R_0= local mean radius of the Earth in the measurement area (Eq. 4.9)
H_m=mean elevation of the surveying area
λ_0= geodetic longitude of point (O)
ϕ_0= geodetic latitude of point (O)
λ_P= geodetic longitude of point (P)
ϕ_P= geodetic latitude of point (P)
e^2= first quadratic eccentricity of the reference ellipsoid (given by Eq. 4.4)
$\rho'' = 206, 264.8062''$ (radian to arc second conversion factor)

When applying these equations, latitudes must be taken as negative values for the southern hemisphere and longitudes as negative values west of the Greenwich meridian, as indicated in Sect. 5.8. Azimuths in the local geodetic system are referenced to the Grid North. Therefore, they are rotated concerning the True North according to the convergence of the meridians. The convergence of the meridians is zero only for those points that are on the central meridian. The value of the convergence of the meridians in this system can be determined using the following equations.

$$\gamma_p = -\left[\Delta\lambda'' * \sin(\phi_P) * \sec\left(\frac{\Delta\phi}{2}\right) + F * (\Delta\lambda'')^3\right] \tag{16.79}$$

where

$$F = \frac{\sin(\phi_P) * \cos(\phi_P)}{12\rho''^2} \tag{16.80}$$

Example 16.12

Given the coordinate values in Table 16.22, calculate the two-dimensional Cartesian coordinates of points (E011) and (E014) in the LMS using the equations given in the previous section. Take point (E010) as the origin of the local system.

Table 16.22 Geodetic and local coordinates (SIRGAS2000)

Point	Latitude (ϕ_g)	Longitude (λ_g)	XL [m]	YL [m]	H [m]
E010	−22°01′52.5229″	−47°51′37.2336″	152,596.0920	254,100.8250	866.868
E011	−22°01′51.6174″	−47°51′43.6780″	?	?	?
E014	−22°01′54.6192″	−47°52′18.3045″	?	?	?

Solution:

Table 16.23 *shows the preliminary calculations used to solve the problem. To calculate the elevation scale factor, the average altitude of the site was set to 866.868 m.*

Table 16.23 Preliminary values for determining the coordinates of the point (E011)

$H_{E010} = 866.868$ m	$M_{E010} = 6,344,401.4869$ m	$N_{E010} = 6,381,143.1030$ m	$R_0 = 6,362,745.7745$ m
$k_{alt} = 0.999863777$	$N_{E011} = 6,381,143.0378$ m	$\Delta\lambda'' = -6.4444''$	$\Delta\lambda_1 = -6.4444''$
$\Delta\phi'' = 0.9055''$	$\Delta\phi_1 = 0.9055''$	$B = 0.032511310$	$C = -1.030854292 * 10^{-9}$
$D = -1.694404851 * 10^{-8}$	$E = 6.103833182 * 10^{-15}$	$\Delta x'_{E011} = -184.8105$ m	$\Delta x_{E011} = -184.8357$ m
$\Delta y'_{E011} = 27.8508$ m	$\Delta y_{E011} = 27.8546$ m		

Given the local coordinates of the point (E010) in Table 13.23, *the local coordinates of the point (E011) are as follows.*

$$X_{LE011} = 152,596.0920 - 184.8357 = 152,411.2563 \text{ m}$$

$$Y_{LE011} = 254,100.8250 + 27.8546 = 254,128.6796 \text{ m}$$

Following the same calculation reasoning for point (E014).

$$X_{LE014} = 152,596.0920 - 1,177.9721 = 151,418.1199 \text{ m}$$

$$Y_{LE014} = 254,100.8250 - 64.5231 = 254,036.3019 \text{ m}$$

16.12.3.2   Transformation from Ground to Geodetic Coordinates

For the transformation from ground (XL_P, YL_P) to geodetic coordinates (ϕ_P, λ_P), provided that the coordinates of the reference point (XL_O, YL_O) are known or assumed, the following equations can be applied:

$$\Delta XL = XL_P - XL_O \tag{16.81}$$

$$\Delta YL = YL_P - YL_O \tag{16.82}$$

Calculation of the latitude of (P):

$$\omega = \frac{k_{alt}}{B} \tag{16.83}$$

$$\varepsilon = \Delta YL - \left(\omega * C * \Delta XL^2\right) - \left(\omega * E * C * \Delta XL^4\right) \tag{16.84}$$

$$\tau = \omega + (D * \varepsilon) + \left(\omega * \Delta XL^2 * E\right) \tag{16.85}$$

$$\Delta\phi'' = \frac{\varepsilon}{\tau * \left[1 - 3,9173^{-12} * \left(\frac{\varepsilon}{\tau}\right)^2\right]} \tag{16.86}$$

$$\phi_P = \phi_0 + \Delta\phi'' \quad \text{(note the need for compatibility between units)} \tag{16.87}$$

Calculation of the longitude of (P):

$$\Delta\lambda_1'' = \frac{\Delta XL * \rho''}{\cos(\phi_P) * N_P * k_{alt}} \tag{16.88}$$

$$\Delta\lambda'' = \frac{\Delta\lambda_1''}{\left(1 - 3,9173^{-12} * \Delta\lambda_1''^2\right)} \tag{16.89}$$

$$\lambda_P = \lambda_0 + \Delta\lambda'' \quad \text{(note the need for compatibility between units)} \tag{16.90}$$

Example 16.13

Given the coordinate values in Table 16.24, calculate the geodetic coordinates of the point (E011) using the equations given in the previous section. Take point (E010) as the origin of the local system.

Table 16.24 Local and geodetic coordinates (SIRGAS2000)

Point	X_L [m]	Y_L [m]	Latitude (ϕ_g)	Longitude (λ_g)	H [m]
E010	152,596.0920	254,100.8250	−22°01′52.5229″	−47°51′37.2336″	866.868
E011	152,411.2562	254,128.6796	?	?	

Solution:

   The first step is to calculate the differences in local coordinates between point (E010) and point (E011).

$$\Delta XL = 152,411.2563 - 152,596.0920 = -184.8357\,\text{m}$$

$$\Delta YL = 254, 128.6796 - 254, 100.8250 = 27.8546 \, \text{m}$$

The latitude and longitude of point (E011) can then be calculated using Eqs. (16.83) to (16.90). *Calculation results are given in* Table 16.25.

Table 16.25 Calculation results

$k_{alt} = 0.999863777$	$\omega = 30.754336393$	$\varepsilon = 27.85564022$	$\tau = 30.754335928$
$\Delta\phi'' = 0.905746763$	$N_{E011} = 6, 381, 143.0378 \, \text{m}$	$\Delta\lambda_1'' = -6.446156104$	$\Delta\lambda'' = -6.446156105$
$\phi_{E011} = -22°01'51.6172''$	$\lambda_{E011} = -47°51'43.6798''$		

The literature suggests that this coordinate transformation method can be applied to areas up to 100 km × 100 km. Experiments carried out by Morais (2019) on an area of 3.7 km × 12 km in the N-S direction gave average results of around ±5 cm at the ends of the area.

16.12.4   Low Distortion Projection (LDP) Method

Another type of solution that can be used to obtain coordinates with distortions compatible with those of a local geodetic plan is the use of a TM system with a scale factor at the central meridian k_0 calculated as a function of the average altitude of the project area, as described below.

In this case, it is assumed that the geodetic coordinates of all the points involved in the project are known. The central meridian and latitude of the local system must then be defined. This defines a point of origin, which can be the approximate centre of the area of interest, a specific physical location, a point belonging to the existing geodetic control network, or others. This point is assigned a pair of plane coordinate values so that negative numbers are not used or confused with the UTM coordinate values of the same points in the project area.

The next step is to define the ellipsoidal height (h_0) for the origin of the system, which must be based on the average of the ellipsoidal heights of the terrain in the project area. A TM projection can then be generated with the projection surface tangent to the reference ellipsoid at the ellipsoidal height (h_0). In these circumstances, the tangent scale factor k_0 for the central meridian must be calculated using Eq. (16.91).

$$k_0 = 1 + \frac{h_0}{R_0} \tag{16.91}$$

Next, the TM coordinates can be determined using the equations presented in Sect. 16.5.1

As described, the Low Distortion TM projection is a tangent projection at the mean ellipsoidal height (h_0) of the project area. See Fig. 16.21. However, it can be lowered slightly to become secant. This can be done by using an additional scale factor, k_r, thus increasing the length of the usable zone, as shown in Fig. 16.22.

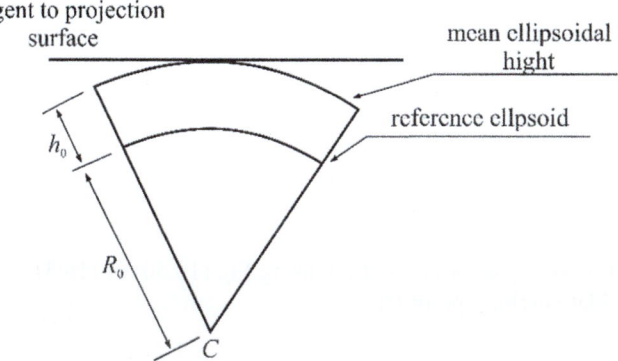

tangent to projection surface

mean ellipsoidal hight

reference ellpsoid

Fig. 16.21 Tangent low distortion projection

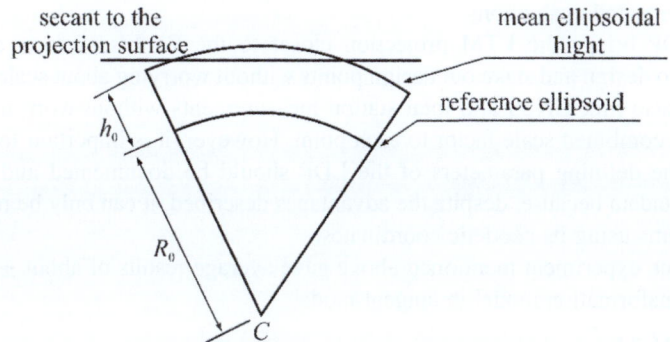

secant to the projection surface

mean ellipsoidal hight

reference ellipsoid

Fig. 16.22 Secant low distortion projection

The additional scale factor k_r can be calculated using Eq. (16.92).

$$k_r = \cos\left[\text{asin}\left(\frac{l}{2R_0}\right)\right] \qquad (16.92)$$

The new scale factor k_{0s} for the central meridian for the secant projection is calculated using Eq. (16.93).

$$k_{0s} = k_0 * k_r \tag{16.93}$$

where

l = greater length of the project area,
R_0 = local mean radius of the Earth.

Then, if desired, the deformation δ_i can be checked for a set of points distributed over the entire design area, according to Eq. (16.94).

$$\delta_i = k_i * \frac{R_0}{R_0 + h_i} - 1 \tag{16.94}$$

where

k_i = projection scale factor at point (i), using Eq. (16.50) or (16.51)
h_i = ellipsoidal height at point (i).

Multiplying the coefficient δ_i by the distance between the points gives the deformation in the unit of distance used. Note that multiplying the coefficient by 1,000,000 gives the deformation in ppm.

As the process of defining a LDP is an optimisation problem, it is necessary to carry out repeated evaluations.

The LDP brings the LTM projection closer to the Earth's surface, allowing engineers to design and stake out design points without worrying about scale factors. They can also mix GNSS and total station measurements without worrying about applying a combined scale factor to each point. However, it is important to emphasise that the defining parameters of the LDP should be documented and kept as project metadata because, despite the advantages described, it can only be related to other systems using its geodetic coordinates.

The same experiment mentioned above gave average results of about ±10.5 cm for this transformation model in tangent mode.

Example 16.14
Given the UTM, local ground and geodetic coordinates of the points shown in Table 16.26, calculate the coordinates of points ($E011$) and ($E014$) in the local ground plane using the LPD method with the coordinates of the point ($E010$) as the origin.

Table 16.26 Coordinate values

Point	UTM coordinate		Geodetic coordinate (SIRGAS 2000)		Local ground coordinate		
	E [m]	N [m]	Latitude	Longitude	X_L [m]	Y_L [m]	H [m]
E007	204,031.2178	7,560,476.9482	−22°02′07.3702″	−47°52′01.8531″	151,889.9871	253,644.0658	867.029
E010	204,729.0177	7,560,947.1156	−22°01′52.5229″	−47°51′37.2336″	152,596.0920	254,100.8250	866.868
E020	204,584.4814	7,559,967.1848	−22°02′24.2665″	−47°51′42.9097″	152,433.3026	253,124.3034	877.881
E031	203,446.7804	7,560,309.7056	−22°02′12.4457″	−47°52′22.3254″	151,302.8407	253,487.8928	867.787
E011	204,543.5922	7,560,971.5163	−22°01′51.6174″	−47°51′43.6780″	?	?	–
E014	203,551.8201	7,560,860.4734	−22°01′54.6192″	−47°52′18.3045″	?	?	–

Solution:

To solve this exercise, it is first necessary to calculate the mean elevation of the project area, which, according to Table 16.26, is 869.891 m. The local mean radius of the Earth at the latitude of the project area is 6.362.746 m.

The next step is to calculate the scale factor at the central meridian of the LTM using Eqs. (16.91), (16.92) and (16.93). Assuming that the maximum width of the project is equal to 1.500 m, the following results are obtained:

$$k_0 = 1 + \frac{869.891}{6,362,746.000} = 1.000136716$$

$$k_r = \cos\left[\text{asin}\left(\frac{1.500}{2 * 6,362,746.000}\right)\right] = 0.999999993$$

$$k_{0s} = 1.000136716 * 0.999999993 = 1.000136709$$

Using the calculated scale factor and taking point (E010) as the centre of the map projection, the coordinates of points (E011) and (E014) can be determined using Eqs. (16.5) to (16.22). Assuming that the central meridian is the meridian passing through the point (E010) and that the scale factor at the central meridian is k_{0s}, the preliminary values shown in Table 16.27 are obtained.

Table 16.27 Preliminary values for calculating the LDP coordinates of points (E010), (E011) and (E014)

	$S = -2,437,261.5097$ m	$\Delta\lambda = 0.0''$	$p = 0.0$	$(I) = -2,437,594.7062$ m
	$(II) = -2608.0085$ m	$(III) = -2.1458$ m	$(IV) = 286,815.6092$ m	$(V) = 81.2970$ m
E010	$A_6 = -0.0016$ m	$B_5 = 0.0205$ m	$N' = -2,437,594.7062$ m	$E' = 0.0$ m
E011	$S = -2,437,233.6579$ m	$\Delta\lambda = -6.4444''$	$p = -0.000644440$	$(I) = -2,437,566.8506$ m
	$(II) = -2607.9848$ m	$(III) = -2.1458$ m	$(IV) = 286,816.1157$ m	$(V) = 81.2978$ m
	$A_6 = -0.0016$ m	$B_5 = 0.0205$ m	$N' = -2,437,566.8517$ m	$E' = -184.8358$ m
E014	$S = -2,437,325.9888$ m	$\Delta\lambda = -41.0709''$	$p = -0.004107090$	$(I) = -2,437,659.1942$ m
	$(II) = -2608.0634$ m	$(III) = -2.1458$ m	$(IV) = 286,814.4364$ m	$(V) = 81.2951$ m
	$A_6 = -0.0016$ m	$B_5 = 0.0205$ m	$N' = -2,437,659.2381$ m	$E' = -1,177.9727$ m

To obtain the coordinates of points (E010), (E011), and (E014) in the same coordinate system as the previous example, it is necessary to translate the calculated LDP coordinates to the local coordinates of the point (E010), as given below.

$$X_{E010} = 152,596.0920 + 0.0000 = 152,596.0920 \text{ m}$$

$$Y_{E010} = 254,100.8250 + 0.0000 = 254,100.8250 \text{ m}$$

$$X_{E011} = 152,596.0920 - 184.8357 = 152,411.2563 \ \text{m}$$

$$Y_{E011} = 254,100.8250 + 27.8560 = 254,128.6810 \ \text{m}$$

$$X_{E014} = 152,596.0920 - 1,177.9721 = 151,418.1199 \ \text{m}$$

$$Y_{E014} = 254,100.8250 - 64.5319 = 254,036.2931 \ \text{m}$$

16.12.5   Local Ground Coordinates by Geometric Reductions

In some specific cases, where high precision is not required, the UTM plane coordinates (N, E) can be transformed into local plane coordinates (XL, YL) by applying a rotation based on the convergence of the meridians and scaling the ground distances based on the combined scale factor, calculated according to the mean elevation of the site and the UTM coordinates of a point taken as a reference. The calculation sequence to use this transformation method is as follows:

1. Select a point in the project area as the origin of the local system (P_0).
2. Calculate the convergence of the meridians and the combined scale factor k_T at that point.
3. Calculate the grid azimuths (Az_G) of each line P_0Pi and rotate them by the convergence of the meridians and the arc-to-chord correction, if necessary.
4. Calculate the projections $\Delta X_{P_0P_i}$ and $\Delta Y_{P_0P_i}$ of each line considering the combined scale factor k_T.
5. Calculate the transformed coordinates for each point (P_i).

The equations to be used are as follows:

$$Az_G = \text{atan} \ \frac{\Delta E}{\Delta N} \pm \alpha_{quad} \quad \text{same as equation (16.63)}$$

$$\gamma'' = (XII) * p + (XIII) * p^3 + (C_5) * p^5 \quad \text{same as equation (16.41)}$$

$$\delta''_{AB} = \left(\frac{2E'_A + E'_B}{6R_0^2} \right) * (N_B - N_A) * \rho'' \quad \text{same as equation (16.47)}$$

$$Az_T = Az_G + \gamma + \delta \quad \text{same as equation (16.64)}$$

$$\Delta X_{P_0P_i} = \frac{d_{UTM(P_0P_i)}}{k_T} * \sin\left(Az_{T(P_0P_i)}\right) \tag{16.95}$$

$$\Delta Y_{P_0 P_i} = \frac{d_{UTM(P_0 P_i)}}{k_T} * \cos\left(Az_{T(P_0 P_i)}\right) \tag{16.96}$$

$$X_{L(P_i)} = X_{L(P_0)} + \Delta X_{P_0 P_i} \tag{16.97}$$

$$Y_{L(P_i)} = Y_{L(P_0)} + \Delta Y_{P_0 P_i} \tag{16.98}$$

This method should be used with caution as it is an approximate method with no geometric consistency with the cartographic projection model. However, it can be useful in cases where the project area is small and has little variation in elevation.

The same experiment mentioned above gave average results of about ±9.5 cm for this transformation model.

Example 16.15

Given the coordinates values in Example 16.12, calculate the coordinates of points (E011) and (E014) by geometric reductions using the coordinates of the point (E010) as the origin point.

Solution:

The results of the preliminary coordinate transformation calculations are shown in Tables 16.28 and 16.29. *Note that some values relating to points (E010) and (E014) are known from previous application examples. Those relating to point (E011) have been calculated for this example. The calculations are not given in this text.*

As in the previous examples, the mean elevation and the local mean radius of the Earth used in the calculation are 869.891 m *and* 6,362,746 m, *respectively.*

Table 16.28 Convergence of the meridians and scale factor

Line	Convergence of the meridians E010	Arc-to-chord	k_{UTM} E010	k_{UTM}	k_{UTM}(mean)	k_{alt}	k_T
E010-E011	1.073729724	−0.01834″	1.000677393	1.000678746	1.000678070	0.999863534	1.000541511
E010-E014		0.0653″		1.000686002	1.000681697		1.000545138

Table 16.29 Distance, azimuth and transformed coordinates

Line	UTM distance [m]	Local ground distance [m]	Az_{UTM}	Az_T	ΔX [m]	ΔY [m]	X_L [m]	Y_L [m]
E010-E011	187.024	186.923	277°29′47.9391″	278°34′13.3478″	−184.8356	27.8560	152,411.256	254,128.681
E010-E014	1180.382	1179.739	265°47′26.1537″	266°51′51.6460″	−1177.9723	−64.5319	151,418.120	254,036.293

16.12.6   Comparing Results of Each Coordinate Transformation Method

Table 16.30 *shows a comparative summary of the local ground coordinates calculated for each transformation model.*

Table 16.30   Local ground coordinates for points (E011) and (E014)

	E011		E014	
Transformation model	X_L [m]	Y_L [m]	X_L [m]	Y_L [m]
Two-dimensional conformal coordinate transformation using the parametric adjustment model (Chap. 5)	152,411.258	254,128.679	151,418.120	254,036.295
Two-dimensional conformal coordinate transformation using the generalized gauss-helmet adjustment model (Chap. 5)	152,411.258	254,128.679	151,418.120	254,036.295
Global geocentric cartesian to topocentric cartesian (Chap. 5)	152,411.257	254,128.681	151,418.124	254,036.293
Tangent plane	152,411.256	254,128.680	151,418.119	254,036.293
Low distortion projection	152,411.256	254,128.680	151,418.119	254,036.293
Geometric reductions	152,411.256	254,128.681	151,418.120	254,036.293

16.13   Problems

1. Using the geodetic coordinates of points (P) and (Q) in Table 16.31, calculate their corresponding UTM coordinates (E, N).

Table 16.31   Geodetic coordinates of points (P) e (Q) (SIRGAS2000)

Point	Latitude(ϕ_g)	Longitude (λ_g)
P	22°00'17.8160" S	47°53'57.0497" W
Q	22°00'56.3027" S	47°51'42.8090" W

2. Using the values given in Table 16.31, calculate the convergence of the meridians for points (P) and (Q) using the rigorous equation.
3. Using the values given in Table 16.31, calculate the convergence of the meridian for points (P) and (Q) using the approximate equation. Compare the results with those obtained in problem 2.
4. Using the UTM coordinates calculated in problem 1, calculate the arc-to-chord correction of the line PQ at point (P), and of the line QP at point (Q).
5. Using the results from the previous problems, calculate the UTM scale factor at points (P) and (Q).
6. Using the geodetic coordinate of point (P) in Table 16.32, calculate its UTM coordinates, the convergence of the meridian for this location and the UTM scale factor at point (P).

Table 16.32 Geodetic coordinates of point (P) (SIRGAS2000)

Point	Latitude (ϕ_g)	Longitude (λ_g)
P10	22°00'17.8160" S	42°06'02.9503" W

7. Using the coordinates given in Table 16.33, calculate the horizontal ground distance d_{PQ} between points (P) and (Q) at the elevation of (P).

Table 16.33 Coordinates of points (P) and (Q) (SIRGAS2000)

Point	UTM coordinates		Latitude	Longitude	Elevation
	E [m]	N [m]	ϕ_g	λ_g	H [m]
P	200,662.024	7,563,785.991	−22°00'17.8160"	−47°53'57.0497"	827.500
Q	204,536.615	7,562,674.235	−22°00'56.3027"	−47°51'42.8090"	918.534

8. Using the field data given in Table 16.34 and the coordinates of point (P) given in Table 16.35, calculate the UTM distance between points (P) and (R).

Table 16.34 Field measurements

Station	Sighting point	Slope distance [m]	Zenith angle
P	R	1562.880	88°32'44"

Table 16.35 Coordinates of point (P)

Point	Elevation (H) [m]	Latitude (ϕ_g)
P	827.500	−22°00'17.8160"

9. Using the coordinates of point (P) given in Table 16.33 and the values calculate in the previous problem, and assuming that the azimuth of the line PR is equal to 112° 35' 44", calculate the UTM coordinates of point (R).

10. Using the coordinates of point (Q) given in Table 16.33 and the field measurements given in Table 16.36, calculate the ellipsoidal and the UTM distances between points (Q) and (S).

Table 16.36 Field measurements

Station	Sighting point	Slope distance [m]	Zenith angle
Q	S	1,300.000	94°15'22"

11. Assuming that the grid azimuth of the line PQ in problem 1 is equal to 106° 00' 36" and that the values of the convergence of the meridian and the arco-to-chord calculated in problems 2 and 4, calculate the geodetic azimuth of the line PQ.

12. Using the geodetic parameters given in Table 16.37, calculate the grid azimuth of the line P_1 - P_2.

Table 16.37 Geodetic parameters

Longitude of (P1)	48°30'40.0000"W
Longitude of (P2)	48°28'45.9141"W
Geodetic azimuth of P_2 - P_1	220°33'42.9"
Convergence of the meridian for point (P_1)	−1°04'18.9"
Arch-to-chord reduction angle of the line P_2 - P_1	−2.4"

13. Using the values for points (P1) and (P2) in the problem 12 problem and the geodetic parameters for point (P3) given in Table 16.38, calculate the angle α_{P1}, as given in Fig. 16.23.

Table 16.38 Geodetic parameters for point (P3)

Longitude (P3)	48°28'56.9080"W
Geodetic azimuth of P1-P3	148°23'08.4"
Arch-to-chord reduction angle of the line P1-P3	−3.0"

Fig. 16.23 Geometric sketch

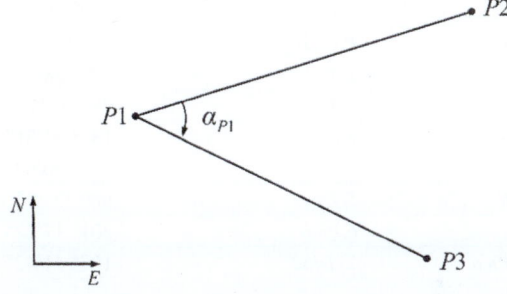

14. Using the coordinates of points (P11), (P12) and (P13) given in Table 16.39, calculate the angle and distance for setting-out point (P13) using a total station set up at point (P12) and backsighted to point (P11). Assume that the local mean radius of the Earth is 6,362,725 m and the UTM scale factor is 0.999865480.

Table 16.39 UTM coordinates of points (P11), (P12) and (P13)

Point	E [m]	N [m]	Elevation [m]
P11	647,121.510	7,562,473.820	–
P12	646,581.360	7,561,597.515	495.700
P13	646,299.408	7,562,963.594	–
SIRGAS20000 (UTM zone 23)			

15. A traversing measurement was carried out using a total station, as shown in Fig. 16.24. The field measurements are given in Table 16.40, and the control point coordinates are given in Table 16.41. Assuming that the mean local radius of the Earth is 6363 km, the mean local elevation is 762.700 m, and the mean UTM factor is 1.00056440, calculate the errors of closure of the traverse and the adjusted coordinates of the traverse stations.

Fig. 16.24 Traverse sketch

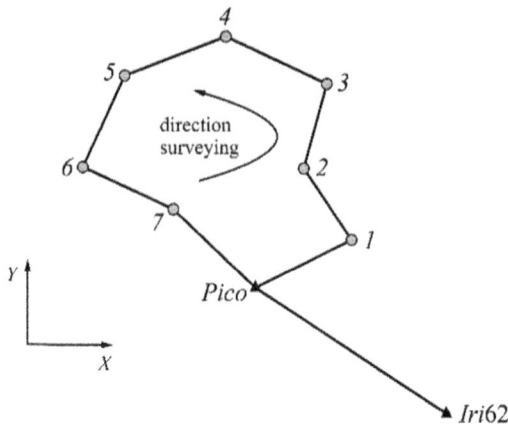

Table 16.40 Field measurements

Station	Back-Fore sights	Horizontal clockwise angle	Horizontal distance [m]
Pico	Iri62,1	300° 59′51″	722.011
1	Pico,2	82° 57′53″	560.600
2	1,3	229° 28′50″	572.694
3	2,3	98° 50′12″	746.715
4	3,5	134° 51′31″	725.027
5	4,6	135° 30′09″	670.221
6	5,7	88° 22′12″	677.849
7	6,Pico	201° 17′22″	743.166
Pico	7, Iri62	167° 41′42″	

Table 16.41 UTM coordinates of points Pico and Iri62

Point	E [m]	N [m]
Pico	220,633.662	7,542,850.539
Iri62	221,937.230	7,542,025.781

SIRGAS20000 (UTM zone 23)

16.14 Review Questions

1. Discuss what a map projection is and its importance in Civil Engineering.
2. Explain what a cylindrical map projection is.
3. Explain the meaning of cylindrical, transverse and secant map projection.
4. Explain what a map projection scale factor is and where the minimum and the maximum UTM scale factors occur in the UTM zones.
5. Explain how the combined factor is calculated for the UTM projection.

6. Explain the calculation steps required to convert the ground distance to the UTM distance.
7. Explain the calculation steps required to convert the UTM distance to the ground distance.
8. Explain what convergence of the meridians is and when it is necessary to use it in Civil Engineering projects.
9. Explain what the arc-to-chord reduction is and when you would recommend using it in a Civil Engineering project.
10. Explain what is the difference between grid and ground coordinates and when it is important to consider the transformation from grid to ground coordinates in a Civil Engineering project.
11. Discuss which grid to ground transformation methods can be used in Civil Engineering and which are the most common.
12. Discuss how to handle grid and ground coordinates in a Civil Engineering project using a total station and a GNSS instrument.

References

A Mira (2000). *Sistema de coordenadas planas LTM aplicada em projetos rodoviários*. Ano XXI. N° 159. Criciúma, SC, Brasil.

Billings, S. (2013a). *Low Distortion Projections: Part 1: Ground versus Grid*. The American Surveyor, v. 10, n. 9.

Billings, S. (2013b). *Low Distortion Projections: Part 2: Ground versus Grid*. The American Surveyor, v. 10, n. 10.

Bugayevskiy, L. M, Snyder, J. P., (2000). *Map Projections—A Reference Manual*. Taylor & Francis Ltd. London, England.

BURKHOLDER, E. F. (2012). *Contrasting a Low Distortion Projection (LDP) With the Global Spatial Data Model (GSDM)*. New Mexico State University. New Mexico, USA.

Deakin, R.E., (2006). *Traverse Computation on the UTM Projection for Surveys of Limited Extent*. School of Mathematical and Geospatial Sciences, RMIT University, March.

Denis, M. L. (2015). *Ground Truth: Design and documentation of Low distortion Projections for Surveying and GIS*. Professional Land Surveyors of Oregon, 2015 Annual conference. Sedona, USA.

IBGE (1995). *Tabelas para cálculos no Sistema de Projeção Universal Transverso de Mercator—UTM*. 2a edição, Rio de Janeiro.

JEKELI, C. (2016). *Geometric reference systems in geodesy*, Division of Geodesy and Geospatial Science. School of Earth Sciences. Ohio State University. 214p.

Ordnance Survey. (1998). *The ellipsoid and the Transverse Mercator projection*. Geodetic information N0 1, V 2.2, England.

Pickford, J., Gibbings, P. (2009). *Local ground-based plane coordinates*. Spatial Science Queensland. Brisbane, Australia.

Pacileo Neto, N.; Tostes, F.; Idoeta, I. V. (2003). *Sistema TM—Sistema topográfico local*. EPUSP. São Paulo.

Panigrahi, N. (2014). *Computing in Geographic Information Systems*. CRC Press. Taylor & Francis Group. ISBN 13: 978-1-4822-2316-3 Boca Raton, USA.

Rapp, R. H. (1991). *Geometric Geodesy—Part I.* The Ohio State University. Department of Geodetic Science and Surveying, Columbus, Ohio, USA.

Snyder, J.P., (1987). *Map Projections—*A Working Manual, U.S. Geological Survey Professional Paper: 1395.

Yang, Q. H., Snyder, J. P, Tobler, W. R (2000). *Map Projection Transformation—Principles and Applications.* Taylor & Francis Inc, UK.

Chapter 17
Global Navigation Satellite System: GNSS

Irineu da Silva, Paulo C. L. Segantine,
and Guilherme Poleszuk dos Santos Rosa

17.1 Introduction

As explained in the previous chapters, the determination of the coordinates (positioning) of a point in space is generally based on terrestrial measurements (horizontal directions, vertical angles and distances) carried out using surveying instruments installed on the Earth's surface and oriented to a frame of control points linked to a predefined local or geodetic reference system. Exceptions to these prerogatives are the determination of coordinates using airborne photogrammetry (Chap. 19), airborne laser scanning (Chap. 18) and artificial satellite positioning, as discussed in this chapter.

As the reader will see in this chapter, the advent of point positioning using artificial satellites has transformed countless technologies in human activities, such as social networks, transport of people and goods, terrestrial and aerial navigation, agriculture, engineering and many others. Although initially developed for military purposes, it quickly became available to civil society, and several positioning systems were developed by different countries, creating what is known as *Global Navigation Satellite Systems* (GNSS). As a result, position and time information are available worldwide 24 h a day, anywhere on the Earth's surface and in all atmospheric conditions. This chapter discusses the details of these systems, the methods for determining coordinates (positioning), and the geodetic and topographic surveying methods currently used for Geomatics applied to Civil engineering.

I. da Silva (✉) · P. C. L. Segantine
São Carlos School of Engineering, University of São Paulo, São Carlos, SP, Brazil
e-mail: irineu@sc.usp.br; pclsegantine@usp.br

G. Poleszuk dos Santos Rosa
Cartographic Engineer, Researcher at the Faculty of Science and Technology of the São Paulo
State University (UNESP), São Paulo, Brazil

17.2   GNSS Structure

The determination of point coordinates in space by tracking signals from artificial satellites began shortly after the launch of the first satellites in the 1960s. Soon after, following these launches, the US Navy realized that by accurately determining a satellite's position relative to a terrestrial reference system, they could calculate the position of a ground-based tracking station. This led to the creation of the *Navy Navigational Satellite System* (NNSS), commonly known as the TRANSIT system, which became operational in 1964. The system was the first of its kind to provide accurate terrestrial navigation with an accuracy of about 400 m.

The encouraging results of this first system prompted the US Air Force to propose a new system in 1973 called *NAVSTAR Global Positioning System* (NAVigation Satellite Timing And Ranging), commonly known as GPS. This system was immediately adopted by the civilian and military communities, making GPS synonymous with satellite-based positioning.

Parallel to the development of GPS, the former Soviet Union began testing a new satellite positioning system in 1982, called the *Globalnaya Navigatsionnaya Sputnikovaya Sistema* (GLONASS). Initially, its use was restricted to the USSR and its allies for military and strategic purposes. However, in 2000, GLONASS became an open system available to the global civilian community.

Given the success of these systems and to reduce its dependence on them, the European Community also decided to develop a global positioning system for its Member States, known as the GALILEO system. By August 2023, GALILEO had 28 satellites in orbit, consisting of four In-Orbit Validation (IOV) satellites and 24 Full Operational Capability (FOC) satellites.

Several other satellite positioning systems are available to different communities. These include the BeiDou system, developed by China and made available to the global civilian community; the Japanese Quasi-Zenith Satellite System (QZSS), serving Japan and Oceania; the Indian Regional Navigation Satellite System (IRNSS), under development by India; and the SLR and DORIS systems.

The term GNSS, therefore, refers to a range of satellite systems used to determine the geographical position, distances, directions, velocity, and local time of a receiver located at or near any point on the Earth's surface.

As shown in Fig. 17.1, for the GNSS to produce geospatial data that meets the needs of Civil Engineering and Geosciences, it comprises three main segments that ensure the quality of GNSS surveying data collection. They are:

Space segment: refers to the constellation of artificial satellites orbiting the Earth along predefined orbital planes at various altitudes that determine the spatial coordinates of each satellite and its trajectory relative to a specific reference system. Each satellite is equipped with time processors for accurate timekeeping, electromagnetic wave emitters for transmitting signals and code generators for generating the codes used in GNSS signals, among other GNSS signals. These signals are transmitted by the satellites' antennas to the user's tracking receiver antennas.

Control segment: responsible for managing and maintaining the constellation of satellites to ensure accurate positioning and timing information. It includes the

Operational Control Segment (OCS), which comprises the Master Control Station (MCS), the Monitor Station (MS), and the Ground Antennas (GA). The OCS's main operational tasks are to track the satellites for orbit and clock determination and prediction, time synchronisation, and to upload the navigation data message to the satellites. The data collected from each satellite is sent to a Master Station. It calculates its positions and the parameter corrections generated for the satellites and retransmits them via the satellites themselves to the user segment.

User segment: refers to all users who, using an antenna and a receiver with a signal processor, can capture the signals emitted by the GNSS satellites and utilize them for various applications. This segment includes all devices, systems, and applications that receive and process GNSS signals to determine the user's position, velocity, and time. Typically, user antennas and receivers are passive devices that only receive satellite signals. They never transmit signals to the satellites. However, there are specific cases where the received signals can be retransmitted to other receiver antennas, such as in surveying using differential positioning methods like Real-Time Kinematic[1] (RTK) mode. Another example is in the case of active stations, such as *Continuously Operating Reference Stations* (CORS), which are fixed satellite tracking stations that continuously provide tracked data to authorised users.

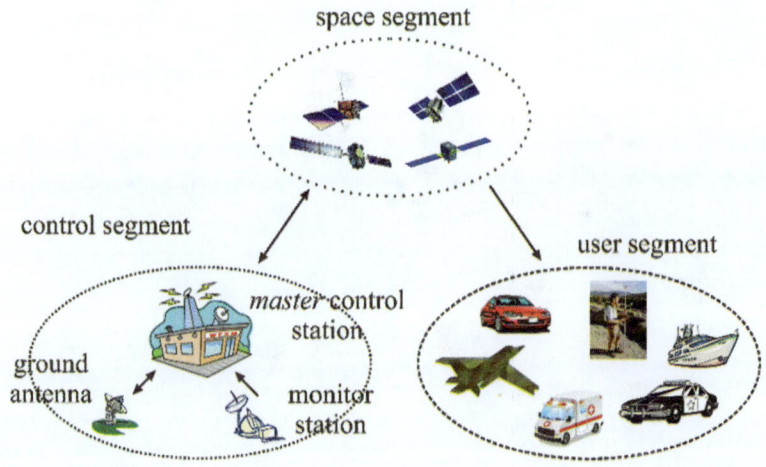

Fig. 17.1 GNSS segments

17.3 Composition and Characteristics of GNSS System

As a general guide for the reader, the following are the main characteristics of the positioning systems that comprise GNSS.

---

[1] See Sect. 17.6.3.2 for more information.

17.3.1   *Global Positioning System - GPS*

The GPS is managed by the Joint Program Office (JPO), a component of the Space and Missile Centre at El Segundo, California, under the US government. It was originally designed to provide a minimum constellation of 24 satellites (maximum 32 satellites). However, in recent years, the number of operational satellites has consistently exceeded this minimum. The satellites are placed in inclined planes describing elliptical orbits with an orbital period of 11 h 57 min 2.0 s (sideral time), at an altitude of approximately 20,200 km above the Earth's surface. These satellites are divided into six orbital planes, each with an inclination of 55° about the equatorial plane, covering the latitudes between 80° N and 80° S, as shown in Fig. 17.2. This configuration ensures that at least four satellites are always visible above the horizon of any GPS antenna located on the Earth's surface at any given time.

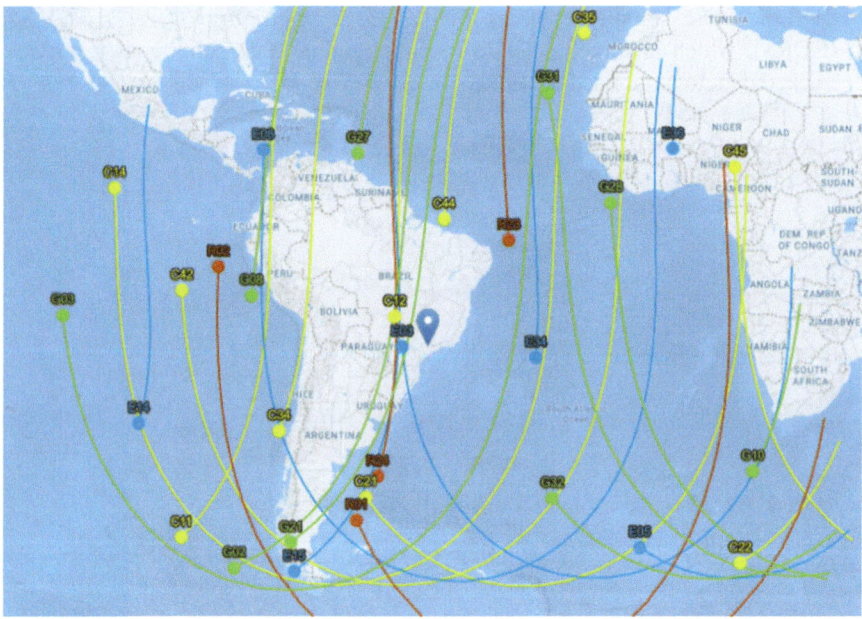

Fig. 17.2 GNSS satellite constellation. (Source: Trimble Planning Online. Location: https://www. gnssplanning.com/#/maps). G: GPS, R: GLONASS, E: Galileo, C: BeiDou, Time: 2023-10-09, UTC +00:00

There are several classes (types) of GPS satellites: Block I, Block II, Block IIA, Block IIR, Block IIR-M, Block IIF and Block III satellites. Detailed information on launch dates, orbital position and operational periods can be found on the USNO website at https://www.navcen.uscg.gov/nanu-abbreviations-and-descriptions.

The signals transmitted by GPS satellites contain pseudo-random codes and information within the transmitted navigation messages, including the Keplerian

elements that allow the calculation of the satellites´ positions at the time of transmission, the ionospheric correction coefficients, satellite clock correction coefficients and the "health" status of the satellite, among other data. Precise coordinates
of GPS satellites at specific times can also be obtained through *precise GPS satellite
ephemerides*, provided by the International GNSS Service (https://igs.org) or
authorised governmental and commercial agencies a few hours or days after the
satellite signals have been tracked. For more detailed information on this subject,
readers are encouraged to visit https://navcen.uscg.gov/gps-precise-ephemeris-
information.

The radio signals transmitted by GPS satellites are referred to as 'carriers'
because they convey the navigation information that GPS receivers use to determine
their position, velocity, and time. GPS satellites from Block IIF onward transmit
three key carrier frequencies for geodetic positioning: L_1, L_2 and L_5. These carriers
and their signals are synthesized from a standard frequency of 10.230 MHz.[2] An
oscillator generates the signal components based on this frequency, as illustrated in
Fig. 17.3.

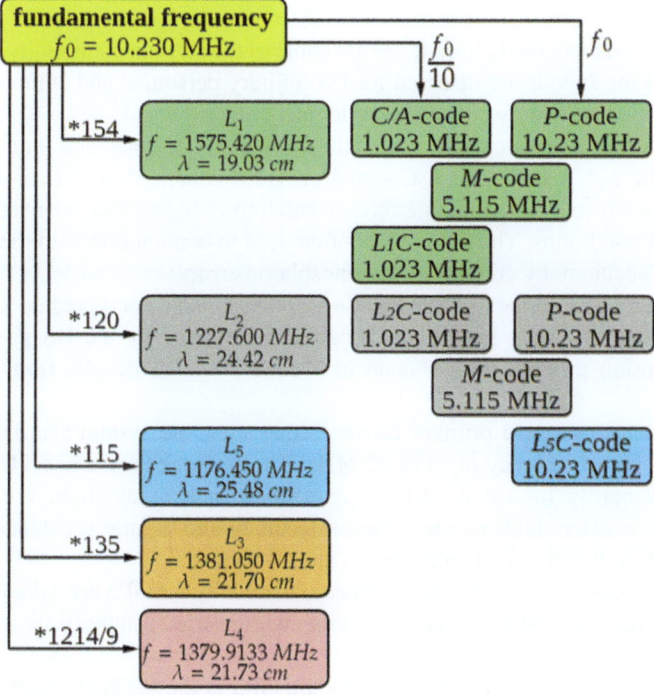

Fig. 17.3 Basic structure of GPS signals

---

[2]The actual internal reference of the satellites is 10.22999999543 MHz to compensate for
relativistic effects that make observers on the Earth perceive a different time reference with respect
to the transmitters in orbit.

The carrier frequency L_1 is modulated with four types of pseudo-random codes: two at a frequency of 1.023 MHz, called C/A-code and L_1C-code; one at a frequency of 5.115 MHz, called M (military) code; and the fourth at a frequency of 10.23 MHz, called P-code. For the reader to better understand the meaning of wave modulation, the effect of modulating the carrier wave with the P-code is shown graphically in Fig. 17.4.

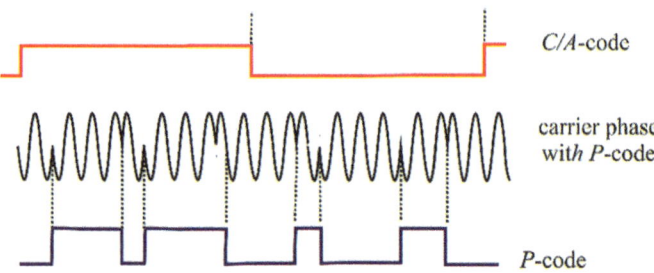

Fig. 17.4 Effect of modulation with the P-code

The C/A-code and the L_1C-code are available to the civilian community, while the M-code and the P-code are reserved for US military personnel and authorised users only. All codes are binary sequences of logical values 0 and 1.[3]

The L_2 carrier frequency also carries the P-code and M-code, and on newer GPS satellites, the L_2C signal. The L_2C signal provides enhanced signal robustness by reducing the effects of external interference and improving measurement accuracy in challenging conditions. The L_2 signal is often used in conjunction with the L_1 signal to enhance accuracy by correcting for ionospheric errors.

By military requirements, the P-code is intentionally encrypted to derive the Y-code, which can only be read by receivers authorised by the US government. This encryption process from P-code to Y-code is called the AS (*anti-spoofing*) effect.

In addition to the two primary carrier frequencies, the system also has a third, called L_5, with a frequency of 1176.45 MHz. The L_5 carrier is the latest GPS signal designed primarily for safety-of-life applications such as aviation. For geodetic positioning, it offers higher power, better accuracy and greater resistance to interference and multipath effects than the L_1 and L_2 signals.

For the civilian community, the services available by the GPS are called *Standard Positioning Services* (SPS), while the services with restricted authorisation are called *Precise Positioning Services* (PPS).

The geodetic reference system used by the GPS is the WGS84 Geodetic Reference System, as described in Sect. 4.3.1.2.

---

[3]Additional information can be found in Interface Control Document IS-GPS-2000 M (https://www.gps.gov/technical/icwg/IS-GPS-200M.pdf)

17.3.2   GLONASS

The Russian GLONASS system, or simply GLONASS, has been operational since 1982 when its first satellite was launched. It was designed to provide a minimum constellation of 21 satellites to transmit signals continuously in two frequency bands, using the same measurement principles as GPS. The GLONASS satellites are distributed in three orbital planes with an inclination of 64.8° about the equatorial plane and orbit at about 19,100 km with a period of approximately 11h 15' 44" (solar time). More information can be reached on the website: https://glonass-iac.ru/en/about_glonass/.

The open GLONASS access signals (for FDMA Signals Centre Frequency Values are:

Frequency band L_1 OF $\rightarrow f_1 = 1602$ MHz
Frequency band L_2 OF $\rightarrow f_2 = 1246$ MHz
Frequency band L_1 OC $\rightarrow f_1 = 1600$ MHz
Frequency band L_2 OC $\rightarrow f_2 = 1246$ MHz

The restricted GLONASS signals are:

Frequency band L_1 SF $\rightarrow f_1 = 1592$ MHz
Frequency band L_2 SF $\rightarrow f_2 = 1237$ MHz
Frequency band L_1 SC $\rightarrow f_1 = 1600$ MHz
Frequency band L_2 SC $\rightarrow f_2 = 1248$ MHz

The frequencies transmitted by the GPS and GLONASS satellites are remarkably close. It is, therefore, possible to use a GNSS antenna and amplifier in the same equipment, allowing the user to benefit from a single receiver to track signals from both systems.

GLONASS provides two levels of accuracy: a high accuracy signal—the *Channel of High Accuracy* (CHA)—and a standard accuracy signal—the *Channel of Standard Accuracy* (CSA). Like GPS, the CHA signal is the prerogative of military users, while the CSA signal is available to civilian users. On the other hand, unlike GPS, the GLONASS system has never had selective signal degradation.

The coordinate system used by GLONASS is the PZ-90 geodetic reference system, based on the PZ90 ellipsoid. See Table 17.1.

Modernisation is also taking place in the GLONASS program. The new series of satellites, GLONASS-M, will have improved onboard clock stability and a civilian code (L_2C) available in the (L_2) frequency band. For further information about the GLONASS system, the reader can refer to the following website: https://www.glonass-iac.ru/en.

Table 17.1 Parameters of the ellipsoid PZ90

Ellipsoid	Semi-major axis [m]	Flattening
PZ90	6,378,136.000	1/298.25784

https://www.unoosa.org/pdf/icg/2014/wg/PZ-90.11_2014.pdf

17.3.3   GALILEO

The European Global Navigation Satellite System, or GALILEO, is fully compatible
with GPS and GLONASS, although it is still in the constellation complementation
phase. When fully deployed, the project is expected to have 24 operational satellites
and six spare satellites placed in three orbital planes with an inclination angle of 56°
about the equatorial plane at altitudes of 23,257 km and 23,222 km and nominal
circular orbits with a semi-major axis of about 29,600 km. The orbital period is
14h 04'45", with repeated passes over the same point on the Earth every 10 days. This
configuration ensures good coverage for all European countries, North Africa, and
regions up to a latitude of up to 75°.

GALILEO relies on a geocentric Cartesian reference frame as defined by the
GALILEO Terrestrial Reference Frame (GTRF), related to the International Terres-
trial Reference Frame (ITRF), which has been established by the International Earth
Rotation Service (IERS). The GTRF is specified to differ from the latest version of
ITRF by more than 3 cm.

The GALILEO system transmits in three frequency bands: (E_5), (E_6), and (E_1).
For this purpose, four carrier frequencies are used: (E_{5A}), (E_{5B}), (E_6), and (E_1), as
shown in Table 17.2. Note that the carrier frequencies (E_{5A}) and (E_1) have the same
frequencies as the GPS carrier frequencies (L_1) and (L_5).

Table 17.2 Carrier frequencies of the GALILEO system

Carrier wave	Average frequency [MHz]
E_5	1191.795
E_{5A}	1176.450
E_{5B}	1207.140
$E6$	1278.750
E_1	1575.420

Europe has adopted a service-oriented approach to the design of GALILEO.
During the definition phase, user requirements were categorized into five service
levels.

1. *Open Access Service* (OS): available worldwide, free of charge and independent
   of other systems.
2. *Commercial Access Service* (CS): the GALILEO concessionaire controls access
   to the data messages and encrypted ranging data.
3. *Safety-of-Life Service* (SoL): it relies on the same signals as the OS but adds
   integrity information to provide a service guarantee to users. The SoL has been
   designed to comply with different standards in the aeronautical, maritime and
   railway domains to maximise the benefit to the user community.
4. *Public Regulated Service* (PRS): provides position and timing to specific users
   requiring a high continuity of service, with controlled access. Two PRS navigation signals with encrypted ranging codes and data will be available.

5. *Search and Rescue Service* (SAR): this service has been defined and built by Russia, USA, France and Canada to provide a means for humanitarian SAR operations worldwide. In the future, GALILEO satellites will also detect emergency signals at 406 MHz and forward the emergency message to the SAR ground segment in the SAR downlink frequency band. Galileo is contributing to the international Cospas–Sarsat system for search and rescue. More information can be found on the website: https://www.esa.int/Applications/Navigation/Galileo_and_EGNOS.

The geodetic reference system of GALILEO is the *GALILEO Terrestrial Reference Frame* (GTRF), which is related to the *International Terrestrial Reference Frame (ITRF)*. More information can be obtained in Sect. 4.3.1.1 and on the website https://itrf.ign.fr.

17.3.4 BeiDou (Compass)

The Chinese *Compass Navigation Satellite System* (CNSS), or *BeiDou*-2, is a partially operational global positioning system that provides positioning information compatible with the GPS, GLONASS and GALILEO systems. The satellite constellation includes geostationary satellites (GEO—at an altitude of 35,786 km) and orbital satellites (MEO, operating at an altitude of 21,528 km; and IGSO, operating at an altitude of 35,786 km), both at an inclination angle of 55° to the equatorial plane. This system uses several frequency bands. For more information, the reader should refer to http://en.beidou.gov.cn/SYSTEMS/System/. When writing this chapter, 47 satellites were operating (refer to: http://www.csno-tarc.cn/en/system/constellation).

The geodetic reference system used is the *China Geodetic Coordinate System 2000* (CGCS2000), which is related to the ITRS.

17.3.5 QZSS

The *Japanese Quasi-Zenith Satellite System (QZSS)* is a regional positioning system (East Asia and Oceania) provided by Japan since 2010, designed to provide positioning services in urban canyons and mountainous environments. It consists of a constellation of four near-zenith satellites orbiting at an altitude of approximately 32,000 km and with an elevation angle of 70°, allowing them to be positioned above the horizon for more than 12 h per day. Table 17.3 shows the carrier frequencies emitted by the system and their frequencies.

The L_1C/A, L_1C, L_2C, and L_5 carrier frequencies are compatible with GNSS receivers, making the QZSS valuable for the Japan region and its neighbouring countries. More information on this system can be found in specific literature.

Table 17.3 QZSS frequency bands

Carrier wave	Average frequency [MHz]
L_1	1575.42
L_2C	1227.60
E6/LEX	1278.75
L_5	1176.45

Although *QZSS* is primarily an augmentation and complement to GPS, it also has the potential to operate in autonomous mode, providing a regional service but with reduced positioning performance. Nevertheless, the system may soon be expanded to a fully operational, high-performance Japanese regional system. The QZSS is based on the Japanese geodetic system, which uses the GRS80 ellipsoid as a reference.

17.3.6   NaviC

NaviC refers to the Navigation Indian Constellation satellite system developed by the Government of India to cover the whole of India's landmass and up to 1500 km from its boundaries. It is designed with a constellation of seven satellites, three geostationary, at 32.5°, 83° and 131.5° longitudes. The remaining four satellites describe predefined orbits, two with an orbital plane inclined at 55° and two with an orbital plane at 111.75° concerning the equatorial plane. This system provides two types of service: Standard Positioning Service (SPS) to all users, and Restricted Service (RS) only to the authorised users, both based on the L_5 and S carrier frequencies compatible with the GPS and GALILEO systems.

17.3.7   SBAS

Satellite-Based Augmentation Systems (SBAS) use geostationary satellites to transmit corrections that enhance the accuracy and reliability of GNSS positioning. They achieve this by correcting signal measurement errors and providing integrity information, allowing user equipment to obtain a highly reliable position estimate. The system relies on a network of CORS reference stations connected to a processing centre, which calculates the corrections and transmits them to GNSS receivers via the SBAS satellite telecommunications link.

The CORS network continuously records GNSS observations and determines correction parameters sent to geostationary satellites and, from there, to users' receivers. As discussed in the following sections, these corrections make achieving GNSS positioning accuracies of a few metres possible, which is helpful for many ground positioning applications.

Several countries have implemented their own SBAS, such as:

- European Union: EGNOS system (*European Geostationary Navigation Overlay Service*)
- USA: WAAS (*Wide Area Augmentation System*), LASS (*Local Area GPS Enhancement*), GBAS (*Ground-based Augmentation System*), WAGE (*Wide Area GPS Enhancement*)
- Canada: CDGPS (*Canada-Wide Differential* GPS)
- Japan: MSAS (*Japan's Space-Based Augmentation System*)
- India: GAGAN system (*Indian GPS Aided Geo Augmented Navigation*)
- China: SNAS (*Satellite Navigation Augmentation System*)
- South Korea: KASS (*Korean Augmentation Satellite System GPS*)
- Russia: SDCM (*System for Differential Correction and Monitoring*)

17.4 Advantages of GNSS for Geomatics

Due to its ease of use and the quality of the results, the technology of determining spatial coordinates by tracking signals from artificial satellites has quickly become one of the main geospatial data collection tools for Civil Engineering projects. This is mainly due to its advantages over other surveying technologies, as described below.

- The stations on which the GNSS antennas are installed do not need to be visible from each other. One of the primary advantages of not requiring visibility between antennas is that the baseline vectors can be extended by hundreds of kilometres.
- Measurements can be taken under all weather conditions 24 h a day.
- The antennas that receive the signals broadcast from the satellites collect the data independently. In other words, a GNSS antenna does not need to be "aware" that other antennas are collecting data at the exact time.
- The system ensures global coverage for the entire Earth's surface or close to it.
- The system allows precision positioning, speed, direction, and time accuracy.
- The system allows for determining the position of points in real-time, which facilitates its use for automating positioning procedures in engineering, such as machine guidance, precision farming, and others.
- User antennas and receivers are generally small, light, and easy to use.

On the other hand, some disadvantages can also be highlighted, such as:

- GNSS does not work correctly when the ground antennas are installed inside buildings, underground locations, under trees with dense foliage, near places with robust radio transmission, or in densely built-up areas, such as urban canyons and others.
- It depends on the quality of the antennas and receivers.
- It also depends on appropriate positioning methods.

Furthermore, it is also important to emphasise that the effective use of GNSS technology requires the user to have a basic understanding of Cartography, Geodesy and the operation of positioning systems.

17.5 Principles of Point Positioning Using GNSS Technology

The determination of the geospatial position of a point through the reception of signals emitted by artificial satellites is achieved by using the principle of *spatial multilateration*,[4] using the satellite coordinates as reference points. In principle, to determine the Three-Dimensional Cartesian coordinates of a GNSS antenna r installed on the Earth's surface, consider the distances d_1, d_2, and d_3 from satellites "Sat1", "Sat2" and "Sat3", respectively, to the GNSS antenna installed over point (P), as shown in Fig. 17.5.

Fig. 17.5 Spatial trilateration in GNSS point positioning

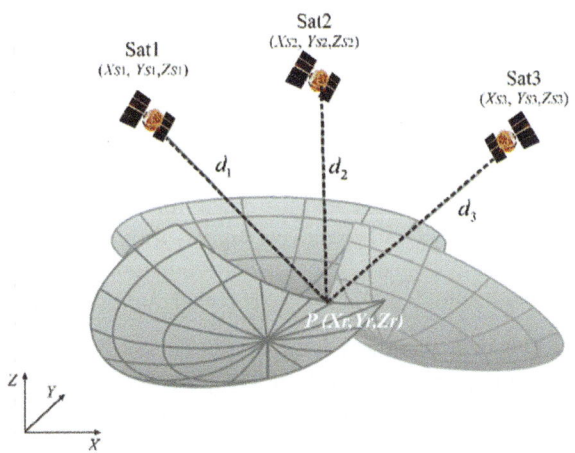

The distances d_1, d_2, and d_3 represent the (unobstructed) "lines of sight" between the satellite antennas and the ground antenna. Thus, knowing the geocentric Cartesian coordinates (X_s, Y_s, Z_s) of the satellites "Sat1", "Sat2", and "Sat3" and the distances d_1, d_2 and d_3, it is, in principle, possible to determine the geocentric Cartesian coordinates (X_r, Y_r, Z_r) of the ground antenna by spatial trilateration, as shown in Fig. 17.5.

Once the satellite positions are known, the decisive factor in determining the Three-Dimensional Cartesian coordinates of the ground antenna installed at point (P) is the precise measurement of distances d_1, d_2, and d_3, carried out using electronic distance measurement methods similar to those used for electronic distance metres, as described in Sect. 8.3.2.

However, there are some significant differences between electronic distance measurement using an EDM instrument and electronic distance measurement using GNSS technology. Firstly, in addition to measuring the phase difference of

---

[4]A method used to determine the location of a point by measuring its distances from multiple known points.

the carrier wave, measurement with GNSS technology also uses code correlation. Another important difference is that it is not possible to transmit a wave from the satellite (or receiver) and have it reflected by the ground antenna (or satellite). In this case, the solution is to consider only the outward path of the electromagnetic wave emitted by the satellite. To do this, it is necessary to replicate the electromagnetic wave and its codes inside the receiver, resulting in one signal being emitted by the satellite and another being generated internally in the receiver.

The electromagnetic waves emitted by the satellites are called *carrier waves*. To calculate the distance between the satellite and the ground antenna, the signals are modulated with information about the satellite (for example, the satellite's position about the reference system) and with some codes, which vary for each GNSS positioning system.

Similar to EDM, distances are measured with GNSS technology based on the propagation time of electromagnetic waves or the phase difference. However, as the signals are generated in different devices, codes are used, and the satellites and receivers must have clocks synchronised in a standard time scale, called *"GNSS time"*.

Satellite clocks are based on atomic frequency standards, and receiver clocks are generally quartz. Therefore, in most cases, it is necessary to include the time variable in the system of equations to determine the geocentric Cartesian coordinates (X_r, Y_r, Z_r) of the ground antenna. Therefore, a system of equations is obtained with four unknowns, which are the geocentric Cartesian coordinates (X_r, Y_r, Z_r) of the ground antenna and the component Δt related to the error between the GNSS time and the time recorded by the receiver's clock. In this way, the position of the ground

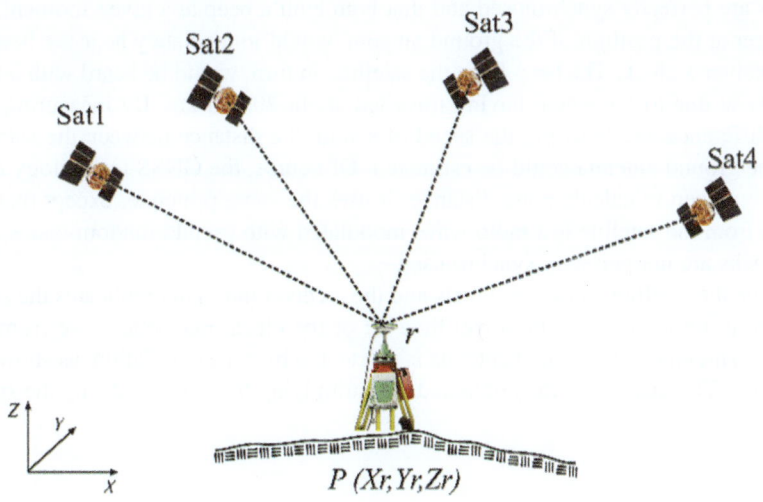

Fig. 17.6 Minimum configuration for determining the geocentric Cartesian coordinates of the GNSS ground antenna

antenna can only be determined by simultaneously tracking signals from at least four satellites, as shown in Fig. 17.6.

After solving the system of equations, the coordinates of the ground antenna are obtained, referring to the same coordinate system as the GNSS geodetic reference system. The following sections provide theoretical details on the determination of these coordinates.

17.5.1   Calculating the Distances Between the Ground Antenna and the Satellites

As mentioned, calculating the distances between the ground antenna and the satellites is based on the propagation time of the electromagnetic wave or the phase difference of the carrier wave. When the distance is determined by the propagation time of the wave, the result of the calculation is called *pseudorange*; otherwise, the distance is said to be determined by *carrier phase ranging*.

The technical details of each distance measurement method are given below.

17.5.1.1   Pseudorange (Code)

The geometric distance between the satellite and the ground antenna is called the pseudorange. This distance is measured from the time difference between the pulses of the *C/A* and *P* codes generated by the satellite and replicated in the receiver. To better understand the measurement process, imagine that the satellite's and receiver's clocks are perfectly synchronised and that both emit a beep at a given moment. An observer at the position of the ground antenna would immediately hear the beep of the receiver's clock. The beep from the satellite, in turn, would be heard with a time difference due to the signal having travelled about 20,000 km. By measuring this time difference and knowing the speed of sound, the distance between the satellite and the ground antenna could be estimated. Of course, the GNSS technology does not use a beep to calculate the distance. It uses the same principle, except that the signal from the satellite is a radio wave modulated with pseudo-random codes, and the clocks are not perfectly synchronised.

Since the satellite emits the signals and the receiver internally replicates the same signals at the same time, the travel time Δt of the electromagnetic wave from the satellite antenna to the ground antenna is calculated by code correlation, as shown in Fig. 17.7. The distance is then obtained by multiplying the value of Δt by the speed of light in a vacuum. Because it is "contaminated" by the clock synchronisation difference error, it is considered to be a *"pseudorange"*, i.e., a *"false distance"*.

Thus, according to Fig. 17.7.

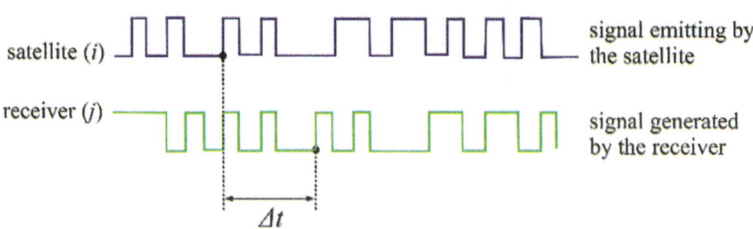

satellite (*i*)   signal emitting by
                                                                            the satellite

receiver (*j*)   signal generated
                                                                            by the receiver

Δt

Fig. 17.7 C/A-code correlation for pseudorange determination

$$\rho = c^* \Delta t \tag{17.1}$$

where

ρ = *pseudorange* between the satellite antenna and the receiver antenna at GNSS system times

c = speed of light in a vacuum

Δt = time travel of the electromagnetic wave from the satellite antenna to the ground antenna in GNSS time

The satellite and receiver clocks have a time difference in relation to the GNSS time t_{GNSS}. The satellites have four atomic particle clocks, with a time difference Δt^s of a few nanoseconds per day. This time difference, also known as the satellite clock error, is determined each time the satellite passes over a ground control station and is retransmitted to the receivers along with the navigation messages. In this way, the time t^s at which the signal was transmitted by the satellite is known precisely on the satellite's time scale t_{GNSS}. Thus, knowing the time t_r of the arrival of the signal at the ground antenna, in the receiver's time and the error Δt_r of the receiver's clock concerning the satellite time t_{GNSS}, the following equations are obtained:

$$\Delta tr = t_{GNSS} - tr \tag{17.2}$$

$$\Delta t^s = t_{GNSS} - t^s \tag{17.3}$$

From this, it can be deduced that the signal propagation time interval, in GNSS time, is given by Eq. (17.4).

$$\Delta t_r^s = (tr - t^s) + (\Delta tr - \Delta t^s) \tag{17.4}$$

Thus, the pseudorange Eq. (17.1) becomes

$$\rho_r^s = c * \Delta t_r^s + c * \Delta t^s - c * \Delta tr \tag{17.5}$$

As the signal emitted by the satellite travels in space, other factors must also be considered when calculating pseudoranges, such as the influence of atmospheric refraction (ionosphere and troposphere), multipath,[5] orbital errors, noise errors in the receiver, noise errors in the satellites and others. Thus, given the geocentric Cartesian coordinates (X_s, Y_s, Z_s) of the satellites at instant i and the geocentric Cartesian coordinates (X_r, Y_r, Z_r) of the ground antenna to be calculated, the final pseudorange equation is obtained as follows:

$$\left(\rho_r^s\right)_i = \sqrt{\left(X_i^s - X_r\right)^2 + \left(Y_i^s - Y_r\right)^2 + \left(Z_i^s - Z_r\right)^2} + c * (\Delta t^s - \Delta tr) + I_r^s$$
$$+ T_r^s + m_r^s + \varepsilon_r^s + \varepsilon_r + \varepsilon^s \tag{17.6}$$

where

$\left(\rho_r^s\right)_i$ = pseudorange measured at time (i)
I_r^s = ionospheric delay
T_r^s = tropospheric delay
m_r^s = multipath effect in the GNSS antenna
$\varepsilon_r^s, \ \varepsilon_r, \ \varepsilon^s$ = orbital errors and noise errors on satellites and receivers

The scientific details and consequences of these delays (ionospheric and tropospheric) and errors, although essential for determining the coordinates of the ground antenna, are not discussed in this book. At the current stage of GNSS technology, these effects are well-known and are corrected during GNSS measurements and data processing.

Assuming that the values of t^s, t_r, Δt^s, ionosphere delay, troposphere delay, satellite and receiver noise errors are known, and multipath can be avoided, the unknowns in Eq. (17.6) are the coordinates (X_r, Y_r, Z_r) of the ground antenna and the correction of the receiver clock error Δt_r concerning GNSS time. These unknowns can be calculated by observing at least four satellites simultaneously. The Cartesian spatial coordinates thus determined are then transformed into geodetic coordinates (latitude, longitude and ellipsoidal height) and, if necessary, into coordinates on the projection plane, as presented in Chap. 16.

17.5.1.2   Carrier Phase Ranging

As mentioned, GNSS carrier phase distance measurement is similar to total station carrier phase distance measurement. The main difference is that with a total station, the EDM instrument emits two waves: a reference wave and a measurement wave, whereas with GNSS, the measurement wave is generated in the satellite, and the reference wave is generated in the receiver.

---

[5]Multipath is the effect resulting from the reception of GNSS signals by the ground antenna that do not come directly from satellites but are reflected or refracted by objects in the surroundings of that antenna.

In this case, distance measurement is based on measuring the phase difference between the electromagnetic wave emitted by the satellite (s) and the reference wave

Fig. 17.8 Principle of carrier phase ranging

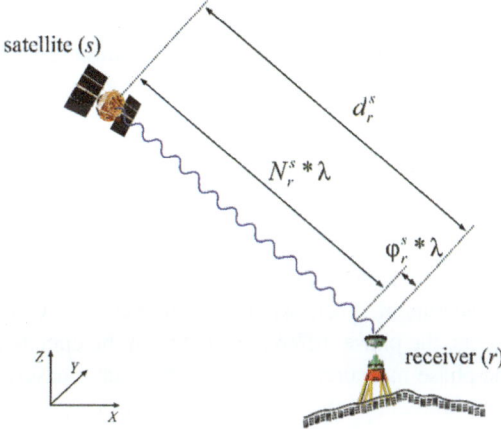

continuously reproduced in the receiver (r) and counting the number N_r^s of integer waves between the satellite and the ground antenna. As in measurements with an EDM instrument, this number of integer waves is called "system ambiguity". The sum of the two variables gives the measured distance, as shown in Fig. 17.8.

The waves used for this measurement are the carrier waves mentioned in the previous sections, which in the case of GPS are the L_1, L_2 and L_5 carrier frequencies or a combination of them such as wide lane, narrow lane, ionospheric free, geometry free and others.

According to Fig. 17.8, the distance between the satellite antenna and the ground antenna is given by Eq. (17.7).

$$\left(d_r^s\right)_i = N_r^s * \lambda + \left(\varphi_r^s\right)_i * \lambda \tag{17.7}$$

where

$\left(d_r^s\right)_i =$ distance between the satellite antenna s and the receiving GNSS antenna r at time i

$N_r^s =$ ambiguity integer cycles between the receiver r and the satellite s

$\left(\varphi_r^s\right)_i =$ the measured phase difference between the wave emitted by the satellite s and the wave replicated inside the receiver r at an instant i

$\lambda =$ wavelength of the carrier frequency

Equation (17.7) is only valid for the ideal case of wave propagation in a vacuum and with perfectly synchronised satellite and receiver clocks. If this is not the case, there will be ionospheric and tropospheric delays and other errors in the system. Thus, the phase difference value measured at the instant i is given as follows:

$$\left(\varphi_r^s\right)_i = \frac{1}{\lambda} * \sqrt{\left(X_i^s - X_r\right)^2 + \left(Y_i^s - Y_r\right)^2 + \left(Z_i^s - Z_r\right)^2} + N_r^s$$
$$+\frac{c}{\lambda} * \left(\Delta t^s - \Delta tr\right) - I_r^s + T_r^s + \text{noises}$$

$$(17.8)$$

The variables involved in the equation are the same as in Eq. (17.6). Note the consideration of signal delay in the troposphere and acceleration in the ionosphere.

Equation (17.8) has four unknowns: the coordinates $(X_r,\ Y_r,\ Z_r)$ and the ambiguity N_r^s. The phase difference is measured, and the other parametres are known, as mentioned in the previous section.

Ambiguity resolution, i.e., the calculation of the value of N_r^s, is performed by specific calculation strategies considering the receiver type and satellite constraints. Because it is an ambiguous value, it can be treated as an integer value on the first measurement. Then, when integer cycles are counted between subsequent measurements, the phase differences between the epochs are unambiguous. This means that the phase measurement is not a distance measurement, as with pseudoranges, but a measurement of the variation in distance as a function of time.

In reality, the value of N_r^s is not an integer. In its calculation, it must be considered as a real number. Its value is calculated by successive adjustments considering Eq. (17.8). In simple terms, it can be understood as if, in the first adjustment, a real value N_r^s plus a standard deviation σ were obtained. In this way, by defining a confidence interval for it, such as $\left(N_r^s - 3\sigma, N_r^s + 3\sigma\right)$, if the interval contains an integer value, this value is adopted as the value of N_r^s, i.e., the *ambiguity is fixed* (ambiguity resolution).

However, when calculating the adjustment, the confidence interval may not contain an integer value or may have several. Thus, there is a non-fixed solution called a *float solution*. In this case, the result for short observation times (tens of minutes) is not considered adequate and should be avoided. For long observation times (hours), the results with a fixed or a float solution for the ambiguity are practically the same.

In practice, the algorithms used to calculate the value of N_r^s are much more complicated than presented. Over the years, several methods and strategies have been developed to calculate the value of ambiguity. Some are widely known in the scientific community; others are proprietary (patented) by GNSS processing software vendors. It is beyond the scope of this book to discuss them in detail, and interested readers are suggested to consult specialised bibliographies on the subject.

Although irrelevant to this book, it is worth mentioning that the phase difference measurement at the receiver is not performed directly on the carrier waves. As already mentioned, satellites transmit their signals using carrier frequencies (in the case of GPS on carrier frequencies L_1, L_2 and L_5). When these carrier frequencies reach the receiver, they differ slightly from those originally generated at the receiver due to the Doppler effect. These two signals create a new wave called the *carrier beat phase*, which measures the phase difference at the receiver. This is why some authors prefer the term *beat-wave phase difference* to refer to phase difference measurement.

17.6 Positioning Methods with GNSS Technology

Considering the methods for measuring the distances between the satellite and the ground antennas presented in the previous section, three methods for positioning the ground antenna have been developed.

- Absolute positioning[6]
- Relative positioning and
- Differential positioning

The most important technical details of each of them are presented below.

17.6.1 Absolute Positioning

The receiver antenna is said to be *"absolutely positioned"* when its coordinates are determined directly in relation to the GNSS satellites and using only one GNSS receiver, as shown in Fig. 17.9. The coordinates are determined concerning the origin of the Geocentric Geodetic System adopted by the GNSS.

Fig. 17.9 Absolute positioning

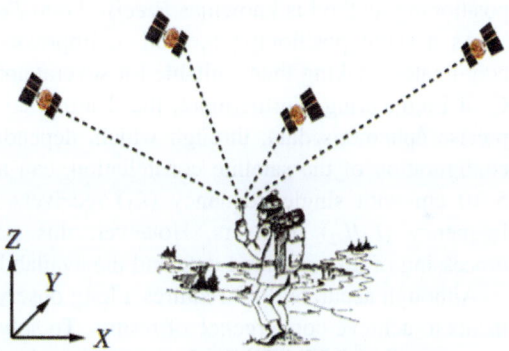

Absolute point positioning can be performed in *kinematic mode* or *static*. It is called kinematic when the GNSS antenna is moved during the measurement period and static the antenna remains stationary over the interest point. The technical details of each of these modes are presented below.

17.6.1.1 Absolute Positioning in Kinematic Mode

For absolute kinematic positioning, the operator requires a single GNSS receiver capable of simultaneously measuring only pseudoranges between the ground

---

[6] Also called autonomous positioning by some authors.

antenna and at least four satellites. Measurements are carried out with the antenna in motion, and the determined coordinate values are displayed in real time on the measuring instrument's screen.

The accuracy achieved by absolute GNSS positioning in kinematic mode depends on several factors, as described in the following sections of this book. In general, for handheld receivers, the user can expect accuracies in the order of 3 to 10 m when using pseudoranges (code) only. It is, therefore, unsuitable for topographic and geodetic surveys where higher accuracy is required.

The instruments used for this type of positioning are small, low cost and simple data processing. Classic examples of this type of GNSS point positioning are navigation through mobile phone applications and vehicle navigation. In some cases, it can be used in the preliminary stage of engineering work or for data collection for low-accuracy GIS applications.

17.6.1.2   Absolute Positioning in Static Mode

For absolute static positioning, the operator uses a single GNSS receiver to track pseudoranges (code) and L_1/L_2 carrier information. The antenna remains stationary over the tracked point for a period ranging from a few minutes to a few hours. This positioning method is known as *Precise Point Positioning* (*PPP*).

As a static positioning method, it improves the accuracy of the determined coordinates, making them suitable for several applications of Geomatics applied to Civil Engineering. Furthermore, the determination of the coordinates is based on precise ephemeris data, through which, depending on the tracking time and the configuration of the satellite constellation, can achieve accuracies of the order of 5–10 cm with single-frequency (L_1) receivers and from 1 to 5 cm with dual-frequency (L_1/L_2) receivers. However, this positioning method requires post-processing of the measured data and the availability of precise ephemeris data.

Although accurate, PPP requires a long observation interval and phase measurements to achieve convergence of results. To solve this problem, a new positioning concept called *PPP-RTK* has been developed, which extends the PPP concept by providing corrections for atmospheric errors, which are calculated by a CORS network, similar to *NRTK positioning*, as described in Sect. 17.9.2.4. The interested reader should consult specialised references on this subject.

17.6.2   Relative Positioning

Relative positioning consists of determining the position of a point in space using two or more GNSS receivers that simultaneously track the C/A-code and carrier phase data of all satellites located above a user-defined elevation angle for a given period of time.

The principle of relative positioning is installing a GNSS receiver at a control point with known coordinates, called the base receiver or reference station, and determining the coordinates of remote points by properly installing another GNSS receiver on them. This determines the spatial vector between the observed points, called the *baseline vector* or simply the *baseline*, as shown in Fig. 17.10.

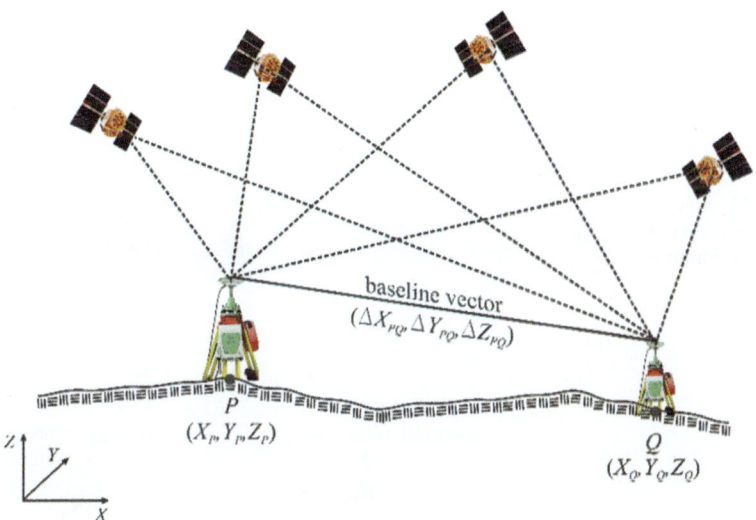

Fig. 17.10 Relative positioning

Mathematically, this can be expressed as follows:

$$X_Q = X_P + \Delta X_{PQ} \tag{17.9}$$

$$Y_Q = Y_P + \Delta Y_{PQ} \tag{17.10}$$

$$Z_Q = Z_P + \Delta Z_{PQ} \tag{17.11}$$

where

X_P, Y_P, Z_P = known geocentric Cartesian coordinates of the reference station
X_Q, Y_Q, Z_Q = geocentric Cartesian coordinates of the remote station
ΔX_{PQ}, ΔY_{PQ}, ΔZ_{PQ} = components of the baseline vector between points (P) and (Q)

The advantage of relative positioning is that it allows better accuracy of the coordinates of remote points. This increase in quality is obtained by processing the observations using linear combinations between the measurements collected by the different GNSS ground antennas. This eliminates specific common errors in measuring the distances between the ground and the satellite antennas due to the phase difference between the carriers.

The linear combinations used are called *Single-, Double-* and *Triple-Difference carrier phases*, in addition to the *linear combinations between frequencies L_1 and L_2*. The analytical details of each of these are given below.

17.6.2.1   Single-Difference Carrier Phase

The *single-difference carrier phase* is a linear combination of observations obtained between receivers and satellites, as shown in Fig. 17.11. In this case, consider Eq. (17.12) of the carrier phase difference for two receivers (P) and (Q) and one satellite S_1 at time i.

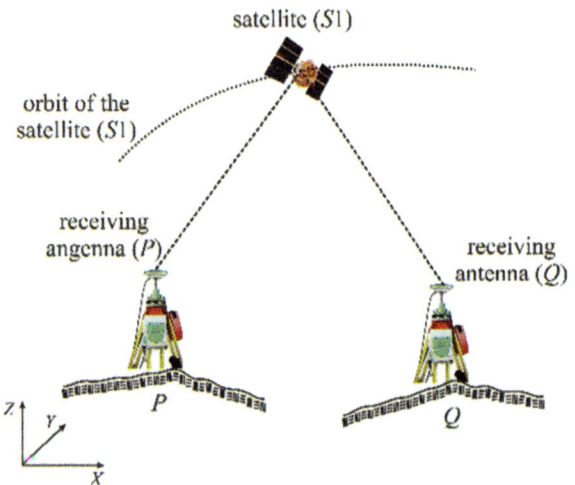

Fig. 17.11 Single-difference carrier phase in relative positioning

$$\left(\varphi_{PQ}^{s1}\right)_i = \left(\varphi_P^{s1}\right)_i - \left(\varphi_Q^{s1}\right)_i \qquad (17.12)$$

Thus, in a simplified way, for the GNSS receivers positioned at points (P) and (Q).

$$\left(\varphi_P^{s1}\right)_i = \frac{1}{\lambda}\left(R_P^{s1}\right)_i + N_P^{s1} - \frac{c}{\lambda}\left(\Delta tr_P - \Delta t^{s1}\right)_i - \left(I_P^{s1}\right)_i + \left(T_P^{s1}\right)_i \qquad (17.13)$$

$$\left(\varphi_Q^{s1}\right)_i = \frac{1}{\lambda}\left(R_Q^{s1}\right)_i + N_Q^{s1} - \frac{c}{\lambda}\left(\Delta tr_Q - \Delta t^{s1}\right)_i - \left(I_Q^{s1}\right)_i + \left(T_Q^{s1}\right)_i \qquad (17.14)$$

where

$$\left(R_r^s\right)_i = \sqrt{\left(X_i^s - X_r\right)^2 + \left(Y_i^s - Y_r\right)^2 + \left(Z_i^s - Z_r\right)^2} \qquad (17.15)$$

Substituting Eqs. (17.13) and (17.14) into Eq. (17.12) gives

$$\left(\varphi_{PQ}^{s1}\right)_i = \frac{1}{\lambda}\left(R_{PQ}^{s1}\right)_i + N_{PQ}^{s1} - \frac{c}{\lambda}(\Delta tr_{PQ})_i - \left(I_{PQ}^{s1}\right)_i + \left(T_{PQ}^{s1}\right)_i \qquad (17.16)$$

From Eq. (17.16), it can be seen that satellite clock errors are eliminated, and orbit and electromagnetic wave propagation errors are reduced by single-differencing.

17.6.2.2   Double-Difference Carrier Phase

When a new satellite is added to the scenario, as shown in Fig. 17.12, using the single-difference carrier phase from two satellites, known as the *double-difference carrier phase* technique, eliminates the receiver clock errors, as given by Eq. (17.18). So, considering the double-difference equation.

$$\left(\varphi_{PQ}^{s1s2}\right)_i = \left(\varphi_{PQ}^{s1}\right)_i - \left(\varphi_{PQ}^{s2}\right)_i \qquad (17.17)$$

Substituting Eq. (17.16) into (17.17) gives

$$\left(\varphi_{PQ}^{s1s2}\right)_i = \frac{1}{\lambda}\left(R_{PQ}^{s1s2}\right)_i + N_{PQ}^{s1s2} - \left(I_{PQ}^{s1S2}\right)_i + \left(T_{PQ}^{s1S2}\right)_i \qquad (17.18)$$

From Eq. (17.18), the terms relating to the receiver clock errors have been eliminated. Since the errors associated with the ionosphere and troposphere can be estimated using mathematical models, the ambiguity values can be fixed to integer values using Least Squares Adjustment Methods such as M-Lambda, FARA, FASF and others, allowing the values of the baseline components to be obtained with high accuracy. For this and other reasons, this is the method commonly used in commercial GNSS data processing applications.

Fig. 17.12 Double-difference carrier phase in relative positioning

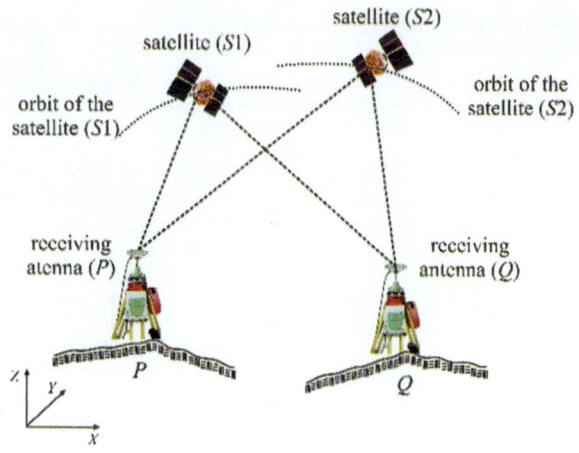

17.6.2.3   Triple-Difference Carrier Phase

Using the same reasoning as in the previous methods, if the double-differencing is performed between the same GNSS receivers and the same satellites at different epochs, i and j, a *triple-difference carrier phase* is obtained, as shown in Fig. 17.13 and given by the following equations.

$$\left(\varphi_{PQ}^{s1s2}\right)_i = \left(\varphi_{PQ}^{s1}\right)_i - \left(\varphi_{PQ}^{s2}\right)_i \qquad (17.19)$$

$$\left(\varphi_{PQ}^{s1s2}\right)_j = \left(\varphi_{PQ}^{s1}\right)_j - \left(\varphi_{PQ}^{s2}\right)_j \qquad (17.20)$$

Solving the double differences results in Eq. (17.21).

$$\left(\varphi_{PQ}^{s1s2}\right)_{i,j} = \frac{1}{\lambda}\left(R_{PQ}^{s1s2}\right)_{i,j} \qquad (17.21)$$

In this case, the ambiguities have been eliminated, simplifying the observations' treatment. Once the number of unknowns has been significantly reduced and the ambiguities, which are the most difficult unknowns to determine, have been eliminated, the solution is quickly obtained. It is, therefore, the solution adopted for calculating the initial adjustment coordinates and allowing the detection of momentary discontinuities in a receiver's phase lock on a satellite signal, known as *cycle slips*.

On the other hand, eliminating the unknowns related to the ambiguity makes it impossible to reach a fixed solution, which reduces the quality of the results and makes it rarely used in the definitive processing of GNSS observations.

Fig. 17.13 Triple-difference carrier phase in relative positioning

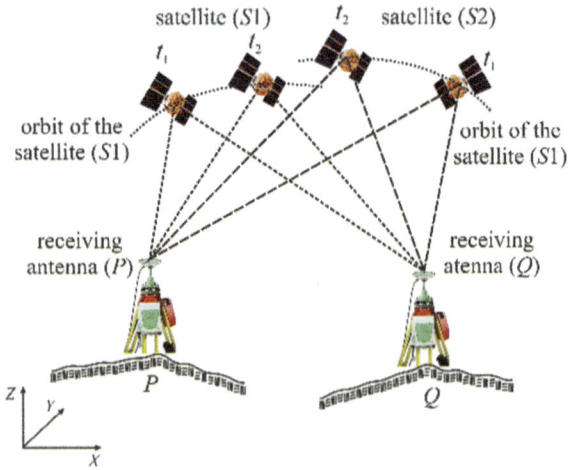

17.6.2.4   Linear Combinations Between Carriers (L_1) and (L_2)

Linear combinations between carriers L_1 and L_2 are used to eliminate the effect of ionosphere refraction or to facilitate ambiguity resolution in the case of long baselines (greater than 50 km).

The most used linear combination between carriers L_1 and L_2 is the combination called L_3, which is given by Eq. (17.22). It is a linear combination free from ionospheric effects and is generally used in GNSS positioning with baselines of a few tens of kilometres.

$$(L_3) = \frac{f_1^2}{f_1^2 - f_2^2} * L_1 - \frac{f_2^2}{f_1^2 - f_2^2} * L_2 \qquad (17.22)$$

where

f_i: frequency of the carrier wave L_i

In addition to the linear combination L_3, there are others that are not discussed in this book.

Based on the linear combinations presented, relative positioning can be performed in static or kinematic modes. In both cases, the positioning results are obtained after post-processing the data in specific application programs, usually provided by the GNSS manufacturers.

17.6.2.5   Positioning in Relative Static Mode

Relative static positioning requires that the GNSS antennas remain static and track the same satellites throughout the positioning period. An observation recording interval[7] (generally a few seconds) guarantees the almost simultaneity of the collected data. The coordinates of the tracked points can be determined by measuring the pseudo-distances and the phase differences of the carrier L_1 or the carriers L_1/L_2. The accuracy of this positioning method depends on the distance between the reference station (receiver) and the remote receiver. Under good measurement conditions, GNSS manufacturers quote centimetre-level accuracy for relative static carrier phase L_1 positioning and ±3 mm + 0.1 ppm horizontally and ±3.5 mm + 0.4 ppm vertically, with a 95% confidence level when using carrier phases L_1/L_2.

---

[7] Data collection rate.

17.6.2.6   Positioning in Relative Kinematic Mode

In relative kinematic positioning, a GNSS antenna is kept static at the base station, i.e., over a point of known coordinates, while one or more remote antennas can be moved over other points of interest while tracking signals from satellites, as schematically shown in Fig. 17.14. This is a highly productive GNSS positioning method under certain operational conditions, as described in Sect. 17.9.

Fig. 17.14 GNSS relative kinematic positioning

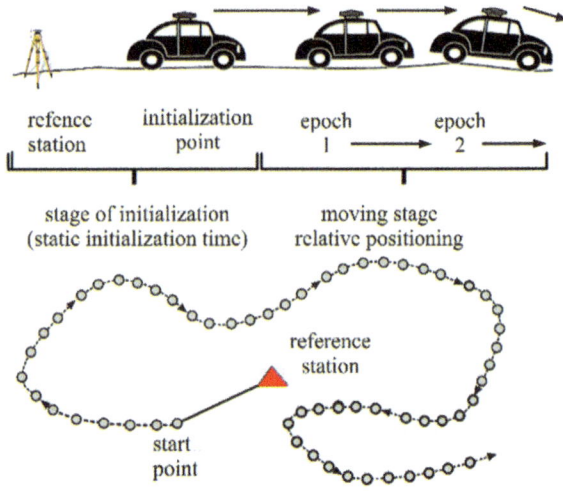

As this is a relative positioning method, all remote antennas must simultaneously track signals from the same satellites tracked by the base station antenna throughout the measurement time.

Distance measurements for positioning can be carried out using either the pseudoranges or the carrier phase. When using pseudoranges, the remote antenna can be moved immediately after the receiver starts tracking the satellite signals. The expected accuracy level, in this case, is sub-metre.

When the carrier phases are used, there are two positioning options: *relative kinematic positioning with initialisation* and *relative kinematic positioning without initialisation* or *PPK*, as described below.

17.6.2.6.1   Kinematic Relative Positioning with Initialisation[8]

In this positioning mode, the remote antenna must remain static for some time, called the *static initialisation time*, to resolve the initial ambiguity. See Fig. 17.14. In this

---

[8] Because of the need to keep the antenna static for a period over the point, some authors prefer the term *semi-kinematic* for this positioning method.

case, the receivers can be either single-frequency L_1 or dual-frequency L_1/L_2. The static initialisation time of the remote antenna and the distance between the reference and remote antennas vary for different instrument manufacturers. Further details about these values are given in Sect. 17.9.2.2. The accuracy obtained with this type of positioning using dual-frequency receivers is in the order of ±5 mm + 0.5 ppm horizontally and ±10 mm + 0.5 ppm vertically.

17.6.2.6.2 Kinematic Relative Positioning with on-the-Fly Ambiguity Resolution: PPK

In this positioning mode, the remote antenna can move during the ambiguity resolution initialisation period, i.e., ambiguity resolution is performed while the remote antenna is in movement. Some authors refer to this type of positioning as *on-the-fly kinematic mode,* while others refer to it as *post-processing kinematic mode.* (PPK).[9] Post-processed kinematic (PPK) is a GNSS correction technology used in surveying that corrects the location data after it is collected and uploaded.

When positioning in PPK mode, the receivers must preferably be dual-frequency L_1/L_2, and the manufacturer specifies the distance between the reference and remote antennas. Note that during the initialisation time for ambiguity resolution, the coordinate values determined with the moving remote antenna are of poor quality. The quality specified by the manufacturer is only obtained after the ambiguity resolution. When using single-frequency L_1 receivers, the ambiguity resolution time is longer than with dual-frequency, and the distance between the base station and the remote receiver should be shorter.

The accuracy specified by the manufacturers for this positioning method using dual-frequency receivers, after the ambiguity resolution, is also in the order of ±5 mm + 0.5 ppm horizontally and ±10 mm + 0.5 ppm vertically.

For both positioning methods, all the antennas involved in the survey must continuously track at least four satellites throughout the measurement period. If the signal is somehow interrupted, the measurement operation must be restarted.

17.6.3 Differential Positioning

This positioning method is similar to relative positioning, as two or more GNSS receivers track satellite data simultaneously, with one of the antennas being held static over a control point. The difference is that, in differential positioning, the differential corrections determined by the reference station receiver are transmitted to the rovers in real time by telecommunication link, as shown in Fig. 17.15.

---

[9] Is an alternative technique to Real-Time Kinematic (RTK). With PPK workflow, accurate positioning does not happen in real-time, all algorithms are applied afterwards.

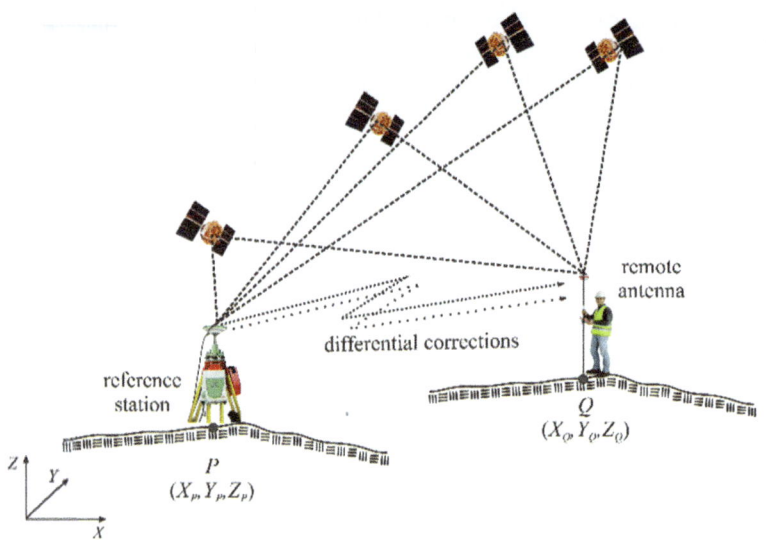

Fig. 17.15 Principle of differential positioning

The great advantage of this positioning method is that the operator at the remote point has real-time information about its coordinates, i.e., there is no need to post-process the collected data. Processing takes place in real-time at the remote receiver.

The disadvantage is that it requires a communication system between the base station and the remote receivers, which can limit the size of the survey area, depending on the quality of the communication system and the level of obstruction.

Differential positioning can be carried out using a *Differential Global Positioning System* (DGPS) or differential *Real-Time Kinematic* (RTK) mode, as described below.

17.6.3.1 DGPS Positioning

DGPS positioning is based on differential corrections performed on the pseudoranges or the coordinates determined by them.

As it is a surveying method based on differential positioning and pseudoranges, the coordinates of the tracked points are displayed in real-time immediately after the receiver starts tracking the satellite signals. Using the known coordinates of the reference station, it is possible, among other things, to calculate the errors of the measured pseudoranges and transmit them to the remote antennas, which in turn use them to determine the coordinates of the remote points. The corrections are transmitted according to the GNSS data exchange protocols described in the next section.

If only the pseudorange corrections are calculated, DGPS positioning can achieve sub-meter accuracy.

17.6.3.2 Positioning in Differential RTK Mode

Differential positioning occurs in *differential RTK mode* when the differential corrections include the values of the carrier phase measurements L_1/L_2. It, therefore, requires the same prerogatives as relative positioning in *post-processed kinematic mode—PPK*, i.e., an initialisation time for the measurement system. The great advantage, in this case, is that as the differential corrections are transmitted from the base station receiver to the remote receiver, and the remote receiver itself indicates when the ambiguity resolution is reached and the accuracy of the calculated values for the tracked point coordinates, which are displayed on the instrument's data collector screen. Furthermore, as with DGPS positioning, corrections are also broadcast using GNSS data exchange protocols, as described in the next section.

According to GNSS instrument manufacturers, the accuracy of baselines determined by GNSS positioning in differential RTK mode is in the order of ± 5 mm + 0.5 ppm horizontally and ± 10 mm + 0.5 ppm vertically.

17.7 GNSS Data Exchange Formats

In addition to data transmission from satellites in the GNSS constellations, there is also data transmission from GNSS receivers to other receivers or measuring devices. Satellite data transmission protocols have been well established since the beginning of the GNSS technology. However, the transmission between receivers is not standardised, and there are different data formats, as shown below.

17.7.1 NMEA 0183 Format

The NMEA 0183 data transmission protocol was developed by the US Navy and is considered a universal GPS information transmission format. It is commonly used to transmit information between the GNSS receiver and a computer or to other navigation devices. Data is transmitted in ASCII format and includes information about position, time, speed, ephemerides, etc. To facilitate the transmission, several standardised sentence types have been created with different types of information, with names such as GGA, GLL, GSA, GSV, VTG, ZDA and others.

The NMEA 0183 format is the data transmission protocol used by most navigation systems based on GNSS information.

17.7.2 RINEX Format

The RINEX (Receiver INdependent Exchange) format was developed by the Astronomical Institute of the University of Bern to facilitate the exchange of raw data collected by receivers from different manufacturers. As all GNSS instrument manufacturers provide data from their receivers in the RINEX format, it is possible to post-process combined data from different receivers in a single operation.

The latest version of the RINEX format is 4.01,[10] which consists of the following three files in ASCII format:

- Observation data file
- Navigation messages file and
- Meteorological data file

Each file consists of a header section and a data section. The header section contains information about the entire file and is located at the beginning of the text. For information on the format and order of the information contained in the header section, the interested reader is encouraged to consult specialised references.

Observation and meteorological data files contain the data of a particular location and an observation session. The Navigation Messages file can contain messages from single or multiple satellites tracked by different receivers. In version 3, all the files now contain messages from GPS, GLONASS, GALILEO, QZSS, *BeiDou*, IRNSS and SBAS satellites.

The types and formats of available observation, navigation and meteorological data vary between the systems. They are presented in format definition tables, and data examples are available in official documents on the RINEX format. It is beyond the scope of this book to discuss these. The reader interested in this information can, for example, consult the tables in the document available at https://files.igs.org/pub/data/format.

17.7.3   RTCM SC1040 Protocol

The RTCM SC104 data transmission protocol was developed by Committee 104 of the *"US Radio Technical Commission for Maritime Services"* to provide a differential correction transmission service for DGPS and differential RTK positioning. Currently, version 2.3 (RTCM 10402.3) is used to transmit differential code corrections (DGPS) for applications using the GPS and GLONASS satellite constellations; version 3.3 (RTCM 10403.3, Differential GNSS Services—April 2020) is used to transmit differential code and phase corrections for high-precision positioning in RTK, NRTK and PPP-RTK modes, for operations using the GPS, GLONASS, GALILEO, QZSS and BeiDou satellite constellations, including SBAS observations.

The messages are encoded in different groups containing different information based on numerical values, including information on stability and parameters of the reference station, satellite constellation stability, differential corrections, etc. Users should consult specific manuals for this protocol to fully understand RTCM messages.

---

[10] For more information: https://igs.org/news/rinex-4-operations-updates

17.7.4   NTRIP Protocol

With the advance in data transmission capabilities in general, such as via mobile telephony and the Internet, a new data transfer protocol called Networked Transport of RTCM via Internet Protocol (NTRIP) has been created. This protocol allows real-time GNSS receivers to operate in DGPS and RTK modes via Internet communication.

The characteristics of the NTRIP protocol are as follows:

- It allows access via HTTP (Hypertext Transfer Protocol)
- It makes all GNSS data continuously available
- It allows simultaneous access by multiple users
- It provides secure data access without the need for the user to be in direct contact with the reference stations
- It can stream data over any TCP/IP (*Transfer Control Protocol / Internet Protocol*) mobile network
- The bandwidth required to transmit GNSS corrections is low

NTRIP comprises four components: the *NTRIP Source*, the *NTRIP server*, the *NTRIP Caster* and the *NTRIP Client*, as shown in Fig. 17.16.

The *NTRIP Source*[11] continuously provides the GNSS data stream, usually represented by a CORS.

The *NTRIP server* acts as an intermediary in the transmission of data between the source to the Caster.

The *NTRIP Caster* consists of an *HTTP* server that collects, organises and checks the quality of the collected data and broadcasts it to the *NTRIP Client*.

The *NTRIP Client* is a device or software that uses the NTRIP protocol to receive RTCM data from the NTRIP Caster.

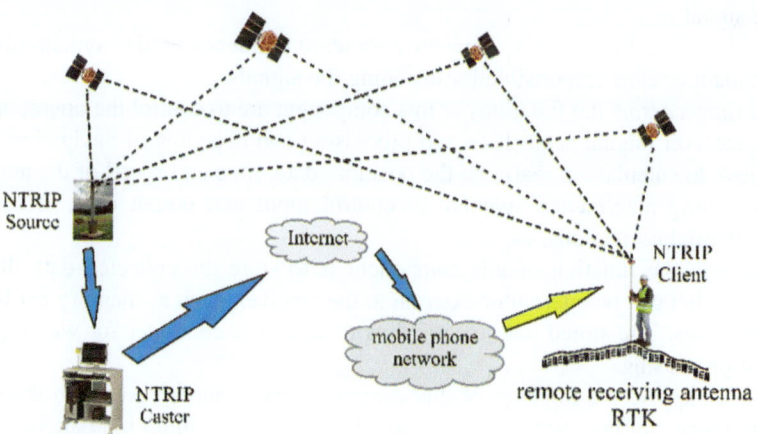

Fig. 17.16 NTRIP protocol components

[11] For further information, the reader is advised to consult the website http://igs.bkg.bund.de/root_ftp/NTRIP/documentation/NtripDocumentation.pdf

17.8   Composition of a GNSS Instrument

The main components and accessories of a GNSS instrument are shown in Fig. 17.17. Their characteristics and functions are described below.

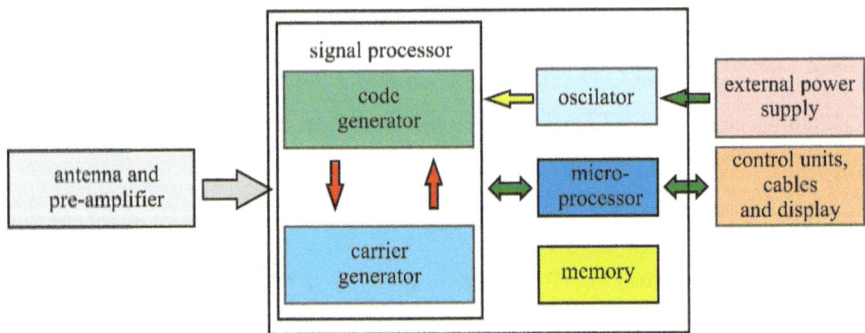

Fig. 17.17   Main components of a GNSS instrument

- *Antenna and pre-amplifier:* the function of the antenna is to detect the electromagnetic waves emitted by satellites, convert them into electrical current, amplify the signal and send it to the receiver. GNSS antennas must be susceptible to receive satellite signals, have phase centre stability and be protected against multipath. There are several types of GNSS antennas, but the most used in topographic or geodetic instruments are the microstrip antennas, as they are the most robust and compact.
- *Signal processor:* the function of this component is to convert the received signals to a lower frequency so that other components of the receiver can process the signal.
- *Oscillator:* its function is to generate a standard frequency used to synchronise all the main circuits responsible for decoding the signals.
- *Microprocessor:* the functions of this component are to control the operations of the receiver (signal acquisition and processing and decoding of navigation messages), to calculate in real-time the position, date, time and speed of the antenna concerning a reference antenna, to control input and output data, to control differential corrections, etc.
- *Memory:* the function of this component is to store the collected data. It is a device that can be internal or external to the receiver, such as memory cards and pen drives. The stored data is later downloaded to a computer for viewing and post-processing.
- *Power supply:* the function of this component is to supply power to the other components of the instrument. It generally consists of lithium-ion batteries, and, in some instances, voltage converters are also used to bring the voltage to a level suitable for the device.

- *Control units (or human communication interface), cables and display*: external components that allow users to communicate with the receiver via an alphanumeric keypad and VGA screen. Communication can also be performed via computers in the office or the field so that the receivers can operate in the field without operator intervention. The connection between the antenna and the receiver and between the control unit and the receiver is usually made by various cable types but can also be wireless.

Figure 17.18 shows the general setup of a GNSS instrument for installation on a tripod or a concrete pillar with a forced centring device. This type of setup is used for reference stations or remote stations when operating in relative static mode. The instruments are mounted on a pole for operations in relative kinematic mode or DGPS/RTK mode, as shown in Fig. 17.19. Other standard setups include those on vehicle roofs, construction machinery and aircraft.

The (receiver + antenna) set of the GNSS instrument can be separated, as shown in Fig. 17.18, or combined into a single (smart antenna) unit, as shown in Fig. 17.19. They can be operated with or without the external control unit in both cases. Each GNSS instrument manufacturer proposes the most suitable combination for their measurement solutions.

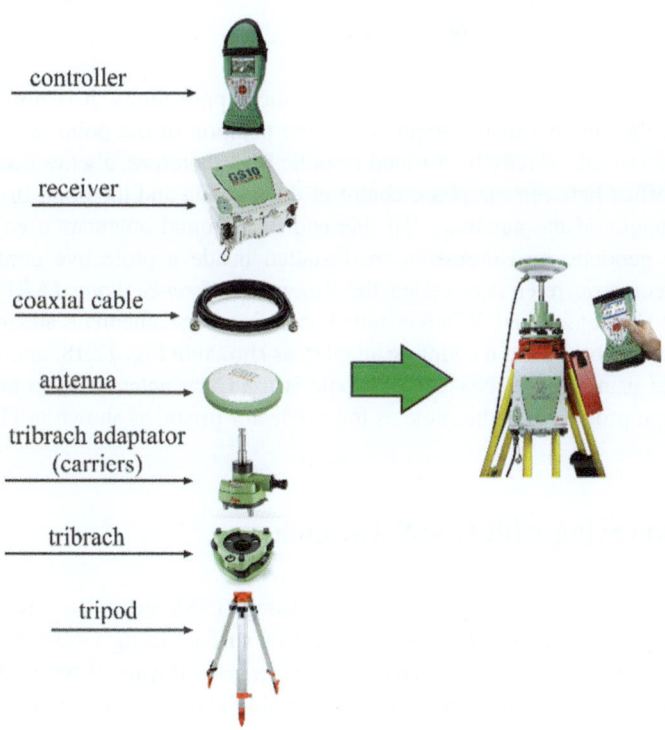

controller

receiver

coaxial cable

antenna

tribrach adaptator
(carriers)

tribrach

tripod

Fig. 17.18 Setting up a GNSS instrument on a tripod

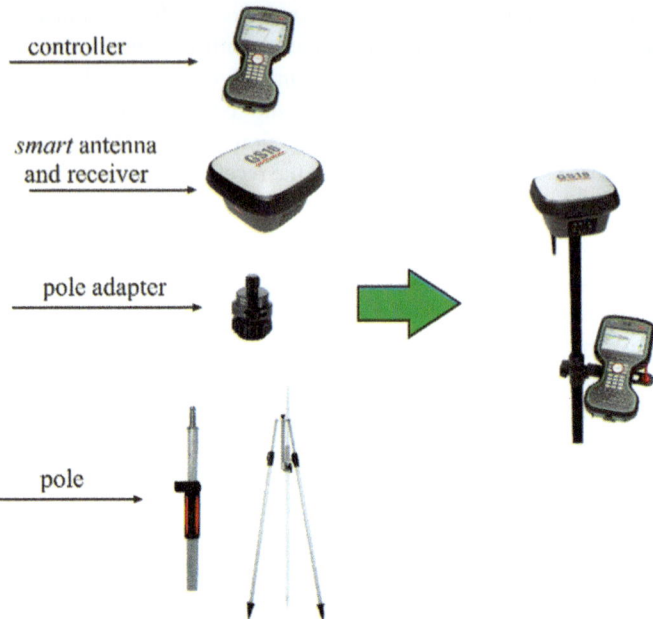

controller

smart antenna
and receiver

pole adapter

pole

Fig. 17.19 Installing a GNSS instrument on a pole

It is important to note that in most Geomatics applications, it is not the exact position of the antenna that is required but the position of the point to which it is fixed or referenced, usually by a tripod or pole. It is, therefore, always necessary to know the offset between the phase centre of the antenna and the point in question, i.e., the "height of the antenna". To this end, the ground antennas used in topographic or geodetic measurements are installed inside a protective container on which a geometric reference, called the *Antenna Reference Point* (ARP), allows this offset to be determined. When installed on a tripod, the antenna is attached to the tripod using a tribrach and a coupling adapter, as shown in Fig. 17.18, and the offset is measured using a tape measure. If a pole is used, the antenna setup and height measurement procedure is the same as for a reflector prism, as shown in Fig. 17.19.

17.9 Surveying with GNSS Technology

Conducting a topographic or geodetic survey using GNSS technology means determining the coordinates of points on the Earth's surface using GNSS positioning methods. In the case of geodetic surveys, they are usually carried out to determine control points or to establish geodetic networks. In the case of topographic surveys, they are carried out for many Geomatics applications, ranging from the establishment of control points for Civil Engineering works to cadastral surveying.

Whatever the purpose of the survey, a number of conditions must be met for the work to succeed. The most important of these are:

- Survey planning
- Selection of the positioning method to be used
- Selection of the appropriate type of instrument
- Planning field operations
- Defining the data processing strategy

17.9.1 Survey Planning

The outcome of a GNSS survey depends on several precautions the user must take before starting the fieldwork to acquire data, particularly the identification and location of points to be used as reference stations. The stations to be used are provided by federal and state agencies, such as NGS-NOAA (USA), which provide information from geodetic databases through a Geodetic Station Report. It is also recommended that a preliminary visit to the survey area be made to check for interference with GNSS signals and to select the best locations for implementing new topographic or geodetic landmarks (if applicable). It is also recommended to check the quality of existing landmarks to be used as reference points.

The GNSS data processing software packages supplied with the measurement system allow for checking the availability and geometric conditions of the constellation, thus allowing the choice of the best locations and time for the survey.

Particular attention should be paid to the availability of continuous monitoring stations. Many countries have networks of continuous monitoring stations maintained by the government with free access and/or private networks with paid access, depending on the time of use. The reader can find, for example, more information on global continuous monitoring stations at https://stdb2.isee.nagoya-u.ac.jp/GPS/GPS-TEC/gnss_provider_list.html. Note that this type of reference station depends on the availability of data from the maintaining agency. Therefore, the user should be aware of data availability problems for critical surveys.

The locations defined for the surveys must guarantee easy access and visibility of the horizon in all directions from a minimum elevation angle of 10° above the horizon of the ground antenna, thus guaranteeing the quality of the tracked signals and minimum influence of multipath, which degrade the accuracy of the determined coordinates. Another critical factor is the choice of the raw observations interval to be stored in the receiver. For surveys in relative static mode, rates between 5 and 15 s are generally used; in relative kinematic mode, rates vary from 1 to 5 s. In differential positioning, rates are always equal to or less than 1 s.

17.9.2   Positioning Method, Instrumentation and Operations for Field Survey

The choice of positioning method and field operations depends on the purpose of the survey, the level of accuracy required by the project, the equipment available, the size of the area to be surveyed, the infrastructure conditions for carrying out the survey and the skills of the field operators. In the case of geodetic surveying, the positioning method is restricted to the relative static mode; in the case of topographic surveys, however, it can be performed in static, kinematic or differential modes. The field operations for each mode and the instruments to be used are presented in the next section.

17.9.2.1   Static Positioning Survey

Field surveys using static positioning are carried out for geodetic or topographic work where it is desired to determine the coordinates of well-defined individual points on the ground, such as geodetic and topographic markers.

This type of survey is generally carried out using relative static positioning. Depending on the accuracy required and operating conditions of the survey, such as in remote areas where reference points are not available, absolute positioning in static mode can be used. The operational details of each of them are described below.

17.9.2.1.1   Relative Static Positioning Survey

This type of survey has long been the most widely used for surveying measurements with GNSS technology, mainly because it directly applies the basic concepts of the technology. In addition, GNSS surveying guarantees the highest accuracy of the determined coordinates, precisely when dual-frequency receivers can track multiple GNSS signals.

In this type of survey, it is necessary to have at least one point with known geodetic coordinates where it is possible to install the GNSS ground antenna to set the reference station or to have a continuous GNSS monitoring station, in both cases, positioned at a distance close to the points to be surveyed (remote points). If a user reference station is to be used, the user must have at least two receivers to install, one over the reference point and the other over the remote point. When starting the survey, the receivers must be set up according to the manufacturer's specifications and collect GNSS raw observations interval simultaneously for a time that varies according to the distance between them.

When using data from a CORS, the operator must also ensure that the remote GNSS receiver is set to collect data at a rate equal to or a multiple of the data collection rate of the continuous monitoring station. Figure 17.20 schematically illustrates a hypothetical survey using a CORS and a reference station.

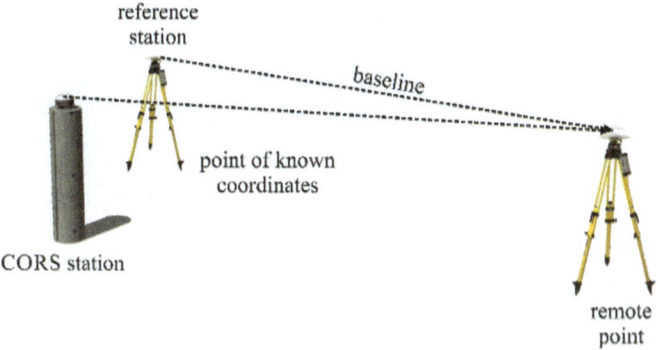

Fig. 17.20 Principle of GNSS relative static survey

Depending on the number of receivers available for the survey, it can be carried out using one or multiple baselines, creating a network of GNSS baselines, as shown in Fig. 17.21.

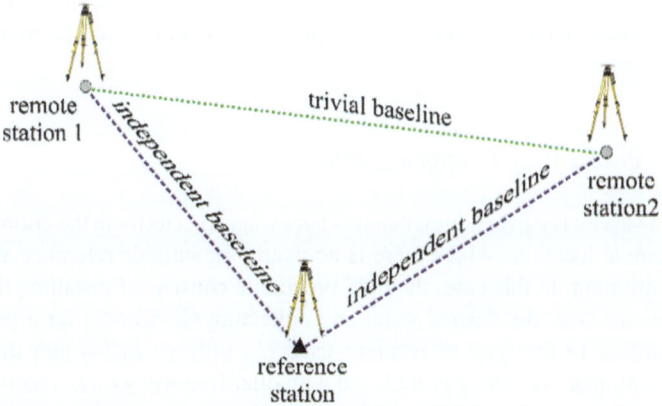

Fig. 17.21 Independent and trivial baselines

Repeating baselines establishes what is known as a *GNSS baseline network*. In this case, the points are occupied more than once, which is why network surveys are used only in relative static mode.

Surveys are usually carried out with only one baseline. However, because of the lack of data redundancy, this procedure can lead to incorrect values in the case of undetected disturbances in the GNSS system components, even in situations where the result indicates ambiguity resolution.

The best field practice is always to survey with at least two baselines from different reference stations. As GNSS measurements are highly correlated, it is also recommended that the baselines be measured at different epochs. The results will be more reliable and allow for more robust statistical analysis.

Surveys can be performed with single-frequency L_1 or dual-frequency L_1/L_2 receivers. The results are obtained by post-processing the data collected by both receivers and by processing the phase differences.

The distance between the reference station and the remote point can vary from a few metres to a few hundred kilometres. With current technology, it is estimated that for distances up to 50 km, available ambiguity resolution techniques can provide a fixed solution with a tracking time of 15 to 20 min for dual-frequency L_1/L_2 receivers. For single-frequency L_1 receivers, it is recommended to follow the manufacturer's recommendations. In these cases, it is said that a rapid static survey is being carried out. For longer distances, where a float solution is commonly used, data collection time should be around 3 h for long baselines (200–300 km). Tracking longer than 3 h is recommended for baselines longer than 300 km. For a better recommendation, the user must consult the User Manual of his instrument.

The GNSS equipment used for this type of survey usually consists of an antenna and a receiver, either as a single unit or as separate units connected by cables. When using a reference station above a control point, both the base and the remote antenna are often mounted on a tripod, as shown in Fig. 17.18. As the survey is limited to collecting raw data in static mode, all the settings for the data collection can be carried out in the office, and the operator only needs to install the antennas over the points and switch on the receiver if the equipment does not have an automatic on/off facility.

17.9.2.1.2   Precise Point Positioning (PPP)

This type of survey is carried out when it is necessary to determine the coordinates of points in remote locations where there is no nearby or suitable reference station for relative positioning. In this case, the field procedure consists of installing the GNSS receiver/antenna over the desired point and collecting GNSS data for a period that varies according to the type of receiver used (L_1 only or L_1/L_2) and the desired accuracy. Centimetre accuracy is achieved with dual-frequency L_1/L_2 receivers and a tracking time between 3 and 4 h. As this is a point positioning in static mode, the antenna can be installed using any static installation methods described in the previous sections.

For this type of survey, the collected data must be post-processed using precise ephemeris data and all corrections involved in PPP GNSS positioning. As precise ephemerides are used, processing can only be carried out a few days after the data collection due to the time required for the ephemeris to be made available. In general, processing is carried out by specialised agencies that offer the service free of charge, such as, for example, the University of New Brunswick in Canada, through the website http://www2.unb.ca/gge/Resources/PPP/OnlinePPPs.html, the National Geodetic Service—NGS, USA, which offers the online service called OPUS (Online Positioning User Service—https://www.ngs.noaa.gov/OPUS/). In Brazil, this type of service is provided by IBGE (https://www.ibge.gov.br/geociencias/informacoes-sobre-posicionamento-geodesico/servicos-para-posicionamento-geodesico/16334-servico-online-para-pos-processamento-de-dados-gnss-ibge-ppp.html?=&t=processar-os-dados), which already provides the results in the reference system used in Brazil (SIRGAS 2000.4).

17.9.2.2 Kinematic Positioning Survey

As mentioned above, in several Geomatics applications, it is desirable to determine the coordinates of successive points along a path. This type of measurement is known as *relative kinematic measurement.*

A typical example of this type of survey application is the work carried out to determine groups of points and the work carried out to position mobile topographic sensors (aerial and terrestrial), as already described in previous chapters.

Figure 17.22 schematically illustrates the situation of a relative kinematic survey consisting of a reference station, a remote antenna on the roof of a vehicle, and a remote antenna in a backpack.

Fig. 17.22 Principle of GNSS relative kinematic survey

Users can benefit from a continuous monitoring station or establish their baseline using a second instrument, like static surveys. Measurements can be made in relative kinematic mode using code, code and phase with initialisation, or relative kinematic PPK mode.

Relative kinematic surveys to measure pseudoranges are not widely used in Geomatics due to their low accuracy. However, this type of survey is suitable for collecting data for GIS or for agricultural applications that do not require high accuracy. Despite its low positional quality, it does not require any system initialisation time at the start of the survey or when the signal is interrupted.

Regarding relative kinematic positioning surveys with code and phase with initialisation, data acquisition must not be interrupted during antenna movement. If this happens, the survey must be restarted from where the signal was lost.

In the case of PPK positioning surveys, the remote antenna can keep moving if the signals are interrupted. However, as mentioned, the determined coordinates are of low precision during the time interval with the float solution.

Interruptions occur when the signals received by the remote antenna are blocked, for example, when it passes through dense vegetation, under an overpass, or when buildings or other obstacles block the signal. Receivers usually emit a warning signal when the number of satellites being tracked falls below a pre-set threshold.

The number of points determined in this type of survey will depend on the data recording interval and the speed of movement of the remote antenna. In the first case, it usually varies between 1 and 5 s but can reach a tenth of a second in the case of aerial surveys.

As this is a relative positioning survey, the data collected must be post-processed to obtain the final coordinates of the surveyed points.

17.9.2.3   Differential Positioning Survey

The ability to know the precise coordinates of a point in real time using GNSS technology has ushered in a new era for GNSS surveying. Currently, most surveys are carried out using this type of positioning. They are used in all kinds of work where it is necessary to know the coordinates of the measured points in real-time, such as Civil Engineering construction work, machine guidance for Civil Engineering, precision farming, etc.

Similar to the field surveys mentioned in the previous sections, in this case, it is also necessary to use two receivers, one operating as a reference antenna and the other as a remote antenna. The difference with this type of survey is that the receivers must have a data transmission system that allows real-time communication between the reference and the remote receivers. The communication system can be UHF radios, Bluetooth communication, GSM mobile communication, Internet communication, geostationary satellite communication, and others. The data exchange formats are those presented in Sect. 17.7.

In this case, field operations consist of installing the reference antenna over a control point, whose coordinates must be stored in the receiver, and sending the differential corrections (DGPS) or GNSS observations (RTK) from the reference to the remote receivers. As soon as the remote receiver receives the corrections, it begins to determine the coordinates of the remote point, which are displayed on the receiver's data logger screen, along with the accuracy information. A typical example of a differential positioning field survey is using a GNSS antenna fixed to a pole, as shown in Fig. 19.19. In this case, the operator positions the pole with the antenna over the desired point, checks the quality of the position displayed in the data logger and stores the coordinates in the instrument's memory.

When positioning in DGPS mode, the values indicated by the receiver can be used immediately after display. In the case of differential RTK positioning, the operator must wait for the system to resolve the ambiguity. This type of survey uses dual-frequency receivers *L1/L2* and multiple GNSS signal tracking.

As the survey is based on differential positioning, post-processing is unnecessary, and the coordinates are available in real-time on the data logger's screen and/or stored in the instrument's internal memory, if necessary.

17.9.2.4 Surveys Assisted by Virtual Reference Stations (VRS)

The Virtual Reference Station (VRS) concept is an efficient way of transmitting GNSS corrections provided by a CORS network through a differential correction interpolation method and using them to process baselines of remote points within the network coverage area. Typically, CORS are spaced 50 to 60 km apart to cover as much of the area of interest as possible.

Operationally, when determining a VRS, the pseudorange and carrier phase information collected by the antenna of each receiver in the network is processed in a control centre. Once a fixed solution to the network ambiguity has been obtained, the differential correction modelling phase begins. From then on, the control centre can interpolate values for specific points within the network area and generate GNSS observations and/or corrections data as if a hypothetical ground antenna.

This generated VRS data is then sent to the user via a wireless connection, often using the Networked Transport of RTCM via Internet Protocol (NTRIP). Finally, as if the VRS data had come from a physical reference station, the rover receiver uses standard single baseline algorithms to determine the coordinates of the remote point in near real-time kinematic or post-processed modes, as shown in Fig. 17.23.

Surveys performed using differential RTK positioning are called *Network Real-Time Kinematic* (NRTK). There are two solutions preferably used for surveys using the differential method: the first one is based directly on the VRS concept, which requires the remote receiver to be able to receive and send communications to and from the control centre, usually via a mobile phone or the Internet. In this case, the remote receiver sends the navigation position to the control centre, which performs the initial calculations and transmits them to the remote receiver via the NMEA protocol. The positioning process is iterative until a final position is obtained. From there, differential corrections are continuously transmitted via the RTCM or proprietary protocol. See Fig. 17.23.

The second type of solution is called the *Master-Auxiliary Concept* (MAC). This solution is based on the concept of creating an area of ambiguity data to be used as differential corrections, which are made available to all remote receivers operating within the CORS network area via the RTCM protocol without the need for duplicate communication. For the system to work, the network of stations is divided into areas formed by clusters and cells, which achieve a common level of ambiguity to be made available to users.

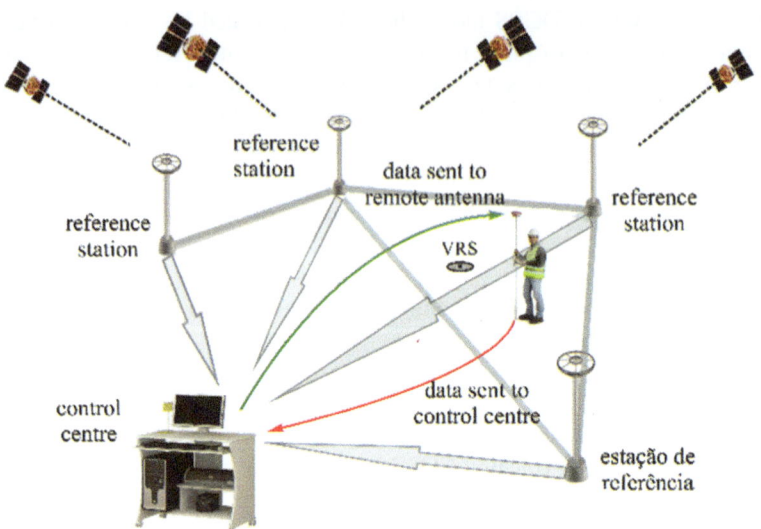

Fig. 17.23 Principle of a VRS-assisted survey

After conducting several field tests, studies have shown that both technologies provide the same level of accuracy, with values varying between 1 to 2 cm horizontally and 1.5 to 3.5 cm vertically, with a confidence level of 68.27% (1σ).

In the case of post-processing data in static or kinematic positioning surveys, the solution can be obtained by using a Virtual RINEX (VRINEX) file with interpolated raw data from each CORS for a hypothetical positioning of a point within the network coverage area. The raw data is the same as a conventional RINEX file, with the advantage of being related to a station close to the remote point. The user then uses the information from this RINEX file to post-process the data.

The advantages of such a solution are huge, which is why it is used in several countries and regions with extensive GNSS measurement activities. These basic considerations have led to reference station networks like the Austrian positioning service (APOS), the German satellite positioning service (SAPOS), and others.

17.10 Quality of Surveys with GNSS Technology

The quality of the surveying measurements carried out with GNSS technology also varies due to instrumental, atmospheric and operational errors, as follows:

17.10.1 Instrumental Errors

The main instrumental errors and their effects on GNSS measurements have been described in detail in the previous sections on positioning methods. In addition to these, there are others that can reduce the quality of the survey, as described below:

Errors between receiver channels: current receivers offer multi-channel models to receive signals from different positioning systems. This can lead to instrument errors as each satellite signal from each system follows a different electronic path. These errors can be eliminated in the differentiation of observations if instruments of the same brand and model are used in the survey.

Receiver noise errors: receiver noise is a complex error generated at the receiver's side while measuring satellite signals. The receivers themselves introduce several sources of error to the measurement of satellite ranges. Thermal noise produced by the environment and the various components within a receiver (antenna, cables and amplifiers) cause small random errors. The received signal-to-noise ratio, quantization of the analog-to-digital converter, and the type of tracking loop a receiver uses are also determining factors in the noise level. To measure these interferences, the instrument uses the *signal-to-noise ratio* (SNR) parameter, which measures the strength of the satellite signal received by the receiver relative to the background noise and the environmental interferences. For this reason, receivers must be checked periodically to ensure that the electronic noise generated is within the tolerances specified by the manufacturer.

Antenna phase errors: the electrical phase centre of the antenna is the virtual point to which the signal measurements are referenced, and it does not coincide with the antenna's geometric centre (ARP). For this reason, it is recommended to calibrate the antennas to know the values of the phase centre variations for high-quality work. In a simplified way, this calibration is done by rotating the antenna around an axis to determine the direction in which it shows the least phase variation. This error can be minimised by using antennas of the same type/model from the same manufacturer and aligning them with True North so that the processing software can correctly use the calibration parameters.

17.10.2 Atmospheric Errors

The main errors related to atmospheric conditions affecting GNSS surveys are as follows:

Errors during signal propagation: when signals are transmitted from the satellites and before reaching the ground antenna, they pass through the Earth's atmosphere and are, therefore, exposed to "atmospheric refraction" effects. Refraction occurs in all layers of the atmosphere, with the most relevant effects occurring in the troposphere and ionosphere. When using multi-frequency receivers, the effects of ionospheric refraction can be removed by mathematical models, but they must be modelled appropriately for single-frequency receivers. These errors cannot be ignored for long-range surveys where the atmosphere varies between points. Refraction errors are removed by signal phase differentiation in the processing for short-range surveys. In general, processing programs provided by GNSS manufacturers include mathematical modelling options for the troposphere and ionosphere. The processing operator must choose the most appropriate processing strategy for the region where the survey will be performed.

Errors due to solar flare occurrences: solar flares are harmful to signals travelling through space before reaching the ground antennas. Therefore, whenever possible, the engineer should consult relevant online portals to check the occurrence of solar flares during the design phase of the survey project and avoid performing the survey during these periods.

Errors due to multipath: this type of error occurs due to reflected signals reaching the GNSS antenna from reflective surfaces. Depending on local conditions, the multipath effect can be significant enough to cause cycle slips and reduce the coordinate accuracy of the measured point as the signal travels longer paths. Some manufacturers offer special types of antennas that can filter out reflected signals, but this does not guarantee that the effect will be eliminated. The best way to reduce it is to choose locations with no reflective surfaces near the ground antenna.

Errors due to loss of cycles: this type of error is detected when, for some reason, the GNSS antenna stops receiving the signal from one or more satellites. This results in a loss of signal, which leads to a loss of count of the integer number of cycles measured by the receiver, thus causing the "loss of ambiguity". If the cycle slips are small (few observation epochs), the software can correct these errors using "cycle slip fixing" techniques. On the other hand, if the cycle slips are large (many observation epochs), this may compromise the final precision of the processed coordinates, leading to the cancellation of the work and the need to return to the field to perform new observations to repeat the work.

Errors due to relativity: these are errors caused by the orbital motion of the satellites around the Earth, which, according to the theory of relativity, causes the clocks of the atomic particles in the satellites to be "delayed" because the clocks of the receivers and the clocks of the satellites are in different gravitational fields and moving at different speeds. These corrections are calculated by the master station and applied to the satellites' clocks.

In addition to the errors listed above, others of lesser significance are beyond the scope of this book. Readers interested in more details should consult specialised references.

17.10.3   Operational Errors

The operational errors to be considered in a GNSS survey are as follows:

Antenna centring error: this error occurs similarly to the centring errors of surveying instruments mentioned in Sect. 10.2.

Error in measuring the antenna height: the user must pay attention to the precise measurement of the antenna height (vertical or inclined), as this measurement is important for determining the geometric height of the measured point.

Receiver setup errors: users must ensure that the receivers (base and remotes) are set up with the same parameters, such as data acquisition rates, elevation (cut-off) angles, positioning methods, coordinate systems and others, to avoid losing measured data.

In addition to the errors described above, it is also important to consider the accuracies of other elements involved in the field survey, as follows:

- Accuracy of the coordinates *(X, Y, Z)* of the reference station, usually provided with the coordinate values.
- Accuracy of the level bubble of the remote antenna pole (if applicable).
- Tripod stability (ISO 12858-2:1999).
- Accuracy of mathematical models used for coordinate transformation and geoidal undulation (if applicable).

Another factor affecting the accuracy of GNSS measurements is the geometric arrangement of the satellites in space over the measurement area at the time of measurement. To consider this effect, a quality factor called *Dilution of Precision* (DOP) has been defined, which relates the geometry of the satellites in space with the precision of the position of the measured point. To this end, two DOP values must be prioritised in GNSS technology measurements: PDOP and GDOP.[12] The PDOP (*Position Dilution of Precision*) value expresses the dilution of coordinate precision at the time of observation, and the GDOP (*Geometric Dilution of Precision*) value represents the influence of satellite geometry and time measurement on the quality of the observations. In general, it is recommended that observations be carried out at times when the PDOP and GDOP are less than 3 and 6, respectively. Users can obtain information about satellite availability, PDOP and GDOP from the mission planning programs that are part of the data processing program that comes with the instruments or from applications available on Internet portals. This information is often displayed in a variety of methods, including graphs, charts and diagrams, such as a skyplot which displays the satellite constellation over a location.

The values of the PDOP and GDOP coefficients are determined as a function of the diagonal elements of the cofactor matrix Q_{XX} obtained in the adjustment of GNSS observations performed to position the tracked points. Thus,

$$\text{GDOP} = \sqrt{q_{XX} + q_{YY} + q_{ZZ} + q_{\Delta t \Delta t}} \qquad (17.23)$$

$$\text{PDOP} = \sqrt{q_{XX} + q_{YY} + q_{ZZ}} \qquad (17.24)$$

These values are calculated a priori based on the approximate coordinates of the observation area and the predicted satellite orbit provided by the satellite almanac files.

The coordinate values obtained in the GNSS survey result from combining all the above errors. It should be noted that the accuracy values given in manufacturers'

---

[12] In addition to PDOP and GDOP, VDOP, HDOP, TDOP and HTDOP are also available

catalogues (manuals) only consider instrumental and atmospheric errors for GNSS surveys. Moreover, they indicate the baseline accuracy and not the absolute accuracy of the measured point. For this reason, it is recommended that the results obtained are continuously evaluated for reliability. The reader should remember that GNSS processing is based on statistical evaluations and may give different results under different conditions.

The most reliable results are obtained from network surveys, which allow absolute and relative precision to be determined according to a confidence interval (due to data redundancy).

In the case of results obtained using a single baseline, the only indication of the quality obtained is the internal processing standard deviation, which is always too optimistic due to the high correlation between the measured values. In general, the accuracies obtained are in the order of millimetres, which do not correspond to the precision values of the measured points. As it is impossible to determine the accuracy of the measured point, an empirical rule applied by GNSS technology users is to consider the accuracy to be ten times the value of the precision obtained in processing.

Note that if the post-processing includes a precise ephemeris, the processed coordinate results may be of better quality.

As a summary of the quality values of a GNSS survey, Fig. 17.24 shows typical values of the accuracy ranges obtained in a GNSS survey depending on the positioning method used.

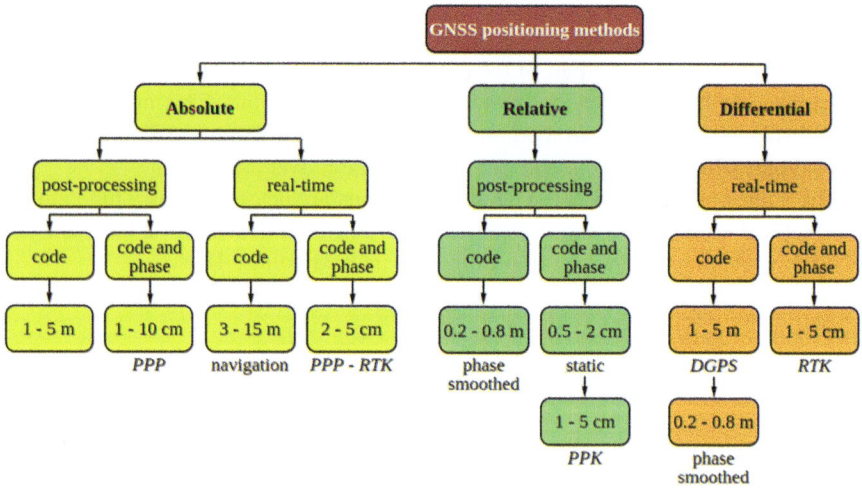

Fig. 17.24 GNSS survey accuracy

17.11 GNSS Data Processing

Processing GNSS data collected in the field means using a computer program developed for this purpose. By processing the data, it is possible to obtain the coordinates of the tracked points and all the information related to the survey. It is important to emphasise that the primary coordinates obtained by post-processing GNSS data are the Three-Dimension Cartesian coordinates (X, Y, Z) of the measured point or the baseline components $(\Delta X, \Delta Y, \Delta Z)$ referenced to the WGS84 geodetic reference system. As explained in Chap. 4, for use in Civil Engineering projects, they must be transformed into geodetic coordinates (latitude, longitude and ellipsoidal height) and then into the desired map projection system.

The first step in processing the data collected in the field is to transfer the data from the receiver to a computer. If the survey is performed in real-time, the coordinates have already been determined in the field, and the program displays them with the other information related to the survey. In the case of relative static and kinematic mode surveys, the coordinates are obtained after combining the observation data from all the baselines involved in the survey, a process known as post-processing.

According to the types of surveys presented in previous sections, data post-processing can be autonomous, radial or network.

Autonomous post-processing strictly follows the prerogatives mentioned in Sect. 17.9.2.1.2.

Data post-processing is considered radial if performed for each independent baseline or by section. In the post-processing section, independent baselines are processed simultaneously for all the survey sections, as shown in Fig. 17.25. In this case, the coordinates of one or more stations are usually (but not necessarily) tightly constrained to their known values, and the coordinates of the new stations are estimated relative to the reference frame of the known stations. As an example, consider the case where there is a set of points whose coordinates need to be determined. If the user has only one pair of receivers, the coordinates of these stations can be determined from pairwise sessions, starting from a reference station.

Fig. 17.25 Principle of radial data post-processing

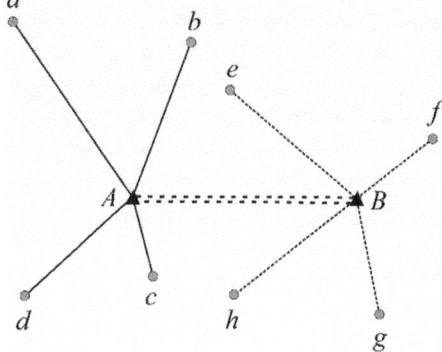

In Fig. 17.25, assuming that point (A) is a point with known coordinates, it can be used as a reference station to determine the coordinates of points (a), (b), (c), (d) and (B). Then, the coordinates of points (e), (f), (g) and (h) are determined from point (B). Although commonly used for GNSS measurements post-processing, this is not the best strategy for determining coordinates because the points are disconnected and do not guarantee rigidity in the adjusted solution. The solution is not considered good even if the points are reoccupied. The best solution is achieved if the measured points are connected through sessions to "geometrically close the sections", as shown in Fig. 17.26.

Fig. 17.26 Principle of closed section post-processing

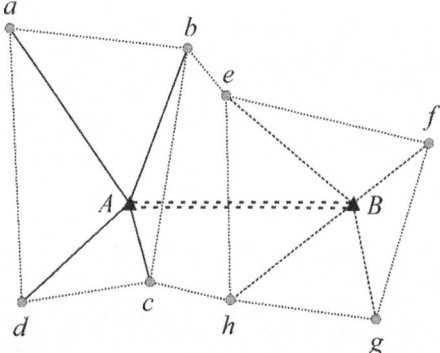

Close the section, as stated above, means determining a network of closed triangles or quadrilaterals of vector lines, which is called a *GNSS network*, as described in the next section.

17.12   Establishment and Adjustment of GNSS Network

The differencing technique used in carrier phase measurements does not directly produce positions for the points occupied by the receivers. Instead, baselines are determined. These baselines are computed concerning their corresponding coordinate differences, ΔX, ΔY and ΔZ. For example, suppose that two stations, (A) and (1), are occupied for an observation session, where station (A) is a control point and station (1) is a point of unknown position, as shown in Fig. 17.27. The session would produce coordinate differences of ΔX_{1A}, ΔY_{1A} and ΔZ_{1A}. The (X, Y, Z) coordinates of station (1) are then obtained by subtracting the baseline components to the coordinates of station (A) as follows:

Fig. 17.27 GNSS network

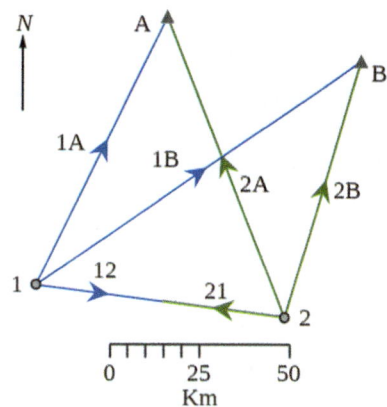

$$X_1 = X_A - \Delta X_{1A}$$
$$Y_1 = Y_A - \Delta Y_{1A} \qquad\qquad (17.28)$$
$$Z_1 = Z_A - \Delta Z_{1A}$$

Adding observation errors to the observation Eqs. (17.28), the following error equations are obtained:

$$\Delta X_{1A} + v X_{1A} = X_A - X_1$$
$$\Delta Y_{1A} + v Y_{1A} = Y_A - Y_1 \qquad\qquad (17.29)$$
$$\Delta Z_{1A} + v Z_{1A} = Z_A - Z_1$$

where

$\Delta X_{1A},\ \Delta Y_{1A},\ \Delta Z_{1A}$ = observations
$v X_{1A},\ v Y_{1A},\ v Z_{1A}$ = observation errors

Suppose at least three independent baselines are measured to form a closed triangle, and a GNSS network with redundant data is created, allowing least squares adjustment for coordinate determination. If more than three independent baselines are created, as in the example in Fig. 17.27, where points (A) and (B) are control points and points (1) and (2) are remote points, there are eighteen error equations. Grouping and rearranging these equations gives the following set of error equations:

$$vX_{1A} = -1X_1 + 0Y_1 + 0Z_1 + 0X_2 + 0Y_2 + 0Z_2 - (\Delta X_{1A} - X_A)$$
$$vY_{1A} = 0X_1 - 1Y_1 + 0Z_A + 0X_2 + 0Y_2 + 0Z_2 - (\Delta Y_{1A} - Y_A)$$
$$vZ_{1A} = 0X_1 + 0Y_1 - 1Z_1 + 0X_2 + 0Y_2 + 0Z_2 - (\Delta Z_{1A} - Z_A)$$
$$vX_{1B} = -1X_1 + 0Y_1 + 0Z_1 + 0X_2 + 0Y_2 + 0Z_2 - (\Delta X_{1B} - X_B)$$
$$vY_{1B} = 0X_1 - 1Y_1 + 0Z_A + 0X_2 + 0Y_2 + 0Z_2 - (\Delta Y_{1B} - Y_B)$$
$$vZ_{1B} = 0X_1 + 0Y_1 - 1Z_1 + 0X_2 + 0Y_2 + 0Z_2 - (\Delta Z_{1B} - Z_B)$$
$$vX_{12} = -1X_1 + 0Y_1 + 0Z_1 + 1X_2 + 0Y_2 + 0Z_2 - \Delta X_{12}$$
$$vY_{12} = 0X_1 - 1Y_1 + 0Z_1 + 0X_2 + 1Y_2 + 0Z_2 - \Delta Y_{12}$$
$$vZ_{12} = 0X_1 + 0Y_1 - 1Z_1 + 0X_2 + 0Y_2 + 1Z_2 - \Delta Z_{12}$$
$$vX_{21} = 1X_1 + 0Y_1 + 0Z_1 - 1X_2 + 0Y_2 + 0Z_2 - \Delta X_{21} \quad (17.30)$$
$$vY_{21} = 0X_1 + 1Y_1 + 0Z_1 + 0X_2 - 1Y_2 + 0Z_2 - \Delta Y_{21}$$
$$vZ_{21} = 0X_1 + 0Y_1 + 1Z_1 + 0X_2 + 0Y_2 - 1Z_2 - \Delta Z_{21}$$
$$vX_{2A} = 0X_1 + 0Y_1 + 0Z_1 - 1X_2 + 0Y_2 + 0Z_2 - (\Delta X_{2A} - X_A)$$
$$vY_{2A} = 0X_1 + 0Y_1 + 0Z + 0X_2 - 1Y_2 + 0Z_2 - (\Delta Y_{2A} - Y_A)$$
$$vZ_{2A} = 0X_1 + 0Y_1 + 0Z + 0X_2 + 0Y_2 - 1Z_2 - (\Delta Z_{2A} - Z_A)$$
$$vX_{2B} = 0X_1 + 0Y_1 + 0Z_1 + 1X_2 + 0Y_2 + 0Z_2 - (\Delta X_{2B} - X_B)$$
$$vY_{2B} = 0X_1 + 0Y_1 + 0Z + 0X_2 + 1Y_2 + 0Z_2 - (\Delta Y_{2B} - Y_B)$$
$$vZ_{2B} = 0X_1 + 0Y_1 + 0Z + 0X_2 + 0Y_2 + 1Z_2 - (\Delta Z_{2B} - Z_B)$$

In Eq. (17.29), the unknown parameters are the coordinates (X_1, Y_1, Z_1) and (X_2, Y_2, Z_2), which can be calculated using the Parametric adjustment model given by the equation $v = Ax - l_0$ (Eq. 3.79), where

$$A = \begin{bmatrix} -1 & 0 & 0 & 0 & 0 & 0 \\ 0 & -1 & 0 & 0 & 0 & 0 \\ 0 & 0 & -1 & 0 & 0 & 0 \\ -1 & 0 & 0 & 0 & 0 & 0 \\ 0 & -1 & 0 & 0 & 0 & 0 \\ 0 & 0 & -1 & 0 & 0 & 0 \\ -1 & 0 & 0 & 1 & 0 & 0 \\ 0 & -1 & 0 & 0 & 1 & 0 \\ 0 & 0 & -1 & 0 & 0 & 1 \\ 1 & 0 & 0 & -1 & 0 & 0 \\ 0 & 1 & 0 & 0 & -1 & 0 \\ 0 & 0 & 1 & 0 & 0 & -1 \\ 0 & 0 & 0 & -1 & 0 & 0 \\ 0 & 0 & 0 & 0 & -1 & 0 \\ 0 & 0 & 0 & 0 & 0 & -1 \\ 0 & 0 & 0 & -1 & 0 & 0 \\ 0 & 0 & 0 & 0 & -1 & 0 \\ 0 & 0 & 0 & 0 & 0 & -1 \end{bmatrix} \quad l_0 = \begin{bmatrix} \Delta X_{1A} - X_A \\ \Delta Y_{1A} - Y_A \\ \Delta Z_{1A} - Z_A \\ \Delta X_{1B} - X_B \\ \Delta Y_{1B} - Y_B \\ \Delta Z_{1B} - Z_B \\ \Delta X_{12} \\ \Delta Y_{12} \\ \Delta Z_{12} \\ \Delta X_{21} \\ \Delta Y_{21} \\ \Delta Z_{21} \\ \Delta X_{2A} - X_A \\ \Delta Y_{2A} - Y_A \\ \Delta Z_{2A} - Z_A \\ \Delta X_{2B} - X_B \\ \Delta Y_{2B} - Y_B \\ \Delta Z_{2B} - Z_B \end{bmatrix} \quad x = \begin{bmatrix} X_1 \\ Y_1 \\ Z_1 \\ X_2 \\ Y_2 \\ Z_2 \end{bmatrix}$$

The GNSS network survey consists of two processing phases. The first is processing the GNSS data to obtain the "n" baseline vectors (sets of adjusted components ΔX, ΔY, ΔZ) created with $n + 1$ points in the network. The second is the adjustment of these baselines to obtain the adjusted coordinates.

As the baseline vectors are the result of the first least squares adjustment, the processing software can provide the variance-covariance matrix (3x3) of each baseline, which can be used to weight the observations. The inverse of the variance-covariance matrix of each baseline is then used as the weight matrix in the network adjustment.

The adjustment computation can be performed using Eqs. (3.84) to (3.87), from which the coordinates of the remote points are obtained. Finally, as each baseline generates three observation equations and each remote point creates three unknowns, the standard deviations of the unit weight can be calculated using Eq. (17.30), and the standard deviation of the unknowns using the variance-covariance matrix given by Eq. (17.31). The remaining statistical parameters can be determined as indicated in Sect. 3.9.1.2.

A total of n observations and two unknowns have been considered.

$$s_0^2 = \frac{v^{\mathrm{T}} P v}{3(\text{baselines-remote points})} \qquad (17.31)$$

$$\Sigma_{xx} = s_0^2 N^{-1} \qquad (17.32)$$

When implementing a GNSS network, some fundamental issues need to be considered to achieve the desired accuracy for the project. The following are some suggestions that may be helpful in this task:

1. The choice of station location must ensure their physical integrity against vandalism and ease of access to carry out work from this point. A suitable alternative is to set out these points in areas of public institutions that have internal security services.
2. The site must provide an unobstructed view of the observer's horizon so that the antenna is immune to obstacles that prevent perfect signal reception.
3. The site must have minimum characteristics to influence multipath signals.
4. Whenever possible, the station should be monumentalised. If possible, a concrete landmark should be built with a foundation that guarantees its stability, and antennas should be installed at this point using forced centring pins.
5. Using forced centring mechanisms is essential to ensure that observation sessions are always carried out under the same installation conditions.
6. The equipment to be used must be suitable for the accuracy required by the project. The user must be aware of the limitations described by the instrument manufacturer. Whenever possible, dual-frequency receivers should be used. This minimises atmospheric effects.

7. Depending on the characteristics of the project and the area to be surveyed, the type of antenna used must be considered. The best thing is to look for a type and model suitable for the project. It is important to know all the information about this accessory.
8. The professionals responsible for processing and adjusting data must have technical and scientific skills and knowledge that help them explore the full capacity of the processing and adjustment software to find the best solution.

Regarding practical and logistical constraints, the best strategy for carrying out observation sessions is based on the experience from previous projects. Given that the accuracy of a local GNSS network depends fundamentally on the length of the vectors involved, the planning and execution of sessions will be an economic (cost) and logistical (equipment, personnel, ease of access, etc.) function. Regardless of the size of the project, some precautionary measures are recommended:

- Each station should be occupied for data collection at least twice under different circumstances, with different geometric configurations of the satellites. The objective of making two or more observations per station is to detect any errors that may occur. The number of redundant observations will depend on logistical and economic factors.
- The closest stations should collect data simultaneously to obtain fixed solutions for ambiguities, as they are more easily resolved and fixed for short baselines.
- All receivers involved should use the same data recording collection rates. Failure to do so may be detrimental to the resolution and fixation of ambiguities and, consequently, will influence the processing results.
- Whenever possible, the number of receivers should be greater than four to obtain higher work rates, greater geometric rigidity, and simplify the operation logistics.
- To check their accuracy, some vectors should be observed more than once (especially those of longer lengths) whenever possible.

During the planning stage for implementing a network, the users should seek information about a network close to their area of interest. In this case, the planners must include at least two observations for each point of the nearby network in their projects. This indication aims to ensure that these networks have common points to form a more extensive network and are not seen as isolated. Otherwise (if there is no nearby network), the network should start from a reference point that is part of a network of points recognised by legal authorities.

The engineer should aim for the best possible accuracy and reliability when implementing a network. The degree of reliability of a network indicates its quality. Reliability means the ability of the network itself to guarantee the checking of systematic errors and the occurrence of blunders. The degree of reliability of a network is related to its rigidity, which is directly related to the number of occupations per station and, consequently, the number of vectors per station.

The processing phase can begin as soon as the data has been collected and downloaded onto computers. This step must be carried out with maximum care, not simply accepting the default conditions of the software used. The user must be able to choose the best processing options to obtain the best results, immune to blunders.

Once processing is complete, the next step is to adjust the network points. In this step, it is necessary to define a point with known coordinates and their respective variances and covariances, which will serve as a reference for the adjustment. This strategy is known as *minimally constrained*. After this preliminary step, it is always advisable to include more than one point with known coordinates for the adjustment whenever possible. This strategy ensures greater reliability of the results.

Most receiver manufacturers include data processing software in their sales packages. The user should check whether the adjustment module is included in this package. In this case, it is recommended that all work with network features be adjusted, as the final adjusted coordinates are of better quality than the processed coordinates. Otherwise, the user should purchase the adjustment module for their projects.

17.12.1 Final Work Report

In all engineering projects, a final report must be prepared at the end of the field and office work to document the work carried out and to allow for future analysis. The report must include at least the following:

- A detailed description of the area where the survey was carried out, if possible, with an image of the location of the point taken with a camera or easily obtained using Google Earth.
- The objectives of the work and the desired and achieved accuracies.
- Description of the geodetic markers with photos and details of their positions to facilitate possible future visits.
- A detailed description of the software used to process and adjust the field data, including name and version.
- A detailed description of all the stages of the field survey, including the names of the field operators.
- A description of all events during the observation session that are considered relevant for data processing and adjustment analysis.
- All files generated by data collection, processing and adjustment are conveniently stored on secure media to ensure data integrity and security.
- A final printed and digital report for presentation to the client.

Example 17.1

Using the GNSS network shown in Fig. 17.28, calculate the adjusted coordinates of points (A) and (B) and their standard deviations. Plot the absolute error ellipses for points (1) and (2). The control point coordinates are shown in Table 17.4. The baseline lengths are given in Table 17.5. The variance-covariance elements for the baselines are given in Table 17.6.

Fig. 17.28 GNSS network
baselines

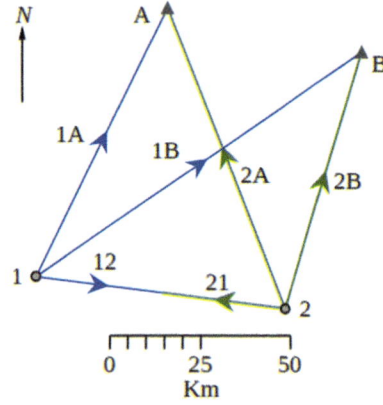

Table 17.4 Control point coordinates (SIRGAS2000)

Control point	Geocentric Cartesian coordinates		
	X [m]	Y [m]	Z [m]
A	4,007,215.1075	−4,306,650.1936	−2,458,220.5598
B	4,043,683.2088	−4,266,290.4477	−2,469,479.8486

Table 17.5 Baseline lengths

Baseline	ΔX [m]	ΔY [m]	ΔZ [m]
1A	46,218.611394	3442.261024	67,514.461922
1B	82,686.714523	43,801.999098	56,255.196407
12	49,103.012747	50,165.158713	−7803.775996
21	−49,103.023432	−50,165.148751	7803.783709
2A	−2884.382438	−46,722.885419	75,318.227307
2B	33,583.679489	−6363.147623	64,058.970535

Table 17.6 Variance-covariance matrices (3x3) for each baseline

Baseline	Variance-covariance elements		
1A	$3.243815 * 10^{-5}$	$2.697051 * 10^{-6}$	$-5.060925 * 10^{-6}$
	$2.697051 * 10^{-6}$	$3.064896 * 10^{-5}$	$7.122440 * 10^{-6}$
	$-5.060925 * 10^{-6}$	$7.122440 * 10^{-6}$	$8.630786 * 10^{-6}$
12	$3.213665 * 10^{-5}$	$3.000032 * 10^{-6}$	$-5.478830 * 10^{-6}$
	$3.000032 * 10^{-6}$	$3.031527 * 10^{-5}$	$7.364385 * 10^{-6}$
	$-5.478830 * 10^{-6}$	$7.364385 * 10^{-6}$	$8.541649 * 10^{-6}$
2A	$3.172201 * 10^{-5}$	$1.716886 * 10^{-6}$	$-5.698628 * 10^{-6}$
	$1.716886 * 10^{-6}$	$2.983454 * 10^{-5}$	$6.701419 * 10^{-6}$
	$-5.698628 * 10^{-6}$	$6.701419 * 10^{-6}$	$8.525115 * 10^{-6}$
1B	$3.714610 * 10^{-5}$	$2.481507 * 10^{-6}$	$-6.385173 * 10^{-6}$
	$2.481507 * 10^{-6}$	$3.804805 * 10^{-5}$	$9.186476 * 10^{-6}$
	$-6.385173 * 10^{-6}$	$9.186476 * 10^{-6}$	$9.834433 * 10^{-6}$
21	$3.103147 * 10^{-5}$	$2.216674 * 10^{-6}$	$-5.868894 * 10^{-6}$
	$2.216674 * 10^{-6}$	$2.957456 * 10^{-5}$	$6.981379 * 10^{-6}$
	$-5.868894 * 10^{-6}$	$6.981379 * 10^{-6}$	$8.341353 * 10^{-6}$
2B	$3.656343 * 10^{-5}$	$3.976674 * 10^{-6}$	$-5.879752 * 10^{-6}$
	$3.976674 * 10^{-6}$	$3.523883 * 10^{-5}$	$8.350366 * 10^{-6}$
	$-5.879752 * 10^{-6}$	$8.350366 * 10^{-6}$	$9.741889 * 10^{-6}$

Solution:

Since the network in this example has the same configuration as in Fig. 17.27, *for the application of the Parametric adjustment model, using* Eq. (3.79), *the following matrices can be written.*

$$
A = \begin{bmatrix}
-1 & 0 & 0 & 0 & 0 & 0 \\
0 & -1 & 0 & 0 & 0 & 0 \\
0 & 0 & -1 & 0 & 0 & 0 \\
-1 & 0 & 0 & 0 & 0 & 0 \\
0 & -1 & 0 & 0 & 0 & 0 \\
0 & 0 & -1 & 0 & 0 & 0 \\
-1 & 0 & 0 & 1 & 0 & 0 \\
0 & -1 & 0 & 0 & 1 & 0 \\
0 & 0 & -1 & 0 & 0 & 1 \\
1 & 0 & 0 & -1 & 0 & 0 \\
0 & 1 & 0 & 0 & -1 & 0 \\
0 & 0 & 1 & 0 & 0 & -1 \\
0 & 0 & 0 & -1 & 0 & 0 \\
0 & 0 & 0 & 0 & -1 & 0 \\
0 & 0 & 0 & 0 & 0 & -1 \\
0 & 0 & 0 & -1 & 0 & 0 \\
0 & 0 & 0 & 0 & -1 & 0 \\
0 & 0 & 0 & 0 & 0 & -1
\end{bmatrix}
\quad
l_0 = \begin{bmatrix}
-3,960,996.4961 \\
4,310,092.4546 \\
2,525,735.0217 \\
-3,960,996.4943 \\
4,310,092.4468 \\
2,525,735.0450 \\
49,103.0127 \\
50,165.1587 \\
-7803.7760 \\
-49,103.0234 \\
-50,165.1488 \\
7803.7837 \\
-4,010,099.4899 \\
4,259,927.3082 \\
2,533,538.7871 \\
-4,010,099.5293 \\
4,259,927.3001 \\
2,533,538.8191
\end{bmatrix}
\quad
x = \begin{bmatrix}
X_1 \\
Y_1 \\
Z_1 \\
X_2 \\
Y_2 \\
Z_2
\end{bmatrix}
$$

Assuming the weighting model equal to the inverse of the standard deviations, the weight matrix **P** *is given as follows:*

$P = s_0^2 * \Sigma_{ll}^{-1}$. *Thus, for an* a-priori *value of s_0 equal to* 0.001 m, *the following weight matrix is obtained:*

$P =$

3.628	−1.007	2.958	0	0	0	0	0	0
−1.007	4.316	−4.152	0	0	0	0	0	0
2.958	−4.152	16.748	0	0	0	0	0	0
0	0	0	3.270	−0.937	2.999	0	0	0
0	0	0	−0.937	3.662	−4.030	0	0	0
0	0	0	2.999	−4.030	15.879	0	0	0
0	0	0	0	0	0	3.827	−1.233	3.518
0	0	0	0	0	0	−1.233	4.570	−4.731
0	0	0	0	0	0	3.518	−4.731	18.043	......
0	0	0	0	0	0	0	0	0
0	0	0	0	0	0	0	0	0
0	0	0	0	0	0	0	0	0
0	0	0	0	0	0	0	0	0
0	0	0	0	0	0	0	0	0
0	0	0	0	0	0	0	0	0
0	0	0	0	0	0	0	0	0
0	0	0	0	0	0	0	0	0
0	0	0	0	0	0	0	0	0

0	0	0	0	0	0	0	0	0
0	0	0	0	0	0	0	0	0
0	0	0	0	0	0	0	0	0
0	0	0	0	0	0	0	0	0
0	0	0	0	0	0	0	0	0
0	0	0	0	0	0	0	0	0
0	0	0	0	0	0	0	0	0
0	0	0	0	0	0	0	0	0
...... 0	0	0	0	0	0	0	0	0
4.039	−1.213	3.857	0	0	0	0	0	0
−1.213	4.578	−4.685	0	0	0	0	0	0
3.857	−4.685	18.624	0	0	0	0	0	0
0	0	0	3.795	−0.957	3.289	0	0	0
0	0	0	−0.957	4.312	−4.029	0	0	0
0	0	0	3.289	−4.029	17.096	0	0	0
0	0	0	0	0	0	3.320	−1.066	2.918
0	0	0	0	0	0	−1.066	3.903	−3.989
0	0	0	0	0	0	2.918	−3.989	15.445

Considering the weight matrix given above and using Eqs. (3.84) *to* (3.87).

$$N = \begin{bmatrix} 14.76337123 & -4.390408595 & 13.33155369 & -7.865197281 & 2.446394738 & -7.374627027 \\ -4.390408595 & 17.1268679 & -17.59840726 & 2.446394738 & -9.148288386 & 9.416575433 \\ 13.33155369 & -17.59840726 & 69.29349534 & -7.374627027 & 9.416575433 & -36.66642995 \\ -7.865197281 & 2.446394738 & -7.374627027 & 14.98039073 & -4.469679394 & 13.58156083 \\ 2.446394738 & -9.148288386 & 9.416575433 & -4.469679394 & 17.36368035 & -17.43526936 \\ -7.374627027 & 9.416575433 & -36.66642995 & 13.58156083 & -17.43526936 & 69.20800547 \end{bmatrix}$$

$$n = \begin{bmatrix} 20,450,972.741471 \\ -21,835,792.949531 \\ -23,151,100.625300 \\ 21,632,102.854114 \\ -22,382,588.348648 \\ -23,792,273.140231 \end{bmatrix} \quad x = \begin{bmatrix} X_1 \\ Y_1 \\ Z_1 \\ X_2 \\ Y_2 \\ Z_2 \end{bmatrix} = \begin{bmatrix} 3,960,996.4935 \\ -4,310,092.4534 \\ -2,525,735.0294 \\ 4,010,099.5099 \\ -4,259,927.3018 \\ -2,533,538.8060 \end{bmatrix}_m$$

Using Eq. (3.79),

$$v = \begin{bmatrix} 0.002626 \\ -0.001229 \\ 0.007712 \\ 0.000797 \\ 0.006597 \\ -0.015573 \\ 0.003675 \\ -0.007084 \\ -0.000548 \\ 0.007010 \\ -0.002878 \\ -0.007165 \\ -0.019963 \\ -0.006415 \\ 0.018871 \\ 0.019410 \\ 0.001689 \\ -0.013157 \end{bmatrix}_m$$

In this example, since the approximate values are equal to zero, the adjusted values of the unknowns are equal to the correction values given in the **x** vector.

The standard deviation of unit weight is calculated using Eq. (17.31) with $n - u = 12$.

$$s_0^2 = 1.2943618784 * 10^{-3} \text{ m}^2$$

$$s_0 = \pm 0.03597 \text{ m} = \pm 3.60 \text{ mm}$$

Using Eq. (17.32), the variance-covariance matrix for the parameters (coordinates X, Y, Z) is given as follows:

$$\Sigma_{xx} = \begin{bmatrix} 1.46561*10^{-4} & 1.13727*10^{-5} & -2.45708*10^{-5} & 7.5961*10^{-5} & 5.76884*10^{-6} & -1.24013*10^{-5} \\ 1.13727*10^{-5} & 1.42118*10^{-4} & 3.35422*10^{-5} & 5.78333*10^{-6} & 7.3728*10^{-5} & 1.70847*10^{-5} \\ -2.45708*10^{-5} & 3.35422*10^{-5} & 3.8975*10^{-5} & -1.24132*10^{-5} & 1.70948*10^{-5} & 2.02096*10^{-5} \\ 7.5961*10^{-5} & 5.78333*10^{-6} & -1.24132*10^{-5} & 1.45206*10^{-4} & 1.14795*10^{-5} & -2.48728*10^{-5} \\ 5.76884*10^{-6} & 7.3728*10^{-5} & 1.70948*10^{-5} & 1.14795*10^{-5} & 1.38731*10^{-4} & 3.2337*10^{-5} \\ -1.24013*10^{-5} & 1.70847*10^{-5} & 2.02096*10^{-5} & -2.48728*10^{-5} & 3.2337*10^{-5} & 3.87911*10^{-5} \end{bmatrix}_{\text{m}^2}$$

Using the diagonal values in the variance-covariance matrix for the parameters, the standard deviations of the adjusted coordinates are given as follows:

$X_1 = 3,960,996.4935$ m ± 12.1 mm $X_2 = 4,010,099.5099$ m ± 12.1 mm

$Y_1 = -4,310,092.4534$ m ± 11.9 mm $Y_2 = -4,259,927.3018$ m ± 11.8 mm

$Z_1 = -2,525,735.0294$ m ± 6.2 mm $Z_2 = -2,533,538.8060$ m ± 6.2 mm

Finally, using Eqs. (3.73) and (3.94) and calculating the precision of the adjusted observations by writing the matrix $\Sigma_{\tilde{l}\tilde{l}}$, the following results are obtained:

$\Delta X_{1A} = 46,218.6140$ m ± 12.1 mm $\Delta X_{1B} = 82,686.7153$ m ± 12.1 mm

$\Delta Y_{1A} = 34,422.5980$ m ± 11.9 mm $\Delta Y_{1B} = 43,802.0057$ m ± 11.9 mm

$\Delta Z_{1A} - 67,514.4696$ m ± 6.2 mm $\Delta Z_{1B} = 56,255.1808$ m $+ 6.2$ mm

$\Delta X_{12} = 49,103.0164$ m ± 11.8 mm $X_{21} = -49,103.0164$ m ± 11.8 mm

$\Delta Y_{12} = 50,165.1516$ m ± 11.5 mm $Y_{21} = -50,165.1516$ m ± 11.5 mm

$\Delta Z_{12} = -7803.7765$ m ± 6.1 mm $Z_{21} = 7803.7765$ m ± 6.1 mm

$X_{2A} = -2884.4024$ m ± 12.1 mm $X_{2B} = 33,583.6989$ m ± 12.1 mm

$Y_{2A} = -46,722.8918$ m ± 11.8 mm $Y_{2B} = -6363.1459$ m ± 11.8 mm

$Z_{2A} = 75,318.2462$ m ± 6.2 mm $Z_{2B} = 64,058.9574$ m ± 6.2 mm

The coordinates of points (1) and (2) presented above are in geocentric Cartesian coordinates; therefore, the precision distribution is best represented by an error ellipsoid rather than an error ellipse. However, in most cases, the error ellipse is preferred because it is the easiest to plot. In these cases, it is necessary to transform the geocentric Cartesian coordinates to local ground coordinates (Sect. 5.9.2.4) and plot the error ellipse in the (X_L, Y_L) plane. This can also be done by calculating the *variance-covariance matrix for the topocentric coordinates (e, n, u), which can be done by applying the general law of propagation of variance-covariance to the observation Eq. (5.98). Thus, using Eq. (3.28),*

$$\mathbf{\Sigma}_{ff(e,n,u)} = F^T \mathbf{\Sigma}_{xx} F \tag{17.32}$$

where

$$
F^T =
\begin{bmatrix}
-\sin\lambda_1 & \cos\lambda_1 & 0 & 0 & 0 & 0 \\
-\cos\lambda_1 * \sin\phi_1 & -\sin\lambda_1 * \sin\phi_1 & \cos\phi_1 & 0 & 0 & 0 \\
\cos\lambda_1 * \cos\phi_1 & \sin\lambda_1 * \cos\phi_1 & \sin\phi_1 & 0 & 0 & 0 \\
0 & 0 & 0 & -\sin\lambda_2 & \cos\lambda_2 & 0 \\
0 & 0 & 0 & -\cos\lambda_2 * \sin\phi_2 & -\sin\lambda_2 * \sin\phi_2 & \cos\phi_2 \\
0 & 0 & 0 & \cos\lambda_2 * \cos\phi_2 & \sin\lambda_2 * \cos\phi_2 & \sin\phi_2
\end{bmatrix}
\tag{17.33}
$$

Then, after transforming the geocentric Cartesian coordinates (X, Y, Z) of points (1) and (2) to geodetic coordinates (λ, ϕ, h), the following results are obtained:

Station 1:	Station 2:
$\phi_1 = -23° 28' 45.089756''$	$\phi_2 = -23° 33' 20.332453''$
$\lambda_1 = -47° 25' 00.607510''$	$\lambda_2 = -46° 43' 49.123033''$
$h_1 = 633.7939$ m	$h_2 = 730.6268$ m

$$
F^T =
\begin{bmatrix}
0.73629594 & 0.67665965 & 0 & 0 & 0 & 0 \\
0.26959202 & -0.29335298 & 0.91720483 & 0 & 0 & 0 \\
0.62063550 & -0.67533420 & -0.39841599 & 0 & 0 & 0 \\
0 & 0 & 0 & 0.72813548 & 0.68543323 & 0 \\
0 & 0 & 0 & 0.27392624 & -0.29099175 & 0.91667236 \\
0 & 0 & 0 & 0.62831770 & -0.66746167 & -0.39963957
\end{bmatrix}
$$

$\mathbf{\Sigma}_{ff(e,n,u)} =$
$$
\begin{bmatrix}
1.558593*10^{-4} & 4.724034*10^{-6} & -6.833312*10^{-7} & 8.068073*10^{-5} & 2.866421*10^{-6} & 4.956570*10^{-7} \\
4.724034*10^{-6} & 2.367027*10^{-5} & 6.089680*10^{-6} & 2.374066*10^{-6} & 1.263741*10^{-5} & 3.511087*10^{-6} \\
-6.833312*10^{-7} & 6.089680*10^{-6} & 1.481251*10^{-4} & -1.258198*10^{-6} & 3.615038*10^{-6} & 7.656942*10^{-5} \\
8.068073*10^{-5} & 2.374066*10^{-6} & -1.258198*10^{-6} & 1.536222*10^{-4} & 4.730863*10^{-6} & 7.070828*10^{-7}
\end{bmatrix}
$$

The cofactors of (e) and (n) are obtained from the 2 x 2 covariance submatrices for the east and north coordinates given in the matrix $\mathbf{\Sigma}_{ff(e,n,u)}$ above. So, given that

$$\mathbf{Q}_{ff(e,n)} = \frac{1}{s_0^2} \mathbf{\Sigma}_{ff(e,n)} \tag{17.34}$$

The following cofactor matrices are obtained:

Station 1:		Station 2:	
$\mathbf{Q}_{ff(e,n)} = \begin{bmatrix} 0.1204139866 & 0.0036497010 \\ 0.0036497010 & 0.0182872110 \end{bmatrix}$		$\mathbf{Q}_{ff(e,n)} = \begin{bmatrix} 0.114438695 & -0.000972060 \\ -0.000972060 & 0.118685666 \end{bmatrix}$	

The up component is also obtained from the diagonal of the $\sum_{ff(e,\,n,\,u)}$. They correspond to the positions 3.3 and 6.6 in the matrix).

Using Eqs. (3.149), (3.150) and (3.151).

Station 1:	Station 2:
Semi-Major Axis (p) = 12.5 mm	Semi-Major Axis (p) = 12.4 mm
Semi-Minor Axis (q) = 4.9 mm	Semi-Minor Axis (q) = 4.8 mm
Azimuth (θ) = 87° 57′ 21″	Azimuth (θ) = 87° 55′ 04″

Figure 17.29 shows the two absolute error ellipses for stations 1 and 2 together with the up vector represented by the vertical rectangles at each station.

Fig. 17.29 Absolute error ellipses and the up vector of the GNSS network for stations 1 and 2

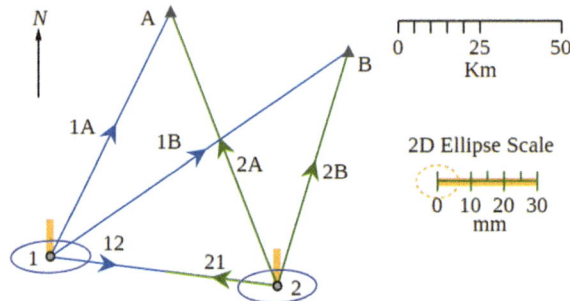

17.13   Review Questions

1. Describe the main GNSS segments, their composition and their importance to the satellite system.
2. Briefly discuss the advantages and disadvantages of GNSS.
3. Discuss the main factors affecting the accuracy of GNSS observations.
4. How important are tropospheric and ionospheric models in GNSS data processing?
5. Briefly discuss the differences between using pseudorange and carrier phase ranging techniques for GNSS point positioning.
6. Define what L_1, L_2 and L_5 frequencies are in GPS.
7. What is ambiguity resolution for GNSS point positioning?
8. What is cycle slip and how does it affect GNSS positioning?
9. Briefly explain how GNSS errors can be corrected for point positioning.
10. Explain the importance of precise ephemerides for GNSS positioning.

11. What are the differences between absolute, relative and differential GNSS positioning?
12. Discuss the basic requirements for using relative static and kinematic positioning.
13. Explain what SBAS and WASS means for GNSS.
14. What is a PPK positioning?
15. What is the difference between DGPS and RTK positioning modes, and what is the expected level of accuracy when using each mode?
16. What is the NMEA 0183 format, and what is it used for?
17. What is the RINEX format, and what is it used for?
18. What is the RTCM protocol, and how is it used for DGPS and RTK differential positioning?
19. What is the NTRIP protocol, and how is it used for DGPS and RTK differential positioning?
20. Describe what are the main components of a GNSS instrument for geodetic point positioning.
21. Briefly discuss which are the point positioning methods that can be used in Civil Engineering works.
22. Briefly explain what a Virtual Reference Station is and how it is used for GNSS positioning.
23. Discuss what accuracy levels are expected for relative static point positioning, PPP, PPK, DGPS and RTK.
24. Briefly explain how the GNSS data post-processing process is carried out for GNSS point positioning.
25. Briefly discuss the advantages of using a GNSS network instead of an independent baseline measurement for GNSS point positioning.
26. Briefly explain how a GNSS network survey is carried out in the field.
27. Discuss the main precautions to be taken when setting up a GNSS geodetic network.
28. Briefly discuss for which types of Civil Engineering works you would recommend the use of GNSS relative static and RTK surveying modes.
29. Which GNSS surveying mode would you recommend for setting out points in Civil Engineering works?
30. Explain how to stake out a point for Civil Engineering work using RTK point positioning and what level of accuracy can be expected.
31. Field measurements were carried out to establish a GNNS geodetic network of four points. The Cartesian coordinates, the baseline post-processing results of the baseline length, and their standard deviations are given in Table 17.7. Calculate the adjusted coordinates of each station using the Parametric adjustment model.

Table 17.7   Processing results (SIRGAS2000)

Reference station		Cartesian coordinates (SIRGAS2000)		
		X [m]	Y [m]	Z [m]
CHUA		4,010,549.1890	−4,470,077.4480	−2,143,179.4240
JABO		3,958,011.6951	−4,440,249.2085	−2,295,903.2570
Baseline		$\Delta X_j \pm s_X$[m]	$\Delta Y_j \pm s_Y$[m]	$\Delta Z_j \pm s_Z$[m]
CHUA	AVER	−179, 380.1646 ± 0.0260	−150, 316.9989 ± 0.0160	−8, 223.9392 ± 0.0170
JABO	AVER	−126, 843.4598 ± 0.0260	−180, 144.2821 ± 0.0160	144, 500.3241 ± 0.0170
CHUA	SJRP	−124, 688.5592 ± 0.0250	−56, 942.2957 ± 0.0150	−106, 126.8658 ± 0.0170
JABO	SJRP	−72, 151.8544 ± 0.0250	−86, 769.5788 ± 0.0150	46, 597.3974 ± 0.0170
SJRP	AVER	−54, 691.6254 ± 0.0260	−93, 374.7032 ± 0.0160	97, 902.9267 ± 0.0170
AVER	SJRP	54, 691.6254 ± 0.0250	93, 374.7032 ± 0.0150	−97, 902.9267 ± 0.0170

References

Cina, Alberto; Dabove, Paolo; Manzino, Ambrogio M.; Piras, Marco (2020). *Network Real Time Kinematic (NRTK) Positioning – Description, Architectures and Performances.* https://www. intechopen.com/books/satellite-positioning-methods-models-and-applications/network-real-time-kinematic-nrtk-positioning-description-architectures-and-performances.

Ghilani, C. D., Wolf, P. R. (2007) - Elementary Surveying – An Introduction to Geomatics. 12 Edition. Pearson Prentice Hall, EUA.

Hofmann-Wellenhof, Bernhard; Lichtenegger, Herbert; Wasle, Elmar (2008) *GNSS. Global Navigation Satellite Systems. GPS, Glonass, Galileo & more.* Springer Wien New York. ISBN: 978-3-211-73012-6.

International GNSS Service (IGS), RINEX Working Group and Radio Technical Commission for Maritime Services Special Committee 104 (RTCM-SC104). RINEX (2018). *The Receiver Independent Exchange Format.* Version 3.04. Access: http://acc.igs.org/misc/rinex304.pdf.

Janssen, V. (2009). *A comparison of the VRS and MAC principles for network RTK.* International Global Navigation Satellite Systems Society IGNSS Symposium 2009. Australia.

Kaplan, Elliott D. (1996). *Understanding GPS. Principles and applications.* Artech House. ISBN: 0-89006-793-7.

Leick, Alfre; Rapoport, Lev; Tatarnikov, Dmitry (2015). *GPS satellite surveying.* 4ª Edition. John Wiley & Sons, New Jersey. ISBN: 978-1-118-67557-1.

Lekkerkerk, H-Jan. (2007). *GPS Handbook for professional GPS users.* Netherlands.

Segantine, Paulo C. L. (1995). *Establishment of geodetic high precision São Paulo GPS network.* Doctoral Thesis, Polytechnic School of the University of São Paulo.

Segantine, Paulo C. L. (2005). *GPS – Sistema de Posicionamento Global.* EESC/USP. ISBN: 85-85205-62-8.

Chapter 18
Terrestrial Laser Scanning

Irineu da Silva

18.1 Introduction

Laser scanning technology has become, in a short time,[1] one of the main innovations in Geomatics for geospatial data collection. Based on LiDAR (*Light Detection and Ranging*) technology, which allows the precise and reflectorless measurement of the distance between a sensor[2] and a measurement object by means of spatial scanning (3D scanning) and at a measurement speed of thousands of points per second, laser scanning has established a new method of 3D geospatial data acquisition based on the vectorisation or spatial modelling of the set of scanned points with known spatial coordinates (X, Y, Z).

Unlike a total station, which measures selected discrete points in space, laser scanners determine 3D coordinates by expanding the one-dimensional distance measurement into a spatial measurement of scattered points called a *point cloud*.[3] See Fig. 18.1. Due to their characteristic of spatially scanning the scene, laser scanning instruments are called *laser scanners*. A laser scanner is then classified as an *active data acquisition system* where the raw data product is a point cloud with coordinates (X, Y, Z) georeferenced to a predefined Cartesian spatial coordinate system.[4]

In Civil Engineering, it can be seen as the process of creating a 3D model of a real-world object for which there is no digital information, which can then be

---

[1] The first laser scanning experiments began in the mid-1990s.

[2] A sensor is an electronic device that allows collecting data from the physical environment that can be interpreted by either a human or a machine.

[3] Point cloud is a collection of points converted from range and angular measurements into a common Cartesian coordinate system.

[4] Note that (X, Y, Z), in this case, refers to a generic Cartesian spatial coordinate system.

I. da Silva (✉)
São Carlos School of Engineering, University of São Paulo, São Carlos, SP, Brazil
e-mail: irineu@sc.usp.br

Fig. 18.1 Laser scanning
point cloud

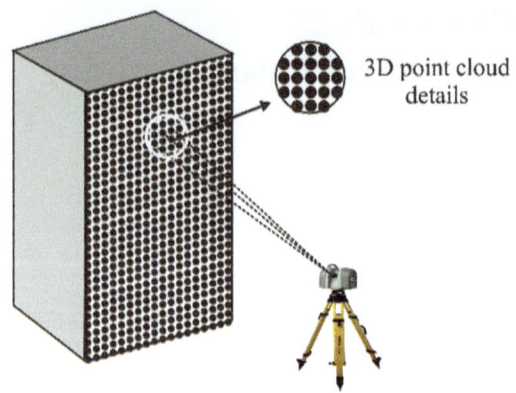

3D point cloud
details

incorporated into existing digital models for further elaboration and use with the information already available.

Compared to other surveying technologies, it has the advantages of high speed, multiple data sampling rate, high data acquisition accuracy, low external impact, non-contact measurement and a high degree of data modelling for mapping purposes. Once the point cloud data is acquired, it is possible to vectorise the scene or reconstruct a three-dimensional model for further analysis and investigation, with greater precision than other tools.

Laser Scanning Measurement Technologies are classified according to the measurement technology used or the platform on which the instrument is installed. Regarding the latter, they can be classified as follows:

- Airborne Laser Scanning (ALS)
- Terrestrial Laser Scanning (TLS)

Laser Scanning Technology is classified as ALS when the measuring system is installed in an aircraft (manned or unmanned) and as TLS when the measuring system is located on the Earth's surface. The latter is further classified as *stationary* and *dynamic*,[5] with dynamic being those where the instrument moves during the measurement process. See Fig. 18.2.

Depending on the measurement technology and the type of instrument used, the measurement distances can vary from a few meters to kilometres, making it possible to use laser scanning technology in areas as diverse as 3D modelling of industrial parts to Digital Terrain Modelling. In Civil Engineering, it has been primarily used in the following applications:

ALS: digital surface modelling, corridor surveys (gas and oil pipelines, transport routes, etc.), geotechnical surveys, environmental management surveys, etc.
Stationary TLS: building or industrial 3D modelling, engineering surveys, as-built surveys, facilities management, mining, geology, archaeological surveys,

---

[5] Also called mobile by some authors.

deformation and monitoring surveys, architectural and historic preservation surveys, earthworks surveys, forensic surveys, erosion surveys and others.

Dynamic vehicular TLS: road/pavement topographic surveys, tunnelling surveys, railways surveys, urban area surveys and others.

Dynamic portable TLS: mapping small areas with difficult access, inside buildings, urban canyons, etc.

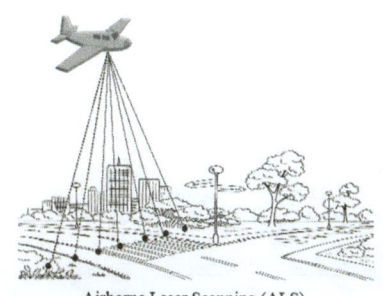

Airborne Laser Scanning (ALS).

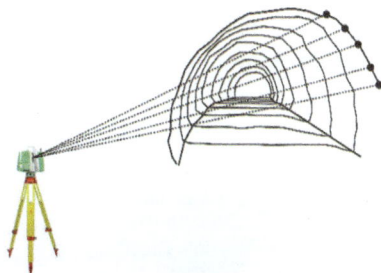

Stationary Terrestrial Laser Scanning (STLS).

Dynamic vehicular Terrestrial Laser Scanning.

Dynamic portable Terrestrial Laser Scanning.

Fig. 18.2 Laser scanning platforms

It is worth noting that despite its great capacity for data capture, laser scanning will not be a solution for all surveying tasks. Field and data post-processing can require significant effort to achieve the required level of results, as discussed later in this chapter. In addition, laser scanning is best suited to scanning surfaces rather than edges. Accurately capturing irregular edges can require an extremely high resolution that may not be justified for the rest of the scene, wasting valuable field and office processing time.

Due to the extension of the subject and the scope of this book, this chapter covers only the most relevant aspects of time-based stationary TLS technology.[6] In other words, this chapter aims to introduce the technology and guide the reader in making decisions about its rational use.

---

[6]TLS can also be performed using triangulation-based technology. For further information, readers are advised to consult specialist literature.

18.2   Terrestrial Laser Scanning

A *laser scanner* is a reflectorless distance measurement instrument consisting of a transmitter of electromagnetic energy, a receiver and a detector. The carrier wave is a laser that contains the modulation information necessary to determine the signal's propagation time between the emitter and the receiver. The detector is formed by a network of highly sensitive photodiodes that can detect the returned signal's intensity (I) and its signature. The TLS thus determines sets of (X, Y, Z, I) for each reflection point of the scanning object.

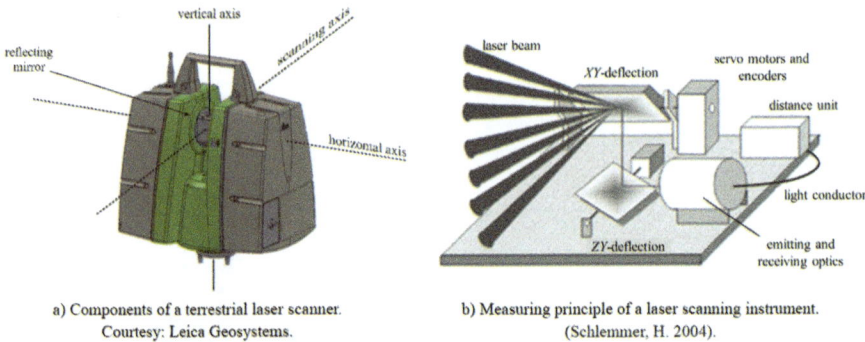

a) Components of a terrestrial laser scanner.   b) Measuring principle of a laser scanning instrument.
Courtesy: Leica Geosystems.   (Schlemmer, H. 2004).

Fig. 18.3   Components of a stationary terrestrial laser scanner

As shown in Fig. 18.3a, the components of a terrestrial laser scanner are similar to those of a total station, except that there is no telescope. Instead, distance measurement and surface scanning are performed by a reflecting mirror, which is moved by a motor coupled to it, deflecting the electromagnetic wave in different directions in space, as shown in Fig. 18.3b. In some instruments, the mirror performs a vertical rotation of 360°; in others, it oscillates 180°. The horizontal rotation is performed by the rotation of the instrument, completing the spatial scan.

Similar to a robotic total station, a laser scanner consists of an alidade containing the electronic components that operate the instrument, the three axes, the angle sensors and the distance measuring system.

The instrument's stability during the measurement is crucial for accurately determining the positions of the measured points. For this reason, many of the laser scanners on the market are equipped with electronic compensators that store information about the direction of gravity before each measurement and adjust the values measured by the instrument according to the inclinations indicated by the compensators.

For field operation, stationary terrestrial laser scanners are mounted on a tripod and operated by an external computer or a VGA monitor attached to the instrument, as shown in Fig. 18.4. They vary in shape, weight, scanning direction, data acquisition speed, distance range and measurement accuracy. Readers interested in more details about the characteristics of the instruments available on the market should consult the technical information provided by the manufacturers.

Data capture	Point cloud

Fig. 18.4 Stationary terrestrial laser scanner in operation. (Courtesy: Select Surveys. Location: https://selectsurveys.com/infrastructure-surveys)

Technical details of the main scanning parameters of a stationary terrestrial laser scanner are presented in the following subsections.

18.2.1 Distance Measurement

The first parameter to consider with a laser scanner is the distance measurement method used by the instrument. The principle of laser scanning distance measurement is the same as described in Sect. 8.3.2. Some instruments use the time-of-flight method, others use the phase-shift method, and the latest ones use WFD technology. They all use laser beam modulation to measure distances without a reflector prism. The measured distance is the slope distance between the instrument and the object. The distance range varies from a few meters to kilometres. In general, short distances use the phase-shift distance measurement method (high accuracy), and medium and long distances use the time-of-flight (low accuracy) or the WFD (high accuracy) methods.

Since laser scanners operate with laser beams, some parameters related to the laser spot and the scanned surfaces must be considered when characterising the survey quality, as described below.

18.2.1.1 Beam Divergence

In simple terms, the geometric shape of the laser beam emitted by a laser scanner can be visualised as concentric circles whose diameter increases as the beam moves away from the instrument. This increase in diameter is called the laser beam divergence. The result is a laser footprint on the surfaces being measured with different dimensions depending on the distance from the instrument. Note that the laser spot may be circular or elliptical depending on the inclination of the surface illuminated by the laser beam. For these reasons, beam divergence greatly influences the resolution of the point cloud and the positional uncertainty of a measured point. Specifications for measuring laser beam diameters, divergence angles, and propagation rates are given ISO/TR11146-3:2004 standard—*Lasers and laser-related equipment—Test methods for laser beam widths, divergence angles, and beam propagation rations. Part 3: Intrinsic and geometrical laser beam classification, propagation, and details of test methods.* As an electromagnetic wave propagating in space, laser beams are also subject to all the effects of atmospheric refraction, which cannot be neglected to avoid systematic errors in the measured values.

Figure 18.5 illustrates the effect of laser beam divergence on laser spot size at different distances. The smaller the laser spot, the greater the accuracy of the measured distance, as the return signal will always be the average of the reflections occurring within the circular (or elliptical) area illuminated by the incident laser beam. See Fig. 18.6.

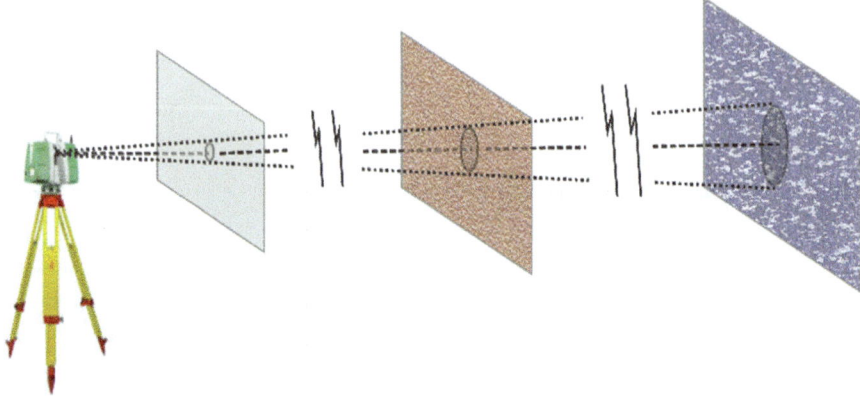

Fig. 18.5 Laser beam divergence

Fig. 18.6 Effect of laser beam size

18.2.1.2   Reflectivity

The reflectivity of the incident surface is another essential factor to consider when measuring the distance with a laser scanner. The greater the reflectivity of the object's surface, the greater the power of the return signal and, therefore, the greater the range of the laser beam and the ability of the instrument to measure the distance correctly. However, reflectivity values vary according to the type of laser radiation used, the type of surface and the angle of incidence of the laser beam, making it difficult to assess correctly. In extreme cases, the distance cannot be measured at all. In this case, the only sources of information are the tests carried out by the manufacturers, who state the reflectivity values in the technical specifications of their instruments.

In addition to the reflectivity of the object's surface, it is also essential to consider the effect of the atmosphere in the space through which the laser beam passes. The existence of dust particles in the air and humid environments deteriorates the quality of the return signal.

18.2.2   Scanning Resolution

The second parameter to consider in a laser scanner is the spacing between the laser points, i.e., the scanning resolution or point density that the instrument can achieve. This parameter is a function of the emission rate of the laser beam, the distance to the object and the rotation increments of the reflecting mirror. See Fig. 18.7. Longer distances may require smaller rotations and vice versa. In the case of the time-of-flight method, for example, the measurement system must wait for the signal to return before sending another signal. Therefore, the greater the distance, the longer the waiting time and the lower the scan rate. On the other hand, in the case of the phase-shift measurement method, the system does not need to wait for the signal to return before sending the next one, allowing for higher scan rates.

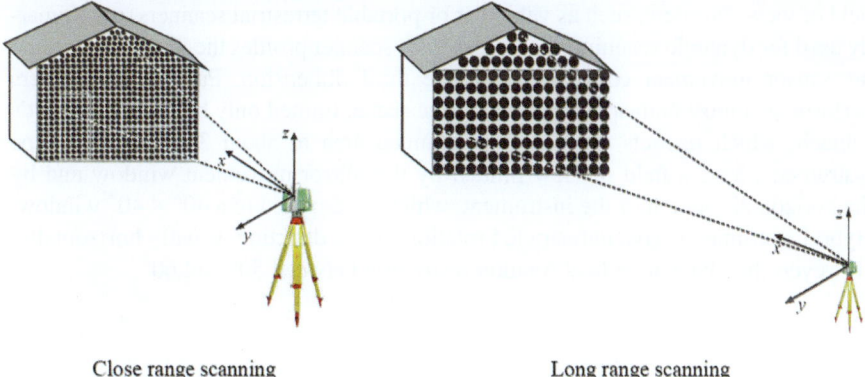

Close range scanning   Long range scanning

Fig. 18.7   Laser scanning resolution as a function of the distance from the instrument to the object

18.2.3   Angular Measurement

The third parameter to consider in a laser scanner is the angular measurement method used by the instrument. The laser beam orientation in a TLS is measured using horizontal and vertical angle encoders (angle measuring circles), similar to those used in total stations and electronic theodolites. See Chap. 9. For low-resolution instruments, angle sensors based on the absolute angle measurement method are more common, whereas, in high-resolution instruments, incremental sensors are required.

18.2.4   Beam Deflection System

The fourth parameter to consider is the beam deflection system. As already mentioned, to scan a scene with a laser scanner, it is necessary to have a laser beam deflection system in the horizontal and vertical directions. This system consists of a set of mirrors that deflect the laser beam emitted by the instrument towards the scene and scan the object to be measured using servomotors (rotors) coupled to the angle encoders.

A grid of points is measured as a function of constant and equal angular increments, thus defining the point cloud. Furthermore, depending on the intensity of the returned laser signal (intensity of the reflection), it is possible to determine the different reflectance values of each measured point and assign a grey scale or a colour image overlay to it, giving the point cloud of the scanned scene an image aspect, as shown in Fig. 18.8.

Deflection systems vary for different laser scanning technologies (terrestrial and airborne). Figure 18.9 illustrates the different types of deflectors used in the different classes. Stationary terrestrial laser scanners use oscillating or rotating mirror deflection systems.

As shown in Fig. 18.10, laser scanners are classified as *profiler* (or linear), *panorama, camera,* and *hybrid,* depending on the type of mirror deflection and field of view. Profilers, such as vehicular or portable terrestrial scanners, are primarily used for dynamic scanning. In this case, the scanner profiles the 2D direction, and the sensor movement completes with the third dimension. Panoramic scanners perform an almost hemispherical scan of the scene, limited only by the instrument's tribrach, which restricts the vertically scanned area to about 310°. Camera-type instruments have a field of view limited by the mirror movement window and by the horizontal rotation of the instrument, which is restricted to a 40° × 40° window. Hybrid instruments have unrestricted rotation in one direction, usually horizontally. However, they have a vertical rotation restriction between 50° and 60°.

Fig. 18.8 Illustration of a coloured point cloud. (Courtesy: Rogue Visual Design, LLC. Location: https://rogue3d.com/lidar-point-cloud-modeling-services)

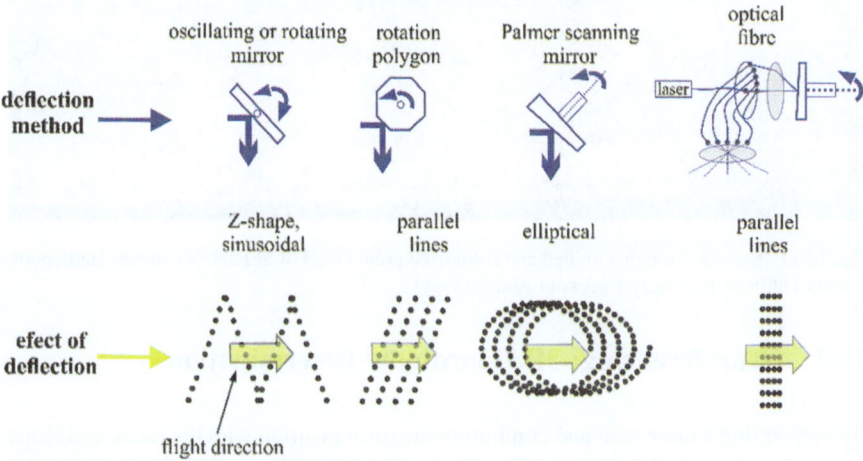

Fig. 18.9 Laser beam deflection systems

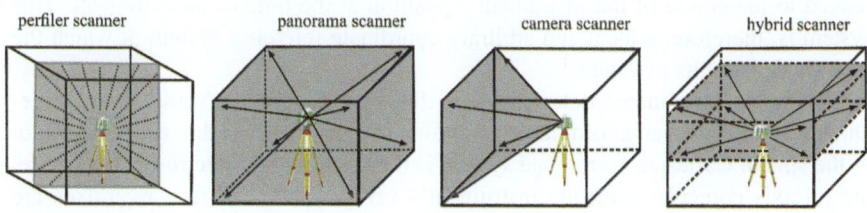

Fig. 18.10 Classification of laser scanners according to the type of mirror deflection and field of view

18.2.5   Internal Digital Camera

Another parameter to consider in a laser scanner is the internal camera. Some instruments have an internal CCD camera installed coaxially with the collimation axis of the instrument. It can photograph the measurement scene so that the resulting images can then be subsequently arranged as a mosaic to be incorporated into the point cloud in the processing stage by adding the RGB information of the pixel to the corresponding point, thus enhancing the imaging aspect of the mapped scene. Therefore, the camera's quality is an essential parameter for a laser scanning instrument. Figure 18.11 shows an example of a section through a unified and coloured point cloud of St Hilda's Church, Hartlepool, County Durham.

Fig. 18.11 Section through a unified and colourised point cloud of St Hilda's Church, Hartlepool, County Durham. (Courtesy Cotswold Archaeology)

18.3   Georeferencing: 3D Coordinate Determination

By performing a laser scan and combining the rotor position with the measured slope distances, it is possible to determine the position of points using the polar spatial coordinates (Az_i, z_i, d'_i) defined by the sensor. The result is a point cloud with coordinates referenced to the *Scanner's Own Coordinate System* (SOCS), which is related to the centre of the instrument's position at the time of measurement. This system is, therefore, a local and arbitrary coordinate reference system in which the scanner delivers its raw data.

The polar coordinates determined in this way can be converted into Three-Dimensional Cartesian coordinates, as shown in Fig. 18.12. In this case, the origin of the spatial Cartesian coordinate system is the instrument's electro-optical centre, the (z) axis coincides with the instrument's vertical axis, and the (x, y) axes are oriented considering the SOCS. Note that different laser scanner manufacturers may use different orientations for the axes of their coordinate system.

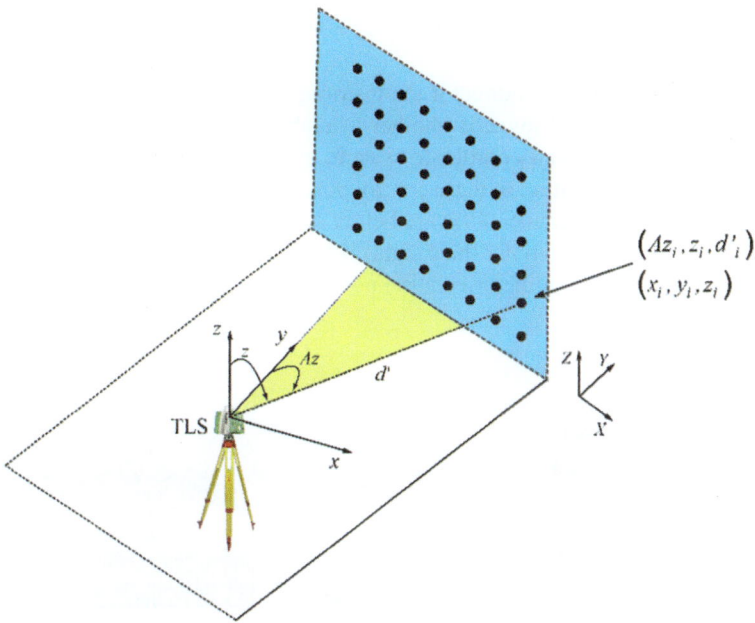

Fig. 18.12 Georeferencing of points surveyed by laser scanning

According to the mathematical theories presented in Chap. 5, the relationship between the polar spatial coordinates measured by the scanner and the local Cartesian spatial coordinates of the measured point are given by Eqs. (18.1) and (18.2).

$$\begin{bmatrix} x \\ y \\ z \end{bmatrix} = \begin{bmatrix} d' * \sin(z) * \sin(Az) \\ d' * \sin(z) * \cos(Az) \\ d' * \cos(Az) \end{bmatrix} \tag{18.1}$$

$$\begin{bmatrix} Az \\ z \\ d' \end{bmatrix} = \begin{bmatrix} \mathrm{atan}\left(\Delta x / \Delta y\right) \\ \mathrm{acos}\left(\Delta z / d'\right) \\ \sqrt{\Delta x^2 + \Delta y^2 + \Delta z^2} \end{bmatrix} \tag{18.2}$$

The (x, y, z) coordinates of the point cloud computed using Eq. (18.1) are referenced to the SOCS and are only valid for a single instrument position. In many cases, however, data acquisition requires multiple scanning positions to cover the object's entire surface. Therefore, all scanned data from different scanner positions must be transformed into a common reference frame (georeferenced) for visualisation and further processing of the scanned scene. There are, therefore, two georeferencing methods that can be used for Geomatics applied to Civil Engineering:

- Direct georeferencing
- Indirect georeferencing

18.3.1   *Direct Georeferencing*

Georeferencing is considered direct if the instrument can be installed over a known station point and oriented towards another available control point referenced to a predefined Cartesian spatial coordinate system, as shown in Fig. 18.13. All information regarding the position of these control points and the height of the instrument can be entered into the scanner software before scanning or during data processing. In this case, the orientation point (backsight) is a specific scanning target that allows determining the position of the orientation point when scanned by the instrument as shown in Fig. 18.13.

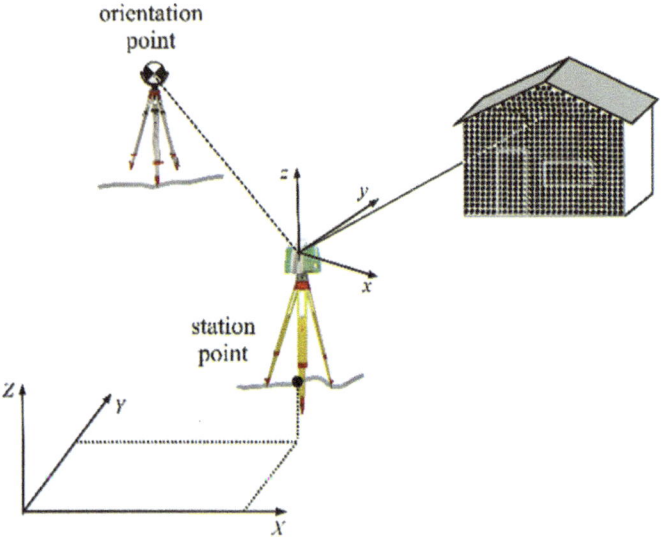

Fig. 18.13   Direct georeferencing

As with total station surveys, a network of control points must be set up whenever scanning objects requires more than one scanner installation. Figure 18.14 shows an example of a network of control points set up for a laser-scanning survey of a basilica in Italy. Note the need to set up control points inside the building to scan its interior.

The network of control points is usually established using a total station or GNSS instruments. However, some laser scanners can establish network points using traversing or free station techniques like modern total stations. For more details, the reader should consult the technical manual of the instrument of interest.

The point cloud thus generated has its 3D coordinates directly referenced to the spatial coordinate system of the control point network. It should be noted, however, that for this type of georeferencing, the laser scanner must be equipped with appropriate resources to operate in a similar way to a total station, such as devices for installing the instrument over ground landmarks, backsight pointing and electronic compensators.

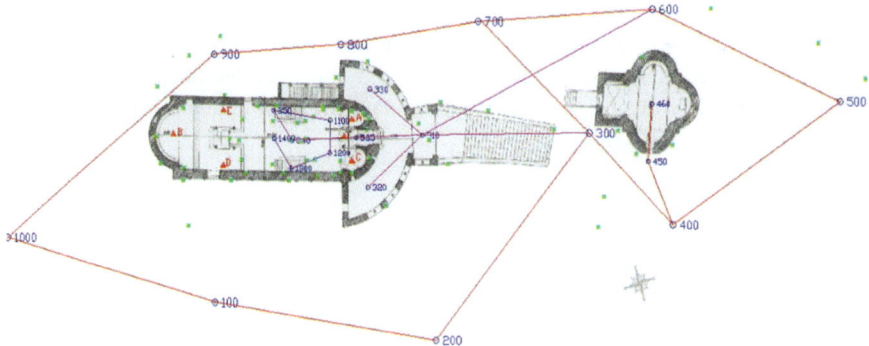

Fig. 18.14 Network of control points deployed for the laser scanning of the Basilica di San Pietro al Monte, Italy. (Alba et al. 2007)

18.3.2   Indirect Georeferencing

Georeferencing is indirect when measurements taken from different scanner positions are combined into a single point cloud and then converted into the coordinate system of the scanned object. This implies the use of target features (artificial or natural) referenced to the coordinate system of the scene.

Combining multiple point clouds into a single database is called *scan registration* or *scan co-registration*. The principle of this georeferencing method is shown in Fig. 18.15. According to the figure, the polar spatial coordinates $\left(Az_{ji}, z_{ji}, d'_{ji}\right)$ of point (i), determined from the instrument position (j), are first transformed into

Fig. 18.15 Principle of indirect georeferencing

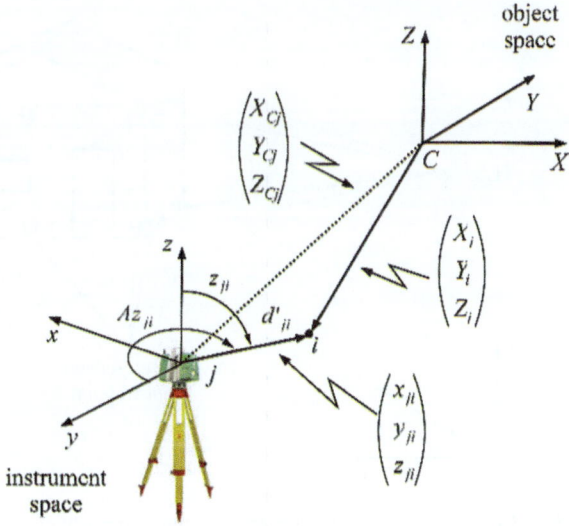

Cartesian spatial coordinates (x_{ji}, y_{ji}, z_{ji}) according to the SOCS. Assuming that the coordinates (X_i, Y_i, Z_i) of point (i) are also known in the object space coordinate system, the different scans (j) can be registered and georeferenced to the object space (X_{Cj}, Y_{Cj}, Z_{Cj}) using a spatial coordinate transformation (Eq. 18.3), as presented in Sect. 5.9.2.2.

$$\begin{bmatrix} X_i \\ Y_i \\ Z_i \end{bmatrix} = \boldsymbol{R} * \begin{bmatrix} x_{ji} \\ y_{ji} \\ z_{ji} \end{bmatrix} + \begin{bmatrix} X_{Cj} \\ Y_{Cj} \\ Z_{Cj} \end{bmatrix} \tag{18.3}$$

where

$\boldsymbol{R}$ = rotation matrix between object and instrument spaces (j). See Eq. (5.79).
(X_i, Y_i, Z_i) = coordinates of the point (i) in object space
(X_{Cj}, Y_{Cj}, Z_{Cj}) = coordinates of the position (j) of the scanner in object space.

As discussed in Chap. 5, the transformation parameters between any two spatial coordinate systems can be determined if there are at least three homologous points with known coordinates. Usually, more than three points are available, and the transformation parameters are determined using the Least Squares method based on the georeferencing targets whose coordinates are measured in instrument space by the laser scanner and in object space by conventional surveying methods. The target object can be an artificial target provided by the laser scanner companies or a target printed on paper by the user. See Fig. 18.16. They can also be natural targets of the structure that can be accurately identified or manually selected in the scan, such as surface edges or cornices. The coordinates are then transformed using the 3D Helmert transformation method. Special attention must be paid to the distribution of

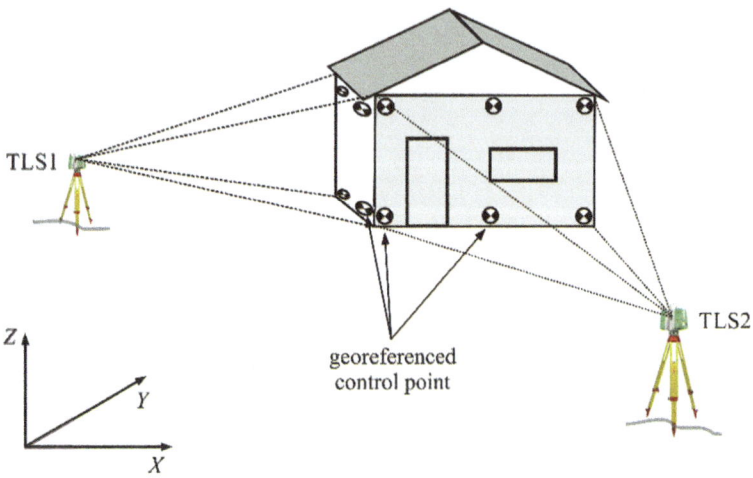

Fig. 18.16 Principle of indirect georeferencing using artificial control point

the targets in order to ensure the best possible georeferencing of the point cloud. Care should be taken when scanning elongated scenes, where co-registration errors quickly accumulate. As with photogrammetry, final verification of the georeferencing must be carried out using checkpoints.

Targets used as control points in laser scanning are usually flat or spherical, as shown in Fig. 18.17. Typically, the spherical is preferred because it is easy to determine its centroid at any scan angle. Target and laser spot size, distance from the scanner, and target scan resolution determine how precisely the target reference positions can be determined. In the data processing phase or even in the field, targets are automatically detected in the point cloud due to their high level of reflectivity.

Fig. 18.17 Examples of targets used as control points in laser scanning

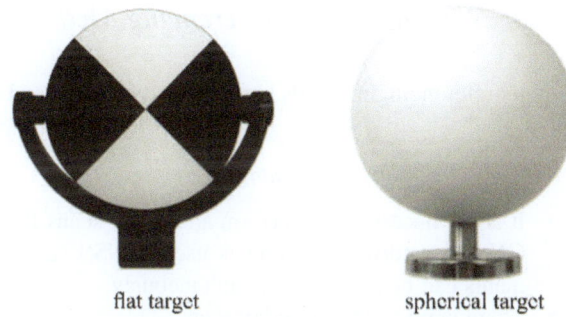

flat target   spherical target

The co-registration can also be done by using point cloud overlaps without external reference targets. This targetless registration approach, known as *cloud-to-cloud*, relies on collecting sufficiently dense point clouds with sufficient overlap between adjacent scans to derive sufficient matching points. Different matching algorithms are used to merge the scans until a complete registration of all scenes is obtained, forming a single point cloud referenced to the instrument's spatial coordinate system, as shown in Fig. 18.18.

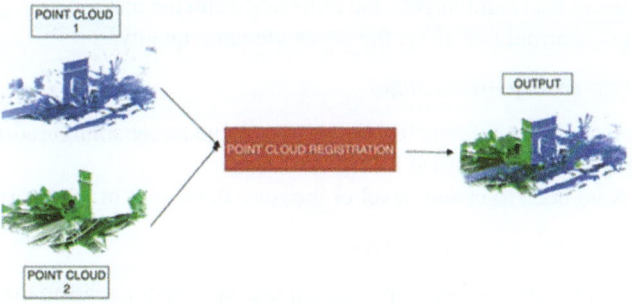

Fig. 18.18 Principle of cloud-to-cloud registration. (P. W. Theiler, 2015)

The matching (or registration) algorithms currently available are diverse. Some are based on the point-to-point iterative nearest point algorithm, and matching between registered objects using geometric properties of the point cloud, such as curvature. Another common type of algorithm is the feature-based or surface geometry algorithm, which relies on feature extraction and matching algorithms. The more sophisticated ones are based on graph matching and stochastic probability distribution models. For more detailed information on registration algorithms, readers are advised to consult specialised references, as this subject is beyond the scope of this book.

18.3.3   Comparing Georeferencing Methods

Both georeferencing methods have their advantages and disadvantages, as indicated below:

Direct georeferencing—advantages

- It is not necessary to perform additional scans if control points are present. If GNSS-supported scanning is used, GNSS scanning and scanning measurements can be performed simultaneously.
- There is no need to overlay scans.
- Point clouds are georeferenced in the field and do not require post-processing in the office.
- Data density and quality can be verified in the field, thus reducing survey time and the need for rescanning.
- It is a well-known procedure that facilitates understanding and integration with traditional surveys.

Direct georeferencing—disadvantages

- Less accurate than the indirect method.
- When using backsight targets, the reflectance and the method used to calculate the target centroid can affect the georeferencing quality.

Indirect georeferencing—advantages

- High-accuracy georeferencing provided a stable target configuration is used.
- Scanning can be performed from any station.
- There is no need to centre, level or measure the height of the instrument.

Indirect georeferencing—disadvantages

- The need to establish control points and use more instruments in the field, such as a total station.
- The need for careful planning to ensure a good target configuration.
- The need for coverage between scans.

- Difficulty in establishing a good target configuration in some cases, such as longitudinal works (roads, tunnels, pipelines and others).
- Difficulty in positioning targets in restricted access locations such as dams, tall buildings, towers and others.
- The need for post-processing in the office.
- The results of registration and georeferencing depend on the reflectivity and the methods used to calculate the centroids of the targets.

18.4 Software Used for TLS Surveys

The software used for TLS surveys consists of several modules for different functions. Looking at the whole process of a scanning project, from data collection to the final model, a rough division can be made as follows:

- Software for scanning control
- Software for noise filtering, target identification, scan registration and georeferencing
- Software for point cloud processing
- Software for 2D and 3D CAD modelling
- Software for texturing and image mapping
- Software for project management
- Software for data sharing and integration into existing software such as CAD, GIS and BIM systems

The scanner manufacturer usually develops its software for scanning control, registration and georeferencing. For data processing, however, third-party processing software packages that can import point clouds in (X, Y, Z) format and process them to produce final products are available on the market.

The processing software is specifically designed to handle large volumes of point cloud data, enabling the data to be rotated, zoomed and panned in real-time. It should also be able to automatically detect control targets, perform automatic cloud-to-cloud registration and classify the points by, for example, distance from the scanner, height above ground and reflection intensity. Additional processing may be available to generate a triangular mesh from the points, onto which the coloured images can be accurately incorporated, improving data interpretation. Some software systems allow direct 3D vectorisation from the georeferenced images in a CAD environment.[7] Some examples of processing software are:

1. Cyclone and Cyclone Cloudworx (Leica)
2. Polyworks (Innovmetric)
3. Riscan Pro (Riegl)
4. Isite Studio (Isite)

---

[7]It is important to note that not all CAD software can handle the large number of points generated by laser scanning without additional plug-ins.

5. LFM-Software (Zoller+Fröhlich)
6. Luposcan (Lupos3D)
7. Split FX (Split Engineering)
8. RealWorks Survey (Trimble)
9. Pointools (Pointools)

In addition to the software, it is also important to consider the capabilities of the computer available for data processing, mainly in terms of 3D graphics acceleration, RAM configuration, data storage and screen resolution.

18.5   Data Modelling

The products generated by laser scanning surveys are the point cloud, vector maps and spatial models. The point cloud is the raw data from which the others are generated. It is, therefore, usually pre-processed to filter out unwanted points, group homogeneous ones, and sometimes segmented to facilitate data modelling. The pre-processed dataset can then be used for 3D visualisation, creation of dimension lines and extraction of 3D coordinates, as shown in Fig. 18.19, or formatted and imported into vectorisation or spatial data modelling software for further interpretation and analysis.

The vectorisation or feature extraction of the point cloud consists of creating a 3D vector map of geographical features detected in the point cloud, such as corner points of buildings and constructions, geomorphological structures, centre points of small objects and many others, depending on the object and map scale. It is a semi-automatic process based on connecting points that refer to the same feature

Fig. 18.19 Example of obtaining geometric values of elements from the point cloud of a laser scanning survey. (Courtesy Beijing LidarTop Technology Co., Ltd.)

until obtaining the desired final map. It is a mapping process usually carried out using CAD software packages capable of operating with point clouds. Depending on the mapping purpose, creating the 3D vector map may require much human intervention due to the high density of the point cloud and the lack of textual and topological information. See Fig. 18.20. Special techniques and automatic processing algorithms are being developed to speed up the extraction process. Readers interested in further details should consult specific software packages and instrument manufacturers.

Point cloud   Vectorization

Fig. 18.20 Illustration of a vectorised image from a point cloud. (Courtesy Beijing LidarTop Technology Co., Ltd.)

Spatial modelling of point clouds consists of mathematically determining the spatial shape of the scanned objects, which can be individual three-dimensional objects or the entire landscape of the scene. In this case, modelling means developing a georeferenced mathematical representation of three-dimensional surfaces using an application program designed for this purpose. As a mathematical representation, the generated spatial model can be viewed, edited, modified, plotted and rotated in space using appropriate computer resources.

There are two ways to perform 3D mathematical modelling of the point cloud from a laser scanning survey. The first is to apply a Digital Surface Modelling (DSM) algorithm, as described in Chap. 15. In this case, the steps to be followed are those described in that chapter. The model thus generated is visually composed of triangular elements that form the image of the modelled surface, which computer graphics resources can process for various purposes. Figure 18.21 shows an example of spatial modelling of a point cloud using DSM resources.

The second way to perform 3D mathematical modelling of the point cloud from a laser scanning survey is to use dedicated structural modelling application programs. These application programs can automatically recognise geometric primitives from the point cloud, such as rectangles, cylinders, structural profiles and many others, which are used to generate spatial geometric shapes of complex structures, as shown in Fig. 18.22.

Point cloud.   Modelled surface.

Fig. 18.21 Spatial modelling of surfaces using DSM. (Courtesy Beijing LidarTop Technology Co., Ltd.)

Fig. 18.22 Structural modelling of a point cloud. (Courtesy: Rogue Visual Design, LLC. Location: https://rogue3d.com/lidar-point-cloud-modeling-services/)

The reader should note the paradigm shift in graphical representation and resources for engineering design due to the possibility of working with point clouds and 3D models of objects and surfaces. Particular attention should be given to professional training in these design resources.

18.6   TLS Measurement Quality

Engineering survey data points collected using the TLS point cloud can be verified in various ways, including comparing scan points to checkpoints, reviewing the digital terrain model or the morphological lines in the plane and profile, and performing redundant measurements. The accuracy of point positioning depends on several factors, such as the physical quality and orientation of the laser beam, the measurement distance, the physical properties of the scanned object, atmospheric and environmental conditions, the quality of the georeferencing, and the stability of the scanner mechanism, among others. In addition, the size of the laser footprint on the measured object is much larger than its image portrayed in the point cloud, and the density of the point mesh (resolution) is so high that the points' confidence ellipsoids to overlap, making it difficult to analyse the positional quality of the points. All these factors make evaluating the quality of a laser scanning measurement a difficult task, mainly due to the lack of specific standards for this purpose.

As a result, the manufacturers determine the quality of laser scanning instruments and indicate it in the technical specifications. In addition, the behaviour of the internal components of the instrument to temperature variations and vibrations during measurements requires repetitive tests in special calibration chambers, which are only available in specialised laboratories, which is why only the manufacturer can carry out the appropriate calibrations and adjustments. An example of a calibration chamber is shown in Fig. 18.23.

Fig. 18.23   Example of calibration chamber for TLS instruments. (Courtesy: Leica Geosystems)

For the assessment of the precision (repeatability) of a laser scanning instrument, it is recommended to use the *ISO 17123-9:2018 (E)* standard—*Field Procedures for Testing Geodetic and Surveying Instruments-Part 9: Terrestrial Laser Scanners*, which provides procedures for testing the suitability of a given instrument for a given intended task. The reader should consult the respective documents for further details. Some manufacturers also provide field procedures that operators can use to check the quality of their instruments and ensure they operate within specified specifications.

18.7   Source of Errors in Laser Scanning Surveys

In practice, as laser scanners are basically angular and linear measuring instruments, they are subject to the same types of errors as a total station, with some inherent characteristics of scanning technology. In this sense, to describe the sources of laser scanning errors, they are grouped into four categories:

1. Instrumental (scanner mechanism)
2. Object properties
3. Atmospheric and environmental conditions
4. Scanning geometry

18.7.1   Instrumental Errors

The instrumental errors of a laser scanner can be both systematic and random. They are related to misalignment of the hardware components, calibration, settings and variations in the properties of the emitted laser beam and its detection process. More specifically, they are:

- Angular errors related to the angle measurement circles
- Angular errors related to the laser beam deflection system
- Angular errors related to the orthogonality of the instrument axes
- Linear errors associated with the distance measurement system

The main physical characteristics of each of these are presented below.

18.7.1.1   Angular Error Related to the Angle Measurement Circles

The horizontal and vertical circles of a laser scanning instrument are practically the same as those of a total station, and, therefore, they behave in the same way concerning the instrumental systematic circle errors, as presented in Sect. 10.2.1.1. For this reason, they are not discussed in this chapter.

18.7.1.2   Angular Errors Related to the Laser Beam Deflection System

As already described, the laser beam deflection system directs the electromagnetic wave towards the object to be measured. The angular error of this system occurs due to the irregularities of its physical and optical components, such as the misalignment of the laser beam emitter and receiver sensors in relation to the vertical axis of the instrument, the deviation of the laser beam incidence on the reflecting mirror, the irregularities of the mirror surface and others. These errors depend exclusively on the construction quality of the instrument, with no possibility of operator interference.

18.7.1.3   Angular Errors Related to the Instrument's Axes Orthogonality

The axes of rotation of a laser scanner, as shown in Fig. 18.3, are the same as those of a total station, namely:

- Vertical axis: also called the instrument rotation axis, around which the instrument rotates horizontally.
- Horizontal axis: also called the rotation axis of the reflecting mirror, around which the laser beam scans vertically.
- Collimation axis: defined by the axis that passes through the centre of the scanning mirror and the centre of the laser spot on the object's surface.

The geometrical relationships between the three axes and the effects of errors due to non-orthogonality between them in angular measurements are the same as those indicated in Sect. 10.2.1.2, including the calibration of the electronic compensator. Therefore, the equations for correcting these errors are the same as those shown in the corresponding texts. The values of the axis errors are determined during the instrument calibration and stored in the instrument memory for appropriate angular correction.

For instruments with an internal digital camera, it is also necessary to calibrate the possible misalignment between the centre of the CCD (or CMOS) sensor array and the laser beam collimation axis to ensure perfect coincidence between the measured point and the camera's optical centre.

18.7.1.4   Linear Errors Related to the Distance Measurement System

Again, the errors related to measuring the distance between the instrument and the measuring object in a laser scanning instrument are the same as those described for the EDM errors presented in Sect. 10.2.3. However, some unique scanner features should be highlighted, such as errors in distance measurements due to the laser beam's divergence and the incidence angle on the reflecting surface, as described in the previous sections.

18.7.2   *Errors Related to the Object Properties*

Errors related to the reflecting surface are mainly related to the reflectivity index of the object's surface, as briefly discussed in Sect. 18.2.1.2. The change in the reflectance index mainly causes an error in the distance measurement. Highly reflective surfaces allow more accurate measurements than less reflective ones because the former reflects a greater amount of energy back to the sensor. Note that excessive reflectivity can also affect or even prevent measurement quality. However, the effects vary for different instruments at different laser beam wavelengths.

In addition to reflectivity, laser beam energy can penetrate some materials, such as wood, polystyrene, marble, etc. In this case, the laser beam is refracted and reflected in the material, resulting in a new addition constant for the measured distance.

The physical characteristics of the reflecting surface that commonly cause errors in distance measurements with a laser scanning instrument are as follows:

- Colour
- Roughness
- Surface temperature
- Surface humidity
- Physical composition of the material

18.7.3   *Errors Related to Atmospheric and Environmental Conditions*

The atmospheric and environmental conditions at the measurement site are often difficult factors to regulate and predict. The user has very little or no control over them. Environmental conditions such as temperature and atmospheric pressure, relative humidity, vibration and others are important sources of errors in laser scanning measurements. They cause distortions in the shape of the reflected laser beam, attenuate the signal intensity and change its propagation speed, as described in Sect. 8.3.2.4. Another substantial effect to consider is false returns due to multiple return signals caused by rain or dust particles in the air.

18.7.4   *Errors Related to the Scanning Geometry*

Scanning geometry errors are mainly related to the scanner's georeferencing and the instrument's orientation concerning the scanned surface, which determines the angle of incidence, range and point density. As far as georeferencing is concerned, the sources of error are the same as those associated with total station georeferencing:

- Accuracy of the station and reference point coordinates in the case of direct georeferencing and of control point coordinates in the case of indirect georeferencing.
- Levelling errors of the instrument in the absence of an electronic compensator
- Instrument centring error.
- Errors in determining the centre of the backsight targets in the case of direct georeferencing.

In addition, processing steps such as point registration, segmentation and filtering can affect the quality of the measured point cloud. These are random errors, the sources and effects of which can be evaluated as described in Chap. 10.

18.8 Scanning Process

A typical scanning process consists of several steps. Some of these are automated, while others require high human intervention. The typical workflow for a scanning process is shown in Fig. 18.24.

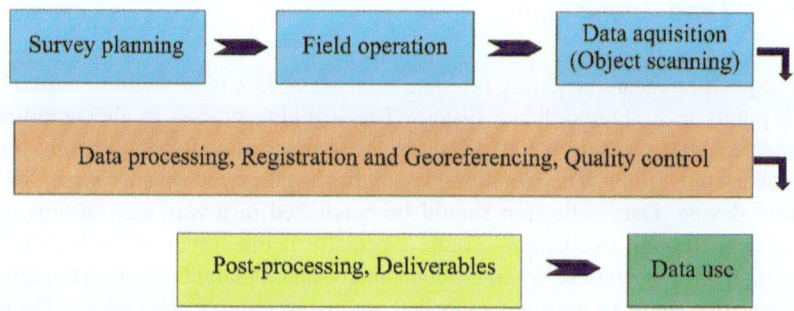

Fig. 18.24 Scanning process workflow

18.8.1 Survey Planning

The survey planning step comprises all the procedures to be taken before the field measurements. It varies according to the surveying work to be carried out. It usually consists of the following topics:

- Define goals, objectives, survey time, deliverables and costs.
- Study the area to be surveyed and define the complexity of the structure to determine the scanning and target locations.

- Determine the measuring techniques and equipment for georeferencing and data scanning according to the required accuracy and level of detail.
- Select one or more software packages to manipulate, edit, register and model the point cloud.

18.8.2 Field Operation

Field operations include all procedures for surveying control points and setting up the instrument and targets. Control point surveying is carried out where necessary using a total station or GNSS instrument to determine a local coordinate system within the scan area. Instrument installation is similar to that of a total station or GNSS antenna. Instrument configuration consists of entering the scanner settings according to the instructions provided by the manufacturer.

The operator must remember that, as with reflectorless total stations, laser scanning measurements perpendicular to a surface give better accuracy than those with a large angle of incidence. The larger the angle of incidence, the more the beam can elongate, producing errors in the measured distance. The point cloud resolution will also become more dispersed as the distance from the scanner increases.

18.8.3 Data Acquisition

Data acquisition means acquiring raw data from the surface to be scanned, which is a set of points in a 3D coordinate system. The scanning process is almost entirely automated. After pressing the start button, the scanner moves to the starting point and collects the points. The cloud points are displayed on the scanner screen or an external device. Data collection should be conducted in a way that ensures data redundancy through overlapping scans, especially if the survey relies heavily on cloud-to-cloud registration. When artificial and/or natural targets are used to register the scans, they must be accurately labelled and measured to ensure good accuracy. Finally, checking the integrity and quality of the scan after completing the survey is essential to avoid rework.

18.8.4 Data Processing

The first step in data processing is performing the registration and georeferencing processes described in Sect. 18.3. After (or during) registration and georeferencing, it is also necessary to perform some processing steps before providing deliverable products, such as cleaning and filtering the data to remove unwanted features such as people, vegetation, obstructions, poor signal returns and others. For some specific applications, data segmentation and classification may also be of interest to facilitate data modelling. In addition, there are different data formats for scan data from different scanner manufacturers, so it is sometimes necessary to convert the scan data into another format, depending on the modelling software used for the post-processing step.

18.8.5   Post-processing

After pre-processing, the point clouds are positioned and georeferenced to a specific coordinate system, allowing the registered raw data point cloud to be transformed into a final deliverable. The work process differs depending on whether the final product is a 2D model, such as plans, profiles, sections, or textured 3D models.

Direct 2D modelling from point clouds is achieved through feature vectorisation using plug-ins for CAD packages such as AutoCAD and MicroStation or dedicated point cloud management applications.

To create a 3D surface model, the point cloud can be converted into geometric shapes or a triangle grid, as described in Sect. 18.5. Once the mesh has been created, a plain texture or coloured images can be added to each triangle face for better visualisation.

18.8.6   Data Use

In Civil Engineering, terrestrial laser scanning is mainly used by architects and project managers in the design and construction phases, often, combined with dynamic portable TLS and UAV data acquisition to photograph and scan outdoors and indoor sites, as shown in Fig. 18.25. This allows the design to be adapted to the specificities of the construction environment and provides data for integration into BIM and GIS models. In the construction phase, it has a wide range of applications, such as helping stakeholders to compare the designed model with the current status of the construction or to map the final as-built. In the maintenance phase, 3D scanning can be combined with UAV data collection to scan hard-to-reach objects such as bridges, tall buildings, transmission towers and others.

Fig. 18.25 Principle of using a portable laser scanner indoors. (Courtesy: Heron)

18.9   Review Questions

1. Briefly explain how the laser scanning technology works for geospatial data collection.
2. Briefly explain the different types of lasers scanning technology available for geospatial data collection, depending on the platform on which the scanning sensor is installed.
3. Briefly explain the main components of a terrestrial laser scanning instrument.
4. What are the products of terrestrial laser scanning?
5. Discuss what are the main advantages and disadvantages of using terrestrial laser scanning in Civil Engineering.
6. Briefly discuss the main measurement parameters that need to be considered when specifying a terrestrial laser scanning instrument for Civil Engineering projects.
7. Explain what is meant by laser scanning georeferencing.
8. Briefly discuss the different types of lasers scanning georeferencing used for geospatial data collection and the advantages and disadvantages of each.
9. Explain why it is necessary to use software when working with laser scanning and what processing modules it must have in order to fulfil the complete workflow of a TLS geospatial data collection.
10. Explain the importance of data modelling when working with laser scanning.
11. Briefly discuss the main sources of error that can occur when working with terrestrial laser scanning.

12. Briefly outline the scanning process workflow of a terrestrial laser scanning job.
13. Briefly discuss the general technical characteristics of terrestrial laser scanning equipment currently available on the market.

References

Alba, M., Giussani, A., Roncoroni, F. and Scaioni, M. (2007). *Review and comparison of techniques for terrestrial 3D-view georeferencing.* In: Proceedings of the fifth international symposium on mobile mapping technology, Padua, Italy, 29–31 May. 8p.

Historic England (2018). *3D Laser Scanning for Heritage: Advice and Guidance on the Use of Laser Scanning in Archaeology and Architecture.* Swindon. Historic England. HistoricEngland.org.uk/advice/technical-advice/recording-heritage/.

Lerma, JL et al. (2008). *Theory and Practice on Terrestrial Laser Scanning.* Training Material Based on Practical Applications. Universidad Politecnica de Valencia Editorial; Valencia, Spain.

Mohd Azwan Abbasa, B, Lau Chong Luha, Halim Setana, Zulkepli Majida, Albert K. Chongc, Anuar Aspuria, Khairulnizam M. Idrisa, Mohd Farid Mohd Ariffa (2014). *Terrestrial Laser Scanners Pre-Processing: Registration and Georeferencing.* Journal Teknologi (Sciences & Engineering) 71:4, pages 115–122.

P. W. Theiler, J. D. Wegner, and K. Schindler (2015). Globally consistent registration of terrestrial laser scans via graph optimization. ISPRS journal of photogrammetry and remote sensing, 109: 126–138.

Schlemmer, H. (2004). *Einfuhrung in die Technologie des Laserscannings.* In 'XIV Kurs fur Ingenieurver messung, Tutorial Laserscanmng', ETH Zurich, Institut fur Geodäsie und Photogrammetrie, Prof. Dr. H. Ingensand.

Schulz, Thorsten (2008). *Calibration of a Terrestrial Laser Scanner for Engineering Geodesy.* Doctoral Thesis, ETH Zurich, https://doi.org/10.3929/ethz-a-005368245.

Shan, J. Toth C. K. Toth (2018) *Topographic Laser Ranging and Scanning: Principles and Processing*, Taylor & Francis Group, LLC. Florida, USA.

Si, H.; Qiu, J.; Li, Y. *A Review of Point Cloud Registration Algorithms for Laser Scanners: Applications in Large-Scale Aircraft Measurement.* Appl. Sci. 2022, 12, 10247. Access: https://doi.org/10.3390/app122010247.

Soudarissanane, S. S. (2016). *The geometry of terrestrial laser scanning—identification of errors, modelling and mitigation of scanning geometry.* Masters Thesys. Technische Universiteit Delft, 13p. http://repository.tudelft.nl/

UNAVCO (2013). *Terrestrial Laser Scanning (TLS) Field Camp Manual.* UNAVCO Boulder, CO, USA, v1.3.

Chapter 19
Airborne Photogrammetry

Irineu da Silva

19.1   Introduction

As seen in previous chapters, surveying measurements for determining the values of geospatial data are most often performed using terrestrial instruments and in direct contact with the measurand. Depending on the size of the site and the number of objects to be measured, this work may be laborious, costly, and, in some cases, even unfeasible with the technology currently available. This type of measurement can include cadastral surveys of large areas, cartographic mapping, mapping of corridors for linear engineering projects, mapping of difficult access areas, data collection for digital terrain model (DTM) or digital surface model (DSM) generation of large areas, and many others.

Due to the difficulty in using conventional surveying methods in data collection for these types of projects, remote measurement techniques with the potential to generate large amounts of data are used, such as the airborne and terrestrial laser scanning techniques described in the previous chapter and the photogrammetric techniques discussed in this chapter.

The term photogrammetry has different meanings for different applications, and its definitions have changed over time depending on the technological stage of its development. Currently, an adequate definition is to consider it the methodology for modelling 3D space from 2D images[1] generated by digital sensors. For Geomatics, such modelling predominantly involves collecting geometric (location and shape), semantic (interpretation) and physical (electromagnetic energy properties) information of the modelled space.[2]

---

[1] Photograph.

[2] Other professionals may be interested in other aspects of 3D modelling, as is the case, for example, of those involved with Computer Graphics or Computer Vision.

I. da Silva (✉)
São Carlos School of Engineering, University of São Paulo, São Carlos, SP, Brazil
e-mail: irineu@sc.usp.br

© The Author(s), under exclusive license to Springer Nature Switzerland AG 2025   769
I. da Silva, P. C. L. Segantine, *Geomatics Applied to Civil Engineering*,
https://doi.org/10.1007/978-3-031-75737-2_19

To achieve these objectives, photogrammetry is considered a science and technique with a strong interdisciplinary aspect, incorporating technical information from disciplines such as Geodesy, Geometry, Physics, Image Processing and Computer Vision, among others. In the early stages, it was just *Photogrammetry,* which became *Analog Photogrammetry, Analytical Photogrammetry* and finally, *Digital Photogrammetry,* which is now replaced by two competing terms: *Conventional Photogrammetry* and *UAV Photogrammetry.*

In this context, this chapter aims to discuss the basic concepts of *Digital Photogrammetry.* It will focus on the technical concepts of collecting geometric information from geospatial data using airborne sensors, defining what is called *Airborne* or *Aerial Photogrammetry.* The interpretative aspects of the images (photointerpretation) are not covered, nor are the concepts concerning *Terrestrial Photogrammetry.* As this is a book chapter, only the relevant aspects of *Digital Airborne Photogrammetry* will be discussed to guide engineering professionals in the rational use of this technology.

19.2   Digital Photogrammetry

Digital photogrammetry is the name given to the geospatial data collection system based on electronic equipment and computer programs that allow the definition of the position, shape, dimension, orientation and semantic information of objects in space through images stored in digital media and mathematical models governing the geometric relationships between the 3D object and its 2D images. *Image sensors* or *digital cameras* are used to collect digital images. The equipment used to extract geospatial information from photogrammetric images is called *Digital Workstations* or *Digital Photogrammetric Systems,* which operate through computer programs and integrated physical devices capable of handling photogrammetric operations. The companies operating in this professional category are called *Aerial Photogrammetry Companies,* and the professionals working in them are called *photogrammetrists.*

Although digital photogrammetry can be categorised in many ways, for the purposes of Geomatics applied to Civil Engineering, it will be considered according to the type of data acquisition platform, as presented below:

Terrestrial photogrammetry: in which the data acquisition platform is located on the surface of the Earth.
Airborne photogrammetry: in which the data acquisition platform is an aircraft. In the case of a manned aircraft, it is called *conventional airborne photogrammetry,* and in the case of an unmanned aircraft, it is called *UAV photogrammetry.*

Engineering applications vary between categories. For the specific case of conventional airborne photogrammetry, the following are highlighted:

(a) Extraction of three-dimensional information for updating the spatial database.
(b) Collection of three-dimensional information for mapping, Civil Engineering, Environmental Engineering, Forestry Engineering, Architecture and Urbanism, rural and urban cadastre, historical and cultural heritage registration, forensic analysis, and others.

(c) Generation of three-dimensional models for study in different phases of the life cycle of an engineering project, such as BIM technology applications, environmental studies, geotechnics, hydrological modelling, monitoring of structures and virtual reality.
(d) Mapping and production control in agriculture.
(e) Mapping for regional and urban planning.
(f) Traffic control and accidents on urban and rural roads.
(g) Military intelligence.
(h) Others.

The applications related to UAV photogrammetry are described in Sect. 19.17.

19.3 Digital Image

The primary source of information used in digital photogrammetry is the photographic image, i.e., the photograph. For its use in photogrammetry, image data are captured by electro-optical sensors (or photodiodes) and recorded in the form of a matrix with values corresponding to the amount of electromagnetic radiation of the light spectrum captured by the array of photodiodes. The amount of radiation captured is then converted into panchromatic (grey tones) or colour tones so that it can be visualised. The resulting image is formed by picture elements or *pixels* numerically represented according to the *radiometric resolution* of the image. The size of the image depends on the number of photodiodes in the sensor, i.e., the size of the array of photosensitive elements. Figure 19.1a shows an example of an original aerial digital image represented in 8-bit greyscale values and resampled with a lower resolution. Figure 19.1b shows a portion of the image at its lowest resolution and the corresponding matrix of greyscale values.

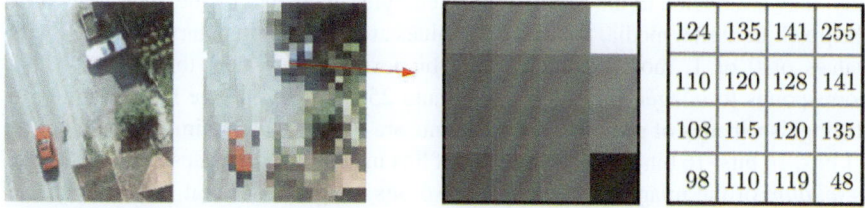

a) Original and resampled aerial 8-bit image.   b) Pixel grey shades and respective greyscale values.

Fig. 19.1 8-bit greyscale digital image. (**a**) Original and resampled aerial 8-bit image. (**b**) Pixel grey shades and respective greyscale values

In addition to the attribute relating to the greyscale or the colour channels of the imaged object, each pixel has a dimension (Δr, Δc) and a position in the image frame corresponding to the row (r) and the column (c) in which it is located in the sensor array. The pixel size is the smallest unit of an image. It is directly related to the resolution of the image, as it corresponds to the discrete grid of pixels that make up the image. The geometric relationship between the pixel size and its position in the image allows for establishing a *two-dimensional pixel coordinate system* (r, c) expressed in pixels. It is usually defined with its origin at the centre of the pixel, located in the upper left corner of the image, as shown in Fig. 19.2b. On the other hand, the mathematical equations of photogrammetry are based on a *two-dimensional image coordinate system* (x, y), expressed in linear units, with the origin at the centre of the sensor array. The geometric relationship between the two coordinate systems is shown in Fig. 19.2b.

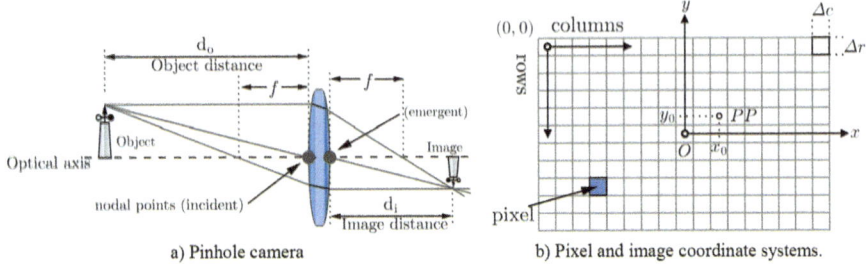

a) Pinhole camera   b) Pixel and image coordinate systems.

Fig. 19.2 Principle of the lens system in a pinhole camera and image coordinate systems. (**a**) Pinhole camera. (**b**) Pixel and image coordinate systems. (Based on Skaloud (2019))

For imaging, the photodiodes are fixed on a substrate, forming a sensor array based on charged-coupled (CCDs) or complementary metal-oxide-semiconductor (CMOS) technology, already described in previous chapters. It has different dimensions for photogrammetric applications, reaching resolutions of about 450 megapixels[3] for frame sensors,[4] with a pixel size of about 4.0 μm.

The values recorded in the digital image depend on the number of photons[5] captured by each photodiode in a given time interval, called *pixel intensity*. Generally, they are related to the average of the energies generated by the area covered by the pixel. In digital media, the intensity values are recorded in binary form, i.e., with values of 0 or 1, most commonly in 8-bit format (equivalent to 1 byte), which corresponds to integer values between 0 and 255, far beyond the human ability to distinguish shades of grey. Bit arrangements are varied; existing images with 8 bits, 11 bits, 14 bits, 16 bits, etc. The number of bits in an image indicates the *radiometric resolution* of the image. For example, a 16 bits image means that there are 65,536 possible digital numbers between 0 and 65,535 for that sensor to record information.

---

[3] The development of CCD and CMOS sensors is dynamic. The values given are indicative only. The reader should consult specialised and updated materials whenever information on this subject is required.

[4] For further details, see Sect. 19.4.1.

[5] Elementary particles of electromagnetic radiation, corresponding to the amount of energy in the respective light spectrum.

Colour images are generally formed by three matrices of intensity values of 8-bit, corresponding to the R (red), G (green) and B (blue) regions of the electromagnetic spectrum that the sensor can detect.

As mentioned, image acquisition occurs through an *image sensor*, which, in its simplest case, is based on the pinhole camera model. The geometric theory of the pinhole optical systems assumes straight light rays reflected from an object illuminated by a light source entering the camera through the pinhole, forming an inverted image on the focal plane of the sensor array. Each measured pixel corresponds to a spatial direction from the projection centre to the object point. The resulting 2D image is an ideal projection of the 3D scene on the sensor array. By construction, a complex lens system is mounted in the camera housing to enlarge the size of the camera opening, defining a *perspective centre* through which the optical ray enters the camera. The optical axis of such a lens system is defined as the central ray passing through the lens system without deviation. See Fig. 19.2a. The *focal length* (f) of the system is the distance between the perspective centre and the focal plane, also called the *principal distance* (c). The point where the optical axis hits the focal plane is called the *principal point* (*PP*). Thus, the principal point, the principal distance and the sensor array configuration are the elementary geometric parameters defining the geometric model of the camera, from which a bundle of spatial rays can be reconstructed from image coordinate measurements. The mathematical link between the object bundle of rays and its correspondence in the image plane is expressed by *collinearity equations*, described in Sect. 19.10.1.

In an ideal central perspective projection image formation event, a bundle of light rays reflected from the object traverses through the atmosphere, intersects at the perspective centre, and reaches a planar image focal plane, where the photodiode array records the intensity of the light rays. Suppose the lens aperture is set to zero so that an *image point* is directly related to an *object point* by the spatial ray passing through the perspective centre O. Consider the distance between the object plane and the image plane as *hf*, the sensor photosensitive element size as *pix*, the camera's focal length as *f* and the Ground Sample Distance as (*GSD*), as shown in Fig 19.3. By similarity of triangles, the four parameters are related as given by Eq. 19.1.

Fig. 19.3 Principle of central perspective image formation

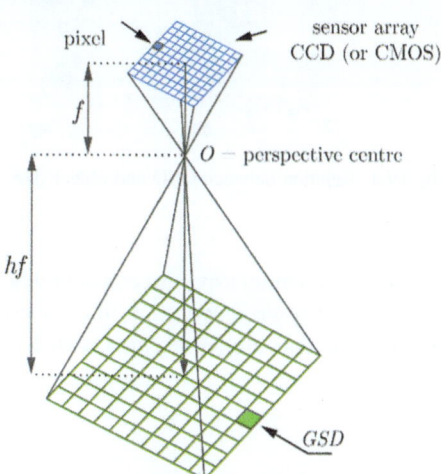

$$GSD = \frac{hf * pix}{f} = m * pix \qquad (19.1)$$

where

$$m = \frac{hf}{f} = \text{image scale} \qquad (19.2)$$

Radiometric resolution indicates the smallest change of intensity detected in an image, and *spatial resolution* indicates the smallest area of terrain that can be represented in the image.

Note that the smaller the *GSD* value, the higher the image resolution and the better the identification of objects in the image. The human brain's ability to detect, recognise and identify an object in a digital image depends on the number of pixels representing the object and its surroundings. This number is not consistently defined. In general, it is accepted that detection occurs when at least 2 pixels represent the object, recognition when there are between 4 and 6 pixels and identification when there are between 8 and 12 pixels (Sandau R. et al., 2010). This means that with a *GSD* of 4 cm, it will only be possible to identify approximately 32 to 48 cm objects. These values depend on the shape of the object and the contrast with neighbouring objects. A typical example of the relationship between *GSD* and object size in the image is shown in Fig. 19.4.

Fig. 19.4 Relation between GSD and object size

Another essential aspect to consider when creating a photographic image is that it is a *perspective projection*, i.e., it does not have a uniform scale since the objects closer to the lens are represented with a larger scale than those further away. This effect causes an image's radial deformation related to the relief of the terrain, as shown in Fig. 19.5a. Similarly, suppose the image has been acquired by an aerial camera inclined concerning the vertical line (oblique image). A deformation related to this inclination will also occur in this case, as shown in Fig. 19.5b.

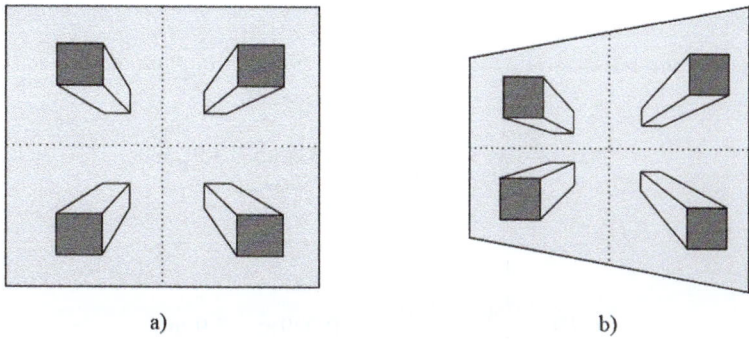

a)   b)

Fig. 19.5 Aerial image deformation. (**a**) Deformation due to the relief of the terrain. (**b**) Deformation due to the camera inclination

At this point, the Civil Engineer needs to note that, due to image deformations, a photograph cannot be considered geometrically equivalent to a map, except in rare circumstances.

As shown in Fig. 19.6, a map is an orthogonal projection, i.e., all objects are represented with the same scale, which is the map's scale. This means that on an aerial photograph, depending on the viewpoint, some building facades, for instance, may be visible, which is not the case on an orthogonal projection map.

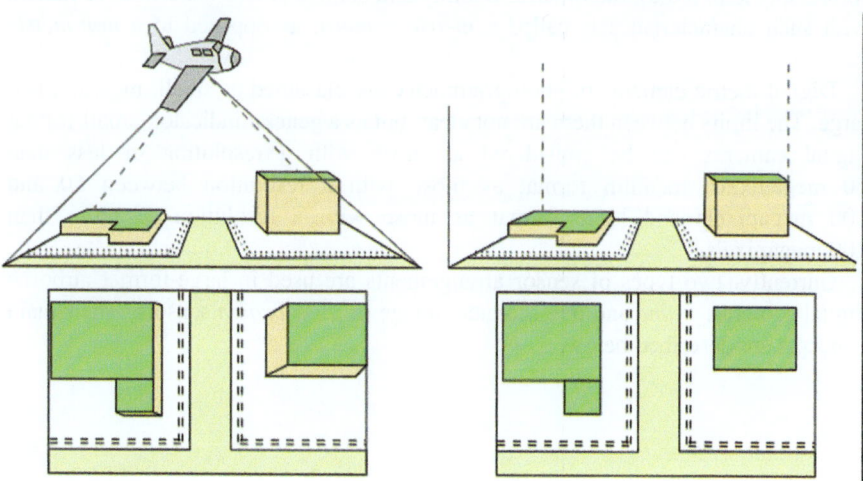

Fig. 19.6 Geometric difference between a map and an aerial image

Example 19.1

Considering a digital camera with principal distance c equal to 3.4 cm, composed of a CMOS sensor array with 6000 × 4000 rectangular photodiodes and dimensions 23.4 × 15.6 mm, respectively, calculate the pixel size pix and the GSD for a photogrammetric flight with flight height hf equal to 250 m.

Solution:

Considering the sensor array dimension and the number of photodiodes.

$$pix = \frac{23.5}{6000} = \frac{15.6}{4000} = 0.0039\,\text{mm} = 3.9\,\mu\text{m}$$

Using Eq. (19.1),

$$GSD = \frac{250^*3.9^*10^{-6}}{0.034} = 0.029\,\text{m} = 2.9\,\text{cm}$$

19.4   Aerial Image Acquisition

Any digital camera can perform the acquisition of photogrammetric aerial images. However, to achieve the quality required for photogrammetric purposes, they must have some essential physical characteristics that guarantee the geometry and the stability of their components so that the images are reproducible when subjected to the same operational conditions. These include the stability of principal distance (preferably kept fixed), radiometric stability and control of lens distortion. A camera with such characteristics is called a *metric camera*, as opposed to a *non-metric camera*.

Digital metric cameras in photogrammetry are classified as small, medium, and large. The limits between them are not clear, but as a general indicator, small-format digital cameras can be considered as those with a resolution of less than 50 megapixels, medium format as those with a resolution between 50 and 100 megapixels and large format as those with a resolution of more than 100 megapixels.[6]

Currently, two types of sensor arrangements are used in large-format airborne digital cameras: *frame* and line scanners image or *push-broom* sensors. Their main concepts are described below.

---

[6]Contextual classification suggested by the author.

19.4.1   Frame Sensor

Two technologies are used for frame sensors: the *single-head* and the *multi-head* systems. The former consists of a single CCD (or CMOS) photodiode array with a few hundred megapixels with dimensions ranging from 3.8 to 12.0 µm, radiometric resolution of 14 bits (R, G, B, NIR[7] or thermal infrared) and focal lengths ranging from 80 to 210 mm, which enables the generation of images with GSD of the order of 5 cm for flight heights of approximately 1000 m. The composition of the frames generated with this type of camera is shown in Fig. 19.7. The geometry of the image produced by this sensor is shown in Fig. 19.8. The multi-head system consists of several small-format single-head sensors acquiring images that are then remapped and radiometrically fused to produce a unique large-format image.

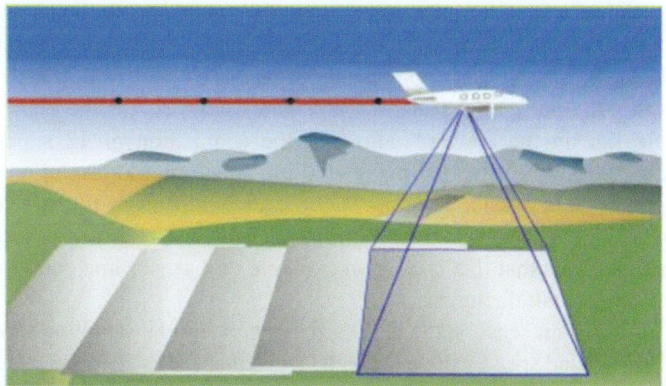

Fig. 19.7 Principle of airborne frame camera imaging. Courtesy Leica Geosystems

Fig. 19.8 Geometry of airborne frame camera data acquisition

Photograph with central perspective

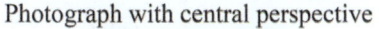

Flight line with overlapping photographs

---

[7] Near InfraRed.

Most current frame sensor-based cameras are rectangular, with the larger dimension transverse to the flight line to minimise the number of flight strips. The resolution of the sensor array is defined by the number of pixels in the image, for example, 26,460 × 17,004 pixels.

19.4.2   Line Cameras (Push-Broom)

Airborne push-broom cameras are based on the principle of spatial scanning performed by linear photodiodes arranged perpendicular to the flight line. The entire image is obtained by the aircraft's motion. In the case of the aerial cameras currently manufactured by Leica Geosystems, the linear photodiode array is arranged in three different viewing directions: backwards, nadir and forward, as shown in Fig. 19.9. Thus, three different stereo angles are possible: forward and nadir, backward and nadir, and forward and backwards. In this type of camera, four lines of multispectral channels (R, G, B, and NIR) and one panchromatic are positioned in the backward and nadir, plus the additional panchromatic line staggered[8] in the nadir and one panchromatic line in the forward, allowing the acquisition of 12 superimposed images, with 20,000 linear pixels (pixel ~5.0 μm), radiometric resolution of 14-bit and focal lengths of 62.5 and 120 mm. The images produced by this type of sensor are available in continuous bands (see Fig. 19.10), which can be cut out according to the user's needs. Note that this is only an example of a sensor line configuration. It varies for different manufacturers.

For the push-broom sensors, the GSD size depends on the sampling time t of the system and the ground speed v of the aircraft. Since the GSD sizes are independent in across- and along-track, the corresponding GSD_{across} in the orthogonal direction to the flight motion is given by Eq. (19.1) and in the direction of the flight (GSD_{along}) by Eq. (19.3).

$$GSD_{along} = v * \Delta t \qquad\qquad (19.3)$$

Airborne sensors can be installed alone or combined with other types of sensors. For example, a manned aircraft may be equipped with a digital camera and a LiDAR

---

[8] Sensor line shifted by a half pixel in the across-track direction.

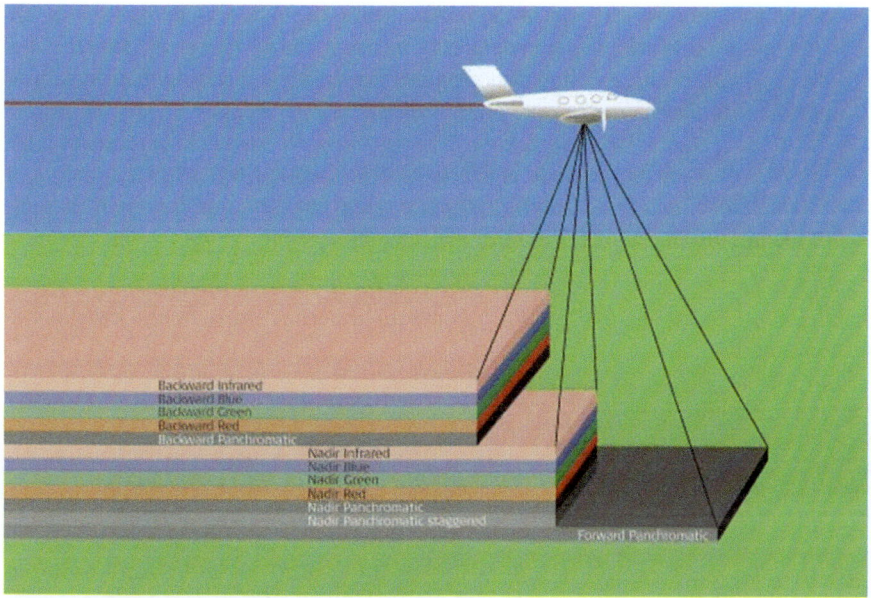

Fig. 19.9 Principle of airborne push broom camera imaging. Courtesy Leica Geosystems

Fig. 19.10 Geometry of airborne push-broom camera data acquisition

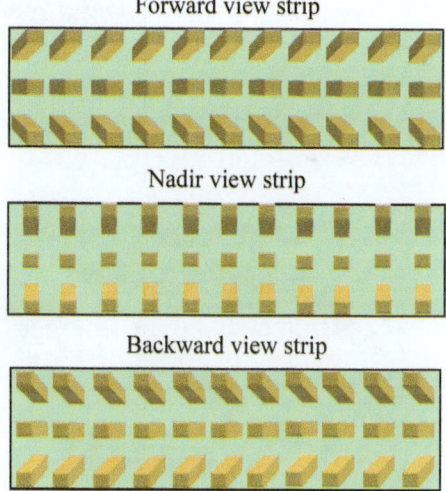

sensor, and an unmanned with an RGB frame camera and a thermal camera. In addition to image sensors, airborne image systems have several other components necessary for producing images of sufficient spatial quality for photogrammetry. These are:

(a) *Gyro-stabilised sensor mount: compensates for angular motion (vibrations and swing) and drift during data acquisition.*
(b) *Time Delay Integrator (TDI): avoids the image blur effect when capturing images at very high aircraft speed or a low light level, compensating the effect*

of the ground displacement in the image during the shutter opening time. Such compensation is achieved by adequately delaying the pixel signals delivered by the pixels of the same column (along-track direction) and adding them synchronously with the optical sensor. The result is a sharper image.

(c) *Positioning (GNSS) and inertial navigation system INS*: assist in determining the sensor's position and attitude in space, as described in Sect. 19.11.2.

(d) *Flight control and image capture devices*: related to the control of the trajectory of the aircraft and the operation of the imaging system.

Figure 19.11 shows an example of an airborne camera mounted on a gyro-stabilised sensor mount. Figure 19.12 shows an airborne camera before installation on the aircraft.

Fig. 19.11 Airborne camera installed on the gyro-stabilised sensor mount. (Courtesy: PHASEONE Industrial)

Fig. 19.12 Airborne camera system to be installed inside the aircraft. (Courtesy: Vexcel Imaging)

19.5 Aerial Photogrammetric Flight Mission

The acquisition of images, as described in the previous section, is performed by a photogrammetric flight, whose purpose is to acquire a set of images covering the entire area of interest. In the case of flight planning for the frame camera, the images must overlap each other to a minimum value along and across the flight line, as shown in Fig. 19.13. A sequence of images acquired along the same line is generally referred to as a *strip*.

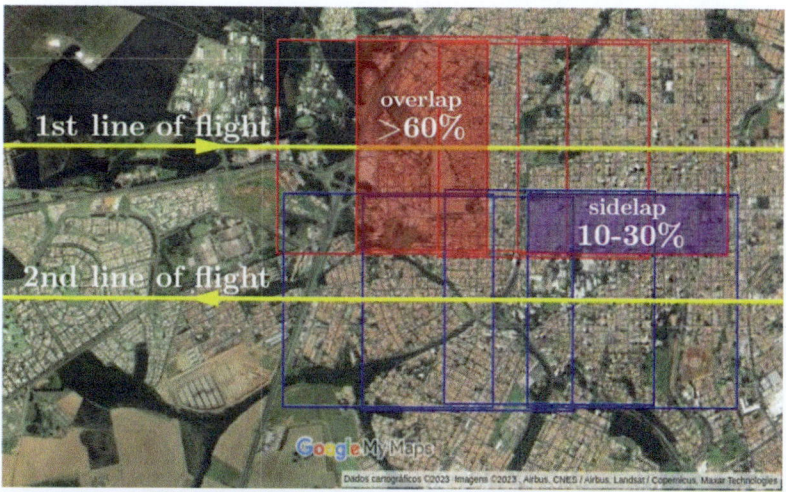

Fig. 19.13 Principle of photogrammetric flight lines for data acquisition. (Source: Google Maps image)

The photogrammetric flight is performed in strips and at predetermined flight heights using flight planning software. To compensate for the effect of aircraft angular motion, the overlaps in the longitudinal direction of flight range from 60% to 80% and the transversal strips sidelaps from 0% to 30%.

The geometric layout of the resulting strips depends on the terrain relief and the shape/extent of the project area. The number of strips and the direction of the flights depend on the configuration of the area to be covered.

The pair of longitudinal overlapping images is called *conjugated images* or *stereo pair*. Figure 19.14 shows an example of a pair of longitudinally overlapping images forming a stereo pair.

Generally, each strip has its altitude regarding the height of the terrain, determining the minimum and maximum GSD values, which must be checked in hilly terrain to ensure that they remain within the values determined for the project. In addition, it is also necessary to analyse the correct stereoscopic coverage of the entire area. Another flight planning solution consists of maintaining a relatively constant altitude over the terrain during the flight. In this case, an accurate terrain model is required for flight planning. The main advantage of this approach is a more uniform spatial resolution along the flight lines. In practice, however, this type of flight mission is more likely to be used for unmanned aircraft flights.

Stereopair.

Longitudinal overlapping.

Fig. 19.14 Example of a longitudinal stereo pair

The number of strips and, consequently, of images depends on the size of the sensor photodiode, the flight height, the desired GSD, the specified overlaps and side laps and the size and shape of the area to be covered. For this purpose, equations and application programs can be used to calculate the quantities of strips and images required to cover a given area. For more information, readers are advised to consult specialised references.

19.6 Aerial Photogrammetric Model

The geometric relationship between the plane image and the spatial object is called the *photogrammetric model. See* Sect. 19.9. It is divided into two geometric spaces: the image and object spaces, each with its coordinate system. The object space is referenced to the object coordinate system, and the image space is referenced to the image coordinate system (see Fig. 19.21). The spatial geometric representation of this model is shown in Fig. 19.15.

The object coordinate system is a 3D Cartesian coordinate system defined by the reference points of the object. The image coordinate system is described in Sect. 19.3.

As shown in Fig. 19.15, the image space can be arranged in a negative or positive position concerning the perspective centre. Generally, the positive position is preferred, i.e., with the image plane facing downwards from the perspective centre.

Fig. 19.15 The photogrammetric model

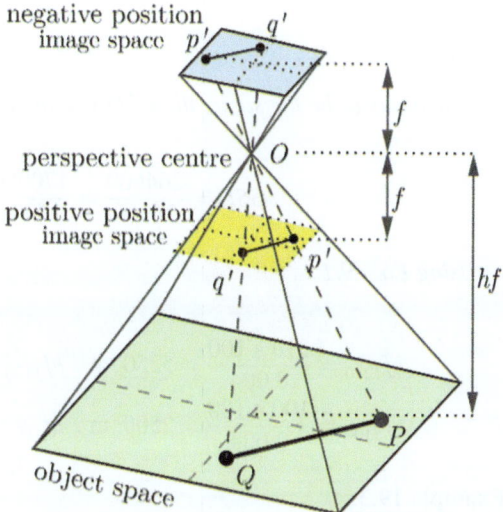

Example 19.2

Consider the situation of four photogrammetric flights with four airborne frame cameras, each with a 4×4 μm photodiode dimension and $26,460 \times 17,004$ pixels, transversal and longitudinal to the flight line, respectively. Given that they have focal lengths of 210, 120, 100 and 80 mm, calculate the flying heights so that the terrain areas covered by each image have dimensions equal to 2646×1700 m, as shown in Fig. 19.16.

Fig. 19.16 Geometry of the flight

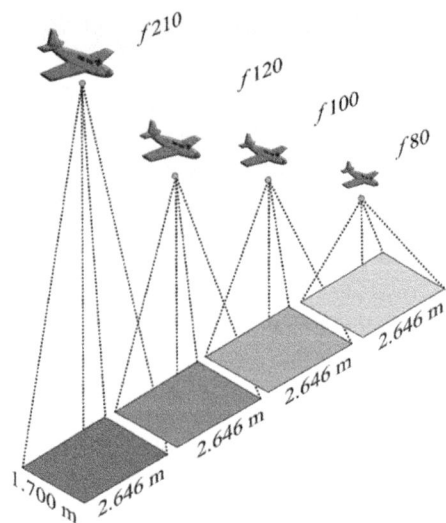

Solution:

For the areas to be the same, the GSD size in each flight must be the same. Thus,

$$GSD = \frac{264600}{26460} = \frac{170000}{17004} = 10 \text{ cm}$$

Using Eq. 19.1.

$$hf_{f210} = \frac{210 * 100}{0.004} = 5250 \text{ m} \quad hf_{f120} = \frac{120 * 100}{0.004} = 3000 \text{ m}$$

$$hf_{f100} = \frac{100 * 100}{0.004} = 2500 \text{ m} \quad hf_{f80} = \frac{80 * 100}{0.004} = 2000 \text{ m}$$

Example 19.3

Using the same airborne sensor as in the previous example, in its configuration with a focal length equal to 100 mm, calculate the terrain area covered by a stereo pair with longitudinal coverage equal to 60% and flying height equal to 1500 m.

Solution:

Using Eq. 19.1.

$$Rectangle \ side \ 1 = \frac{1500 * 26460 * 4 * 10^{-6}}{0.1} = 1.59 \text{ km}$$

$$Rectangle \ side \ 2 = \frac{1500 * 17004 * 4 * 10^{-6}}{0.1} = 1.02 \text{ km}$$

The terrain area covered by each image is given as follows:

$$area_{img} = 1.59 * 1.02 = 1.62 \ \text{km}^2$$

The terrain area covered by the stereo pair is given as follows:

$$area_{pair} = 0.6 * 1.59 * 1.02 = 0.97 \ \text{km}^2$$

19.7 Stereoscopy

Stereoscopy is the visual effect that allows human beings to virtually reconstruct in their brains the 3D view of a scene from two-dimensional visualisations of a pair of images of the same object taken from different viewpoints, known as conjugate images. The physical principle of the stereoscopic effect is based, among other things, on the ability of human beings to detect changes in the angle of convergence of their optical axes of sight and thus detect depth differences in a scene, as shown schematically in Fig. 19.17.

Fig. 19.17 Angle of convergence

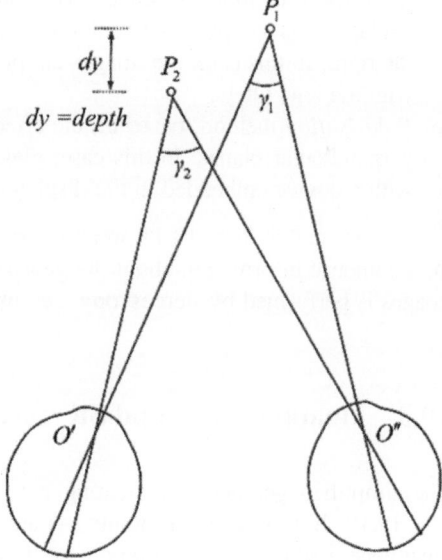

In the case of two overlapping images, the observer can only reconstruct the virtual 3D space (stereoscopic model) if each eye views the respective images separately and simultaneously, as shown in Fig. 19.18. Several devices are created to make this possible, as shown below.

Fig. 19.18 Principle of stereoscopy view

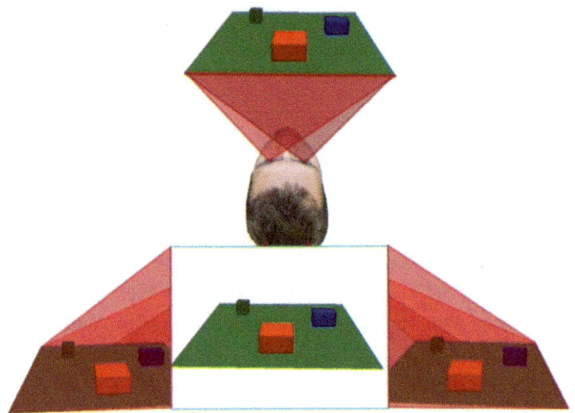

(a) *Images split* using diverging lenses, as, for example, in *pocket stereoscopes* and *mirror stereoscopes*.
(b) *Anaglyphic* method, which is based on the effect of viewing images through ocular lenses equipped with coloured filters, e.g. red for the left lens and blue for the right. In this case, the images are projected or printed in colours compatible with the lens filters.
(c) *Polarisation* method, based on the effect of the oscillation of the image in two perpendicular planes. In this case, glasses with polarised lenses are combined with a device embedded in the display to polarise two views differently.

As presented in Sect. 19.11, in photogrammetry, the process of manually capturing geometric information about the geospatial data represented in a pair of digital images is performed by stereoscopy, i.e., using a stereoscopic model.

19.8 Floating Mark and Stereoscopic Parallax

The simplified geometry of creating two overlapping aerial images is shown in Fig. 19.19. In this case, the points (A) and (B) in the terrain, called *object points*, have generated two *image points* in each photo, which are the points (a', b') in the left picture and the points (a'', b'') in the right. The image points (a') and (a'') represent the images of point (A) in each photograph and are, therefore, homologous points. Similarly, points (b') and (b'') are also homologous. Taking the homologous points as a reference, each has undergone an apparent displacement in the plane of the images, with a variable value, depending on the elevation at which they are located on the terrain. This apparent shift is called *stereoscopic parallax*.

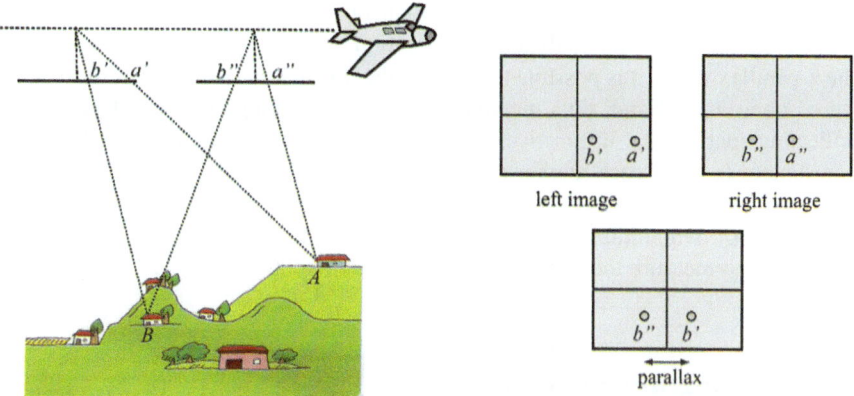

Fig. 19.19 Stereoscopic parallax

Considering the effect of the relief on the apparent distance between homologous points, as shown in Fig. 19.19, it can be seen that the closer the object point is to the photograph, the smaller the distance between its images in the two pictures, that is, the smaller the stereoscopic parallax. For example, in Fig. 19.20, object point (P_1) has a lower elevation than object point (P_2). Therefore, the horizontal distance between the homologous points (p_2') and (p_2'') is smaller than the horizontal distance between the homologous points (p_1') and (p_1'').

Fig. 19.20 Principle of the floating mark

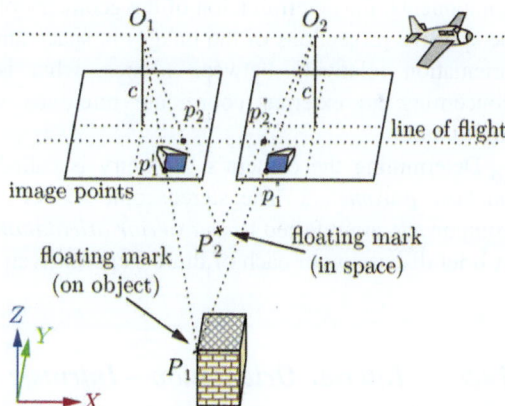

As shown in Fig. 19.20, object point (P_2) is not in the terrain. It will appear in the stereoscopic model space if, for instance, two identical artificial marks are placed on the photographs at the intersection of two rays from the optical centres (O_1) and (O_2). In this situation, it will appear to float above the ground in the stereoscopic model, and because it is a virtual mark, it is called the *floating mark*.

The reader should note that by simultaneously moving the two marks over the photographs, keeping them at the intersection of homologous rays, and changing the x-parallax value, it is possible to position the floating mark over any object in the stereoscopic model and, thus, use the marks in the photographs as references for collecting spatial data. In practice, this means determining the values of the image coordinates (x, y) of each marker on each overlapping photograph.

If the homologous rays do not intersect in space, the observer will have the impression of two points in the images. In this case, the x and y parallaxes must be eliminated to measure the image coordinates of the point correctly.

19.9   Photogrammetric Model

The brilliant idea of photogrammetry is recreating the spatial geometrical situation of image taking through a mathematical model, which represents the geometrical relationship between image and object spaces, allowing its exploration through a stereoscopic model (3D view). Physically, this means creating the conditions to have a reduced model of the object space in which geometrical and semantic information of spatial data can be collected. Such a model is called a *photogrammetric model.*

The creation of a photogrammetric model requires the determination of the geometric relationship between the airborne sensor and the real world. In the literature, this process is called *georeferencing* or *camera orientation* in photogrammetry and *Structure from Motion* (SfM) in computer vision. It is based on two components: the determination of the geometry of the camera to recreate the central perspective projections of the images in space and the determination of the camera orientation relatively between scenes, what is called *relative orientation*, or concerning an external coordinate reference system, what is called *absolute orientation.*

Determining the camera's geometry is called *interior orientation*[9] (IO) or its *intrinsic parameters.* The stereoscopic model reconstruction and georeferencing components are referred to as *exterior orientation* (EO) or its *extrinsic parameters.* A brief discussion of each of these steps is given below.

19.9.1   Interior Orientation—Intrinsic Parameters

In practice, the interior orientation (IO) of a camera means determining the image geometry, i.e., the size of the camera sensor, the calibrated focal length c of the lens, the position of the principal point (*PP*) in the image plane, and the lens distortion. These parameters are called *interior orientation parameters* or *intrinsic parameters.*

---

[9] Also inner orientation.

They are available from camera calibration and are considered fixed for the corresponding physical camera.

In addition to determining the interior orientation parameters, the (x, y) image coordinate system is also established in this orientation stage, as described in Sect. 19.3. This means that a coordinate transformation is performed to adapt the pixel coordinate system to the metric image coordinate system. This transformation from the pixel coordinate system to the image-centred coordinate system is given in matrix form by Eq. (19.4).

$$
\begin{bmatrix} x \\ y \end{bmatrix} = \begin{bmatrix} \Delta r & 0 \\ 0 & -\Delta c \end{bmatrix} * \begin{bmatrix} r - rc \\ c - cc \end{bmatrix} \tag{19.4}
$$

where

x, y = image coordinate
$\Delta r, \Delta r$ = pixel size
r, c = pixel coordinate (row and column)
r_c, c_c = pixel coordinate of the central pixel

Considering that the number of rows and columns of the image are H and W, respectively, the coordinates (r_c, c_c) are given as follows:

$$
r_c = (H - 1)/2 \tag{19.5}
$$

$$
c_c = (W - 1)/2 \tag{19.6}
$$

19.9.2   Image Coordinates Refining

As previously stated, the determination of geospatial data in airborne photogrammetry is based on the measurements of the image coordinates of homologous points, assuming that the ray connecting the object point, the perspective centre, and the image point is linear, as shown in Fig. 19.3. The main phenomena invalidating this linearity are atmospheric refraction, lens system distortion, geometric inaccuracies of the photodiode array and errors in the platform calibration. It is beyond the scope of this book to discuss atmospheric refraction and geometric inaccuracies of the photodiode array. Therefore, only the nonlinearities caused by the lens distortions and the platform calibration will be discussed in the following subsections.

19.9.2.1   Optical Distortion Models

Current metric cameras have excellent optical and physical qualities. Nevertheless, the user must be aware of the optical problems that can occur in the images produced by a digital camera, such as *image blurring, lens distortion, lens aberration, lens diffraction, depth of field* and *depth of focus*. In photogrammetry, only image blurring and lens distortion are considered. *Image blurring* is an image aberration that occurs as a result of the forward motion, rotation and vibrations of the camera during data collection, causing small structures in the scene to appear unsharp. It is minimised if the camera is equipped with a time-delayed integrator feature, as previously described. On the other hand, lens distortions are the physical deviations of pixels between the mathematical model of perspective projection and the physical formation model. These distortions cause, for example, straight lines in the object space not to be straight in the image, i.e., they distort the image. These problems become even more relevant with non-metric cameras, where the quality of the optical system may be inferior to that of metric cameras. For photogrammetric purposes, lens distortion is characterised by *radial* and *decentring* distortions. *Radial symmetric distortion* is the most common and varies radially from the principal point. The mathematical model used to correct it is generally expressed via the high-order polynomials given by the following equations:

$$\Delta x_r = (x - x_0) * \left(k_1 * r^2 + k_2 * r^4 + k_3 * r^6 \right) \tag{19.7}$$

$$\Delta y_r = (y - y_0) * \left(k_1 * r^2 + k_2 * r^4 + k_3 * r^6 \right) \tag{19.8}$$

$$r^2 = (x - x_0)^2 + (y - y_0)^2 \tag{19.9}$$

where

$\Delta x_r, \Delta y_r$ = radial symmetric distortion at (x, y)
x, y = image coordinate
x_0, y_0 = image coordinate of the principal point
k_i = coefficient of radial distortion
r = radial distance from the principal point

Decentring distortion is due to a lack of centring of lens elements along the optical axis, which causes radial non-symmetric and tangential distortions in the image. The mathematical model used to correct it is generally expressed as given by the following equations:

$$\Delta x_d = p_1 * \left[r^2 + 2(x - x_0)^2 \right] + 2p_2 * (x - x_0) * (y - y_0) \tag{19.10}$$

$$\Delta y_d = p_2 * \left[r^2 + 2(y - y_0)^2 \right] + 2p_1 * (x - x_0) * (y - y_0) \tag{19.11}$$

where

$\Delta x_d, \Delta y_d$ = component of the decentring distortion at (x, y)
x, y = image coordinate
x_0, y_0 = image coordinate of the principal point

p_i = coefficient of decentring distortion
r = radial distance from the principal point

Generally, decentring distortion has little influence on the result of the coordinates measured on the images compared to symmetric radial distortion. For this reason, it is often disregarded in photogrammetric models.

19.9.3   Exterior Orientation—Extrinsic Parameters

Exterior Orientation (EO) of the photogrammetric model means mathematically establishing the interrelation between image and object spaces, i.e., determining the mathematical equations that relate the coordinate systems of both spaces. In practice, this means georeferencing the sensor concerning the coordinate system of object space. Such an interrelation is expressed by determining the six exterior orientation parameters (e.g. spatial position X_0, Y_0, Z_0 and attitude angles ω, ϕ, κ of an image at the time of exposure) concerning the object coordinate system, as described in Sect. 19.11.

19.10   Mathematical Concepts of Photogrammetry

The mathematical concepts of photogrammetry are based on the principles of descriptive geometry with the following assumptions:

(a) Object invariable during data acquisition.
(b) Object consisting of a set of points in space.
(c) Image formed by a set of light rays considered as straight lines connecting the image point to the object point passing through a perspective centre O (principle of collinearity).
(d) Only a single ray from a point object forms the corresponding image point.
(e) Mathematical model based on the equations of the perspective projection.
(f) Object represented in two or more images.
(g) Image considered to be flat.

Based on these assumptions, two Cartesian coordinate systems are involved in the spatial geometry of the photogrammetric model, as shown in Fig. 19.21. The coordinates (X, Y, Z) of the object space are related to the coordinate system adopted for the project. In this case, they are generic coordinates that can be related to either a local 3D Cartesian coordinate system, a cartographic projection system, or a local-based plane, plus the orthometric altitude. The image coordinates (x, y, z) refer to the 3D Cartesian coordinate system with the origin at the optical centre O of the camera. The z-axis completes the Cartesian coordinate system, where the distance from the optical centre to the image plane along the optical axis is the camera's principal distance c. The angles (κ, ϕ, ω) or *(roll, pitch, yaw)* are the rotation angles (attitude angles) of the camera in space.

In Fig. 19.21, point (O) has coordinates (X_0, Y_0, Z_0). Point (P) is the object point on the terrain with coordinates (X_P, Y_P, Z_P). Point (p) is the image point with coordinates $(x_P, y_P, -c)$. Point (PP) is the principal point, which does not necessarily coincide with the central pixel of the image and, therefore, has image coordinates (x_0, y_0).

The (x, y, z) image coordinate system is rotated in space by the angles $(\omega, \varphi, \kappa)$ concerning the (X, Y, Z) object coordinate system and its origins are shifted in space by (X_0, Y_0, Z_0).

The geometric relationship between these parameters is given by a mathematical model based on a set of equations called *Collinearity Equations*, as shown below.

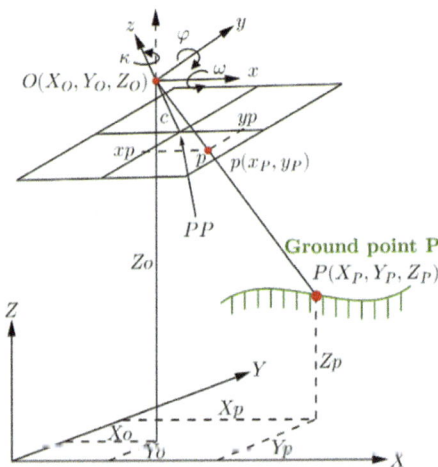

Fig. 19.21   Photogrammetric model geometry

19.10.1   The Collinearity Equations

The mathematical model used to represent the physical model shown in Fig. 19.21 is based on the projective equations model, through which the collinearity equations are obtained. They make it possible to determine the (X, Y, Z) Cartesian coordinates of an object point on the terrain by measuring its image coordinates (x, y) in a pair of photographs or the inverse. In short, the collinearity equations relate the three-dimensional coordinates of a point in the object space to those of its counterparts in the image space.

The parameters of the collinearity equations are determined by applying a 3D Cartesian coordinate transformation with seven parameters between the object coordinate system (X, Y, Z) and the image coordinate system (x, y, z). In this geometric configuration, the perspective centre O of the camera has the following coordinates:

Concerning the object coordinate system:

$$\begin{bmatrix} X \\ Y \\ Z \end{bmatrix} = \begin{bmatrix} X_0 \\ Y_0 \\ Z_0 \end{bmatrix} \tag{19.12}$$

Concerning the image coordinate system:

$$\begin{bmatrix} x \\ y \\ z \end{bmatrix} = \begin{bmatrix} x_0 \\ y_0 \\ -c \end{bmatrix} \tag{19.13}$$

To apply the coordinate transformation, it is first necessary to make the two coordinate systems parallel, which can be done by rotating the (X, Y, Z) object coordinate system. Thus, assuming rotations $(\kappa, \varphi, \omega)$ gives

$$\begin{bmatrix} x \\ y \\ z \end{bmatrix} = R(\kappa, \varphi, \omega) * \begin{bmatrix} X \\ Y \\ Z \end{bmatrix} \tag{19.14}$$

Next, the origins of both coordinate systems must coincide so that the origin of the (X, Y, Z) coordinate system is placed in the perspective centre. Finally, the difference in magnitude between the two must be eliminated by introducing a scale factor k into the equations. Thus,

$$\begin{bmatrix} x - x_0 \\ y - y_0 \\ -c \end{bmatrix} = k * R(\kappa, \varphi, \omega) * \begin{bmatrix} X - X_0 \\ Y - Y_0 \\ Z - Z_0 \end{bmatrix} \tag{19.15}$$

Considering the elements of rotation matrix R $(\kappa, \varphi, \omega)$ expressed according to Eq. (5.81), Eq. (19.15) can be rewritten to generate Eq. (19.16).

$$\begin{bmatrix} x - x_0 \\ y - y_0 \\ -c \end{bmatrix} = k * \begin{bmatrix} r_{11} & r_{12} & r_{13} \\ r_{21} & r_{22} & r_{23} \\ r_{31} & r_{32} & r_{33} \end{bmatrix} * \begin{bmatrix} X - X_0 \\ Y - Y_0 \\ Z - Z_0 \end{bmatrix} \tag{19.16}$$

Multiplying the matrix and the vector on the right side of the equation gives the following set of algebraic equations:

$$x = x_0 + k * [r_{11}(X - X_0) + r_{12}(Y - Y_0) + r_{13}(Z - Z_0)] \tag{19.17}$$

$$y = y_0 + k * [r_{21}(X - X_0) + r_{22}(Y - Y_0) + r_{23}(Z - Z_0)] \tag{19.18}$$

$$-c = k * [r_{31}(X - X_0) + r_{32}(Y - Y_0) + r_{33}(Z - Z_0)] \tag{19.19}$$

Dividing Eqs. (19.17) and (19.18) by Eq. (19.19) and adding noise parameters $(\Delta x, \Delta y)$ gives the classical form of the collinearity equations as follows:

$$
\begin{aligned}
x &= x_0 - c * \frac{r_{11}(X - X_0) + r_{12}(Y - Y_0) + r_{13}(Z - Z_0)}{r_{31}(X - X_0) + r_{32}(Y - Y_0) + r_{33}(Z - Z_0)} + \Delta x \\
y &= y_0 - c * \frac{r_{21}(X - X_0) + r_{22}(Y - Y_0) + r_{23}(Z - Z_0)}{r_{31}(X - X_0) + r_{32}(Y - Y_0) + r_{33}(Z - Z_0)} + \Delta y
\end{aligned} \tag{19.20}
$$

Note that the scale factor k disappears in the equation system (19.20).

Assuming that the interior orientation and noise parameters are known, the system of Eq. (19.20) has six transformation parameters to calculate, which are the three rotation angles $(\kappa, \varphi, \omega)$ and the three coordinates (X_0, Y_0, Z_0) of the perspective centre. Therefore, considering that each point with known coordinates in the image and object coordinate system generates two equations of the type (19.20), it is necessary to have at least three points to calculate the six unknowns $(\kappa, \varphi, \omega, X_0, Y_0, Z_0)$.

Equations (19.20) relate the object coordinates to image coordinates. To perform the inverse, i.e., reconstruct the 3D space from the images, the inverse o matrix R can be used, and since it is an orthogonal matrix, the new set of equations can be written as follows:

$$
\begin{bmatrix} X - X_0 \\ Y - Y_0 \\ Z - Z_0 \end{bmatrix} = k * R^{\mathrm{T}}(\kappa, \varphi, \omega) * \begin{bmatrix} x - x0 \\ y - y0 \\ -c \end{bmatrix} \tag{19.21}
$$

And then,

$$
\begin{aligned}
X &= X_0 + (Z - Z_0) * \frac{r_{11}(x - x_0) + r_{21}(y - y_0) - r_{31} * c}{r_{13}(x - x_0) + r_{23}(y - y_0) - r_{33} * c} \\
Y &= Y_0 + (Z - Z_0) * \frac{r_{12}(x - x_0) + r_{22}(y - y_0) - r_{32} * c}{r_{13}(x - x_0) + r_{23}(y - y_0) - r_{33} * c}
\end{aligned} \tag{19.22}
$$

From Eq. (19.22), it is possible to calculate the coordinates (X, Y) of any point, provided that the IO parameters, the six EO parameters and the corresponding Z-coordinate are known. Therefore, if the Z-coordinate is an unknown value, there are infinite object points for each image coordinate (x, y). This means that for an unambiguous solution, a second photograph of the same object with known orientation parameters or additional information on the Z-coordinate is required, for example from a DSM.

Considering that the interior orientation parameters are known from camera calibration, and the Z-coordinate is not available, the solution of Eq. (19.22) requires at least one stereo pair of images and three control points in the terrain (object space) identifiable i6n each conjugate image. Each control point generates four Eq. (19.22), producing 12 equations for three control points. Each conjugate image has 6 unknowns, resulting in 12 unknowns for the stereo pair. Solving this system of equations gives the values of the 12 transformation parameters. As a result, the 2D to 3D coordinate transformation can be performed by measuring the (x, y) coordinates in each conjugate image to obtain their respective (X, Y, Z) object coordinates. When there are more than three homologous control points, which is often the case, the solution to the system of equations is obtained by applying an adjustment computation model, as presented in Chap. 3.

Example 19.4

Considering the orientation parameters of the airborne camera in positions (1) and (2) for a given aerial photogrammetric flight, as shown in Table 19.1, calculate the ground coordinate values of point (P), whose image coordinates are given in Table 19.2.

Table 19.1 Exterior orientation parameters

Perspective centre	Coordinate [m]	Rotation angle	Principal distance [mm]
X_{01}	201,961.524	$\omega_1 = -0.013197159$ rad	$c = 153.5$
Y_{01}	7,564,499.471	$\varphi_1 = 0.023878410$ rad	
Z_{01}	2038.068	$\kappa_1 = -1.592169455$ rad	
X_{02}	201,948.101	$\omega_2 = -0.002147713$ rad	
Y_{02}	7,563,826.864	$\varphi_2 = 0.030440581$ rad	
Z_{02}	2036.582	$\kappa_2 = -1.573313763$ rad	

Table 19.2 Image coordinates of the point (P)

Point	Image 1		Image 2	
P	$x_{p'}$ [mm]	$y_{p'}$ [mm]	$x_{p''}$ [mm]	$y_{p''}$ [mm]
	97.722	−37.373	14.082	−37.044

Solution:

Considering Eq. (5.80), the rotation matrices of images 1 and 2 are as follows:

$$R_1(\kappa, \varphi, \omega) = \begin{bmatrix} -0.021365409 & -0.999677808 & 0.013703986 \\ 0.999486593 & -0.021684656 & -0.023586574 \\ 0.023876141 & 0.013193014 & 0.999627868 \end{bmatrix}$$

$$R_2(\kappa, \varphi, \omega) = \begin{bmatrix} -0.002516267 & -0.999994360 & 0.002224325 \\ 0.999533554 & -0.002582795 & -0.030430307 \\ 0.030435880 & 0.002146716 & 0.999534416 \end{bmatrix}$$

For simplification, consider the second term of Eq. (19.22) as follows:

$$X = X_0 + (Z - Z_0) * kx_1$$
$$Y = Y_0 + (Z - Z_0) * ky_1$$

where

$$kx_i = \frac{r_{11}(x - x_0) + r_{21}(y - y_0) - r_{31} * c}{r_{13}(x - x_0) + r_{23}(y - y_0) - r_{33} * c}$$

$$ky_i = \frac{r_{12}(x - x_0) + r_{22}(y - y_0) - r_{32} * c}{r_{13}(x - x_0) + r_{23}(y - y_0) - r_{33} * c}$$

$$r_{ij} = elements\ of\ rotation\ matrix\ R\ (\kappa, \varphi, \omega)$$

Using data from Tables 19.1 and 19.2.

$$kX_1 = 0.285055182$$
$$kY_1 = 0.654039053$$
$$kX_2 = 0.274079441$$
$$kY_2 = 0.094015691$$

As there are four equations and three unknowns, the solution to this problem can be achieved using the Parametric adjustment model for the following observation equations:

$$X = X_0 + (Z - Z_0) * kx_1$$
$$Y = Y_0 + (Z - Z_0) * ky_1$$
$$X = X_0 + (Z - Z_0) * kx_2$$
$$Y = Y_0 + (Z - Z_0) * ky_2$$

Or,

$$X + 0Y - Z * kx_1 = X_{01} - Z_{01} * kx_1$$
$$0X + Y - Z * ky_1 = Y_{01} - Z_{01} * ky_1$$
$$X + 0Y - Z * kx_2 = X_{02} - Z_{02} * kx_2$$
$$0X + Y - Z * ky_2 = Y_{02} - Z_{02} * ky_2$$

Using Eq. (3.79), the following matrices can be written as

$$A = \begin{bmatrix} 1 & 0 & -0,285055182 \\ 0 & 1 & -0.654039053 \\ 1 & 0 & -0.274079441 \\ 0 & 1 & -0.094015691 \end{bmatrix} \quad l_0 = \begin{bmatrix} 201,380.56216 \\ 7,563,166.49493 \\ 201,389.91574 \\ 7,563,635.39334 \end{bmatrix} \quad x = \begin{bmatrix} X \\ Y \\ Z \end{bmatrix}$$

*Using Eqs. (3.84) to (3.87) and assuming that the weight matrix **P** is equal the identity.*

$$N = \begin{bmatrix} 2.0 & 0 & -0.559135 \\ 0 & 2.0 & -0.748055 \\ -0.559135 & -0.748055 & 0.592982 \end{bmatrix} \quad n = \begin{bmatrix} 402,770.4779 \\ 15,126,801.8883 \\ -5,770,308.0683 \end{bmatrix}$$

$$x = \begin{bmatrix} 201,619.3177 \\ 7,563,714.1133 \\ 837.2894 \end{bmatrix}_m$$

Using Eq. (3.79),

$$v = \begin{bmatrix} 0.081859 \\ -0.001604 \\ -0.081858 \\ 0.001604 \end{bmatrix}_m$$

Finally, for n = 4 and u = 3, the standard deviations of the unit weight and the variance-covariance matrix for the unknowns (coordinates) are given as follows:

$$s_0^2 = 0.01341 \, m^2$$
$$s_0 = 0.116 \, m = \pm 11.6 \, cm$$

$$\Sigma_{xx} = \begin{bmatrix} 0.013383026 & 0.008936501 & 0.023892638 \\ 0.008936501 & 0.018659386 & 0.031965470 \\ 0.023892638 & 0.031965470 & 0.085462917 \end{bmatrix}$$

Using the diagonal values in the variance-covariance matrix for the parameters, the standard deviations of the adjusted angles are given as follows:

$$X_P = 201,619.318 \pm 0.116 \text{ m}$$

$$Y_P = 7,563,714.113 \pm 0.137 \text{ m}$$

$$Z_P = 837.289 \pm 0.292 \text{ m}$$

19.11   Determination of Exterior Orientation Parameters

As described in the previous section, the collinearity equations relate the coordinates (X, Y, Z) of the object space to the coordinates (x, y) of the image space. Thus, provided that the (EO) parameters $(\kappa, \varphi, \omega, X_0, Y_0, Z_0)$ of each image as well as the IO parameters $(c, x_0, y_0, \ldots)$ of the camera are known, it is possible to determine the value of the coordinates of any point on the ground (object space) by measuring its image coordinates in the conjugate photographs.

If the values of the *EO* parameters $(\kappa, \varphi, \omega, X_0, Y_0, Z_0)$ of a pair of conjugate images are known, a stereoscopic pair is immediately formed, i.e. it is possible to visualise the 3D model (stereoscopic model) by positioning the images on the computer screen and by using an image separation device, as described in Sect. 19.7.

Measuring the coordinates of a point on an image means determining the pixel position of the point in the photodiode array. This can be done monoscopically, stereoscopically or by image correlation as described in Sect. 19.13. In the first case, the position of the point in each image is determined by visual recognition. In this case, the (x, y) image coordinates are measured separately. In the second case, the location of the point on the images is done using the floating mark as described in Sect. 19.8. In the third case, the coordinates are determined automatically using an image correlation algorithm. In the case of the stereoscopic measurement, when the operator visually considers that the floating mark is positioned over the representation of the point on the stereoscopic model, he confirms the position, and the system directly measures the values of the (x, y) image coordinates in both conjugate photographs.

The whole photogrammetric process is based on the knowledge of the *EO* parameters $(\kappa, \varphi, \omega, X_0, Y_0, Z_0)$ of the images. They can be derived *indirectly* from overlapping images and ground control points distributed over the image block, *directly* by integrating GNSS and *Inertial Navigation System* (INS) into the image data acquisition system or even by combining both approaches.

19.11.1 Indirect Method of Exterior Orientation

In the *indirect method*, the *EO* parameters of each image are considered unknowns and calculated by aerial triangulation, where adjacent images are connected by automatic or manual measurement of homologous tie points. A reduced number of ground control points is sufficient to calculate the EO parameters when using such a method.

A control point for aerial photogrammetry is a marked point on the ground and identifiable in the image, whose coordinates are known concerning a previously selected 3D coordinate system. This means they must be determined by survey measurements using GNSS technology or total stations. They can be Geocentric, UTM or local topographic coordinates with orthometric altitudes. Usually, a ground control point is usually abbreviated to GCP.

A control point can be planimetric, altimetric or planialtimetric. To calculate the exterior orientation parameters of conjugate images, it is necessary to have at least two points with known (X, Y, Z) coordinates and one point with Z-coordinate. Using more than three control points is recommended for better accuracy and to know the precision of the calculated values.

There are two types of control points used in aerial photogrammetry.

19.11.1.1 Natural Control Point (Photo Identifiable)

Generally, a natural control point is visually selected on the images after flying the project. Its identification depends on the GSD of the project, provided it is sharp and well-defined horizontally. A natural feature control point can be any feature on the ground, such as a manhole, parking stripe, etc. Once identified in the image, it must be identified and measured in the field. Figure 19.22 shows an example of a natural control point that can be used in airborne photogrammetry.

Fig. 19.22 Example of the
natural control point

19.11.1.2   Artificial Control Point (Pre-signalised)

An artificial control point is established in the field by marking or painting figures on
the ground before flying the project. The shape, type and size of these points vary
according to the flight's height and the image's geometric resolution, provided they
are large enough to be identified from the particular flight altitude. The points thus
marked may be measured in the field before or after the flight. It is more laborious to
set up than a natural one. However, its identification and the quality of the results can
be significantly improved as its digital identification is facilitated by using automatic
identification tools in photogrammetry programs. Figure 19.23 shows an example of
an artificial control point used in aerial photogrammetry.

Fig. 19.23 Example of the artificial control point

19.11.2   Direct Georeferencing Using GNSS/INS Systems

Direct measurement of EO parameters typically relies on integrating the GNSS receiver and antenna with an inertial navigation system (INS), as shown in Fig. 19.24. In such an integrated system, the GNSS data provide absolute orientation, velocity information, and error control of the inertial measurements. On the other hand, the INS contributes to attitude estimation by interpolating the trajectory between GNSS positions and by mitigating the GNSS loss of satellite tracking. In this case, it is said that a *direct georeferencing* of the photogrammetric model is performed.

Fig. 19.24 Typical installation of GNSS and INS sensors in the aircraft

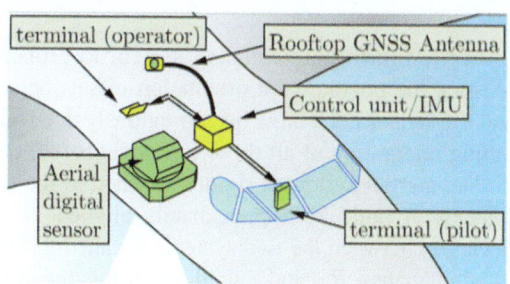

GNSS positioning is achieved by applying different surveying techniques, depending on the quality required. Post-processed kinematics (PPK) and Real-Time Kinematics RTK positioning have been used, as described in Chap. 17.

The inertial navigation system derives position, velocity and attitude from the integration of measurements of accelerations and angular velocities along the flight path at an output rate of approximately 100–500 Hz. These observations are typically obtained from a minimum of three gyroscopes and three accelerometers mounted orthogonally in an inertial measurement unit (IMU). An IMU coupled with a navigation computer forms an INS. The performance of a standalone INS is characterised by a time-dependent drift in its measurements, which can vary between 3 and 10 cm/s in position and between 0.1 and 0. 2$''$ per second, depending on the quality of the sensor. However, it is essential to consider that it is not worth dwelling on specific data due to the technological advances in this field. It is recommended that the reader consult recent specialised references and manufacturers' catalogues whenever necessary.

To overcome the problem of inertial instability, inertial and GNSS data are integrated. In short, this means that the deviation from the trajectory determined by the INS is corrected by integration with GNSS positioning. Considering that inertial systems provide accurate relative position and attitude values at a frequency of the order of 200 Hz when a GNSS correction is applied every second, for example, the orientation of the aircraft in space is obtained at distances suitable for airborne photogrammetry, as shown in Fig. 19.25. The accuracy of GNSS limits the absolute accuracy in such a combined system. The absolute roll and pitch accuracy is limited to accelerometer bias stability and accuracy of the gravity model. Absolute heading accuracy is limited to the gyro bias and scale factor stability, and the accelerometer and gyro noises limit the relative position and orientation accuracy.

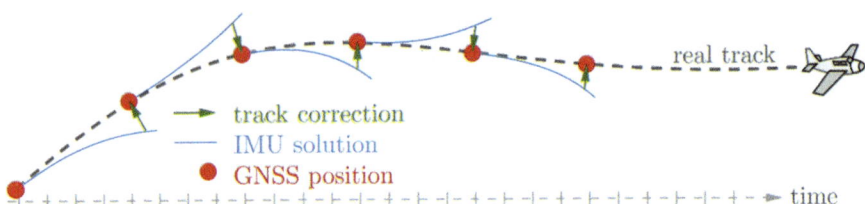

Fig. 19.25 Principle of trajectory correction by GNSS/INS integration

For a direct georeferencing system to properly integrate position and attitude sensors, the position and orientation offsets between the GNSS antenna, the IMU, and the sensor must have been previously determined. Only accurate geometric and timing integration of all these components will ensure valuable results.

The inertial system is rigidly attached to the camera body so that the coordinate systems of both are approximately aligned. Referring to Fig. 19.26, the angular deviation between the two systems is called the *boresight* angles represented by the rotation matrix $\boldsymbol{R}_b^c$ and calculated using the image rotation matrix computed by photogrammetry and the IMU-derived rotation matrix.[10] The GNSS antenna is rigidly mounted to the aircraft body, and the deviation vector between the centre of the antenna and the IMU centre is called the *lever arm from IMU to GNSS antenna* a_{GNSS}^b, and between the IMU centre and the camera perspective centre is called the *lever arm from IMU to camera perspective centre* a_c^b. The lever arms are transformed to the object coordinate system using a rotation matrix computed by the inertial navigator R_b^m. The scale s_i is determined by stereo processing or from an existing DEM. The position of each pixel relative to the perspective centre is computed using the EO represented by the rotation matrix $R_b^m(t)$ and the IO described by $r_i^c(t)$. The lever arms values are usually provided by the camera manufacturers or measured. Even then, they must be determined by field calibrations. Considering all the above parameters, the equation for determining the coordinate vector r_i^m of point (i) in the object coordinate system (m) is given by Eq. (19.23).

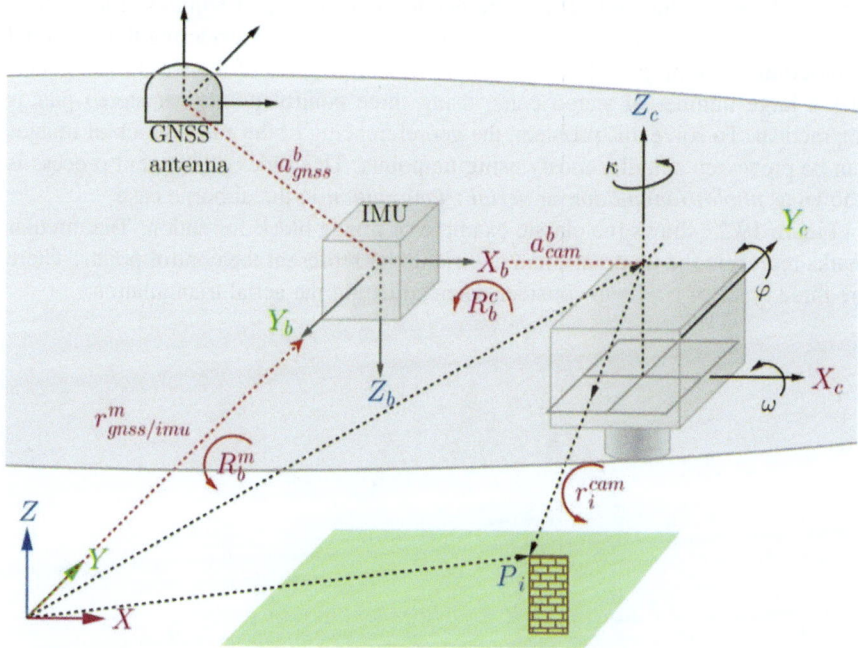

Fig. 19.26 Principle of direct georeferencing

$$r_i^m = r_{GNSS/IMU^{(t)}}^m + R_b^m(t) * \left[s_i * R_c^b * r_i^c(t) + a_E^b \right] \qquad (19.23)$$

---

[10]Figure 19.26 shows the IMU and the camera separately for easy of understanding.

Another essential aspect to consider when integrating GNSS-INS systems into airborne camera positioning is the choice of the project coordinate system. Generally, a map projection system related to the national geodetic reference system is used, such as the UTM projection and orthometric altitude. See Chap. 16. However, the photogrammetric model is based on the Cartesian spatial coordinate system. It is, therefore, necessary to perform coordinate transformations to combine the two systems. In addition, it must be considered that the elevations used in engineering projects are orthometric or normal, as described in Sect. 4.4.1. On the other hand, those generated by the GNSS system are ellipsoidal, so it is essential to consider the geoid undulation and the deviation from the vertical in the final mapping product.

19.12   Aerial Triangulation

As already stated, the indirect georeferencing of a stereo pair requires at least three control points distributed over the overlapped area. Thus, considering that an aerial photogrammetric project almost always covers large areas of the terrain and generates a large number of stereo pairs, using three control points per stereo pair is impractical. To solve this problem, the georeferencing of the entire block of images can be processed simultaneously using tie points. This block adjustment process is known as *phototriangulation* or *aerial triangulation* in the airborne case.

Figure 19.27 shows the classic example of image block formation. The circular marks represent the tie points, and the triangular represent the control points. There are three types of points to consider when adjusting the aerial triangulation.

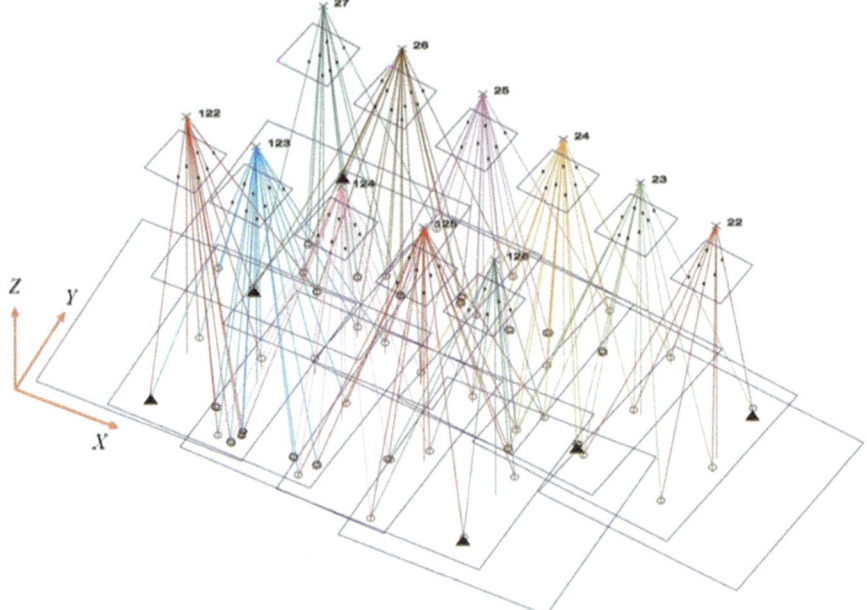

Fig. 19.27 Principle of bundle block adjustment

Control point: as described in Sect. 19.11.1.

Tie point: Those measured on the overlapping areas of the adjacent images of the photogrammetric block, whose function is to guarantee the longitudinal and lateral connection between the models. They are detected on the image and do not need to be identified on the terrain, i.e., they do not need to have (X, Y, Z) coordinates. There is no fixed rule to indicate which type of object to use as a tie point. In general, the only requirement is that they are easily identifiable in the adjacent images, avoiding shaded areas, especially between different strips, due to the changing position of the sun. See the example in Fig. 19.28. Photogrammetry programs usually have algorithms to automatically detect these points and generate hundreds of them in each stereo pair.

Fig. 19.28 Example of tie point

Checkpoint: the one used to check the quality of the aerial triangulation adjustment. It is a control point not used in the adjustment but to check the accuracy of the block adjustment by comparing the checkpoint known coordinates with those calculated by the adjustment computation model.

Several methods of phototriangulation have been developed in the literature. The most flexible and accurate among them, available in almost all photogrammetry software, use the bundle block adjustment method, which can be understood as a mathematical model that simultaneously relates the image space coordinates of the entire block to the sensor parameters and the object space coordinates. In short, it is the simultaneous application of the collinearity equations for the whole block of images. In this case, the values of the orientation parameters are determined by an adjustment computation model.

Regarding the number of control points, some authors recommend using at least one point in every three to six pairs of images distributed in the corners and lateral overlapping areas. Vertical control points (points of known elevation) can also be included in the lateral coverage areas in every four models.

As the tie points are automatically detected by the image correlation algorithm, their number is much higher than the minimum required.

The accuracy of the adjustment computation is determined by examining the variance-covariance matrix of the adjustment parameters, which indicates the value of the standard deviation with which each parameter was specified. Accuracy, in turn, is determined using checkpoints. The RMSE of the coordinate values measured in the adjusted model and the checkpoints indicates the accuracy of the adjustment. It is essential to note that the control points already used in the adjustment computation cannot be used as checkpoints.

There is little information in the literature on the accuracy to be expected when verifying the adjustment computation. Some authors give values from 0.5 to one pixel for planimetry and one to two for altimetry in the case of natural control points, assuming that there are no gross errors. These values should be treated with caution and taken only as comparative indicators since the level of accuracy to achieved must consider the purpose of the project.

19.13   Image Correlation

Searching for tie points in photogrammetry has always been a tedious process, especially when done manually for large image blocks. Due to the slowness of this process, since the beginning of Digital Photogrammetry, solutions have been sought to automate the search processes. The algorithms developed for this automation are called *image correlation* or *image matching*.

In recent years, quite a number of algorithms have been proposed, and significant progress has been made in image correlation. The basic principle of these algorithms is to determine the correspondence between two grayscale templates (area-based matching) or feature-based matching of physical shapes. In either case, the problem to be solved is automatically identifying them in the conjugate images. In this respect, several algorithms have been developed over time and are very well described in specialised references.

Area-based correlation algorithms are very popular and widely used in image template matching. They are based on the assumption that the grey value of the neighbourhood of a given pixel in two adjacent images is similar. Thus, instead of searching for the location of the homologous pixel, one looks for the location of a set of pixels determined by a predefined pixel template. In this case, the pixel template consists of an array of pixel greyscale with dimensions ranging from 5×5 pixels to 11×11 pixels. The central pixel of this template is the homologous pixel.

The matching is carried out by positioning the pixel template on a predefined and fixed position of the first image, called the reference template. Once the reference template has been selected in one of the images, a search region must be selected in the other image, and the same pixel template must be positioned over it (now called the search template).

The search template is then moved over the search region to carry out all possible combinations of the reference template with the different positions of the search template to determine the position with the most significant similarity (or correspondence) to the reference template. The positions of the selected central pixels of both templates define the coordinates of the homologous points, as shown in Fig. 19.29.

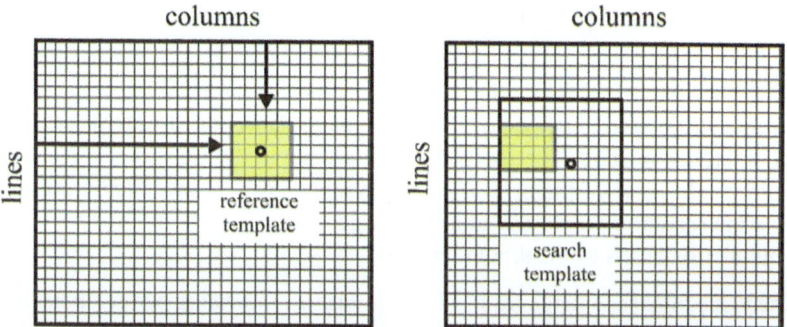

Fig. 19.29   Example of image matching templates

Feature matching algorithms are based on the detection, description and matching of salient and distinctive objects of conjugate images such as points, edges or objects. They are more complicated and involve the extraction of features at different scales and rotations, with a subsequent comparison based on feature characteristics such as size and shape. Several algorithms have been developed on this subject over the years, many of which are widely used in image processing software, such as the Scale-Invariant Feature Transform (SIFT) algorithm. A discussion of these algorithms is beyond the scope of this book. Readers interested in more details are advised to consult specialised references.

To help the reader to better understand the conceptual meaning of image correlation, the following section presents the basic concepts of the image correlation algorithm using the *normalised cross-correlation* method, which is a very simple and powerful image correlator used in many image processing packages.

19.13.1   Normalised Cross-Correlation Method

This image correlation method is based on maximising the correlation coefficient (ρ) between the greyscales of the two correlation templates, as given by Eq. (19.24).

$$\rho = \frac{(F^{\mathrm{T}}G) - \frac{1}{n}(e^{\mathrm{T}}F)(e^{\mathrm{T}}G)}{\sqrt{\left[(F^{\mathrm{T}}F) - \frac{1}{n}(e^{\mathrm{T}}F)^2\right] * \left[(G^{\mathrm{T}}G) - \frac{1}{n}(e^{\mathrm{T}}G)^2\right]}} \tag{19.24}$$

where

$$F = \begin{bmatrix} f_1 \\ f_2 \\ \vdots \\ f_n \end{bmatrix} \qquad G = \begin{bmatrix} g_1 \\ g_2 \\ \vdots \\ g_n \end{bmatrix} \qquad e^T = \begin{bmatrix} 1 & 1 & \cdots & 1 \end{bmatrix}$$

where

f_i = greyscale value of the pixel (i) of the template (f) – conveniently positioned on the reference image

g_i = greyscale value of the pixel (i) of the template (g) – conveniently positioned over the search region of the adjacent image

n = number of pixels of the search template

The value of (ρ) can range from -1 to $+1$, with $+1$ indicating perfect matching. A coefficient of -1 indicates a negative correlation, which would occur if identical images from a photographic negative and positive were compared. Thus, the closer to $+1$, the more significant the correlation between the greyscales of the chosen template. Figure 19.30 shows an example of the distribution of the correlation coefficients in a search region. The peak in the graph indicates the position of the template with a more significant correlation. Figure 19.31 shows an example of a point mesh created by image correlation.

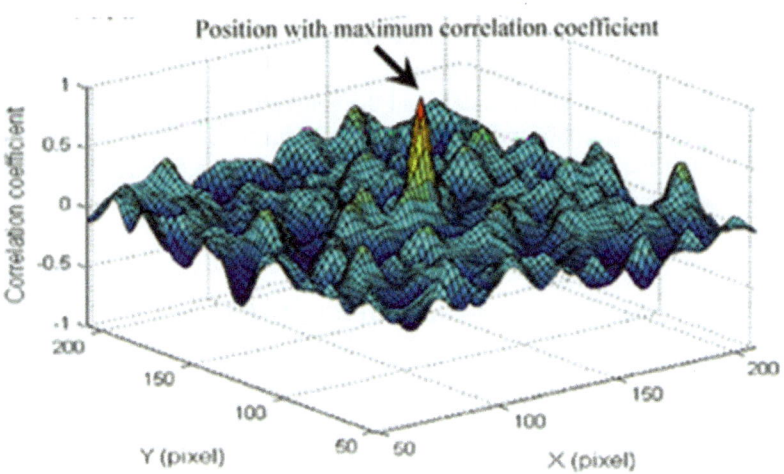

Fig. 19.30 Example of correlation coefficient distribution in an image correlation search of 200×200 pixels

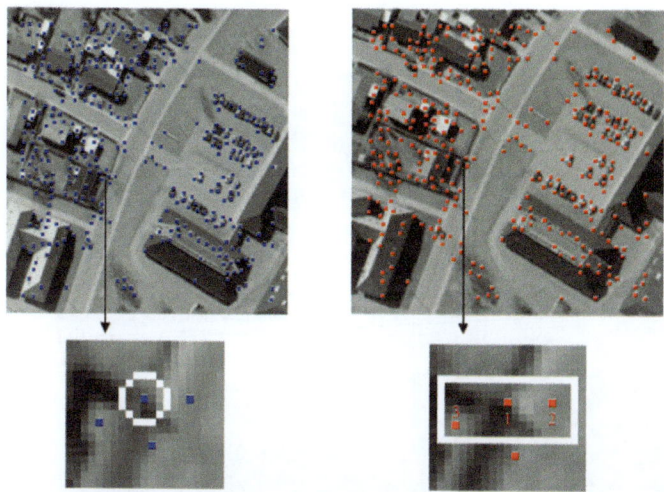

Fig. 19.31 Example of point mesh created by image correlation

19.13.1.1   Search Area Definition

Before starting the search for homologous points by any correlation method, it is necessary to define the most probable geometric location of the points to be correlated, i.e., the region of the image where the search will be performed. The correct definition of this region is of utmost importance to avoid high computational costs and to reduce the ambiguous correlation results. In this respect, all correlation algorithms use some techniques to restrict the search area, as described below.

Epipolar line: is based on the assumption that if an object appears in one of the images, it may or may not appear in the other, but if it does, it will undoubtedly be on a zero-parallax line, called the *epipolar line*. The epipolar constraint thus reduces the search space for matching points from two dimensions to one.

Hierarchical process: consists of using a pyramid of images, as shown in Fig. 19.32. In this case, the pyramid is formed by a set of identical images with different geometric resolutions, reduced by a factor of two. The highest level corresponds to the image with the lower resolution, and the lowest level corresponds to the one with the higher resolution, usually the original image.

The image correlation procedure along the pyramid starts at the top of the pyramid, i.e., in the lower-resolution images. The results obtained at each upper level are projected onto the lower image until the original image is reached.

Regarding image correlation methods, it is essential to note that this area of image processing has been extensively explored by researchers in many different fields related to computer vision and has been implemented with excellent results in

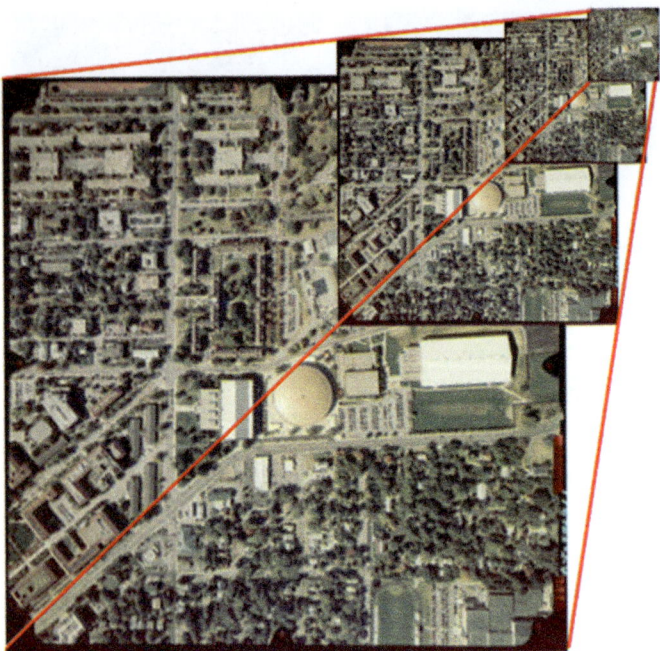

Fig. 19.32 Principle of hierarchical area search procedure

specific programmes and systems for photogrammetric data processing. In this section of the book, the objective is to introduce the subject to the readers to help them understand the fundamentals of the correlation methods used in photogrammetry so that they can conveniently evaluate the results obtained in their practical work.

19.14   Camera Calibration

A photogrammetric camera is ultimately a measuring instrument. As such, it must be calibrated. Calibration, in this case, is performed to determine the values of the IO parameters (c, x_0, y_0), the radiometric values of the CCD (or CMOS) sensor array and the lens distortion parameters $(k_1, k_2, k_3, P_1, P_2)$. In the case of cameras with resources for direct orientation, i.e., integrated with GNSS/IMU sensors, the calibration also includes information on the geometry of this integration, which is related to the offsets of the GNSS antenna and the inertial system (IMU), concerning the perspective centre of the camera, and the misalignment angles between the IMU and camera reference systems, as described in Sect. 19.11.2.

The values of the calibrated parameters and their accuracies must be presented in a calibration certificate with a graphical indication of the results. The parameters to be calibrated and the format of the calibration certificate vary according to the type of camera, especially for small-format cameras. The frequency with which calibration must be performed depends on the type of camera and governmental regulations. Generally, for large and medium format cameras, it is recommended to calibrate them at least once every 2 years it is recommended that large and medium format cameras be calibrated at least once every 2 years, or as recommended by the manufacturer.

The procedures for calibrating a photogrammetric camera are based on three classical calibration techniques: laboratory calibration, ground calibration (test field) and self-calibration.

19.14.1 Laboratory Calibration

This calibration technique consists of physically relating the geometry of the sensor array to a controlled optical beam generated by a set of collimators mounted at a known angular arrangement. Each collimator then projects an image of a cross-mark (or the image of a coded plate) onto the CCD (or CMOS) sensor array. The coordinates of the intersections are subsequently determined, and thus, the geometric parameters of the interior orientation are calculated.

Another parameter determined in this calibration step is the spectral response of the camera, i.e., the radiometric calibration, which determines the functional relationship between the input radiation and the digital value displayed by the camera. An example of such a calibration is the use of a high-quality spectrometer that generates an isotropic and homogeneous light source, taken as a standard, which illuminates the camera lens system and determines the spectral response of the sensor array.

In addition to the parameters mentioned, several others are calibrated in the laboratory. It is beyond the scope of this book to discuss them all. The reader interested in further information is advised to consult specialised references or the user manual of airborne camera manufacturers.

19.14.2 Ground Calibration (Test Field)

This type of calibration is based on a suitable targeted field of object points with known coordinates and distances. A classical network of calibration points consists of a grid of targets with known spatial coordinates painted over a stable bulkhead. The calibration parameters are determined by taking overlapping photographs from different positions and directions from several camera stations and applying a bundle adjustment model using the expanded collinearity equations with additional parameters, which are: the calibrated focal length c, the coordinates of the principal point

(x_0, y_0), the three coefficients of the radial symmetric distortion and the two coefficients of the decentring distortion. These are the most common ones.

The solution is obtained by applying an adjustment computation model, from which the EO parameters are also obtained. This type of calibration is usually performed using dedicated software.

19.14.3   Self-calibration (Auto-calibration)

This calibration technique is similar to the previous one, except that the observations are performed using control points over the actual surveyed object measurement, as described in Sect. 19.12. To ensure the quality of the results, pre-marked control points are used in most cases, although natural control points can also be used. Self-calibration can also be performed without known control points. In this case, the IO parameters are calculated by the photogrammetric determination of the object shape, such as small-format camera calibration using checkboard patterns.

After calibration, it is assumed that the camera's intrinsic parameters will remain stable for some time. In reality, no camera is ever stable. Therefore, in cases where non-metric cameras are used, it is advisable to check the calibration parameters before and after the flight to ensure that there are no significant changes during image acquisition.

The best way to determine the geometric parameters of a camera is to perform laboratory calibrations, where all measurements are made under controlled environmental conditions. However, this procedure requires the camera to be sent to a certified laboratory according to the manufacturer's requirements. Field calibration is used almost exclusively for small-format cameras. Self-calibration is used for all types of cameras. In some cases, even lab-calibrated cameras are self-calibrated during airborne photogrammetric projects to verify the stability of the parameters. For this purpose, most photogrammetric programmes have tools to determine the calibration parameters of aerial cameras.

19.14.4   Boresight and Lever Arms Calibration

Calibration of the boresight angles and offsets of the GNSS antenna and the IMU sensor is similar to auto-calibration, described earlier. In this case, an aerial photogrammetric flight is performed over a flat terrain with well-distributed and well-defined control points. An aerial triangulation process is then performed to determine the position of the optical centre of the camera and the corresponding rotation matrix of the model concerning a predefined coordinate system. The measured position of the optical centre and the rotation matrix for each IMU position are obtained using the GNSS antenna and the IMU system processors. The values of the offsets and the boresight angles are obtained by comparing the two positions and the rotation matrices. Finally, the adjusted values and the associated variance-covariance matrix are obtained by adjusting the entire block of images. The process is better controlled, and the results are more reliable if there is already preliminary information on these parameters, which is generally provided by the manufacturers.

As mentioned, the mathematical model used for this calibration consists of adding the calibration parameters directly to the collinearity equations and performing the bundle adjustment computation method.

19.15   Digital Photogrammetric Workstation

Digital Photogrammetric Workstation (DPW) is the name given to the set of equipment and computer application software designed to generate cartographic products using Digital Photogrammetry concepts. In this case, the equipment includes a computer with at least two monitors, a 3D mouse, an image separation device for 3D visualisation, glasses for stereoscopic vision and accessories for computer manipulation. Figure 19.33 shows an example of a complete digital photogrammetric workstation with its equipment and accessories. Figure 19.34 illustrates the principle of image separation of the photogrammetric workstation shown in Fig. 19.33. Figure 19.35 shows examples of 3D mouse.

Fig. 19.33 Digital photogrammetric workstation. (Courtesy 3D Pluralview)

Fig. 19.34 Principle of image separation for 3D view

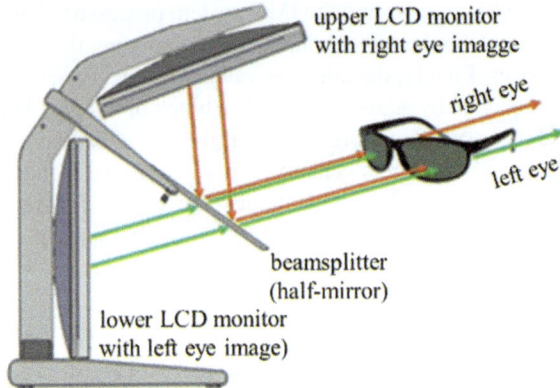

upper LCD monitor
with right eye imagge

right eye

left eye

beamsplitter
(half-mirror)

lower LCD monitor
with left eye image)

Fig. 19.35 3D mouse

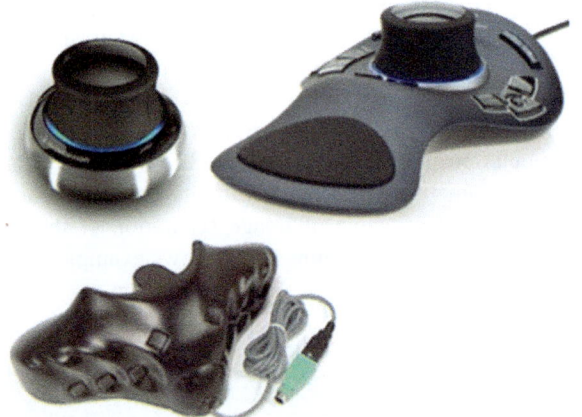

The image separation device shown in Fig. 19.34 is an example of a device based on the polarised light method, as presented in Sect. 19.7. Its principle of operation is self-explanatory in the figure. The reader interested in other examples may find several available on Internet websites.

The 3D mouse is the accessory of the photogrammetric workstation with the function of moving the cursor in the (X, Y, Z) directions of the stereoscopic model on the computer display screen. The movements in (X, Y) are performed by moving the 3D mouse on the working table in the corresponding directions. Movement in (Z) is performed using a sphere inserted into the 3D mouse body, as shown in Fig. 19.35. By revolving this sphere, the operator separates or joins the floating marks, which means moving it up and down the stereoscopic model, allowing it to be placed on the terrain. For ease of use, the 3D mouse has configurable buttons that enable functions such as parallax elimination, zooming, and the creation of geometric figures in a CAD environment.

A photogrammetric workstation is managed by a computer program generically called a *photogrammetric program*. It is through this program that the photogrammetric workstation communicates with the operator. Its function is to carry out all the routines for importing images, editing and storing control points, carrying out the camera orientations, managing and displaying stereoscopic models and generating cartographic products by collecting spatial data represented in the images. In this case, the data is collected by measuring the coordinates of the homologous points representing the objects to be measured. The photogrammetric program calculates the coordinates of the homologous points indicated by the operator, transforms them into object coordinates, as mentioned in the previous sections, and displays the point on the screen of the graphic monitor through a CAD program integrated into the photogrammetric workstation. The cartographic product is then generated by graphically combining the measured points. For this purpose, the photogrammetric programme has modules dedicated to the development of different products, which are described in more detail in the following sections.

19.16   Products Generated by a Photogrammetric Workstation

The products that Digital Photogrammetry can generate are many and varied. For Geomatics applied to Civil Engineering, there are three products of interest, which are:

- Photogrammetric restitution
- Point mesh generation for digital terrain or surface models
- Orthophoto

They are all based on the fundamental principle of measuring the coordinates of homologous points in the stereoscopic model. The technical details of each of them are presented below.

19.16.1   Photogrammetric Restitution

In photogrammetry, the process of line map creation through the vectorisation of geographic elements, represented in a photogrammetric stereoscopic model, is called *photogrammetric restitution*. This process produces a vector map with a planimetric or planialtimetric representation of the geospatial data, as shown in Fig. 19.36. Figure 19.36a shows the digital aerial image of the site, and Fig. 19.36b shows the vector map.

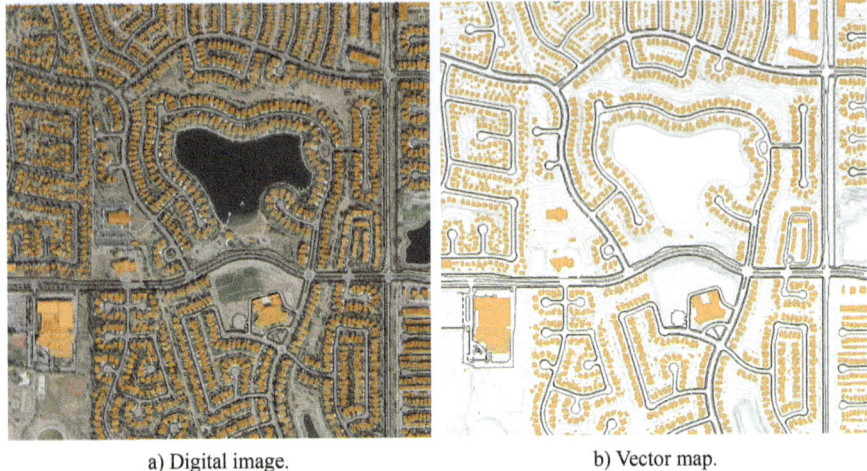

a) Digital image.   b) Vector map.

Fig. 19.36 Example of vector map created by photogrammetric restitution. (**a**) Digital image. (**b**) Vector map. (Source: WGS Aerial survey. Location: https://www.wgsair.com/)

As vectorisation, the reader must understand the transformation process of an object represented on the image (raster format) to an object represented in vector format, formed by points, lines and curves in space.

The type of geospatial data to be extracted depends on the purpose of generating the vector map. In any case, what is obtained is a map with geometric elements in scale represented in several layers, which can be manipulated by the user. Furthermore, as a vector model, it does not have a fixed scale and can be zoomed up or down according to the user's needs.

Photogrammetric restitution is a manual process that requires operator experience. It is based on positioning the floating mark on each object's vertex to be vectorised in the stereoscopic model and connecting them using lines, polylines, and other geometric compositions. To make this possible, the photogrammetric software has an integrated CAD program that helps to perform the desired vectorisation functions. The primary geometric element plotted in CAD drawing space is the point with coordinates (X, Y, Z) defined by the floating mark. The geometric figures are then composed of lines and curves using the CAD functions available in the application.

Another essential type of photogrammetric restitution is the generation of contour lines from the stereoscopic model. Although this process is increasingly being carried out using digital terrain modelling, drawing contour lines directly on the stereoscopic model has been regularly used in digital mapping projects. In this case, the procedure for generating contour lines is based on positioning the floating mark at the desired elevation, say 700 m, and moving it across the terrain surface, represented in the stereoscopic model, keeping the height of the mark fixed as it moves. As the operator moves the floating mark across the model, the photogrammetric program creates points at predetermined intervals, and the CAD application draws a polyline connecting them to create a contour line, as shown in Fig. 19.37.

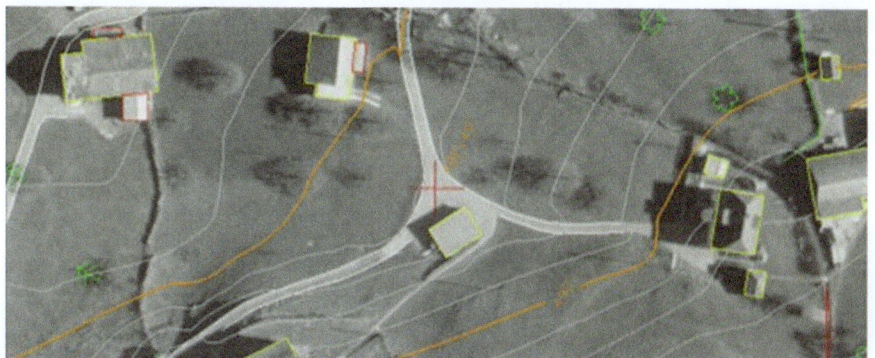

Fig. 19.37   Example of contour lines drawn over a stereoscopic model

19.16.2   Point Mesh

Point mesh generation is a photogrammetry application that has gained prominence with the development of image correlation methods. The objective is to create a cloud of 3D points distributed over the terrain and the objects represented in the stereoscopic model, with (X, Y, Z) coordinates automatically generated. The operator specifies the spacing or the number of points that may contain millions of points. The resulting mesh is often used as input data for generating digital terrain or surface models, as presented in Chap. 15. The use of the digital model depends on the application for which it was created. In this context, the creation of digital surface models, or 3D models for short, has gained prominence due to the number of computer applications developed to use these models in various engineering fields.

Figure 19.38 shows an example of a 3D model created using image correlation. Note that it is a hybrid image, where a part of the original image and a part of the mesh of triangulated points are shown for ease of visualisation. Each vertex of the triangles corresponds to a mesh point, determined by image correlation.

Fig. 19.38 Illustration of a 3D TIN model generated by image correlation

19.16.3   Orthophoto

An orthophoto is a digital image represented in orthogonal projection, similar to a conventional map, i.e., it has the geometric characteristics of a map and the graphical qualities of a photograph. This means that in an orthophoto, the measured distances, areas and angles correspond to their terrain values, corrected for scale and distortions due to cartographic projection. For this to be possible, it must undergo a photogrammetric processing called orthorectification, which removes image distortions due to terrain relief and camera inclination, among others associated with acquiring a photographic image.

Distortions associated with image acquisition are removed during the aerial triangulation process and by sensor calibration. Distortions due to terrain relief are removed using digital terrain modelling in the orthorectification process.

Several mathematical models can be used for orthorectification. Among them are those based on polynomial transformations, finite elements and collinearity equations. Experts consider that the mathematical model based on the collinearity equations gives better results. In this case, the orthorectification process is performed according to the operational flow shown in Fig. 19.39. To do so, it is necessary to have an oriented image and a digital terrain model of the region to be orthorectified, available in the same coordinate system.

Fig. 19.39 Principle of
image orthorectification

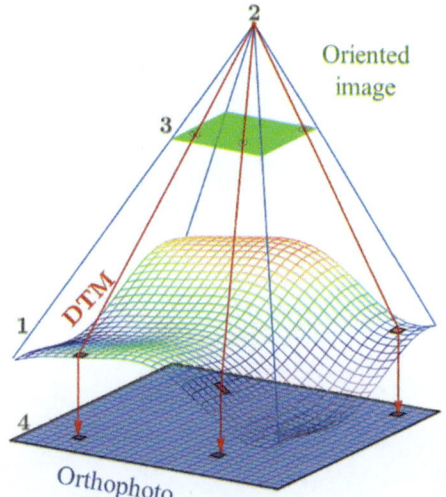

The operational flow consists of assigning the grey scale value of the oriented image to the corresponding pixel of the numerical terrain model and then calculating a new rectified image.

The process of image orthorectification is as follows:

(a) Define the desired pixel size for the orthorectified image.
(b) Define the area of the object space to be orthorectified, which must match that of the image to be orthorectified and the digital terrain model.
(c) Based on the coordinates (X, Y, Z) of the first pixel, obtained using the digital terrain model, the coordinates (x, y) of the corresponding point on the image is determined using the collinearity equations.
(d) In most cases, the (x, y) image coordinates do not correspond precisely to the position of a pixel on the image. It is, therefore, necessary to interpolate the grey values of its neighbours to determine the grey value to be assigned to the pixel in the orthorectified image. This interpolation is performed using digital image processing, called *resampling*. Several resampling methods can be used to produce orthophotos. Among them, the following stand out: nearest neighbour, bi-linear interpolation and cubic convolution. It is beyond the scope of this book to discuss them. Interested readers are advised to consult specialised references.

Because orthophotos are usually generated using a digital terrain model, objects not belonging to the terrain, such as buildings, viaducts and others, are not orthorectified and appear inclined in the image, as shown in Fig. 19.40. Note the presence of the building's facades in the orthophoto. This means that the objects are represented in the image with their upper parts displaced about the lower parts. The key to solving this problem is to use a digital surface model (DSM) instead of the digital terrain model (DTM). As the (X, Y, Z) object coordinates are also available, buildings can be orthorectified to minimise the effects of their vertical displacements. The result is an orthorectified image known as a *true orthophoto*. Figure 19.41 shows the true orthophoto of the same scene as in Fig. 19.40.

Fig. 19.40 Conventional orthophoto

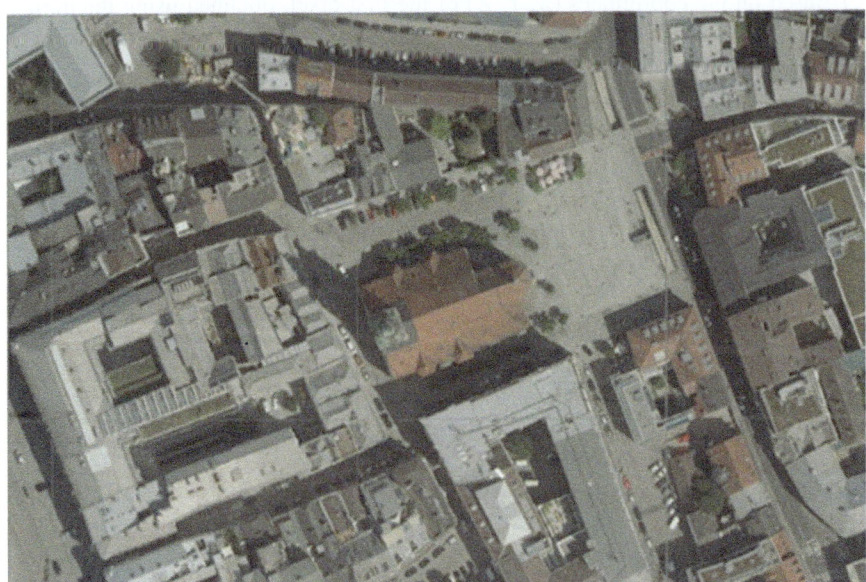

Fig. 19.41 True orthophoto

Using true orthophotos solves the problem of object inclination. However, it creates the problem of colouring the occluded areas of the original image. These areas correspond to those covered by the inclined portion of the objects that become visible with the orthorectification. This problem is solved if other images showing views of the occluded areas are available. See the example of a true orthophoto in Fig. 19.42.

Fig. 19.42 True orthophoto with the reconstructed occluded area

Another critical issue to consider is that digital surface modelling rarely models the edges of objects accurately, causing their orthorectified images to appear distorted or blurred.

Since an orthophoto can be considered equivalent to a map, it can be used for the same purposes as maps, with the advantage of being more detailed. Furthermore, if necessary, they can be complemented with the overlay of the geometric elements of the photogrammetric restitution, as shown in Fig. 19.43.

Finally, an orthophoto is cheaper to produce, easier to update than a digital map and can be used by non-technical users.

In large project areas, successive individual orthophotos are connected to form a block of orthorectified images, called an *orthomosaic*.

Fig. 19.43 Orthophoto with superposed vector map

19.17 UAV Photogrammetry

As seen throughout this chapter, the resources of Digital Photogrammetry have greatly simplified digital data processing, reduced costs and opened up an infinite horizon of applications for photogrammetry. New users have been co-opted by the sheer power of the digital age and the ease with which geoinformation can be generated using automated computer vision resources. At the same time, with the advent of unmanned aerial vehicles (UAV), with the ability to integrate miniaturised sensors, a new era was opened for airborne photogrammetry, facilitating geospatial data collection for small engineering projects, where the use of conventional airborne photogrammetry is impossible or inefficient. This new airborne platform for photogrammetric data collection is called *UAV Photogrammetry*.

At first glance, the main difference between conventional and UAV photogrammetry is the type of aircraft used for data collection. In practice, however, from the end-user's point of view, UAV photogrammetry operation is much simpler, more accessible, cheaper, faster and more multitasking than conventional photogrammetry. The fact here is that in conventional photogrammetry, images are obtained using large-format digital metric cameras taken perpendicular to the object, in a well-behaved environment, with well-defined longitudinal and lateral overlaps, creating regular image blocks that are processed in high-performance photogrammetric workstations, developed by photogrammetrists. On the other hand, UAV photogrammetry reduces all these constraints by allowing the use of low-cost non-metric digital cameras and oblique images collected at different heights. They can also be distributed unevenly in space, as long as the necessary overlaps are guaranteed. Finally, data processing, in general, is performed in programs developed by computer vision specialists with little experience in photogrammetric mapping. As a result, data processing and user interfaces are quite different when comparing the two technical approaches. Conventional airborne photogrammetry relies on well-known collinearity equations, image matching and aerial triangulation by block adjustment. On the other hand, the integration of computer vision and image processing in UAV photogrammetry has resulted in a technique called *Structure from Motion* (SfM), which, coupled with *Multi-View Stereo* (MVS) algorithms and *Scale-Invariant Feature Transform* (SIFT) operators, automatically solves the geometric problem of camera positioning and orientation with very little user interaction, making UAV photogrammetry an easy-to-use tool for geospatial data collecting by non-photogrammetrists. In the end, even if all the processing steps are the same for both, which include camera calibration, interior orientation (or intrinsic orientation), tie point (or key-point) detection, ground control point (GCP) measurements or direct GNSS/INS georeferencing for exterior orientation (or extrinsic orientation), the UAV photogrammetry users hardly notice. However, it is important for the reader to understand that using unmanned aircraft as a platform for photogrammetric data collection does not constitute a new form of photogrammetry. The theoretical basis and the basic mathematical and operational concepts of photogrammetry remain the same.

The potential of UAV photogrammetry for Geomatics is evident in cost and efficiency. In practice, it fills the gap between airborne and terrestrial photogrammetry, allowing its use in project areas such as:

(a) Mapping and collection of selective geographic data on small corridors (highways, railways, pipelines, transmission lines and other infrastructure elements).
(b) Mapping areas for energy production, such as inspection of electric power generating plants, wind power, solar panels, etc.
(c) Precision farming.
(d) Localised environmental studies.
(e) Natural disaster monitoring.
(f) Inspection and monitoring of Civil Engineering works.
(g) Mineral exploration and volume calculation support.
(h) Topographic and cadastral mapping of small areas.
(i) Data collection for updating Geographical Information Systems.
(j) Forensic photogrammetry.
(k) Architectural and archaeological photogrammetry.
(l) Others.

In this sense, the following sections briefly describe the main components of UAV photogrammetric data acquisition and processing in the context of Geomatics applied to Civil Engineering.

19.17.1   Definition of Unmanned Aircraft System (UAS)

For the purposes of this book, an unmanned aircraft system (UAS) is defined as a set of mechanical and electronic devices interconnected to perform flights and collect data by unmanned aerial vehicles (UAVs) capable of carrying various types of payloads, which vary according to type, functionality, operational characteristics and mission objective. As a system, it comprises several subsystems, of which the aircraft is only one. In order not to confuse the aircraft with the system, the following terms are used internationally:

- Unmanned Aerial Vehicles (UAV)
- Unmanned Aircraft Systems (UAS)
- Remotely Piloted Aircraft (RPA)
- Remotely Piloted Aircraft Systems (RPAS)

DRONE is another term currently used to replace the term Unmanned Aerial Vehicle (UAV). However, some authors consider it a generic term for an unmanned aerial vehicle for recreational or military use. At the same time, a UAV would be an unmanned aerial vehicle with onboard intelligence to carry out programmed missions.

Figure 19.44 shows the general operational composition of a UAS for photogrammetry. As shown in the figure, the main components of this system are the unmanned aerial vehicle (UAV) with navigation, georeferencing and data acquisition sensors, the *ground control station* (GCS) and the communication data link.

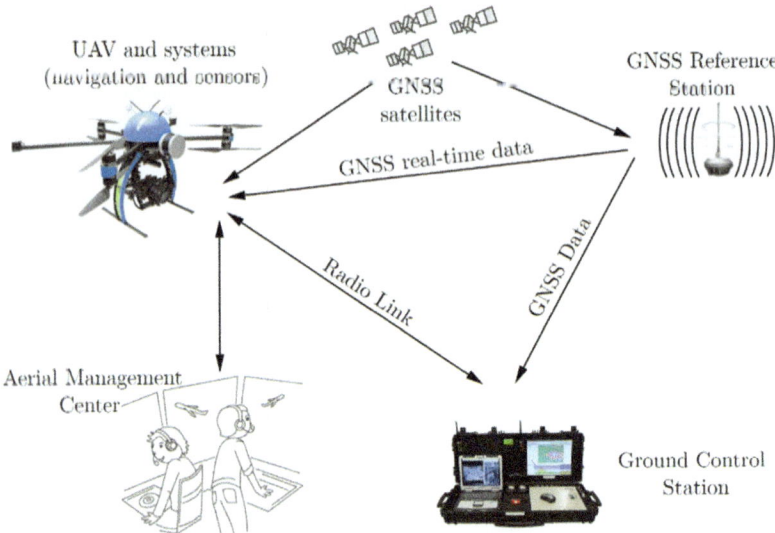

Fig. 19.44 UAV control system

19.17.2 UAV Platforms and Categories

Unmanned aerial vehicles can be classified according to different aspects related to their physical components, flight heights, Maximum Take-Off Weight (MTOW) and others. In the absence of an official classification accepted worldwide, it is proposed in this book to categorise them according to the platform type, flight height and MTOW, as presented below.

19.17.2.1 Platform Construction

UAV platforms are many in shape and size, and their structure is mainly based on the type of wing, which can be *Fixed-wing*, *Rotary-wing* and *Airships* platforms, as shown in Fig. 19.45.

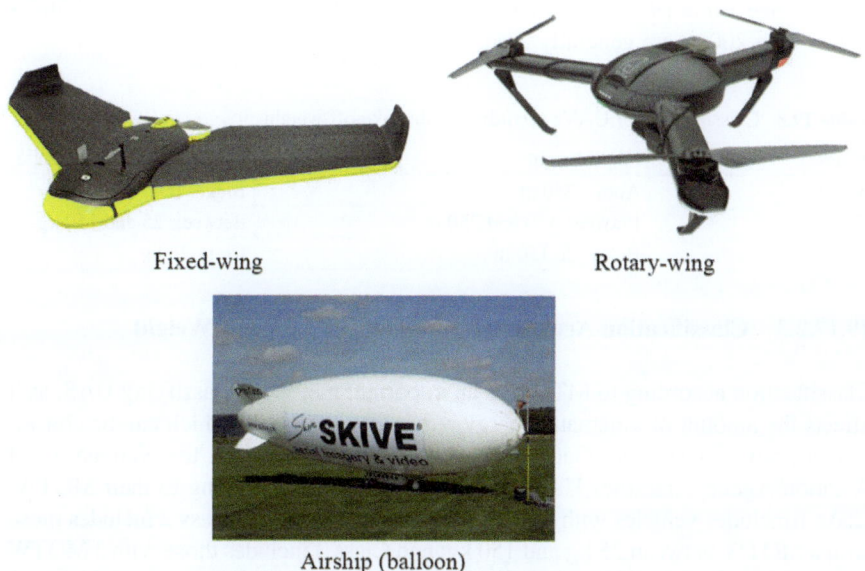

Fixed-wing   Rotary-wing

Airship (balloon)

Fig. 19.45 UAV platforms

In general, most UAVs used in Geomatics have aerodynamic characteristics of fixed and rotary wings. They are built using lightweight materials, such as EPP (Expanded Poly Propylene) for fixed-wing and carbon fibre for rotary-wing.

Fixed-wing aircraft have a single structure and efficient aerodynamics, allowing them to fly longer distances than other aircraft. However, they have limitations in take-off and landing procedures.

Rotary-wing aircraft are more complex than Fixed-wing and are, therefore, slower, less stable and have a shorter flight range. However, they can take off, land vertically, and perform agile manoeuvres suitable for asset inspections, close-range photogrammetry, and 3D object modelling. They can also carry larger payloads and are more affordable.

Airship platforms vary from miniature to large systems and can be equipped with small to large data acquisition systems. The main advantages of this platform are its stability, long endurance and safety. In Geomatics applications, they are used for 3D modelling and as a complementary image source for ground imagery (Martin, 2017).

19.17.2.2   Flight Height

Regarding flight height, UAVs can be divided into low and high altitudes. Low-altitude UAVs are authorised to fly below air traffic, at an altitude of 150–200 m above the ground, and generally only within the operator's visual line of sight. See Table 19.3. For Geomatics applied to Civil Engineering, flight altitudes of less than 200 m are generally used.

Table 19.3 Classification of UAVs according to the take-off weight

Class	Flight height	Maximum take-off weight
1	Above 250 m	Bigger than 150 kg
2	Between 150 and 250 m	Between 25 and 150 kg
3	Below de 150 m	Up to 25 kg

19.17.2.3   Classification According to Maximum Take-Off Weight

Classification according to MTOW is an important metric for classifying UAS, as it affects the amount of kinematic energy required for take-off, which can be considered a primary factor in aviation safety. In the case of Brazil, the National Civil Aviation Agency classifies UAVs into three categories according to their MOTW. Class 1 includes vehicles with a MOTW greater than 150 kg; Class 2 includes those with a MOTW between 25 kg and 150 kg; and Class 3 includes those with a MOTW less than or equal to 25 kg. For Geomatics applied to Civil Engineering, aircraft classified as Class 3 are generally used. See Table 19.3.

Some regulatory agencies also classify UAVs in terms of flight autonomy. However, this classification is of little interest to Geomatics and will not be discussed in this book.

19.17.3   Devices for Flight Control

For the aircraft to orient itself and indicate its position to the control station, it needs a low-cost, lightweight navigation system that combines, among other sensors, a GNSS module and an inertial navigation system. For navigation, these elements do not need to be high-performance and are perfectly suitable for the indirect orientation of photogrammetric models using control points. For direct georeferencing, however, they must have high-quality positioning accuracy that reaches the centimetre range. In this case, depending on the desired survey quality, the GNSS georeferencing must be carried out in RTK or PPK modes, and the IMU must guarantee a high signal processing rate (around 200 Hz) and a drifting quality inferior to 6 ° /h. It should, however, be noted that the navigation and georeferencing systems, although composed of the same components, are independent.

19.17.4   GNSS Base Station

For aircraft positioning in differential RTK or PPK mode, at least one reference station must output RTK data or store GNSS raw data. These can be ground stations installed by the user or a network of SBAS stations. In the case of RTK or SBAS positioning, they require an onboard communication system to receive differential corrections.

19.17.5   Data Collection Devices

The primary objective of a UAV platform for Geomatics applications is to collect high-quality remote sensing data for mapping purposes. To do this, it is commonly equipped with optical sensors, GNSS devices and inertial navigation systems. As shown below, optical sensors can be passive (digital cameras) or active (laser scanners or radars).

19.17.5.1   Digital Camera

The digital cameras currently used in a UAS for Geomatics applied to Civil Engineering are of small format and can be metric or non-metric, weighing less than 500 grams, with a geometric resolution that can reach 40 MP (pixel 4 μm and GSD varying from 1 to 10 cm depending on the flight height) and radiometric resolution in the RGB, multispectral and thermal spectral bands. Although the technology for manufacturing digital cameras for photogrammetric UAS is advancing rapidly, an example of a commercially available digital camera is shown in Fig. 19.46.

Fig. 19.46 Low format
camera for UAV. (Courtesy:
PHASEONE Industrial)

19.17.5.2   GNSS/INS Modules

GNSS receiver chips used in UAS are miniaturised GNSS processing modules. They
can be single frequency (L_1) or dual frequency (L_1/L_2). They can also collect data for
operations in post-processed kinematic mode or differential RTK mode.

The INS modules for UAV photogrammetric applications have accelerometers,
gyroscopes, magnetometers and barometers. They are built around
microelectromechanical systems (MEMS) technology, making them lightweight
and suitable for UAS.

Figure 19.47 shows an example of a single board GNSS-Inertial module from
Applanix. It is typically mounted directly onto the body of the digital camera or laser
scanning sensor to facilitate their calibration.

Fig. 19.47 GNSS and IMU
board. (Courtesy: Applanix)

19.17.5.3 Laser Scanning Sensor

In recent years, airborne laser scanning sensors have undergone a technological process that has reduced their size and weight, enabling them to be used on UAV platforms. The characteristics of this type of sensor vary considerably from manufacturer to manufacturer, and the technological advances are enormous, making it difficult to classify them in the current state of the art.

To give the reader an idea of the state of the art, Fig. 19.48 shows an example of a UAV lidar solution with an integrated image sensor and inertial navigation system, courtesy of SatLab Geosolutions[@]. The sensor characteristics quoted by the manufacturer, in this case, are as shown in Table 19.4.

Fig. 19.48 Example of UAV platform with LiDAR and image sensor onboard. (Courtesy: SatLab Geosolutions)

Table 19.4 UAV Lidar solution with an integrated image sensor and inertial navigation system

Equipment specification					
LiDAR unit	System accuracy	5 cm to 100 m	Pos unit	Position accuracy (pp)	Hz = 0.01 m V = 0.02 m
	Range accuracy	0.5 cm to 100 m		Heading accuracy (pp)	0.04°
	Measuring range	100 m		Rolling/pitch accuracy	0.01°
	Data	640,000 points/s (single echo)			
		1,280,000 points/s (dual echo)	Camera unit	Effective pixel	26 mega pixel (6252 × 4168)
		1,920,000 points/s (triple echo)		Focal length	16 mm

19.17.6   Communication System

The communication and flight control systems of a UAV are semi-autonomous. They operate via a radio link between the ground station and the aircraft, generally in the UHF, microwave (5 GHz) or long-range Wi-Fi bands. The ground station, consisting of a radio control unit, monitors the vehicle via telemetry and allows the operator to intervene at any time. The autopilot unit consists of a processor connected to all the UAV's devices and components, executing the flight plan and monitoring the aircraft's status. Some autopilots can automatically return the aircraft to the launch site in the event of a control system failure.

The radio link allows information such as ground speed, estimated wind speed, flight direction and altitude, and battery charge level to be transmitted to the ground station in real time. With this information, the pilot controls the aircraft and aborts the flight if necessary.

19.17.7   Ground Control Station

The Ground Control Station (GCS) of a UAS is most often located on the ground, close to the flight area. In the case of Geomatics applied to Civil Engineering, it is a local control station based on computer devices through which the flight mission is programmed and executed. Its function is to monitor the status of the flight and to send commands and receive information to/from the aircraft to ensure that the mission is carried out as planned. For special applications other than those performed for geospatial data acquisition, ground control stations may be more elaborate and have more advanced features, which are beyond the scope of this book. It is important to note that there is also a growing development of software for mobile applications and proprietary operating systems that avoid using large computerised equipment in the field.

19.18   UAS for Geomatics

For Geomatics applied to Civil engineering, UAS can be classified according to the type of application, as follows:

- Aerial inspection of infrastructure
- Remote Sensing
- Photogrammetry

19.18.1   UAS for Inspection of Infrastructure

UAS for aerial reconnaissance and assessment is primarily used to capture images (or video) for remote verification of events related to the maintenance of Civil Engineering infrastructures. In most cases, they are used for inspection operations in places of difficult access, where a UAV is advantageous. Numerous examples of this type of application include the inspection of bridges and viaducts, high-tension and wind towers, oil and gas pipelines, open-pit mining areas, water dams and others.

19.18.2   UAS for Remote Sensing

UAS for remote sensing is primarily used to capture images for studies related to the semantics and physical conditions of the object being imaged. Classic examples of this application are the use of UAS in areas as diverse as archaeology, agriculture, Forestry Engineering, and road traffic.

19.18.3   UAS for Photogrammetry

UAS for photogrammetry is used for image data acquisition or laser scanning point clouds. The types and characteristics of the photogrammetric products and point clouds are the same as those of conventional airborne photogrammetry and airborne laser scanning. They differ in the dimension of the project area, the type of features that can be collected and the conditions under which they are collected.

19.19   UAV Photogrammetric Workflow

The workflow for UAV mapping is similar to conventional airborne photogrammetry. However, some are different or new. Figure 19.49 shows the operational workflow of these steps.

It is important to note that because some processing programs specific to UAV photogrammetry has been developed by non-photogrammetrists, some terms may differ from those commonly used in conventional photogrammetry.

Below are some relevant details for each operational workflow step shown in Fig. 19.49.

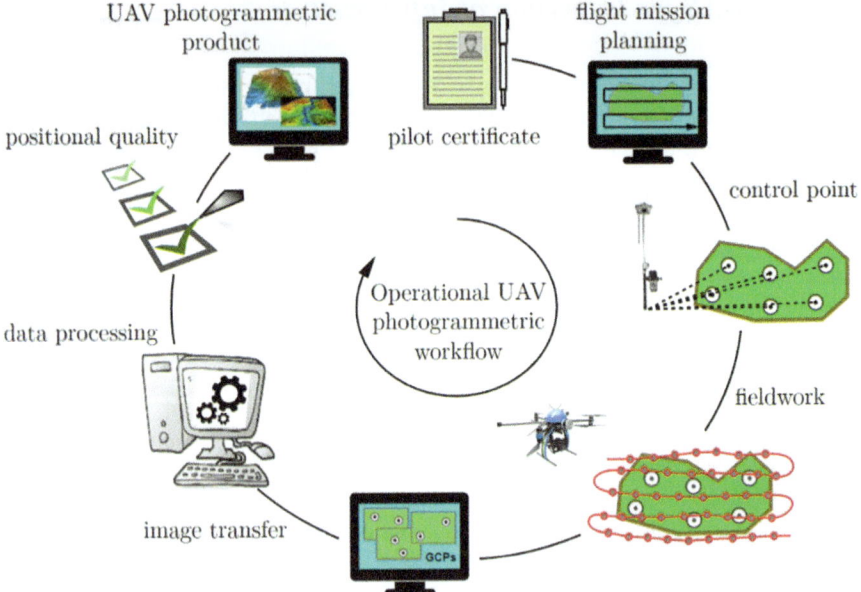

Fig. 19.49   Operational UAV photogrammetric workflow

19.19.1   *Project Preparation and Field Reconnaissance*

The operational procedures involved in this phase are as follows:

(a) Project feasibility analysis, including verification of legal and safety procedures determined by regulatory authorities in each country.
(b) Analysis of project parameters such as output data, GSD and accuracy requirements.
(c) Site reconnaissance through site visits or third-party documentation such as Google Maps or similar.
(d) Considerations on the type of UAV, onboard sensors and environmental conditions.

19.19.2   *Flight Mission Planning*

The photogrammetric flight mission planning software used in UAS is usually supplied with the purchased aircraft as it must be fully compatible with the vehicle's characteristics and the onboard sensors. Typically, the flight plan is created using Google Earth® imagery, over which the aircraft trajectory is defined considering waypoints, strips, speed, flight heights, image overlaps and sidelaps and GSD dimension, among other flight parameters.

When planning a photogrammetric mission, it is essential to consider logistics, weather conditions and the purpose of the survey. As a minimum, the following points need to be defined:

(a) Data redundancy, with high overlap rates between conjugate images and flight strips, around 80% and 60% and 80%, respectively. Cross-flights can also increase the quality of the block adjustment.

(b) In flat terrain, regular standard flights (80% longitudinal and 60% lateral) are possible with only one flight height, provided there are sufficient control points, a good navigation system and a camera with calibrated IO parameters.

(c) For hilly terrain or urban areas, it is recommended to perform cross-flights at different heights. Special attention should be paid to the uniformity of the GSD. Some flight planning software includes digital surface models to assist in determining flight height.

(d) For flights in corridors, it is recommended to fly at least three to four flight lines along the corridor, with control points distributed along strips to avoid excessive model deformation and to ensure quality in case of camera self-calibration.

(e) For 3D modelling, flying at an oblique angle of up to 45° in a cross pattern helps to overcome shading effects.

(f) For 3D models with facades, it is recommended to add images perpendicular to the object.

(g) Wind speed must be less than 12 m/s in all cases. As a general guideline, when the wind speed is less than 6 m/s, it is possible to fly in either the direction of the wind or perpendicular to it. When the wind speed is between 7 m/s and 12 m/s, it is recommended to fly perpendicular to the wind.

(h) Adequately define the landing and take-off areas.

(i) Civil engineering projects generally require a GSD between 1 and 30 cm and flight height between 50 and 800 m. Flights above 120 m generally require special permission from regulatory agencies.

(j) Low flights over areas with trees, tall buildings, or significant height variation are difficult to process. A rule of conventional airborne photogrammetry, which can also be applied to UAV photogrammetry, is that object height should exceed 20% of the flight height.

(k) Ensure flight duration is commensurate with the autonomy of the vehicle. With current technology, this is between 20 and 50 min.

19.19.3   Control Point

Depending on the type of camera georeferencing, there may be a need for control points. The type of control point, quantity, measurement methods and distribution must follow the same prerogatives indicated for conventional photogrammetry. In the case of UAS, however, there may be situations where control points must be set up in an unconventional way to meet specific UAV photogrammetric survey purposes. The reader should pay particular attention to this subject, as in most cases, the quality of the control point, in most cases, will determine the quality of the results.

19.19.4   Fieldwork

The fieldwork step includes the following:

(a) Establishment and measurement of GCPs.
(b) Flight execution and data collection.
(c) Quality checks of collected data, such as overlaps, sidelaps and sensor signal quality.

As stated, UAS images are obtained using small-format digital cameras, which, in many cases, may be non-metric. For this reason, the images may contain significant distortions and be blurred. It is therefore recommended to pay close attention to the type of camera and to carry out regular calibrations before each flight. The minimum geometric resolution must be equal to 6 MP, recommended 10–20 MP, with pixel dimensions of 5 μm. The camera's focal length f should preferably be fixed, ranging between 18 and 80 mm. Note that the greater the value of f, the greater the number of images required to cover the same area.

Most programs work with images in JPEG format; others recommend using RAW images and converting them to Tag Image File Format or Tagged Image File Format, commonly known by the abbreviations TIFF or TIF. Remember that there is always a risk of loss of quality in image compression.

In the specific case of topographic mapping, it is recommended that the images be vertical or nearly vertical, all have the same dimensions, such as 4000 × 3000 pixels, and that the flight be performed in strips similar to those of conventional airborne photogrammetry, i.e., in sequential parallel lines. However, images can be oblique and irregularly distributed in space for particular applications, as shown in Fig. 19.50.

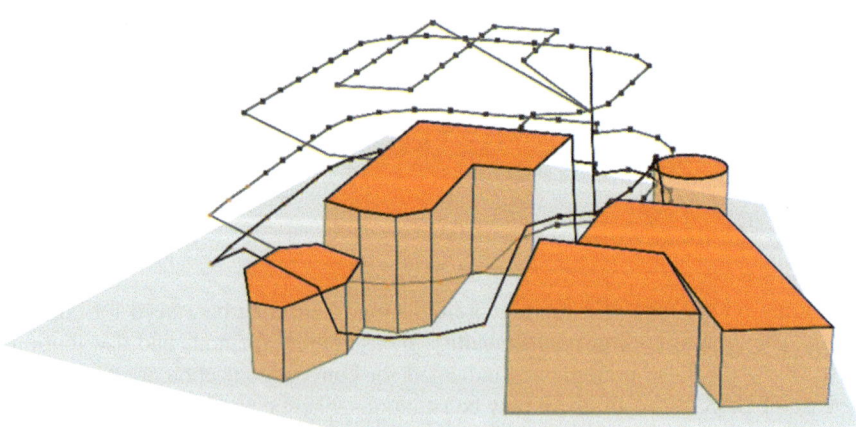

Fig. 19.50 Oblique image acquisition using UAS

19.19.5   Image Transfer

This step consists of importing all the images acquired during the airborne photogrammetric flight into the computer for processing using specific computer programs. Special care must be taken when handling the imported files to ensure the correct processing of the images, as any change in their structure may cause difficulties in processing. Image exposure corrections and quality checks are carried out at this stage of the workflow.

19.19.6   Data Processing

UAS data processing is similar to conventional airborne photogrammetry. It includes the GNSS/IMU data processing, if available, interior orientation, tie points determination and the photogrammetric block adjustment relying on Structure from Motion (SfM) algorithms. Figure 19.51 shows a typical image block acquired using the UAV platform.

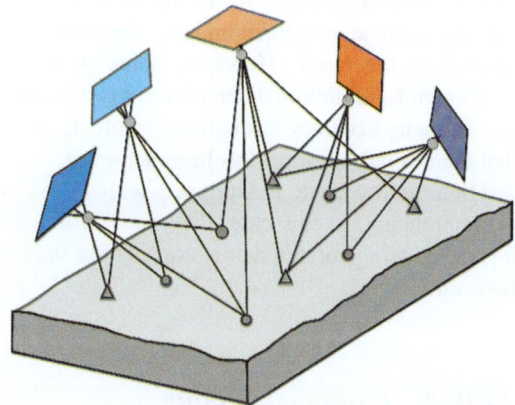

Fig. 19.51 Typical image block acquired using UAV platform

As mentioned, SfM algorithms allow non-specialists to generate 3D coordinate point clouds from digital images relatively cheaply and using non-metric digital cameras. Some of the key features of these algorithms are:

(a) Can be used with any commercially available digital camera
(b) Are implemented in low-cost application programs, some of them free of charge and highly automated,
(c) They require, however, a high degree of coverage between images
(d) Easy to use by users with little knowledge of photogrammetry

(e) However, they are generally implemented as black boxes and can generate solutions with low accuracy and repeatability
(f) Subject to dubious solutions available on blogs and generic websites

Self-calibration and block adjustment in the SfM algorithms is performed using control points and the automatic tie points (or key points) detection by image correlation algorithms. It is important to consider that the image correlation methods used in UAV photogrammetric processing programs are based on high image overlaps. Therefore, the results can be disastrous if this condition is not met.

The processing software determines the number of key points, requiring little user interaction. However, it is advisable to check that the distribution is consistent with the flight lines.

The number of control points required for aerial triangulation depends on several factors, including:

(a) Configuration of the project area: rectangular, corridor or 3D object.
(b) Camera quality: metric or non-metric.
(c) Terrain relief: flat, mountainous, built up, other.
(d) surface texture: qualified for image correlation.

Similar to conventional airborne photogrammetry processing, if the project requires control points (GCPs), they should be positioned between tracks so that they appear in at least three images. The minimum configuration is four points at the corners of the block. However, some photogrammetric software manufacturers recommend a minimum of ten points. In any case, the GCP should ensure planimetry and altimetry accuracy throughout the block, and checkpoints must be used to verify that the desired accuracy has been achieved.

High-quality UAS is equipped with GNSS and INS to perform direct image georeferencing. In this case, they need to be calibrated in the same way as large-format camera georeferencing, considering the GNSS antenna and IMU offsets and boresight angles.

19.19.7 Positional Quality

Assessing the positional quality of the products generated by photogrammetric processes using UAV is difficult due to the multiple components of the system. Similar to conventional airborne photogrammetry, it varies according to the type of aircraft, the quality of the onboard sensors (image, GNSS, IMU), the quality of camera self-calibration and the positional quality of the control points and key points. At the current state of the art, research indicates that the positional quality achieved by UAV photogrammetry for data collection in Geomatics is around 0.5 GSD in planimetry and 1.5 GSD in altimetry under favourable conditions. However, these values should not be taken as a rule, as data acquisition and processing techniques are diverse and evolving rapidly. It should be noted that these accuracies are reported for the positional quality of the photogrammetric model points and do not apply to the photogrammetric products, which will depend on other factors, such as the quality of the image correlation, the quality of the interpolation method used to generate the digital surface model and others. Studies indicate accuracy between 0.5 and 2.5 GSD in planimetry and between 2.5 and 4.0 GSD in altimetry. In the case of laser scanning, the most sophisticated systems have achieved accuracies ranging from a few centimetres to decimetres, depending on the quality of the onboard sensors and the flight height.

It is also important to emphasise that the high degree of automation of the operation may lead inexperienced users to believe that the results will always be positive. In practice, however, it has been observed that a lack of knowledge of photogrammetry, or a failure to follow its basic procedures, has led to results incompatible with the correct use of the technology.

19.19.8 UAV Photogrammetric Product

The main products generated by UAV photogrammetry are the same as those of conventional photogrammetry: the orientation parameters and the grid of points with known 3D coordinates. The point grid is then used to generate digital surface models. Some programs have algorithms for filtering artificial terrain elements to obtain the digital terrain model. The products generated as a function of the digital surface or terrain models are diverse and depend on the applications included in the processing program. They include the generation of three-dimensional models of objects, contour lines, slope maps, volume calculations, terrain profiling, and others.

In general, UAV photogrammetry programs do not have photogrammetric restitution capabilities. However, they can export the orientation parameters in the formats of the most popular conventional photogrammetry programs so that they can be used in sophisticated photogrammetric workstations to extract three-dimensional features.

Other essential products generated by UAV photogrammetry are orthophotos and orthomosaics. In these cases, it is necessary to correctly evaluate the GSD value, which varies with image resolution, camera focal length and flight height. The definition of this value has implications for the details that can be identified in the image and then in the final map, which is reproduced at a specific graphical scale. In the case of orthophoto production, the American Society of Photogrammetry and Remote Sensing (ASPRS) recommends that the GSD of the raw image should be at most 95% of the final GSD of the ortho-image. This means that one can have a raw GSD of 5 cm and generate an orthophoto with a GSD of 10 cm, but not the opposite.

19.20   Review Questions

1. Discuss what makes photogrammetry a particular technology for geospatial data collection when compared to total station and GNSS technologies.
2. What is digital photogrammetry?
3. What is the difference between terrestrial and airborne photogrammetry?
4. Briefly explain what a digital image is.
5. Briefly explain what CCD and CMOS sensors are.
6. Briefly explain how a digital image is acquired for photogrammetric purposes.
7. What is a GSD in the context of photogrammetry?
8. What GSD would you expect from an aerial image is taken with a digital camera with a pixel size of 6 μm, a focal length of 6 cm and a flight height of 1000 metres?
9. Briefly outline the principle of the central perspective image in the context of aerial photogrammetry.
10. Briefly explain the difference between geometric and radiometric digital image resolutions.
11. Explain why a digital aerial photograph cannot be considered a map.
12. Explain the image deformations that occur in a digital aerial photograph.
13. Briefly compare a frame sensor and a push-broom sensor in the context of digital photogrammetry.
14. Briefly explain how an aerial photogrammetric flight mission is carried out for geospatial data acquisition.
15. What is a stereo pair (conjugate images) in the context of photogrammetry, and are they obtained in the process of image acquisition?
16. Explain the meaning of the aerial photogrammetric model.
17. Briefly explain why stereoscopy is important in photogrammetry.
18. Briefly explain what the principle of the floating mark is and how it is used in the photogrammetric data acquisition process.
19. Explain what camera orientation (camera georeferencing) means in the context of photogrammetry.
20. Explain the importance of internal and external orientation in the process of creating a photogrammetric model.
21. Explain what are the geometric parameters to be considered in the interior orientation process.
22. Briefly discuss how the collinearity equations are used in the process of creating a photogrammetric model.
23. Briefly explain why it is necessary to have a stereo pair for geospatial data acquisition on a photogrammetric model.
24. Explain which parameters are determined during a photogrammetric exterior orientation process.
25. Explain the difference between direct and indirect exterior orientation methods.
26. Briefly discuss the advantages and disadvantages of using direct and indirect exterior orientation methods.

27. What is aerial triangulation? Why is it important to create a photogrammetric model?
28. What are the types of points to consider when adjusting an aerial triangulation model?
29. What is image correlation in the context of digital photogrammetry?
30. Briefly discuss how the image correlation process is carried out for point matching in the context of digital photogrammetry.
31. Discuss the importance of camera calibration in photogrammetry.
32. Briefly discuss the types of camera calibration currently used in aerial photogrammetry.
33. What is a digital photogrammetric workstation?
34. What are the main products that can be generated from a photogrammetric workstation for use in Civil Engineering projects?
35. What is an orthophoto?
36. Briefly explain how an orthophoto can be produced from digital aerial images.
37. What is a true orthophoto?
38. What makes UAV photogrammetry a particular type of airborne photogrammetry?
39. Discuss the applications of UAV photogrammetry in Civil Engineering that cannot be done with conventional photogrammetry.
40. Briefly outline the operation workflow of the UAV photogrammetric process.

References

Austin, Reg., (2010). *Unmanned aircraft systems.* John Wiley & Sons Ltd, The Atrium, Southern Gate, Chichester, West Sussex, PO19 8SQ, United Kingdom. ISBN 9780470058190 (H/B).

Bing Pan, Huimin Xie, and Zhaoyang Wang (2010). *Equivalence of digital image correlation criteria for pattern matching.* Vol. 49, No. 28 / Applied Optics.

Colomina, I., Molina P., (2014). *Unmanned aerial systems for photogrammetry and remote sensing: A review.* ISPRS Journal of Photogrammetry and Remote Sensing, 92, pages 79–97.

Everaerts, J., (2008). *The use of unmanned aerial vehicles (UAVS) for Remote Sensing and Mapping.* The International Archives of the Photogrammetry, Remote Sensing and Spatial Information Sciences. Vol. XXXVII. Part B1. Beijing.

Fraser, Clive, S. (1997). *Digital camera self-calibration.* ISPRS Journal of Photogrammetry & Remote Sensing 52. Pages 149-159.

Hutton, J. Mostafa, M. M. R. (2005) - *10 Years of Direct Georeferencing for Airborne Photogrammetry.* Photogrammetric Week. Stuttgart, Germany. 16p.

Kasser, M. Egels, Y. (2002). *Digital Photogrammetry.* Taylor & Francis, London, UK.

Krauss, K. (1993). *Photogrammetry: Fundamentals and Standard Processes*, Vol 1. Fer. Dümmler Verlag. ISBN 3-427-78684-6. Bonn. Germany.

Krauss, K. (1997). *Photogrammetry: Advanced Methods and Applications*, Vol 2. Fer. Dümmler Verlag. ISBN 3-427-78694-3. Bonn. Germany.

Luhmann, T., Robson, S., Kyle S., Boehm, J. (2020). *Close-Range Photogrammetry and 3D Imaging.* 3 Edition. De Gruyter. Berlin, Germany. 822p.

M. Cramer (2009). *EUROSDR Network on digital camera calibration.* Official Publication N. 55. EuroSDR. Web: www.eurosdr.net.

Massimiliano Pepe, Luigi Fregonese & Marco Scaioni (2018). *Planning airborne photogrammetry and remote-sensing missions with modern platforms and sensors.* European Journal of Remote Sensing, 51:1, 412–436, https://doi.org/10.1080/22797254.2018.1444945.

Mohamed M.R. Mostafa, Joseph Hutton (2001). *Direct positioning and orientation systems. How do they work? What is the attainable accuracy?* Proceedings, American Society of Photogrammetry and Remote Sensing Annual Meeting. St. Louis, MO, USA, 11p.

Mostafa, M.M.R. (2001). *Boresight Calibration of Integrated Inertial/Camera Systems.* Proceed. Int. Symposium on Kinematic Systems in Geodesy, Geomatics and Navigation—KIS 2001, Banff, Canada, June 5-8.

Novak, K. (1992). *Rectification of Digital Imagery.* Photogrammetric Engineering & Remote Sensing. Vol. 58, No.3, March, pages 339-344.

Potucková, M. (2018). *Image matching and its applications in photogrammetry.* Aalborg: Institut for Samfundsudvikling og Planlægning, Aalborg Universitet. ISP-Skriftserie, No. 314. Denmark, 2004. Downloaded from vbn.aau.dk on: november 25.

Rehak, M. (2017). *Integrated Sensor Orientation on Micro Aerial Vehicles.* Geodäticsh-geophysikalische Arbeiten in der Schweiz. Achtundneunzigster Band, Volume 98. Switerland.

Sandau R. et al (2010). *Digital Airborne Camera—Introduction and Technology.* Springer Verlag GmbH, Heidelberg, Germany.

Schenk, T. (1999). *Digital Photogrammetry.* Vol 1. TerraScience. Laurelville, USA. 428p.

Skaloud, J. (2019). *Geomonitoring & Mapping.* Geomatics engineering. EPFL. Switzerland.

Wolf, P. R. et Dewitt, B.A. (2000). *Elements of Photogrammetrwy with Applications in GIS.* 3[a] ed. MC Graw Hill, New York.

Chapter 20
Detail Survey

Irineu da Silva and Paulo C. L. Segantine

20.1 Introduction

A detailed survey is the name given for the engineering techniques and methods used to assess and record the 3D location, orientation, dimensions, and shape of geospatial data for land management. In the case of Civil Engineering, this means collecting information such as building corners, fences, benches, curbs or even a tree, which can be used to help engineers design plans for buildings, roads, bridges and various other construction and infrastructure projects. The data collécted is then processed in the office to produce topographic maps.

The survey techniques and the types of features to be surveyed will depend on the specific objectives for which the survey is being carried out, and the instruments and field procedures to be used will be determined. In general, total stations, topographic levels, GNSS instruments and terrestrial laser scanners are used. Airborne photogrammetry and airborne scanning techniques are used for large areas where such instruments are not cost-effective. The purpose of this chapter is to discuss the specifications, techniques and surveying methods available for collecting such information.

20.2 Specifications for Detail Survey

Before requesting a detailed survey, the project manager should prepare the technical specifications indicating the surveying objective, techniques, methods, accuracies, the type of feature to be surveyed, the method of data processing and storage, and the

I. da Silva (✉) · P. C. L. Segantine
São Carlos School of Engineering, University of São Paulo, São Carlos, SP, Brazil
e-mail: irineu@sc.usp.br; pclsegantine@usp.br

© The Author(s), under exclusive license to Springer Nature Switzerland AG 2025
I. da Silva, P. C. L. Segantine, *Geomatics Applied to Civil Engineering*,
https://doi.org/10.1007/978-3-031-75737-2_20

scale of graphical representation, if applicable. Based on these specifications, the surveyor shall establish the surveying parameters described in Sect. 20.3. In addition, the recommended survey instruments and fieldwork techniques shall also comply with the technical standards and regulations of the project to ensure that its objectives can be achieved.

20.2.1   Surveying Objective

Regarding the surveying objectives, the project manager should clarify the primary uses of the features to be surveyed so that the surveyor can decide on the level of detail to be collected. Categorising them into levels of importance is a resource that can facilitate and speed up the fieldwork. At the same time, it should also indicate the deadline for completion of the work and the quality control procedures to be followed for final acceptance.

20.2.2   Measurement Accuracy

Regarding accuracies, it has long been defined as a function of the size of the smallest detail to be measured and the scale of the final graphical representation. Today, however, with the computer resources available, it is possible to produce a drawing at any graphic scale, i.e., with the "zoom" resources available in computer programs, the scale of a digital map is no longer fixed. For this reason, the graphical representation only becomes an essential component in defining the accuracy of the survey when it is specified that the project will be delivered in paper form, which is rare nowadays. In cases where the survey will be manipulated in databases, the required accuracies must be specified according to technical standards or defined in the project mandate. For example, the accuracy required for a tunnel project is different from that required for a road, dam or housing project. The accuracy specification must, therefore, ensure that the position of the surveyed geospatial data is within the limits accepted by the construction or project management methods. In addition, the engineer must consider the accuracy of the instruments commercially available. For this purpose, Table 20.1 lists the surveying instruments commonly used in detail surveys and their respective nominal accuracies.

Table 20.1 Nominal accuracy of surveying instruments

Instrument	Manufacturer nominal accuracy	
	Lower accuracy	Higher accuracy
Electronic theodolite—Angular	9″	2. 0″
Total Station—Angular	7″	0. 5″
Distance	2.5 mm + 2 ppm	0.6 mm + 1 ppm
Optical level—double run	5 mm/1 km	0.2 mm/1 km
Digital level—double run	2 mm/1 km	0.2 mm/1 km
GNSS navigation (absolute)	3 – 15 m	2 – 5 cm (PPP-RTK)
GNSS DGPS	meter	submeter
GNSS RTK	10 mm + 1 ppm (horizontal)	8 mm + 1 ppm (horizontal)
	20 mm + 1 ppm (vertical)	15 mm + 1 ppm (vertical)
GNSS Post-processing static	5 mm + 0.5 ppm (horizontal)	3 mm + 0.1 ppm (horizontal)
	10 mm + 0.5 ppm (vertical)	3.5 mm + 0.4 ppm (vertical)
Terrestrial laser scanner	Position = 8 mm/20 m	Position = 3.2 mm/50 m

Instrument accuracy is only one component of the position quality of geospatial data surveying. The position error of the control point, the centring error of the instrument and reflector prism, or the position error of the GNSS receiver antenna must also be considered, in addition to other random errors discussed in Chap. 10. Therefore, for each new point measured, it is necessary to apply the law of error propagation to determine the precision of its coordinates. Another common practice to ensure the quality of point positioning using total station measurements is to determine the point coordinates from more than one reference station or to perform double-face readings.

20.2.3   Type of Detail to Be Surveyed

Regarding the type of detail to be surveyed, it is necessary to know whether the survey will be *planimetric, altimetric* or *planialtimetric*. It is planimetric when the relief of the terrain is not considered, and only the coordinates (*X, Y*) need to be determined. It is altimetric when the objective is only to determine the difference in elevation between points on the terrain. Finally, it is planialtimetric when altimetry and planimetry are combined, i.e., when, in addition to horizontal positioning, it is also required to include information on the relief of the terrain or the elevations of the survey details.

Having defined the type of survey, the project manager needs to list the details above and below the ground surface that are to be surveyed. In this regard, they can be divided into four groups as shown below:

- *Well-defined terrain details*, which are those that are clearly defined on the ground, such as buildings, road systems, boundary structures and natural features such as slopes, embankments and others.

- *Poorly defined terrain details*, which are those that are difficult to define precisely on the ground, such as the banks of a river or lake and areas of vegetation.
- *Aerial details*, which, as the name suggests, are those that are above the ground, such as power and telephone lines.
- *Underground details*, which, as the name suggests, are those below the ground, such as power and telephone lines.

The surveying specifications should also indicate which features are to be represented by their full geometric shapes and which can only be represented by symbols. If available, a list of symbols should be provided.

For planialtimetric surveys, the specifications should list the recommended levelling methods and the type of relief representation. The maximum acceptable equidistance between the grid vertices should be specified if the altimetric survey requires regular grid data collection. In the case of spot elevation measurements (irregular grid), the type of feature and the maximum acceptable spacing between them shall be specified.

20.2.4   Data Processing and Storage

Whatever the purpose of the data collected in the field, it will always need to be mathematically processed to produce geospatial data, which will be associated with an attribute to provide geospatial information appropriate to the project. To be used orderly and efficient, the information needs to be stored in databases and made available graphically in CAD/BIM environments. For these processes to be effective, the way in which the surveyed features are handled and stored must be part of the technical specifications of the surveying work.

20.3   Field Reconnaissance and Work Planning

It is recommended that a thorough reconnaissance of the project site is carried out before starting the detailed survey. This can be done by visiting the site or using available documents, such as existing topographical maps, aerial photographs and Google Earth® images. The objective of the field reconnaissance is to identify parameters that can assist in the preparation and definition of the survey, such as access roads, vegetation cover, type of relief, maximum sight distances, need or not for densification of the control point network, sky visibility for GNSS observations, availability of electrical power for battery charging and others.

For the detailed survey to be carried out, it is necessary to have a network of control points in the project area. As emphasised throughout this book, this network can be referenced to a local or a national geodetic reference system. In either case, there must be a set of landmarks with known coordinates on the site to which the surveying instruments can be set up and referenced. These landmarks should be visited and their physical condition checked.

After the field reconnaissance, the surveyor should be able to define the following measurement parameters:

- The preliminary work required to start the fieldwork.
- Which surveying instruments will be used, and under what conditions.
- The measurement methods to be used.
- The composition and qualifications of the field team.
- Field support material, such as survey accessories, stakes and support material.
- Authorisation and documentation to carry out the work.
- Budgetary planning.
- Schedule of activities.

Once the specified parameters have been defined, the field operations are prepared. This phase can determine the degree of success of the surveying work and must, therefore, be very well conducted. During this phase, the following auxiliary elements for the surveying work must be defined:

- The type of coding to be used for the different types of details to be collected.
- Preparation of the code list and attributes to be used in the field data collection.
- Field measurement instructions.
- Methodology for assessing the quality of the measurements.
- Data storage and backup procedures.
- Data processing procedures.
- Software packages for technical reports and drawings.
- Instructions and metadata, where relevant.

20.4   Fieldwork

The fieldwork described below is based on total stations, GNSS receivers and terrestrial laser scanners.

20.4.1   Detail Survey Using Total Station

Assuming that a network of control points adequately covers the area to be surveyed, the field procedure for a detailed survey using a total station is to set up the instrument at a control point, orient it with a backsight to another point on the network, and then measure the details by taking a series of side-shot (or irradiation) measurements (directions, angles and distances), as shown in Fig. 20.1. As a result, all measured points are referenced to the control point reference system.

Fig. 20.1 Principle of detail
survey using a total station

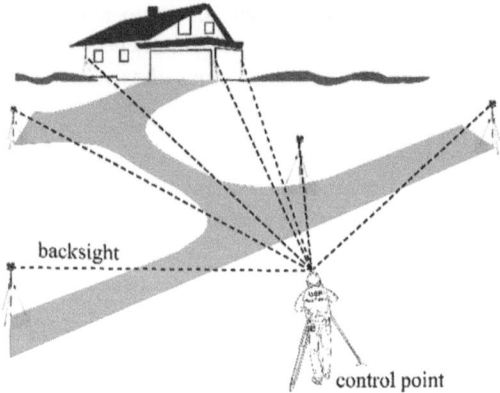

backsight

control point

Whenever the measurement capabilities of the occupied station point are exceeded, the instrument is moved to another control point, and the setup, orientation and other preliminary procedures are repeated.

If the surveyor is used to working with code lists, he can set the coding on the instrument using a predefined code list or enter the code manually. See Sect. 9.3.7 for more details on the code list. There is no need for booking sheets when using code lists, as the instrument's field software will help the operator define the features and point numbering as the measurements are taken in the field. It is important to remember that a well-structured point numbering facilitates data collection in the field and data handling in the office.

In cases where field books are used to record measurements, the observer should catalogue all the information necessary for the office staff to manipulate the field data to correctly create the project database and for drawing management. Typically, the field notes also include sketches of the surveyed details in 2D view, which will be used for the final CAD drawing manipulation. It is also important to note that some total stations with coding capabilities may not have graphical resources for data collection. In this case, it is recommended that the observer produces a sketch of the measurement to assist with data management in the office.

Current total stations have an internal memory to store all the data information collected in the field, such as point identifiers, instrument and reflector prism height, horizontal directions, vertical angles, distances, coordinates, attributes and observations assigned to the measured data, as well as code lists and feature related points. Therefore, the sequence of field operations for a detailed survey with a total station is as follows:

1. Set the atmospheric parameters in the total station for distance measurement.
2. Check that the prism constant is correctly displayed and set on the instrument.
3. Enter the correction parameters for distance reduction in case of cartographic projection, if the instrument allows it.
4. Upload and store the list of control point coordinates in the instrument's memory.

5. Set up the total station on the most convenient control point or location (in case of free station measurements).[1]
6. Enter the instrument and the reflector prism heights if the survey is planialtimetric with trigonometric levelling.
7. In the case of a control point, enter the point identifier into the total station. The instrument will display its coordinates for verification. In the case of free station, take measurements to calculate the station's position. The instrument will display the calculated coordinates and their accuracies.
8. Select the backsight point for instrument orientation and aim at the target accurately.
9. Specify the point identifier and record the backsight direction readings.
10. Select the type of detail to be surveyed and apply the appropriate coding and point identifier.
11. Guide the field assistant to the point to be surveyed.
12. Observe the slope distance, horizontal direction, and vertical angle, store them in the instrument's memory, check the increment of the point identifier.
13. Repeat steps 10 to 12 to measure the other features.

The feature point is measured by positioning the prism pole vertically (or inclined depending on the type of prism pole) over the feature point. Therefore, the field assistant must keep the prism pole plumb during the measurement to minimise the prism centring error. It is recommended that, wherever possible, the prism be placed at the bottom of the pole, as close as possible to the point on the ground. In cases where the centre of the prism cannot coincide with the axis of the measured object, such as trees, poles, and building corners, observations must be carried out in two steps. First, measure the distance and then the direction to the feature point, as shown in Fig. 20.2.

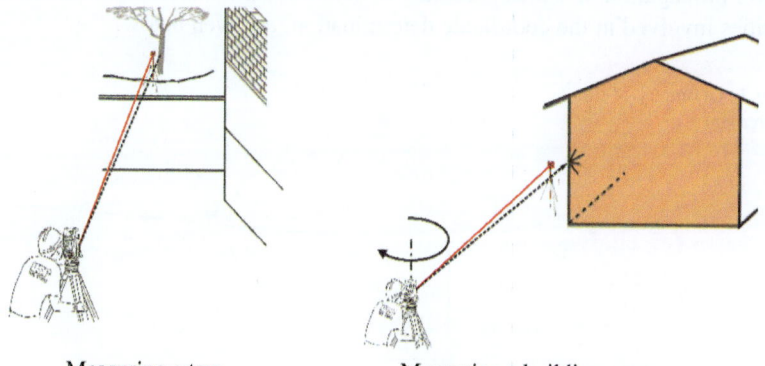

Measuring a tree   Measuring a building corner

Fig. 20.2 Principle of detail survey in two steps (distance and angle)

---

[1] See Sect. 11.10.5.

In the case of cadastral surveys using a total station, and in places where the control point network is not available, surveyors often carry out the control point (traverse) determination and cadastral data collection simultaneously, i.e., they measure the horizontal directions, vertical angles and slope distances of the traverse lines and feature points sequentially.

There are also situations where the quality and location of the existing control points are insufficient to survey all features. As mentioned in other chapters, in these cases, it will be necessary to densify the control point network with one or more secondary traverses linked to the existing control points.

20.4.1.1 Accuracy of Point Positioning Using a Total Station

The geometric elements involved in determining the coordinates of a point using a total station are the slope distance, the forward and backward horizontal directions and the vertical angle. To calculate the precisions of the coordinates of a detail point, it is, therefore, necessary to consider the precision with which each of these elements can be measured. Furthermore, since the coordinates of the new points depend on the values of the coordinates of the station and the reference points, their precision must also be considered. For a complete error analysis, it is also advisable to consider the centring errors of the instrument and the target, as well as the angular and linear standard deviations of the instrument. The precisions of the new coordinates are then calculated by error propagation, as described in Sect. 11.3.1.

To assess the problem, suppose that the coordinates of point (Q) are to be determined using a total station positioned at point (P) and backsighted at point (A), as shown in Fig. 20.3. In this case, the field observations are the backsight and foresight horizontal direction readings, the slope distance and the vertical angle. For the error propagation, it is then necessary to determine the standard deviations of the quantities involved in the coordinate determination, as given below.

Fig. 20.3 Geometry of the measurement

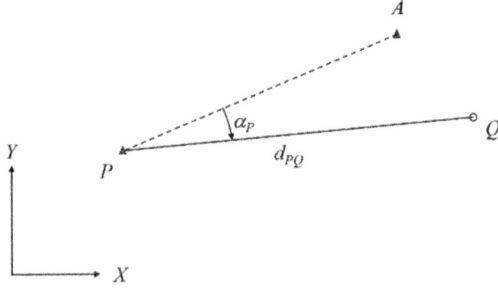

Standard deviation sd'_{PQ} of the slope distance using Eq. (20.1) (additive method), where a and b are given by t
he instrument manufacturer or from instrument calibration.

$$sd'_{PQ} = \pm (a + b * d[\text{km}])\,\text{mm} \qquad (20.1)$$

Standard deviation sd_{PQ} of the horizontal distance using Eq. (7.2), where the standard deviation of the vertical angle is given by the manufacturer or from instrument calibration.

Standard deviation sAz_{PA} of the reference azimuth using Eqs. (11.6), (11.7), (11.8), (11.9), (11.10) and (11.12) or Eq. (11.13).

Combined standard deviation $s_{\alpha P}$ of the angle α_P using Eq. (10.23).

Standard deviation sAz_{PQ} of the azimuth Az_{PQ} using Eq. (20.2).

$$sAz_{PQ} = \pm \sqrt{sAz_{PA}^2 + s_{\alpha P}^2} \qquad (20.2)$$

With all standard deviations determined, the standard deviation of the coordinates of the point (Q) can be calculated using Eqs. (11.22), (11.23), (11.24), (11.25), (11.26), (11.27), (11.28), (11.29), (11.30), (11.31) and (11.32).

Example 20.1 Given the coordinates of points (B) and (A) in Table 20.2 and the field observations in Table 20.3, calculate the coordinates of point (C) and their standard deviation. Assume that the total station used for data collection has a horizontal direction and vertical angle standard deviation (ISO 17123-3) of $\pm 5''$ and linear standard deviation (ISO 17123-4) of $\pm(3\,\text{mm} + 2\,\text{ppm})$ and that the centring errors of the instrument and target are ± 1.5 mm and ± 5 mm, respectively.

Table 20.2 Cartesian coordinates of points (A) and (B)

Ponto	X [m]	Y [m]
B	$5,000.000 \pm 5.0$ mm	$10,000.000 \pm 8.0$ mm
A	$5,044.169 \pm 10.0$ mm	$10,385.381 \pm 12.0$ mm

Table 20.3 Field observations

Station	Sighting point	Horizontal direction	Zenith angle	Slope distance [m]
B	A	0° 00′ 00″	–	–
	C	35°42′ 56″	89°44′ 12″	323.112

Solution:

   The horizontal distance can be calculated using Eq. (7.1).

$$d_{BC} = 323.112 * \sin(89°\,44'12'') = 323.109\,m$$

The standard deviation of the measured slope distance by the additive method can be calculated using Eq. (20.1).

$$sd'_{BC} = \pm(3 \text{ mm} + 2 * 0.313112) = 3.7 \text{ mm}$$

Since all field measurements were carried out on a single face reading, according to the explanation in Sect. 6.2.1.2, the standard deviation of the zenith angle is equal
$$szBC = \pm 5''\sqrt{2} = \pm 7.1'' = \pm 3.42815 * 10^{-5} \text{ rad.}$$
The standard deviation of the clockwise horizontal angle is equal
$$s\alpha_B = \pm 2 * 5'' = \pm 10.0'' = \pm 4.84814 * 10^{-5} \text{ rad.}$$

Using Eq. (7.2),

$$sd_{BC} =$$

$$\pm\sqrt{[\sin(89°44'12'') * 3.7]^2 + [323.112 * \cos(89°44'12'') * 3.42815 * 10^{-5}]^2}$$

$$= \pm 3.7 \text{ mm}$$

Using the coordinates of points (A) and (B),

$$d_{BA} = \sqrt{44.169^2 + 385.381^2} = 387.904 \text{ m}$$

The error in angle due to the reading error and the centring errors (instrument and target) can be calculated using Eqs. (10.16), (10.18) and (10.23).

$$s\alpha B_i = \pm \ 0.56''$$
$$s\alpha B_r = \pm \ 4.19''$$
$$s\alpha B_T = \pm \ 10.84''$$

The azimuth of the line BA can be calculated using Eq. (11.4).

$$Az_{BA} = \text{atan}\left(\frac{5,044.169 - 5,000.000}{10,385.381 - 10,000.000}\right) = 6°32'17.56''$$

The standard deviation of the azimuth Az_{BA} can be calculated using Eqs. (11.6), (11.7), (11.8), (11.9), (11.10), (11.11), (11.12) and (11.13).

$$s^2_{Az_{BA}} = J^T \Sigma_{ll} J$$

$$\Sigma_{ll} = \begin{bmatrix} 10.0^2 & 0 & 0 & 0 \\ 0 & 12.0^2 & 0 & 0 \\ 0 & 0 & 5.0^2 & 0 \\ 0 & 0 & 0 & 8.0^2 \end{bmatrix}$$

$$J^{T} = \frac{1}{150,469.4160} * [385.3810 \quad -44.1690 \quad -385.3810 \quad 44.1690]$$

$$s_{AZ_{BA}}^{2} = \pm 8.378854268 * 10^{-10} \ \text{rad}^{2}$$

$$s_{AZ_{BA}} = \pm 6.0''$$

The azimuth of the line BC can be calculated using Eq. (11.45).

$$Az_{BC} = Az_{BA} + \alpha_{B} = 6°\,32'18'' + 35°\,42'56'' = 42°\,15'14''$$

The standard deviation sAz_{BC} of the azimuth BC can be calculated using the general law of propagation of variances to the equation above.

$$s_{AZ_{BC}} = \pm \sqrt{6.0^{2} + 10.8^{2}} = \pm 12.4''$$

The coordinates of the point (C) can be calculated using Eq. (11.21).

$$X_{C} = 5,000.000 + 323.109 * \sin(42°\,15'14'') = 5,217.263 \ \text{m}$$
$$Y_{C} = 10,000.000 + 323.109 * \cos(42°\,15'14'') = 10,239.157 \ \text{m}$$

The variance-covariance matrix of the coordinates of point (C) can be calculated using Eqs. (11.23), (11.24), (11.25), (11.26), (11.27), (11.28), (11.29), (11.30), (11.31) and (11.32).

$$\Sigma_{ff} = F^{T}\Sigma_{ll}F$$

$$\Sigma_{ll} = \begin{bmatrix} 5^{2} & 0 & 0 & 0 \\ 0 & 8^{2} & 0 & 0 \\ 0 & 0 & 3.7^{2} & 0 \\ 0 & 0 & 0 & (6.001079 * 10^{-5}\text{rad})^{2} \end{bmatrix}$$

$$F^{T} = \begin{bmatrix} 1 & 0 & 0.672415 & 239.156660 \\ 0 & 1 & 0.740174 & -217.263111 \end{bmatrix}$$

$$\Sigma_{ff} = \begin{bmatrix} 2.36991 * 10^{-4} & -1.80505 * 10^{-4} \\ -1.80505 * 10^{-4} & 2.41278 * 10^{-4} \end{bmatrix}$$

Finally,

$$s_{X_c} = \pm 15.4\,\text{mm}$$
$$s_{Y_c} = \pm 15.5\,\text{mm}$$

And then,

$$X_C = 5,217.263\,\text{m}\ \pm 15.4\,\text{mm}$$
$$Y_C = 10,239.157\,\text{m}\ \pm 15.5\,\text{mm}$$

It is also possible to determine the error ellipse of point (C) based on the covariance matrix using Eqs. (3.149), (3.150) and (3.151) or (3.154), (3.155) and (3.156). Figure 20.4 shows the error ellipse configuration.

Fig. 20.4 Error ellipse for point (C)

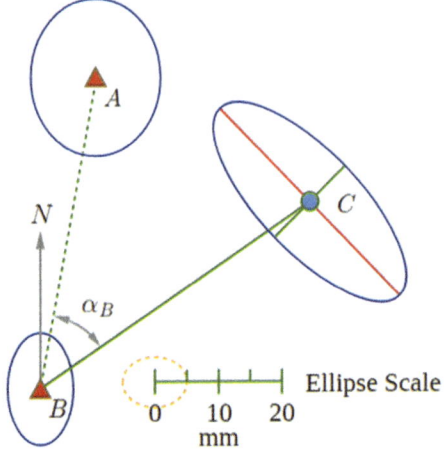

20.4.2   *Detail Survey Using GNSS Instrument*

As presented in Chap. 17, a detailed survey using GNSS technology can be carried out using one of the following surveying methods:

- Static post-processing
- Post-processing kinematics (PPK)
- Real-time kinematics (RTK)

Although the first two methods can produce consistent results, they are only partially suitable for feature surveying because they do not allow the quality of the survey to be checked at the time of measurement, as the GNSS observations have to be post-processed for coordinate determination. RTK measurements are then the most common method used for feature surveying with GNSS instruments. See Fig. 20.5.

Fig. 20.5   Detail survey using RTK GNSS measurement

The sequence of field operations for the detailed survey using GNSS technology is as follows:

1. Upload the list of coordinates of the control points to be used as base stations to the GNSS receiver.
2. Install the GNSS base station antenna over a suitable control point.
3. Set up the receiver according to the manufacturer's instructions, measure the height of the antenna, and enter the identifier of the control point to be used as the base station.
4. Start transmitting the real-time raw data from the base station.
5. Set up the rover GNSS receiver according to the manufacturer's instructions and measure the height of the rover antenna.
6. Select the type of detail to be surveyed and apply the appropriate coding and numbering to the feature point to be measured.
7. Move to the rover point to be measured, check the RTK data quality and record the coordinate values.
8. Move to the next feature point and repeat step 7 for the remaining feature points.

RTK GNSS data collection is generally faster than using a total station. However, it requires an open sky view for satellite tracking and may be less accurate than total station measurements.

In the traditional setup, the RTK GNSS data collection requires the GNSS antenna to be installed at a base station and transmit RTK data via radio link to remote receivers operating around the base. Therefore, if the project area is larger than the range of the radio link, it is necessary to move the base station to another available control point, which requires time and availability of control points in the area. One solution to this problem is using Continuous Observation Stations (CORS) with the ability to generate RTK data.

20.4.3   Detail Survey Using Terrestrial Laser Scanning

The technical details of terrestrial laser scanning systems are presented in Chap. 18. The georeferencing and point cloud surveying methods are also described in detail in this chapter. This section briefly discusses the field operations required for terrestrial laser scanning data collection.

As with any surveying project, collecting and processing terrestrial laser scanning data involves field measurements and office processing. However, field measurements are simplified by the high degree of automation in data collection. On the other hand, data processing and product generation still require many manual operations. Figure 20.6 briefly illustrates the operational workflow of a terrestrial laser scanning data collection.

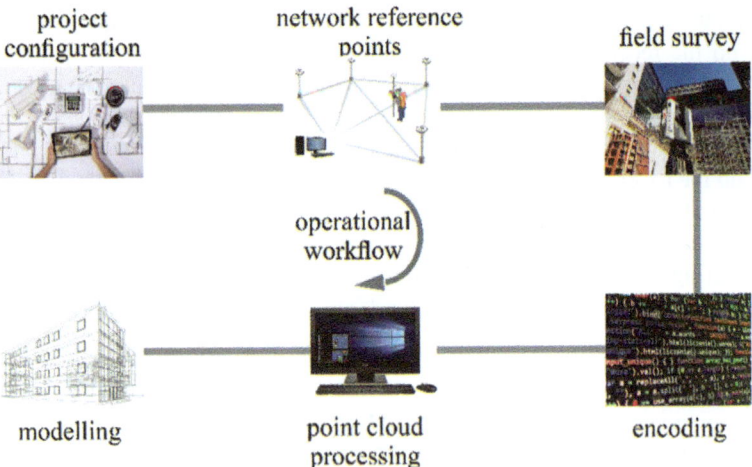

Fig. 20.6 Terrestrial laser scanning workflow

20.4.3.1 Project Configuration

For a 3D terrestrial laser scanning mapping or modelling to be effective, field operations must be planned in detail to avoid unnecessary scanning or occlusion of essential areas. It is, therefore, necessary to consider the following aspects of the project to avoid unnecessary rework:

1. Specify in detail the objectives of the project.
2. Carefully outline the tasks and responsibilities of each member of the professional team involved in the project.
3. Study in advance the geometric details of the areas and objects that will be scanned.
4. Evaluate the overall dimensions of the area to be surveyed.
5. Check the coordinate systems involved in the project.
6. Carefully select the type of instrument and accessories to be used.
7. Carefully select the georeferencing methods to be used in the project. For example, Google Earth® images, existing maps, or photographs to determine the best locations for the control points.
8. Establish field and office work schedules.
9. Determine the formats and forms of raw data storage.
10. Define the application programs to be used for data processing.
11. Carefully evaluate the hardware available for data processing in the office, considering the amount of data normally handled in laser scanning surveying.
12. Define the final deliverables and how they will be delivered to the client.

20.4.3.2 Control Point Network

As described in Chap. 18, the first step in the detailed survey using a terrestrial laser scanner is to establish a control point network. The details of this network will depend on the type of georeferencing to be used, as already described in that section. Once the georeferencing has been defined, the field survey workflow consists of the following steps:

1. Configure the instrument according to the type of measurement, georeferencing and target type to be used.
2. Install the instrument at a suitable control point, in the case of direct georeferencing, or at an appropriate point on the terrain, in the case of indirect georeferencing.
3. If necessary, measure the instrument and the target heights.
4. Indicate the atmospheric corrections and, if necessary, the geometric parameters according to the type of reference system adopted.

5. In the case of direct georeferencing, position the targets over control points and perform the necessary instrument orientation sightings.
6. For indirect georeferencing, position the targets at specific points in the scene and verify their capabilities for indirect georeferencing.
7. Configure scan parameters, such as scan window and type, resolution, camera parameters, filters and others.
8. Execute the scanning of the scenes.

Some laser scanners come with application software that allows programming the field data collection, for example, on georeferenced maps or Google Maps®. In this case, if the instrument is equipped with a GNSS instrument, it can determine its position, and by pre-programming the scans, it will show to the operator the sequence of operations required for each scanning session.

20.4.3.3   Office Work

Once the fieldwork is complete, the office work begins. This is a laborious step that requires skilled and experienced professionals. Because scanning generates millions of points, processing this point cloud is hard work that should not be underestimated. In summary, the main operational steps at this stage of the scanning work are as follows:

1. Prepare the code and attribute lists and load them into the processing software.
2. Download the data collected in the field into the processing software.
3. Using specialised software, code the points according to the predefined code list.
4. Generate lists of geocoded points in formats suitable for insertion into 3D modelling or mapping software.
5. Import the field data into the software.
6. Perform the modelling or mapping as specified in the project.
7. Generate the final products.

Data processed in this way can be used to produce engineering reports, plan views and profiles, digital surface and terrain models, perspective views and graphical animations. Figure 20.7 shows an example of a detailed survey using terrestrial laser scanning.

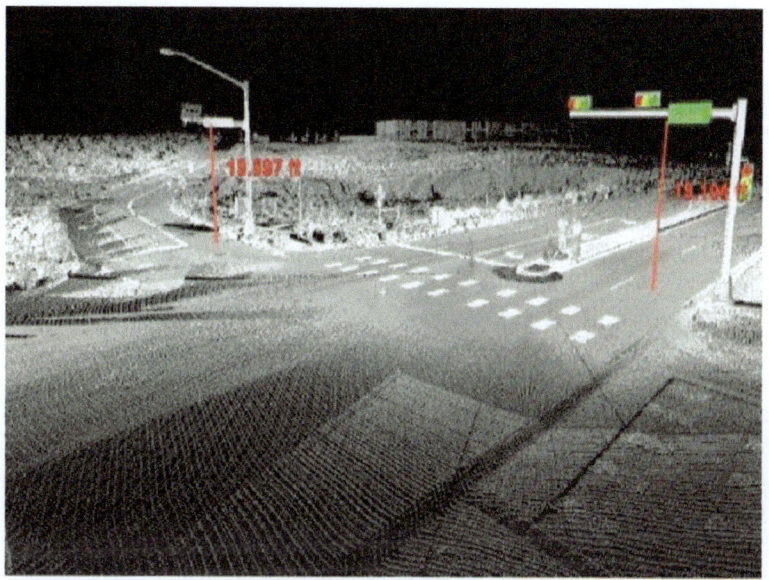

Point cloud perspective view

Topographic mapping from terrestrial laser scanning

Fig. 20.7 Example of point cloud and topographic mapping using terrestrial laser scanning

20.5   Technical Documentation

The main product of a detailed survey is the map, which for Civil Engineering purposes, has the finality to provide a graphical representation of feature elements positioning to facilitate the design of infrastructure projects and once designed, to provide layout facilities. In a topographic vector map, feature elements are represented by points, lines and polygons, depending on the scale of the map or how each feature is to be represented. Points are identified by their coordinates; lines are identified by the coordinates of their endpoints and polygons are described by a series of connected lines in a closed loop.

The drawing of a map is always carried out using a design program, that can work with stored coordinates allowing the user to graphically represent field data and then edit, create and design feature data using points, lines and polygons. The designed features are defined by their coordinates which, when the project is complete, allow the surveyor to take the designed information and carry out a construction set-out survey. Depending on the data structure, the design data can also be transferred directly to the onboard computers of construction machines for machine automation and control functions.

The type of feature data to be plotted and the data structure management to be carried out depends on the finality of the map. In terms of Geomatics applied to Civil Engineering, it can be a CAD map, a GIS map or a BIM map. The type of program to be used and the data management will be different for each application. It is beyond the scope of this book to discuss the details of these applications. The reader is advised to consult specialised literature for further information.

Generally, the basic geospatial data plotted on a CAD topographic map are the horizontal and vertical positions of the feature data collected, with attributes describing the feature where applicable. Therefore, the reader must be aware of the type of CAD application, as they have different capabilities. Depending on the type of project, the program must be able to work with map projection, have capabilities for COGO functions, digital terrain modelling, 3D object modelling, earthworks calculations, land division and design tools for various Civil Engineering applications such as road design and many others.

The final map processing is carried out by a professional designer and generally includes the following information:

- Title
- Scales
- Date of the survey
- Location of the surveying area
- Limits
- Date of the survey
- Location of control points
- Graphical representation of each feature element measured in the field, including contour lines and spot elevations, where appropriate
- Point IDs and attributes about particular feature points or objects

- Geodetic reference system
- Coordinate reference system
- Altimetric reference system
- Central meridian and UTM zone in case of UTM projection
- Grid and Geodetic North directions and the Meridian Convergence value, in case of UTM projection
- Coordinate grid
- Legend describing the symbols, line types and other abbreviations used in the drawing
- A list of the coordinates of the control points, if required
- Azimuth and distances of special feature lines, if required
- Name of the engineer in charge of the project

Figure 20.8 shows a typical example of a topographic mapping drawing for a detailed survey project.

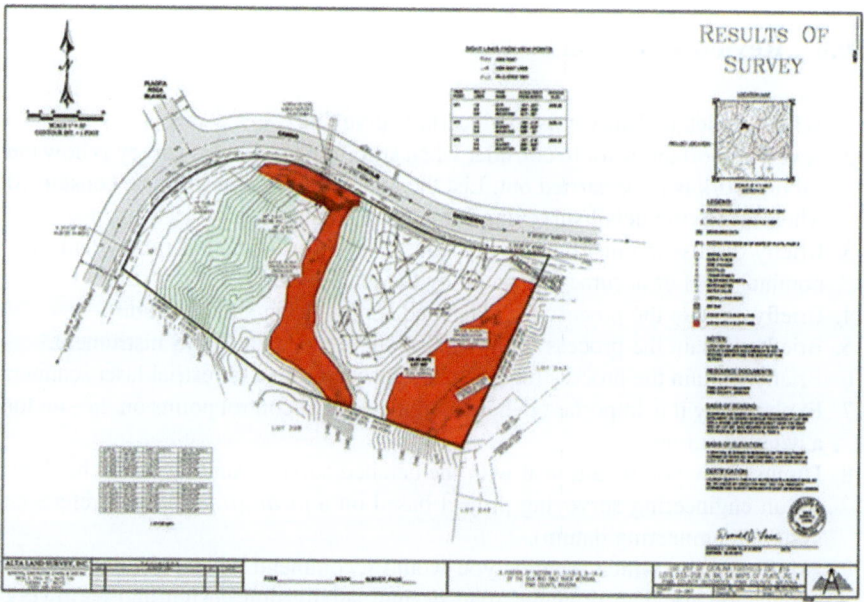

Fig. 20.8 Example of a topographic mapping drawing. (Courtesy: ALTA Survey. Location: https://www.altasouthwest.com/alta-arizona-land-surveying/topographic-mapping/)

If 3D modelling, DTM or DSM are also part of the project, as these are computer application models, they are supplied separately and georeferenced to the main map. Some applications can also link the model to the map, allowing online navigation in both graphical representations.

After the final map processing, deliverables must include technical documents describing working procedures, data processing parameters and engineering reports. The nature of the deliverables will depend on the nature of the project. In general, the following information should be included:

- Equipment used for field data collection and calibration certificates.
- Standards and technical specifications used for the project.
- Software package used for data processing.
- Topographic survey methods used for control point determination and data collection.
- Raw data files.
- Technical report containing information about computation methods, quality control, and coordinate listing.
- Descriptive memorandum of feature data included in the mapping representation.
- Monument description and location (sketch or image).

20.6   Review Questions

1. What is a detailed survey in engineering surveying?
2. A very important factor to consider when specifying a detailed survey is how the survey work is to be carried out. List the main points that need to be considered when specifying detail surveying for a Civil Engineering project.
3. Briefly discuss the instruments that can be used for a detailed survey and their nominal level of accuracy as stated by the manufacturer.
4. Briefly explain the process for a detailed survey using a total station.
5. Briefly explain the process for a detailed survey using a GNSS instrument.
6. Briefly explain the process for a detailed survey using a terrestrial laser scanner.
7. Explain why it is important to have a framework of control points on the site for a detailed survey.
8. Discuss how you would deal with the detailed survey using GNSS technology for an engineering surveying project based on a *local ground-based* reference system (engineering datum).
9. Briefly discuss which features you would recommend when selecting a CAD application for map drawing, road design, subdivision design and earthwork calculations.
10. Briefly discuss why an application software needs to have different capabilities when working with CAD drawings, GIS modelling and BIM modelling.

Chapter 21
Setting Out

Irineu da Silva

21.1 Introduction

For a Civil Engineering project to be completed, it needs to be *set out* (*layout, stakeout*) on the job site so that the designed construction elements are outlined on the ground to guide workers in carrying out the construction work. Setting out is necessary for all construction projects, regardless of size, especially for small projects where its importance is sometimes neglected. Typically, the following construction elements are set out for a Civil Engineering project:

- Site boundaries.
- Earthworks (cut and fill) sites.
- Axis of structural elements (foundations, columns, walls).
- Centre lines and offsets of roads, tunnels and other types of linear projects.
- Profile boards for road construction.
- Levels and slope of drainage runs.
- Other necessary structural parts of a construction site.

According to ISO 7.078:2020, setting out Civil Engineering works is defined as the *"establishment of marks and lines to define the position and level of the elements for the construction work so that work can proceed with reference to them"*.

The type of markers, reference lines, and field procedures vary according to the type of project. Setting out the elements of a road, the blocks and lots of a subdivision, a bridge or a building requires different field procedures, thus making it difficult to establish generic procedures. The ideal solution will always be specific to each case, and the criteria that will guide the engineer in selecting one or another

I. da Silva (✉)
São Carlos School of Engineering, University of São Paulo, São Carlos, SP, Brazil
e-mail: irineu@sc.usp.br

© The Author(s), under exclusive license to Springer Nature Switzerland AG 2025 861
I. da Silva, P. C. L. Segantine, *Geomatics Applied to Civil Engineering*,
https://doi.org/10.1007/978-3-031-75737-2_21

setting-out technique or surveying instrument will be dictated by the characteristics of the work to be done, such as:

- The accuracy required.
- The terrain relief and the physical conditions of the work site.
- The dimensions of the project.
- The tools and instruments available.
- The possibilities of quality control.
- Others.

Engineers responsible for setting-out works on a construction site must consider their interventions as part of a work plan. For the result of the work to be effective, they must, therefore, cooperate with the other professionals involved in the construction project to:

- Obtain all the information necessary for georeferencing and calculation of the setting-out elements.
- Ensure that the latest construction documents are used. Many project errors occur due to the use of outdated information.
- Establish and implement a work plan agreed upon with the other sectors of the site. This plan should consist of a schedule and a detailed setting-out project with the definition of control points, benchmarks, axis numbering, distances, and angles to be set out.
- Understand how setting-out elements will be used during construction to deploy them according to the users' needs.
- Ensure that the users can easily understand the identification of the control points, benchmarks and the outlined elements.
- Indicate the location of the stakeout points using a marking process that enables them to be identified and accessed quickly and accurately.
- Carry out a complete final inspection of all outlined elements with the construction foreman after each phase of work.

In a large construction project, a network of permanent control points is usually established during field data collection. It is, therefore, essential to use the same control points and to check their consistency throughout the staking operation to ensure homogeneity between the different operations. An important recommendation is not to use design points already staked as control points or reference lines for setting new stakes.

Considering these recommendations, for the purposes of this book, the methods of setting out are divided into two categories.

- Geometric methods
- Analytical methods

The details of each are presented below.

21.2   Geometric Stakeout Methods

A construction stakeout method is said to be geometric if the establishment of the landmarks and reference lines is performed by directly applying geometric concepts. Generally, such methods are preferred for small projects, or site works where the geometric positions are referenced to geometric elements of the building site, such as parallels and orthogonal lines, rather than coordinates.

In these cases, setting out is usually done by measuring distances directly with a tape measure or a laser distance metre. However, if angles are to be laid out or the accuracy requires sophisticated instruments, a theodolite or total station should be used.

The advantage of geometric methods is that they are more intuitive than analytical methods but require more practice by the engineer and his/her assistants.

21.2.1   Batter Board

The most common geometric method used in almost all small building constructions is the setting-out method using batter boards along the sides of the construction site. The assembly of such a batter board is very simple, consisting only of horizontal battens nailed to stakes fixed on the ground, forming a wooden frame, as shown in Fig. 21.1. Such a frame then becomes the geometric reference structure for setting out procedures.

Fig. 21.1 Example of a batter board on a construction site

For its practical use, the batten board must be constructed according to specific rules, such as:

- Cover the entire perimeter of the building.
- Stay firmly in the ground.
- Have at least one side referenced to the main alignment of the building.

- Be level.
- Have a height above the ground that facilitates its visualisation and use as a reference frame.
- Be set out so that construction machinery can access the working site.

Taking the battens as a reference, the engineer establishes the baselines of the construction elements by nails driven into opposite battens. The baselines are then marked by wires stretched between the nails. The points of intersection of the wires are then plumbed down using an ordinary plumb bob, as illustrated in Fig. 21.2.

Fig. 21.2 Use of the plumb bob at the intersection of string lines

As shown in Fig. 21.1, lines can also be established using a theodolite or total station. In these cases, instruments with a laser plummet and the capability for reflectorless and visible laser measurements can increase the efficiency and quality of the work.

Levelling works at the construction site should be performed using the surveyor´s levels and differential levelling methods or laser levels, as described in Chap. 13.

21.2.2 Drawing Perpendicular Lines

On a construction site, it is often necessary to define right angles or lines perpendicular to existing alignments. In such cases, the following procedures can be helpful, provided the required accuracy of the setting out allows.

(a) *The 3-4-5 method.* According to Pythagoras, every right triangle whose cathetus measures 3 and 4 (whatever unit of measurement is used) has a hypotenuse measuring 5. Therefore, this property of right triangles can be used to find perpendiculars on a construction site.

As shown in Fig. 21.3, the measurement procedure for establishing the perpendicular is as follows:

Fig. 21.3 Principle of
drawing perpendicular lines
using the 3-4-5 method

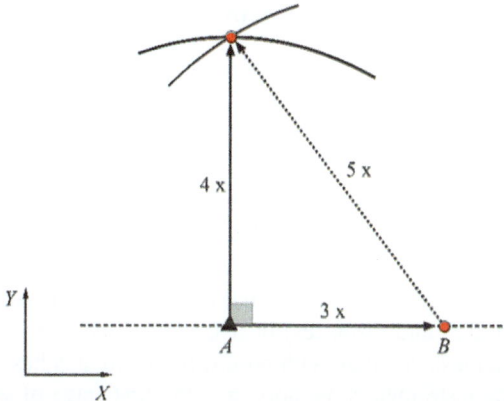

Using a tape measure, mark a point (B) on the reference line 3 units from the base (A) of the future perpendicular. From point (A), draw an arc with a radius of four units. From point (B), draw an arc with a radius of five units. The intersection of the two arcs is the point that defines the perpendicular to the line AB through the point (A). Obviously, all the measurements must be in the same plane.

(b) *The isosceles triangle method.* It is well known that the height of an isosceles triangle intersects the midpoint of the side opposite the isosceles sides. This property can, therefore, be used to draw perpendiculars on a construction site.

As shown in Fig. 21.4 the measurement procedure for establishing the perpendicular is as follows:

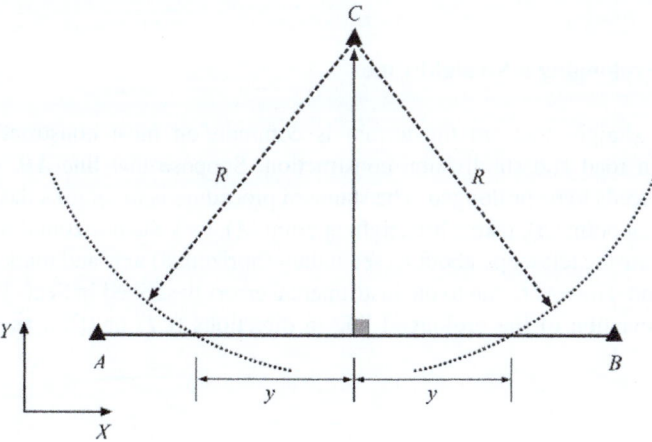

Fig. 21.4 Principle of drawing perpendicular lines using the isosceles triangle method

Using point (C) as a centre point, draw an arc with a radius greater than the distance from point (C) to line AB. Then, measure the distance between the arc intersections on line AB and find the midpoint. The position of the midpoint defines the perpendicular to line AB through the point (C).

21.2.3 Geometric Setting-out Using Total Station

Using *construction total stations* is a suitable solution for large projects where a batter board is not cost-effective. As briefly described in Sect. 9.3.8, this type of instrument is designed for detailed surveys and setting-out works on a construction site. In this case, the geometric setting-out can be performed by polar measurements (angles and distances) or using reference lines and offsets. In practice, this means that a sketch plan with points, lines and arcs has already been created and stored in the instrument's memory, and the first stage of setting out control points and lines has already been carried out. The total station then uses this information to guide the surveyor in setting-out procedures according to a previously developed setting-out plan. In addition, some construction total stations have onboard software that drives the users in setting out cut & fill points against the heights of a DTM, checking defined slopes, setting out batter boards and placing marks in the field along predetermined road lines and cross-sections.

In the case of a standard total station, the surveyor can carry out much the same procedures as for a construction total station, except that in this case, he must prepare the sketch plan, the setting out parameters and the procedures for stakeout points, lines and arcs, using external devices and spreadsheets. It is beyond the scope of this book to discuss the details of these procedures, mainly because they vary for different projects. However, some standard techniques may be helpful in setting out works and are, therefore, presented in the following subsections.

21.2.3.1 Prolonging a Straight Line

Prolonging straight lines in the terrain is common on most construction sites, especially in road and subdivision construction. Suppose that line AB, shown in Fig. 21.5a, needs to be prolonged. The standard procedure is to set up a theodolite or total station at point (B), take a backsight at point (A), lock the horizontal instrument rotation, rotate the telescope about its secondary (horizontal) axis and mark point (C) on the ground. However, due to the instrumental errors described in Sect. 10.2, there will be a deviation of the prolonged line in directions (C1) or (C2), as shown in Fig. 21.5b.

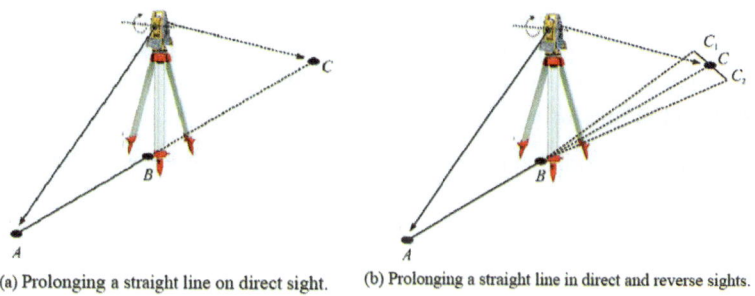

(a) Prolonging a straight line on direct sight.   (b) Prolonging a straight line in direct and reverse sights.

Fig. 21.5 Principle of prolonging straight lines. (**a**) Prolonging a straight line on direct sight. (**b**) Prolonging a straight line in direct and reverse sights

Therefore, point (C) must be marked with the telescope in direct and reversed positions to compensate for the instrumental error. The result will be two points on the ground (C1 and C2). The midpoint between them is in the prolongation axis.

The same procedure can be used for tracing vertical lines, as shown in Fig. 21.6.

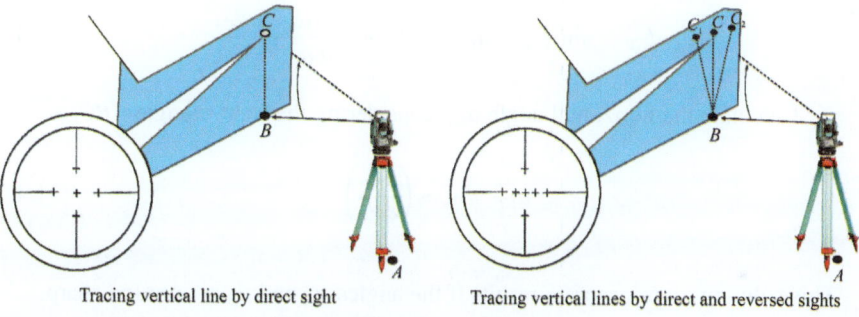

Tracing vertical line by direct sight   Tracing vertical lines by direct and reversed sights

Fig. 21.6 Tracing vertical lines by direct and reversed sights

21.2.3.2 Running a Straight Line Between Two Points

In everyday work on a construction site, it may be necessary to align a theodolite or total station between two points (A) and (B) marked on the ground. This is a difficult operation because there are no references to position the instrument. The solution to this problem is to set up the instrument at point (C) at a distance d_{CP} from line AB so that points (A) and (B) are visible for direction and distance observations, as described below.

(a) **Measurement of directions and distances to points (A) and (B)**. If a prism
   pole can be positioned at points (A) and (B) for distance and direction observations from point (C), as shown in Fig. 21.7, the procedure for positioning the
   instrument in line AB is as follows:

   1. Measure the horizontal directions L_{CA} and L_{CB} and horizontal distances d_{CA}
      and d_{CB}.
   2. Calculate $\alpha_C = L_{CB} - L_{CA}$.
   3. Calculate the horizontal distance d_{AB} using the law of cosines.

$$d_{AB}^2 = d_{CA}^2 + d_{CB}^2 - 2d_{CA} * d_{CB} * \cos(\alpha_C) \tag{21.1}$$

   4. Calculate horizontal angle α_A using the law of sines.

$$\frac{\sin(\alpha A)}{d_{CB}} = \frac{\sin(\alpha C)}{d_{AB}} \rightarrow \alpha_A = \mathrm{asin}\left[\frac{d_{CB}}{d_{AB}} * \sin(\alpha_C)\right] \tag{21.2}$$

   5. Calculate the horizontal distance d_{CP} using Eq. (21.3).

$$d_{CP} = \sin(\alpha_A) * d_{CA} = \frac{d_{CB} * \sin(\alpha_C) * d_{CA}}{d_{AB}} \tag{21.3}$$

   6. Calculate the horizontal angle α_{CP} considering the right triangle APC.

$$\alpha_{CP} = \mathrm{acos}\left(\frac{d_{CP}}{d_{CA}}\right) \tag{21.4}$$

This solution gives accurate results if the angles α_A and α_B are not too sharp.

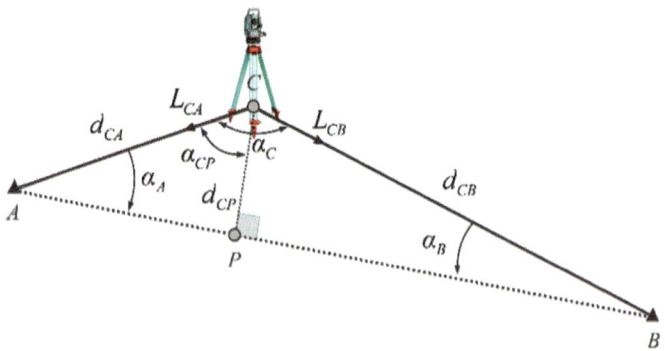

Fig. 21.7 Principle of running a straight line between two points measuring directions and
distances

(b). **Measurement of directions to points (A) and (B)**. If the endpoints of the line cannot be occupied but can be sighted with the instrument at a remote location, as shown in Fig. 21.8, the procedure for positioning the instrument in line AB is as follows:

Fig. 21.8 Principle of running a straight line between two points measuring directions

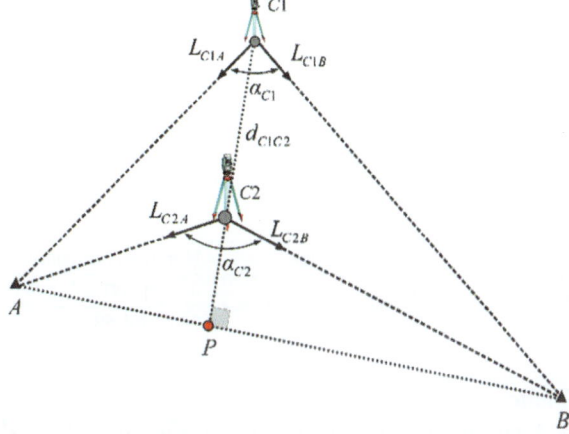

1. Position the instrument at a suitable point $(C1)$, in the terrain and observe the horizontal directions L_{C1A} and L_{C1B}.
2. Calculate the angle $\alpha_{C1} = L_{C1B} - L_{C1A}$.
3. Move the instrument to another convenient point $(C2)$, measure the horizontal distance d_{C1C2} and observe the horizontal directions L_{C2A} and L_{C2B}.
4. Calculate the angle $\alpha_{C2} = L_{C2B} - L_{C2A}$.

The horizontal distance d_{C2P} can be calculated using Eq. (21.5).

$$d_{C2P} = d_{C1C2} * \left[\frac{\tan\left(\frac{180° - \alpha_{C1}}{2}\right)}{\tan\left(\frac{180° - \alpha_{C2}}{2}\right)} - 1 \right]^{-1} \tag{21.5}$$

This method is less accurate than the previous one and should only be used when a low level of accuracy is acceptable. The closer the lines $C1$-P and AB are to the perpendicular, the greater the accuracy. If the position of the calculated point (P) is not satisfactory, the process can be repeated, taking into account point $(C2)$ and the calculated point (P).

Example 21.1 Suppose a total station is to be installed between points (A) and (B), as shown in Fig. 21.9. For this task, the instrument was first installed at point (C). Given the observation values in Table 21.1, calculate the angle and distance the instrument must be moved from point (C) to be positioned at the calculated point (P) in line AB.

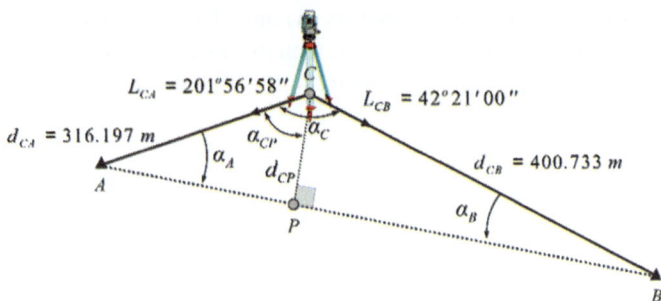

Fig. 21.9 Geometry of the measurement

Table 21.1 Field measurements

Station	Sighting point	Horizontal direction	Horizontal distance [m]
C	B	42° 21′ 00″	400.733
	A	201° 56′ 58″	316.197

Solution:

According to the calculation procedure described above:

$$\alpha_C = 201°56'58'' - 42°21'00'' = 159°35'58''$$

$$d_{AB} = \sqrt{316.197^2 + 400.733^2 - 2 * 316.197 * 400.733 * \cos(159°35'58'')}$$

$$= 705.758 \, \text{m}$$

$$d_{CP} = \frac{316.197 * 400.733 * \sin(159°35'58'')}{707.758} = 62.584 \, \text{m}$$

$$\alpha_{CP} = \text{acos}\left(\frac{62.584}{316.197}\right) = 78°35'03.4''$$

Example 21.2 Suppose that points (*A*) and (*B*) of the previous example cannot be occupied by a prism pole but can be sighted for observations of horizontal directions. Given readings in Table 21.2, calculate the distance d_{C2P}, as shown in Fig. 21.8, that must be measured to place the instrument on line *AB*.

Table 21.2 Field measurements

Station	Sighting point	Horizontal direction
C1	B	42° 21′ 00″
	A	201° 56′ 58″
C2	B	0° 00′ 00″
	A	169° 31′ 42″
Distance C1-C2 = 30.721 m		

Solution:

According to the calculation procedure described above:

$$\alpha_{C1} = 201°56'58'' - 42°21'00'' = 159°35'58''$$

$$\alpha_{C2} = 169°31'42'' - 0°00'00'' = 169°31'42''$$

$$d_{C2P} = 30.721 * \left[\frac{\tan\left(\frac{180° - 159°35'58''}{2}\right)}{\tan\left(\frac{180° - 169°31'42''}{2}\right)} - 1 \right]^{-1} = 31.884\,\text{m}$$

21.2.4 Trenches Excavation for Pipeline Setting Out

Another type of work that often requires surveying intervention is the excavation of trenches for setting out pipelines, mainly for grade control. This is often the case for drainage, sewerage, oil and gas projects, where precise levelling instruments are required because most of these projects are designed with low gradients to minimise costs and fit the project to the relief of the site.

The first operation in this work is to determine the excavation axes using alignments and grades from the coordinates specified in the design CAD drawing. Depths and centrelines are determined using batter boards at appropriate intervals along the alignments. Nails are then driven into the top of the battens and connected by stretched string lines to define the centrelines and grades, as shown in Fig. 21.10. The centreline determines the excavation offsets, and the grades are set out by differential or laser levelling. In both cases, the values to be read from the levelling staff are calculated as a function of the slope ($i\%$) and the horizontal distance d between levelling points, as shown in Fig. 21.11.

Fig. 21.10 Principle of trench excavation for pipeline setting out

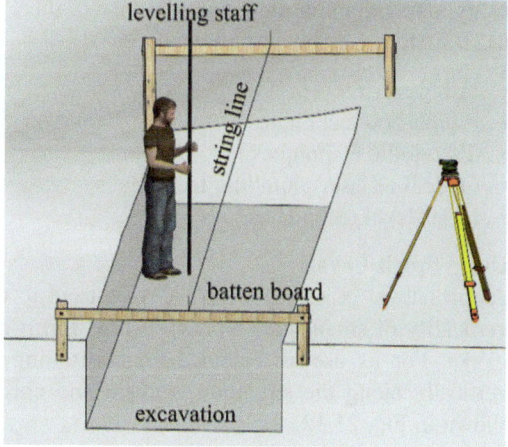

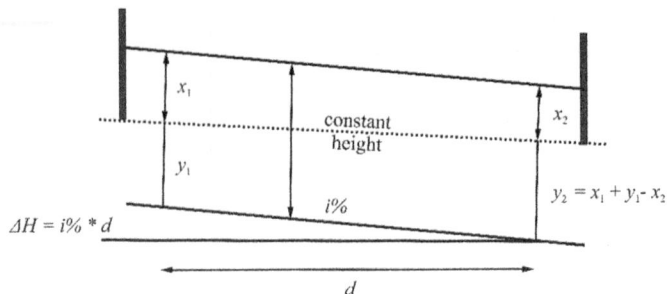

Fig. 21.11 Geometric relationship of the excavation profile

It is important to note that in the case of long pipelines, once the bottom of the trench has been levelled, it is still necessary to level the pipes to ensure they are laid at the correct slope. As indicated in Sect. 9.5.3, there are pipe laser instruments with horizontal beams specifically designed for this purpose. For information on the operation of this type of instrument, the reader is suggested to refer to the catalogue or to the user manual of the instrument of his/her choice.

21.2.5   Controlling Verticality

In addition to the horizontal positioning of design points, many projects require determining their elevations and controlling the verticality of structures, such as columns, lift shafts, walls and others. The accuracy of the verticality of these elements is an essential factor not only for structural stability but also for the economy and aesthetics of the construction. As stated in Chap. 13, vertical control points must be established on the site to determine the elevations of design points, which are used to level the required structural elements. For controlling the verticality, depending on the heights involved, there are several techniques that can be used, such as:

- Using spirit levels
- Plumb-bob techniques
- Theodolite techniques
- Optical or laser plumbing techniques
- Laser-level techniques

Using Spirit Levels
A spirit level is a very simple device used to check the level of surfaces or the verticality of small-scale works, such as formworks, door frames and bricklayer's works. For its use as a device for controlling the verticality, it must be placed vertically along the structural element and checked for the bubble's position, as shown in Fig. 21.12.

Fig. 21.12 Principle of using a spirit level. (Courtesy of Stabila. Location https://www.stabila.com/en/products/list/spirit-levels.html)

Plumb-Bob Technique

The plumb-bob technique is a common technique used to check or control the verticality of structural elements, such as lift shafts, foundations, walls, and columns, as well as for controlling the verticality of the full height of a building. A standard plumb-bob, as shown in Fig. 21.13, consists of a weight with a pointed tip on the bottom attached to the end of a string, which can define a precise vertical line (plumb line) under gravity. It can be purchased at building retail shops and is commonly used for controlling the verticality of small-height structures.

Fig. 21.13 Standard plumb bob. (Courtesy iStock-477856221)

A steel wire with a weight at its end is used for medium-height structural elements, as illustrated in Fig. 21.14. In this case, to avoid loss of accuracy due to wind force, the weight may be dropped in a viscous material to prevent its movement. A plumb bob under this condition with a weight of 3 kg is considered to have verticality accuracy of between 0.5 mm and 1.0 mm per metre. Although reasonably accurate, its use is not recommended for tall buildings.

Fig. 21.14 Principle of using steel plumb bob for building verticality control

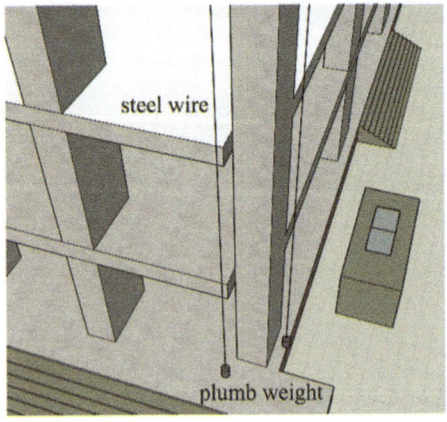

Theodolite Technique

A theodolite can control the verticality of towers, walls, foundations and columns, as shown in Fig. 21.15, specifically along a reference line. In this case, the instrument is set up on an extension of the reference line, away from the structural element and a mark or a steel tape is attached to the formwork. Verticality can be checked or controlled by turning the instrument telescope up and down and checking the top and bottom readings on the formwork.

Fig. 21.15 Principle of verticality control using theodolite

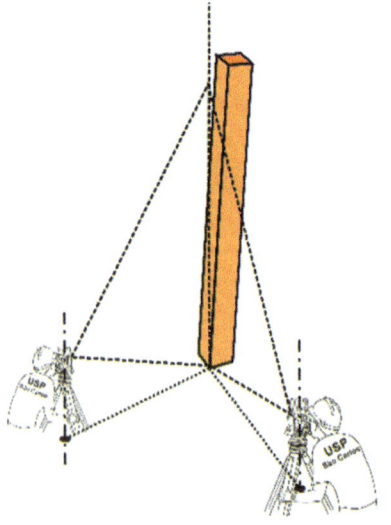

The instrument should then be moved perpendicularly to the reference line, and the procedure should be repeated until the verticalities of the two sides are adjusted. If the work can be performed with two instruments simultaneously, the efficiency of the operations is much higher.

As explained in Sect. 21.2.3.1, the measurement must be carried out using double-face readings to avoid instrumental errors.

Optical or Laser Plumbing Techniques

For controlling the verticality of high-rise buildings, bridge columns, chimneys, crane structures, oil platforms and cooling towers, as well as for tunnelling works, to connect control points of the surface network with the underground network through tunnel holes, it is recommended to use an optical or laser plumbing technique, which can be performed using a *zenith plummet* device designed exclusively for this purpose.

A zenith plummet instrument allows vertical transfer using an optical telescope or laser beam. Several variations of this instrument are available in the market, and most of them incorporate a coaxial telescope for both laser and optical observations and have an automatic compensator, which increases its accuracy significantly compared to other methods used for verticality control. The accuracy of the zenith plummet is expressed in mm/100 m. The usual accuracy ranges from ±1 mm/100 m to ±2.5 mm/100 m. See Fig. 21.16.

Fig. 21.16 Laser Zenith plummet. (Courtesy of Geo Funnel. Location: https://geo-fennel.de/en/products/surveying-instruments/laser-zenith-plummet/flp-1 50-green?)

According to the manufacturer:
Accuracy: ±1 mm/1.5 m; ±2.5 mm/100m
Maximum visible distance: ≥ 120 m (day); ≥ 120 m (night)

In general, using a zenith plummet is carried out by placing the instrument on a surveying tripod, levelling it, and aiming a target on the ground. Once such an operation is accomplished, the instrument is set up, and the plumb points upwards, the plumb points to the ground, and the optical axes are concentrical. In the case of high-rise buildings, the control or transfer of verticality is carried out through openings left on the building floor and using a particular target device appropriately installed over the openings, which can be moved horizontally on the upper or lower floor until the transfer point is set. Fig. 21.17 illustrates an example of such an installation.

Fig. 21.17 Principle of zenith laser plummet installation for verticality control on building construction

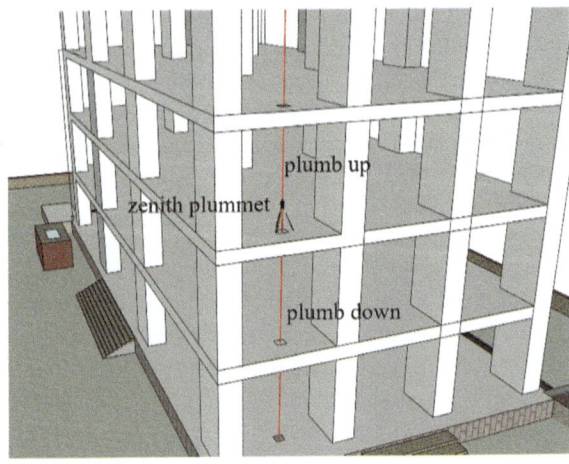

In the case of high-rise buildings, it is essential to note that as the height of the structure increases, some factors, such as wind and vibrations, can cause difficulties in controlling the verticality using the previous techniques. Alternative technologies are then being tested, such as combining or individually using GNSS instruments and total stations. Fig. 21.18 shows an example of an installation combining a GNSS antenna and 360° reflector prism.

Fig. 21.18 Principle of GNSS antenna and reflector prism installation for controlling the verticality of a high-rise building

In addition to GNSS measurements and laser plummets, electronic inclinometers can also be used for verticality control. In this case, the variation in the verticality of the whole building or parts of it can be monitored by installing the inclinometer at appropriate points in the structure. For more information on this subject, the reader is advised to consult specialised literature.

21.3 Analytical Setting-out Methods

A setting-out procedure is considered analytical when the coordinates of design points on a CAD drawing are used to position landmarks and reference lines on the construction site. A total station or GNSS RTK instrument is commonly used to stake out the design points. Particular attention must be paid to the type of coordinates available. They can be referenced to local or UTM projection planes. No scale factor is required when setting out with a total station if they are referenced to the local plane. On the other hand, if they are referenced to the UTM projection plane, setting out with a total station will require the application of scale factors, and with GNSS instrument only requires the correct receiver configuration to work with UTM coordinates. In both cases, the engineer must fully control the transformation models and the scale factors to be applied. See Chap. 16, for more information.

21.3.1 Setting-out with Total Station

The field procedures for analytical stakeout with a total station are the reverse of the detail survey procedures described in Chap. 20. An important element to consider is the availability of a network of control points on the site. In this case, the field procedure consists of setting up the total station over a control point, orienting it with a backsight to another control point, as shown in Fig. 21.19, and setting out the design points using their coordinates previously loaded into the instrument's memory.

Field operations are carried out using the setting-out program embedded in the total station. Firstly, the operator must store the coordinates of all the control and design points to be used in the setting-out process in the instrument's memory. The operator then needs to set up the instrument over a control point and indicate its number to the program. The next step is to select the backsight point in the terrain and identify it in the program menu. Using this information, the instrument calculates the direction and distance to any design point stored in the instrument. Once the desired design point has been selected, the operator must rotate the instrument in the direction indicated and instruct the field assistant to position the reflector prism pole in the direction indicated and at the approximate distance from the design point. An initial horizontal distance measurement is taken to the reflector prism, from which the instrument indicates how far the field assistant should move towards the desired position. The process is iterative, and successive distance measurements are repeated until the total station indicates a displacement of approximately zero.

Whenever the possibilities of setting out from the ground station occupied are exhausted, the instrument is moved to another control point, a new backsight is performed, and the process is repeated. This type of work is easy to carry out and, if well planned, can be highly satisfactory. However, the results will depend on the practice and skill of field personnel.

If the engineer needs to manually calculate the setting out elements, refer to Sect. 11.7.

To ensure the quality of the stakeout, it is recommended that the engineer establishes rules for field checking, such as staking out the same point twice from different ground stations and measuring the distance to the stakeout point to check against the design data records. Another simple rule is to check the squareness of the figures and the distances between design points set out in the field.

Fig. 21.19 Setting-out with total station

Depending on the nature of the site, e.g., construction in densely built-up areas, it may not always be possible to have a network of control points that can be occupied by a reflector prism. The solution may be to establish a network of secondary control points distributed around neighbouring buildings and determined by reflectorless total station measurements or by the point positioning by resection method described in Chap. 11. See Fig. 21.20. With this network of secondary control points at his disposal, the engineer can carry out the survey by positioning the total station at the most appropriate location for the measurements and determining its position and orientation using the *free station method.* Most instruments already have this application built into their operating system, which can guide the operator through field procedures or automatically search for reflecting prisms and interactively perform the free station calculation. In this case, it is important to have redundant observations and check that the residuals obtained after positioning the instrument are within the limits previously set for the project. It is also recommended that at least one of the secondary control points be used as a checkpoint to verify the quality of the instrument positioning.

Fig. 21.20 Free station

In cases where the number of design points to be set out is significant, using *robotic total stations* may be a cost-effective solution.

Another resource that can be used for setting out works in small construction sites is to combine batter boards with total station measurements. The field procedures, in this case, are as follows:

1. Verify that the project is georeferenced to a local control point network. If not, create a local control point network and georeference the design elements to it.
2. Lay out a batter board on the construction site following to the procedures presented in Sect. 21.2.1.
3. Georeference of the horizontal batten axes to the control point network.
4. Create a CAD project drawing combining the horizontal batten axes and the design baselines of the construction elements.
5. Using CAD resources, determine the coordinates of the intersections of the baselines with the horizontal batten axes.
6. Stake out the intersection points in the horizontal batten using the coordinates and a total station referenced to the control point network. Drive a nail into the intersection points.
7. Finish the job by laying the baselines with wires stretched between the corresponding nails.

The advantage of this procedure is that it can be accurately repeated, if necessary, to re-establish the position of points that have been moved or lost during construction. In addition, the position of the points can be checked by observing them from different control points.

21.3.2 Setting-out Accuracy Using Total Stations

To calculate the accuracy of a point set out by a total station, it is necessary to consider all the systematic and random errors described in Chap. 10. The propagation of all these errors gives the standard deviation of the position of the point. Thus, in the case of setting out points from a single instrument station, the positional standard deviation can be calculated as follows:

The positional standard deviation of the design point in the direction transversal to the alignment, ignoring the standard deviation of the azimuth of the reference line, is given by Eq. (21.6).

$$s_{tran} = d * s_\alpha \tag{21.6}$$

s_{tran} = positional standard deviation in the transversal direction
d = horizontal distance between the total station and the design point
s_α = precision of the horizontal angle in rad

In the longitudinal direction, the standard deviation is calculated according to Eq. (21.7), considering the centring error ε_i of the instrument and the centring error ε_r of the reflector prism.

$$s_{long} = \pm \sqrt{a^2 + (b * d[km])^2 + \varepsilon_i^3 + \varepsilon_r^2} \tag{21.7}$$

- a, b = EDM parameters, as presented in Sect. 8.3.2.5.

The precision of the resultant s_p can be calculated using Eq. (21.8).

$$s_p = \sqrt{s_{tran}^2 + s_{long}^2} \tag{21.8}$$

A practical recommendation for determining the precision with which a point should be set out in a building site is to always set it out with half the accuracy required for the work to be carried out. For example, if the construction methods for a bridge column allow it to be built with an accuracy of 1.0 cm, its centre should be set out on the ground with a precision equal to 0.5 cm.

Example 21.3 Suppose a point (Q) is to be set out at an angle α_P of 43° 22′ 15″ and a horizontal distance d_{PQ} of 340.950 m from a control point (P) using a total station with horizontal direction standard deviations (ISO 17123-3) of ±5″ and linear standard deviation (ISO 17123-4) of ±(3 mm + 2 ppm). The backsight control point is at 288.151 m from point (P). Considering that the centring error of the instrument and reflector prism are ±1.5 mm and ±5.0 mm, respectively, calculate the precision with which point (Q) will be set out.

Solution:

Transversal precision:

Considering that the horizontal direction reading was performed on a single face of the telescope, the transversal precision is given as follows:

$$s_l = \pm 5'' * \sqrt{2} = \pm 7.1''$$

The horizontal angle standard deviation is then calculated as follows:

$$s_\alpha = \pm 7.1'' * \sqrt{2} = \pm 10.0''$$

Using Eqs. 10.16 and 10.18.

$$s\alpha_i = \pm \frac{0.0015 * 206,264.8062''}{340.950 * 288.151} *$$

$$\sqrt{340.950^2 + 288.151^2 - 2 * 340.950 * 288.151 * \cos(43\,22'15'')} = \pm 0.7''$$

$$s\alpha_r = \pm \frac{0.005 * 206.264.8062'' \sqrt{340.950^2 + 288.151^2}}{340.950 * 288.151} = \pm 4.7''$$

Using Eq. (10.23),

$$s\alpha_T = \pm \sqrt{10^2 + 0,7^2 + 4.7^2} = \pm 11.1'' = \pm 5,36633 * 10^{-5} \text{ rad}$$

Using Eq. 21.6.

$$s_{tran} = \pm 340.950 * 5,36633 * 10^{-5} = \pm 18.3 \text{ mm}$$

Longitudinal precision:
Using Eq. 21.7.

$$s_{long} = \pm \sqrt{3^2 + (2 * 0.34095)^2 + 1.5^2 + 5^2} = \pm 6.1 \text{ mm}$$

Using Eq. 21.8.

$$s_p = \pm \sqrt{18.3^2 + 6.1^2} = \pm 19.3 \text{ mm}$$

21.3.3   Setting-out with GNSS

Setting out design points using GNSS instruments can be carried out by RTK-GNSS measurements. The field surveying procedures are similar to those described in Sect. 20.4.2. Similarly to the total station setting-out, RTK-GNSS measurements require local or geodetic control points distributed over the site. The advantages, in this case, are that the control point occupied by the GNSS base station does not need to be visible to the GNSS rover antenna, and several RTK remote antennas can operate simultaneously with only one GNSS base station. It is only required that the location and distance between them ensure continuous transmission links for RTK processing. However, many obstacles or tall buildings between the base station and the remote antenna can interfere with the RTK communication link or satellite tracking and prevent the GNSS system from working correctly.

As described in the previous sections, there are also solutions for GNSS relative positioning in RTK mode for construction site setting out using the CORS system. For more information on this subject, the reader is advised to consult the details of the CORS systems available in the region of the worksite.

Again, it is important to emphasise the care the operator must take when matching the GNSS coordinate system with the project coordinates.

The field procedures for point setting-out with GNSS receivers are as follows:

1. Load a data file containing the coordinates of all control points that can be used as GNSS base stations into the data logger of the GNSS receiver.
2. Load a data file containing the coordinates of all design points to be set out into the data logger of the GNSS receiver.
3. Set up the GNSS base station over a suitable control point according to the manufacturer's instructions, measure the antenna height and enter the occupied point number.
4. Start transmitting the RTK data from the GNSS base station.
5. Select the design point to be set out and move the rover antenna to it according to the offsets on the data logger screen, as shown in Fig. 21.21.
6. Repeat step 5 for each design point to be set out.

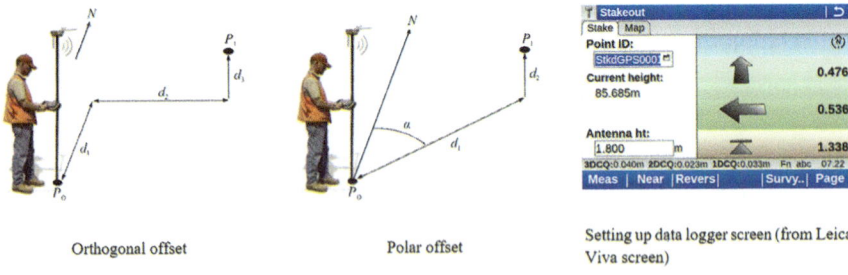

Orthogonal offset   Polar offset   Setting up data logger screen (from Leica Viva screen)

Fig. 21.21 Setting-out procedure using GNSS measurement

As shown in Fig. 21.21, data loggers operating with GNSS receivers have graphical displays to guide the operator to the stakeout point. In this case, the procedure is to move the antenna pole towards the point until the data logger on-screen graph indicates that it has been reached. Typically, the datalogger will emit an audible beep to indicate the position has been reached.

The reader should refer to the instrument's user manual for information on the positional quality of points set with the GNSS instrument being used. In addition to the nominal precision specified by the manufacturer, the operator must consider other factors, such as the precision of the bubble level used on the antenna pole and the instrument and target centring errors. In general, it is expected to achieve precisions of 10–15 mm in position and twice in elevation for conventional setting-out operations.

21.3.4   Machine Control Setting-out

Considering that today's Civil Engineering projects are completely digital and that the standardisation of geospatial data in non-proprietary formats and with the capacity to store information related to Civil Engineering features is a reality, a new perspective opens up for Civil Engineering design work through *machine automation*, generally referred to in Geomatics as *machine control*.

Based on the definition of XML (Extensible Markup Language), professionals in the field of Civil Engineering, in particular road construction, have developed the LandXML format, whose purpose is to allow the precise and unambiguous exchange of geospatial data in the field of Geomatics applied to Civil Engineering. This new standardised markup language thus provides a reliable tool for a construction setting-out operation, allowing the automation of fieldwork in 1D, 2D and 3D space, as shown in Fig. 21.22.

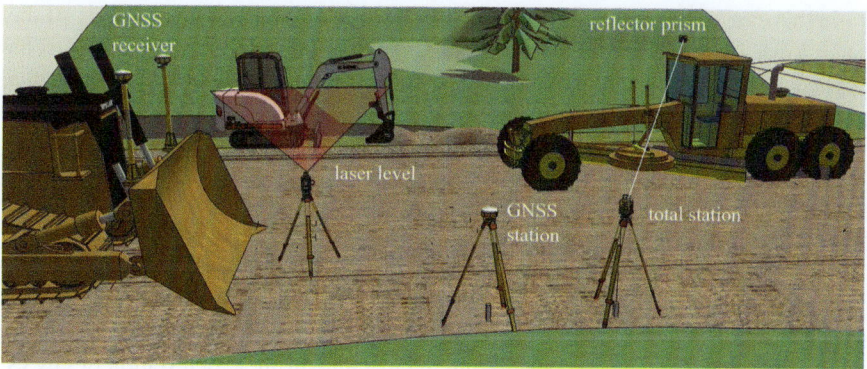

Fig. 21.22 Machine control systems

21.3.4.1   1D Machine Control System

For cases where the construction machine is only controlled in height, a 1D machine control system is defined. Typically, this type of machine control is performed using laser levelling, as described in Sect. 13.10. In this case, the types of machines commonly automated are excavators and bulldozer tractors. See laser levelling in Fig. 21.22.

21.3.4.2   2D Machine Control System

In cases where the machine can be controlled in height, and the inclinations and rotations of its tools are also taken into account, a 2D machine control system is defined, as shown in Fig. 21.23.

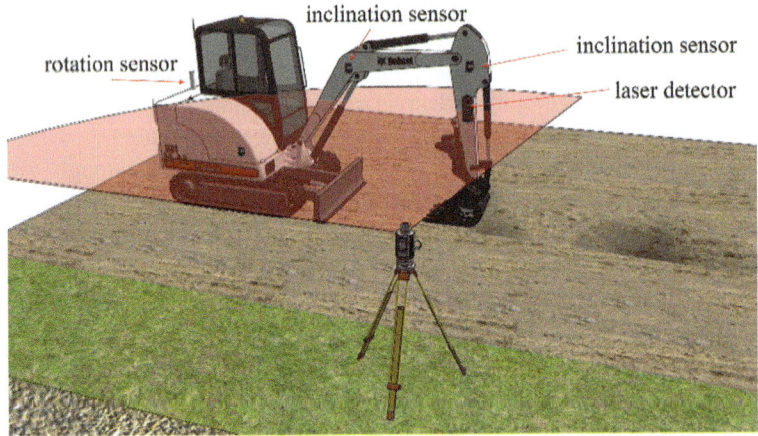

Fig. 21.23 2D Machine control system

In this case, the laser level controls the vertical positioning of the machine. At the same time, inertial sensors, discussed in Chap. 19, indicate the corrections to be applied to the machine components to perform their movements according to the project. In this case, the machines that are often automated are excavators and bulldozers.

21.3.4.3   3D Machine Control System

In cases where the machine can be controlled in its three axes, a 3D machine control system is defined. In this case, automation is achieved by installing inertial sensors in the machine components and positioning the machine using a total station or GNSS receivers, as shown in Fig. 21.22. Typically, all earthmoving and paving machines can be automated using a 3D machine control system.

Using the above-mentioned machine control systems requires experience on the part of the user and perfect synchronisation between the geometric elements of the project and the geographical elements of the site, i.e., between the two coordinate systems. A detailed discussion of this subject is beyond the scope of this book. For further information, the reader is advised to consult specialised literature on the subject.

21.4 Quality Control on Setting-Out

As mentioned, the quality requirements for setting-out Civil Engineering works vary according to the type of work. There are, however, three ISO Standards that deal with this subject, which are:

- ISO 4.463-1 (1989)—*Measurement methods for building – Setting out and measurement. Part 1: Planning and organisation, measuring procedures, acceptance criteria.*
- ISO 4.463-2 (1995)—*Measurement methods for building – Setting out and measurement. Part 2: Measuring stations and targets.*
- ISO 4.463-3 (1995)—*Measurement methods for building – Setting out and measurement. Part 3: Checklists for the procurement of surveys and measurement services.*

Different countries and contractors may have different standards on this subject. It is beyond the scope of this book to discuss the technical details of the applicable standards. Readers interested in further information should consult other references and the texts of specific standards.

21.5 Review Questions

1. Explain what a setting out process is in the context of a Civil Engineering construction site.
2. What is the difference between a setting out survey and a detailed survey?
3. Briefly explain what a batter board is in the context of Civil Engineering stakeout work and how it is assembled on site.
4. Explain how to extend a straight line in the field using a total station.
5. Briefly explain the procedures for setting out a trench in a Civil Engineering context using a laser level.
6. Discuss what are the steps required to provide the resources for the horizontal and vertical setting out of a building.
7. Discuss what vertical control is in the context of Civil Engineering.
8. Explain the conditions under which a spirit level is used on a construction site.

9. Explain how the plumb-bob technique is used to control verticality on a building site.

10. Explain how the laser plumbing technique is used to control verticality on a construction site.

11. Briefly explain the field operations required to set out design points using a total station.

12. Briefly explain the field operations required to set out design points using a GNSS RTK instrument.

13. Briefly discuss how to handle coordinate systems when staking out elements of a Civil Engineering site using a total station and GNSS receivers.

14. Explain how to carry out a stakeout using the free station method.

15. Discuss what are the advantages of using a rotating laser rather than a surveyor´s level on a building construction.

16. Discuss which stakeout technique you would recommend for road construction, building construction and subdivision.

17. Explain what machine control setting-out is.

18. Briefly discuss the difference between 1D, 2D and 3D machine control systems.

19. Briefly discuss under what conditions you would recommend using 1D, 2D and 3D machine construction systems.

References

Baykal, O., Tari, E., Coskun, M. Z., Erden, T, (2005). *Accuracy of Point Layout with Polar coordinates*. Journal of Surveying Engineering, Vol. 131, No. 3. https://doi.org/10.1061/(ASCE)0733-9453~2005!131:3(87). USA.

Barras, V (2015). *Les Implantations. heig-vd – Haute Ecole d´Ingénierie et de Gestion du Canton de Vaud*. Yverdon-les-Bains. Suisse.

Crawford, W.G. (1995). *Construction surveying and layout*. 2nd Edition. Creative Construction Publishing Inc., Indiana, USA.

Irvine, W. (1980). *Surveying for construction*. McGraw-Hill Book Co. (UK) Limited. 2nd Edition, London.

ISO 4.463-1 (1989). *Measurement methods for building – Setting out and measurement. Part 1: Planning and organisation, measuring procedures, acceptance criteria*;

ISO 4.463-2 (1995). *Measurement methods for building – Setting out and measurement. Part 2: Measuring stations and targets*;

ISO 4463-3:1995. *Measurement methods for building — Setting-out and measurement — Part 3: Check-lists for the procurement of surveys and measurement services.*

ISO 7.7078:2020. *Buildings and civil engineering works – procedures for setting out, measurement and surveying – Vocabulary.*

Milles, S., Lagofun, J. (2011). *Topographie et topométrie modernes*. Eyrolles, Paris.

Ragab Khalil, (2015). *Alternative Solutions for RTK-GPS Applications in Building and Road Constructions*. Open Journal of Civil Engineering, 5, pages 312–321.

Van den Berg, J., Lindberg A. (1983). *Measuring practice on the building site*. The National Swedish Institute for Building Research. Bulletin M83:16. Gävle, Sweden.

Chapter 22
Areas and Volumes

Irineu da Silva, Paulo C. L. Segantine, and Marcelo Monari

22.1 Areas

Determining the area of geometric figures is an important task performed by professionals involved with Geomatics. Almost all Civil Engineering projects require
calculating the area of a land surface or a geographical or designed feature element.
In this sense, although almost all current software applied to Geomatics have
modules dedicated to the calculation and division of areas in a CAD environment,
this chapter presents the best-known methods of estimating and dividing areas.
However, it is important to note that, for legal and administrative purposes, the
area of a land surface is generally calculated according to the horizontal projections
of the boundary lines that delimit it on the local horizontal ground plane, which is
different from the area generated from slope distances or map projections. However,
there are design situations where it is necessary to calculate the surface areas
represented in a map projection. This chapter only deals with the area calculation
of flat surfaces in the local horizontal ground plane.

Existing methods for calculating areas can be classified into:

- Geometric methods
- Analytical methods
- Mechanical method

I. da Silva (✉) · P. C. L. Segantine
São Carlos School of Engineering, University of São Paulo, São Carlos, SP, Brazil
e-mail: irineu@sc.usp.br

M. Monari
Civil Engineer, Assistant Professor in the Department of Civil Engineering at the Federal
University of São Carlos (UFScar), São Paulo, Brazil
e-mail: marcelo.monari@ufscar.br

22.2   Geometric Methods for Calculating Areas

The name *geometric method* is given to the area calculation method based on the division of the surface into elementary geometric figures such as triangles, rectangles, or trapezoids, with simple and well-known mathematical formulations for calculating their areas. In this section, only the areas of triangles and quadrilaterals are considered.

22.2.1   Area of a Triangle

The most elementary geometric figure of all is the triangle, whose geometric relationships are shown in Fig. 22.1. In this case, the area can be calculated using the following mathematical formulation:

$$A = \frac{1}{2}a * h_a = \frac{1}{2}b * h_b = \frac{1}{2}c * h_c \tag{22.1}$$

$$A = \frac{1}{2}a * b * \sin(\alpha_C) = \frac{1}{2}b * c * \sin(\alpha_A) = \frac{1}{2}a * c * \sin(\alpha_B) \tag{22.2}$$

$$A = \sqrt{p * (p - a) * (p - b) * (p - c)} \tag{22.3}$$

where

A = area of the triangle

$$p - \frac{a + b + c}{2} \tag{22.4}$$

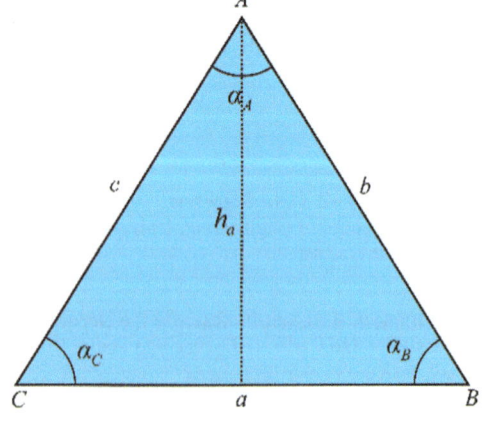

Fig. 22.1 Geometric elements of the triangle

Example 22.1

Given the data in Table 22.1 and the geometric elements in Fig. 22.2, calculate the area of the corresponding triangle using the equations shown in the previous section.

Table 22.1 Values of the geometric elements of the triangle

Geometric element	Value
c	1250.684 m
α_B	52° 45′ 32″
α_C	56° 23′ 10″

Fig. 22.2 Geometric elements of the triangle

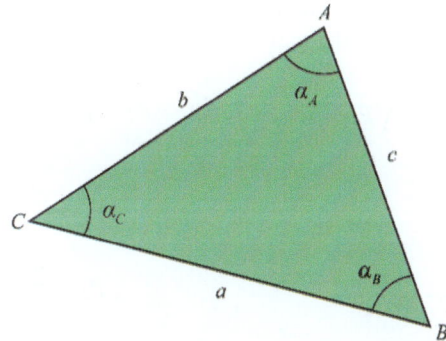

Solution:

  Before using Eq. (22.1), it is necessary to calculate the supplementary angle at the vertex (A), the height of the triangle in relation to the point (A) and the length of the side (a).

$$\alpha_A = 180° - 52°45'32'' - 56°23'10'' = 70°51'18''$$

$$h_A = \sin(52°45'32'') * 1,250.684 = 995.664 \text{ m}$$

$$a = 1250.684 * \frac{\sin(70°51'18'')}{\sin(56°23'10'')} = 1418.743 \text{ m}$$

$$\text{Area} = \frac{1}{2} * 1418.743 * 995.664 = 706,295.875 \text{ m}^2 = 70.630 \text{ ha}$$

To use Eq. (22.2), it is necessary to calculate the value of (b) as follows:

$$b = \frac{\sin(52°45'32'')}{\sin(56°23'10'')} * 1250.684 = 1195.581 \text{ m}$$

$$\text{Area} = \frac{1}{2} * 1195.581 * 1250.684 * \sin(70°51'18'') = 706,295.875 \text{ m}^2$$
$$= 70.630 \text{ ha}$$

Finally, using Eq. (22.3),

$$p = \frac{1418.743 + 1195.581 + 1250.684}{2} = 1932.504 \ \text{m}$$

$$\text{Area} = \sqrt{1932.504 * 513.761 * 736.923 * 681.820} = 706,295.875 \ \text{m}^2 = 70.630 \ \text{ha}$$

22.2.2   Area of a Quadrilateral

Using the geometric elements shown in Fig. 22.3, the area of a quadrilateral can be calculated by dividing it into several triangles. The total area can then be calculated by adding the partial areas as follows:

$$A = \sqrt{p_1 * (p_1 - a) * (p_1 - d) * (p_1 - e)} + \sqrt{p_2 * (p_2 - b) * (p_2 - c) * (p_2 - e)} \tag{22.5}$$

$$p_1 = \frac{a + d + e}{2} \tag{22.6}$$

$$p_2 = \frac{b + c + e}{2} \tag{22.7}$$

Fig. 22.3 Geometric relations of the quadrilateral

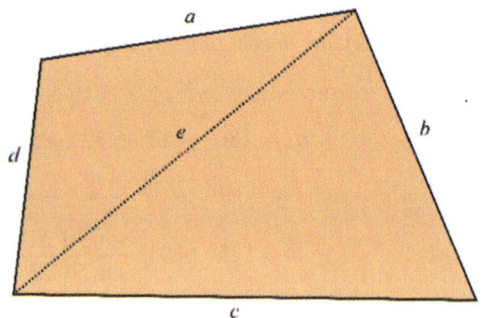

Example 22.2
Using the data in Table 22.2 and Fig. 22.3, calculate the area of the corresponding quadrilateral.

Table 22.2 Geometric values

Geometric element	Value [m]
a	808.679
b	1224.791
c	906.421
d	825.571
e	1455.023

Solution:

Using Eqs. (22.5, 22.6 and 22.7)

$$p_1 = \frac{808.679 + 825.571 + 1455.023}{2} = 1544.637 \ \text{m}$$

$$p_2 = \frac{1224.791 + 906.421 + 1455.023}{2} = 1793.118 \ \text{m}$$

$$\text{Area} = \left[\begin{array}{c} \sqrt{1544.636 * (1544.637 - 808.6789) * (1544.636 - 825.571) * (1544.636 - 1455.023)} + \cdots \\ \sqrt{1793.118 * (1793.118 - 1224.791) * (1793.118 - 906.421) * (1793.118 - 1455.023)} \end{array} \right]$$
$$= 823{,}377.493 \ \text{m}^2$$

22.3   Analytical Methods for Calculating Areas

An area calculation method is analytical if, instead of using known equations to calculate the areas of elementary geometric figures, field survey data (directions and distances) or known coordinates of the vertices of the polygon representing the surface are used. The most relevant of these methods are presented below.

22.3.1   Area Calculation by Radial Surveys

Consider the case where a surveying instrument is installed at any point (*C*) within the terrain whose area is to be determined, and the values of the horizontal directions L_i, zenith angles z_i and slope distances d'_i are taken from point (*C*) to each vertex (*i*) of the polygon, as shown in Fig. 22.4. The area measured in this way can be calculated using Eq. (22.8). Note that the slope distances must be reduced to horizontal distances d_i before calculating the area.

$$A = \frac{1}{2} \sum [d_i * d_{i+1} * \sin(L_{i+1} - L_i)] \qquad (22.8)$$

where

d_i = horizontal distance
L_i = observed directions

Fig. 22.4 Radial surveys

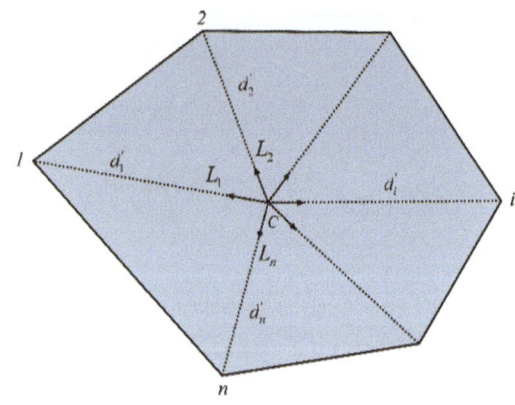

Example 22.3

Using the field data given in Table 22.3, calculate the area of the corresponding geometric figure.

Table 22.3 Field data and calculated values

Station	Sighting point	Horizontal direction	Horizontal distance [m]	$\sin(L_{i+1} - L_i)$ (1)	$d_i * d_{i+1}$ [m^2] (2)	(1) * (2) [m^2]
C	1	0 ° 00′ 00″	882.371	0.774573591	713,554.898	552,700.780
	2	129 ° 14′ 01″	808.679	0.964609850	936,038.664	902,912.116
	3	203 ° 56′ 42″	1157.491	0.778410566	955,591.002	743,842.133
	4	255 ° 03′ 37″	825.571	0.966197581	728,459.909	703,836.202
					Sum	2,903,291.231
					Area (m^2)	1,451,645.615

Solution:

To solve this example, it is recommended to follow the calculation steps given in Table 22.3.

22.3.2 Area Calculation by Polar Coordinate Surveys

Consider the case where a surveying instrument is installed at any point (C) outside the terrain whose area is to be determined, and the horizontal directions L_i, the zenith angles z_i and the slope distances d_i' are measured from point (C) to each vertex (i) of the polygon, as shown in Fig. 22.5. In this case, the area can also be calculated using Eq. (22.8).

Fig. 22.5 Polar coordinate
surveys

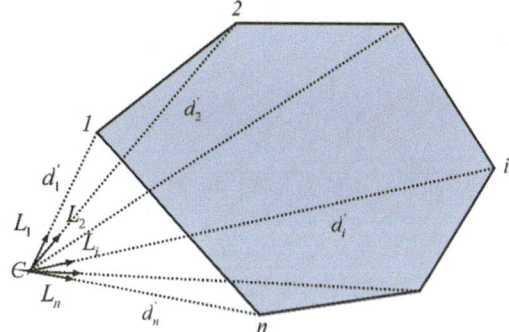

22.3.3   Area Calculation by Coordinates (Gaussian Method)

This method is based on the (X, Y) Cartesian coordinates of the vertices of a polygon, as shown in Fig. 22.6. In this figure, the area of the trapezoid A_{12}, containing the side 1–2 of the polygon, is given by Eq. (22.9).

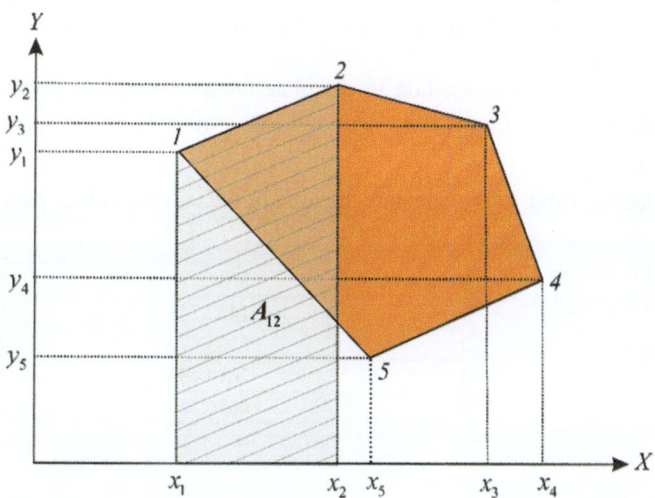

Fig. 22.6 Gaussian method for calculating areas

$$A_{12} = \frac{y_1 + y_2}{2} * (x_2 - x_1) \qquad (22.9)$$

If all sides are considered, Eq. (22.10) is obtained, which allows the area of any polygon to be calculated from its known coordinates.

$$A = \frac{y_1+y_2}{2} * (x_2 - x_1) + \frac{y_2+y_3}{2} * (x_3 - x_2) + \frac{y_3+y_4}{2} * (x_4 - x_3) + \frac{y_4+y_5}{2} * (x_5 - x_4)$$
$$+ \frac{y_5+y_1}{2} * (x_1 - x_5)$$

$$(22.10)$$

For an n-sided polygon, Eq. (22.10) can be rewritten as

$$A = \frac{1}{2} \sum [y_i * (x_{i+1} - x_{i-1})] \qquad (22.11)$$

Similarly, exchanging the (x) and (y) coordinates gives

$$A = \frac{1}{2} \sum [x_i * (y_{i+1} - y_{i-1})] \qquad (22.12)$$

As shown in Eq. (22.10), the double area of the polygon is the algebraic sum of each Y (or X) coordinate multiplied by the difference between the X (or Y) coordinates of the adjacent vertices. As shown in Eqs. (22.11) and (22.12), the double area divided by 2 gives the final area. The final area can be positive or negative, the algebraic sign only reflecting the direction of the calculation approach (clockwise or counterclockwise).

CAD software uses this method to calculate areas of irregular polygons, even those with curved sides, by dividing them into line segments.

Example 22.4
Using the data in Table 22.4, calculate the area of the corresponding polygon.

Table 22.4 Coordinates of the vertices of the polygon

Vertex	X [m]	Y [m]
1	7453.743	12,743.125
2	8105.479	11,360.951
3	7019.484	10,794.624
4	6676.216	11,633.531

Solution:
Using Eq. (22.11) or (22.12),

$$\text{Area} = 1,451,645.351 \ \text{m}^2$$

Another way of using the Gaussian method is to cross-multiply the corresponding coordinates of the different vertices of a polygon to find its area (as shown below), which is why the method is also known as the *shoelace method*. The absolute value of the difference between the sums of the ascending and descending products, divided by 2, gives the area of the polygon. Note that the coordinates of the first vertex are repeated at the end of the sequence to close the polygon.

Vertex	Rectangular coordinates	
1	X_1	Y_1
2	X_2	Y_2
3	X_3	Y_3
4	X_4	Y_4
1	X_1	Y_1

Example 22.5
Using the data from Example 22.4, calculate the area of the polygon using the *shoelace method*.

Solution:
   Computing the sums of ascending and descending products.

$$\sum \text{descending products} = 338,914,446.867 \text{ m}^2$$
$$\sum \text{ascending products} = -341,817,737.570 \text{ m}^2$$

The absolute value of the difference between the sums, divided by 2, gives an area of 1,451,645.351 m^2.

22.4 Mechanical Method for Calculating Areas

The most common instrument used to calculate areas mechanically is a *polar planimeter* or simply a *planimeter*. The planimeter can be used to measure the area bounded by any line drawn on a map of known scale. Although computational methods for calculating areas have become widespread, the planimeter still provides an adequate measuring technique due to its ease of use, speed and efficiency in its application.

The earliest planimeters were analogue, consisting of two rods (articulated arms), a set of graduated discs and a vernier. The area was then measured from the readings on the graduated discs and the vernier. Figure 22.7 shows an example of a mechanical polar planimeter.

Fig. 22.7 Example of a
mechanical planimeter

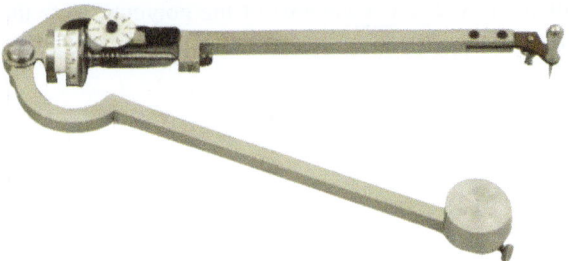

The latest planimeters, however, are digital and allow easier and more convenient evaluation of the measured area than their predecessors. Figure 22.8 shows an example of a digital polar planimeter.

Fig. 22.8 Example of a
digital planimeter

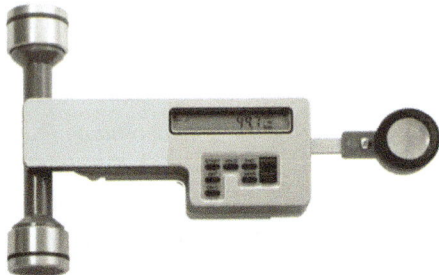

Planimeter procedures are not covered in this book. The interested reader should refer to the instruction manual supplied with the instrument.

The accuracy of the area measured by a mechanical or digital polar planimeter is estimated to be about 0.2%, i.e., 2 m^2 per 1000 m^2. However, this estimate depends on the shape of the measured figure and the operator's practice.

22.5   Computational Methods for Calculating Areas

As mentioned at the beginning of this chapter, most polygon area calculations are performed using computer application programs that are easy to use and give consistent results. Usually, the polygon whose area is to be calculated is already drawn on the computer screen, and the calculation is performed by moving the cursor along the edges of the polygon. However, in cases where the polygon is drawn on a topographic map, the map must first be digitised using a calibrated scanner. It is also necessary to use a computer program to convert the shades of grey of the line representing the perimeter of the figure into vectors.

There are highly efficient software packages that can be used to vectorise the perimeter of polygons. This requires little operator interaction. Once the figure is digitised, calculating the area becomes a simple process. The user should be aware of the need to scale the figure on the computer screen properly so that the estimated area matches the paper drawing.

22.6 Division of Areas

In some situations, once the polygon area has been calculated, the engineer may need to divide it into smaller portions according to the designed features in the project. Several methods can be used to do this. They are usually based on a dividing line of a known direction or a line through a known point, as shown in the following subsections.

22.6.1 Dividing Triangular Areas

Equations for three different situations for dividing triangular areas are presented below.

1. Dividing a triangle into successive areas A_1, A_2, A_3, etc., which are proportional to the ratios m, n, p, etc., and whose dividing lines start from a common vertex, as shown in Fig. 22.9. In this case, the distances forming the base of each triangle are given as follows:

Fig. 22.9 Dividing triangular areas by lines from a common vertex

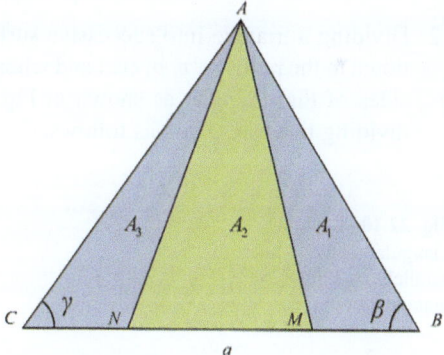

$$BM = \left(\frac{m * a}{m + n + p} \right) \qquad (22.13)$$

$$MN = \left(\frac{n * a}{m + n + p} \right) \qquad (22.14)$$

$$NC = \left(\frac{p * a}{m + n + p}\right) \qquad (22.15)$$

Example 22.6

• Suppose that the triangular surface in Fig. 22.9 has an area of 383, 047.780 m²
  and that $CB = 917.870$ m. This area must be divided into three parts so that A_1 is
  equal to A_3 and A_2 is equal to the sum of $(A_1 + A_3)$. Calculate the values of the
  distances BM, MN and NC and the values of the areas A_1, A_2 and A_3.

  If the reader wants to check the calculated areas, the following data should also be
  considered:

  $AB = 931.184$ m $AC = 975.522$ m $\beta = 63°40'\,45''$ $\gamma = 58°\,49'\,29''$.

  Solution:

  According to the data in this example, the ratios between the areas are as follows:
  $m = 1, n = 2, p = 1$
  The results are given in Table 22.5.

Table 22.5 Field data and calculation results

Ratio	Value	a [m]	Line segment	Distance [m]	Total area [m²]	Partial area [m²]
m	1		BM	229.468		95,761.945
n	2	917.870	MN	458.935	383,047.780	191,523.890
p	1		NC	229.468		95,761.945
Sum	4					383,047.780

2. Dividing a triangle into successive surfaces A_1, A_2, A_3, etc., which are proportional to the ratios m, n, p, etc., and whose dividing lines are parallel to one of the
   sides of that triangle, as shown in Fig. 22.10. In this case, the lengths of the
   dividing lines are given as follows:

Fig. 22.10 Dividing
triangular areas by lines
parallel to a side of the
triangle

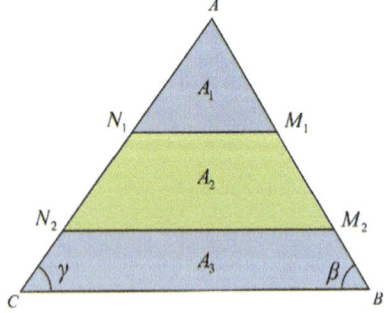

$$AM_1 = AB * \sqrt{\frac{m}{(m+n+p)}} \qquad (22.16)$$

$$AM_2 = AB * \sqrt{\frac{m+n}{(m+n+p)}} \qquad (22.17)$$

$$AN_1 = AC * \sqrt{\frac{m}{(m+n+p)}} \qquad (22.18)$$

$$AN_2 = AC * \sqrt{\frac{m+n}{(m+n+p)}} \qquad (22.19)$$

Example 22.7
Suppose that the triangular surface in Fig. 22.10 has an area of $383,047.780 \text{ m}^2$ and that $AB = 931.184$ m and $AC = 975.522$ m. This area must be divided into three parts so that A_1 is equal to A_3 and A_2 is equal to the sum of $A_1 + A_3$. Calculate the values of the distances AM_1, AM_2, AN_1, and AN_2, and the values of the areas A_1, A_2, and A_3.

Solution:
As in the previous example, the ratios between the areas are as follows: $m = 1$, $n = 2$, $p = 1$
The results are shown in Table 22.6.

Table 22.6 Field data and calculation results

Ratio	Value	Line segment	Distance [m]	Line segment	Distance [m]	Total area [m²]	Partial area [m²]
m	1	AB	931.184	AC	975.522		95,761.945
n	2	AM_1	465.592	AN_1	487.761	383,047.780	191,523.890
p	1	AM_2	806.429	AN_2	844.827		95,761.945
Sum	4						383,047.780

3. Dividing a triangle into successive surfaces A_1, A_2, A_3, etc., which are proportional to the ratios m, n, p, etc., and whose dividing lines depart from a common point on one of the sides of that triangle, as shown in Fig. 22.11. In this case, the lengths of the dividing lines are given as follows:

$$BM = \frac{2A_1}{BQ * \sin(\beta)} \qquad (22.20)$$

$$CN = \frac{2A_3}{CQ * \sin(\gamma)} \qquad (22.21)$$

Fig. 22.11 Dividing triangular areas by lines departing from a common point on one of the sides of the triangle

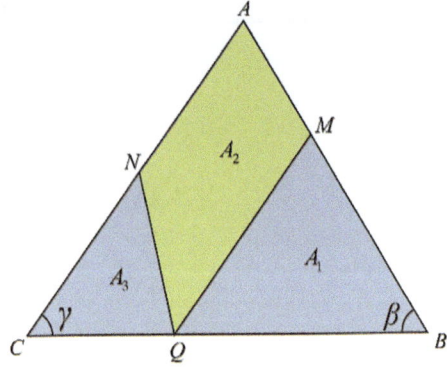

Example 22.8

Suppose that the triangular surface in Fig. 22.11 has an area of 383,047.780 m^2. This area must be divided into three parts so that A_1 is equal to A_3 and A_2 is equal to the sum of $A_1 + A_3$. Given that $BC = 917.870$ m and that the point (Q) is 400.000 m from the vertex (C), calculate the values of the distances BM and CN and the values of the areas (A_1), (A_2), (A_3). Given: $\beta = 63°\,40'45''$ and $\gamma = 58°\,49'29''$.

Solution:

According to the data above:

$A_1 = 95,761.945 \ m^2$
$A_2 = 191,523.890 \ m^2$
$A_3 = 95,761.945 \ m^2$.

Thus,

$$BQ = 917.870 - 400.000 = 517.870 \ m$$

$$BM = \frac{2 * 95,761.945}{517.870 * \sin(63°40'45'')} = 412.607 \ m \quad CN = \frac{2 * 95,761.945}{400.000 * \sin(58°49'29'')} = 559.627 \ m$$

22.6.2 Dividing Quadrilateral Areas

Quadrilaterals can be divided into successive areas A_1, A_2, A_3, etc., which are proportional to the ratios m, n, p, etc., and whose dividing lines start from a common vertex, as in Fig. 22.12. In this case, the distances BN and DM are calculated as follows:

Fig. 22.12 Dividing
quadrilaterals by lines
starting from a common
vertex

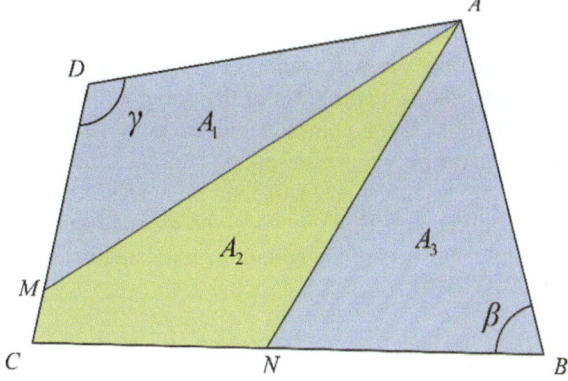

$$BN = \frac{2A_3}{AB * \sin(\beta)} \qquad\qquad (22.22)$$

$$DM = \frac{2A_1}{DA * \sin(\gamma)} \qquad\qquad (22.23)$$

Example 22.9
The quadrilateral in Fig. 22.12 has an area of 599, 813.880 m². This area must be
divided into three parts so that A_1 is equal to A_3 and A_2 is equal to the sum of $A_1 + A_3$.
Knowing that $AB = 931.180$ m, $DA = 697.880$ m, $\beta = 63°40'45''$ and
$\gamma = 94°59'36''$, calculate the values of the distances BN and DM and the values of
the areas A_1, A_2, and A_3.

Solution:
   According to the data above:

$A_1 = 149, 953.470$ m², $A_2 = 299, 906.940$ m² and $A_3 = 149, 953.470$ m².

   Thus,
$BN = \frac{2*149,953.470}{931.180* \sin(63°40'45'')} = 359.325$ m $DM = \frac{2*149,953.470}{697.880* \sin(94°59'36'')} = 431.377$ m

22.6.3   Dividing a Polygon by a Line Through a Known Point

In the case of dividing a polygon into two areas (A_1) and (A_2) by a line through a
known point, as shown in Fig. 22.13, there are two possible solutions. The first is to
apply the principle of the analytical method for calculating areas. Thus, given that
the boundary of the two areas begins at point (B) with known coordinates, the

coordinates of point (F) can be calculated using Eqs. (22.11) or (22.12) separately for each area to obtain two equations as a function of (X_F, Y_F). Solving the system of linear equations gives the values of (X_F, Y_F). Note that point (B) can be either a vertex or a point on one of the sides of the polygon.

In Fig. 22.13, the point (1) and the point (n) are, respectively, the first and last vertices of the polygon representing area (A_1), the point $(n + m)$ is the end point of the polygon representing area (A_2), and point (F) is located on the $(1\text{-}n + m)$ side. Based on this point numbering sequence, the second solution to calculate the coordinates of the point (F) is to use the equations indicated below:

Fig. 22.13 Dividing a polygon by a line through a known point

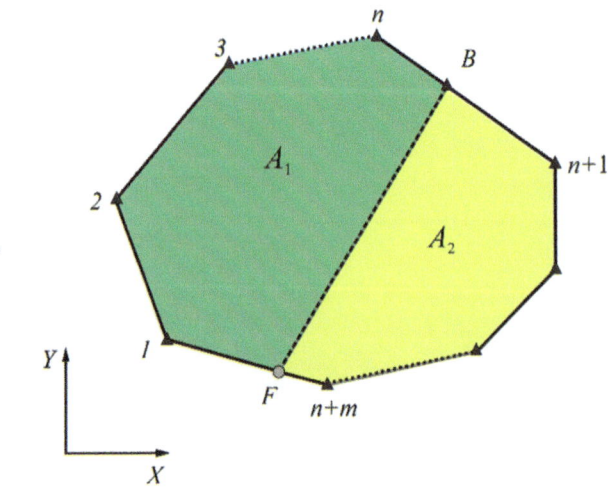

$$K_1 = \left[\sum_{i=1}^{n} (X_{i+1} * Y_i - X_i * Y_{i+1}) \right] - 2A_1 \qquad (22.24)$$

$$X_F = \frac{-K_1 - X_B * Y_n + X_n * Y_B + (X_B - X_1) * [Y_1 - X_1 * \cot(Az_{1-n+m})]}{Y_B - Y_1 - (X_B - X_1) * \cot(Az_{1-n+m})}$$
$$(22.25)$$

$$Y_F = Y_1 + (X_F - X_1) * \cot(Az_{1-n+m}) \qquad (22.26)$$

Note that Az_{1-n+m} corresponds to the azimuth of the alignment on which the point (F) is located.

Example 22.10

The polygon in Fig. 22.14 has an area of $772, 700.069$ m^2, and the coordinates of its vertices are given in Table 22.7. Calculate the coordinates of a point (F) so that the line segment FB divides the polygon into two parts of equal area. Use the two calculation methods presented in the previous section.

Fig. 22.14 Geometric situation of the polygon division

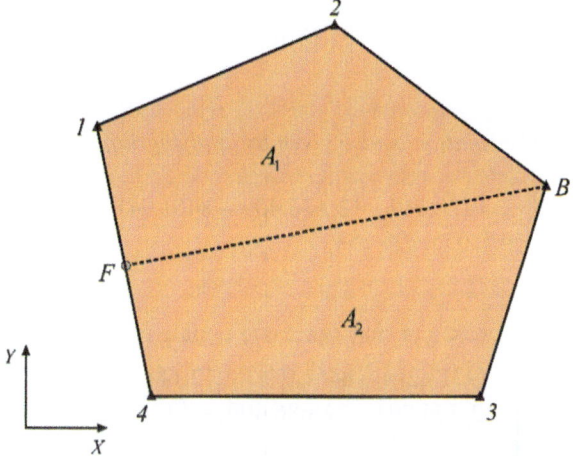

Table 22.7 Coordinates of the polygon vertices

Point	X [m]	Y [m]
1	8461.941	12,686.409
2 = n	9120.717	12,916.735
B	9713.294	12,556.294
3	9533.600	12,082.090
4 = n + m	8615.729	12,082.090

Solution:

The calculation results using the Gaussian method are shown in Table 22.8.

Table 22.8 Gaussian method for calculating areas (A_1) and (A_2)

Area	Vertex	X [m]	Y [m]	Area	Vertex	X [m]	Y [m]
A_1	1	8461.941	12,686.409	A_2	B	9713.294	12,556.294
	2	9120.717	12,916.735		3	9533.600	12,082.090
	B	9713.294	12,556.294		4	8615.729	12,082.090
	F	X_F	Y_F		F	X_F	Y_F
	1	8461.941	12,686.409		B	9713.294	12,556.294

The following system of linear equations with 2 unknowns is obtained, which are precisely the coordinates (X_F, Y_F) of the point (F) to be determined.

$$X_F + 9.617 Y_F = -127,406.042$$
$$X_F - 2.315 Y_F = 20,060.370$$

Solving the system of equations above gives the following results:

$$X_F = 8545.240 \text{ m} \qquad Y_F = 12,359.081 \text{ m}.$$

Using Eq. (11.4),

$$Az_{1-n+m} = \text{atan}\left(\frac{8,615.729 - 8,461.941}{12,082.090 - 12,686.409}\right) + 180° = 165°43'20.4''$$

The solution is also given by applying Eqs. (22.24) to (22.26) as follows:

$$K_1 = 9120.717 * 12,686.409 - 8461.941 * 12,916.735 - 2 * 386,350.034$$
$$= 5,635,796.684$$

X_F

$$= \left\{ \frac{-5,635,796.684 - 9713.294 * 12,916.735 + 9120.717 * 12,556.294 +}{(9713.294 - 8461.941) * [12,686.409 - 8461.941 * \cot(165°43'20.4'')]}{(12,556.294 - 12,686.409) - (9713.294 - 8461.941) * \cot(165°43'20,4'')} \right\}$$

$$= 8545.240 \text{ m}$$

$$Y_F = 12,686.409 + (8545.240 - 8461.941) * \cot(165°43'20.4'') = 12,359.081 \text{ m}$$

22.6.4   Dividing a Polygon by a Line of Known Direction

In many cases, instead of dividing the polygon by the known coordinates of a point, it is desirable to divide it by an alignment of known azimuth, which intersects two sides of the polygon, as shown in Fig. 22.15. Again, there are two points of intersection with the polygon, called points (B) and (F).

Fig. 22.15 Dividing a polygon by a line of known direction

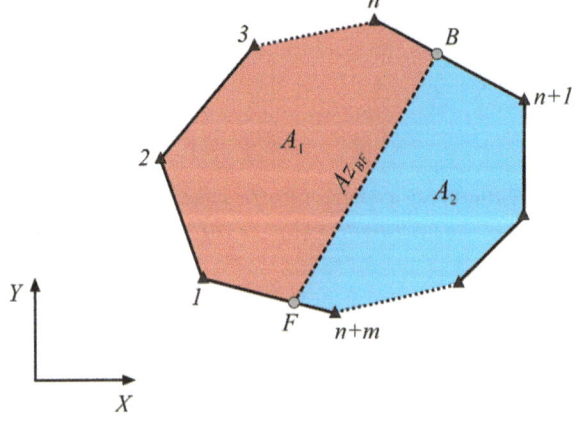

Based on the point numbering sequence in Fig. 22.15, the coordinates of points (B) and (F) can be determined as follows:

$$K_2 = -\left[\frac{Y_n - Y_1 + X_1 * \cot(Az_{1-n+m}) - X_n * \cot(Az_{n-n+1})}{\cot(Az_{BF}) - \cot(Az_{1-n+m})}\right] \quad (22.27)$$

$$K_3 = \frac{\cot(Az_{BF}) - \cot(Az_{n-n+1})}{\cot(Az_{BF}) - \cot(Az_{1-n+m})} \quad (22.28)$$

$$K_4 = K_1 + X_1 * Y_1 - X_n * Y_n - \cot(Az_{1-n+m}) * X_1^2 + \cot(Az_{n-n+1}) * X_n^2 \quad (22.29)$$

$$K_5 = Y_n - Y_1 + \cot(Az_{1-n+m}) * X_1 - \cot(Az_{n-n+1}) * X_n \quad (22.30)$$

$$K_6 = \cot(Az_{n-n+1}) - \cot(Az_{1-n+m}) \quad (22.31)$$

For point (B): The value of (X_B) is obtained by solving the following quadratic equation:

$$(K_3 * K_6) * X_B^2 + [K_5 * (K_3 + 1) + K_2 * K_6] * X_B + (K_4 + K_2 * K_5) = 0 \quad (22.32)$$

The value of (Y_B) is obtained using Eq. (22.33).

$$Y_B = Y_n + (X_B - X_n) * \cot(Az_{n-n+1}) \quad (22.33)$$

For point (F):
$$X_F = K_2 + K_3 * X_B \quad (22.34)$$

$$Y_F = Y_1 + (X_F - X_1) * \cot(Az_{1-n+m})$$

Note Due to the intervening cotangents, the equations will produce undefined values if the direction of the dividing line is at or near azimuth $0°$ or $180°$. In this case, the solution may be to rotate the figure to have a different azimuth for the dividing line and return to the starting position after calculating the coordinates of the intersection points.

Example 22.11
Given the same area as in Example 22.10, divide the polygon so that the areas (A_1) and (A_2) are equal and the dividing line of intersection between the two areas is in the azimuth direction $Az_{BF} = 260° \, 25'00''$. Calculate the coordinates of points (B) and (F) of the intersection of the dividing line with the respective sides of the polygon. Table 22.9 gives the coordinates of the vertices of the polygon, and Fig. 22.16 shows the geometric situation of its division.

Table 22.9 Coordinates of
the vertices of the polygon

Point	X [m]	Y [m]
1	8461.941	12,686.409
2 = n	9120.717	12,916.735
3 = n + 1	9713.294	12,556.294
4	9533.600	12,082.090
5 = n + m	8615.729	12,082.090

Fig. 22.16 Geometric
situation of the polygon
division

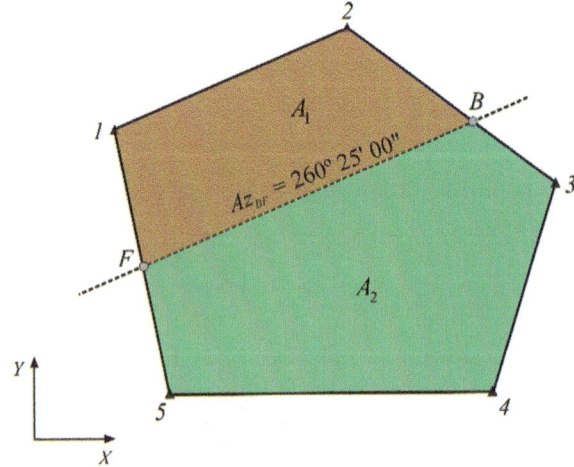

Solution:

The first step in solving this example is to assess which sides of the polygon will be intersected by the dividing line, which can be done visually. If the calculation results indicate that points (B) and (F) are not on the sides of the polygon, the process is repeated using the calculated position. As shown in Fig. 22.16, point (B) is on line 2–3 and point (F) is on line 1–5. Thus,

$$Az_{1-n+m} = 165°43'20.4'' \quad \text{(calculated in Example 22.10)}$$

$$Az_{n-n+1} = \text{atan}\left(\frac{9713.294 - 9120.717}{12,556.294 - 12,916.735}\right) + 180° = 121°18'37.7''$$

$$K_2 = -\left[\frac{12,916.735 - 12,686.409 + 8461.941 * \cot(165°43'20.4'') - 9120.717 * \cot(121°18'37.7'')}{\cot(260°25'00'') - \cot(165°43'20,4'')}\right]$$
$$= 6703.499$$

$$K_3 = \frac{\cot(260°25'00'') - \cot(121°18'37.7'')}{\cot(260°25'00'') - \cot(165°43'20.4'')} = 0.190$$

$$K_4 = \begin{bmatrix} 5,635,796.684 + 8461.941 * 12,686.409 - 9120.717 * 12,916.735 \\ - \cot(165°43'20.4'') * 8461.941^2 + \cot(121°18'37.7'') * 9120.717^2 \end{bmatrix}$$
$$= 225,951,809.044$$

$$K_5 = 12,916.735 - 12,686.409 + \cot(165°43'20.4'') * 8461.941$$
$$- \cot(121°18'37.7'') * 9120.717 = -27,473.600$$

$$K_6 = \cot(121°18'37.7'') - \cot(165°43'20.4'') = 3.321$$

Coordinates of the point (B):
Using Eqs. (22.32) and (22.33),

$$0,630 * X_B^2 - 10,418.556^* X_B + 41,782,557.455 = 0$$

$$X_B = 9713.295 \text{ m}$$

$$Y_B = 12,916.735 + (9713.294 - 9120.717) * \cot(121°18'37.7'') = 12,556.293 \text{ m}$$

Note that the calculated point (B) practically coincides with the vertex (3) of the polygon. This is a very particular situation. However, this geometric situation has been created so that the readers can check their calculations against the results of the previous example.

Coordinates of the point (F):
Using Eqs. (22.34),

$$X_F = 6703.498 + 0.190 * 9713.294 = 8545.240 \text{ m}$$

$$Y_F = 12,686.409 + (8545.240 - 8461.941) * \cot(165°43'20.4'') = 12,359.081 \text{ m}$$

22.7 Volume Calculation

Volume calculations are required for many Civil Engineering projects that involve earthworks, such as building sites, swage systems surveys and road design. In many cases, this involves a series of operations to move large quantities of earth (using heavy machinery) from the project site to a dump area (cut), from a borrow pit to the project site (fill), or to compensate between cuts (excavations[1]) and fills (embankments[2]) within the project site itself. In this context, four methods are commonly used in Civil Engineering to calculate earth volumes.

- *Volume calculation from cross-sectional areas:* used to calculate earth movements in linear projects such as roads, railways and canals.

---

[1] The amount of material that needs to be removed from the grade.
[2] The amount of material that needs to be added to the grade.

- *Volume calculation from grid spot elevations:* used to calculate earth movements in large open excavation areas such as borrow pits, building sites, tanks and others.
- *Volume calculation from contour lines:* useful for calculating volumes in preliminary projects.
- *Volume calculation using Digital Terrain Models:* used in all cases.

Details of each of these calculation methods are presented below.

22.7.1 Volume Calculation from Cross-Sectional Areas

For linear engineering projects, the earthworks calculation is based on the areas of cut-and-fill cross-sections required to match the natural terrain to the design surface. In this case, the cross-sections are generated from a topographic survey of the excavation area before and after excavation or from a design cross-section comparing the natural and the design surfaces. Volume calculations are then made from the cross-sectional areas taken at regular intervals (usually every 20 m) along the axis (centreline) of the proposed works, as shown in the cross-sections in cut and fill in Fig. 22.17.

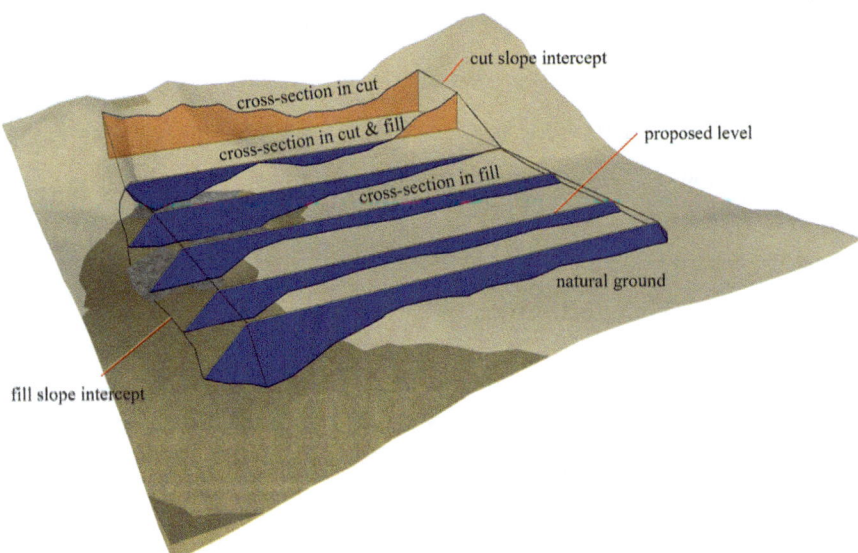

Fig. 22.17 Cut cross-sections, fill cross-sections and side-hill section

The shape of the cross-section varies according to the natural surface of the terrain and the position of the axis of the longitudinal profile of the structure relative to the ground surface. Profile levelling is the first step to be carried out, followed by cross-sectional levelling at regular intervals. Each section is a flat vertical surface approximated by a series of straight lines defining the cut cross-sections, the fill cross-sections, and the side-hill sections,[3] in the latter case both cut and fill, as shown in Fig. 22.17. The area of each section can be calculated using any of the area calculation methods presented in this chapter.

If the earthwork section is entirely cut or entirely embankment, the total area of the section is calculated and added to the volume. If the section is mixed, the cut and embankment areas must be calculated separately before volume calculation. Therefore, it is first necessary to determine the *zero-point* in each cross-section, as shown in Fig. 22.18. The location of the zero point can be calculated using Eqs. (22.35) and (22.36).

Fig. 22.18 Zero-point calculation

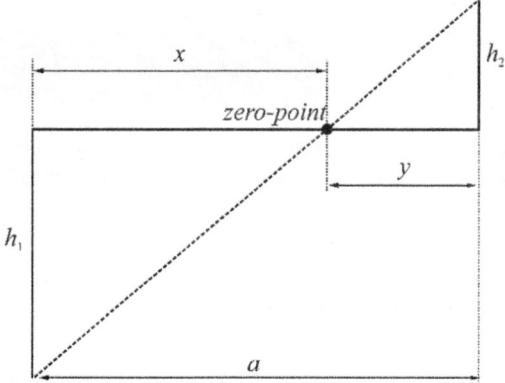

$$x = \frac{h_1}{(h_1 + h_2)} * a \qquad (22.35)$$

$$y = a - x \quad or \quad y = \frac{h_2}{(h_1 + h_2)} * a \qquad (22.36)$$

Once the cross-sectional area at different sections is calculated, the volumes between any number of cross-sections can be calculated using one of the following methods:

- Trapezoidal method
- Prismoidal method

---

[3] Some authors also refer to them as *mixed sections*.

22.7.1.1 Trapezoidal Method

As shown in Fig. 22.19, the trapezoidal method for calculating volumes consists of multiplying the distance (spacing) between two successive cross-sections by the average of their areas A_1 and A_2, as given by Eq. (22.37).

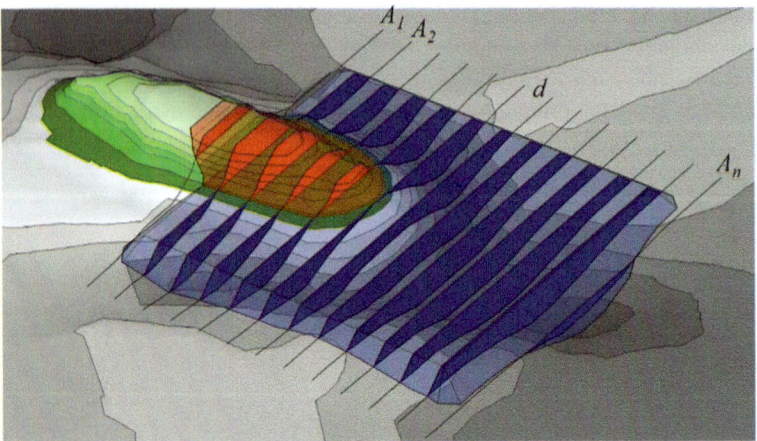

Fig. 22.19 Successive cross-sections. (Courtesy of Kubla Ltd. Location: https://www.kublasoftware.com/how-to-calculate-cut-and-fill/)

$$V = d * \left(\frac{A_1 + A_2}{2}\right) \qquad (22.37)$$

where

V = volume between the cross-sections of areas A_1 and A_2
d = distance between cross-sections

In the case of several cross-sections, the total volume is given by the sum of the partial volumes. Therefore, if the distance between successive sections is always equal to d, the total volume can be calculated using Eq. (22.38) or (22.39).

$$V = \frac{d}{2} * [(A_1 + A_2) + (A_2 + A_3) + (A_3 + A_4) + \ldots\ldots + (A_{n-1} + A_n)] \qquad (22.38)$$

Or,

$$V = d * \left[\frac{A_1 + A_n}{2} + A_2 + A_3 + \ldots\ldots + A_{n-1}\right] \qquad (22.39)$$

This volume calculation method can be applied to any number of equally spaced cross-sections and gives good results in situations where the cross-sectional areas are similar. The accuracy of the calculations depends on the sections' spacing and shape. It is therefore recommended that the distances do not exceed 20 m and that the areas are calculated as accurately as possible.

22.7.1.2 Prismoidal Method

Another way to calculate the volume of earthworks is to consider successive cross-sections, two by two, and calculate the volume of the prism they form by adding a cross-section A_m halfway between them with a weight equal to 4. Given that the distance between the two successive sections is equal to d, the prism's volume can be calculated using Eq. (22.40).

$$V = \frac{d}{6} * [A_1 + 4A_m + A_2] \tag{22.40}$$

The area of the intermediate cross-section A_m must not be taken as an average of the areas of the end sections; otherwise, there will be no difference between the results obtained by the trapezoidal and prismoidal methods. It should be determined from the average heights and widths of the end sections or measured in the field.

In general, the calculation is performed considering a sequence of three cross-sections. Thus, if Eq. (22.40) is developed for a sequence of three sections, e.g., with areas A_3, A_4 and A_5 equally spaced by a distance d, the volume between them can be calculated using Eq. (22.41).

$$V = \frac{d}{3} * [A_3 + 4A_4 + A_5] \tag{22.41}$$

Considering all the cross-sections, provided they add up to an odd number, the final volume can be calculated using Eq. (22.42).

$$V = \frac{d}{3} * (A_1 + 4A_2 + 2A_3 + 4A_4 + A_5) \tag{22.42}$$

Or generalising

$$V = \frac{d}{3} * \left(A_1 + A_n + 4 \sum \text{even areas} + 2 \sum \text{odd areas} \right) \tag{22.43}$$

This volume calculation method is considered more accurate than the previous one, particularly in cases where the cross-sections are perpendicular to the alignment axis and where the areas of successive cross-sections are sufficiently different.

22.7.1.3 Volume Calculation Between Transition Sections

In cases where the successive cross-sections are of the same type (cut only or fill only), the calculation of volumes is simplified, and both methods presented above give good results. However, as shown in Fig. 22.20, if the sections are different (fill or cut), it is necessary to calculate the position of the intersection line (transition line) at the "zero position" of the surface.

Fig. 22.20 Transition sections and intersection line position

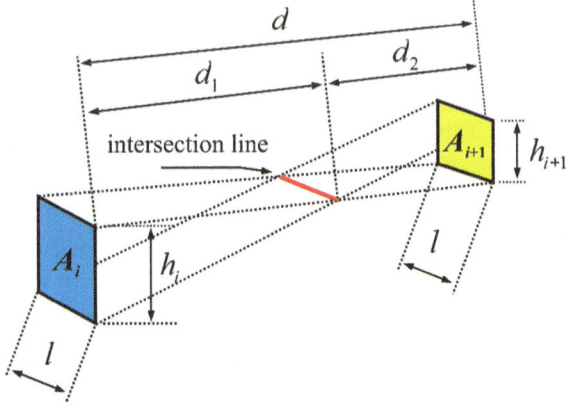

Considering the sequence of fill and cut shown in Fig. 22.20 and that the areas of the sections A_i and A_{i+1} and the distance d between them are known, the volumes can be calculated using the following equations.

$$d_1 = d * \frac{A_i}{A_i + A_{i+1}} \tag{22.44}$$

$$d_2 = d * \frac{A_{i+1}}{A_i + A_{i+1}} \tag{22.45}$$

$$V_{fill} = d_1 * \frac{A_i}{2} \tag{22.46}$$

$$V_{cut} = d_2 * \frac{A_{i+1}}{2} \tag{22.47}$$

If the sections themselves are mixed (cut and fill), the areas of each cross-section must be calculated separately, and the volume between them must be calculated considering each intersection line.

In some cases, volume calculations are performed without considering the positions of the intersection line, i.e., they are calculated by directly adding the cut or fill areas and multiplying by the distances between the cross-sections.

To calculate the volumes of successive sections, the reader can use spreadsheets such as those shown in Tables 22.10 and 22.11. In Table 22.10, the cut volume and fill volume columns give the geometric volumes of each section. The last column shows the cumulative volumes from the initial section used for longitudinal compensation. This means that only the excess volume of a section (the volume not used for transversal compensation) is cumulated from the initial section with its respective algebraic sign. Therefore, the last column of the table indicates how much volume is left over (if positive) or missing (if negative) from the initial section to the section of interest. Thus, if the distances to move the material from the excavations to the embankments do not exceed the limit of economical haul (*freehaul distance*[4] plus the maximum *overhaul*[5]), the cumulative volumes of the last section will indicate how much material is left or missing in the entire project. On the other hand, wasting and borrowing material would be more economical if this limit is reached.

In road and railway construction, a well-designed grade line should seek to balance the total volumes of cut and fill. In addition, the engineer must never forget that the excavation of materials increases their volume due to the destruction of their natural structure and the increase in their void ratio. The swelling factor depends on the type of material being excavated and can never be accurately estimated. On the other hand, when the embankment is compacted, a given excavated material reduces its natural volume. Therefore, fill volumes must be increased or cut volumes reduced to achieve a balance. For example, Table 22.10 lists expanded fills using a 20% factor.

Another way of obtaining the final column of cumulative volumes is to accumulate the total cut volume and the volume required for embankments, as in Table 22.11, instead of deducting the transversal compensation and accumulating the surplus volume. In this case, the last column of cumulative volumes is obtained by the direct algebraic sum of the two previous volumes.

---

[4]The distance contractors can haul a cubic unit of excavated material and place it in fill without extra charge above the cost for excavation.
[5]The haul beyond the freehaul distance.

Table 22.10 Spreadsheet 1 for earthworks calculation

Section	Area [m²]		Volume [m³]		Expanded fills	Transversal compensation [m³]	Longitudinal compensation [m³]	Cumulative volume [m³]
	Cut	Fill	Cut	Fill				
0	1.75	0.00	0	0	0	0	0	0
1	19.93	0.00	216.80	0	0	0	216.80	216.80
2	32.01	0.00	519.40	0	0	0	519.40	736.20
3	64.09	0.00	961.00	0	0	0	961.00	1697.20
4	90.77	0.00	1548.60	0	0	0	1548.60	3245.80
5	70.27	0.00	1610.40	0	0	0	1610.40	4856.20
6	26.66	0.00	969.30	0	0	0	969.30	5825.50
7	3.88	24.69	305.40	246.90	296.28	296.28	9.12	5834.62
8	0.00	64.14	38.80	888.30	1065.96	38.80	−1027.16	4807.46
9	0.00	136.50	0	2006.40	2407.68	0	−2407.68	2399.78
10	0.00	124.22	0	2607.20	3128.64	0	−3128.64	−728.86
11	0.00	44.87	0	1690.90	2029.08	0	−2029.08	−2757.94
12	37.44	10.37	374.40	552.40	662.88	374.40	−288.48	−3046.42
13	101.93	0.00	1393.70	103.70	124.44	124.44	1269.26	−1777.16
14	90.36	0.00	1922.90	0	0	0	1922.90	145.74
15	54.37	0.00	1447.30	0	0	0	1447.30	1593.04
16	1.75	0.00	561.20	0	0	0	561.20	2154.24

Table 22.11 Spreadsheet 2 for earthworks calculation

Section	Area [m²] Cut	Fill	Volume [m³] Cut	Fill	Cumulative cut volume [m³]	Cumulative expanded fill volume [m³]	Cumulative total volume [m³]
0	1.75	0.00	0	0	0	0	0
1	19.93	0.00	216.80	0	216.80	0	216.80
2	32.01	0.00	519.40	0	736.20	0	736.20
3	64.09	0.00	961.00	0	1697.20	0	1697.20
4	90.77	0.00	1548.60	0	3245.80	0	3245.80
5	70.27	0.00	1610.40	0	4856.20	0	4856.20
6	26.66	0.00	969.30	0	5825.50	0	5825.50
7	3.88	24.69	305.40	246.90	6130.90	296.28	5834.62
8	0.00	64.14	38.80	888.30	6169.70	1362.24	4807.46
9	0.00	136.50	0	2006.40	6169.70	3769.92	2399.78
10	0.00	124.22	0	2607.20	6169.70	6898.56	−728.86
11	0.00	44.87	0	1690.90	6169.70	8927.64	−2757.94
12	37.44	10.37	374.40	552.40	6544.10	9590.52	−3046.94
13	101.93	0.00	1393.70	103.70	7937.80	9714.96	−1777.16
14	90.36	0.00	1922.90	0	9860.70	9714.96	145.74
15	54.37	0.00	1447.30	0	11,308.00	9714.96	1593.04
16	1.75	0.00	561.20	0	11,869.20	9714.96	2154.24

Example 22.12

Calculate the volume of earthworks between stakes 92 to 97 of a road design, whose cross-sections are shown in Figs. 22.21, 22.22, 22.23, 22.24, 22.25 and 22.26, using the two calculation methods presented in the previous section. The stakes are spaced at regular intervals of 20 m. Consider a slope of 2:1 for the cut cross-sections (stakes 92 to 95) and a slope of 3:1 for the fill cross-sections (stakes 96 and 97). All values are given in metres. Present the volume calculation results using a spreadsheet, as shown in Table 22.11.

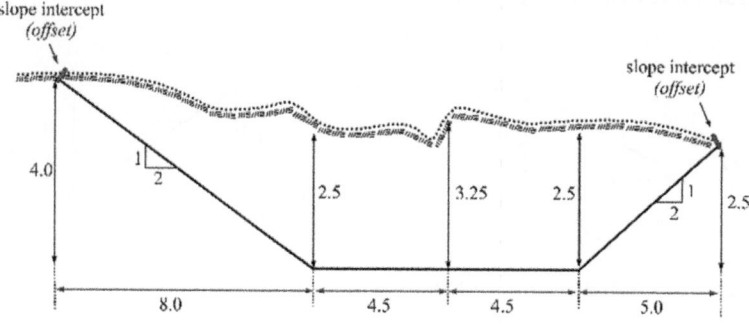

Fig. 22.21 Cross-section of stake 92 (cut)

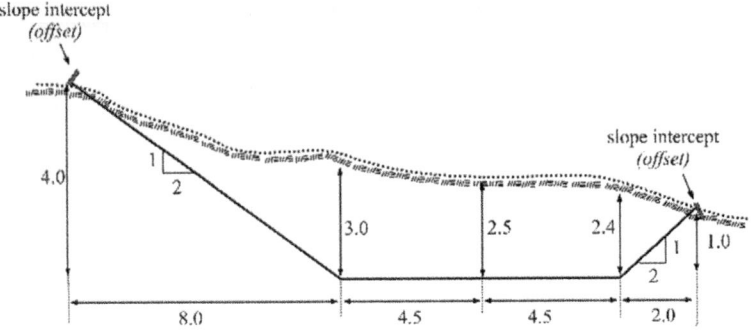

Fig. 22.22 Cross-section of stake 93 (cut)

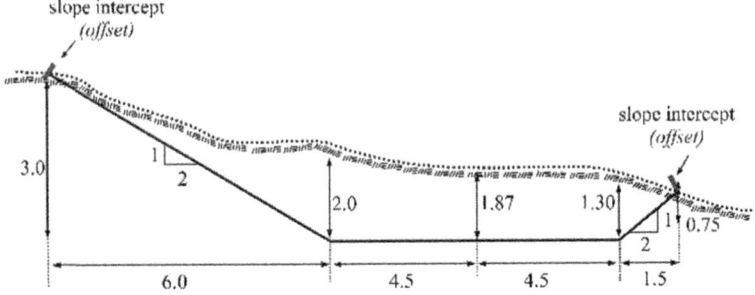

Fig. 22.23 Cross-section of stake 94 (cut)

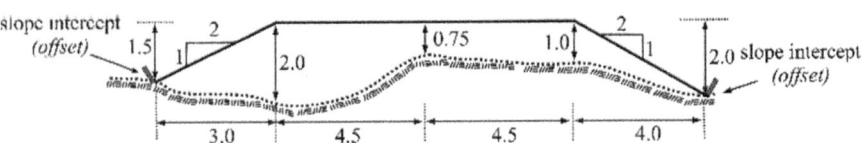

Fig. 22.24 Cross-section of stake 95 (fill)

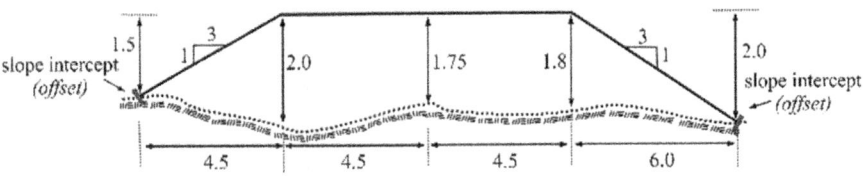

Fig. 22.25 Cross-section of stake 96 (fill)

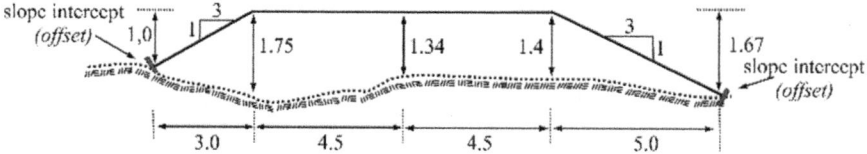

Fig. 22.26   Cross-section of stake 97 (fill)

Solution:

Cross-sectional area of each stake:

$$Stake~92 : A_{cut-92} = \left[(2.5+3.25)*\frac{4.5}{2}+(3.25+2.5)*\frac{4.5}{2}\right]$$
$$+\left(\frac{8*4.0}{2}-\frac{8*1.5}{2}\right)+\left(\frac{5*2.5}{2}\right)=42.13~\text{m}^2$$

$$Stake~93 : A_{cut-93} = \left[(3+2.5)*\frac{4.5}{2}+(2.5+2.4)*\frac{4.5}{2}\right]+\left(\frac{8*4.0}{2}-\frac{8*1.0}{2}\right)$$
$$+\left(\frac{2*2.4}{2}-\frac{2*1.4}{2}\right)=36.40~\text{m}^2$$

$$Stake~94 : A_{cut-94} = \left[(2+1.87)*\frac{4.5}{2}+(1.87+1.5)*\frac{4.5}{2}\right]$$
$$+\left(\frac{6*3.0}{2}-\frac{6*1.0}{2}\right)+\left(\frac{1.5*1.3}{2}-\frac{1.5*0.55}{2}\right)=22.40~\text{m}^2$$

$$Stake~95 : A_{fill-95} = \left[(2+0.75)*\frac{4.5}{2}+(0.75+1.0)*\frac{4.5}{2}\right]+\left(\frac{3*2}{2}-\frac{3*0.5}{2}\right)$$
$$+\left(\frac{4*2}{2}-\frac{4*1}{2}\right)=14.38~\text{m}^2$$

$$Stake~96 : A_{fill-96} = \left[(2+1.75)*\frac{4.5}{2}+(1.75+1.8)*\frac{4.5}{2}\right]$$
$$+\left(\frac{4.5*2}{2}-\frac{4.5*0.5}{2}\right)+\left(\frac{6*2.0}{2}-\frac{6*0.2}{2}\right)=25.20~\text{m}^2$$

$$Stake~97 : A_{fill-97} = \left[(1.75+1.34)*\frac{4.5}{2}+(1.34+1.4)*\frac{4.5}{2}\right]$$
$$+\left(\frac{3*1.75}{2}-\frac{3*0.75}{2}\right)+\left(\frac{5*1.67}{2}-\frac{5*0.27}{2}\right)=18.12~\text{m}^2$$

Volumes by the trapezoidal method:
Using Eq. (22.38),

$$V_{cut} = \frac{20}{2} * [42.13 + 22.40 + 2 * (36.40)] = 1373.28 \text{ m}^3$$

$$V_{fill} = \frac{20}{2} * [14.38 + 18.12 + 2 * (25.20)] = 828.93 \text{ m}^3$$

The cut-and-fill volumes between the stakes 94 and 95 can be calculated using Eqs. (22.44) to (22.47).

$$d_1 = 20 * \left(\frac{22.40}{14.38 + 22.40}\right) = 12.18 \text{ m} \qquad d_2 = 20 * \left(\frac{14.38}{14.38 + 22.40}\right) = 7.82 \text{ m}$$

$$Vcut = 12.18 * \left(\frac{22.40}{2}\right) = 136.46 \text{ m}^3 \qquad V_{fill} = 7.82 * \left(\frac{14.38}{2}\right) = 56.19 \text{ m}^3$$

Finally, the total volume of earthworks is as follows:

$$V_T = 1373.28 - 828.93 + 136.46 - 56.19 = 624.63 \text{ m}^3$$

Volumes by the prismoidal method:

To calculate the volumes using the prismoidal method, it is suggested to organise the data according to Table 22.12. Consider a 20% factor for the expansion of fill volumes.

Table 22.12 Areas and volumes between stakes 92 to 97

Stakes	Area [m²] Cut	Fill	Volume [m³] Cut	Fill	Cumulative cut volume [m³]	Cumulative expanded Fill volume [m³]	Cumulative total volume [m³]
92	42.13						
$A_{M(92\text{-}93)}$	38.70		777.75		777.75	0	777.75
93	36.40						
$A_{M(93\text{-}94)}$	29.14		548.54		1362.29	0	1362.29
94	22.40						
			136.46	56.19	1498.75	67.43	1431.32
95		14.38					
$A_{M(95\text{-}96)}$		19.59		393.12	1498.75	539.17	959.58
96		25.20					
$A_{M(96\text{-}97)}$		20.41		416.53	1498.75	1039.00	459.75[a]
97		18.12					

[a]This result refers to the total volume of earthworks considering a 20% expansion factor for the fill volumes. If the reader wants to compare the results of using the trapezoidal and prismoidal methods, i.e., not considering the 20% expansion factor for the fill volumes in the latter case, the total volume of the earthworks would be 632.92 m³

In road design, a volume balance function based on the mass diagram or Brückner curve theory is used to assess the earthworks along the road. The details of this diagram are beyond the scope of this book. Interested readers are advised to consult specialised literature on road design.

22.7.1.4   Volume Calculation for Curved Sections

The equations presented in the previous section assume that the section for which the volume is being calculated is part of the tangent, i.e., a straight section of the line. In the case of a curved road section, however, the cross-sections are not parallel to each other. Therefore, the volume calculation must consider the radius of the curve.

In this case, the solution is to consider that the volume of a constant cross-section moving along a curved axis is equal to the cross-sectional area multiplied by the distance travelled along the axis of the centre of gravity of the cross-section. Therefore, the axis of the centre of gravity of the cross-section is used instead of the alignment axis, as shown in Fig. 22.27.

Considering an eccentricity c of the centre of gravity axis of the cross-section from the alignment axis, the geometric condition shown in Fig. 22.28 is obtained, for which the volume calculation is given by Eq. (22.48).

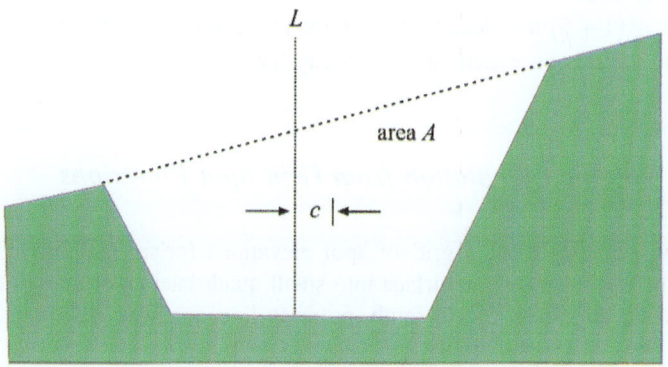

Fig. 22.27   Eccentricity (c) of the centre of gravity axis of the cross-section

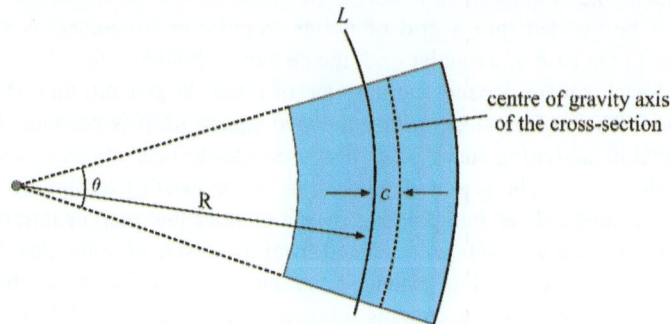

Fig. 22.28   Geometry of cross-section eccentricity in a curved section

$$V = A * \theta * (R + c) \tag{22.48}$$

where

V = volume of the cross-section along the curved road section
A = cross-sectional area
R = radius of the curve
c = eccentricity of the centre of gravity axis of the cross-section
θ = angle at the centre of the curve (in radians)

Considering L as the length of the curve,

$$\theta = \frac{L}{R} \tag{22.49}$$

Substituting Eq. (22.49) into (22.48),

$$V = \frac{L * A * (R + c)}{R} = L * \left(A + \frac{A * c}{R}\right) = L * A * \left(1 + \frac{c}{R}\right) \tag{22.50}$$

From Eq. (22.50), it is sufficient to correct the cross-sectional area by the coefficient $\pm \left(1 + \frac{c}{R}\right)$ to calculate the volume in a curved section, where the sign $\pm$ depends on the side on which the eccentricity lies.

22.7.2   Volume Calculation from Grid Spot Elevations

Calculating volumes from a grid of spot elevations (prism method) consists of dividing the entire earthwork surface into small quadrilaterals or triangles oriented to a particular alignment (north-south, property line, roadway alignment, etc.), as shown in Figs. 22.29 and 22.30.

In Fig. 22.29, a horizontal earthwork platform has been created, resulting in a cut-and-fill area. In the cut area, the base (platform) intersection with the ground creates a cut slope. In the fill area, the intersection creates a fill slope.

To calculate the volume of earthworks, the entire surface of the platform and its slopes must be divided into a grid of points (regular or irregular), as shown in Fig. 22.30a. In the case of a regular grid, the elements should preferably be squares. The size of each square depends on the type of relief. In general, they should not exceed 20 m. The surface slope within the grid square itself is accounted for and approximated by assigning surveyed or proposed (design) elevations to each of the corners of the square. The grid node elevations of the existing terrain are obtained using levelling methods or interpolating over a contour line map or terrain model. The "*proposed ground*" levels are obtained from the designed point elevations. To calculate volumes, each cell is extended vertically until it intersects the natural terrain or the designed surface, creating a prism as shown in Fig. 22.30b. Measurements can then be taken to determine the depth of cut or fill at each corner of the cell.

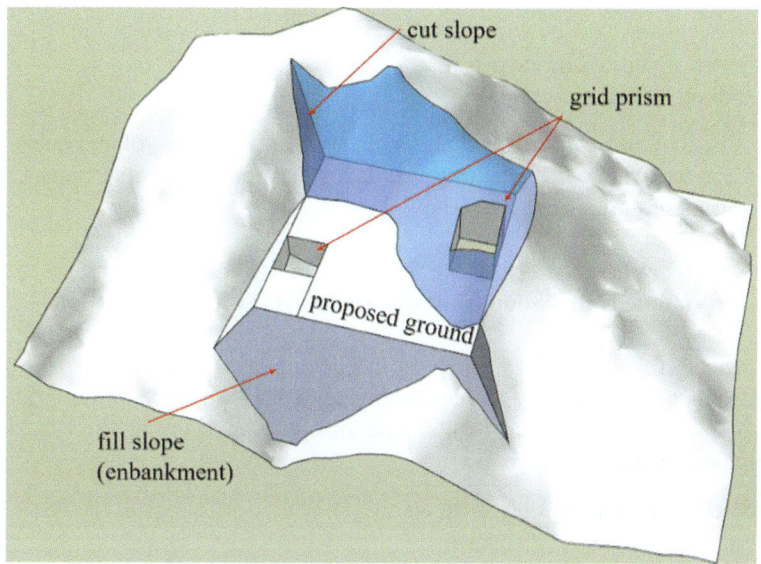

Fig. 22.29   Illustration of cut-and-fill areas on an earthwork

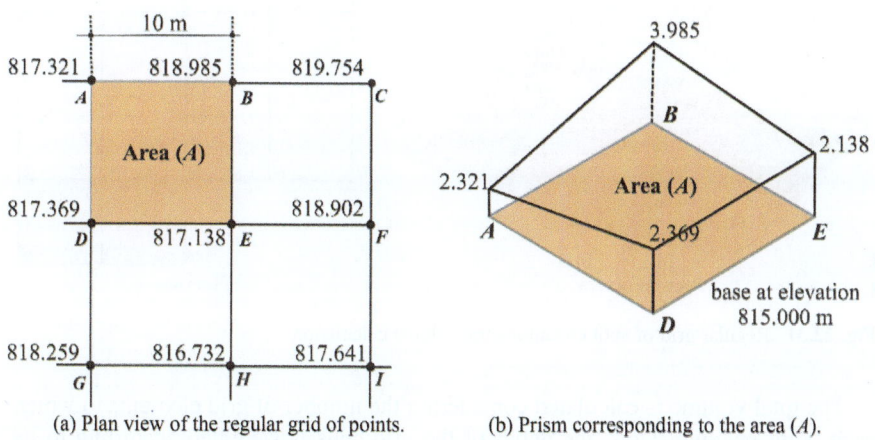

(a) Plan view of the regular grid of points.   (b) Prism corresponding to the area (A).

Fig. 22.30   Volume of a quadrangular prism

Once the depths of the square grid cells have been determined, they are averaged by adding them and dividing by four. This gives the average depth for the grid cell, which can then be multiplied by the area A of the square to give the volume V of the prism, as given by Eq. (22.51).

$$V = \frac{\sum h_i}{n} * A \qquad (22.51)$$

where

n = number of vertices of the prism
h_i = depth of cut/fill at each grid vertex (i)

The distribution of points over the surface determines the volume calculation method for the whole surface. The calculation methods for regular squares and irregular triangles grids are given below.

22.7.2.1   Volume Calculation from a Regular Grid of Spot Elevations

In the case of a regular grid of squares, the calculation of the total volume depends on the grid layout. There may be vertices belonging to one, two, three or four grid cells. For example, considering the regular grid of spot elevations in Fig. 22.31, the distribution of vertices is as follows:

- 5 vertices belonging to only 1 square
- 8 vertices belonging to 2 squares
- 1 vertex belonging to 3 squares
- 4 vertices belonging to 4 squares

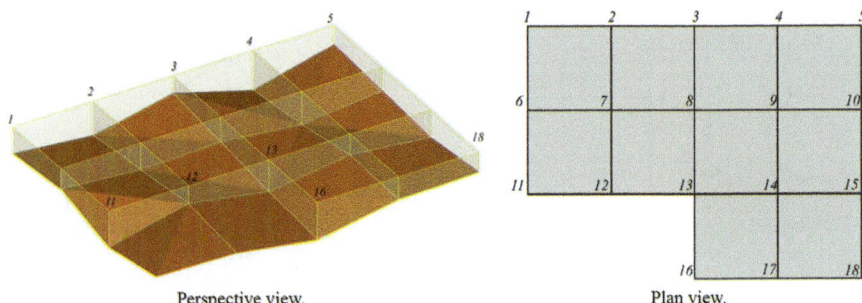

Perspective view.   Plan view.

Fig. 22.31 Regular grid of spot elevations for volume calculation

The total volume is calculated considering the number of grid elements to which each point belongs. Thus, the depth of the grid cells is weighted according to its position within the whole grid, as given by Eq. (22.52).

$$V_T = \frac{A_T}{n} \left(\sum h_1 + 2 \sum h_2 + 3 \sum h_3 + 4 \sum h_4 \right) \qquad (22.52)$$

where

A_T = total grid area
h_i = depth of point (i)
n = sum of the number of times the vertex is repeated, considering its respective weight

$i = 1$, for vertices belonging to only 1 square
$i = 2$, for vertices belonging to two squares and so on....

The most obvious factor affecting the accuracy of the grid method is the size of the grid squares used. The smaller the squares, the greater the accuracy of the method. However, when measuring the elevations of points in the field, a balance must be achieved between accuracy and field work.

The volume can also be calculated by drawing the cross-sections of the regular grid of spot elevations and using one of the volume calculation methods given in Sect. 22.7.1. This is the recommended method when earthworks involve cut and fill areas.

Another common situation in a regular grid of spot elevations is the occurrence of irregular areas at the edges of the grid, as shown in Fig. 22.32. In this case, the solution is to calculate the volumes of each irregular area separately or to use Eq. (22.52), assigning individual weights to the vertices of the irregular boundaries according to the size of the contributing area. In the case of cut-and-fill volumes, it is recommended to calculate them using the cross-sectional area method in the most appropriate direction to include irregular areas, as shown in Fig. 22.32.

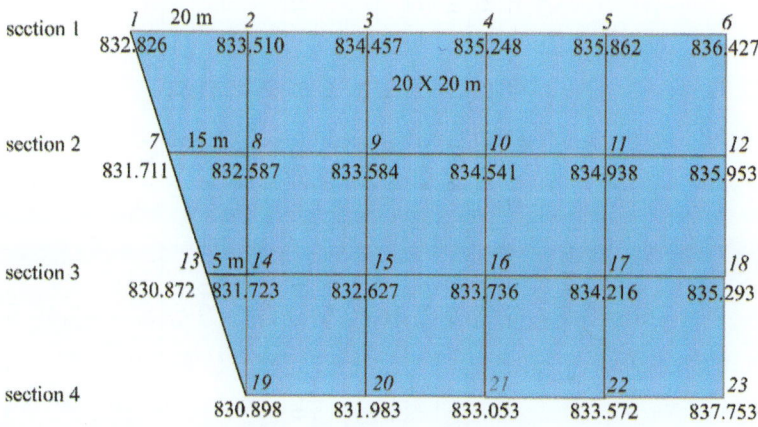

Fig. 22.32 Example of a regular grid of points with irregular boundaries

Example 22.13
Using the grid of spot elevations shown in Fig. 22.32, calculate the cut volume using Eq. (22.52) so that the final platform has an elevation of 830.000 m. Also, calculate the cut volume for the same grid layout if the final platform should have an elevation of 834.000 m. In the latter case, use the cross-sectional area method,

Solution:
  Volume calculation using grid cells.
  To calculate the cut volume so that the platform has an elevation of 830.000 m, it is first necessary to assign weights (according to the contributing area) and calculate the depth of each grid vertex, as shown in Table 22.13.

Table 22.13 Weights and depth of the grid vertices

Vertex	Contributing area in the irregular grid [m²]	Weight (1)	Vertex depth (h_i) [m] (2)	(1) * (2)	Vertex	Contributing area in the irregular grid [m²]	Weight (3)	Vertex depth (h_i) [m] (4)	(3) * (4)
1	350	0.875	2.826	2.473	13	250	0.5 + 0.125	0.872	0.545
2	750	1 + 0.875	3.510	6.581	14	1050	2 + 0.5 + 0.125	1.723	4.523
3		2	4.457	8.914	15		4	2.627	10.508
4		2	5.248	10.496	16		4	3.736	14.944
5		2	5.862	11.724	17		4	4.216	16.864
6		1	6.427	6.427	18		2	5.293	10.586
7	550	0.875 + 0.5	1.711	2.353	19	450	1 + 0.125	0.898	1.010
8	1350	2 + 0.875 + 0.5	2.587	8.731	20		2	1.983	3.966
9		4	3.584	14.336	21		2	3.053	6.106
10		4	4.541	18.164	22		2	3.572	7.144
11		4	4.938	19.752	23		1	7.753	7.753
12		2	5.953	11.906					

section 1

$$x = \frac{20*0.490}{0.947} = 10.348\text{m} \quad \rightarrow \quad y = 9.652\text{m}$$

$$A_{cut} = \left(\frac{0.457 * 9.652}{2}\right) + \frac{20}{2} * \left[\frac{0.457 + 2.427 +}{2(1.248 + 1.862)}\right]$$

$$= 93.245\,\text{m}^2$$

$$A_{fill} = \frac{20}{2} * [1.174 + 0.490] + \left(\frac{0.490 * 10.348}{2}\right)$$

$$= 19.175\,\text{m}^2$$

section 2

$$x = \frac{20*0.416}{0.957} = 8.694\text{m} \quad \rightarrow \quad y = 11.306\text{m}$$

$$A_{cut} = \left(\frac{11.306 * 0.541}{2}\right) + \frac{20}{2} * \left(\begin{array}{c}0.541 + 1.953 \\ +2 * 0.938\end{array}\right)$$

$$= 46.758\,\text{m}^2$$

$$A_{fill} = \frac{15}{2} * (2.289 + 1.413) + \frac{20}{2} * (1.413 + 0.416) +$$

$$\left(\frac{0.416 * 8.694}{2}\right)$$

$$= 47.863\,\text{m}^2$$

(continued)

(continued)

section 3

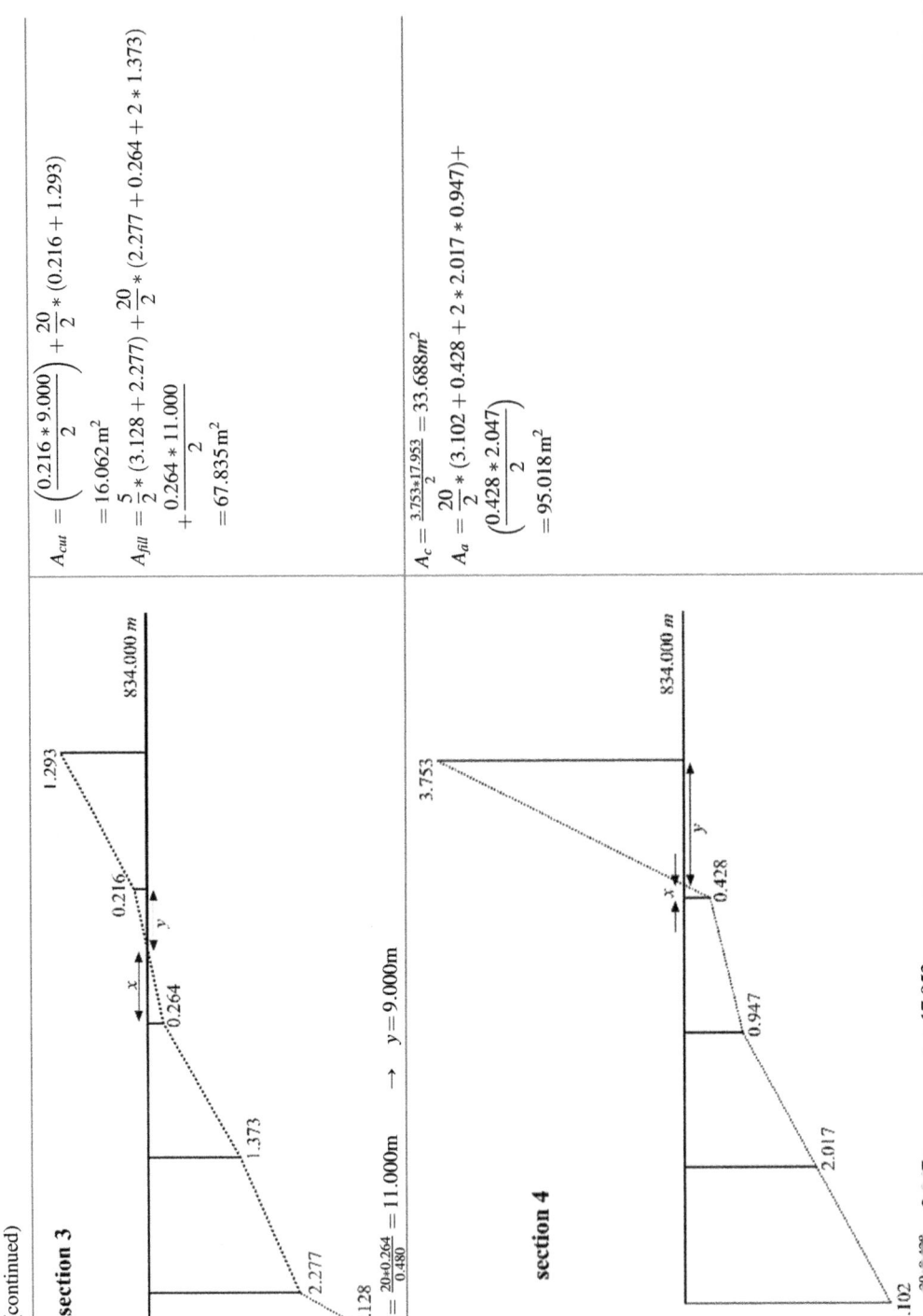

$$A_{cut} = \left(\frac{0.216 * 9.000}{2}\right) + \frac{20}{2} * (0.216 + 1.293)$$

$$= 16.062\,\text{m}^2$$

$$A_{fill} = \frac{5}{2} * (3.128 + 2.277) + \frac{20}{2} * (2.277 + 0.264 + 2 * 1.373)$$

$$+ \frac{0.264 * 11.000}{2}$$

$$= 67.835\,\text{m}^2$$

$$x = \frac{20*0.264}{0.480} = 11.000\text{m} \quad \rightarrow \quad y = 9.000\text{m}$$

section 4

$$A_c = \frac{3.753*17.953}{2} = 33.688\,\text{m}^2$$

$$A_a = \frac{20}{2} * (3.102 + 0.428 + 2 * 2.017 * 0.947) +$$

$$\left(\frac{0.428 * 2.047}{2}\right)$$

$$= 95.018\,\text{m}^2$$

The cut volume can be calculated using Eq. (22.52).

$$A_T = (12^*400) + \frac{20}{2}(20 + 15) + \frac{20}{2}(15 + 5) + \frac{1}{2}(5^*20) = 5400\,\text{m}^2$$

$$n = 53.875$$

$$V_T = \frac{5,400}{53.875}$$

$$* \begin{pmatrix} 2.473 + 6.581 + 8.914 + 10.496 + 11.724 + 6.427 + 2.353 + 8.731 + \\ 14.336 + 18.164 + 19.752 + 11.906 + 0.545 + 4.523 + 10.508 + 14.944 + \\ 16.864 + 10.586 + 1.010 + 3.966 + 6.106 + 7.144 + 7.753 \end{pmatrix}$$

$$= 20,628.338\,\text{m}^3$$

Volume calculation using cross-sections.

Using the cross-sectional area method, the cut-and-fill volumes for a platform at an elevation of 834 metres are given as follows. Since there will be cut-and-fill areas, it is necessary to determine the zero point in each cross-section using Eqs. (22.35) and (22.36).

Finally, using the trapezoidal method, the following cut and fill volumes are obtained:

$$V_{cut} = \frac{20}{2} * [93.245 + 33.688 + 2(46.758 + 16.062)] = 2525.741\,\text{m}^3$$

$$V_{fill} = \frac{20}{2} * [19.175 + 95.018 + 2(47.863 + 67.835)] = 3455.891\,\text{m}^3$$

22.7.2.2   Volume Calculation from an Irregular Grid of Points

The principle of calculating the volume of prisms based on an irregular grid of points is similar to that of a regular grid. The difference is that the irregular grid is formed by a network of triangles, as shown in Fig. 22.33. This is by far the most technically difficult method but also the most accurate.

This method starts by triangulating the existing points with known coordinates (X_i, Y_i, H_i) in the terrain using the Triangulated Irregular Network (TIN), as described in Sect. 15.3.2. This step is repeated for the design surface.

The next step is to merge these two triangulations to create a third triangulation containing all the edges from the original triangulations, as shown in Fig. 22.33. Merging the two input triangulations means that every detail of both the existing and the proposed triangulation is included in the calculations. This is the basis of the accuracy of this method.

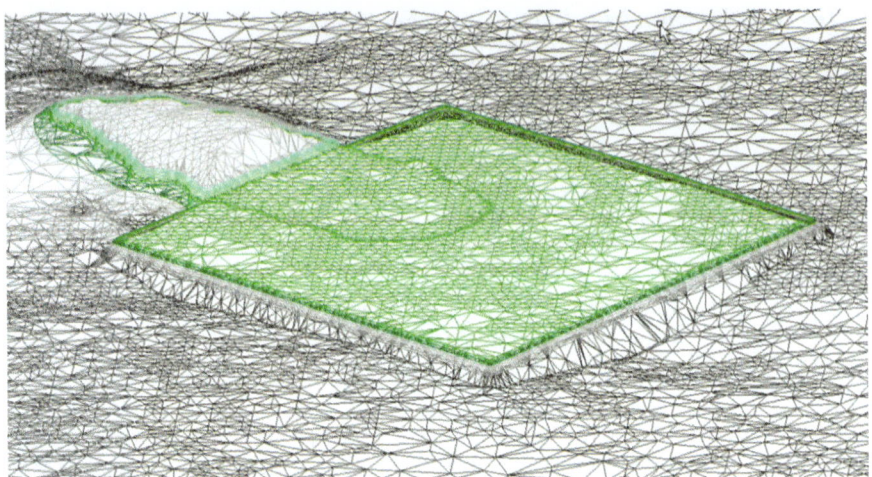

Fig. 22.33 Principle of volume calculation using a triangular grid. (Courtesy of Kubla Ltd. Location: https://www.kublasoftware.com/how-to-calculate-cut-and-fill/)

The final step is to calculate the cut and fill of each vertex on the calculation TIN. These values can be used to calculate the cut and fill for each triangle, as shown in Fig. 22.34, and the total volumes are obtained by adding all the triangles together.

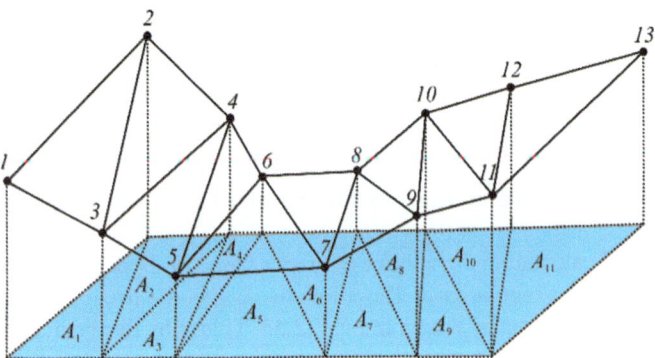

Fig. 22.34 Irregular grid of triangles

For the triangle formed by the vertices (1-2-3), for example, the volume V_1 of the triangular prism is given by Eq. (22.53).

$$V_1 = \frac{A_1}{3}(h_1 + h_2 + h_3) \tag{22.53}$$

$$A_1 = abs \left[det \begin{vmatrix} X_1 & X_2 & X_3 \\ Y_1 & Y_2 & Y_3 \\ 1 & 1 & 1 \end{vmatrix} \right] \qquad (22.54)$$

where (X, Y) are the planimetric coordinates of the vertices of the triangle formed by the vertices (1-2-3) and A_1 is the area of the base of this triangle.

The total volume within the grid area is given by Eq. (22.55).

$$V_T = \frac{\sum (hi_{1,2,3} * Ai)}{3} \qquad (22.55)$$

where

$hi_{1, 2, 3}$ = depth of the vertices of the triangle (i) referring to the calculation platform
A_i = horizontal area of the triangle (i).

Due to the complexity of these calculations and the thousands of triangles that can be generated depending on the number of points available, it is impractical to calculate triangular prisms manually. Instead, these calculations are carried out using specialised software. It should be noted, however, that not all earthworks' software uses this method; some software calculations are based on automated high-density square grid calculations, or the cross-section method used with TIN.

Example 22.14
A levelling work was carried out for the irregular grid of points shown in Fig. 22.35. Table 22.14 gives the values of the known coordinates of the vertices. Calculate the cut-and-fill volumes if a platform is to be constructed at an elevation of 815.000 m.

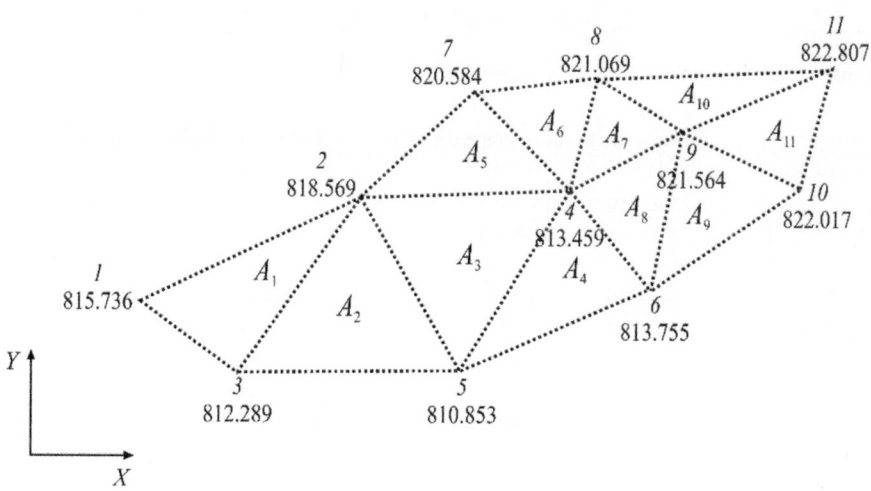

Fig. 22.35 Illustration of the irregular grid of points

Table 22.14 Coordinates of the vertices

Vertex	X [m]	Y [m]	H [m]
1	5236.258	10,480.369	815.736
2	5245.159	10,495.148	818.569
3	5248.456	10,465.965	812.289
4	5265.152	10,490.753	813.459
5	5268.481	10,463.486	810.853
6	5285.342	10,478.147	813.755
7	5255.257	10,510.981	820.584
8	5270.854	10,513.108	821.069
9	5285.357	10,502.015	821.564
10	5303.875	10,495.096	822.017
11	5295.753	10,510.481	822.807

Solution:

To solve this example, it is recommended to prepare a calculation table as shown in Table 22.15 and use Eqs. (22.53) to (22.55).

The volume calculation for the triangle A_1 is shown below.

$$A_1 = \det \begin{vmatrix} 5236.258 & 5245.159 & 5258.456 \\ 10,480.369 & 10,495.148 & 10,465.965 \\ 1 & 1 & 1 \end{vmatrix}$$

$$= 308.484\,\text{m}^2$$

$$V_1 = \frac{308.484 * (0.736 + 3.569 - 2.711)}{3} = 163.908\,\text{m}^3$$

Table 22.15 Calculation of areas and volumes

Triangle	Area [m²]	Average depth [m]	Cut volume [m³]	Fill volume [m³]
A_1 (1 - 3 - 2)	308.484	0.531	163.908	
A_2 (2 - 3 - 5)	576.216	-1.096		631.725
A_3 (2 - 5 - 4)	530.518	-0.706		374.723
A_4 (4 - 5 - 6)	508.555	-2.311		1175.271
A_5 (7 - 2 - 4)	360.930	2.537	915.799	
A_6 (7 - 4 - 8)	336.543	3.371	1134.374	
A_7 (8 - 4 - 9)	387.467	3.697	1432.594	
A_8 (4 - 6 - 9)	482.084	1.259	607.104	
A_9 (9 - 6 - 10)	442.091	4.112	1817.880	
A_{10} (8 - 9 - 11)	238.105	6.813	1622.290	
A_{11} (9 - 10 - 11)	228.703	7.129	1630.502	
		Total	9324.452	2181.719

22.7.3   Volume Calculation Based on Contour Lines

In some specific cases where an approximate volume calculation is acceptable, e.g., for labour and material costs in preliminary projects or to determine the capacity of reservoirs, the volume calculation can be done using a contour map, as shown in Fig. 22.36. This method is simpler to calculate volumes than the precedents, as no additional drawings and sections are required.

This calculation method is similar to the case of calculating volumes from cross-sectional areas, where the areas between each contour line are taken as a horizontal cross-section, and the distance d is replaced by the vertical equidistance h between the contour lines. Since contours do not predict the full irregularities of the ground and the contour intervals are not small, the volume calculated from contours is likely to be an estimate.

Fig. 22.36 Principle of volume calculation using contour lines

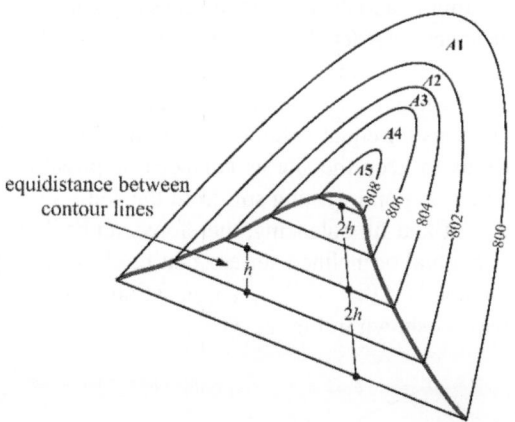

To calculate volume using this method, general recommendations for the contour interval are a maximum of 2 m for a regular ground surface and 0.5 m for irregular topography. It should also be noted that, in many cases, a residual solid above or below the last contour line cannot be included in the cross-sectional calculation. It is, therefore, suggested that it be calculated separately using the most appropriate volume calculation method for its geometric shape.

The area of the surface enclosed by each contour line can be calculated using a planimeter if the contour map is drawn on paper or using a software application if it is digital. For more information on calculating volume using this method, see Example 22.15.

Example 22.15
Using Fig. 22.36 and the data below, calculate the earthwork volume for an elevation of 802 m.

$A_1 = 4280\,\text{m}^2$ $A_2 = 3250\,\text{m}^2$ $A_3 = 2180\,\text{m}^2$ $A_4 = 1940\,\text{m}^2$ $A_5 = 930\,\text{m}^2$.

Solution:

In Fig. 22.36, *it can be seen that the vertical equidistance of the contour lines is*
2 m. *Thus, using Eq. (22.38).*

$$V = \frac{2}{2} * [3250 + 930 + 2 * (2180 + 1940)] = 12,420 \text{ m}^3$$

The volume above the 808.000 m elevation was disregarded.

22.7.4   Cut-and-Fill Balancing

The ultimate goal of cut and fill is to conserve energy and maximise the use of existing materials to avoid unnecessary earthwork. To avoid such problems, project designers use detailed and intelligent cut-and-fill maps, which provide comprehensive plans to guide excavation teams to the most efficient use of mass and labour. However, the best solution is to have equal cut-and-fill volumes. A site design that provides for equal volumes of cut and fill is referred to as 'balanced'. It is advantageous to keep the site as balanced as possible, as there are costs associated with borrowing and wasting soil from the site.

Cut-and-fill balancing, therefore, can be done by calculating the elevation of the horizontal or inclined surface that produces an equilibrium between the cut-and-fill volumes ($V_{cut} = V_{fill}$), at an elevation called the *transition* or *economic elevation* (H_t), as shown in Fig. 22.37.

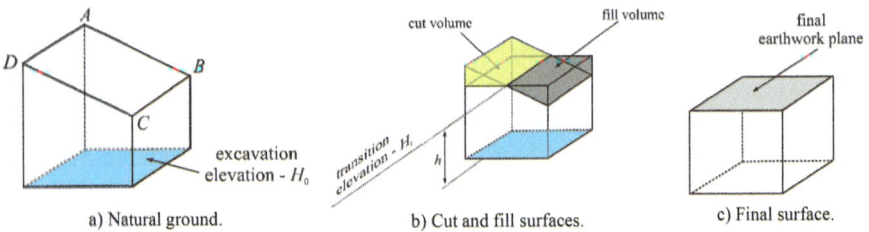

a) Natural ground.   b) Cut and fill surfaces.   c) Final surface.

Fig. 22.37 Transition or economic elevation

Considering Fig. 22.37, the problem is to calculate the transition elevation (H_t) at which the volume of the solid $ABCD$ (Fig. 22.37a) is equal to the volume of the final surface (Fig. 22.37c). To solve this problem, an excavation elevation (H_0) is adopted, and the cut volume (V_0) above this elevation is calculated. In this case, volume calculation is performed using one of the methods presented in Sect. 22.7.2.1. Then, an average height (h) is calculated by dividing that cut volume by the total platform area (A_T). Finally, the average height added to the value of the excavation elevation (H_0) results in the final value of the transition elevation (H_t), as given by Eq. (22.58)

$$V_0 = A_T * h \tag{22.56}$$

$$h = \frac{V_0}{A_T} \tag{22.57}$$

$$H_t = H_0 + h = H_0 + \frac{V_0}{A_T} \tag{22.58}$$

where

V_0 = total cut volume above the excavation elevation
A_T = total platform area
h = average height
H_0 = reference excavation elevation
H_t = transition elevation

The calculated transition elevation can be plotted on a terrain contour map, indicating the transition elevation contour as shown in Fig. 22.38.

Fig. 22.38 Transition elevation

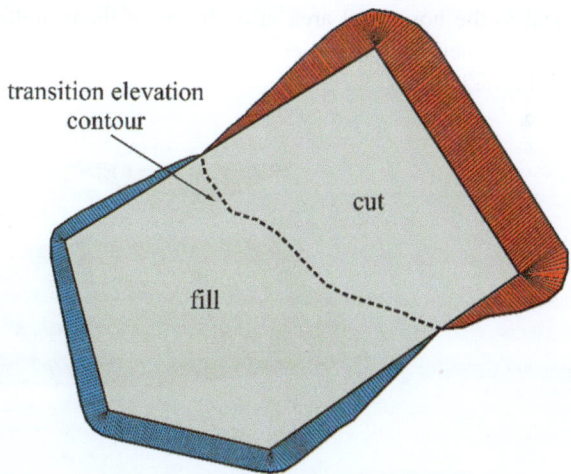

For this purpose, as discussed in the previous sections, the cut-and-fill volumes can be calculated using either the cross-sectional area method or the prism method.

Example 22.16
Given the results of Example 22.13, calculate the transition elevation for this platform, i.e., the elevation at which the cut-and-fill volumes are equal.

Solution:
From Example 22.13:

$$A_T = 5400.000 \text{ m}^2 \qquad V_T = 20,628.338 \text{ m}^3$$

Using Eq. (22.58),
$$H_t = 830.000 + \left(\frac{20,628.338}{5400.000}\right) = 833.820 \text{ m}$$

22.7.5 Volume Calculation Using Digital Terrain Models

One of the main applications of a DTM is volume calculation. This simple calculation process requires the creation of two surfaces: one to be considered as the base
surface and the other as the earthwork surface. For example, the two surfaces could
come from two surveying measurements at two different stages of a construction
project or from a situation where the base surface is the natural ground surface, and
the earthwork surface is the design surface. Regardless of the type of surfaces being
compared, the calculation methods to be applied are the same, as shown below.

22.7.5.1 Triangulation Method

The simplest method of calculating volumes using a DTM is to define a series of
prisms with triangular bases on the earthwork surface and consider the volume to be
equal to the horizontal area of each one of them multiplied by the average of the
depths of their vertices projected onto the base surface, as presented in Sect. 22.7.2.2.
See Fig. 22.39.

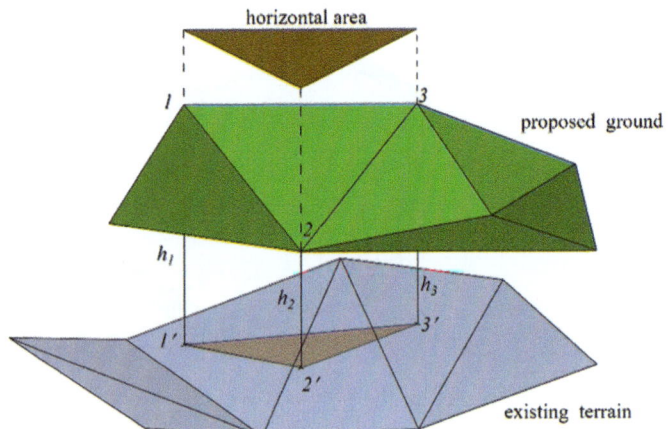

Fig. 22.39 Volume calculation by horizontal area

This solution gives reliable results for the two surfaces with equivalent triangles.
Otherwise, the results can be inaccurate.

A variant of the previous solution is to project the ground surface triangle onto the
base surface and calculate the prism's volume bounded by the two non-horizontal
surfaces. See Fig. 22.40.

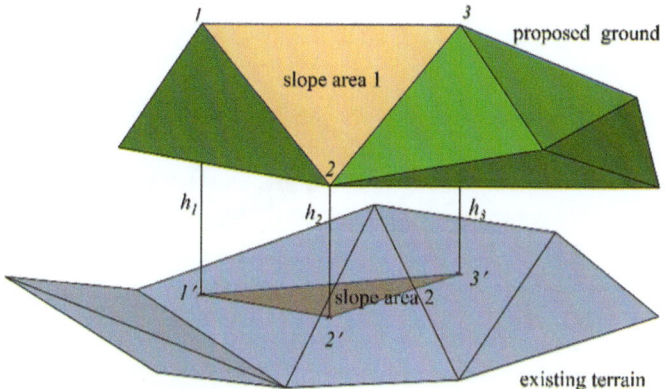

Fig. 22.40 Volume calculation by sloping area

The disadvantage of this method is the need to calculate the areas of the two non-horizontal triangular surfaces. As in the previous case, this solution only gives reliable results if the two surfaces have equivalent triangles.

The solution that gives the best results is to project the earthwork surface onto the base surface and calculate the volume using the areas of the projected triangles. This solution has the disadvantage of increasing the number of interpolation points and, therefore, the processing time, but it does not require the triangular surfaces to be equivalent. See Fig. 22.41.

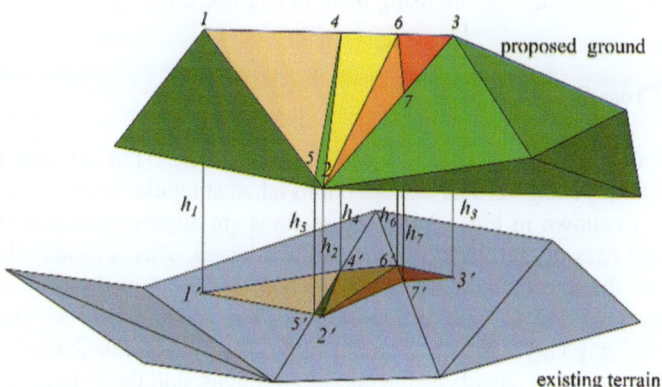

Fig. 22.41 Volume calculation by projected area

22.7.5.2   Regular Grid Method

Another solution also available in DTM software is to define a rectangular grid on the earthwork surface, project the grid edges onto the base surface and calculate the volume by taking the average of the depths of the vertices between the two surfaces, as described in Sect. 22.7.2.1. See Fig. 22.42.

Fig. 22.42 Volume
calculation by regular grid
interpolation

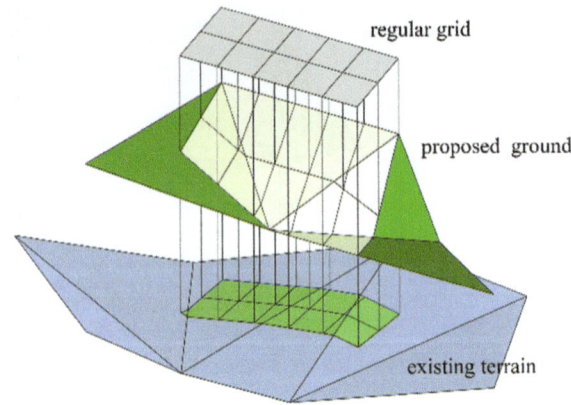

regular grid

proposed ground

existing terrain

In this case, the calculated volume's quality depends on the triangles' dimensions on the two surfaces and the regular grid. The smaller the grid, the higher the quality obtained, but the higher the computational cost.

Whichever method is used, the engineer must remember that the earthwork volume calculation is always an estimate. The accuracy of the excavation, the rate of expansion and shrinkage of the soil, the nature of the materials and the accuracy of the topographic survey can all significantly affect the results.

It is also important to note that knowing which volume calculation method is used in DTM software is not always possible. For this reason, tests are suggested to verify the quality of the models before using them in engineering projects.

22.7.5.3   Cross-Sectional Method

Instead of calculating volumes using grid methods, some DTM software packages calculate volumes using vertical cross-sections taken at regular intervals through the fills or cuts, as shown in Fig. 22.19. The sections are aligned perpendicularly to a centreline that runs the full length of the earthwork area. This is usually the longest dimension of the site to increase accuracy, but can be aligned along a property feature, surveying stakes, road centreline, etc. The spacing between the parallel cross-sections may vary depending on the site's size and the calculation's intended accuracy. The software calculates the cut-and-fill areas, and the volume calculation is then performed using the trapezoidal or prismoidal methods.

22.8   Problems

1. Calculate the area of the quadrilateral shown in Fig. 22.43 using the values given in Table 22.16.

Fig. 22.43 Geometry of the measured quadrilateral

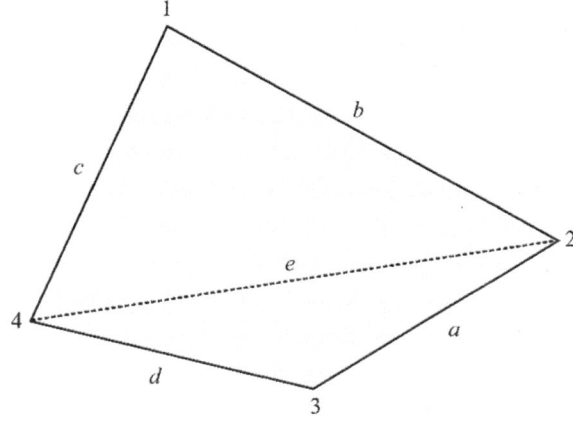

Table 22.16 Field measurements

Geometric elements	Value [m]
a	736.117
b	1152.229
c	833.859
d	753.009
e	1,382,461

2. To calculate the plot area, a total station has been installed within the area at point (*P*), as shown in Fig. 22.44. Using the field measurements given in Table 22.17, calculate the measured area.

Fig. 22.44 Geometric elements of the surveyed area

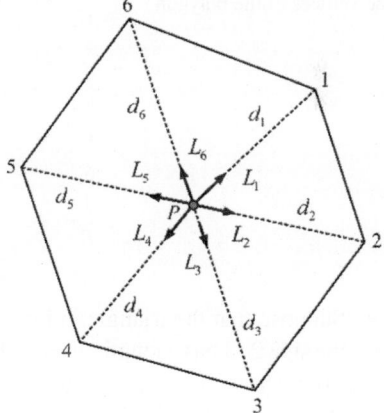

Table 22.17 Field measurements

Station	Sighting point	Horizontal direction	Horizontal distance [m]
P	1	0° 00′ 00″	815.369
	2	55° 24′ 11″	845.257
	3	115° 06′ 46″	958.147
	4	173° 25′ 25″	874.123
	5	236° 00′ 52″	862.258
	6	294° 10′ 24″	951.486

3. Calculate the area of the polygon shown in Fig. 22.45 from the coordinates of its vertices given in Table 22.18.

Fig. 22.45 Geometry of the polygon

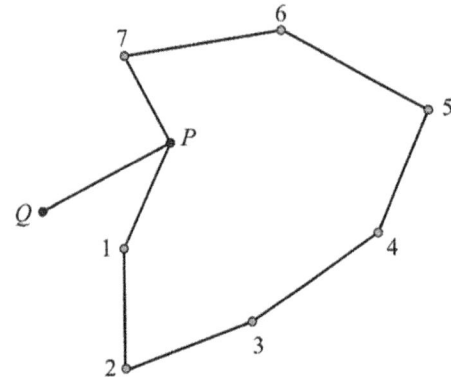

Table 22.18 Coordinates of the vertices of the polygon

Station	X [m]	Y [m]
1	14,320.712	10,672.640
2	14,320.699	10,120.913
3	14,913.263	10,345.044
4	15,511.571	10,758.842
5	15,741.686	11,310.600
6	15,051.300	11,672.655
7	14,320.336	11,556.696
P	14,550.798	11,159.290

4. Suppose that the triangle in Fig. 22.46 has an area of 670, 848.810 m^2 and that the side 2–3 has a length of 915.884 m. This area must be divided into three parts from vertex (1) so that A_1 (on the right side) is equal to $A_3/2$ and A_3 (on the left side) is equal to $1.5A_2$. Calculate the values of the lengths of the intersection points on side 2–3.

Fig. 22.46 Geometry of the area

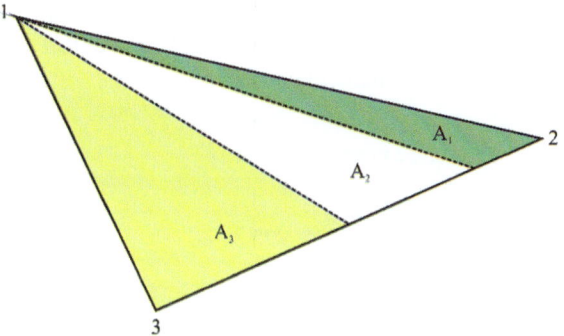

5. Suppose the quadrilateral shown in Fig. 22.47 has a total area of 670, 848.810 m², the side 1–2 has a length of 1152.229 m, the side 4–1 has a length of 833.859 m, the angle $\alpha_2 = 59°06'12''$, and the angle $\alpha_4 = 77°52'09''$. This area must be divided into three equal parts from the vertex (1) and intersecting sides 2–3 and 3–4. Under these conditions, calculate the lengths of the points of intersection on the corresponding sides of the polygon.

Fig. 22.47 Geometry of the divided area

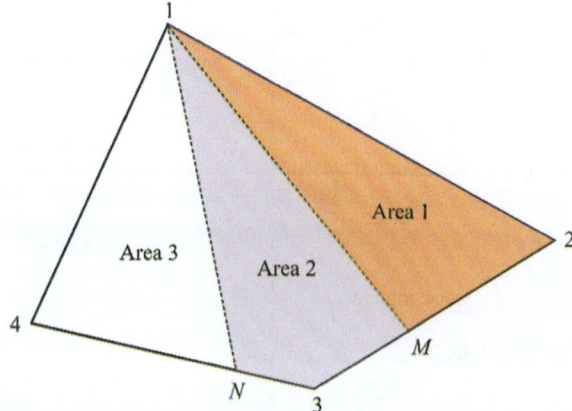

6. Using the coordinates from Problem 3, calculate the coordinates of the point (F) of the intersection of the line that starts at point (P) and divide the area of the polygon into two equal parts.
7. Using the coordinate from Problem 3, calculate the coordinates of the point (F) of the intersection of the line that starts at point (P) and is parallel to the side 6–5 of the polygon and divides the polygon into two equal parts.
8. Using the coordinates from Problem 3, suppose that the polygon has to be divided into three parts. In this case, calculate the coordinates of the intersection points of the two dividing lines parallel to the side 6–5 so that the first is 500 m from it and the second is 500 m from the first.

9. Using the cross-sections shown in Fig. 22.48, calculate the volume of earth-works between stakes 41 and 43. Consider that the distance between the stakes is 50 m in this particular case.

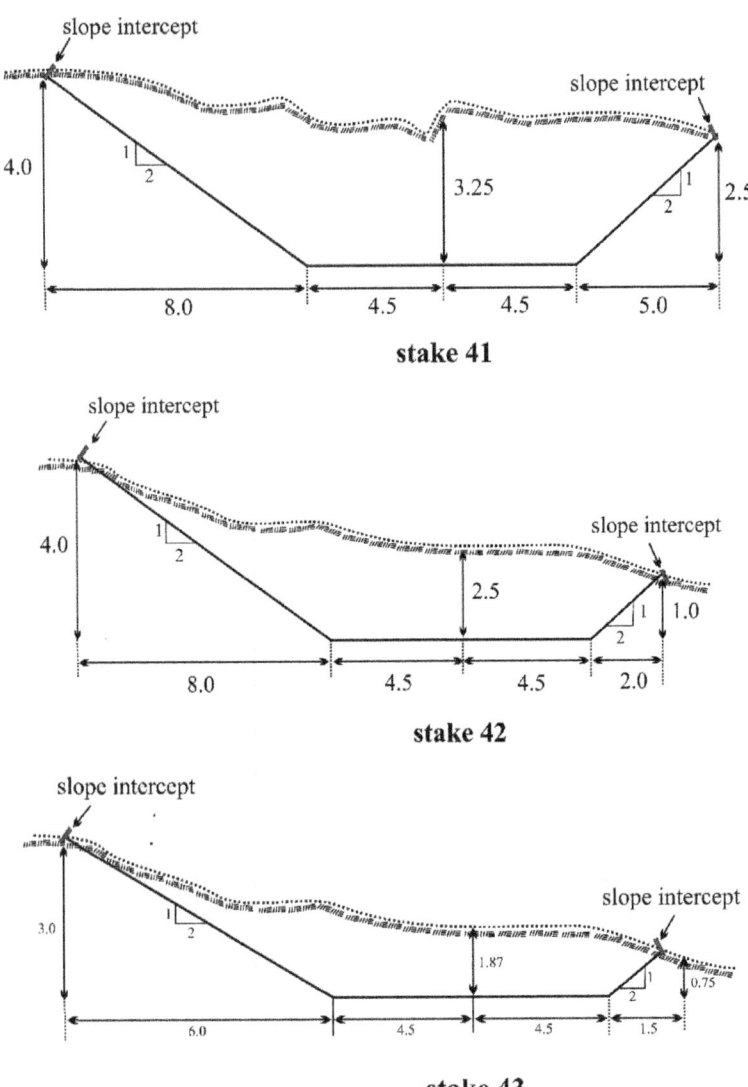

Fig. 22.48 Cross-sections of the stakes 41 to 43

10. Given the regular grid of spot elevations shown in Fig. 22.49, calculate the total cut volume generated for an elevation of 750.000 m by assigning individual weights to the grid vertices. Note that the sides of the grid are 20 m long.
11. Using the data from Problem 10, calculate the total cut volume generated for an elevation of 750.000 m using the cross-sectional area method.

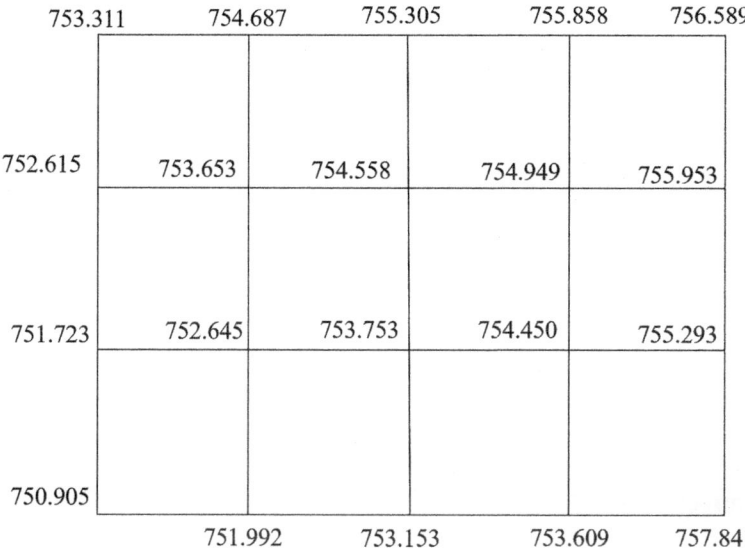

Fig. 22.49 Regular grid of spot elevations

12. Calculate the transition elevation value using the data from Problem 10.
13. Using the data from Problem 10, calculate the total cut-and-fill volumes, considering that the reference elevation is 753.000 m.
14. Calculate the cut volume for the grid layout shown in Fig. 22.50, assuming the final platform should have an elevation of 560.000 m.

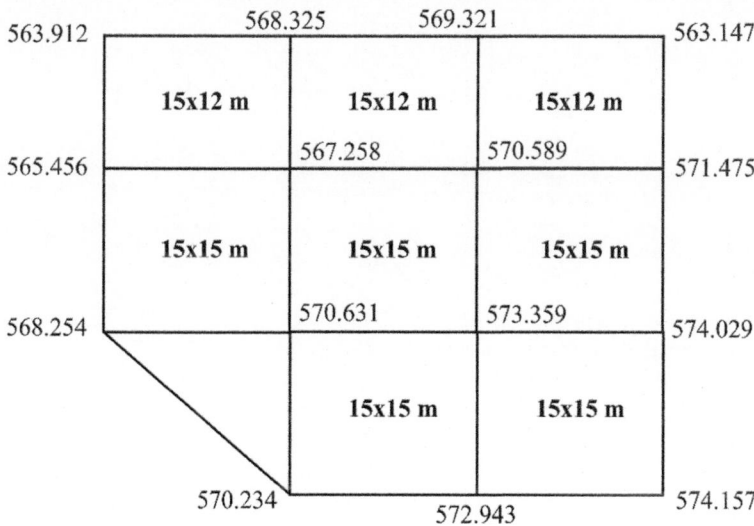

Fig. 22.50 Geometry of the grid area

15. Calculate the cut-and-fill volumes between each pair of consecutive stakes whose areas are given in Table 22.19 and the cumulative volume along the road section. Consider that the distance between the stakes is 20 m.

Table 22.19 Cross-sectional areas

Stake	Areas [m^2]	
	cut	fill
101	85.63	
102	68.52	
103	72.45	
104	95.63	112.36
105		98.15
106		196.32

References

Barras, V (2015). *Les relevés de terrain en 3D, Les profils, les plans topographiques, les MNTs.* heig-vd—Haute Ecole d'Ingénierie et de Gestion du Canton de Vaud. Yverdon-les-Bains. Switzerland.

Easa S. M.; (1989). General direct method for land subdivision. Journal of Surveying Engineering, Vol 115, no. 4, Nov, USA. pages 402–411.

Yanalak, M.; Baykal, O. (2003). *Digital elevation model-based volume calculation using topographic data.* Journal of Surveying

Chapter 23
Horizontal and Vertical Curves

Irineu da Silva, Paulo C. L. Segantine, and Marcelo Monari

23.1 Introduction

In Civil Engineering, many projects are designed based on geometric alignments, such as highways, railways, urban roads, tunnels, gas and oil pipelines, among many others. In these cases, alignments are composed of straight lines and curved segments arranged in space to fit the project to the existing terrain, as shown in Fig. 23.1. The linear elements, as shown in this figure, determine the geometric layout of the project and, consequently, the design guideline, which is composed of straight sections, defined by *tangents* or *grades (gradients)*, and curved sections, defined by *horizontal* and *vertical curves*.

Since linear elements are spatial structures, in Civil Engineering, tangents and horizontal curves are represented on the horizontal plane and gradients and vertical curves are represented on the vertical plane, as shown in Fig. 23.2. A series of connected horizontal straight lines (tangents) and curves is called the *horizontal alignment*. A series of straight grades (gradients) connecting vertical curves is called the *vertical alignment*. Therefore, in this chapter, the study of curves used in Civil Engineering is divided into horizontal and vertical. It should be noted that this chapter does not discuss the applications of curves in Civil Engineering but only their analytical and geometric details. Readers interested in using curves in Civil Engineering should consult specialised references.

I. da Silva (✉) · P. C. L. Segantine
São Carlos School of Engineering, University of São Paulo, São Carlos, SP, Brazil
e-mail: irineu@sc.usp.br

M. Monari
Civil Engineer, Assistant Professor in the Department of Civil Engineering at the Federal University of São Carlos (UFScar), São Paulo, Brazil
e-mail: marcelo.monari@ufscar.br

© The Author(s), under exclusive license to Springer Nature Switzerland AG 2025
I. da Silva, P. C. L. Segantine, *Geomatics Applied to Civil Engineering*,
https://doi.org/10.1007/978-3-031-75737-2_23

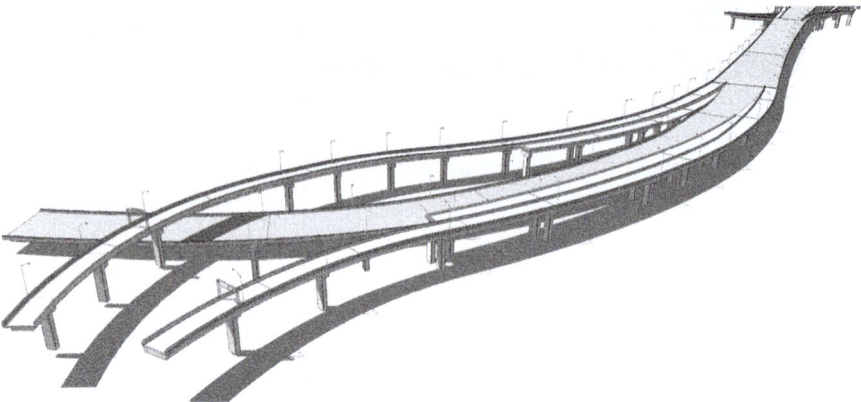

Fig. 23.1 Example of the spatial composition of a linear engineering project

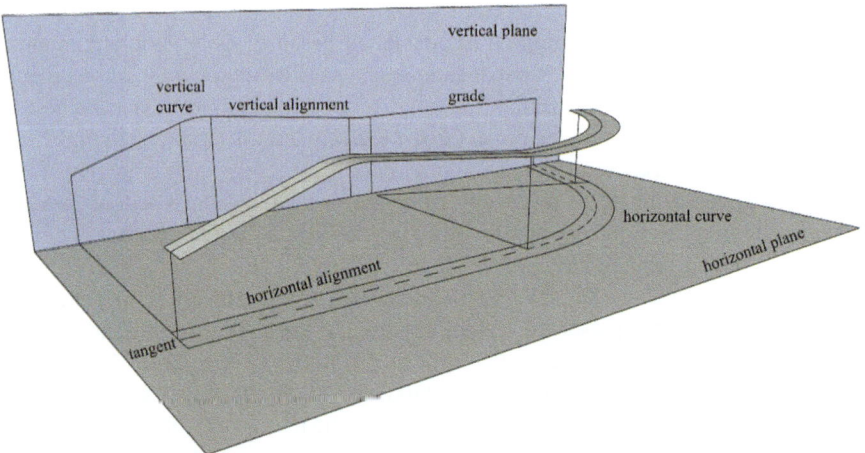

Fig. 23.2 Plan and profile view of a road design with horizontal and vertical alignments

The curves commonly used in Civil Engineering are:

- Horizontal circular curves
- Horizontal transition curves (spiral curves)
- Vertical curves

23.2   Horizontal Circular Curves

A horizontal curve is a circular arc drawn on a horizontal plane to provide a change of direction to the centreline of a road, railway, etc. It usually lies between two tangents also drawn on that plane.

There are three main types of horizontal circular curves to consider.

- Simple horizontal circular curve
- Compound horizontal circular curve
- Reverse circular curve

The geometric details of each of them are presented below.

23.2.1   Simple Horizontal Circular Curves

A simple horizontal circular curve has a constant radius of curvature R, as shown in Fig. 23.3. It is a geometrically well-defined curve that is easy to design and layout and is therefore the most used type of curve in Civil Engineering. It connects two tangents in such a way as to allow continuity of the alignment without vertices, i.e., it begins and ends at the two points of tangency with the tangents.

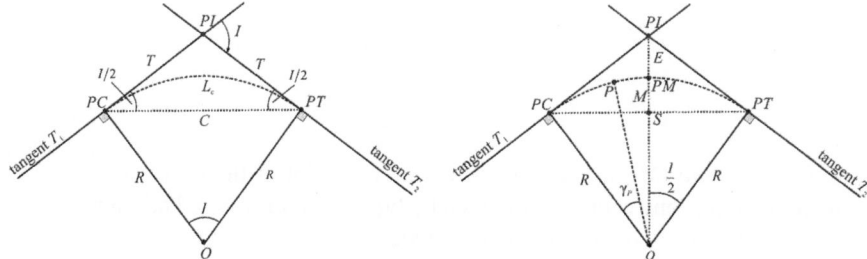

Fig. 23.3 Geometric elements of a simple horizontal circular curve

As shown in Fig. 23.3, the simple horizontal circular curve has the following geometric elements:

PI = point of intersection between the tangents T_1 and T_2
PC = point of curvature, which is the point where the circular curve begins (the back tangent is tangent to the curve at this point)
PT = point of tangency, which is the endpoint of the curve (the forward tangent is tangent to the curve at this point)
PM = midpoint of the curve
C = long chord
S = midpoint of the long chord
M = middle ordinate
P = random point on the curve
E = external distance
I = deflection angle between tangents T_1 and T_2 = central angle of the curve = deflection angle at (PI)
γ_P = central angle of the point (P)
R = radius of the curve

T = tangent distance
L_c = length of the curve from (PC) to (PT)
O = centre of the curve

The algebraic relationships between these elements are defined by Euclidean geometry, as described below.

Tangent distance (T): the distance from the (PI) to the (PC) or (PT), as given by Eq. (23.1). These distances are equal on a simple curve.

$$T = R * \tan\left(\frac{I}{2}\right)$$ (23.1)

External distance (E): the distance from the (PI) to the midpoint of the curve (PM), as given by Eqs. (23.2) or (23.3). The external distance bisects the central angle at the (PI).

$$\cos\left(\frac{I}{2}\right) = \frac{R}{R+E} \rightarrow E = R\left[\frac{1}{\cos(I/2)} - 1\right]$$ (23.2)

$$E = T * \tan\left(\frac{I}{4}\right)$$ (23.3)

Middle ordinate (M): the distance from the midpoint of the curve (PM) to the midpoint of the long chord (S), as given by Eqs. (23.4) or (23.5). The extension of the middle ordinate bisects the central angle.

$$\cos\left(\frac{I}{2}\right) = \frac{R-M}{R} \rightarrow M = R * \left[1 - \cos\left(\frac{I}{2}\right)\right]$$ (23.4)

$$M = E * \cos\left(\frac{I}{2}\right)$$ (23.5)

Long chord (C): the chord from the (PC) to the (PT), whose length is given by Eqs. (23.6) or (23.7).

$$\sin\left(\frac{I}{2}\right) = \frac{C}{2R} \rightarrow C = 2R * \sin\left(\frac{I}{2}\right)$$ (23.6)

$$C = 2T * \cos\left(\frac{I}{2}\right)$$ (23.7)

Length of the curve (L_c): the distance from (PC) to (PT) measured along the curve, which can be calculated using Eqs. (23.9), (23.10) or (23.11).

$$\frac{L_c}{2\pi R} = \frac{I}{360°}$$ (23.8)

where

$$L_c = \frac{\pi * R * I^\circ}{180^\circ} \quad I \text{ in degrees} \tag{23.9}$$

$$L_c = I * R \quad I \text{ in radians} \tag{23.10}$$

$$L_c = \frac{\pi * R * I}{200^g} \quad I \text{ in gons} \tag{23.11}$$

Degree of the curve (G): the central angle subtended by an arc or chord of one station.

(a) *Arc definition:* the arc definition states that the degree of a curve is the angle formed by two radii drawn from the circle's centre to the ends of an arc of 100 units. This definition is primarily used for roadways. Note that the greater the degree of the curve, the "sharper" the curve and the smaller the radius. It is calculated by Eq. (23.12).

(b) *Chord definition:* the chord definition states that the degree of a curve is the angle formed by two radii drawn from the centre of a circle to the ends of a chord of 100 units. The chord definition is primarily used for railroad construction and by the military for road and rail construction. It is calculated by Eq. (23.13).

$$\text{For roadways :} \quad \frac{G}{100} = \frac{360^\circ}{2\pi R} \quad G \cong \frac{5,729.578^\circ}{R} \tag{23.12}$$

In this case, the radius of a 1° curve is 5729.578 m in the metric system and 5729.578 ft. in the foot system, and a 3° curve would have a radius of 1909.859 units in the respective system.

$$\text{For railroads :} \quad \sin\left(\frac{G}{2}\right) = \frac{50}{R} \quad G = 2a \sin\left(\frac{50}{R}\right) \tag{23.13}$$

According to Eq. (23.13) the radius of a 1° curve is 5729.651 feet or metres, and the radius of a 10° curve is 573.686 feet or metres.

23.2.2   Compound Horizontal Circular Curves

A compound circular curve is formed by two or more consecutive horizontal circular curves with different radii of curvature, as shown in Fig. 23.4. In general, this type of curve is used when it is necessary to avoid obstacles on the terrain, which cannot be done using simple horizontal circular curves with a larger radius.

In Fig. 23.4, line *AB* is the common tangent to the two curves at point T_C, and lines T_1 and T_2 are the tangents to both curves. In this way, because they are different curves, they produce two different deflection angles (α and β), so that ($I = \alpha + \beta$). The other elements are the same as the simple horizontal circular curve.

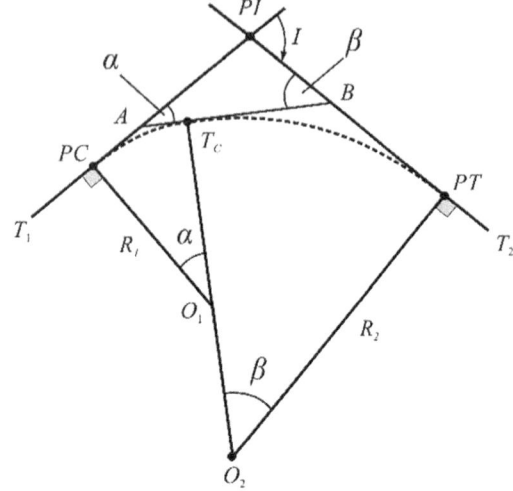

Fig. 23.4 Compound horizontal circular curve

23.2.3 Reverse Horizontal Circular Curves

A reverse horizontal circular curve is formed by two consecutive simple horizontal circular curves with the same radius but opposite centres of curvature, as shown in Fig. 23.5. The geometric design of this type of horizontal curve is usually carried out by considering them as two independent simple horizontal circular curves.

In Fig. 23.5, the lines T_1 and T_2 are the tangents to the two curves. The other elements are the same as for the simple horizontal circular curve.

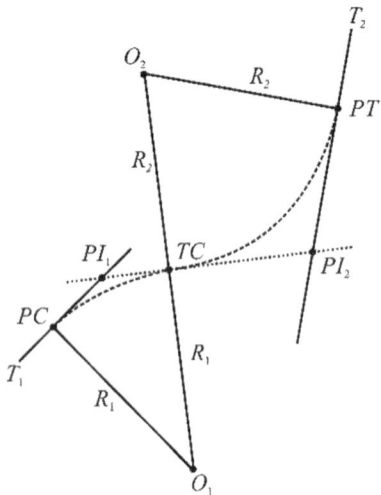

Fig. 23.5 Reverse horizontal circular curves

23.2.4   *Staking of the Horizontal Circular Curve*

Once the alignment is defined, with its tangents intersected by horizontal curves, it is discretised into equidistant points, commonly called *stakes*. The equidistance between the stakes varies depending on the type of project. Depending on the project conditions, they are usually placed at 20, 30 or 40 m intervals. Staking begins at a point specified by the designer and continues according to the adopted equidistance, producing stakes numbered according to their position in the staking sequence. Key points such as intersections, angle points, and the start and end of curves are usually numbered with specific identifiers. If the discretisation of the designed feature point does not correspond to an integer number of stakes, its position is defined by the previous integer stake number plus the distance in metres between the previous stake and the feature point stake.

In road design, the point positioning along the horizontal alignment is defined by a staking system called *stationing*. By convention, stationing is the process of defining locations along the project by station numbers given by a pair of values between brackets of the type [A + B],[1] where A indicates a full station and B is a plus station that indicates a feature point location between full stations. For example, for a 20-metre stationing, the station identifying a point 1272.135 m from the starting point would be [63 + 12.135] m.

Among the various feature points of a road design, the *(PI)*, *(PC)* and *(PT)* stations are always calculated and displayed in the graphical representation of the project. These positions are calculated using the following mathematical relationships:

(PI) station: as it lies on the tangent, it is calculated directly from the length of the tangent, not requiring any equation.

(PC) station: calculated from the station position *PI* and the tangent length *T* as follows:

$$(PC)station = (PI)station - T \tag{23.14}$$

(PT) station: calculated from the station position *PC* and the length of the curve from *PC* to *PT* as follows:

$$(PT)station = (PC)station + L_c \tag{23.15}$$

Example 23.1

On a given section of the road, a change will be made to the geometric design of a simple horizontal circular curve. In this new project, the curve shall meet the following technical characteristics:

---

[1] The use of brackets is optional.

Radius of curvature = 800 m	Interval between stations: 20 m

The degree of the curve (G) should be calculated for an arc of 20 m.

$(PI)station = [65 + 12.498\,]$m	Deflection angle $(I) = 32° \, 10' \, 22''$

Considering all the information above, calculate the geometric elements of the new curve.

Solution:

Using the equations presented in the previous sections.

$$T = 800 * \tan\left(\frac{32° 10' 22''}{2}\right) = 230.702\ m$$

$$E = 800 * \left[\frac{1}{\cos(32° 10' 22''/2)} - 1\right] = 32.600\ m \ \text{ or } \ E = 230.702 * \tan\left(\frac{32° 10' 22''}{4}\right) = 32.600 m$$

$$M = 800 * [1 - \cos(32° 10' 22''/2)] = 31.324 m \ \text{ or } \ M = 32.600 * \cos(32° 10' 22''/2) = 31.324 m$$

$$C = 2 * 800 * \sin(32° 10' 22''/2) = 443.338 m \ \text{ or } \ C = 2 * 230.702 * \cos(32° 10' 22''/2) = 443.338 m$$

$$L_c = \frac{\pi * 800 * 32° 10' 22''}{180°} = 449.217\ m$$

$$G = \frac{20 * 360°}{2\pi * 800} = 1°25'56.6''$$

$$(PC)station = [65 + 12.498]\ m - [11 + 10.702]\ m = [54 + 1.796]\ m$$

$$(PT)station = [54 + 1.796]\ m + [22 + 9.217]\ m = [76 + 11.012]\ m$$

23.3 Horizontal Transition Curves (Spiral Curves)

In engineering construction, the simple definition of a horizontal alignment consisting of straight lines and horizontal curves is not always considered a good solution as it creates a discontinuity of curvature at the points of curvature (PC) and tangency (PT) of the alignment. This is particularly critical in road or railway design as it causes skidding and steering difficulties for the driver when entering the curve. Therefore, a "transition" section of progressive curvature is required at the entry and exit of the circular curve to meet a series of geometric, safety and user comfort requirements. This section of variable curvature is called a *transition horizontal curve*.

There are several types of transition curves in the literature. In this book, only the geometrical details of the *spiral*[2] will be discussed, i.e., a plane curve whose radius of curvature R increases or decreases monotonically with distance along the curve, as given by Eq. (23.16).

$$R * L = K \qquad\qquad (23.16)$$

where

$R=$ radius of curvature at a point (P) on the spiral
$L=$ length of the spiral to a random point (P)
$K=$ constant related to the length of the spiral and the radius of the circular section. This parameter is also called the *spiral parameter*. Different values of K imply different types of spirals

23.3.1   Coordinates of a Random Point on a Spiral (Transition Curve)

The best way to locate a point on a transition curve is to establish a local 2D Cartesian coordinate system (X, Y), where the X-axis coincides with the direction of the tangent and the Y-axis is perpendicular to it, as shown in Fig. 23.6. Taking the starting point of the spiral (*TS*—tangent to the spiral) as the origin of the coordinate system, the following geometric elements must be considered when studying the spiral:

Fig. 23.6 Geometry of a spiral connecting a tangent to a circular curve

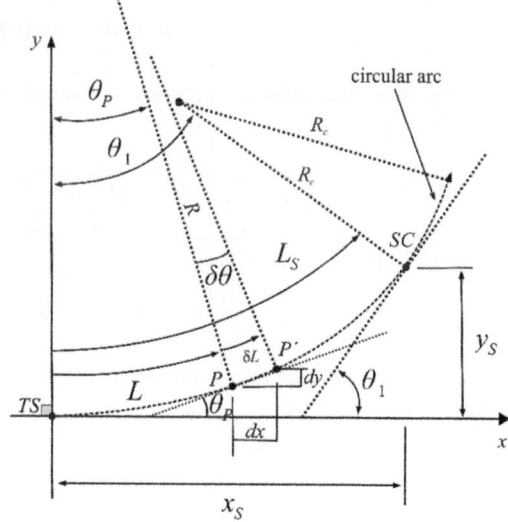

$R=$ radius of curvature at any point (P) on the spiral
$R_c=$ radius of curvature of the horizontal circular curve
$TS=$ starting point of the spiral (tangent to spiral)
$SC=$ endpoint of the spiral (spiral to curve)
$L=$ length of the spiral from (TS) to any given point (P) on the spiral
$L_S=$ total length of the spiral from (TS) to (SC)
$\theta_P=$ angle between the tangent to the spiral at a random point (P) and the X-axis
$x, y=$ axes of the 2D Cartesian coordinate system

The point (P'), infinitesimal of (P), is located at a distance $L + \delta L$ from the (TS). Thus,

$$R * \delta\theta = \delta L \tag{23.17}$$

Substituting Eqs. (23.16) into (23.17) gives the following differential equation of the spiral:

$$\delta\theta = \frac{L * \delta L}{K} \tag{23.18}$$

Integrating Eq. (23.18) gives

$$\theta = \frac{L^2}{2K} = \frac{L}{2R} [\text{rad}] \tag{23.19}$$

Similarly,

$$dx = \delta L * \cos(\theta) \tag{23.20}$$
$$dy = \delta L * \sin(\theta) \tag{23.21}$$

If the sine and cosine functions are developed in series and integrated, the following equations are obtained:

$$x = L * \left(1 - \frac{\theta^2}{10} + \frac{\theta^4}{216} - \frac{\theta^6}{9360} + \frac{\theta^8}{685,440} \cdots \right) \tag{23.22}$$

$$y = L * \left(\frac{\theta}{3} + \frac{\theta^3}{42} - \frac{\theta^5}{1320} + \frac{\theta^7}{75,600} \cdots \right) \tag{23.23}$$

The geometric elements considered for the transition curve (spiral curve) are shown in Fig. 23.7. where

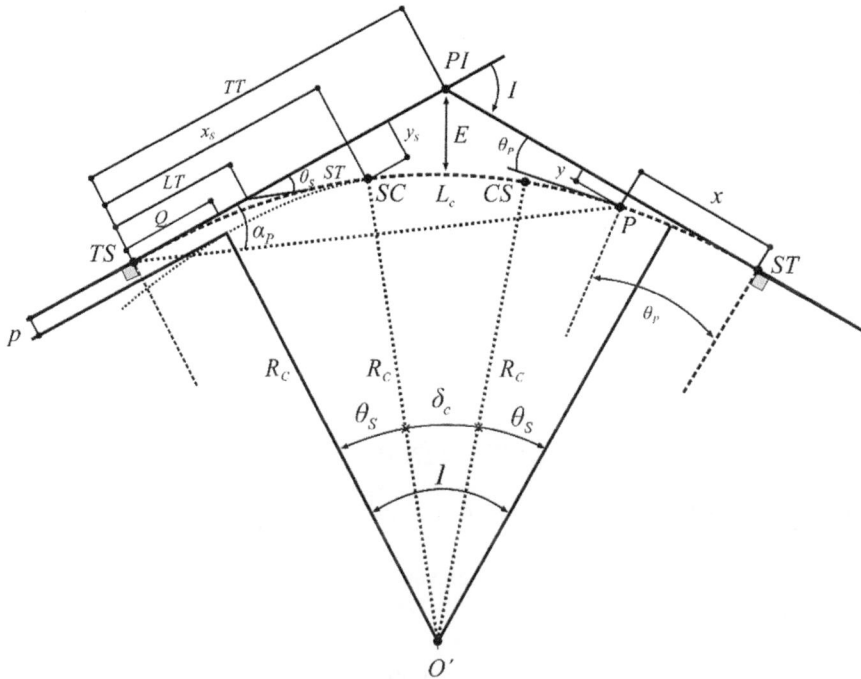

Fig. 23.7 Geometric elements of a horizontal circular curve with spiral curve

I = angle of intersection
TT = total tangent
LT = long tangent
ST = short tangent
p = offset distance perpendicular to the back tangent
E = external distance
δ_c= central angle of the horizontal circular curve
L_c= length of the horizontal circular curve
θ_S= central angle of the spiral
θ_P= angle to the random point (P)
SC= transition station from spiral to circular curve
CS= transition station from circular curve to spiral
TS= transition station from tangent to spiral
ST= transition station from spiral to tangent
x_S= x-coordinate of the (SC) and (CS) stations
y_S= y-coordinate of the (SC) and (CS) stations
Q = x-coordinate of the centre of the offset circle (O')
O'= centre of the offset circle
R_c= radius of curvature of the horizontal circular curve
x, y= coordinates of a random point (P)

Considering that at point (SC), the geometric relations $R = R_c$ and $L = L_s$ are valid for both the spiral and the horizontal circular curve, Eq. (23.19) can be written as follows:

$$\theta_S = \frac{L_S}{2R_c} \text{ [rad]} \tag{23.24}$$

The designer usually chooses R_c and a spiral's length L_S. From there, the remaining parts of the spiral can be determined.

In road design, the values of R_c and L_S are usually determined using parameters based on the centrifugal ratio and the super-elevation of the curve. Based on the centrifugal ratio, the minimum safe radius R_c can be calculated using Eq. (23.25).

$$R_c = \frac{V^2}{127CR} \tag{23.25}$$

where V is in km/h, R_c in metres and CR = centrifugal ratio = 0.21 to 0.25 on roads and 0.125 on railways.

The value of L_S, based on the super-elevation rate, can be calculated considering the interval given by Eqs. (23.26) and (23.27), i.e., $L_{S_{min}} \leq L_S \leq L_{S_{max}}$.

$$L_{S_{min}} = \frac{0.036V^3}{R_c} \tag{23.26}$$

$$L_{S_{max}} = I * R_c \tag{23.27}$$

where V is in km/h and R_c and L_S in metres.

A more detailed discussion of the calculation of R_c and L_S is beyond the scope of this book. The reader interested in further information on this subject is advised to consult specialised literature on road design.

The other geometric parameters in Fig. 23.7 are calculated according to the equations below.

$$x_S = L_S * \left(1 - \frac{\theta_S^2}{10} + \frac{\theta_S^4}{216} - \frac{\theta_S^6}{9360} + \frac{\theta_S^8}{685,440} - \cdots\right) \tag{23.28}$$

$$y_S = L_S * \left(\frac{\theta_S}{3} - \frac{\theta_S^3}{42} + \frac{\theta_S^5}{1320} - \frac{\theta_S^7}{75,600} + \cdots\right) \tag{23.29}$$

$$Q = x_S - R_c * \sin(\theta_S) \tag{23.30}$$

$$p = y_S - R_c * [1 - \cos(\theta_S)] \tag{23.31}$$

$$TT = Q + (R_c + p) * \tan\left(\frac{I}{2}\right) \tag{23.32}$$

$$L_c = (I - 2\theta_S) * R_c \tag{23.33}$$

$$E = \frac{(R_c + p)}{\cos\left(\frac{I}{2}\right)} - R_c \tag{23.34}$$

$$LT = x_S - y_S * \cot(\theta_S) \tag{23.35}$$

$$ST = \frac{y_S}{\sin(\theta_S)} \tag{23.36}$$

$$\theta_P = \left(\frac{L_P}{L_S}\right)^2 * \theta_S[\text{rad}] \tag{23.37}$$

$$x_P = L_P * \left(1 - \frac{\theta_P^2}{10} + \frac{\theta_P^4}{216} - \frac{\theta_P^6}{9360} + \frac{\theta_P^8}{685,440} - \cdots\right) \tag{23.38}$$

$$y_P = L_P * \left(\frac{\theta_P}{3} - \frac{\theta_P^3}{42} + \frac{\theta_P^5}{1320} - \frac{\theta_P^7}{75,600} + \cdots\right) \tag{23.39}$$

$$\alpha_P = \left(\frac{L_P}{L_S}\right)^2 * \frac{\theta_S}{3} = \arctan\left(\frac{y_P}{x_P}\right) \tag{23.40}$$

The value of (TT) positions the (TS) and (ST) stations in relation to (PI). The value of the central abscissa Q positions the centre (O') in relation to the (TS) or (ST) stations. The value of p corresponds to the offset distance from the circular curve to the tangents.

Example 23.2

Given a spiral with $I = 35°18'42''$, $L_s = 90$ m, $R_c = 500$ m, stationing interval of 20 m and (PI) station $= [228 + 3.674]$ m, calculate the geometric elements of the transition curve.

Solution:

The geometric elements of the spiral can be calculated using Eqs. (23.28) to (23.36).

$\theta_S = \frac{90}{2*500} = 0.0900\,\text{rad} = 5°09'23.8''$	$x_S = 90 * \left(1 - \frac{0.0900^2}{10} + \frac{0.0900^4}{216} - \frac{0.0900^6}{9360}\right) = 89.927\,\text{m}$
$y_S = 90 * \left(\frac{0.0900}{3} - \frac{0.0900^3}{42} + \frac{0.0900^5}{1320}\right) = 2.698\,\text{m}$	$Q = 89.927 - 500 * \sin(5°09'23.8'')$ $= 44.988\,\text{m}$
$p = 2.698 - \{500 - [1 - \cos(5°09'23.8'')]\}$ $= 0.675\,\text{m}$	$TT = 44.988 + (500 + 0.675) * \tan\left(\frac{35°18'42''}{2}\right)$
	$TT = 204.348\,\text{m}$
$L_c = [0.616304848 - (2 * 0.0900)] * 500$ $L_c = 218.152\,\text{m}$	$E = \frac{500 + 0.675}{\cos\left(\frac{35°18'42''}{2}\right)} - 500 = 25.425\,\text{m}$
$LT = 89.927 - 2.698 * \cot(5°09'23.8'')$ $LT = 60.025\,\text{m}$	$ST = \frac{2.698}{\sin(5°09'23.8'')} = 30.023\,\text{m}$

23.4   Vertical Curves

To adapt a linear structure design to the variations in relief, once the horizontal alignment has been defined, it is necessary to know the profile of the terrain along the route. This work is carried out by profile levelling, as described in Sect. 14.2. With the terrain profile surveyed, it is possible to design grades (rates of slope) and vertical curves so that they are in accordance with the relief conditions, to meet the safety requirements of the project and to achieve the best cost/benefit ratio for the earthworks.

The first step in designing vertical curves is to define the grade lines. Vertical curves are then defined to provide a transition between tangent grades (straight lines) in the vertical plane, as shown in Fig. 23.8. A vertical curve that connects a descending grade with an ascending grade is called a *sag vertical curve* (convexity downwards). An ascending grade followed by a descending grade defines a *crest vertical curve* (convexity upwards).

In general, in road design, the vertical curves connecting two grade lines are *arcs of circles* and *parabolas*. In the first case, the same rules apply as for simple horizontal circular curves. Therefore, this section discusses the geometric and analytical details of simple second degree parabolic vertical curves. Given the geometry shown in Fig. 23.8, the following geometric elements are of interest for the parabolic curve between tangent grades 1 (i_1) and 2 (i_2):

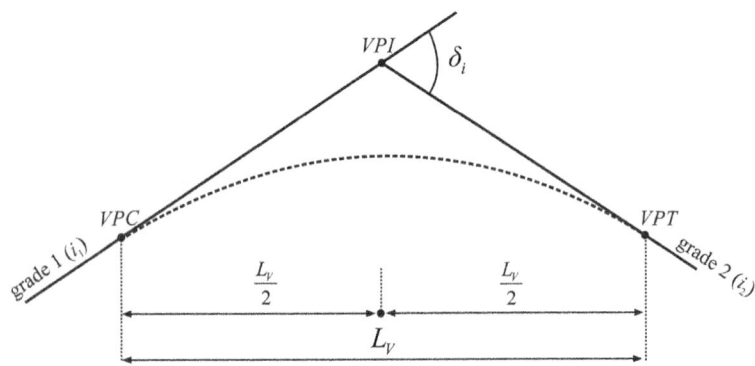

Fig. 23.8 Parabolic vertical curve geometry

VPI = point of vertical intersection
VPC = beginning of the vertical curve
VPT = end of the vertical curve
L_V = length of the vertical curve (horizontal distance from VPC to VPT)
i_1 = grade of tangent 1 (positive for ascending slope in the direction of the survey, given in percentage)

i_2= grade of tangent 2 (negative for descending slope in the direction of the survey, given in percentage)

δ_i= algebraic difference in slope ($\delta_i = i_2 - i_1$)

The parabolic curve is geometrically similar to the circle, therefore, for calculation purposes, it is common to refer to the vertical radius of curvature R_V as the radius of the circle equivalent to the parabola. The minimum length of the vertical curve can therefore be calculated using Eq. (23.41).

$$L_V = |R_V * \delta_i| \tag{23.41}$$

In road projects, it is common to define a value for the vertical radius and calculate the length of the vertical curve.

The following are the algebraic formulations used to study the parabolic vertical curve.

23.4.1 Equation of the Parabola

The equation of the parabola as a function of the (x) and (y) axes, as shown in Fig. 23.9, is given by Eq. (23.42).

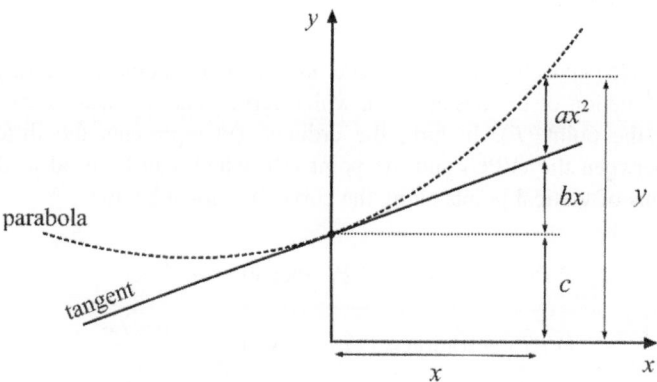

Fig. 23.9 Terms of the parabolic equation

$$y = ax^2 + bx + c \tag{23.42}$$

Note that the (x, y) system, in this case, determines a local coordinate system where the Y-axis is in the vertical plane.

Since the origin of the parabola coincides with the beginning of the vertical curve (VPC), the term c is equal to zero.

Assuming that the parabola meets the tangent grades at the points (VPC) and (VPT), the following algebraic relationship can be written as

$$\frac{\partial y}{\partial x} = i_1 = 2ax + b \tag{23.43}$$

As $x = 0$, at this point, $b = i_1$.

At the end of the vertical curve (VPT), $x = L_V$. Thus,

$$\frac{\partial y}{\partial x} = i_2 = 2a * L_V + b \rightarrow a = \frac{i_2 - i_1}{2L_V} = \frac{\delta_i}{2L_V} \tag{23.44}$$

And therefore,

$$y = \frac{\delta_i}{2L_V} * x^2 + i_1 * x \tag{23.45}$$

23.4.2   Elevations and Positions of the (VPC) and (VPT) Stations

Equation (23.45) gives the value of the ordinate (y) of any point (P) on the parabolic curve as a function of the abscissa (x), which represents the distance between the (VPC) and the point (P). In turn, the ordinate (y) represents the difference in elevation between the (VPC) and the point (P), which can be used to determine the elevations of vertical points along the curve, as shown below:

$$(VPC) \text{ elevation} = (VPI) \text{ elevation} - \frac{i_1 * L_V}{2} \tag{23.46}$$

$$(VPT) \text{ elevation} = (VPI) \text{ elevation} + \frac{i_2 * L_V}{2} \tag{23.47}$$

$$(VPC) \text{ station} = (VPI) \text{ station} - \frac{L_V}{2} \tag{23.48}$$

$$(VPT) \text{ station} = (VPI) \text{ station} + \frac{L_V}{2} \tag{23.49}$$

23.4.3   Highest (or Lowest) Point of the Parabolic Vertical Curve

The highest (or lowest)[3] point of a vertical curve is an important geometric element for designing the curve, which can be defined by setting the derivative of the parabola equation with respect to (x) equal to zero, as given by Eq. (23.50).

$$\frac{\partial y}{\partial x} = \frac{\delta_i}{L_V} * x + i_1 = 0 \rightarrow \frac{\delta_i * L_0}{L_V} + i_1 = 0 \rightarrow L_0 = -\frac{i_1 * L_V}{\delta_i} \tag{23.50}$$

From Eq. (23.50), it can be deduced that the ordinate (y_0) of the highest (or lowest) point of the vertical curve is given by Eq. (23.51).

$$y_0 = -\frac{i_1^2 * L_V}{2\delta_i} \tag{23.51}$$

23.4.4   Calculating Elevations and Offsets from a Tangent to the Parabolic Curve

Considering the parabolic crest vertical curve in Fig. 23.10, the following elements are of interest when designing a vertical curve:

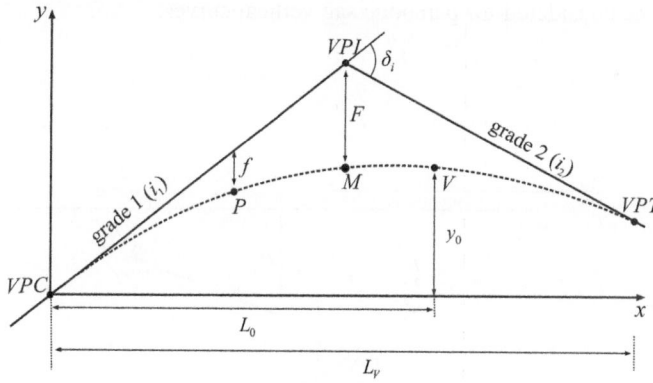

Fig. 23.10   Singular points of a parabolic crest vertical curve

---

[3] If the parabola opens down, the vertex represents the highest point on the graph (maximum value of the quadratic function); otherwise, the vertex represents the lowest point on the graph.

$f=$ offset at a random point (P)
$F=$ maximum offset at VPI (or $VM =$ vertical maximum)
$M=$ midpoint of the vertical curve
$V=$ highest point of the vertical curve

The offset value f at any point on the curve can be calculated using Eq. (23.52).

$$f = \frac{\delta_i}{2L_V} * x^2 \tag{23.52}$$

At point (VPI), where $x = \frac{L_V}{2}$

$$F = \frac{\delta_i}{8} * L_V \tag{23.53}$$

The coordinates of the other singular vertical points can be calculated using the equations in Table 23.1.

Table 23.1 Coordinates of the singular points of a vertical curve

Station	x	y
VPC	0	0
VPT	L_V	$(i_1 + i_2) * L_V/2$
VPI	$L_V/2$	$i_1 * L_V/2$
M	$L_V/2$	$\delta_i * L_V/8 + i_1 * L_V/2$
V	$-i_1 * L_V/\delta_i$	$-i_1^2 * L_V/2\delta_i$

Similarly to Fig. 23.10, Fig. 23.11 illustrates the positioning of the geometric elements to be considered for parabolic sag vertical curves.

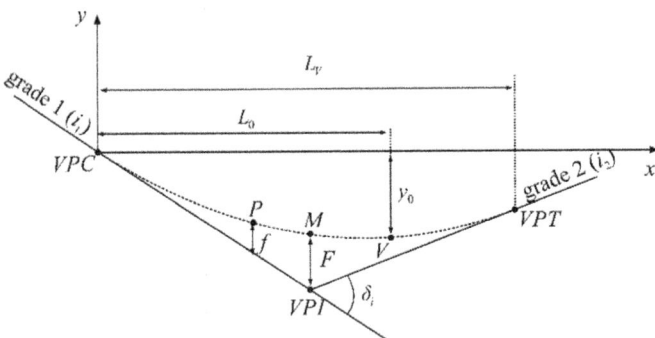

Fig. 23.11 Singular points of a parabolic sag vertical curve

The surveyor uses the highest or the lowest point of a vertical curve to determine the direction and amount of *runoff*,[4] in the case of crest curves, and to locate the low point for drainage.

If the tangent grades are equal, the highest or lowest point will be at the centre of the curve. If both tangent grades are positive, the lowest point is at the *VPC* and the highest point at the *VPT*. If both tangent grades are negative, the highest point is at the *VPC* and the lowest point at the *VPT*. If there are unequal plus and minus tangent grades, the highest or lowest point will be on the side of the curve with the flatter slope.

Example 23.3

Given the values of the geometric elements of the vertical alignment shown in Fig. 23.12, calculate the geometric elements of the parabolic vertical curve that provides a transition between the two tangent grades. Assume a vertical radius of curvature R_V equal to 3000.000 m. In addition, calculate the elevation of point a (P) at station $[32 + 3.475]$ m.

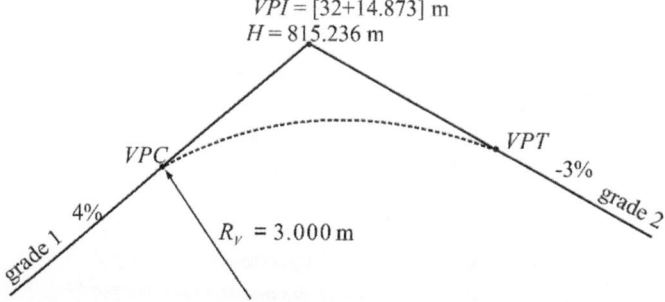

Fig. 23.12 Geometry of the parabolic vertical curve

Solution:

From the values shown in Fig. 23.12.

$$(\delta_i = i_2 - i_1) = -0.03 - 0.04 = -0.07$$
$$L_V = 3000 * |-0.07| = 210.000\,\text{m}$$
$$Elevation\ VPC = 815.236 - \left(\frac{0.04 * 210}{2}\right)$$
$$Elevation\ VPC = 811.036\,\text{m}$$

---

[4]The draining away of water from the surface of an area of land.

$$VPI\ station = [32 + 14.873] - \left(\frac{210}{2}\right)$$

$$VPI\ station = [27 + 9.873]\ m$$

$$Elevation\ VPT = 815.236 + \left(\frac{-0.03 * 210}{2}\right) = 812.086\ m$$

$$VPT\ station = \left[32 + 14.873 + \left(\frac{210}{2}\right)\right] = [37 + 19.873]m$$

$$L_0 = -\left(\frac{0.04 * 210}{-0.07}\right) = 120.000\ m$$

$$y_0 = -\left[\frac{0.04^2 * 210}{2 * (-0.07)}\right] = 2.400\ m$$

$$F = \frac{(-0.03 - 0.04)}{8} * 210 = -1.838\ m$$

Table 23.2 *summarises the calculated values for the singular points of the vertical curve.*

Table 23.2 Local coordinates of the singular points of the vertical curve

	Local coordinates		
Station	x [m]	y [m]	H [m]
VPC	0.000	0.000	811.036
VPI	105.000	4.200	815.236
VPT	210.000	1.050	812.086
M	105.000	2.363	813.399
V	120.000	1.350	813.436
P	93.602	2.284	813.320

The local coordinates of the point (P) are calculated as follows:

$$x_P = [32 + 3.475]\ m - [27 + 9.873]\ m$$

$$x_P = 643.475 - 549.873 = 93.602\ m$$

$$y_P = \frac{-0.07}{2 * 210} * 93.602^2 + 0.04 * 93.602$$

$$y_P = 2.284\ m$$

23.5   Setting-out of Curves

Setting-out the geometric elements of a curve means positioning points along the route where it is to be built using stakes to indicate the linear and elevation references for its construction.

The distance between the stakes varies according to the work type and the geometric element being set out. For example, on a highway, stakes are usually set and levelled every 20 m, but they can also be set every 10, 30, 50 or 100 m. In the case of urban roads, stakes may be set every 1, 2 or 5 m. If grading is very steep or if the site has special soil characteristics, the stakes can be set at closer intervals.

In cases where the project has a platform, as shown in Fig. 23.13, in addition to the centreline of the proposed route, setting out includes the edges of the platform and the points at which the cut and fill slopes intersect the natural ground surface (*offset point* or *catch point*).

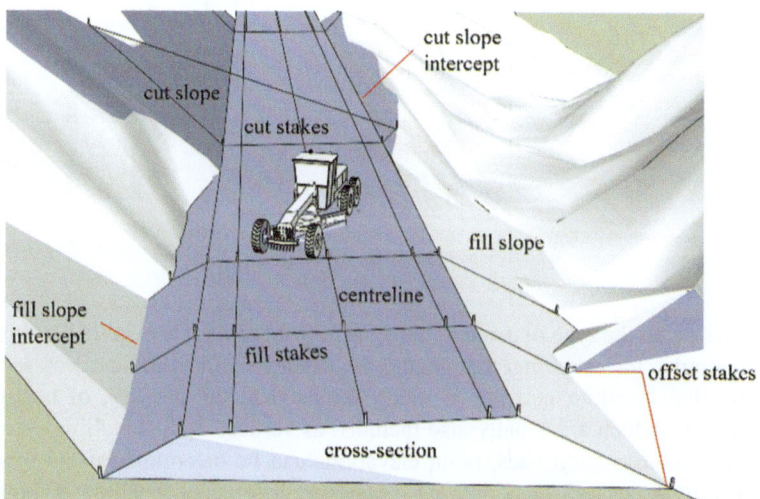

Fig. 23.13 Setting-out stakes

Once the stakes have been positioned to define the route in the horizontal plane, levelling is carried out to define the cut and fill levels, which are the indicators for the subsequent earthworks.

Setting-out the tangents does not present any major difficulties. For this reason, this section only presents the geometric and technical details of the stakeout of horizontal and vertical curves, with an emphasis on the geometric layout of transportation routes.

23.5.1   *Establishing a Network of Control Points*

As with any major engineering project, the first step in setting out a horizontal and vertical alignment is establishing a control point network. This can be done using any of point positioning methods described in previous chapters. Typically, a survey traverse or GNSS surveying methods are used. Whichever method is used, the result should be the implementation of markers with known coordinates along the project section. As they are reference points that must be maintained throughout the construction period, and in many cases, even afterwards, the markers of the control point network must be built in concrete, clearly identified, documented and protected, as shown in Fig. 23.14.

Fig. 23.14 Illustration of a control point network marker

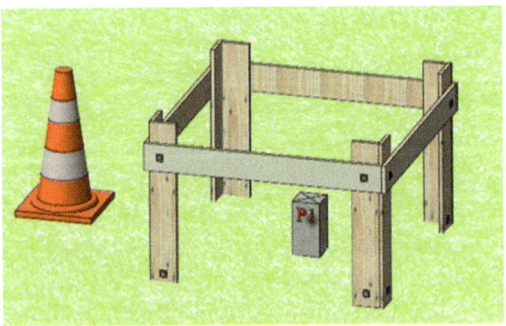

Depending on the type of project, it may also be necessary to establish a network of benchmarks for levelling the geometric elements along the project section. Typically, the levelling network is much denser than the network of horizontal control points, which are usually also included as vertices in the levelling network.

On some construction sites, point elevations can be determined using trigonometric levelling or GNSS technology; otherwise, differential levelling is required. The choice of the most appropriate levelling method is determined according to the accuracy required for the vertical positioning of the construction elements.

Particular attention must be paid to the reference system and vertical datum chosen for the design and setting-out of the project. This choice will depend on the field measurements for detailed surveys and construction setting-out.

23.5.2 Setting-out Horizontal Circular Curves

Except in special cases, the radii of horizontal curves used in Civil Engineering are too large to locate the centre of the circle and swing the arc with a total station. Therefore, some field procedures are recommended for this purpose, which are presented below. They all assume that the tangents have been previously staked out and that the (PC), (PI) and (PT) stations of the curves are known. It should be noted that the (PI) stations are only additional references and are, therefore, not mandatory, mainly because, in some cases, they may be located in places that are difficult to access.

Among the various methods of setting horizontal curves available in the literature, the following are discussed in this chapter:

- *Total coordinates method*, which is based on the positioning of points along the curve according to their plane-rectangular coordinates known in the project and the coordinates of previously set out control points.
- *Deflection angle method*,[5] which is based on the deflection angles to each stake along the curve. A given deflection angle is measured from the tangent at the (PC) or (PT) station to the point of interest on the curve.
- *Offset from the tangent method*, which is based on a local plane-rectangular coordinate system of the curve. A given offset is taken perpendicularly from the tangent line to the point of interest on the curve.

23.5.2.1   Total Coordinates Method

With the growing popularity of total stations and GNSS technology operating in RTK mode, the most commonly used method for setting-out horizontal curves today is the total coordinates method. It is a simple method to use, especially if RTK-GNSS instruments are used. It only requires that control points are located along the design section and that the coordinates of the points to be set out are known in relation to this network of control points. As it is based on known coordinates, it does not necessarily require the prior implementation of (PC), (PT) and (PI) stations.

If total stations are used, setting out is carried as presented in Sect. 21.3.1. Figure 23.15 shows the application of this method. In this case, the point (PS) is a control points and the points (Pi) are station points along the curve. Depending on the length of the curve, points can be set out from more than one control point. When using RTK-GNSS instruments, the stakeout is performed as presented in Sect. 21.3.3.

The coordinates of the station points to be set out are obtained from their positions in the CAD graphics. In cases where it is necessary to calculate these values algebraically, one of the methods presented below can be used. See Fig. 23.16.

---

[5] Called the tangential method by some authors.

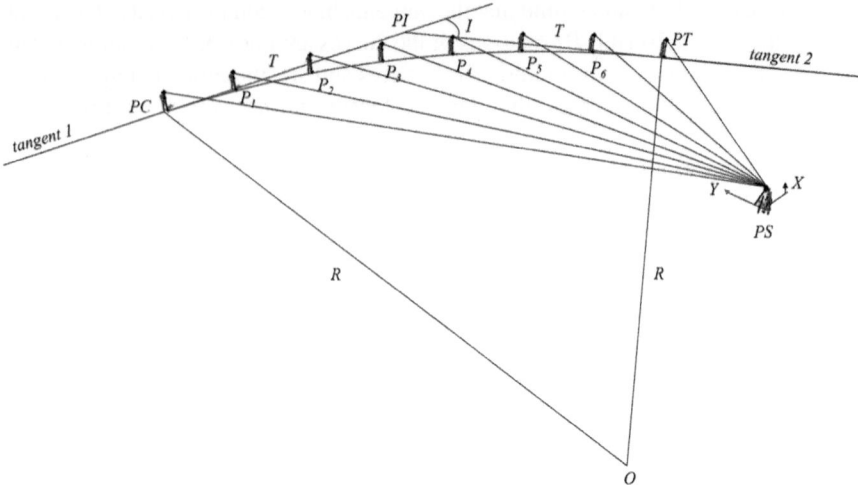

Fig. 23.15 Principle of setting-out a horizontal circular curve using a total station

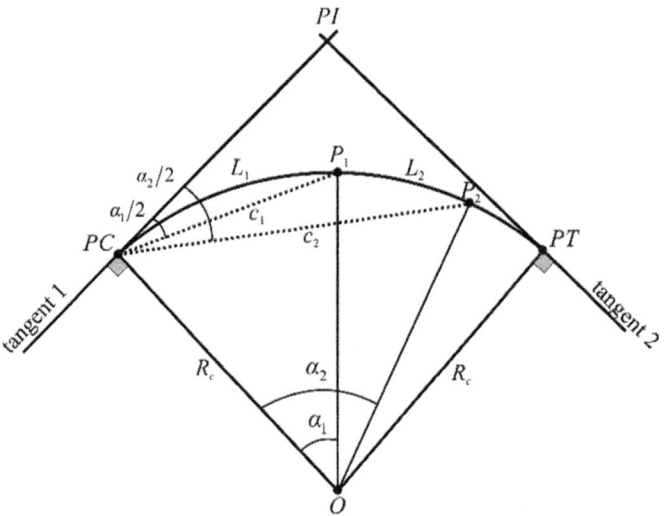

Fig. 23.16 Geometric elements for calculating the total coordinates of a horizontal circular curve

Data:
(PC) station: (X_{PC}, Y_{PC})
(PI) station: (X_{PI}, Y_{PI})
(PT) station: (X_{PT}, Y_{PT})
R_c (Radius)
Arc lengths (L_1) and (L_2)

Calculate:
(P_1) station: (X_{P_1}, Y_{P_1})
(P_2) station: (X_{P_2}, Y_{P_2})

Calculation by deflections:
1. From the coordinates of (PC) and (PI) stations, calculate the azimuth Az_{PCPI} using Eq. (11.4);
2. Calculate the deflection angle $(\alpha_1/2)$ at point (PC) as a function of the arc length L_1 and the radius of curvature R_c.

$$\frac{\alpha_1}{2} = \frac{L_1}{2R} \text{ [rad]} \quad \text{or} \quad \frac{\alpha_1}{2} = \frac{90° * L_1}{\pi * R} \text{ [degrees]} \tag{23.54}$$

3. Calculate the azimuth Az_{PCP_1} of the alignment PCP_1 by adding the azimuth Az_{PCPI} to the deflection angle $(\alpha_1/2)$, considering the direction of the curve in relation to the point (PC).

$$Az_{PCP_1} = Az_{PCPI} \pm \frac{\alpha_1}{2} \tag{23.55}$$

4. Calculate the length of the chord PCP_1 using Eq. (23.56).

$$c_{PCP_1} = 2R * \sin\left(\frac{\alpha 1}{2}\right) \tag{23.56}$$

5. Finally, calculate the coordinates of the point (P_1) using Eq. (23.57).

$$XP_1 = X_{PC} + c_{PCP_1} * \sin(Az_{PCP_1})$$
$$YP_1 = Y_{PC} + c_{PCP_1} * \cos(Az_{PCP_1}) \tag{23.57}$$

Calculation from the coordinates of the centre of the curve (O):

1. Calculate the azimuth Az_{PCPI} as in the previous case.
2. Using the azimuth Az_{PC-PI} and knowing that the centre of the curve (O) is in the perpendicular alignment to the PC-PI tangent, calculate the azimuth of the PC-O alignment, considering the direction of the curve in relation to the point (PC);

$$Az_{PC-O} = Az_{PC-PI} \pm 90° \tag{23.58}$$

3. Calculate the coordinates of the centre of the curve using Eq. (23.59).

$$X_O = X_{PC} + R * \sin(Az_{PC-O})$$
$$Y_O = Y_{PC} + R * \cos(Az_{PC-O}) \tag{23.59}$$

Note that the coordinates of the centre of the curve are often already known, so it is not necessary to consider step 3.

4. Calculate the value of the central angle of the corresponding curved section as a function of the arc length L_1 and the radius of curvature R_c, according to Eq. (23.54):

$$\alpha_1 = \frac{L_1}{R} \text{ [rad]} \quad \text{or} \quad \alpha_1 = \left(\frac{180°}{\pi * R}\right) * L_1 \text{ [degrees]}$$

5. Calculate the azimuth Az_{OP_1} of the O-P_1 alignment by adding the azimuth $Az_{PC-O} \pm 180°$ to the calculated value of the central angle, considering the direction of the curve with respect to the (PC) station.

$$Az_{O-P_1} = Az_{PC-O} \pm \alpha_1 \pm 180° \tag{23.60}$$

6. Finally, calculate the coordinates of the point (P_1) using Eq. (23.61).

$$X_{P_1} = X_O + R * \sin(Az_{O-P_1})$$
$$Y_{P_1} = Y_O + R * \cos(Az_{O-P_1}) \tag{23.61}$$

The same steps can be used to calculate the coordinates of the point (P_2) or any other point on the curve.

Example 23.4

Given the coordinates in Table 23.3 for a simple horizontal circular curve to the right, whose radius of curvature $R_c = 500$ m, calculate the coordinates of the point (P_1) located 50 m from the (PC) along the curve. Use the two calculation methods described above.

Station	X [m]	Y [m]
PC	5172.023	10,520.960
PI	5248.357	10,598.369

Table 23.3 Known coordinates

Solution:

Using the calculation by deflections, the coordinates of the point (P_1) can be obtained as follows:

$$Az_{PCPI} = a\tan\left(\frac{5248.357 - 5172.023}{10,598.369 - 10,598.369}\right) = 44°35'57.8''$$

$$\frac{\alpha_1}{2} = \left(\frac{90°}{\pi * 500}\right) * 50 = 2°51'53.2''$$

$$Az_{PCP_1} = 44°35'57.8'' + 2°51'53.2'' = 47°27'51.0''$$

$$c_{PCP_1} = 2 * 500 + \sin(2°51'53.2'') = 49.979\,\text{m}$$

$$X_{P_1} = 5172.023 + 49.979 * \sin(47°27'51.0'') = 5208.850\,\text{m}$$

$$Y_{P_1} = 10,520.96 + 49.979 * \cos(47°27'51.0'') = 10,554.748\,\text{m}$$

Using the calculation from the coordinates of the centre of the curve, the coordinates of the point (P_1) can be obtained as follows:

$$Az_{PC-O} = 44°35'57.8'' + 90° = 134°35'57.8''$$

$$X_O = 5172.023 + 500 * \sin(134°35'57.8'') = 5528.040\,\text{m}$$

$$Y_O = 10,520.960 + 500 * \cos(134°35'57.8'') = 10,169.887\,\text{m}$$

$$\alpha_1 = \frac{50}{500} = 0.100\,\text{rad} = 5°43'46.5''$$

$$Az_{O-P_1} = 134°35'57.8'' + 5°43'46.5'' + 180° = 320°19'44.3''$$

$$X_{P_1} = 5528.040 + 500 * \sin(320°19'44.3'') = 5208.850\,\text{m}$$

$$Y_{P_1} = 10,169.887 + 500 * \cos(320°19'44.3'') = 10,554.748\,\text{m}$$

23.5.2.2 Deflection Angle Method

As seen in the previous method, deflection angles and chord length values are essential to the setting-out of a horizontal curve. There is, therefore, a method of setting-out based on these geometric elements and the position of the tangents, as shown in Fig. 23.17. It is usually carried out using a total station, although it can also be performed using a theodolite and a tape measure.

In this method of setting-out horizontal curves, as shown in Fig. 23.17, staking starts at the first integer stake that follows the (PC) station, which, in the case of Fig. 23.17, is the point (P_1) located at a distance (L_1) from the (PC) station along the curve.

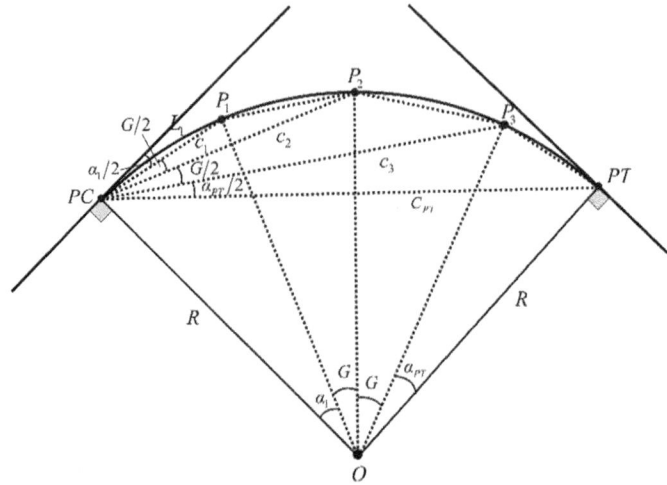

Fig. 23.17 Principle of horizontal circular curve setting-out using the deflection angle method

Using the back tangent of the curve as a reference, the point (P_1) can be set out using a deflection angle and the length of the chord PCP_1. Thus, according to the geometrical relations shown in Fig. 23.17.

$$\frac{\alpha_1}{2} = \frac{L_1}{2R} \text{[rad]} \quad \text{or} \quad \frac{\alpha_1}{2} = \frac{90° * L_1}{\pi * R} \text{[degrees]} \quad \text{according to Eq. (23.54)}$$
$$c_1 = 2R * \sin\left(\frac{\alpha_1}{2}\right) \qquad\qquad\qquad\qquad\qquad \text{according to Eq. (23.56)}$$

Once the first integer stake has been set out, the remaining integer stakes on the curve (in Fig. 23.17, these are the stakes (P_2) and (P_3)) will have the same deflection angle in relation to the direction of the previous chord. Thus, if all the stakes on the curve are visible from the (PC) station, the staking continues regularly until the last integer stake before the (PT) station is set out. From this last integer stake, the position of the (PT) station is verified using the angle ($\alpha_{PT}/2$) and the chord c_{PT}. See Fig. 23.17. This verification is an indicator of the quality of the work carried out. Thus, for a 20 m staking, the following deflection angle value is obtained for successive integer stakes:

$$\frac{\alpha_i}{2} = \frac{G}{2} = \frac{20 * 180°}{2\pi * R} \qquad (23.62)$$

The accumulated chord length for each point on the curve is calculated from the sum of the accumulated deflection angles up to the point of interest, using the following equation:

$$c_i = 2R * \sin\left(\sum \frac{\alpha_i}{2}\right) \quad \text{according to Eq. (23.56)}$$

When setting-out with a total station, the following steps can be followed in the field:

1. Check that all stakes on the curve can be set out from the (PC) station.
2. Prepare a table of angle and distance values for the stakeout. Some total stations have applications that allow this table to be stored in the instrument's memory and used during stakeout.
3. Install the total station on the (PC) station.
4. Orient the instrument at one of the back or forward stakes of the tangent. Set the instrument to $H_z = 0$ in this direction.
5. Set out each stake on the curve according to the angle of deflection and the length of the chord.
6. At the last point, check the position of the (PT) station.

Example 23.5

Given a simple horizontal circular curve (to the right) of a road in a subdivision project, whose known geometric elements are given in Table 23.4, prepare the setting-out table for this curve, considering that a total station will be installed at the (PC) station. Also consider a stationing interval of 20 m.

Table 23.4 Geometric elements of the curve

Element	Value
(PC) station	[340 + 3.000] m
Tangent distance (T)	128.512 m
Central angle of the curve (I)	81°10′34″

Solution:

Before preparing the setting-out table, it is necessary to calculate the elements of the curve as shown below:

$$R = \frac{128.512}{\tan(81°10′34″/2)} = 150.001 \text{ m}$$

$$G = \frac{20 * 180°}{\pi * 150.001} = 7°38'21.8''$$

$$L_c = \frac{\pi * 150.001 * 81°10'34''}{180°} = 212.520 \text{ m}$$

$$(PT) \, station = [340 + 3.000] \text{ m} + 212.520 \text{ m}$$

$$(PT) \, station = [350 + 15.520] \text{ m}$$

$$\frac{\alpha_{station[341]}}{2} = \frac{90° * 17.000}{\pi * 150.001} = 3°14'48.3''$$

$$\frac{\alpha_{PT}}{2} = \frac{90° * 15.520}{\pi * 150.001} = 2°57'50.5''$$

The values of deflection angles and chord lengths for setting-out the curve from the (PC)station are shown in Table 23.5.

Table 23.5 Setting-out table

Station	Arc length [m]	Chord length [m]	Deflection angle
PC = [340 + 3.000] m	0	0	00°00'00''
[341]	17.000	16.991	3°14'48''
[342]	37.000	36.906	7°03'59''
[343]	57.000	56.658	10°53'10''
[344]	77.000	76.157	14°42'21''
[345]	97.000	95.319	18°31'32''
[346]	117.000	114.057	22°20'43''
[347]	137.000	132.288	26°09'54''
[348]	157.000	149.931	29°59'05''
[349]	177.000	166.908	33°48'16''
[350]	197.000	183.144	37°37'27''
PT = [350 + 15.520] m	212.520	195.186	40°35'17''

In Table 23.5, the angular values of the deflections were rounded to the nearest arcsecond.

The calculations performed for stations [341] and [342] are presented below to assist the reader in their studies.

Deflection angle at the station $[341] = \dfrac{\alpha_{station[341]}}{2} = \dfrac{90° * 17.000}{\pi * 150.001} = 3°14'48.3''$

Deflection angle at the station $[342] = \dfrac{\alpha_{station[342]}}{2} = 3°14'48.3'' + \dfrac{7°38'21.8''}{2}$

$$= 7°03'59.2''$$

Deflection angle at the (PT) station $= [350 + 15.520] = \dfrac{\alpha_{station[PT]}}{2} = 37°37'26.5''$

$+ 2°57'50.5'' = 40°35'17.0''$

In some cases, depending on the length of the curve, it may not be possible to set out all the stakes on the curve from a single instrument setup, so the surveying instrument will need to be installed at different points along the curve.

Consider Fig. 23.18, where the total station had to be installed at point (P_2) and backsighted at (PC) station. Under these conditions, to set out the stake (P_3), the operator must rotate the instrument by a clockwise horizontal angle equal to $\left(180° + \frac{\alpha_2}{2} + \frac{G}{2}\right)$ observing the direction of the curve, i.e., whether to the right or to the left. The operator must add half the corresponding central angle to set out all the other stakes from the point (P_2) backsighted towards the (PC) station. The procedure is repeated if the total station is moved to another point on the curve.

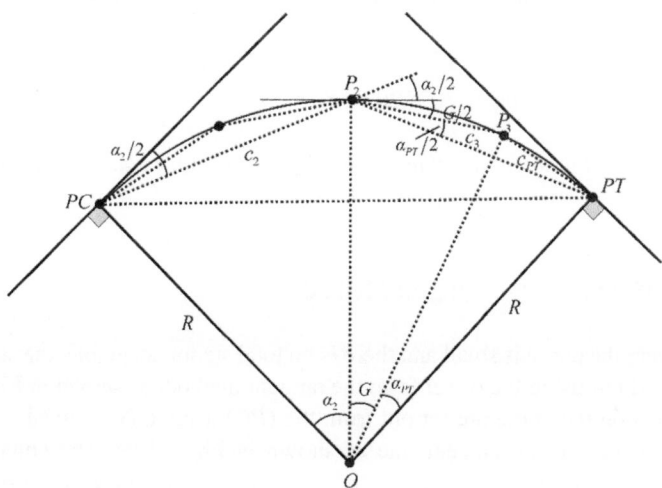

Fig. 23.18 Horizontal circular curve setting-out from a given point on the curve

Example 23.6
Given the data in example 23.5, assume that all integer stakes up to the stake [345 + 0.00] m of the horizontal curve must be set out by installing the total station at the (PC) station. Then, the remaining integer stakes of the curve must be set up from the stake [345 + 0.00] m. Prepare the setting-out table for the curve.

Solution:

Table 23.6 gives the deflection angles and chord lengths for setting-out the curve. The table repeats the values already calculated in example 23.5 for the integer stakes up to stake [345 + 0.00] m.

Table 23.6 Setting out table

Station	Arc length [m]	Chord length [m]	Deflection angle
PC = [340 + 3.000] m	0	0	00°00'00"
[341]	17	16.991	3°14'48"
[342]	37	36.906	7°03'59"
[343]	57	56.658	10°53'10"
[344]	77	76.157	14°42'21"
[345]	97	95.319	18°31'32"
[346]	20	19.985	202°20'43"
[347]	40	39.882	206°09'54"
[348]	60	59.601	209°59'05"
[349]	80	79.055	213°48'16"
[350]	100	98.158	217°37'27"
PT = [350 + 15.520] m	115.520	112.686	220°35'17"

The calculation for stake [346] is given below to assist the reader.

$$Deflection = 180° + 18°31'31.9'' + 3°49'10.9''$$
$$Deflection = 202°20'42.8''$$

In Table 23.6, the angular values of the deflections were rounded to the nearest arcsecond.

23.5.2.3   Offset from the Tangent Method

In cases where the curve is small and there is no total station available, the setting out can be carried out using the offset from the tangent method, as shown in Fig. 23.19.

The stakes on the curve are set out from the (PC) using a (x, y) local coordinate system referenced to the tangent line as shown in Fig. 23.19. The equations for calculating the (x, y) coordinates using the chord length c are given by Eqs. (23.63) and (23.64).

$$x_{P_i} = c_i * \cos\left(\frac{\alpha_i}{2}\right) \tag{23.63}$$

$$y_{P_i} = -c_i * \sin\left(\frac{\alpha_i}{2}\right) \tag{23.64}$$

The (x, y) coordinates can also be calculated using the radius of curvature R, as given by Eqs. (23.65) and (23.66).

$$x_{P_i} = R * \sin(\alpha_i) \tag{23.65}$$

$$y_{P_i} = -R * [1 - \cos(\alpha_i)] \tag{23.66}$$

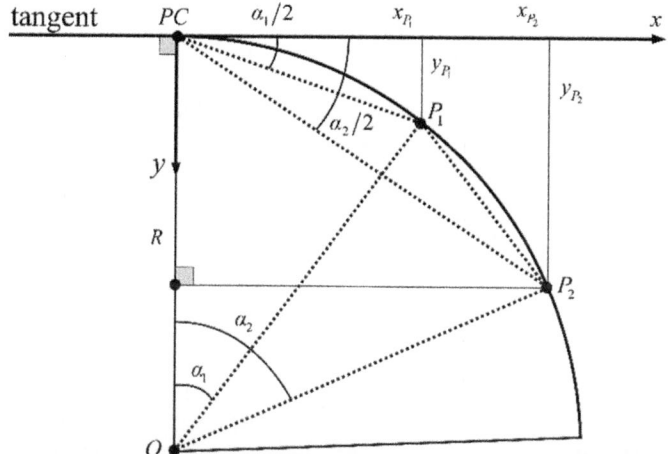

Fig. 23.19 Principle of horizontal circular curve setting-out using the offset from the tangent method

This method can be modified to offset from the long chord and a similar (x, y) local coordinate system.

Example 23.7
Using the offset from the tangent method, prepare the setting-out table for the same curve as in example 23.5.

Solution:

Table 23.7 shows the setting-out table using Eqs. (23.63) and (23.64).

Calculations for stake [341] are given below using Eqs. (23.63) to (23.66).

$$x_{station \ [341]} = 16{,}991 * \cos\left(\frac{3°14'48.3''}{2}\right) = 16.964 \, \text{m}$$

$$y_{station \ [341]} = -16.991 * \sin\left(\frac{3°14'48.3''}{2}\right) = -0.962 \, \text{m}$$

or

$$x_{station \ [341]} = 150 * \sin(6°29'36.6'') = 16.964 \, \text{m}$$

$$y_{station \ [341]} = -16.991 * \sin\left(\frac{3°14'48.3''}{2}\right) = -0.962 \, \text{m}$$

Table 23.7 Setting out table

Station	Arc length [m]	Chord length [m]	Deflection angle	x [m]	y [m]
PC = [340 + 3.000] m	**0**	**0**	00°00′00″		
[341]	17.000	16.991	3°14′48″	16.964	−0.962
[342]	37.000	36.906	7°03′59″	36.626	−4.540
[343]	57.000	56.658	10°53′10″	55.638	−10.700
[344]	77.000	76.157	14°42′21″	73.663	−19.333
[345]	97.000	95.319	18°31′32″	90.379	−30.285
[346]	117.000	114.057	22°20′43″	105.492	−43.363
[347]	137.000	132.288	26°09′54″	118.732	−58.333
[348]	157.000	149.931	29°59′05″	129.864	−74.931
[349]	177.000	166.908	33°48′16″	138.691	−92.861
[350]	197.000	183.144	37°37′27″	145.056	−111.805
PT = [350 + 15.520] m	**212.520**	**195.186**	40°35′17″	**148.226**	**−126.991**

23.5.3 Setting Out Transition Curves

Figure 23.20 illustrates the principle of setting out a point on the transition spiral before the horizontal circular curve. This can be done using any of the horizontal curve setting-out methods described in the previous sections.

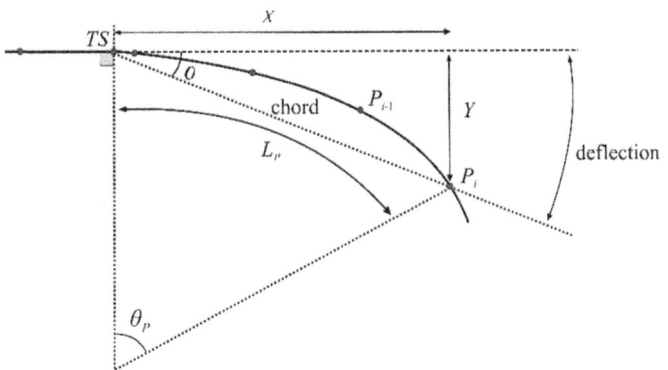

Fig. 23.20 Setting-out a horizontal circular curve with spiral curve

(a) *Setting out using the offset from the tangent method:* in this case, the equations given in Sect. 23.3 must be used.

Referring to Fig. 23.20 and Eqs. (23.37), (23.22) and (23.23) the following equations can be used to determine the setting-out elements:

(b) *Setting out using the deflection angle method:* in this case, the value of the deflection angle is given by Eq. (23.40) and the chord length is given by Eq. (23.64).

$$\theta_P = \left(\frac{L_P}{L_S}\right)^2 * \theta_S \, [\text{rad}] \qquad\qquad\qquad \text{according to Eq. (23.37)}$$

$$x_P = L_P * \left(1 - \frac{\theta_P^2}{10} + \frac{\theta_P^4}{216} - \frac{\theta_P^6}{9360} + \frac{\theta_P^8}{685,440} \cdots\right) \qquad \text{according to Eq. (23.22)}$$

$$y_P = L_P * \left(\frac{\theta_P}{3} - \frac{\theta_P^3}{42} + \frac{\theta_P^5}{1320} - \frac{\theta_P^7}{75,600} \cdots\right) \qquad \text{according to Eq. (23.23)}$$

$$\alpha_P = \left(\frac{L_P}{L_S}\right)^2 * \frac{\theta_S}{3} = \arctan\left(\frac{y_P}{x_P}\right) \qquad \text{according to Eq. (23.40)}$$

$$chord_P = \sqrt{x_P^2 + y_P^2} \qquad\qquad \text{Eq. (23.67)}$$

(c) *Setting out using the total coordinates method:* in this case, the procedure is similar to that for setting out a simple horizontal circular curve.

Example 23.8
Using the geometric elements given in Table 23.8, calculate the setting out elements of the first spiral section of the curve using the three methods presented in the previous sections. Consider staking out every 20 metres.

Table 23.8   Geometric elements of the spiral

[PI] station	[92 + 12.560] m
Total length of the spiral (L_S)	80 m
Radius of the horizontal circular curve (R_c)	500 m
Central angle of the horizontal curve (I)	35°12′47″
Azimuth of the alignment TS-PI	23°45′18″
Coordinates of the (TS) station	X = 5791.164 m
	Y = 11,404.340 m

Solution:

The calculation sequence required to solve this example is shown below.

$$\theta_S = \frac{80}{2 * 500} = 0.080 \, \text{rad}$$

$$x_S = 80 * \left(1 - \frac{0.080^2}{10} + \frac{0.080^4}{216} - \frac{0.08^6}{9360} + \frac{0.080^8}{685,440} \cdots\right) = 79.949 \, \text{m}$$

$$y_S = 80 * \left(\frac{0.080}{3} - \frac{0.080^3}{42} + \frac{0.080^5}{1320} - \frac{0.080^7}{75,600} \cdots \right) = 2.132\,\text{m}$$

$$Q = 79.949 - 500 * \sin\,(4°35'01.2'') = 39.991\,\text{m}$$

$$p = 2.132 - 500 * [1 - \cos(4°35'01.2'')] = 0.533\,\text{m}$$

$$TT = 39.991 + (500 + 0.533) * \tan\left(\frac{35°12'47''}{2} \right) = 198.833\,\text{m}$$

$$(TS)\,Station = [92 + 12.560] - 198.833 = [82 + 13.727]\,\text{m}$$

Table 23.9 *shows the setting out elements from the (TS) station to the (SC) station, using the offset from the tangent method. The calculation of the setting out elements for stake [83] is given below.*

Table 23.9 Setting out table for the transition horizontal curve using the offset from the tangent method

Station	L [m]	θ[rad]	x [m]	y [m]
TS = 82 + 13.727 m	0.000	0	0.000	0.000
83	6.273	0.00049184	6.273	0.001
84	26.273	0.00862820	26.273	0.076
85	46.273	0.02676456	46.269	0.413
86	66.273	0.05490092	66.253	1.213
SC = 86 + 13.727 m	80.000	0.07999945	79.949	2.132

$$L_{station[83]} = 20 - 13.727 = 6.273\,\text{m}$$

$$\theta_{station[83]} = \left(\frac{6.273}{80} \right)^2 * 0.080$$

$$\theta_{station[83]} = 0.00049184\ \text{rad}$$

$$x_{station[83]} = 6.273 * \left(1 - \frac{0.000491882^2}{10} + \frac{0.000491882^4}{216} - \frac{0.000491882^6}{9360} \right.$$

$$\left. + \frac{0.000491882^8}{685,440} \right) = 6.273\,\text{m}$$

$y_{station[83]}$

$$= 6.273 * \left(\frac{0.000491882}{3} - \frac{0.000491882^3}{42} + \frac{0.000491882^5}{1320} - \frac{0.000491882^7}{75,600} \right)$$

$$= 0.001 \, m$$

Table 23.10 *shows the setting-out table for the deflection angle method. The calculation of the setting out elements for stake [83] is given below.*

Table 23.10 Setting out table for the transition horizontal curve using the deflection angle method

Station	x [m]	y [m]	Deflection angle	Chord length [m]
TS = 82 + 13.727 m	0.000	0.000	0°00′00″	0.000
83	6.273	0.001	0°00′34″	6.273
84	26.273	0.076	0°09′53″	26.273
85	46.269	0.413	0°30′40″	46.271
86	66.253	1.213	1°02′55″	66.264
SC = 86 + 13.727 m	79.949	2.132	1°31′40″	79.977

Deflection angle to the stake [83]:

$$\alpha_{station\,[83]} = atan\left(\frac{0.001}{6.273}\right) = 0.000163946 \, rad = 0°00′34″$$

Chord length to the stake [83]:

$$L_{station\,[83]} = \sqrt{6.273^2 + 0.001^2} = 6.273 \, m$$

Table 23.11 *shows the setting-out table for the total coordinates method. Using this method requires the calculation the azimuths of the directions of each stake to be set out, as well as the chord length for each point. The calculation for the coordinates of the stake [83] is given below.*

Table 23.11 Setting-out table for the transition horizontal curve using the total coordinates method

Station	Deflection angle	Azimuths	Chord length [m]	X [m]	Y [m]
TS = 82 + 13.727 m	0°00′00″	23°45′18″	0.000	5791.164	11,404.340
83	0°00′34″	23°45′52″	6.273	5793.692	11,410.081
84	0°09′53″	23°55′11″	26.273	5801.816	11,428.356
85	0°30′40″	24°15′58″	46.271	5810.180	11,446.523
86	1°02′55″	24°48′13″	66.264	5818.962	11,464.491
SC = 86 + 13.727 m	1°31′40″	25°16′58″	79.977	5825.321	11,476.656

$$X_{station[83]} = 5791.164 + [6.273 * \sin(23°45'52'')] = 5793.692 \text{ m}$$

$$Y_{station[83]} = 11,404.340 + [6.273 * \cos(23°45'52'')] = 11,410.081 \text{ m}$$

23.5.4   Setting Out Vertical Curves

Setting out vertical curves basically consists of marking the final elevation of the vertical alignment in the field to guide the construction team in placing the vertical elements. The method of setting out a grade stake is the same, whether it is on a tangent or a curve. In general, once the horizontal alignment is in place, tangent grades, vertical curves, and cut and fill slopes can be set out by coordinates using total stations or RTK-GNSS receivers, or differential levelling where high-accuracy is required.

Assuming that the intermediate elevations at regular intervals along the vertical curve have been calculated, the field procedure is to physically mark each intermediate stake in the ground with the corresponding cut or fill height, i.e., how much the earthmoving machine operator has to cut or fill at that point to reach the level required by the vertical curve design. At the same time, if a cut and fill platform is to be set out, the stakes of the offset points (cut and fill intercepts) are set out by their (X, Y) coordinates specified by the designer.

In the same way as in horizontal curves, the elevations of the stakes on a vertical curve are calculated and stakeout on the terrain considering the spacing between the horizontal stations already set out and, in some cases, fractions of them (depending on the roughness of the terrain), as shown in Fig. 23.21. The rougher the terrain, the smaller the station interval.

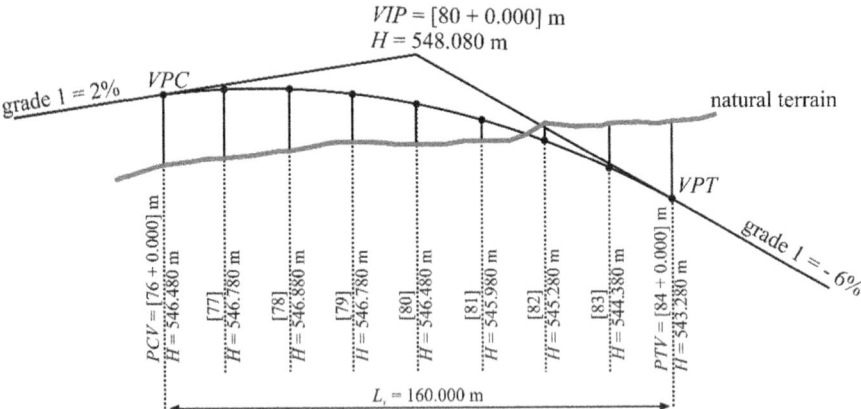

Fig. 23.21 Elevation values and stakes of a vertical curve

The elevations of the stakes on the vertical curve are calculated using Eq. (23.45), taking the elevation of the (*VPC*) station as a reference. The setting out parameters can be calculated as shown in Table 23.12. To facilitate the work of the earthmoving machine operator, it is usual to indicate the difference in elevation using a grade stake, generally consisting of a wooden peg with a mark indicating the height from the reference point to the top elevation, as shown in Fig. 23.22.

Table 23.12 Difference in elevation for setting out the vertical curve

			Difference in elevation (ΔH)	
Station	H_{ground} [m]	H_{design} [m]	Cut [m]	Fill [m]
VPC = [76]	542.325	546.480		4.155
77	542.489	546.780		4.291
78	543.014	546.880		3.866
79	543.873	546.780		2.907
VPI = [80]	544.118	546.480		2.362
81	544.981	545.980		0.999
82	546.257	545.280	0.977	
83	547.148	544.380	2.768	
VPT = [84]	548.053	543.280	4.773	

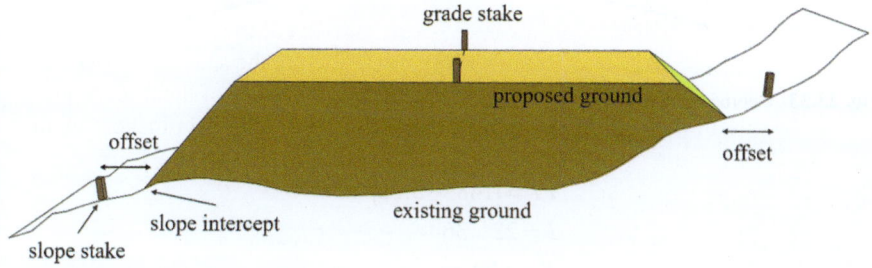

Fig. 23.22 Slope and grade stakes

Where cut and fill sections are required, slope stakes are used to control the construction of earthworks slopes. They are set to mark the designed offset points (catch points), generally, at a constant offset distance from the slope intercept line to facilitate earthworks. See Fig. 23.22.

Table 23.12 shows the cut or fill heights for setting each stake on the vertical curve in Fig. 23.21.

It must be emphasised that the setting out of horizontal and vertical alignments requires the supervision of experienced professionals and a well-trained site team. Due to the constant movement of people, machinery and earth on the work site, it is often necessary to set out the same stake repeatedly, so there must be a constant harmony between the machine operators and the engineer in charge of the project.

Another important factor to consider when settng-out a road project is the increasing use of machine control systems, which Geomatics professionals must be familiar with.

23.6   Problems

1. Calculate the geometric elements of a simple horizontal circular curve with a radius of curvature of 200 m and a central angle of $32°\ 12'\ 35''$.
2. From the values calculated in the previous problem and assuming a (PI) station $= [325 + 16.37]$ m, calculate the values of the (PC) and (PT) stations and prepare the settng-out table using the deflection angle method. Assume a 20-m stationing.
3. From the geometry of the horizontal curve shown in Fig. 23.23 and the known geometric elements presented below, calculate the tangent distance T, the length of the curve L_c and the values of the (PC) and (PT) stations.

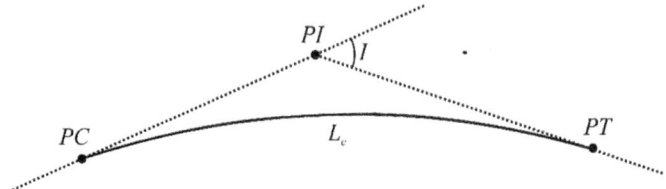

Fig. 23.23   Geometry of the horizontal curve

$$PI = [148 + 5.60]\ \text{m}$$
$$I = 22°\ 36'$$
$$R = 600\ \text{m}$$

4. Prepare the setting out table of the curve in the previous problem using the deflection angle method. Assume a 20-m stationing.
5. From the data shown in Fig. 23.24 and assuming that the two curves are horizontal circular curves, calculate the value of the final stake.

Fig. 23.24   Geometry of the horizontal circular curves

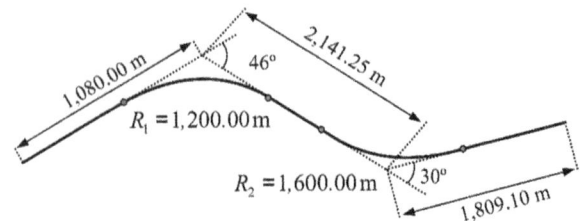

6. From the data shown in Fig. 23.25 and assuming that the tangents must be connected by horizontal circular curves, calculate:

(a) The maximum radius of each curve.
(b) The maximum radii if a tangent of 80.00 m is required between curves.

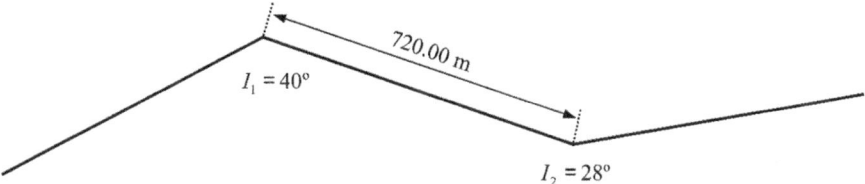

Fig. 23.25 Geometry of the horizontal alignment

7. From the geometric elements shown in Fig. 23.26, calculate the distance between points (PI_1) and (PI_2).

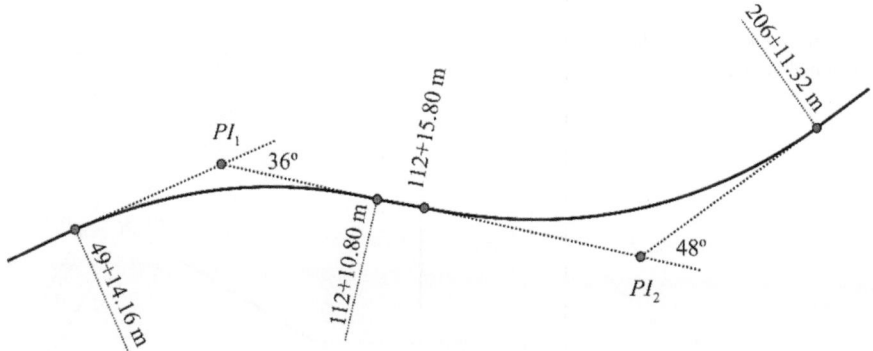

Fig. 23.26 Geometry of the horizontal circular curves

8. The engineer wants to design a simple horizontal circular curve according to the values shown in Table 23.13. Using these values and assuming a 20-m stationing, calculate the geometric parameters shown in Table 23.14.

Table 23.13 Geometric data of the horizontal circular curve

Radius of the curve	1000.000 m
Central angle of the curve (I)	28° 40′ 52″
(PI) station	$PI = [132 + 8.357]$ m

Table 23.14 Parameters to be calculated

Tangent of the curve (T)
External distance (E)
Midpoint of the long chord (S-PM)
Long chord (C)
Length of curve (L_c)
Degree of the curve (G_{20})
(PC) station
(PT) station

9. Consider a simple horizontal circular curve with a central angle of 22° 55′ 14″ and an external distance (E) of 21.523 m. From these values, calculate the curve's geometric elements R, T, and L_C.

10. The engineer needs to change the position of the ($PI1$) station of the horizontal circular curve shown in Fig. 23.27 so that it remains aligned with the back tangent and moves 100.000 m away parallel to the forward tangent. Using the data in Table 23.15 and assuming a 20-m stationing, calculate the new values for the $PI2$, $PC2$ and $PT2$ stations.

Fig. 23.27 Geometry of the horizontal circular curve

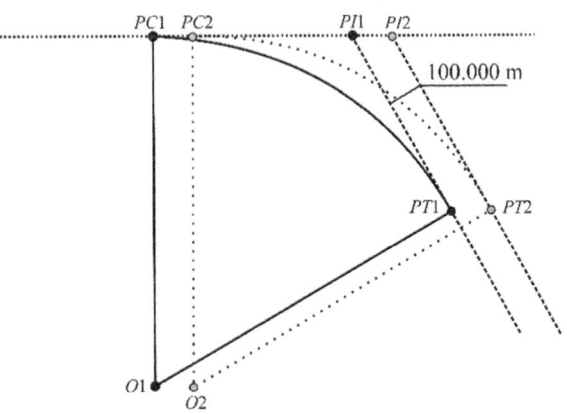

Table 23.15 Geometric data of the horizontal circular curve

(PI) station	[110 + 1.000] m
Central angle of the curve (I)	60° 00′ 00″
Length of circular curve (L_c)	1000.000 m

11. The engineer must set out a transition curve (spiral curve) using the total coordinates method. The geometric elements of the transition curve are given in Table 23.16. Calculate the (X) and (Y) coordinates for the stakes on the first spiral assuming a 20-m stationing.

Table 23.16 Transition curve
geometric data

(PI) station	[89 + 8.147] m
Total length of the spiral (L_s)	100.000 m
Radius of the circular curve (R_c)	500.000 m
Intersection angle (I)	30°

12. Using the results of the previous problem, prepare the settng-out the table for the transition curve using the deflection angle method.

13. Fig. 23.28 shows the geometric situation of the grade between two stakes in the road profile. Using the values shown in the figure, calculate the slope of the ramp.

Fig. 23.28 Geometric
situation of the ramp

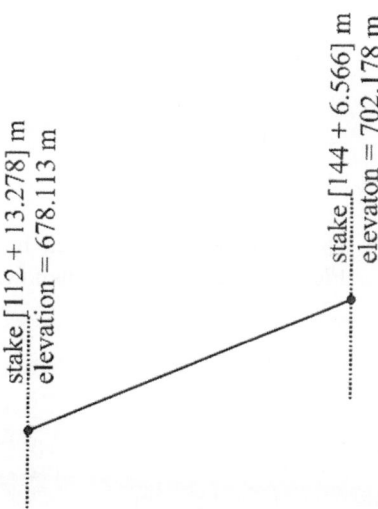

14. Fig. 23.29 shows the continuation of the road gradient from the previous problem. Using the values shown in the figure, calculate the differences in slope (δ_1) and (δ_2) between the grades.

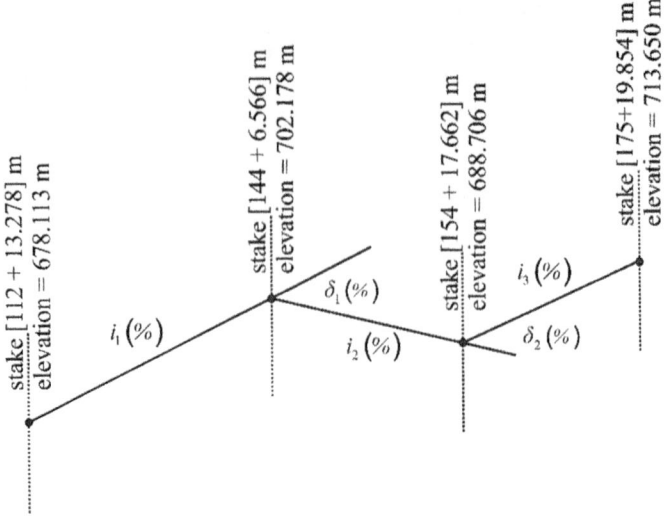

Fig. 23.29 Geometric situation of the grade

15. Using the values calculated in the previous problems and those shown in Fig. 23.30, calculate the unknown elevations of the stakes.

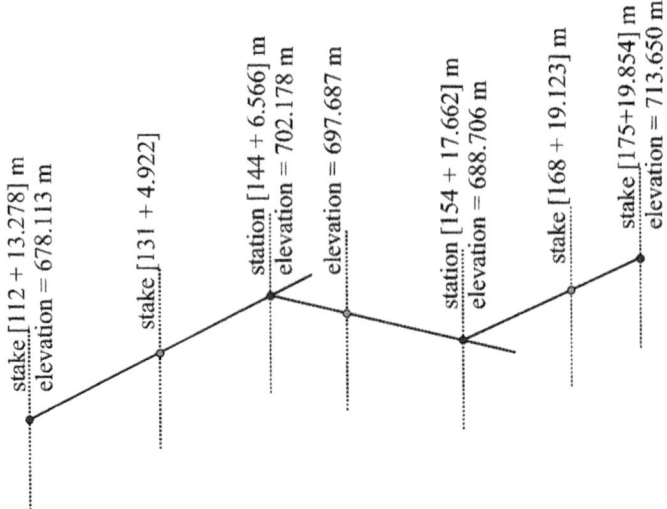

Fig. 23.30 Geometric situation of the grade

16. Using the values calculated in the previous problems and the data given in Fig. 23.31, calculate the elements of the vertical curve shown in Table 23.17,

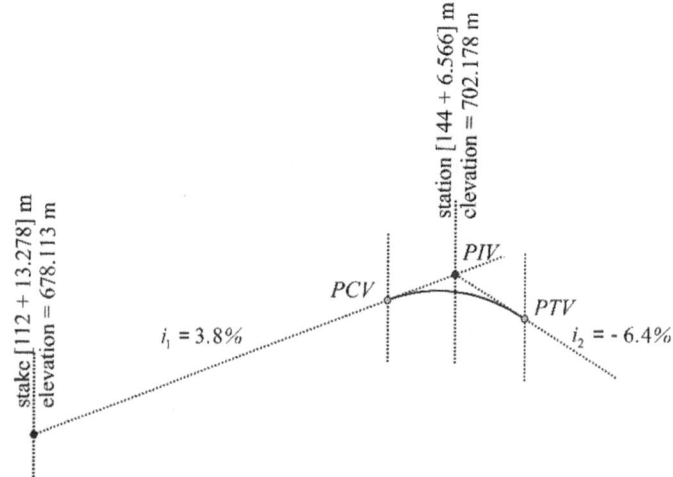

station [144 + 6.566] m
elevation = 702.178 m

stake.[112 + 13.278] m
elevation = 678.113 m

$i_1 = 3.8\%$

PCV

PIV

PTV

$i_2 = -6.4\%$

Fig. 23.31 Geometric situation of the grade

Table 23.17 Geometric elements to be calculated

Length of the vertical curve(L_v)
(VPC) station
(VPC) elevation
(VPT) station
(VPT) elevation
Maximum offset at VPI (F)
Midpoint of the vertical curve (M)
Highest point of vertical curve (V)

assuming that it is to be designed with a radius of curvature equivalent to the circular radius $R_V = 2000$ m.

17. Using the offset from the tangent settng-out method, calculate the (x, y) coordinates to set out the (PC) and (PT) stations of the horizontal circular curve with the following geometric elements:

$$R = 275.000 \text{ m}$$

$$(PI) \text{ station} = [147 + 12.40] \text{ m}$$

$$\text{Deflection angle } (I) = 28° 36'00''$$

References

AASHTO. (2011). A Policy on Geometric Design of Highways and Streets. 6th Edition. Washington, D.C.

Pimenta, Carlos R.T; Silva, I.; Oliveira, Marcio P.; Segantine, Paulo C. L. (2017). *Projeto geométrico de rodovias*. Editora Elsevier. Rio de Janeiro. ISBN:978-85-352-8621-2.

Chapter 24
Basics of Building Information Modelling for Geomatics

Júlio Franco and Irineu da Silva

24.1 Introduction

BIM, or *Building Information Modelling*, is a Civil Engineering project management process that helps to digitally create an accurate virtual model of a building known as a "building information model". It is a set of technologies, practices and collaborative processes for developing and managing projects based on geometric and semantic data, parametric graphical representation, oriented towards three-dimensional objects, and progressively developed and integrated throughout the lifecycle of a building or infrastructure project. This model can be used for the planning, design, construction, and operation processes. It provides greater visibility, better decision-making and more sustainable options to identify potential design, construction or operational issues in Architecture, Engineering and Construction (AEC) projects.

Introduced in the 1980s, the concept of BIM has gained prominence in recent years due to the technological evolution of design tools and the need for digital transformation in the construction industry in favour of greater efficiency, effectiveness and sustainability. The use of BIM aims to facilitate the coordination and compatibility of disciplines and to bring forward the resolution of problems to phases where the ability to influence project quality and cost is greater and the financial impact of changes is lower. These features make it possible to improve

J. Franco
Civil Engineer, Researcher at the Institute of Architecture and Urbanism of the University of São Paulo (USP), São Paulo, Brazil
e-mail: julio.franco@usp.br

I. da Silva (✉)
São Carlos School of Engineering, University of São Paulo, São Carlos, SP, Brazil
e-mail: irineu@sc.usp.br

© The Author(s), under exclusive license to Springer Nature Switzerland AG 2025
I. da Silva, P. C. L. Segantine, *Geomatics Applied to Civil Engineering*,
https://doi.org/10.1007/978-3-031-75737-2_24

information management and reduce waste and rework throughout the construction lifecycle.

As an integration and management tool, BIM technology allows for n-uses approaches, including scheduling, cost analysis and others, according to the project's objectives. Due to its management characteristics, it has been systematically adopted and institutionalized by public and private entities around the world. These characteristics emphasise its importance in Civil Engineering projects, so this chapter is dedicated exclusively to the relationship between this topic and Geomatics applied to Civil Engineering.

24.2   The Basics of BIM

24.2.1   BIM Concepts

To understand the concepts of BIM, it is important to emphasise that BIM is not just about software, hardware or a project management process but a combination of both, configured as a methodological and technological framework for construction project information in the light of the new paradigms of Construction 4.0. It can be roughly understood as a database for the integration of construction project information, including not only 3D visualisation but also semantic (non-geometric) data. This information ranges from component geometry, technical specifications, material properties, quantities, spatial relationships, costs, schedules and other information from the design disciplines throughout the lifecycle of a building. Because BIM tools use 3D parametric and relational modelling, data analysis and other computational resources to create digital models as a collection of linked data, they enable designers, contractors and owners to construct design solutions to analyse feasibility virtually, check compatibility between systems, simulate costs and identify potential problems before construction. These models enable the physical and functional characteristics of a building to be digitally represented to support the technical and management processes of many Civil Engineering projects.

BIM is currently the most developed and used digital technology in the construction sector. To better understand its applications, consider the typical AEC industry, where the entire lifecycle of the planning and realisation of a building or infrastructure project has four phases, which are *planning*, *design*, *building* and *operation*, as shown in Fig. 24.1.

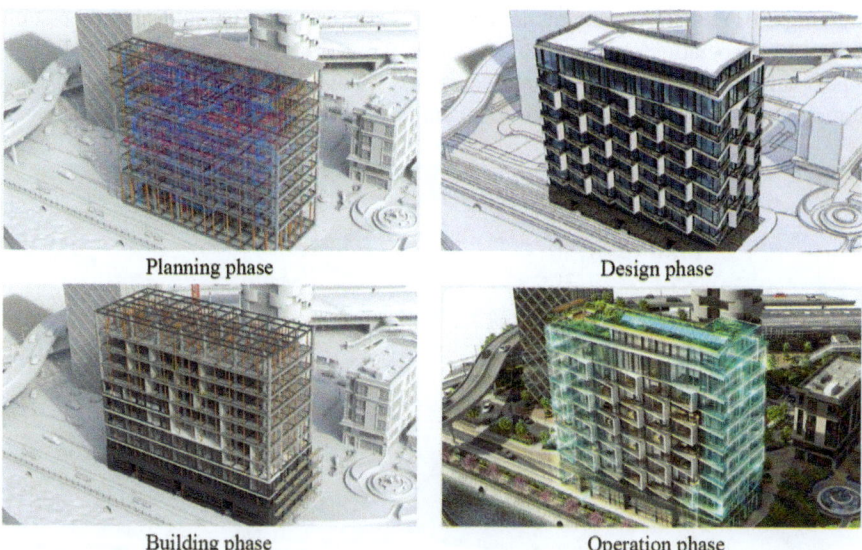

Planning phase   Design phase

Building phase   Operation phase

Fig. 24.1 Lifecycle phases of AEC management processes. (Courtesy of Autodesk. Location: https://www.autodesk.com/solutions/aec/bim/benefits-of-bim)

The *planning phase* consists of analysing the suitability of the project to the natural environment and to the project's needs. This usually consists of georeferenced geospatial information on the terrain and existing infrastructure, provided by digital context models of existing conditions, drawings and other information from which the designer proposes the general guidelines for the project.

The *design phase* consists of conceptualization, analysis, simulations, specification and technical documentation of the construction project in the form of sections drawings, views, details, calculation notes, and quantitative memorials.

The *building phase* consists of the setting out and construction of structural elements and facilities based on the design information provided by the designer.

Finally, in the *operation phase*, the as-built and maintenance projects of the finished assets are prepared and programmed for the designed uses of the building throughout its lifecycle.

However, in most cases, each phase is developed separately and the information they contain can only be partially interpreted and processed by computer methods. The consistency of the various technical drawings can, therefore, only be checked manually, which is a source of error, especially when the drawings are produced by specialists from different design disciplines and in different companies. The same applies to the handover of information to the owner at the end of construction. He/she must invest considerable effort in extracting from the project models and documents the information needed to operate the building and enter it into a facility management system.

This is where Building Information Modelling comes in. Using the resources of BIM, a much deeper use of computer technology is realised in the planning, design, construction and operation of built assets. Instead of recording information in

drawings, BIM stores maintains and exchanges information using digital representations by associating objects with a set of semantic information, such as component
type, materials, technical properties or costs, as well as the relationship between
components and other physical or local entities, as shown in Fig. 24.2.

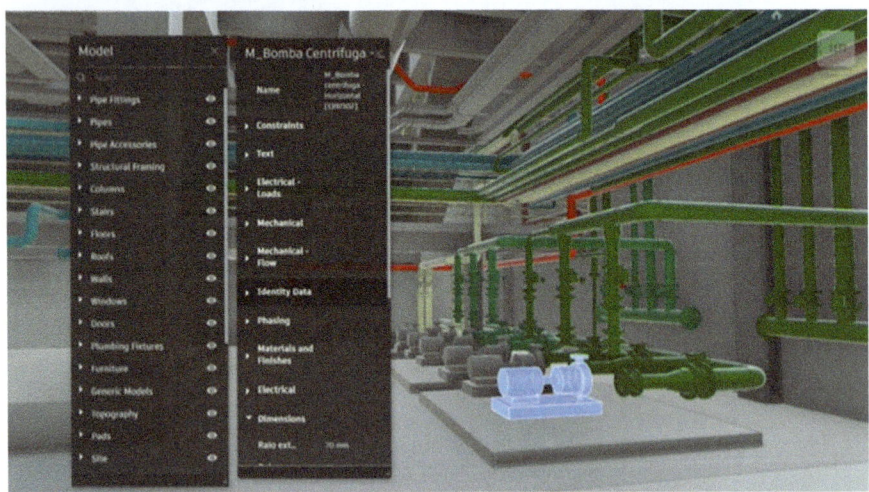

Fig. 24.2 Example of the 3D geometry and semantic information of a BIM model

Among the application examples, BIM has been used worldwide in different
types of construction, whether in new and existing buildings, buildings of historical
value, urban infrastructure, small buildings or complex structures such as the CERN
and Sirius particle accelerators. A famous example is the monumental construction
of Antoni Gaudi's La Sagrada Familia church in Barcelona, whose completion was
made possible using digital modelling technologies based on parametric approaches
to design, manufacture and assembly options.

24.2.2 BIM in Practice

For its practical application, BIM has several technical characteristics as described in
the following subsections.

24.2.2.1 Geometry

The geometric nature of BIM refers to the graphical representation of the project
using parametric Three-Dimensional objects in a virtual environment, including their
spatiality, shapes, dimensions and orientations. This ability to create and visualise
geometry in three dimensions is intuitive to human vision and makes it easier for the
team involved to understand and communicate. It also allows distances, angles, areas

and volumes to be determined virtually in their true magnitude, without the direct need to calculate projections and scales. Figure 24.3 shows an example of the visualisation of the geometries of a case study of an industrial project.

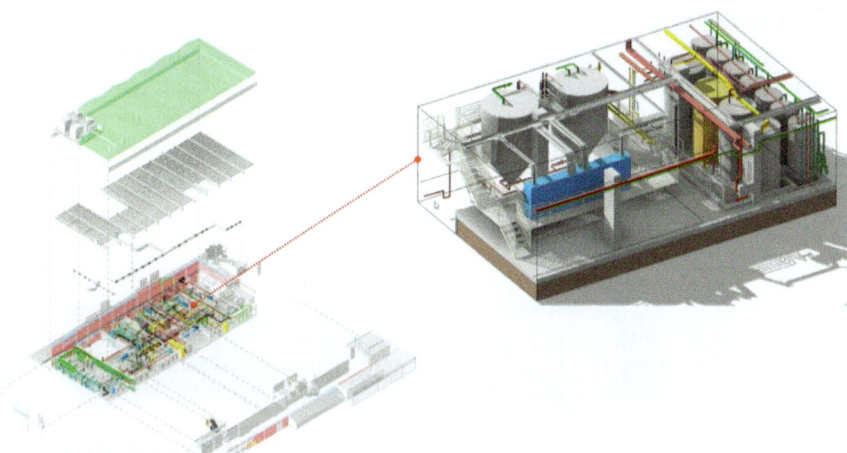

Fig. 24.3 Exploded view of project geometries and internal perspective of an industrial project BIM case

24.2.2.2 Information

In addition to geometric attributes, BIM models can contain a wide range of non-geometric (semantic) data related to the functional characteristics of the project. This includes technical specifications, object classification data, work scheduling and cost data, performance data, and other data relevant to the BIM application being addressed.

It's worth noting that there are different software and design tools to deal with the different uses and information contained in BIM models. This data is not necessarily centralised and stored in a single location, software or file, but may be structured in several parts that are integrated under a common protocol.

In a broad view of project development, BIM information can therefore cover different areas related to the AEC processes, policies and technologies employed. It can also be structured according to the different phases of the project, with applications focused on each discipline or technical purpose. For the reader interested in more information, *Succar's BIM Framework* is an insightful reference for further study of this topic.

24.2.2.3 Parametry

BIM modelling is based on parametric objects, i.e., objects whose geometric and non-geometric properties are determined by parameters and rules.

The concept of parametry makes it possible to define how the objects in the model interact and associate to express a physical or functional characteristic in the project. Furthermore, the parameters and rules are not necessarily fixed, allowing geometries and semantic attributes to be generated dynamically. In other words, objects can be updated according to changes in context and user actions.

For example, complex shape models can be generated from programmable rules so that their objects meet parametric criteria, as shown in Fig. 24.4. It is also possible to link objects so that changes to one are reflected in the other. In addition, it allows automated modelling tasks to save time on revisions and reduce rework.

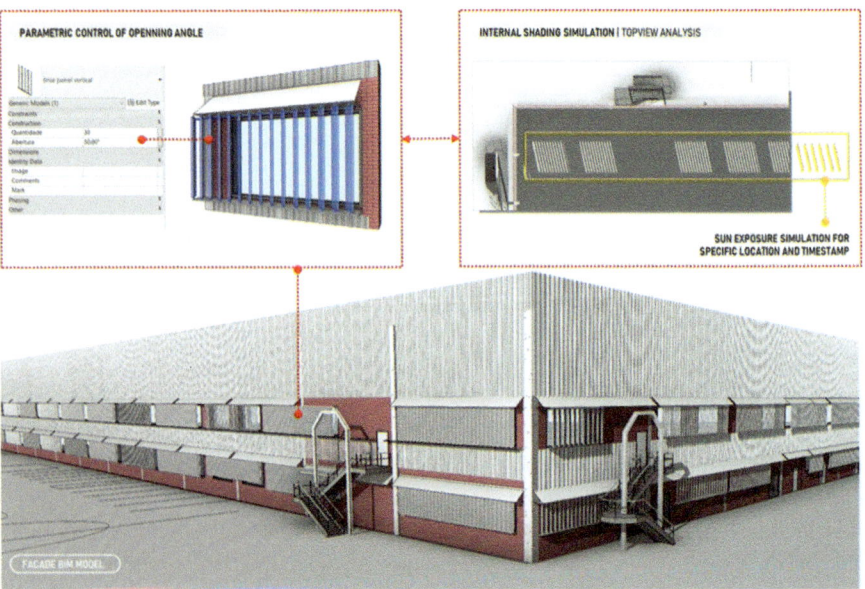

Fig. 24.4 Parametric brise-soleil orientation for shading and daylighting analysis

In short, parametric modelling makes it possible to systematise the creation and handling of objects in a BIM model and to establish relational criteria between geometric and semantic attributes and vice versa.

24.2.2.4 Multiple Visualisation

In a BIM model, the geometries, information and parametric relationships of objects can be visualised in multiple ways, or views.

Generally, views are visualisations of subsets of objects generated from the main model. They can be three-dimensional, two-dimensional, lists, data tables, etc. These multiple ways of visualising and filtering a model are used both in the development of the project itself and in the preparation and export of technical documents.

Multiple views can be understood as 'cameras' on the model: make a change to the model and these cameras automatically update what they see. Similarly, changes to objects in one view are automatically replicated in the others, eliminating redundant transfers between drawings, errors and rework. For example, if a window is deleted in a 3D view, the corresponding 2D view and billing of quantities will also change.

24.2.2.5 Detailed Visual and Information

Depending on the project's development stage, the BIM model and its objects may have different levels of granularity and completeness of their visual and informational details, or more broadly, different levels of development. Therefore, BIM models are usually classified progressively into levels of Detail / Design / Development / Information / etc., so the term Level of Development, abbreviated LOD, is used here as a classification standard.

Although there are different classifications both in the literature and in the standards, the definition of LODs is fundamental for clarifying the deliverables expected for each discipline and project phase, and for establishing coherent guidelines between work teams, both from the point of view of detail and accuracy of the model, as well as reliability for decision-making at each stage of the project.

There are usually several parties involved in the project (client, designers, contractors, consultants, etc.), so it is necessary to determine the level of precision required for each project model, what information it should contain, how reliable this information should be, what its purpose is, and what each stakeholder can expect from other models and disciplines.

To this end, there are different systems for classifying the LOD of models, depending on the geographical location or reference institution. Among the LOD classification standards, there are two that are widely used and referenced: the UK and the US. In the UK there is the BSI—British Standard Institution Specification whose classification of LODs is determined in BS EN ISO 19650. In the USA, there is the American Institute of Architects (AIA), whose classification is found in the BIM Forum LOD Standard.

24.2.2.6 Integrated Approaches

Beyond the combination of digital tools and organisational processes, integrated project approaches deal with the interaction and coordination between the people involved. In the context of BIM, the concept of Integrated Project Delivery (IPD) is a theoretical-practical foundation based on involving stakeholders from the earliest stages of a project, sharing risks and making decisions together.

These approaches are incorporated into strategies for integrating processes, technologies and stakeholders to improve collaboration and efficiency in project development. In this context, BIM models such as Fig. 24.5 are key elements as they provide an integrated virtual representation of the project, enabling improved coordination of disciplines, collaboration between designers, interference analysis, design and construction simulation.

It should be noted that in addition to the integration of BIM models, other organisational processes should be considered to facilitate collaboration between stakeholders. For example, collaboration protocols should be defined to standardise the exchange of information between different teams and disciplines throughout the project phases. The BIM Execution Plan (BEP) is fundamental in defining, among other things, the guidelines for model sharing, delivery dates, expected levels of detail (LoD) and the responsibilities of each stakeholder. This contrasts with the traditional project delivery model where stakeholders work in isolation, often in conflict, leading to delays, additional costs and quality issues.

For more information on this subject, the reader is suggested to consult the Integrated Project Delivery: A Guide provided by the American Institute of Architects (2007) to assist owners, designers, and builders to move toward integrated models and improved design, construction, and operations processes.

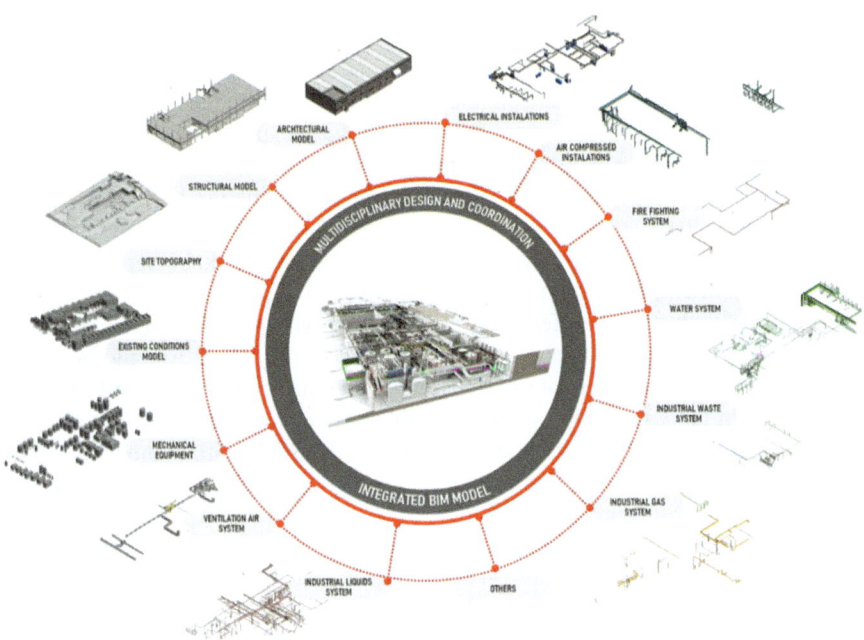

Fig. 24.5 Example of Integrated BIM model of an industrial case

24.2.2.7 Project Compatibility

The compatibility process is a critical stage in the design process as it aims to identify and resolve conflicts or incompatibilities between the different disciplines and avoid errors during and after the project. By identifying issues early in a virtual environment, project compatibility with BIM allows stakeholders to work together to anticipate the identification and resolution of issues before they occur in the real world. This saves time and resources and improves construction quality, as shown by Fig. 24.6.

Interference analysis and structured communication tools are fundamental to this process. In this context, the BIM Collaboration Format (BCF) has emerged as the standard for communicating and sharing information related to project models in an XML-based format. It is used to document and track issues and problems encountered during the development process, allowing interested parties to record and share comments, markups and change requests for review and compatibility.

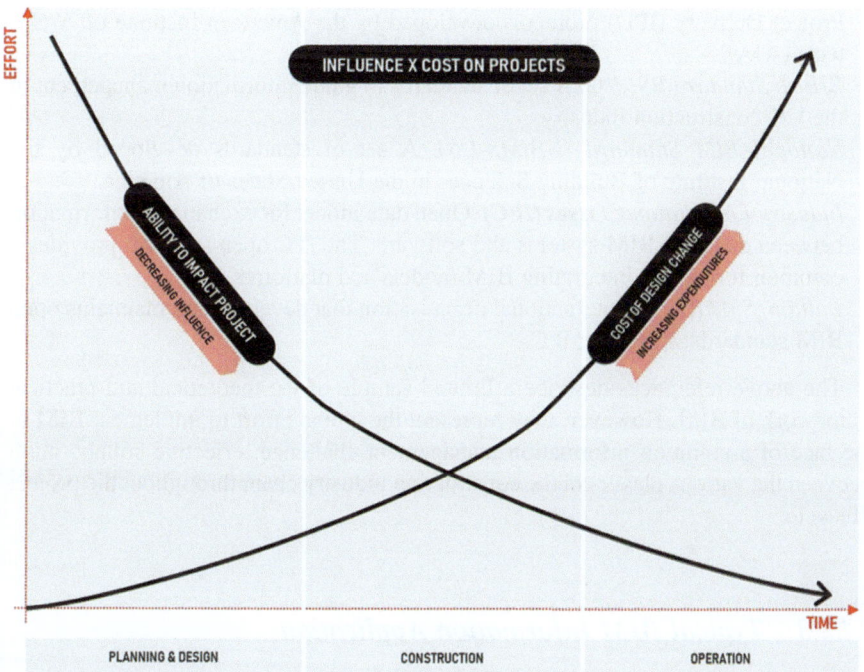

Fig. 24.6 Influence X Cost on projects. Based on Paulson'curve, MacLeamy curve and Sitter's law of evolution costs

24.2.2.8 Standards and Norms

BIM has been standardised by various public and private institutions around the world. The protocols and standards developed consume the knowledge generated and promote a common understanding of the technology in terms of interoperability between software and systems, workflows and collaboration. Some of the main references on this topic are presented below.

- *ISO 19650*. A set of international standards for the management of construction and built environment information. It provides a general framework for the development, management, integration and exchange of information throughout the lifecycle of a construction project.
- *CEN/TC 442*. European committee that develops standards related to information management and BIM for the construction industry.
- *International Alliance for Interoperability (IAI)*. An international organisation promoting the use of open standards in Building Information Modelling.
- *BuildingSMART Data Dictionary (bSDD)*. Dictionary of terms and definitions commonly used in BIM. Repository to support effective communication between different systems through a common set of definitions.
- *Building Information Modelling Protocol Exhibit (BIM-PE)*. A set of Integrated Project Delivery (IPD) protocols developed by the American Institute of Architects (AIA).
- *British Standard BS1192*. A set of standards to guide information management in the UK construction industry.
- *National BIM Standard (NBIMS-US)*. A set of standards developed by the National Institute of Building Sciences in the United States of America.
- *Industry Foundation Classes (IFC)*. Open data model for exchanging information between different BIM systems and software. The IFC open standard provides a common format for integrating BIM models and platforms.
- *BuildingSMART*. An international organisation that develops and maintains open BIM standards, including IFC.

The above references describe a limited sample of the theoretical and practical framework of BIM. However, they represent the global effort to implement BIM in the face of a common information management challenge: effective collaboration between the various players in the construction industry chain throughout the project lifecycle.

24.2.3   Typical BIM Information Application

There is no universal definition of what information a Building Information Model should provide. However, some typical applications can be identified by studying some successful projects. In this sense, using the workflow proposed in Fig. 1.1, the most common application of BIM information in the different phases of a construction project is shown in Fig. 24.7.

Highlighting some BIM uses on the building lifecycle, according to Fig. 24.7.

1. *Project definition*: This includes the initial discussion when designers and managers define the project and the owner's requirements, as well as carrying out site assessments.
2. *Existing conditions model*: BIM helps designers start their projects more efficiently by creating large-scale, intelligent 3D models of their project's real-

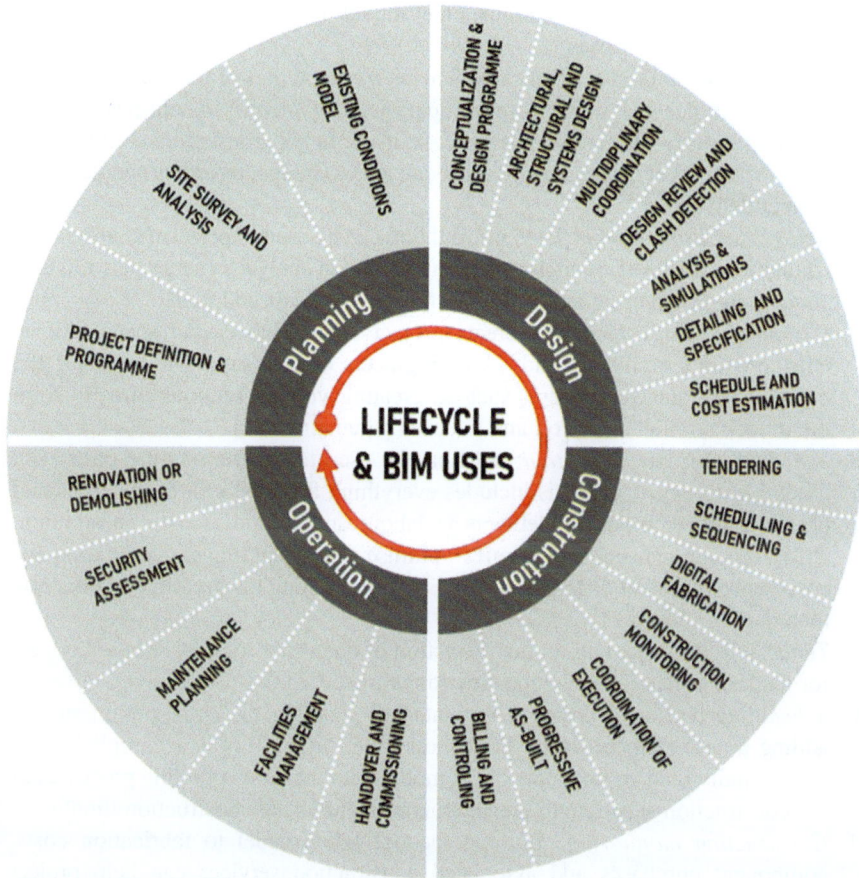

Fig. 24.7 Construction project lifecycle and main BIM uses

world environment. With BIM, they can easily aggregate large amounts of existing data, including reality capture data, 2D CAD, raster data and GIS (Geographic Information System) data, to improve the accuracy of the 3D model of existing sites.

3. *Conceptualization*: 3D modelling represents the three geographical dimensions (*X, Y, Z*) of a building structure. Using an intelligent 3D project model, architects and Civil Engineers can quickly create conceptual designs of their infrastructure and evaluate different options in the preliminary design phase. This helps them to move more quickly to a detailed design stage by adding greater accuracy to their design model.

4. *Multidisciplinary coordination*: BIM helps all stakeholders in a construction project to collaborate on a shared model. This helps them stay involved in the

process, and contractors can use the model to prepare an informed bid, minimising cost surprises.

5. *Design review and clash detection*: This is a critical part of the integrated BIM modelling process. It helps to speed up projects by identifying clashes between different models during the design phase itself, helping designers eliminate the risk of multi-level design changes that can lead to budget overruns and delays in project delivery.

6. *Analysis and simulations*: BIM information can be used to perform analyses and simulations on the 3D models, including structural analysis, energy performance simulation, daylighting analysis, dynamic site analysis and more.

7. *Detailing and specification:* Create project views, details and specifications. BIM provides architects and Civil Engineers with a range of tools for the detailing and design process, such as creating vertical sections directly from the model, conflict detection and material parametrization.

8. *Schedule and cost estimation*: BIM can be used to calculate future costs with much greater accuracy. This includes everything from the cost of materials and prefabricated or modular elements to labour and shipping costs. In addition, BIM software can help companies optimise their spending by comparing the cost-effectiveness of different materials and suggesting the best time to purchase materials at a low market price.

9. *Tendering*: BIM can help in the extraction of quantities to prepare base budgets for tendering materials and construction works.

10. *Scheduling and sequencing*: it is related to planning the construction site by adding a new element to the BIM model, i.e., time. This is accomplished by linking individual components of the model with the corresponding processes of the construction schedule to create an intelligent visual construction timeline.

11. *Construction monitoring:* Linking the 4D BIM model to fabrication costs, component purchases and associated construction services can help project promoters and owners to analyse the costs that will be incurred over time in the project activities. The BIM model also can be used to document construction issues and monitor quality control on site.

12. *Progressive As-built*: Register construction progress by creating and updating BIM models as built.

13. *Facility management*: BIM Facility Management is the application for process management and maintenance of the built environment after the building is completed. It is the final step in the process of designing and completing a building project and includes all those activities necessary to maintain an adequate level of efficiency over time to fulfil the functions for which the building was designed.

14. *Maintenance planning*: Transfer of BIM data to the owner/operator and subsequent incorporation into facility management systems for operation and maintenance planning.

24.3 Geomatics in the BIM Design Process

For BIM to be effective from the outset, it must be based on accurate 3D geospatial data provided by a Geomatics Engineer. The best way to understand how surveying fits into the BIM process is to look at the complete lifecycle of a project, from conception to delivery and operation. At all stages of this cycle, there is a need for fit-for-purpose survey information, that can only be provided through the use of geomatics data acquisition techniques. It is, therefore, important that Geomatics engineers are involved in the BIM conception from the beginning of the project.

To highlight the importance of Geomatics in the BIM design process, the role of the Geomatics Engineer throughout the BIM lifecycle is outlined below.

In the *planning phase*, the Geomatics Engineer may be asked to provide geospatial survey data and information as a background to the project. At this stage, he/she will usually provide a general identification and assessment of the site characteristics based on geographical, topographical and geological data, as well as information on the condition of infrastructure such as public services, transport network, environmental impact, and environmental and urban constraints to the project.

The Geomatics Engineer will also be required to advise on surveying technical requirements such as coordinate systems, measurement methods and instruments, data formats and data management, and quality control methods to be used throughout the project lifecycle.

During the *preliminary design phase*, the studies from the previous phase are extended to other disciplines and may include:

- Use of geospatial data to produce Digital Terrain Models (DTMs) and Geographic Information Systems (GIS) that can be used as a basis for project development.
- Analysis of topographic and drainage data to optimise the positioning and sizing of structures and infrastructure.
- Integration of geospatial data to identify potential conflicts with existing infrastructure and plan adjustments to the project.

In the *detailed design phase*, geospatial information is used as a reference for specifying and detailing design solutions and making systems compatible. Some examples of applications in this phase are described below:

- Incorporating geospatial data to create BIM models with accurate terrain information, such as contours, elevations and geological features.
- Using reality capture methods (laser scanning, photogrammetry, etc.) and geospatial information to optimise the design of infrastructure, such as cut and fill, drainage networks and utility systems, considering geographic and topographical aspects.
- Analysis of geospatial data to identify construction constraints such as risk zones, environmental restrictions, etc.

At this stage, the Geomatics Engineer is also responsible for checking and confirming that the engineering data provided can be used to update the engineering design during the project lifecycle. This allows for a continuous as-built view as the site is prepared, meaning that site control can be managed and maintained, and design changes can be verified at any time.

During the *construction phase*, the Geomatics Engineer is required to use geospatial information to monitor construction in real-time, typically using control points and topographic measurements to check the accuracy of key point location and scanning or photogrammetric techniques to compare the BIM model with as-built cloud points. At the same time, surveyors will supervise contractors on site to ensure that correct coordination is being used and will regularly feed as-built data back into the shared data environment to minimise design error creep.

Finally, in the *management phase*, once the construction contract is completed and the structure is handed over, the Geomatics Engineer will use the geospatial information for asset management and maintenance, such as locating underground infrastructure components, monitoring conditions and updating maintenance records, like a digital twin.[1] While the building is in use, surveyors will be able to locate services in the site coordinate system and could be a party to a service contract, providing data updates to keep the site data up to date.

These examples succinctly illustrate the role of geoinformation in project development processes as an interface between the physical and digital environments, which is fundamental for better-informed decision-making and thus mitigating project risks and errors. Although they come from different domains, BIM and Geomatics are complementary tools in the design process, both in understanding the natural and man-made environment and in developing design solutions.

This relationship has become increasingly closer technologically, unfolding into concepts and technologies such as GIM (Geospatial Information Modelling), GeoBIM, BIM-GIS, and others. Given its relevance, and for further details on the state of the art in Geospatial Information Research, readers are suggested to consult Bill (2022).

In terms of technology, conventional terrestrial surveying instruments are the most used for data acquisition in BIM applications. Typically, imaging and scanning total stations or terrestrial laser scanners are best suited for indoor and outdoor mapping of small building sites. For large areas, UAV photogrammetry and terrestrial scanners are the most used. Figure 24.8 illustrates the integration of geomatic data from an airborne laser scanning survey with the building information model.

---

[1] The real-time digital representation of the physical building or infrastructure.

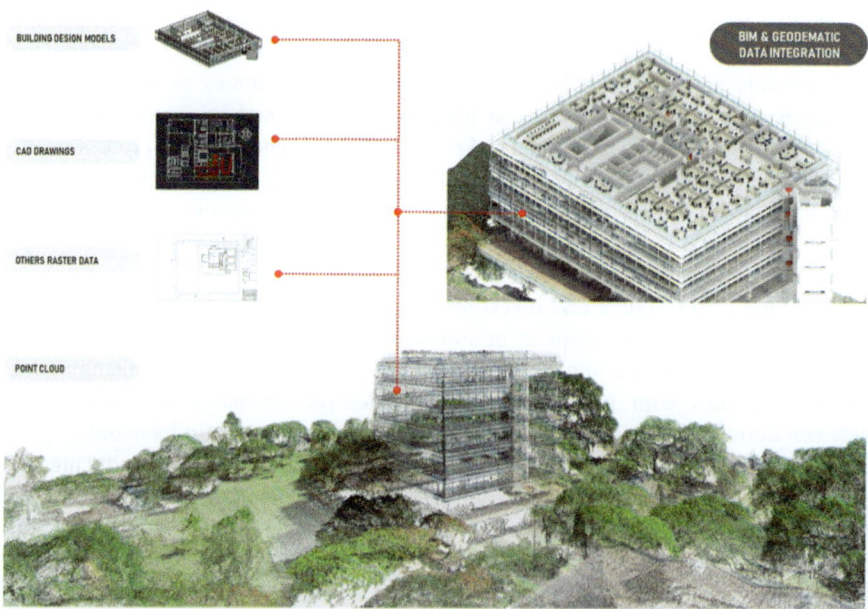

Fig. 24.8 BIM conceptual design from an airborne laser scanning survey

24.4   Final Considerations

This chapter aimed to present general aspects of BIM to provide a basis and reference for professionals wishing to go further into the subject, whether they are beginners or more experienced. In addition, the approach focused on the relationship between BIM and geotechnologies applied to the design process, located at the intersection of these domains.

In this sense, these advances have fuelled the continued evolution of BIM technology, making it increasingly valuable to the construction industry. Such integration could play a key role in optimising project delivery, reducing waste and errors, improving collaboration and providing more resources for decision-making, as well as supporting sustainable practices to reduce the environmental impact of construction.

It should also be noted that, as emerging technologies, there are latent challenges to their use. The first is the level of standardisation. Although these technologies are becoming more widespread, there is still a need for greater standardisation of processes and interoperability between tools, platforms and sectors of the construction industry. The lack of standardisation makes it difficult to share and exchange information between different systems and stakeholders, hindering collaboration and increasing the risk of error. This challenge is an iterative and incremental one as the use of BIM becomes better established and expanded.

Another issue to consider is the resistance to change associated with innovative technologies in the construction industry. BIM can be a significant change for professionals and organisations, requiring time and resources to establish new processes and workflows. This can lead to resistance and reluctance to adopt, particularly among more traditional and conservative stakeholders and professionals.

Add to this the uncertainty of not knowing something relatively new, and the need to establish new processes, adopt new tools and train people. This horizon of digital transformation in construction brings with it new paradigms and barriers to overcome, both technological and human.

There are also resource and investment constraints. Adopting new technologies such as BIM requires investment in software and hardware, as well as time and resources for training and implementing new workflows. This can be challenging for small companies with limited budgets. In addition, exchange rate differences between currencies can limit the choice of BIM systems in certain locations.

In this scenario, the financial viability of BIM must be focused on reducing the risk of error and on opportunities to add value to design and business solutions. To this end, it is necessary to look at the entire project lifecycle so that the benefits, not just the financial ones, can be effectively considered and the limitations can be better balanced, especially in the context of the bioclimatic emergency and our responsibilities.

In a historically conservative and artisanal sector, the search for greater sustainability requires innovations that make planning more robust, facilitate design and construction processes, optimise the use of resources and, a priori, promote the transformation of its "manpower" into "brainpower". Despite the historical and conservative inertia based solely on immediate profit.

The future potential of BIM is huge, and its adoption is expected to increase as its capabilities are improved and integrated with other (geo)technologies.

24.5 Review Questions

1. Explain what Building Information Modelling is and how it relates to Civil Engineering of the future.
2. Briefly explain what the BIM concept is.
3. Briefly explain the differences between BIM, CAD and GIS.
4. Explain under what circumstances you would recommend using BIM modelling for a Civil Engineering project.
5. Explain the difference between the geometric and the semantic concepts of a BIM design.
6. Explain the importance of parametry and multiple visualisation in BIM modelling.
7. Briefly explain the concept of LOD and its importance in BIM modelling. How does it interact with geospatial data?

8. Briefly explain what are the main uses of BIM and geospatial data in the lifecycle of a building and which ones you think will have the most impact on the AEC industry.
9. Explain why Geomatics is an important tool in the BIM process and briefly describe its impact on each BIM design phase.
10. Briefly discuss the benefits you see for implementing BIM modelling in your organisation.
11. Briefly discuss the obstacles you see to implementing BIM modelling in your company in terms of organisational processes, technology, people and knowledge.
12. Briefly discuss the impact and importance of BIM and Integrated Project Delivery on the digital transformation of the construction sector.
13. Explain how BIM can improve the efficiency and effectiveness of construction project development processes.

References

Abbasnejad B, Nepal MP, Ahankoob A, et al (2020). *Building Information Modelling (BIM) adoption and implementation enablers in AEC firms: a systematic literature review.* Archit Eng Des Manag 1–23.

AIA (2007). *Integrated Project Delivery: a Guide.*

Andre Borrmann, Markus König, Christian Koch, Jakob Beetz (2018). *Building Information Modeling: Why? What? How?* Technology Foundations and Industry Practice. Springer, DOI: https://doi.org/10.1007/978-3-319-92862-3

Arup Completing La Sagrada Familia: a collaboration in digital and stone. Access: https://www.arup.com/projects/sagrada-familia#.

Azhar S (2011). *Building Information Modelling (BIM): Trends, Benefits, Risks, and Challenges for the AEC Industry.* Leadersh Manag Eng 11:241–252.

CNPEM SIRIUS: *Accelerating the future of Brazilian science.* Access: https://lnls.cnpem.br/sirius-en/

Eastman C, Teicholz P, Sacks R, et al (2008). *BIM handbook: a guide to building information modeling for owners, managers, designers, engineers, and contractors.* John Wiley & Sons, Inc., Hoboken, New Jersey.

Ghaffarianhoseini A, Tookey J, Ghaffarianhoseini A, et al (2017). *Building Information Modelling (BIM) uptake: Clear benefits, understanding its implementation, risks and challenges.* Renew Sustain Energy Rev 75:1046–1053.

Herle S, Becker R, Wollenberg R, et al (2020). *GIM and BIM: How to Obtain Interoperability Between Geospatial and Building Information Modelling?* PFG—J Photogramm Remote Sens Geoinf Sci 88:33–42.

ISO 12911:2023. *Organization and digitization of information about buildings and civil engineering works, including building information modelling (BIM) — Framework for specification of BIM implementation.*

ISO 19650-1:2018. *Organization and digitization of information about buildings and civil engineering works, including building information modelling (BIM) — Information management using building information modelling — Part 1: Concepts and principles.*

ISO 29481-1:2016. *Building information models — Information delivery manual — Part 1: Methodology and format.*

Loo Y, Sykes M, Sturzaker C, et al (2015). *Early-stage BIM for CERN's future circular collider studies.* Struct Eng 93:12–18.

Noardo F, Ellul C, Harrie L, et al (2020a). *Opportunities and challenges for GeoBIM in Europe: developing a building permits use-case to raise awareness and examine technical interoperability challenges.* J Spat Sci 65:209–233.

Noardo F, Harrie L, Ohori KA, et al (2020b). *Tools for BIM-GIS integration (IFC georeferencing and conversions): Results from the GeoBIM benchmark 2019.* ISPRS Int J Geo-Information 9.

Pezeshki Z and Ivari SAS (2018). *Applications of BIM: A Brief Review and Future Outline.* Arch Comput Methods Eng 25:273–312.

Succar B (2009). *Building information modelling framework: A research and delivery foundation for industry stakeholders.* Autom Constr 18:357–375.

Succar B and Kassem M (2015). *Macro-BIM adoption: Conceptual structures.* Autom Constr 57: 64–79.

U.S. GSA General Services Administration BIM Level of Detail. Access: https://www.gsa.gov/ real-estate/design-and-construction/3d4d-building-information-modeling/bim-software-guide lines/document-guides/level-of-detail#:~:text=Level of Design %2F Development %2F Detail, data associated with the objects.

Yalcinkaya M and Singh V (2015). *Patterns and trends in Building Information Modelling (BIM) research: A Latent Semantic Analysis.* Autom Constr 59:68–80.

Index

A

Abscissa, 121
Absolute, 483
Absolute error ellipse, 83
Absolute orientation, 788
Absolute positioning, 695–696
Accuracy, 616
Additive, 259
Additive constant, 250
Adhesive prism, 295
Adhesive reflector, 300
Adjusted observation vector, 65
Adjusted observations, 76
Adjusted values, 64
Adjustment computation, 37
Aerial triangulation, 804
Affine, 147
Affine coordinate transformation, 133
Airborne Laser Scanning (ALS), 740
Airborne photogrammetry, 770
Alidade, 269
Altimetric, 843
Altitude angle, 179
Ambiguity, 693, 694
Angle adjustment, 452
Angle of refraction, 224
Angle to the right, 448
Angular errors, 317
Angular misclosure, 451, 487
Antenna Reference Point, 710
Anti-reflective coating, 298
À posteriori, 70
Approximate values, 66
Arc-to-chord, 639

Arithmetic mean, 58
Artificial control point, 800
Atmospheric refraction, 200, 226, 546, 692
Auto-calibration, 812
Automatic Target Recognition (ATR), 285
Axial errors, 321
Azimuth, 182, 357
Azimuthal, 623

B

Backsight, 181
Bar code, 350
Barcode levelling staff, 310
Baseline, 697
Baseline network, 713
Batter board, 863–864
Beam deflection, 746
Beam deviation, 298
Beam divergence, 744
Bearing, 182
BeiDou, 685
Benchmarks (BM), 108, 500, 512
Bilateration, 381
BIM Collaboration Format, 997
BIMP, 15
Blunders, 39, 534
Boresight, 812
Bowditch rule, 457
Building Information Modelling (BIM), 7, 989
Bundle Block Adjustment, 805
Bursa-Wolf, 160–162

C

Cadastral management, 5
Cadastre, 4
Calibrations, 16, 336, 343, 810
Carrier wave, 254
Cartesian, 153
Cartesian coordinate system, 120
Cartography, 3
Central meridian, 620, 628
Centring errors, 330, 484
Change point, 508
Charged Coupled Device (CCD), 275, 288
Checkpoint, 805
Chi-square, 90
Chord, 214
Circuit levelling, 512
Circular bubble, 272
Circular curve, 945
Circular prism, 299
Clarke, 102
Clockwise angle, 365
Closed-circuit line, 521
Closing the horizon, 196
Closure accuracy, 488
Cloud-to-cloud, 753
Coefficient of refraction, 224
Cofactor matrix, 65
Cofactors, 63
Collimation error, 322
Collinearity equations, 792–798
Combined direction, 193
Combined scale factor, 644
Compass Navigation Satellite System (CNSS),
    685
Complementary Metal Oxide Semiconductor
    (CMOS), 275, 288
Computer-aided design (CAD), 6
Conditional equations, 73, 79
Confidence ellipse, 91
Confidence level, 485, 538
Conformal, 137, 148, 621
Conformal coordinate transformation, 133
Conical, 623
Conjugated images, 781
Construction total stations, 292, 866
Continuously Operating Reference Stations
    (CORS), 717
Contour, 931
Contouring, 574–583, 608
Contour lines, 572
Control points, 97, 439
Convergence of the meridians, 636
Conversion of coordinates, 129

Coordinate, 97
Coordinate reference system, 97, 119
Coordinate transformation, 129
Co-registration, 751
Correction value, 66
Correlation coefficient, 46, 58, 90
Covariances, 45, 63
Crest, 576, 956
Crosshairs, 278, 343
Cross-sectional, 570
Cross-sectional areas, 907
Cross-sectional profile, 570
Cyclic error, 341
Cylindrical, 622

D

Data modelling, 756
Data structuring, 597
Datum, 97
Dead zones, 576
Decentring, 790
Deflection of the vertical, 108, 110, 225
Degrees, 21
Degrees of freedom, 67, 91
Departure, 455
Detail survey, 846
Differential Global Positioning System
    (DGPS), 704
Differential levelling, 502
Differential RTK, 705
Digital Elevation Model (DEM), 589
Digital elevation modelling, 594
Digital levels, 303, 350
Digital Photogrammetric Workstation, 813
Digital photogrammetry, 770
Digital Surface Model (DSM), 589, 757
Digital Terrain Model (DTM), 589, 934
Digital Workstations, 770
Dilution Of Precision, 721
Direct and reversed, 187
Direct georeferencing, 750, 801
Direct problem, 361
Direct reading, 188
Double-difference, 699
Double-face, 190
Double-face reading, 188
Double-run levelling, 515
DRONE, 824
Dual-frequency, 714
Dynamic portable TLS, 741
Dynamic vehicular TLS, 741

E
Earthwork, 909
Eccentric station, 420
Economic elevation, 932
Electromagnetic, 252
Electromagnetic wave, 254
Electronic compensator, 182, 199, 274
Electronic Distance Measurement (EDM), 251,
    258, 280, 282, 336
Electronic level, 274, 276
Electronic theodolite, 280
Elevation, 499
Elevation scale factor, 216, 644
Ellipse, 27, 83
Ellipsoidal distances, 215, 219
Ellipsoidal height, 108, 125, 128
Ellipsoid of revolution, 27–28, 101
Embankment, 913
Engineering (local) datum, 99, 100
Engineering surveying, 2
Epipolar line, 809
Equation of the parabola, 957
Equidistant, 621
Equipotential surface, 99
Equivalent, 621
Error distribution, 521–523
Error ellipse, 84
Error ellipse confidence level, 89–93
Error ellipse expansion factor, 92
Error equations, 66, 67
Error propagation, 51
Errors of closure, 451, 512, 538, 539
Extensible Markup Language (XML), 997
Exterior orientation, 788
Extrinsic parameters, 788

F
False easting, 620
False northing, 620
F-distribution, 91
Feature matching, 807
Field of view, 279
First eccentricity, 103
First eccentricity squared, 103
Fixed-wing, 825
Flattening, 103
Flight heights, 781
Floating mark, 787
Footscrews, 275
Forced centring, 278, 484
Force of gravity, 108
Foresight, 181
Free code, 291
Free station, 878

Functional, 65

G
GALILEO, 684
Gauss-Helmert adjustment model, 65
Gauss-Helmert model, 79
Gauss-Krüger projection, 648
General adjustment, 79
General adjustment model, 81
General Conference for Weights and Measures
    (CGPM), 17
Generalized Gauss-Helmert, 141
General law of error propagation, 229
General law of propagation of variance-
    covariances, 53
General law of propagation of variances, 69,
    548
Geocentric, 165, 169
Geocentric Cartesian, 162
Geocentric Cartesian coordinates, 109
Geocentric global ellipsoid, 101
Geodesic, 639
Geodesy, 3
Geodetic, 170, 633
Geodetic Altimetry Reference Systems, 107
Geodetic Azimuth, 646
Geodetic coordinates, 162
Geodetic Coordinate System, 128
Geodetic datum, 99, 101
Geodetic distance, 215
Geodetic latitudes, 128
Geodetic longitudes, 129
Geodetic north, 636
Geodetic Reference System, 105
Geodetic Reference System for the Americas
    (SIRGAS), 107
Geodetic Surveying, 101, 209
Geodetic to ground, 660–662
Geographic coordinates, 127
Geographic Information Systems (GIS), 6
Geoidal model, 110
Geoid model, 99
Geoid surface, 100
Geoid undulation, 108, 110
Geomatics Engineering, 2
Geometric Dilution of Precision (GDOP), 721
Geometric height, 108
Georeferenced loop traverse, 448
Global Geocentric, 125
Global Geocentric Geodetic Reference System,
    106
Global interpolation method, 603
Global Navigation Satellite System (GNSS), 6

Globalnaya Navigatsionnaya Sputnikovaya
    Systema (GLONASS), 683
Global positioning system, 680–682
GN, 636
Gnomonic, 622
GNSS Network, 724–736
GNSS time, 689, 691
Gons, 19, 21
Grades, 943
Graduated circle errors, 318
Graduated circles, 186, 280, 318
Gravitational field, 99
Gravity, 99
Greyscale, 771
Grid Azimuth, 646
Grid-based, 603
Grid distance, 230
Grid North, 636
Gross errors, 39
Ground-based horizontal plane, 356
Ground-based plane, 208
Ground control point (GCP), 799
Ground Sample Distance (GSD), 773
Ground surface, 99
Ground to geodetic, 662–664
GRS80, 102
GRS80 ellipsoid, 101
Gyro-stabilised, 779

H
Handheld laser distance meter, 247
Haul, 913
Hayford, 102
Height anomaly, 110
Height of collimation, 504
Height of the instrument, 505, 543
Height of the reflector prism, 543
Helmert 3D, 160–162
Homologous, 149
Homologous points, 132, 138
Horizontal alignment, 943
Horizontal angle, 178
Horizontal circle, 186
Horizontal datum, 99
Horizontal direction, 180, 186
Horizontal distance, 208, 219
Horizontal line, 500
Horizontal plane, 179
Horizontal reference plane, 208, 213
HPC, 505
Hysteresis, 272

I
Identity matrix, 68
Image-assisted total stations, 287
Image coordinate, 772
Image correlation, 806
Image matching, 806
Image sensor, 773
Independent loop traverse, 448
Independent observations, 57
Index error, 337–341
Indirect georeferencing, 751–754
Inertial measurement unit (IMU), 802
Inertial Navigation System (INS), 799
Instrument adjustment, 16
Instrument centring, 267
Instrument levelling, 267
Instrument setup, 267
Interior-angle, 448
Interior orientation, 788
Intermediate sight, 511
International Earth Rotation and Reference
    Systems Service (IERS), 106
International Electrotechnical Commission
    (IEC), 268
International Organization for Standardization
    (ISO), 316
International System of Units (SI), 17
International Terrestrial Reference Frame, 106
International Terrestrial Reference System
    (ITRS), 106
International Vocabulary of Metrology (VIM),
    15
Interpolation, 600
Interpolation functions, 601
Intersection, 378, 384
Intrinsic parameters, 788
Invar, 310
Inverse problem, 356
Ionosphere, 692
IP code, 268
Irregular grid, 578
Irregular triangulation, 597
Isoline, 574

J
Jacobian matrix, 57

K
Kinematic, 695

L

Lagrange multipliers, 75, 80
Laser beam, 260
Laser/built-in video plummet, 276
Laser levelling, 306, 559
Laser plumbing, 872
Laser pointer, 329
Laser scanners, 739, 742
Lateral atmospheric refraction, 201–202
Latitude, 455
Law of propagation of true errors, 52
Law of propagation of variances, 52
Leap-frog, 551
Least-squares, 64, 472, 485
Least squares adjustment, 64
Least squares principle, 66
Level bubble, 272
Level line, 500
Levelling, 499
Levelling error, 334
Levelling misclosure, 512
Levelling network, 520
Levelling section, 521, 524
Levelling span, 525
Levelling staff, 249, 273, 308, 502
Level of Development (LoD), 995
Level surfaces, 500, 502
Light Amplification by Stimulated Emission of
    Radiation (LASER), 276, 306
Light Detection and Ranging (LiDAR), 267,
    739
Line, 25
Linear functions, 56
Linearisation, 56, 66, 79
Linear misclosure, 451, 457, 458
Linear Referencing System (LRS), 120
Linear standard deviation, 82
Line of sight, 267
Line of sight axis, 270
Link levelling, 514
Link traverse, 447
Local ground-based coordinates, 658
Local ground-based horizontal plane, 100
Local mean radius, 103
Local TM, 649
Longitude, 635
Longitudinal profile, 569
Loop levelling, 512
Loop traverse, 447, 454
Low distortion projection, 659, 664–669

M

Machine automation, 883
Machine control, 294, 883

Map projections, 619
Mathematical Cartesian Coordinate System,
    121
Measured value, 40
Measuring instrument, 16
Meridian of Greenwich, 127
Meridians, 127, 182
Metre, 17
Metric camera, 776
Mini-prism, 299
Mistakes, 39
Modulated, 254
Modulated wave, 254
Molodensky, 158
Moving Average, 601
Multilateration, 396
Multipath, 692

N

National Marine Electronics Association
    (NMEA), 705
Natural control point, 799
Navigation Indian Constellation Satellite
    System (NaviC), 686
Networked Transport of RTCM via Internet
    Protocol (NTRIP), 707, 717
Network levelling, 524
Network of triangles, 927
Network Real-Time Kinematic, 717
Network Real-Time Kinematic Positioning
    (NRTK), 717
Normal, 623
Normal elevation, 108
Normal equations, 68
Normal line, 108, 112
Normal vertical line, 109

O

Oblique, 624
Observation, 39
Observational errors, 38
Observation vector, 41, 65
1D machine control, 884
Open traverse, 441
Optical, 248–251
Optical distortion, 789–791
Optical level, 302
Optical plummet, 276
Ordinate, 121
Orientation, 177
Orthographic, 622
Orthometric altitude, 108
Orthomosaic, 821

Orthophoto, 815, 818
Orthorectified, 821

P
Pacing, 238
Parabolas, 956
Parallax, 279, 350, 786
Parallels, 127
Parametric adjustment, 138, 384, 472, 526
Parametry, 994
Parts per million (ppm), 157, 216
Perspective centre, 773
Perspective projection, 774
Perspective view, 609
Phase-shift, 252, 743
Photodiodes, 771
Photogrammetric model, 783, 788
Photogrammetric restitution, 815, 816
Photogrammetry, 4
Phototriangulation, 804
Picosecond, 253
Pixel coordinate, 772
Pixels, 267, 771
Plane, 26–27
Plane Azimuth, 646
Plane surveying, 100, 208
Planialtimetric, 843
Planimeter, 895
Planimetric, 843
Plumb-bob, 242, 872
Plumb line, 99, 168, 500
Plummet, 276, 330
Point, 25
Point cloud, 267, 739
Pointing error, 333, 349
Point mesh, 817
Polar, 153
Polar coordinate, 129
Polar coordinate system, 122
Polygon, 26
Position Dilution of Precision (PDOP), 721
Post-processed kinematic (PPK), 703, 715
Post-processing, 723
PPP Real-Time Kinematic Positioning
    (PPP-RTK), 696
Precise Point Positioning (PPP), 696
Precise Positioning Services, 682
Precision, 210, 211, 259, 336, 542
Precision level, 519
Principal point, 773
Prism method, 920
Prismoidal, 909

Prism pole, 295, 300
Probability distribution, 89
Profiling, 614
Propagation of cofactors, 86
Pseudorange, 690
Pulse, 252
Push-broom, 778

Q
Quantity, 16
Quasi-geoid, 108
Quasi-Zenith Satellite System (QZSS), 685

R
Radial, 790
Radial symmetric distortion, 790
Radians, 18, 21
Radiometric resolution, 771, 774
Radio Technical Commission for Maritime
    Services (RTCM), 706
Radius of curvature, 103
Random errors, 39
Range pole, 245
Raster, 591
Real-Time Kinematic (RTK), 704
Reciprocal, 226, 517, 555
Rectangular coordinate, 121
Red, 216
Reduced level, 500
Reduction factor, 216
Redundancy, 42, 67
Redundant data, 384
Reference direction, 187
Reference ellipsoid, 101
Reference line, 183
Reference plane, 500
Reflecting prism, 212
Reflective coating, 298
Reflectivity, 745
Reflectorless, 252
Reflector prism, 295
Regional TM, 649
Regular grid, 578, 597, 605
Relative, 483
Relative closure, 485
Relative closure ratio, 486–488
Relative error ellipses, 86–89
Relative humidity, 257
Relative kinematic, 702, 715
Relative orientation, 788
Relative precision, 457

Relative static, 701
Remote sensing, 4
Resection, 375, 390
Residual errors, 40, 76
Residual error vector, 41
Residuals, 64
Residual vector, 65
Resolution, 42
Retaining walls, 576
Reverse, 948
Reversed reading, 188
Rise and fall, 504
Robotic total station, 284
Root mean square (RMS), 44
Root mean square error (RMSE), 44–45
Rotary-wing, 825
Rotation, 133, 155
Rotation matrix, 134, 137, 156

S
SAD69, 102
Sag, 956
Satellite time, 691
Scale, 29
Scale error, 341
Scale factor, 134, 157, 620, 641
Scale-Invariant Feature Transform (SIFT), 807
Scanner's Own Coordinate System, 748
Scanning resolution, 745
Scanning total station, 290
Scan registration, 751
Secant, 625
Second eccentricity, 103
Second eccentricity squared, 103
Set out, 861
Setting out, 372, 964–975
Sexagesimal degree, 19
Shading, 608
Shuttle Radar Topography Mission (SRTM),
    591
Sideshot, 514
Sighting point, 267
Signal path, 298
Significant figures, 30–33
Simple levelling, 506
Simultaneous, 226, 517, 555
Single-difference, 698
Single-face, 190
Single-face reading, 188
Single-frequency, 714
Single-point, 267
Skyplot, 721

Slope distance, 208, 211, 213, 230, 247, 317
South American Datum, 102
Spatial resolution, 774
Sphere, 28–29
Spherical, 215
Spherical distance, 213, 219
Spherical polar coordinate, 124
Spiral, 951
Spirit levels, 872
Spot elevations, 573, 920
Stadia, 248
Stadia rod, 249
Staff bubble, 242
Stakeout, 658
Standard deviation of unit weight, 69, 75
Standard error, 43
Standard error ellipse, 91, 92, 485
Standard positioning services, 682
Static, 695
Station, 267
Stationary TLS, 740
Steel turning plate, 311
Stereographic, 622
Stereo pair, 781
Stereoscopy, 785
Stochastic, 65
Straight line, 26
Strip, 781
Structure from motion (SfM), 788, 823
Successive azimuths, 366
Successive coordinate, 368
Surveying instrument, 265
Surveying pole, 295
Surveying Two-Dimensional Cartesian
    Coordinate System, 121
Surveyor's level, 301, 502
Survey standards, 538–541
Survey traverse, 440
Swelling factor, 913
System of units, 16

T
Tacheometry, 247
Tangents, 214, 624, 943
Tape measure, 239
Target recognition, 285
Taylor series, 56
Telescope, 278
Template matching, 806
Temporary Benchmark, 500
Terrestrial frame, 97
Terrestrial Laser Scanning (TLS), 740

Terrestrial photogrammetry, 770
Thalweg, 576
Thematic code, 291
Theodolite, 279
Theory of Errors, 37
Three-Dimensional Cartesian, 155
Three-Dimensional Cartesian Coordinate, 124
3D machine control, 884
3D model, 739
3D mouse, 814
Three-Dimensional point positioning, 401, 406
360° prism, 299
Tie point, 805
Tilting axis, 270
Tilting axis error, 324
Time delay integrator, 779
Time-of-flight, 252, 743
Toe of slope, 576
Tolerance, 43
Topocentric, 126, 165, 169
Total station, 212, 280, 283
Transformation parameters, 132
Transition elevation, 933
Transition horizontal curve, 950
Translations, 135, 157
Transverse, 624
Transverse Mercator (TM), 648
Trapezoidal, 909
Traverse, 440
Traverse stations, 441, 449, 470
Triangle-based, 604, 605
Triangular Irregular Network (TIN), 598, 929
Triangulated Irregular Network, 927
Triangulateration, 440
Triangulation, 439
Tribrach, 275
Trigonometric component, 213
Trigonometric levelling, 542
Trilateration, 440
Triple-difference, 700
Tripod, 271
Tripod drift, 335
Troposphere, 692
True Azimuth, 646
True error, 40
True North (TN), 636
True orthophoto, 820
True value, 40
Tubular, 272

Tubular level, 274
Tubular level bubble, 273
Turning point, 508
Two-Dimensional, 120, 132
Two-Dimensional Cartesian, 155
2D Cartesian coordinate, 121
2D machine control, 884

U

UAV photogrammetry, 770, 822
Unidirectional, 543
Unit of measurement, 16
Units of area, 24
Units of volume, 24–25
Unit vector, 61
Universal Transverse Mercator (UTM), 625
Unweighted, 68
UTM coordinate system, 627

V

Variance, 43, 45
Variance-covariance matrix, 63, 86, 87
Vector data, 591
Vectorisation, 756, 816
Verification, 16
Vertical alignment, 943
Vertical angle, 178
Vertical atmospheric refraction, 536–538
Vertical axis, 186, 270
Vertical axis error, 327
Vertical circle, 186, 199
Vertical collimation error, 344
Vertical control, 499
Vertical curves, 956
Vertical datum, 99, 500
Vertical distance, 211, 213
Vertical index error, 320
Vertical line, 108, 112, 501
Vertical refraction, 228
Virtual Reference Station (VRS), 717
Virtual RINEX (VRINEX), 718

W

Watershed, 576, 584
Wave Form Digitiser (WFD), 252, 256, 743
Weight, 60

Weighted, 68
Weighted mean, 61
Weighted Moving Average, 602
Weight matrix, 64, 68, 74
Wheel odometer, 238
World Geodetic System 1984 (WGS84), 102,
    106–107

Z
Zenith angle, 179
Zero error, 337
Zero-point, 909
Zone, 655
Zone numbering, 627
Zone width, 627

The manufacturer's authorised representative in the EU is Springer
Nature Customer Service Centre GmbH, Europaplatz 3, 69115 Heidelberg,
Germany. If you have any concerns regarding our products, please
contact ProductSafety@springernature.com

Printed and bound by CPI Group (UK) Ltd, Croydon, CR0 4YY
10/06/2025
01898343-0005